AF685598

THE PHILOSOPHY OF SCIENCE

OXFORD STUDIES IN PHILOSOPHY OF SCIENCE

General Editor:
Paul Humphreys, University of Virginia

Advisory Board
Anouk Barberousse (European Editor)
Robert W. Batterman
Jeremy Butterfield
Peter Galison
Philip Kitcher
Margaret Morrison
James Woodward

The Book of Evidence
Peter Achinstein

Science, Truth, and Democracy
Philip Kitcher

Inconsistency, Asymmetery, and Non-Locality: A Philosophical Investigation of Classical Electrodynamics
Mathias Frisch

The Devil in the Details: Asymptotic Reasoning in Explanation, Reduction, and Emergence
Robert W. Batterman

Science and Partial Truth: A Unitary Approach to Models and Scientific Reasoning
Newton C. A. da Costa and
Steven French

Inventing Temperature: Measurement and Scientific Progress
Hasok Chang

The Reign of Relativity: Philosophy in Physics 1915–1925
Thomas Ryckman

Making Things Happen: A Theory of Causal Explanation
James Woodward

Mathematics and Scientific Representation
Christopher Pincock

Simulation and Similarity: Using Models to Understand the World
Michael Weisberg

Systemacity: The Nature of Science
Paul Hoyningen-Huene

Causation and Its Basis in Fundamental Physics
Douglas Kutach

Reconstructing Reality: Models, Mathematics, and Simulations
Margaret Morrison

The Ant Trap: Rebuilding the Foundations of the Social Sciences
Brian Epstein

Understanding Scientific Understanding
Henk de Regt

The Philosophy of Science

A COMPANION

Edited by
Anouk Barberousse, Denis Bonnay, and Mikaël Cozic

OXFORD
UNIVERSITY PRESS

OXFORD
UNIVERSITY PRESS

Oxford University Press is a department of the University of Oxford. It furthers the University's objective of excellence in research, scholarship, and education by publishing worldwide. Oxford is a registered trade mark of Oxford University Press in the UK and certain other countries.

Published in the United States of America by Oxford University Press
198 Madison Avenue, New York, NY 10016, United States of America.

© Editions Vuibert-Paris 2011 as *Précis de Philosophie des Sciences*
© English translation Oxford University Press 2018
© Oxford University Press 2018

All rights reserved. No part of this publication may be reproduced, stored in a retrieval system, or transmitted, in any form or by any means, without the prior permission in writing of Oxford University Press, or as expressly permitted by law, by license, or under terms agreed with the appropriate reproduction rights organization. Inquiries concerning reproduction outside the scope of the above should be sent to the Rights Department, Oxford University Press, at the address above.

You must not circulate this work in any other form
and you must impose this same condition on any acquirer.

CIP data is on file at the Library of Congress
ISBN 978-0-19-069064-9

9 8 7 6 5 4 3 2 1

Printed by Sheridan Books, Inc., United States of America

Contents

Preface

GENERAL INTRODUCTION

Philosophy of science has the aim of answering those questions raised by scientific activity that are not directly addressed by science itself. Among such questions, we can mention: What are the overall goals of science, as well as the specific goals of its various branches? By what means are these goals pursued? What basic principles does it put into practice? Philosophy of science also tries to understand the relationships that exist between the scientific disciplines. To what extent, and in what sense, are they, and should they be, unified? Also belonging to its domain is the relationship between science and reality. What, if anything, does science tell us about reality? And to what extent is it justified in making the claims it does?

Just like the sciences themselves, current philosophy of science is multifaceted and specialized. A philosopher of science may embark on projects as diverse as the development of a formal analysis of the concept of confirmation using probability theory and the study of the potential contribution neuroscience may bring to our understanding of consciousness. Thus, it becomes difficult for both students and researchers within a given domain to be aware of the advances and challenges arising in any specific area in philosophy of science.

The aim of the present book is to expose the main questions, as well as some of the answers, being discussed in today's philosophy of science. We view it as the "missing link" between introductions and research, and our own goals will have been met if this book successfully bridges the gap between introductions to the philosophy of science meant for a general audience on the one hand, and research articles and monographs on the other. It is therefore primarily intended for the use of advanced undergraduate or graduate students who, after a first introduction to the area, may now wish to deepen their knowledge. We also hope that *The Philosophy of Science: A Companion* will be useful to both junior and senior researchers in philosophy of science wishing to familiarize themselves with areas outside of their own.

Philosophy of science has become too specialized for this goal to be achieved by any one person. Thus, our book is a collective effort. We have nevertheless endeavored to present the basic problems that shape contemporary philosophy of science in a coherent way. In contrast with encyclopedias, where contributions tend to simply coexist and thus lack organic unity, we have tried to maximize complementarity and cross-referencing between the chapters. Our hope is that this has favored a strong sense of unity, something that is always hard to attain in such collective undertakings.

Part I: General Philosophy of Science

The two parts of *The Philosophy of Science* mirror the traditional distinction *between general philosophy of science* and *philosophy of the special sciences*. General Philosophy of Science (Part I) deals with generic issues raised by scientific activity, independent of specific disciplines. General philosophy of science was the very core of philosophy of science up to the middle of the twentieth century. Philosophy of science itself has dramatically evolved over the last several decades, becoming increasingly devoted to issues raised by specific scientific disciplines. The study of general problems nevertheless remains a highly active element of philosophy of science. Moreover, it is our opinion that the study of these general problems is indispensable to those who focus on the philosophy of some particular scientific discipline or area, since they represent a set of tools invaluable to understanding their own, specific objects of study.

The objective of the first part of the book is twofold. We intend to both take stock of the traditional questions which have shaped analytic philosophy of science and to introduce certain problems that have been raised more recently. Thus the first two chapters, bearing upon explanation and confirmation, respectively, tackle issues that were the subject of intense debate in the middle of the twentieth century—notably among philosophers of science influenced by logical empiricism—and which, as we shall see, are still much studied today. With causality, chapter 3 also focuses on a traditional concept, though one to which logical empiricism has been rather hostile. Causality is now at the epicenter of a very vibrant area, straddling the borders of philosophy of science and metaphysics. Metaphysics is also at the heart of chapter 4, which deals with scientific realism (an issue that underwent a thorough overhaul during the 1980s) and the metaphysics of science, constituting a topic that is much discussed today. Chapter 5 addresses the issue of knowing how best to analyze some of science's primary products, namely theories and models. Starting from the "received view" of scientific theories, inherited from logical empiricism, it discusses the objections that have been raised against this view while also looking at alternative conceptions. Lastly, chapter 8 deals with issues surrounding the reduction and emergence of properties and/or theories coming from distinct scientific disciplines. Logical empiricism also contributed greatly to this research area. We shall see that current reflection on the matter is closely related to metaphysics, philosophy of knowledge, and sometimes also to the philosophy of the special sciences (particularly the philosophy of mind).

In our view, these six topics—explanation, confirmation, causality, scientific realism, the nature of theories and models, and reduction—constitute the core of general philosophy of science, even if they do not exhaust it. This latter consideration in mind, two further issues are also touched on in Part I. Chapter 6 studies the *diachronic* dimensions of scientific activity, a topic made famous by Kuhn's much celebrated book (*The Structure of Scientific Revolutions*, 1962/1970). Chapter 7 is more meta-philosophical in character: it reviews the relations between philosophy of science and other approaches (notably historical and sociological) which share in the aim of analyzing scientific activity and which are currently referred to as *sciences studies*. Although comprehensive, this does not cover all topics having a justifiable claim to the label of general philosophy of science. For instance, the growing literature on statistics and statistical reasoning is not represented. But it is our contention that Part I of *The Philosophy of Science* will provide the reader with a satisfyingly complete survey of contemporary general philosophy of science.

Part II: Philosophy of the Special Sciences

For several decades, philosophers of science have increasingly directed their attention toward the finer details of scientific activity, in particular to issues exclusive to specific disciplines. These issues are the object of the philosophy of the special sciences, to which the second part of *The Philosophy of Science* is devoted.

Compared with general philosophy of science, philosophy of the special sciences appears two-sided. Certain problems are essentially instances or applications of issues belonging to general philosophy of science. In this case, more often than not, the targeted area of knowledge requires some reconsideration of the issue on the part of the philosopher. For instance, the issue of justification or confirmation of theories raises specific problems when one studies, let's say, economic or mathematical theories, as opposed to theories from physics, which often serve to illustrate confirmation theories. By contrast, certain other issues in the philosophy of the special sciences are entirely generated by the specific concepts and methods of a given field. The discussions on the concept of function (in biology) or on the nature of linguistic universals (in linguistics) are two cases in point. The main objective of the second part of this volume is to introduce the reader to a representative sample of the issues that currently structure the philosophy of the special sciences. We have done our best to respect this two-sided character, i.e., to show how some of the issues are very closely linked to the "big" issues in general philosophy of science while others are specific to certain specialized domains of science.

The first two chapters of Part II are devoted to the philosophy of the formal sciences. More precisely, chapter 9 is concerned with logic and chapter 10 with mathematics. The philosophy of the formal sciences has often been left out of handbooks or textbooks on the philosophy of science. One of the reasons that implicitly underpins this state of affairs is that the issues raised by these formal sciences can seem remote from those raised by *bona fide* empirical sciences. But there are other reasons that speak in favor of integrating philosophical discussion on these disciplines. First, there is some interesting convergence between certain issues in the philosophy of the formal sciences and other issues in general philosophy of science, for example, those related to the nature of explanation. Second, there are certain other issues which call for a unified and coordinated answer from both the philosophy of the formal sciences and other branches within philosophy of science. For example, understanding why mathematics fits into the physical world so well—an issue that lies at the border between the philosophy of mathematics and the philosophy of physics. Or the problem of understanding mathematical cognition, which is of interest to both philosophy of mathematics and cognitive science.

Chapters 11 and 12 are devoted to the philosophy of physics and the philosophy of biology, respectively. These two areas have a special status in philosophy of science. Philosophy of physics is considered basic because physics is viewed as the fundamental scientific discipline. This means at least two things. First, that physics is an area where scientific reasoning is supposed to reach its zenith, and thus, in particular, that it is indispensable to be at least minimally familiar with it if one wishes to gain an understanding of scientific reasoning in general. And, second, that it is crucial to clarify the picture of the world as it is depicted by the physical sciences. Philosophy of biology has become an extremely active field, such that there is probably no other area in the philosophy of the special sciences whose importance has grown more over the last two decades.

An entire chapter is devoted to the philosophy of medicine. Our main reason for this is that philosophy of medicine is an area where philosophy of science overlaps with normative and practical philosophy. This reveals itself with respect to the question of whether the concepts of health and illness have an essential normative dimension, and also as regards the study of clinical reasoning. In both cases, the discussion goes beyond the purely epistemic point of view dominant in the philosophy of the natural sciences.

Another particular feature of Part II is the space we have devoted to philosophy of the human and social sciences (chapters 14 to 17). Interestingly, in these areas the philosopher's stance and corresponding expectations may differ from those that are generally endorsed in the philosophy of the natural sciences. In the former area, philosophers often assume that there is nothing wrong with the way science is done and thus refrain from making recommendations to scientists or from criticizing their methods. Not so in the latter case, and this is to be expected, since there are far more methodological uncertainties, debates, and disagreements involved in the human and social sciences.

Chapters 14 and 15 broach the social sciences. Chapter 14 deals with general issues in the philosophy of the social sciences, for example, methodological individualism and the relations between social sciences and cognitive sciences. Chapter 15 focuses on one specific social science, economics. This emphasis is to be welcomed, in light of the scientific and social impact of economics, and all the more so since it currently constitutes a particularly active field of study for philosophers.

The last two chapters are organized in a similar way. Both are devoted to disciplines that study human cognition. Chapter 16 is a general presentation of the issues raised by cognitive science from the point of view of philosophy of science. Chapter 17, on the other hand, bears on one specific discipline—linguistics. While philosophy of language is a well-structured and well-known area in philosophy, there are relatively few philosophical discussions on linguistics as a science. Both for this reason and for the fact that the philosophy of cognitive science focuses more on disciplines like psychology and neuroscience, we deemed it fitting to devote a whole chapter to linguistics.

Acknowledgments

We are grateful to the contributors to this volume, to Thierry Martin (the editor of the series in which an earlier version of *The Philosophy of Science* was published, under the title "Précis de Philosophie des Sciences" in 2011), to Daniel Andler who brought financial support to this initial version through his Senior Fellowship of the Institut Universitaire de France and to Christopher Robertson, who translated many of the chapters' earlier versions. The current version has benefited from the comments of two anonymous referees. It was notably supported by the Institute of Cognitive Studies at Ecole Normale Supérieure (Paris) under grant ANR-10-LABEX-0087 IEC and by Mikaël Cozic's Junior Fellowship of the Institut Universitaire de France. Lastly, we wish to thank the Institut d'histoire et de philosophie des sciences et des techniques (UMR 8590, Paris I—ENS Ulm—CNRS), which has provided us with a highly stimulating scientific environment for 15 years.

Anouk Barberousse, Denis Bonnay, and Mikaël Cozic, Paris, January 2018.

About the Contributors

Daniel Andler is a philosopher of science focusing on cognitive science. His interests include the philosophical issue of naturalism, the impact of cognitive science on the social sciences and their relevance for societal issues like education and public policy, and artificial intelligence. He is professor emeritus at the Sorbonne and a member of the Académie des Sciences Morales et Politiques.

Anouk Barberousse is a professor of philosophy of science at Sorbonne Université, Paris, where she teaches general philosophy of science, philosophy of physics and philosophy of scientific expertise. She has recently written on the epistemology of computer simulation, the philosophy of probability, and the role of databanks in our knowledge of biodiversity.

Denis Bonnay is an assistant professor in philosophy at Université Paris Nanterre, working in logic, philosophy of science and philosophy of cognitive science. His research ranges from works on the nature of logic and the boundaries between logic and mathematics to studies on judgment aggregation and group beliefs.

Mikaël Cozic (Paris-Sorbonne University, PhD, 2005) is an assistant professor at the Paris-Est University, a researcher and head of the group "Decision, rationality and interaction" at the Institut d'Histoire et de Philosophie des Sciences et des Techniques and a member of the Institut d'Universitaire de France. He studied philosophy (Ecole Normale Supérieure de Paris, 1997–2002), logic (University Paris-Diderot, MSc, 2002), and cognitive science (Ecole des Hautes Etudes en Sciences Sociales, MA, 2001) in Paris. Professor Cozic's research focuses primarily on philosophy of economics and formal theories of rationality. His current research concerns the relationship between cognitive science and positive and normative economics, as well as several issues in Bayesian epistemology, including the revision of one's beliefs upon learning the opinion of others.

Jacques Dubucs is a senior scientist at the Centre National de la Recherche Scientifique and the head of the Social Sciences and Humanities Department at the French Ministry of Higher Education, Research, and Innovation. His scientific work deals with logic and philosophy of science.

Paul Égré (born 1975; PhD, 2004) is directeur de recherche at Institut Jean-Nicod (CNRS) and an associate professor in the Philosophy Department of Ecole Normale Supérieure in Paris. Besides work in formal semantics and on the epistemology of linguistic theory, a large part of Paul Egré's work over the last decade has been on the topic of vagueness in language and in perception, dealing with semantic, logical, and psychological aspects of the phenomenon. Since 2012, Egré is also the editor-in-chief of the *Review* of *Philosophy and Psychology*.

Jon Elster is the Robert K. Merton Professor of Social Science at Columbia University. He is the author or editor of more than thirty-five books translated into more than seventeen languages on the philosophy of social sciences, the theory of rational choice, political psychology, deliberative democracy, and the history of political thought (Marx and Tocqueville), to name a few of their subjects. He is currently working on a comparative study of the Federal Convention (1787) and the first French constituent assembly (1789–1791).

Michael Esfeld is full professor of science at the University of Lausanne. His research is in the metaphysics of science, the philosophy of physics, and the philosophy of mind. His latest book publication is *A Minimalist Ontology of the Natural World*, with Dirk-André Deckert (New York: Routledge, 2017).

Élodie Giroux is an assistant professor at Jean Moulin Lyon 3 University, where she teaches philosophy of science and philosophy of medicine. She is director of the master's in "Culture and Health." Her main research interests are the history and epistemology of "risk factor epidemiology"; causation in medicine and public health; and risk, health, and disease concepts. She is currently working on precision medicine. Besides several papers on modern epidemiology, she published *Après Canguilhem, définir la santé et la maladie* (Paris: PUF, 2010) and *Naturalism in the Philosophy of Health* (Cham: Springer, 2016), and she edited a special issue on the history of risk factor epidemiology in *Revue d'Histoire des Sciences* (2011) and on precision medicine in *Lato Sensu* (2018).

Max Kistler is professor at the Department of Philosophy at University Paris 1 Panthéon–Sorbonne and head of IHPST (Institut d'Histoire et de Philosophie des Sciences et des Techniques). His research topics include causation, powers and dispositions, laws of nature, natural kinds, and reduction. He is the author of *Causation* *and Laws of Nature* (Routledge, 2006), *L'esprit matériel. Réduction et émergence* (Ithaque, 2016), and coeditor (with B. Gnassounou) of *Dispositions and Causal Powers* (Ashgate, 2007).

Hélène Landemore is an associate professor of political science at Yale University. She is a political theorist interested in democratic theory, theories of justice, Enlightenment thinkers, and the philosophy of social sciences. Her book *Democratic Reason* (Princteon, NJ: Princeton University Press, 2013) was awarded the 2015 David and Elaine Spitz Prize for best book in liberal and/or democratic theory published two years earlier. She is currently writing a new book on postrepresentative or "open" democracy.

Maël Lemoine is a professor at the University of Bordeaux, France, where he teaches philosophy of medical science. He published an introductory essay in the philosophy of medical science in 2017 and has recently published various articles on biological research in psychiatry, animal models, and precision medicine.

Pascal Ludwig is an associate professor in the Department of Philosophy, Sorbonne Université, Paris. He has coauthered several books on the philosophy of science and the philosophy of the mind.

Thomas Pradeu is a CNRS senior investigator in philosophy of science (permanent position) at ImmunoConcept (CNRS and University of Bordeaux), and associated member at the Institut d'Histoire et des Philosophie des Sciences et des Techniques (CNRS and University Pantheon–Sorbonne). His research focuses on biological individuality, immunology, the microbiota, and the interactions between philosophy and science.

Philippe de Rouilhan is a senior researcher emeritus at the CNRS and a member of the Institut d'Histoire et de Philosophie des Sciences et des Techniques (CNRS and Université Panthéon–Sorbonne), of which he was the director for a long time. His work pertains to logic lato sensu or, more specifically, to formal ontology, formal semantics, philosophy of logic, philosophy of mathematics, and philosophy of language. He is currently preparing a book on truth, logical consequence, and logical universalism.

Marion Vorms is a lecturer (*maître de conférences*) in philosophy at University Paris 1 Panthéon–Sorbonne and a Marie Curie fellow at Birkbeck College, London, psychology department. Her past work in philosophy of science concerns the nature of scientific theories and representations. Her new project, which is at the crossroads of epistemology and the psychology of reasoning, bears on the notion of reasonable doubt; she is particularly interested in judicial reasoning and decision-making.

1

General Philosophy of Science

1

SCIENTIFIC EXPLANATION

Denis Bonnay (Université Paris Nanterre, IRePh & IHPST)

WHY IS NICOLAS angry? Because he thinks Dominique wanted to play a nasty trick on him. Why was Gomorrah destroyed? Because God wanted to punish its inhabitants. Why did the dinosaurs disappear? Because a giant asteroid crashed into the earth. In asking the question "why?" we bring a real or reputed fact—Nicolas's anger, the destruction of Gomorrah, dinosaur extinction—to the attention of our interlocutor, and we ask for an explanation of that fact. These explanations may rely on simple everyday knowledge: it is well known that people do not like having nasty tricks played on them. Explanations can be of the religious sort: the biblical account tells not only of Gomorrah's existence but also of the sins of its people, going on to explain the destruction of the city by an act of divine retribution. And then there are the explanations offered to us by science: thus, the extinction of the dinosaurs being one of the enigmas that paleontology faces, an asteroid strike is one of the explanations put forward.[1]

More than just a simple side issue of scientific activity, explanation takes its place as one of the specific goals of science. Of course, as we have just seen, it is not just science that claims to offer explanations. And, conversely, science certainly has goals other than explanation too. Science enables us to describe and classify phenomena, as well as

[1] I thank Anouk Barberousse, Mikaël Cozic, Henri Galinon, Marion Vorms, and Kenneth Waters for various discussions, comments, and re-readings, which were of help to me. I also wish to thank Christopher Robertson, who translated the French version. This work received funding from the ANR (The IHPST's Logiscience program) and from the Institut de Recherches Philosophiques (Université Paris Nanterre). The survey on theories of explanation is also obviously indebted to some other, similar enterprises, in particular the surveys by Salmon (1989) and Woodward (2009).

enabling us to predict and control them. Nevertheless, one of the motivations, be they individual or collective, to "do science" in the first place seems to be to find explanations that cannot be found elsewhere—for example, research on electricity and magnetism, and also work on the electromagnetic theory, that is developed to explain a group of mysterious phenomena such as static electricity, the properties of Magnesia stones, or lightning and its effects. In contrast, it is not easy to imagine what sort of thing a scientific theory that explained nothing would be. A strict typology, say a botanical classification of different plant species according to their phenotype for example, doesn't strike us as being a bona fide scientific theory, insofar as it lacks any explanatory power.

Not lacking, however, are opponents to the idea that the aim of science is to provide explanations. Pierre Duhem, in *The Aim and Structure of Physical Theory*, opposes the idea that the object of a scientific theory is to explain a set of observable regularities, an opinion shared by other physicists of his time such as Ernst Mach. But this refusal is primarily grounded in Duhem's own concept of explanation. To explain would be "to strip reality of the appearances that envelop it like a veil, in order to see the bare reality itself" (Duhem, 1908); Duhem considers that attaching an explanatory ambition to science makes it subservient to metaphysics, the only domain to claim possession of the keys to the ultimate essence of things.[2] The approach that we will follow here is not quite the same. In determining whether science provides explanations or not, we will not start out with some overly demanding concept of explanation. We will set out from the intuition that science provides explanations, and we will try to identify a concept of explanation such that this concept would enable us to account for the explanatory power of science.

What can be expected from this line of enquiry? What goals are we pursuing? In a good concept of explanation, we expect first of all that it be adequate; that is, that it will allow us to understand which elements provided by science constitute explanations and by what virtue they come to possess their explanatory power. For example, if an explanation has some epistemological virtue, in that it allows us to "understand what is happening," then a good concept of explanation must tell us how scientific explanations allow us to "understand what is happening." We would hope then, off the back of this, to be in a position to evaluate explanations, that is to say, to have the capacity to distinguish between good and bad explanations. An analysis of the concept of explanation will obviously not tell us if the explanation is right, in the sense of its expressing truth, but it should be able to tell, or at least indicate to us, whether some explanation would be a good explanation, presuming that it does express the truth. And lastly, we would like some insight regarding the relationship between the explanatory aim of science and its other aims—prediction, control, and so on.

We will begin by looking in detail, during the first section, at the theory of scientific explanation proposed by Hempel and Oppenheim known as the deductive-nomological model (DN). The importance of place we give it here is justified conceptually by the rigor of the analysis it proposes and historically by the role of cardinal reference it

---

[2] On the question of realism—does science give us access to the very nature of things or not?—and on the metaphysical scope of science, see chapter 4 of the present volume.

continues to play in contemporary debates on explanation, despite its no longer being the dominant model. In the second section, and in light of the DN model, we will revisit the general properties of explanation, discussing the link between explanation and prediction, the temporal conditions that weigh, or do not weigh, on explanation, as well as the characterization of the laws of nature. The third section is devoted to an examination of the classic objections brought against the DN model, these taking the form of a list of counter-examples. The main rival theories that have emerged to resolve these problematic examples in the DN model's stead—causal theory and unificationist theory—are presented and discussed in the fourth section. In the closing section, we will sketch out some other approaches toward contemporary reflection on explanation.

1. The Deductive-Nomological Model

1.1 TO EXPLAIN IS TO DEDUCE FROM A LAW

Let us begin then by looking at the inaugural example given by Hempel and Oppenheim (1948). A mercury thermometer is rapidly immersed in a basin of hot water. The level of the mercury column falls slightly at first before rising swiftly. Why? Here we have a little puzzle to solve—we were expecting that the level of the mercury would simply rise, though this is not exactly what has happened. In fact the explanation is quite simple. The rise in temperature, at first, affects only the standard quality glass tube which contains the mercury. Expanding, the tube leaves more room for the mercury, whose level promptly drops. Then, rapidly, the heat spreads out and the mercury expands in turn. As its coefficient of expansion is much higher than that of glass, the mercury level rises and exceeds its own initial level.

Analyzing this example makes the distinction between the *explanandum*, what is to be explained, namely the slight decrease followed by rapid rise in the level of the mercury, and the *explanans*, which does the explaining, immediately clear. Under *explanans* we see, first, the initial conditions, the particular facts reported in the explanation, such as the set-up involved—the glass tube, the mercury column, the bowl of hot water—and the act of immersing the tube in hot water itself. Then too, the general laws come into effect, such as the laws governing the thermal expansion of glass and mercury, and a statement regarding the relatively low thermal conductivity of glass. The *explanandum* is subsumed under the general laws, in the sense that it can be deduced from these laws and the initial conditions.

Hempel and Oppenheim's theory is that the full generality of scientific explanation can be read in this particular case. To explain, one need not do anything other than deduce the phenomenon to be explained by using general laws and the initial conditions, which justifies the labeling of their model as the deductive-nomological (DN) model of explanation. Thus, the general form for scientific explanation that we draw from Hempel and Oppenheim is as follows:[3]

[3] The double-lined bar ==== indicates that the statement below follows on logically from those statements above it.

$C_1, \ldots, C_k$	Initial conditions	*Explanans*
$L_1, \ldots, L_l$	General laws	
=============		
E	Empirical phenomenon to be explained	*Explanandum*

For there to be explanation, certain conditions must be met by the *explanans* and by the *explanandum* (the *explanandum* is a statement describing the phenomenon to be explained, the *explanans* is a set of statements describing the initial conditions and the laws involved):

Logical Conditions of Adequacy

(R1) The *explanandum* must be a logical consequence of the *explanans*.

(R2) The *explanans* must contain general laws whose presence is necessary for the *explanandum* to be a logical consequence of the *explanans*.

(R3) The *explanans* must have empirical content.

Condition of Empirical Adequacy

(R4) The statements making up the explanans are true.

The logical conditions of adequacy are purely formal. They specify the properties of the *explanans* and of the *explanandum*, which do not depend on the actual state of the world. This is not the case with the condition of empirical adequacy, which states that a supposed explanation is not truly an explanation unless one additional condition is satisfied: the statements contained in the *explanans* must be true. (R1) and (R4) together imply that the statement, which is the *explanandum*, is also true.

Condition (R1) carries the full weight of the analysis. When we are given the explanation of a phenomenon, we understand why this phenomenon occurred, in the sense that we have an argument that shows that it was to be expected that the phenomenon would occur (see Hempel, 1965b, p. 337). Salmon (1989) summarizes this point by saying that the essence of scientific explanation, according to Hempel, lies in *nomic expectability*.[4] The initial conditions being in place, the phenomenon could only but occur, since it follows on logically from the initial conditions using general laws.

Note that Hempel's model does not leave room for the common idea that to explain is to explain surprising or unfamiliar phenomena by reducing them to facts and principles with which we are already familiar (Hempel, 1966). To explain is to bring everything back to laws. If these laws are familiar, then the explanation will equal reduction to the familiar, but this is not necessarily the case. An example of the first sort of explanation would be the kinetic theory of gases: the behavior of the molecules of a gas, with which we are not familiar, is explained by subsumption under laws that also apply to the movements of things with which we are familiar, such as billiard balls. But science is overflowing with examples of the second sort. Very often, familiar

[4] In this context, *nomological* simply means "relative to the laws of nature."

phenomena are explained by less familiar things, such as when we explain the range of colors of the rainbow, with which we are very familiar, using the laws of reflection and refraction of light, with which we are certainly less familiar. That the proposed model of what a scientific explanation is does not imply that these explanations work by reduction to the familiar is a good thing if it is simply not true that all scientific explanations work by reduction to the familiar.

Condition (R2) enables the distinction of scientific explanations from pseudo-explanations. Carnap (1966) explores the example of the vitalist theories of German biologist and philosopher Hans Driesch. Driesch proposed explaining the various phenomena of life by means of the notion of entelechy. The *entelechy* is "some specific force that makes living beings behave in the way they behave." The various levels of complexity in organisms correspond to various types of entelechies. What we call the spirit of a human being is nothing other than a part of its entelechy. It is this same entelechy, the vital force, that explains, for example, that skin heals over after an injury. To those who criticize the mysterious nature of the concept of entelechy, Driesch replies that it is no more mysterious than the concept of force used in physical theory. Entelechies are not visible to the naked eye, but electromagnetic force is no more observable—in both cases, we see only the effects. But, as Carnap highlights, there is a crucial difference between Driesch's entelechies and the forces of physics. The concept of force used by physical theories is called on from within a set of laws, whether this be the general laws of motion and the law of gravitation in regards to gravitational force, or Coulomb's law when regarding electrical force. If the concept of force has explanatory virtue, in the sense that it can be included in scientific explanations, such as the explanation of an eclipse based on the antecedent position of celestial bodies, the laws of motion, and the law of gravitation, then it is precisely because it plays a crucial part in the formulation of these general laws. No such thing occurs in the case of the entelechy: there are no laws of the entelechy. Driesch offers many zoological laws that are indeed bona fide laws, but the concept of the entelechy is nowhere to be seen, it appears at the end as something of a *deus ex machina* expected to explain away the mystery of life. For Carnap this firmly establishes that entelechy explanations are mere pseudo-explanations, so that a virtue of Hempel's analysis of scientific explanation is precisely that it allows us to establish this.

Condition (R3) means that the statements in the *explanans* can be tested, at least in principle. It is redundant if the *explanandum* is indeed an empirical fact, since in that case the very fact that the *explanandum* is a consequence of the *explanans* enables it to be tested. Its inclusion alongside (R1) and (R2) is no doubt a sign of Hempel and Oppenheim's resolutely empiricist mindset.

Condition (R4) makes the concept of explanation an objective one. Without (R4), the concept of explanation is relative to a theoretical framework. The flaming of a match can be deduced from the presence of phlogiston[5] and the law dictating that phlogiston

[5] In the chemical theory preceding Lavoisier's modern theory, phlogiston was a hypothetical substance supposedly found in all flammable materials and would dissipate into the air during combustion, thus explaining the decrease in mass observed subsequent to combustion.

is released under certain circumstances, causing the phenomenon of combustion. The modern theory of combustion, which explains the same phenomenon from the recombination of various elements with oxygen, provides another explanation. In a relativistic perspective, we would say that these are two explanations for one and the same phenomenon: two explanations existing in two distinct theoretical frameworks, one where the laws of combustion grant pride of place to phlogiston, and another where the laws of combustion accord this honor to oxygen. But if what we want from the concept of explanation is that it be an objective one, then this is clearly not satisfactory. The explanation proposed by Lavoisier is not merely some other explanation for combustion, rather it replaces the phlogistic explanation, the latter no longer to be considered a genuine explanation. Subscribing to this way of seeing things, which is undoubtedly the way of seeing things that would come naturally to scientists, implies having an objective concept of explanation. It is just such a concept that the addition of condition (R4) provides.

The deductive-nomological model is generalized out in two directions. First, the *explanandum* need not necessarily be a particular event, it can also be a law, explained by means of more general laws from which it is derived. This possibility is brought about by the characterization given by Hempel and Oppenheim, since, although the inclusion of initial conditions in the *explanans* may not be strictly required, the inclusion of laws is. The canonical example of this kind of explanation is the derivation of Kepler's laws of planetary motion from the general laws of motion and the law of universal gravitation. A thorough examination of this kind of explanation nevertheless uncovers a set of problems of its own, hidden in the requirement that the laws contained in the *explanans* be more general than the law to be explained.[6] Note that, as before, this explanation clearly shows us that it was to be expected that the planets would move according to the laws set forth by Kepler, since these laws are in fact a consequence of the law of gravitation, by way of the general laws of motion.

1.2 GENERALIZING OUT TO PROBABILISTIC EXPLANATIONS

Second, certain scientific laws liable to arise within explanations are statistical laws,[7] which do not enable us to deduce a particular phenomenon with absolute certainty,

[6] Hempel and Oppenheim (1948, note 28) make the following remark. From the conjunction *K* & *B* of Kepler's laws and Boyle's law, one can derive both Kepler's laws *K* and Boyle's law *B*. But this derivation is not explanatory. Subsuming *K* and *B* under the simple conjunction *K* & *B* does not in any way constitute an advancement in regards to explanation, as opposed to the derivation of Kepler's laws from Newtonian principles. The formulation of the unificationist theory of explanation given in section 4.2 aims, among other things, at resolving this problem.

[7] A statistical law does not tell us that an event will always occur under certain conditions but that under certain conditions an event has a certain probability of occurring. For example, the law that the nucleus of a tritium atom has a three in four chance of disintegrating after 24.6 years is a statistical law. A probabilistic explanation is the explanation of a phenomenon that is based on the probability that is ascribed to this phenomenon.

but simply enable us to ascribe it a high probability. Here is an example taken from Salmon (1989). The ratio of carbon 14 to other carbon isotopes in a piece of wood found on an excavation site is equal to half the same ratio in the atmosphere. Why? Because this piece of wood comes from a tree that was cut down about 5730 years ago and the half-life of carbon 14 is 5730 years. The proportion of carbon 14 in the atmosphere remains constant due to cosmic radiation. The tree absorbs carbon from the atmosphere while it is alive, but the chopped timber does not, and so the percentage of carbon-14 decreases due to radioactive decay. The general form of this kind of explanation is as follows:

$C_1, \ldots, C_k$	Initial conditions	*Explanans*
$L_1, \ldots, L_l$	Laws (including statistical laws)[*r*]	
=============		
E	Empirical phenomenon to be explained	*Explanandum*

where the laws $L_1, \ldots, L_l$ (notably, in our example, the law establishing the half-life of carbon-14) and the initial conditions $C_1, \ldots, C_k$ (notably, in our example, the date on which the wood was cut) enable us to infer E (in our example, that the ratio of carbon-14 isotopes in the wood sample is equal to half the atmospheric ratio) with probability r which must be high. Note that here the probability is assigned to the inductive inference, and not to the *explanandum*. What is explained is that the ratio has been halved, which is neither probable nor improbable—it is quite simply true. The explanation given is a statistical explanation insofar as the phenomenon to be explained is not a logical consequence of the *explanans*, it doesn't "definitely" result from it, but only with a certain probability. It seems natural to demand that this probability be high since, otherwise, the *explanans* wouldn't provide us reason to expect that things should have occurred as they did; that is to say that it wouldn't have provided us reason to expect that the *explanandum* be true. Based on this, it is tempting to modify the conditions of adequacy for the deductive-nomological explanation to the explanation Hempel calls inductive-statistical (IS) in the following manner:

Logical conditions of adequacy	
(R′1)	The *explanandum* must follow on from the *explanans* with strong inductive probability.
(R′2)	The *explanans* must contain at least one statistical law whose inclusion is necessary if we are to be able to derive the *explanandum*.
(R′3)	The *explanans* must have empirical content.
Condition of empirical adequacy	
(R′4)	The statements making up the *explanans* are true.

In light of conditions (R1) and (R′1), the common point between the two types of explanation appears clearly. In both cases, nomic expectability is at the heart of the explanation. As Hempel puts it,

> Any rationally acceptable answer to the question 'Why did event X occur?' must offer information which shows that X was to be expected—if not definitely, as in the case of DN explanation, then at least with reasonable probability. Thus the explanatory information must provide good grounds for believing that X did in fact occur; otherwise that information would give us no adequate reason for saying, "That explains it—that does show why X occurred." (1965b, pp. 367–368)

However, inductive-statistical explanation poses some problems of its own. Let's consider another simple example, taken from Hempel (1965b). John Jones is suffering from a strep infection, he is treated with penicillin and he recovers. Let's imagine that 95% of strep infections are cured by penicillin. We can then explain John Jones's swift recovery in the following manner:

P(G | S and P) = 0.95	Statistical law	*Explanans*
Sa and Pa	Particular fact[0.95]	
=============		
Ga	Empirical phenomenon to be explained	*Explanandum*

where S stands for 'suffering from a strep infection,' P for 'treated with penicillin,' a for 'John Jones,' and G for 'get better.' P(G | S and P) is a conditional probability; it's the probability of G knowing that S and P (thus, in this instance, the probability of getting better knowing that the patient is suffering from a strep infection and is being treated with penicillin). Now, here's the problem. Certain strains of streptococcus are resistant to penicillin; in these cases the probability of getting better if treated with penicillin is very low. So if the specific strain that has made John Jones ill is a resistant strain, we can explain that John Jones doesn't get better in the following manner:

P(~G | S and P and R) = 0.95	Statistical law	*Explanans*
Sa and Pa and Ra	Particular fact[0.95]	
=============		
~Ga	Empirical phenomenon to be explained	*Explanandum*

where R means "infected by a resistant strain."

So it seems just as possible to explain Jones's getting better, if he got better, as his not getting better, if he didn't. We're confronted here with what Hempel calls the ambiguity of inductive-statistical explanations. Two logically compatible *explanans*—which can both be true at the same time—can be used to infer, with a very high probability, one thing and its contrary (in our example, Ga and ~Ga). This problem is unique to statistical explanations. It doesn't arise with the deductive-nomological explanations, since if two sets of statements are such that one allows the deduction of one statement and the other the negation of that statement, then the two sets in question are not logically compatible. But as we have just seen, this is not the case for probabilistic inferences.

The problem cannot be ignored. Of course, only one of the two statements "Ga" and "~Ga" is true, so that one would never be in a situation where Ga and ~Ga had to be explained simultaneously. But in the case where "Ga" is true, the counterfactual possibility of explaining ~Ga (had Jones not gotten better, we could have explained this by saying that the strain of bacteria must have been resistant) enters into direct conflict with the idea of 'nomic expectability.' Clearly it doesn't make sense to speak of a situation where we should simultaneously expect that Jones get better and that Jones not get better.

What should we make of these scenarios? If we know that Jones has a strep infection, and we don't have any other information regarding the nature of the infection, we must expect that Jones will get better, even if we can't completely rule out the possibility that he not get better, in the improbable case that he be unlucky enough to have picked up a resistant strain. If we know not only that Jones has a strep infection but also that he is carrying a resistant strain—because, for instance, an antibiogram has been carried out—then it must be expected that Jones will not get better if he is treated with penicillin. Whether the strain is resistant or not makes a difference to the outcome of the treatment. So since it is relevant, the information that the strain is resistant must, if available to us, be taken into consideration in determining what must be expected. Hempel's solution to the ambiguity problem in IS explanation takes pointed advantage of the intuition that it is necessary for all available relevant information to be taken into consideration. In the case of a statistical explanation of the form

$P(G \mid F) = r$	Statistical law	*Explanans*
Fb	Particular fact[*r*]	
============		
Gb	Empirical phenomenon to be explained	*Explanandum*

Hempel introduces what he calls the requirement of maximal specificity (RMS),[8] which can be stated in the following manner. Let S be the set of statements contained in the *explanans* and K the set of statements accepted at the time of the explanation,

> If the conjunction of S and K implies that b belongs to a certain class F_1 and that F_1 is a subclass of F, then the conjunction of S and K must also imply a statement specifying the statistical probability of G in F_1, say

[8] In inductive logic, Carnap (1950) introduced the *requirement of total evidence* according to which, "in the application of inductive logic to a given knowledge situation, the total evidence available must be taken as a basis for determining the degree of confirmation" (Carnap, 1950, p. 211).

$P(G \mid F_1)=r_1$

here r_1 must equal r, unless the probability statement just cited is simply a theorem of mathematical probability theory. (Hempel, 1965b, p. 400)

If r_1 does not equal r, this means that available and relevant information was not taken into account, since it is from here that the even more precise characterization of b being an F_1 ensues, a characterization that alters the situation regarding the probability of G's occurring. Conversely, when the requirement of maximal specificity is met, we know that all the available and relevant information has been taken into account, since the deployment of all our background knowledge S can tell us no more about the probability of b's being G.

We obtain the conditions of adequacy for IS explanations by adding a condition of empirical adequacy to the conditions (R′1) to (R′4) we already have:[9]

(R′5) The statistical law contained in the *explanans* satisfies the requirement of maximal specificity.

Coming back to the example of John Jones and the strep infection, "P(G | S and P) = 0.95" can be contained in the *explanans* only if we do not know that Jones is carrying a resistant strain. Indeed, since P(G | S and P) and P(G | S and P and R) are, for empirical reasons, completely different values, the requirement of maximal specificity is violated if the statements that we accept imply that Jones belongs to the subclass "S and P and R" of "S and P." Note that P(G | S and P and G) = 1—this is an elementary law of probability calculation. So in the case where we know that Jones got better, without knowing that he was carrying a resistant strain, the requirement of maximal specificity would nevertheless risk not being satisfied since "S and P and G" is a subclass of "S and P" and P(G | S and P) and P(G | S and P and G) have different values. The function of the final clause, "unless the probability statement just cited is simply a theorem of mathematical probability theory," is precisely to eliminate trivial counter-examples of this sort.

Finally, note also that the addition of the condition of adequacy (R′5), in which the set K of statements accepted at the time of explanation appears as a parameter, introduces an important difference between DN explanation and IS explanation. While DN explanation is purely objective—the conditions of adequacy make no reference to our knowledge state—IS explanation has an irreducibly subjective element—since the fact that the *explanans* satisfies or doesn't satisfy the requirement of maximal specificity depends on what we know. In this regard Hempel speaks of an epistemic relativity of statistical explanation.

---

9 This condition of adequacy is genuinely empirical, since it depends on our knowledge state, and thus on the state of the world insofar as the fact that our knowing or not knowing something is, in a broad sense, a fact of the world. To highlight that the only facts on which that condition depends are facts about what we know, we could speak, as Salmon does (1989), about an *epistemic* condition of adequacy.

We can sum up all of the above by drawing out the four types of explanations identified by Hempel in the following table, once again from Salmon (1989, p. 9):

TABLE 1

Types of explanations

Explananda Laws	Particular Facts	General Regularities
Universal laws	DN explanation (deductive-nomological)	DN explanation (deductive-nomological)
Statistical laws	IS explanation (inductive-statistical)	DS explanation (deductive-statistical)

The deductive-statistical explanations, of which we have not explicitly spoken, correspond to those cases where a general statement is derived from laws (like in the DN explanations of general statements), but where the statement in question concerns a statistical regularity.

2. The Properties of Explanation (Following the DN Model)

2.1 A GENERAL MODEL OF SCIENTIFIC EXPLANATION

Let's return, to complete this presentation of the deductive-nomological theory of explanation, to some of its stand-out characteristics. First, this is a general model of what a scientific explanation is. When we answer the question asking why Nicolas is angry by saying it's because Dominique wanted to play a nasty trick on him, we don't give any law to support what we are saying. Such an explanation, measured against the deductive-nomological approach, is at best incomplete and at worst incorrect. Incomplete if it is possible to complete it with some general law, in this instance a statistical law of human psychology according to which people very probably get angry when others attempt to do them wrong. Incorrect if no such law exists, for example because a scientific categorization of mental states would not recognize anger as being a homogeneous psychological state. The DN model is thus truly a model of scientific explanation, insofar as discovering the laws of nature is a properly scientific activity. Further, this model is general to the extent that, as Hempel and Oppenheim (1948, §4) first highlighted, it is called on to be applied not only to the physical sciences, from which its first examples are admittedly taken, but to the empirical sciences in total, thus also including the social sciences. A science can be said to produce explanations only to the extent that it be able to subsume phenomena under certain laws. For example, in psychology, it is possible to explain why an individual may not be able to distinguish, in terms of weight, between two objects, one weighing 10 kg and the other 11 kg, by calling, first, on the fact that this same individual is not able to distinguish, by their weight, between an object weighing 1 kg and another weighing 1.1 kg, and second, on the Weber-Fechner law that links sensation felt to the logarithm of the stimulus's intensity, this implying that the relative differential threshold is a constant. Of course, it could just be that it is particularly difficult to state psychological laws with all the precision and generality required, meaning that explanations in psychology are more often approximate or partial than their counterparts in physics.[10] Nevertheless, the benchmark for explanation, subsumption under laws, remains the same.

Yet it certainly seems that the sciences differ in the types of explanation they produce. There are mechanical explanations in physics, for example the explanation of the movement of billiard balls. There are no mechanical explanations—not of that type at any rate—in economics. Conversely, there are teleological explanations (explanations that call on the ends pursued by agents) in psychology and in economics. For example, in economics, the behavior of companies in a monopoly situation or in a competitive situation is explained by their drive to maximize profit. There are no teleological explanations in physics. But if Hempel and Oppenheim are right, these differences can be entirely understood as differences concerning the laws of the sciences in question. The DN model does not exclude teleological explanations, no more than it favors mechanical explanations or indeed any other type of explanation. Simply put, the DN model dictates that we cannot explain the behavior of an agent by appeal to the goals they are pursuing unless some general laws exist linking goals and behavior. As long as such general laws exist, teleological explanations in economics or in psychology are explanations in the DN model sense. Let's go back to the example of monopolies to see how a teleological explanation can constitute a bona fide explanation. The *explanandum* is that when a competitive industry is replaced by a monopoly, the prices increase and the production decreases. In a competitive situation, the equilibrium price corresponds to the intersection of the demand curve, which gives the sale price as a function of the quantity sold, and the marginal cost curve (aggregated for the industry), which gives the cost of the last unit produced as a function of the quantity produced. In a monopoly situation, the company is not subordinated to the market price, and is thus free to fix its price and act directly on the demand curve, meaning it can increase its profits by selling less but at a higher price. The equilibrium situation corresponds to the intersection of the marginal revenue curve, which gives the difference in total revenue as a function of the quantity sold, and the marginal cost curve, since as long as the company continues to produce at a cost lower than the revenue taken from sales, it increases its profit. The marginal revenue curve decreases faster than the average revenue curve, so that, at equilibrium, prices are higher and production quantity lower in monopolistic cases than in cases of competition. This is

[10] The Weber-Fechner law, the formulation of which is contemporary to the birth of psychophysics, is itself a law whose validity is considered as being only approximate. It is generalized by Stevens's law, according to which sensation is related to stimulation by a power law.

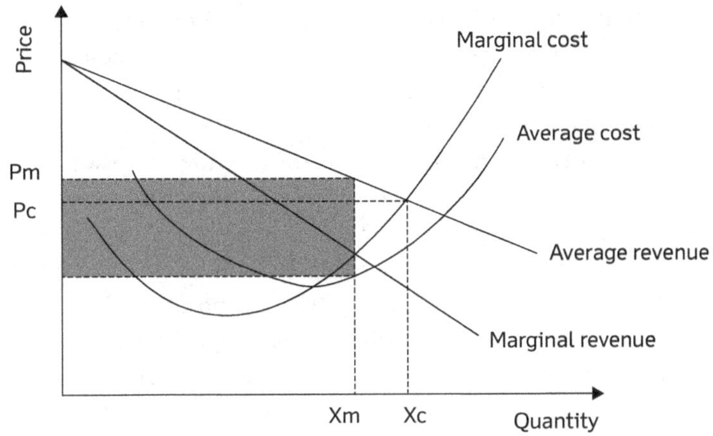

FIGURE 1 Price determination in a monopoly and in a competitive market[1]

[1] At equilibrium, the price P_m in a monopoly situation is higher than the price P_c in a competitive situation, and the quantity produced X_m in a monopoly situation is lower than the quantity produced in a competitive situation. The shaded surface represents profit.

Source: Wikipedia, License Creative Commons Attribution ShareAlike 3.0

where the hypothesis that companies seek to maximize their profits comes into play in determining the equilibrium: the quantity of goods produced by the monopoly is the quantity at the intersection of the curves of marginal revenue and marginal cost, since any other level of production would lead to reduced profits, and the company wants to maximize its profits. This is quite clearly a teleological explanation. The explanation is teleological because the principle of profit maximization informs us on what the economic agents want to do. And it is indeed an explanation because this principle is used as a law that enables, along with other laws, the derivation of a phenomenon to be explained, in this instance the negative effect monopolies have on price and production.

2.2 EXPLANATION AND PREDICTION

The DN model is a general model for scientific explanation based on, as we have seen, the idea of nomic expectability. A phenomenon is explained in so far as it has been shown that it was to be expected that it occur. This brings us to a second important property of the DN model, the symmetry between explanation and prediction. There is symmetry to the extent that the difference between explanation and prediction appears as being purely relative to our epistemic state. If a fact F is already known, its derivation from particular laws and circumstances is an explanation. If a fact F is not known, but the particular laws and circumstances are, the same derivation is a prediction. This symmetry leads to what Hempel calls the thesis of structural identity (Hempel et Oppenheim, 1948, Hempel, 1965b) which can be presented as two sub-theses. On the one hand, every adequate explanation is potentially a prediction, and on the other, every adequate prediction is potentially an explanation.

Hempel (1965b) discusses an objection Scriven (1962) brings against the thesis of structural identity, an objection which more specifically targets the first sub-thesis.[11] Scriven considers the example of a metal bridge which collapses. The collapse could have been brought about by overloading, by external damage, or by metal fatigue. The load weighing on the bridge at the moment of its collapse was normal, and a meticulous inspection revealed that no external damage had been caused to the bridge's structure. The investigators reached a conclusion of fracture by fatigue. Yet even though metal fatigue explained the collapse of the bridge, it couldn't have been used to predict this collapse. By assumption, there is no other sign of the excessive weakening of the metal than the collapsing of the bridge. When, as is the case here, the only reason we have to subscribe to one of the elements of the explanans resides in our acceptance of the *explanandum*, an adequate explanation does not, Scriven explains, have any value for potential prediction. Hempel's response is simple and, it seems to us, convincing. An adequate explanation is a good prediction only when certain epistemic conditions are satisfied—that is, when the statements in the explanans are known and the *explanandum* is not. In Scriven's bridge scenario, these conditions are far from being met, since one of the statements in the explanans cannot be known unless the statement making up the *explanandum* is. The thesis of structural identity has the following counterfactual consequence: had we known, independently, that the metal had been weakened to breaking point, then we would have been in a position to predict that the bridge was going to collapse. However, this counterfactual conditional is indeed true, to the extent that, by assumption, laws of physics assure us that excessive metal fatigue is sufficient to cause the collapse of the bridge. So Scriven's example is not in fact a counter-example to the thesis of structural identity. This response is illuminating in that it brings precision to the relationships between explanation and confirmation.[12] Explanation and confirmation do not generally go in the same direction. The function of explanation is not to assure us of what is to be explained: the phenomenon to be explained is supposed to be known. Very often the *explanandum* can, on the contrary, contribute to confirming the elements contained in the explanans, particularly the general laws. Scriven's bridge scenario is simply a borderline case where an element of the explanans—in this instance a specific circumstance, the fatigue in the metal the bridge is made of—has only the *explanandum* as empirical support.

[11] The second subthesis is only correct if every prediction is based on a law, which is not entirely evident. We can predict that the sixth egg out of the box will turn out to be rotten if the first five were ruined without it seeming necessary to call on a law and without that prediction potentially constituting an explanation for why the sixth egg is rotten. Hempel (1965b) suggests that, for cases such as this, the prediction is correct only if we can present statistical laws that would validate the probabilistic inference that the sixth egg is rotten. Otherwise, Hempel concedes the problematic nature of the second subthesis, which is not, contrary to the first, inseparable from his theory of explanation.

[12] The next chapter of the present volume is dedicated precisely to an analysis of the concept of confirmation.

2.3 THE TEMPORALITY OF EXPLANATION

Whether we consider our general discussion of the criteria of adequacy or the more focused discussion on the difference between explanation and prediction, the issue of temporal conditions was never brought to bear. That might seem strange. When a certain phenomenon has occurred, we can try to explain why it has occurred. Conversely, we can try to predict that a phenomenon which has not yet occurred is going to occur. A prominent difference between explanation and prediction thus seems to be of a purely temporal nature. In Hempel's model this difference is not primitive, it is uniquely the result of an epistemic parameter. When we explain, we explain something we know to be true, and, in the majority of cases, we know this thing to be true because we have seen it happening in the past. Conversely, we predict things that we do not yet know, and our ignorance is quite often related to future events. But nothing prevents our predicting that a certain event of which we have no direct knowledge must have happened in the past, on the basis of other facts. Another potentially relevant temporal condition concerns not the chronological relationships between the particular fact that is the *explanandum* (in cases where the *explanandum* is indeed a particular fact) and the time of the explanation, but rather the chronological relationships between the particular fact that is the *explanandum* and the particular facts contained in the explanans. In the example of the column of mercury thrust into a basin of boiling water, the prominent particular facts of the explanans are prior to the phenomenon to be explained: a certain set-up is described (the column of mercury in a glass tube, at a certain temperature, the water in the basin at a certain temperature) and what will happen next is explained on the basis of these antecedent conditions. The anteriority of the explanans is a natural candidate for the title of condition of adequacy of the explanation. And so, Hempel and Oppenheim (1948, §3) do indeed speak, regarding statements describing the particular facts of the explanans, of statements "stating specific *antecedent* conditions" (the emphasis is ours). All the same, the anteriority of the explanans is not explicitly mentioned in the conditions of adequacy.

What must be made of this situation? Two remarks to start off with. First, we can distinguish, as Hempel does, between *laws of succession*, which describe the evolution of a system, and *laws of coexistence*, which describe the state of a system. The law of universal gravitation and the laws of movement can be used to describe the evolution of the solar system (the movements of the planets). Boyle's law, which relates the pressure, volume and temperature of a real gas, describes the state of a gaseous system. Boyle's law can be used to explain the volume of a gas using its temperature and its pressure. In this particular case, and in all cases where laws of coexistence are used, the particular circumstances contained in the *explanans* are not strictly prior to the *explanandum*, they are concomitant to it. Second, it is sometimes possible to use laws of succession "backwards," when the processes described are reversible. The particular facts described by the statements $C_1, \dots, C_k$ take place at instants $t_1, \dots, t_k$ which are posterior to the instant t when the particular fact F took place and which we derive from laws and also from $C_1, \dots, C_k$. For example, we can deduce the position of the planets at an instant t using the laws of celestial mechanics and the position of the planets at a time t'>t. The deductive-nomological structure is the same as for the explanations or the "genuine" predictions for which the anteriority of the particular circumstances described in the *explanans* is confirmed. Hempel (1962, p. 116) speaks of "retrodiction" to name the counterpart of a prediction where the *explanans* is prior to the time of the explanation. But the introduction of the term does not resolve the problem. If we have retrodiction when the epistemic situation is one of prediction (F was not known ahead of time), is there, yes or no, explanation, admittedly of quite a particular type, the retrodictive type, when the epistemic situation is one of explanation (F was already known)? Here is Hempel's response:

> Any uneasiness in explaining an event with reference to factors that include later occurrences might spring from the idea that explanations of the more familiar sort, such as our earlier examples, seem to exhibit the *explanandum* event as having been brought about by earlier occurrences; whereas no event can be said to have been brought about by factors some of which were not even realized at the time of its occurrence. Perhaps this idea also seems to cast doubt upon purported explanations by reference to simultaneous circumstances. But, while such considerations may well make our earlier examples of explanation, and all causal explanations, seem more natural or plausible, it is not clear what precise construal could be given to the notion of factors "bringing about" a given event, and what reason there would be for denying the status of explanation to all accounts invoking occurrences that temporally succeed the event to be explained. (1965, pp. 353–354)

So yes, the "retrodictive" explanations do indeed have a counter-intuitive character. But however much this counter-intuitive character may be related to a causal notion of explanation, and however much the deductive-nomological model is not an essentially causal model, since subsumption under laws may or may not correspond to the description of a causal history, it seems it is the conflict with our intuitions that we must temper and not the model that must be modified. Another diagnostic is possible, as we shall see in the next section, which uses this sort of disagreement between the DN model and our intuitions as a starting point for a challenge to the DN model. For now, let us just grant credit to the coherence of the DN model. Its central idea is to put, to employ an expression of Hempel's (1962, p. 99), "nomological systematization" at the heart of a certain number of the products of scientific activity, these being explanation, prediction and retrodiction. These things differ among themselves in a purely inessential way, due to either epistemic parameters (prediction vs explanation) or chronological ones (prediction and explanation vs retrodiction and retrodictive explanation). Considering these parameters, one of the reasons not to grant too much importance to our intuitions is precisely the unifying virtue of the DN model, which reveals the essential contribution laws make to science when responding to a certain number of our expectations—whether these expectations correspond to demands for explanation, for prediction or for retrodiction.

2.4 WHAT IS A LAW OF NATURE?

If the full weight of the analysis is carried by the concept of laws, the analysis will only be complete when that concept itself is clear and precise. Following on from Hempel, let us begin by distinguishing laws and nomological statements, a nomological statement being a statement that is a law provided that it be true. It is not for us to decide which nomological statements are true—it is to science itself that it falls to say which nomological statements are confirmed to a high enough degree and are to be accepted as true. Our task, in completing Hempel's analysis, is then to characterize nomological statements, which account for the nomic expectability of the *explanandum* in the DN model.

Nomological statements are typically universal, conditional statements, such as "all metals are conductors" (Hempel and Oppenheim, 1948, §6, entitled "Problems of the concept of general law"). The general form of nomological statements, in logical notation, is $\forall x\ (\varphi(x) \rightarrow \psi(x))$;[13] that is, for every x, if x is a φ then x is a ψ. The putative law thus establishes the relationship between the fact of being φ (for example, the fact of being a metal) and the fact of being ψ (for example, the fact of being a conductor of electricity). By contrast, a particular statement, such as "certain metals are present in nature in a non-oxidized state" clearly doesn't claim the status of general law, and thus does not constitute a nomological statement. A universal statement whose scope is artificially restricted will not count as a nomological statement either. Saying that, on earth, the bodies of all living organisms contain carbon is not stating a general law about living organisms.[14] There is still another way in which a nomological statement is general: it must not make reference to specific individuals. The general unrestricted universal statement, "all of Napoleon's brothers-in-law became kings" is not a candidate to be a law, because it makes reference to a very specific individual, Napoleon. Neither should the generality of the statement be compromised by reference, implicit or explicit, to specific times or places. The statement, "all boats which navigate beyond

[13] Hempel and Oppenheim point out that in reality only the universal form is necessary since, syntactically speaking, the conditional statements can be transformed into equivalent statements that are not conditional. For example, the universal conditional statement, "all metals are conductors," is logically equivalent to the statement, "all things are not metals or are conductors," which is universal but not conditional. Nevertheless, it is possible to make the same remark regarding universal quantification, since "all metals are conductors" is equivalent to "it is false that some metals are not conductors." It is thus necessary to provide a definition of the concept of universal statement that is not purely syntactic (see 1948, §7).

[14] The exclusion of restrictions on scope poses its own problems. Many laws apply *ceteris paribus*. For example, the law establishing the thermal expansion coefficient of a metal only applies all other things being equal: the length of a heated metal bar will not increase by the proportions predicted by the law if somebody hammers at one of the ends of that bar (Lange, 1993). For a discussion of *ceteris paribus* laws in relation to economics, see chapter 15 in this volume.

the 75th degree of northern latitude risk being trapped in the ice" is universal, unrestricted, and doesn't make reference to individuals. Its generality is nevertheless limited by reference to a particular location (the 75th degree of northern latitude) so that it cannot claim to be a law either.[15] Having reached the end of the analysis, it appears that a nomological statement must be a universal statement, without restriction of scope and containing no purely qualitative terms. Are these necessary conditions also sufficient?[16] Consider the following statements:

(1) All signals travel at speed less than or equal to the speed of light.
(2) All solid spheres of gold have a diameter of less than one mile.
(3) All solid spheres of uranium-235 have a diameter of less than one mile.

(1), (2), and (3) satisfy the conditions we have just set forth. However, only (1) and (3) are nomological statements. (1) is one of the fundamental principles of the theory of general relativity, and (3) comes from the laws which govern nuclear fission. The critical mass of uranium-235, the mass beyond which a chain reaction of nuclear fission spontaneously occurs, is well below the mass of a one mile sphere of that isotope. Even if (2) is probably just as true as (1) and (3), it is still not a law of nature. That there is not a gigantic golden sphere in the universe is merely an accidental generalization. Correlatively, (2) does not seem to have any explanatory power. Saying that some metallic sphere has a diameter of less than one mile *because it is made of gold* does not in any way seem to constitute a good explanation. On the contrary, we could explain that the speed of a given signal transmission is inferior or equal to the speed of light by reference to (1).[17] Further, there is no difference between (2) and (3) in terms of the logical form of the statement or in terms of the nature of the expressions contained therein, so that it seems pointless to try and separate them by recourse to conditions like the necessary conditions which have been given thus far.

We can nevertheless point out the differences between (2) and (3). A first difference concerns what happens when certain fictional situations are envisaged. Consider the following counterfactual statements:

(4) If this sphere were made of gold, its diameter would be less than one mile.
(5) If this sphere were made of uranium, its diameter would be less than one mile.

[15] We omit the difficulties relative to the ideas of unrestricted scope and purely qualitative terms. Only a certain number of them are discussed by Hempel and Oppenheim (1948).

[16] This short introduction to the problem of characterizing laws of nature follows the classics van Fraassen (1989, part 1) and Salmon (1989, pp. 14–19). See Carroll (2012) for a more thorough survey.

[17] That the distinction between nomological statements and accidental generalizations seems to intuitively overlap with the distinction between universal statements having explanatory power, and universal statements not having explanatory power, corroborates the importance the DN model ascribes to the laws of nature.

Let's imagine that (4) and (5) are stated in front of an enormous bronze sphere which could well have a diameter of more than one mile. Intuitively, in that context, (4) is false. If the bronze sphere has a diameter of more than one mile, had it been made of gold, it would still have a diameter of more than one mile. Intuitively, in the same context but also in all other contexts, (5) remains true. Had the sphere been made of uranium, then it couldn't have had a diameter of more than one mile since it would have exploded before reaching that mass. Nomological statements support counterfactuals—they remain true when they are reworded counterfactually, like when (3) becomes (5) —while accidental generalizations do not support counterfactuals: (2) may well be true, (4) certainly is not.

Another similar difference is related to modal contexts.[18] So, let's compare the following:

(6) Necessarily, all solid spheres of gold have a diameter of less than one mile.
(7) Necessarily, all solid spheres of uranium-235 have a diameter of less than one mile.

(6) is true to the extent that the existence of such a sphere would defy the laws of physics which apply in all possible worlds (or at least in all the physically possible worlds, were we to posit the existence of logically possible but physically impossible worlds). By contrast, (7) is certainly not true: an enormous solid gold sphere, patiently put together by generations of goldsmiths or present in a natural state thanks to some exceptional conditions, and having a diameter of more than one mile could well exist. Nomological statements have modal import—(6), which is the modalized version of (2), is true—while accidental generalizations have no modal import: (7), the modalized version of (3), is not true, even if (3) is true.

Perhaps we will hold on to these conditions, adding them to the previous ones to characterize nomological statements in a necessary and sufficient manner. A nomological statement would then be defined as a universal statement without restriction of scope, containing only qualitative expressions, that support counterfactuals and have modal import. Less the adequacy of this characterization, it is rather its analytical virtue which is now problematic. We can give account for the notion of nomological statements in either modal or counterfactual terms. But the fact of having modal import or of supporting counterfactuals seems at least as mysterious as the fact of being able to claim the status of a law. It could even be tempting to turn the order of the analysis around and to say that (2), for example, supports counterfactuals because (2) is a law and not simply an accidental generalization. In the same way, it could be tempting to clarify the notion of necessity by saying that anything is possible that doesn't defy the laws of nature. Problems of conceptual priority like this arise with any

[18] By "modal context" we mean a subclause taking on the role of a modal operator, such as "necessarily," "it is necessary that," "it is possible that," etc.

attempt at conceptual analysis, and it could be just as tempting to accept the circularity of these notions as an insurmountable fact. Nevertheless, this circularity poses a particular problem in the case at hand. In fact everything depends on the methodological constraints that we place on the analysis of the concept of explanation. If that analysis has to be acceptable from an empiricist point of view, then the only conditions it should contain are those which can be satisfied by recourse to empirical observations. Experience can refute or confirm a general statement to a certain extent. But how could it tell us whether a statement supported counterfactuals, or had modal import? As Hume would put it, experience can teach us that something is this or that, but not that it is necessarily this or that. Our experience is always only our experience of our world, never the experience of other possible worlds where golden spheres would have or not have a diameter of more than one mile.

The problem with the characterization of nomological statements has become a completely distinct problem for the philosophy of science. Attempts have been made to respond both in a Humean framework and by renouncing empiricist constraints. Under this first category come the holistic conceptions that characterize laws by their attachment to our best scientific theory—to the supporters of this view then, such as Lewis (1973) or Earman (1984), to share the burden of defining what "best" means in this context. Under the second category we find the solutions proposed notably by Dretske (1977) and by Armstrong (1983), which call on the notion of universals, laws expressing relationships of necessitation between universals. We will not open a more in depth discussion of this problem here. From the perspective of the analysis of the concept of scientific explanation, we retain only that the DN model must be completed by a characterization of the concept of laws, and if it is indeed an issue of completing an empiricist model of explanation, this characterization too must be acceptable from that perspective, and also that to propose an acceptable characterization of what a law is from an empiricist point of view is a largely open problem.[19]

3. The Limits of the Deductive Model and How to Get Beyond Them

3.1 COUNTER-EXAMPLES

Even in the absence of a satisfactory characterization of nomological statements, it is possible to agree about the fact that some statements, such as Boyle's law, do seem to be good candidates for the status of nomological statement, while some other statements, such as the affirmation that all of Napoleon's brothers-in-law became kings, do not. In this way, the DN model can be applied without pre-empting the possibility of giving a fully satisfactory characterization of nomological statements. But the question arises whether, as it stands, the DN model provides an extensionally

[19] Salmon (1989), taking stock of the theories of explanation, remarks that the problem of the characterization of nomological statements has not disappeared. This is undoubtedly still true today.

correct account of our naive notion of explanation. It will be correct if something is an explanation in an intuitive sense if and only if that thing is an explanation according to the model's sense. It is surely not reasonable to demand perfection in this matter. Sometimes our intuitions are fuzzy and don't return a definite verdict, sometimes the proposed model contains a sufficient amount of good general properties to prompt us to legitimately revise our intuitions. This, according to Hempel, is the case with explanations which call on occurrences which are posterior to the fact to be explained, which are not clearly explanations in the intuitive sense, if they are at all, and which are nevertheless considered as such by the DN model. Such discrepancies can be accepted now and again. But, in general, when our intuitions are particularly sturdy, when reasons to oppose them are lacking, we do expect that the DN model conform itself to our intuitions regarding the presence, or otherwise, of an explanation.

Criticisms of the DN model and of its probabilistic variant were thus developed on the basis of a series of now standard counter-examples.[20] These counter-examples are of two sorts. Either we have an explanation in the DN model sense without it intuitively seeming like we have an explanation. Or else we intuitively seem to have an explanation without there being an explanation in the DN model sense. We'll start with some cases of the first kind.

Counter-example 1: Shadow of the Empire

On a certain day of the year, at a certain time of the day, at a certain spot on Fifth Avenue, a ray of sunlight hits the ground. The impact is located at a distance of x m from the base of the Empire State Building.[21] The ray brushes past the summit of the building and, there at the spot where it hits the ground, makes an angle of α degrees with the horizontal. Using the laws of geometrical optics, it is possible to deduce the height h of the Empire State Building, that is $h = \tan(\alpha) \cdot x$. This derivation satisfies all of the DN model's conditions of adequacy. It contains, essentially, laws of nature, in this instance the laws of geometrical optics, and the *explanandum* is derived using these laws and certain initial conditions such as the trajectory of the ray, the distance x and the angle α. And yet it seems absolutely counter-intuitive to go about explaining the height of a building by the length of the shadow it casts. Many elements come into play in explaining the height of the Empire State Building, among which the financiers' desires, the architects' decisions, the construction procedure, but certainly not, it would seem, the length of the shadow the sky-scraper would cast at a certain hour of the day at a certain time of year.

[20] Notably these counter-examples, and others besides, are to be found presented by Salmon (1989, pp. 46–50) and by Woodward (2009). The present rendering is much indebted to their clear presentation and insightful discussions.

[21] Various versions of this example, attributed to Bromberger, are in circulation. The object casting the measured shadow is sometimes the Empire State Building, sometimes an anonymous tower, sometimes a mast. The Empire State Building version is found in Bromberger (1966).

Note that it is possible, by reasoning analogous to the previous, to deduce the length of the shadow from the height, the angle between the ray and the ground, and the same optical laws. This would once again be an explanation in the DN model sense and, as for this explanation, things seem to be legitimate: the height of a building does provide an explanation for the length of the shadow it casts. Inferences made from a law are not directional, in the sense that they can be made in "several directions." The same functional law indifferently gives h using α and x or x using α and h. Explanation, unlike nomological inferences in general, seems to be directional: the inference from α and h to x is an explanation, but not the inference from α and x to h.

Counter-example 2: A storm in the air

A sudden drop in a properly functioning barometer's level is (generally) followed by a storm. Let's suppose that this is a law. From the observation of such a drop and from that law, we can deduce that a storm is coming. If we didn't yet know that the storm had happened, or was going to happen, this would be a legitimate prediction. If we already knew that the storm had happened, or was going to happen, this would be, according to the DN model, an explanation for the storm. But it seems absolutely counter-intuitive to consider that the drop in the barometer explains the storm. Many atmospheric phenomena come into play in explaining a storm's arrival, but what happens to barometers certainly does not feature among these phenomena. The drop in the barometer is a secondary effect, if you will, of these phenomena, but it does not contribute to the scientific explanation of the onset of a storm.

This counter-example in particular seems to knock a hole in the theory of symmetry between explanation and prediction since, if it is in fact a possible, or even typical, case of prediction, it doesn't however constitute a possible case of explanation.

Counter-example 3: Contraception for Men[22]

The example doesn't posit the invention of male contraceptive methods which would prevent a female partner becoming pregnant, but the somewhat less medically promising situation of a man taking the female contraceptive pill and not becoming pregnant himself. We consider the following argument:

(P) No man who takes the pill will bear child.
(M) Jean Dupont is a man who takes the pill.
==
(E) Jean Dupont will not bear child.

[22] This example is found in Salmon (1971), who uses it as one of the starting points in presenting a model for statistical explanation to rival the IS model.

Once again, if we accept considering (P) as a law of nature, (M) is an initial condition allowing the derivation of (E) from (P). According to the DN model this derivation constitutes an explanation that Jean Dupont will not bear child. And, once again, this seems absolutely counter-intuitive, since the correct explanation of (E) is simply that Jean Dupont is a man and that men do not bear children. Whether Jean Dupont takes the pill or not has nothing to do with it.

This counter-example indicates a problem of relevance. Logical validity is indifferent in regards to relevance. We can deduce that Jean Dupont will not bear child using the information that Jean Dupont is a man (M′) and the biological law according to which men do not bear children (P′). It is possible to deduce this using the same law and the original (M), since (M) implies (M′. It is also possible to deduce it using (M) and (P). Taking irrelevant supplementary information (strengthening (M′) into (M) and weakening (P′) into (P)) into account does not reduce the validity of the reasoning. However it certainly seems to reduce the explanatory validity.

Counter-example 4: Magic Salt[23]

This case is analogous to the previous one. We consider the following argument:

(S) Magic salt dissolves in water.
(W) These grains of salt have had a spell cast on them.
====================================
(D) These grains of salt dissolve in water.

(D) is a logical consequence of (S) and (W) but, once again, it seems that what we have here is not a bona fide explanation because some of the information contained in (S) and (W) are irrelevant to the phenomenon to be explained, that is the dissolution of salt in water.

Now we will see cases of the second type, where intuitively speaking there is an explanation without there being an explanation in the DN model sense.

Counter-example 5: The Spilled Ink-Bottle[24]

There is a huge fresh ink stain on the carpet. Why? I can explain it by saying that I bumped my desk with my knees and that this knocked the ink-bottle over. Intuitively this indeed seems like a possible explanation for the ink stain on the carpet. However, there is no general law contained in this explanation. So it can't be an explanation in the DN model sense. This case suggests that to explain an event it may be sufficient to "tell a story" which leads to this event, although according to the DN model relating a series of facts is never sufficient in providing an explanation.

---

[23] This example is from Kyburg (1965).

[24] This example is used by Scriven (1962) as an example of singular causal explanation.

Counter-example 6: The Mayor's Syphilis[25]

The town mayor suffers from a motor deficiency, characterized by the limitation of certain movements and a loss in muscular strength, which is called paresis. We know that roughly a quarter of patients with untreated, latent syphilis are victim to paresis, and we know too that the mayor has precisely such latent syphilis, a condition he was not aware of and was consequently not having treated. Intuitively what we have here is an explanation for the mayor's paresis. But the law linking syphilis and paresis, together with the mayor's untreated syphilis, only brings the probability of the mayor's developing paresis to 25 percent. According to the IS model, we only have an explanation if the explanans makes the *explanandum* highly probably. What counts as being highly probable is not determined with precision, but one chance in four certainly doesn't count as highly probable. So we don't have an explanation in the IS model sense, even though, intuitively, we in fact do.

It is the requisite of high probability that seems to be causing trouble here. Given the mentioned statistical medical law, having syphilis is enough to explain the mayor's paresis because this massively increases the chances of falling victim to paresis, even if the chances, globally, remain low. By demanding in absolute terms that the chances be high, we prevent ourselves from understanding how an argument containing a statistical law can be explanatory when the probability conferred onto the *explanandum*, even if it remains low, has been increased considerably.

The counter-examples we have just looked at, and others of the same sort, are, in part, at the root of the progressive abandonment of the deductive-nomological model, criticized by many philosophers of science since the 1960s. Indeed, they are not the only cause of this. Historically speaking, the DN model has been one of the pillars of the "received view" in philosophy of science which developed around logical empiricism, and the challenges to it are to be viewed in the perspective of more general challenges sustained by this received view. In particular, the DN model is connected to the syntactic concept of scientific theories, according to which scientific theories can be seen as axiomatic theories (see chapter 5 of this volume). Indeed, the precise formulation of the DN model is a logical formulation,[26] which presupposes that the elements of explanation can be written down as statements in a formal language. In return, this supposes the possibility of formalizing the scientific theories to be used in

[25] Another example from Scriven (1959).

[26] This logical formalization is employed by Hempel and Oppenheim to clarify, in an admittedly incomplete way, what initial conditions and laws, or, more generally, which theoretical aspects, are contained in the *explanans*, as well as the various associated formal adequacy conditions (in particular the fact that the laws be indispensable to the *explanandum*'s derivation). We have not copied this analysis out in detail insofar as, by the authors' own admission, it does not manage to resolve the major problem, which is the characterization of nomological statements. Curious readers will find its original formulation in "Logical Analysis of Law and Explanation," the third part of their 1948 article, and also a more recent study in Salmon (1989). This logical analysis has been criticized in itself. Eberle, Kaplan, and Montague (1961) point out a technical default that has the infuriating consequence of making any fact explicable from any theory. Satisfactory technical solutions are proposed by Kaplan (1961) and Kim (1963).

describing initial conditions and in stating laws of nature. But, since they constitute the most direct reason for throwing the DN model into doubt, let us now come back to our counter-examples and see what modifications to this model, or indeed what other account, they seem to demand.

3.2 LESSONS FROM COUNTER-EXAMPLES

This is our situation: Counter-examples 1 to 4, if they are accepted as given, show that the adequacy conditions of the DN model are not sufficient for there to be explanation. One possible answer would be to complete the DN model: an explanation would then be a deduction from general laws and initial conditions also satisfying certain supplementary conditions. The question is, of course, what would these supplementary conditions be? Another answer would be to abandon the DN model for some other model of explanation. Counter-examples 5 to 6 pose a potentially more serious problem for the DN model. In so far as they show that its adequacy conditions are not necessary conditions for there to be explanation, they tempt us to reject the DN model of explanation and replace it with another, or at the very least to supplement it with a second model which would account for these counter-examples. Given the existence of counter-examples 1 to 4, replacement is a more tempting option than supplementation, were it possible to find an alternative model which would simultaneously resolve all six counter-examples.[27]

But then, must counter-examples 1 to 6 be in fact accepted as presented? Let's look at what a defender of the DN model might say in objection to some of them. In the case of number 4 (the magic salt), it is possible to turn to the concept of a law of nature, or to its further clarification, to reject the counter-example. Indeed, it is possible to contest that the statement, "Magic salt dissolves in water" is a nomological statement at all. It is a reasonable opinion to hold that nomological statements should have well-defined empirical content, and the absence of an established procedure allowing the determination of whether salt has been hexed or not substantiates doubts that this be the case. If the statement is not nomological then the proposed argument is not an explanation and the DN model falls in line with intuitive thinking.

The problem is that counter-example 3 (contraception for men) seems completely analogous, although here the same strategy cannot apply since the statement (P), "No man who takes the pill will bear child" seems just as testable as any other general statement containing only terms with definite empirical content. Nevertheless, there is a definite sense in which this statement appears a far less apt candidate for the part of general statement playing an essential role in the explanation than the statement

[27] In the case of counter-example 5 (the spilt ink-bottle), supplementation is a strategy worth considering, insofar as we could consider that "telling a story" could be a genuine mode of explanation, perhaps a non- or pre-scientific mode of explanation. Counter-example 6 (the mayor's syphilis) argues more in favor of replacement, insofar as, *prima facie* at least, the example seems analogous to other examples of facts being explained in a probabilistic way using a statistical law and covered by the IS model.

(P′), "No man bears children," does. Any and every good biological theory on human reproduction seems behoven to include (P′) among its principles, primary or derived, and a biological theory on human reproduction containing (P) can be a good theory only in so far as (P) appears as a principle derived from (P′). In other words, it is not clear that the preliminary demand, to accept that (P) is a nomological statement, is an innocent one. On the contrary, the whole problem may be born right there and the best answer might be to just refuse that demand. Unificationist theories of explanation, which we will present in the next section, take advantage of that possibility: they distance (P) and keep (P′) by advancing that (good) explanations are those which contain the most unifying theoretical principles. (P′), albeit in a mundane way, allows for the unification of a whole collection of observations, while (P) is simply a redundant addition, so that it just wouldn't be fitting to call on (P) in an explanation when we could call on the more general principle (P′).

Counter-examples 1 (shadow of the Empire) and 2 (a storm in the air) seem to pose problems of a different sort. It is not so simple to challenge the nomological statement status of the general statements involved in the derivation of the *explanandum*. Not content to merely suggest that irrelevant information may find its way into a DN explanation, the counter-examples suggest that DN explanation is indifferent to a crucial dimension of ordinary explanations, that is the fact that they concern the manner in which the *explanandum* is produced, what made things such that the fact to be explained occurred, that the Empire State Building measures 381 m or that the storm swept in. The problem here is similar to the case of explaining an eclipse using initial conditions which are posterior to the *explanandum*. As it stands, nothing guarantees that the deductive argument of which DN explanation consists relates to what made the *explanandum* actually happen: any deductive argument containing nomological statements is "good to go" from the perspective of the DN model orthodoxy.

An adherent of the DN model could take inspiration from Hempel's response in the case of the eclipse, simply advocating that we must revise our intuitions, in so far as the DN explanation still satisfies the nomic expectability criterion. DN explanation does not necessarily say what makes an event or fact happen, it tells us that it was bound to happen, saying what makes an event happen being just one of the possible ways to say that it was bound to happen. The problem with that response is the persistent feeling that explaining the length of the building's shadow from the building's height is a better explanation than the reverse, and that explaining the storm using meteorological conditions is better than explaining it by changes in the barometer. Even if abandoning 1 and 2 as counter-examples by revising our intuitions were the thing to do, it would still seem like we have to demand that a good theory of explanation tell us why certain explanations are clearly better than others. The most direct manner to resolve this problem is to abandon the DN model by proposing an explanation model centered on the idea that an explanation tell us what made things such that the *explanandum* happened. The causal theory of explanation, which will also be presented in the next section and according to which providing an explanation is to provide the causes of the *explanandum*, constitutes such a theory.

Taking another step back, we can assess the difficulties encountered by the DN model with respect to the distinction, put forward by Salmon (1989) between a theory's descriptive power and its explanatory power.[28] Say that the descriptive power of a theory resides in its ability to "save phenomena," in keeping with the turn of phrase so dear to Duhem, or in other words, in the adequacy linking observations and predictions. Say, in contrast, that the explanatory power of a theory resides in its ability to explain phenomena, in a sense that we seek to clarify. Duhem (1908) rejects the idea that explaining is one of the goals of science because he thinks that descriptive power is the only scientific measure of a theory's success (A) and that explanatory power cannot be reduced to descriptive power (B). The merit of Hempel and Oppenheim's theory of explanation is that, by refusing (B), it makes the idea that explaining is indeed one of the goals of science compatible with (A). Indeed, if the difference between prediction and explanation is only a matter of a subject's epistemic state, then explanatory power does not differ from descriptive power. But the counter-examples to the DN model, on the contrary, seem to speak in favor of (B), in so far as they establish either that it is not sufficient or that it is not necessary to account for known phenomena on the basis of laws to explain them. If we accept this conclusion, several solutions are possible. We can accept (B) and come back on (A), running the risk of passing from science to metaphysics denounced by Duhem.[29] We can also accept (A) and reject (B), though in the name of a more liberal conception of what is meant by the "descriptive power" of a theory rather than, as Hempel and Oppenheim did, on the basis of an overly liberal conception of what an explanation is. This is the strategy which corresponds notably to Salmon's causal theory of explanation. So it is a matter of simultaneously defending that to explain is to give the causes and that science describes the causes of phenomena.

Another kind of approach to the problem is possible. We would begin by accepting (A) and (B), implying the presence of an extra-scientific dimension in explanation. But we would then try to "positively" account for that extra-scientific dimension of explanation on the basis of our discursive practices (what is it to ask "why?"). This is the path taken by the pragmatic theorists of explanation, in particular van Fraassen (1980).[30] It consists in understanding the extra-scientific dimension of explanation as a product of dependence in regard to contextual factors, and not as an irreducibly metaphysical aspect inescapably leading to a Duhem-esque rejection of

[28] As Salmon highlights, the expression "descriptive power" takes on different meanings according to whether, in particular, we reckon it to mean describing only observable phenomena or, more broadly and from a realist perspective, the "workings of nature," whether this involves directly observable phenomena or not.

[29] The analysis of the concept of law of nature in terms of universals, which Armstrong proposed, can be seen as an example of this strategy (see Armstrong, 1983).

[30] In van Fraassen's terms, "The discussion of explanation went wrong at the very beginning when explanation was conceived of as a relationship like description: a relation between theory and fact. Really it is a three-term relation, between theory, fact, and context" (1980, p. 153).

the explanatory demands addressed at science. The importance of contextual factors in demands of explanation is illustrated by van Fraassen through the following example.[31]

Example 7: The Knight at the Tower

A man comes to visit a knight who lives in a secluded tower. The shadow of the tower completely covers the terrace situated out front. The guest wonders why the knight built the tower so high. The knight gives him one explanation. A little later, a maid provides him with another. She explains that the tower was built on the spot where the knight had declared his passion for the woman he had loved and then killed out of jealousy. The knight wished for the tower to be so high that, at sunset, its shadow would cover the terrace where he had first declared his love.

The whole point of this example is obviously that it perfectly offsets the shadow of the Empire example. In the case of the Empire State Building, it seemed clear that the height of the building explained the length of the shadow and not the reverse. The scenario with the knight at the tower is such that the length of the edifice's shadow explains its height. Regarding counter-example 1, it is irrelevant to call on the length of the shadow to explain the height of the building. There is no reason to imagine that the height of the building in any way depends on the length of its shadow. The only pertinent explanations, in this context, would involve the ambition to construct the tallest building in the world and the means that rendered the realization of this ambition possible. It seems difficult to imagine a scenario in which the explanation would move in the opposite direction. Yet example 7 provides precisely such a scenario. In the context of this example, it would not be pertinent to explain the length of the shadow from the height of the tower, because, on the contrary, it was a calculation based on the length of the shadow which determined the desired height of the tower. The fact that things sometimes work in this way has consequences on the lessons we can take from counter-example 1. First, we thought that a good theory of explanation would have to account for a certain objective asymmetry, what we called the "directional character of explanation." But if van Fraassen's analysis of the knight's tower example is correct, there is no objective asymmetry. In certain contexts explanation moves in one direction, in different contexts this direction may change. Thus a theory of explanation need not account for this objective asymmetry that simply does not exist, rather it needs to account for the role played by context. Thus, responding to counter-example 1 with a theory of explanation which accords it a directional character would not be a good idea. In particular, responding to counter-example 1 by adopting a (uniquely) causal theory of explanation would seem completely inadequate in so far as

[31] The example takes the form of a short but charming tale of which the résumé offered here is but a pale reflection.

the relationship of causality concerned is not explanatory in all contexts (in counter-example 1 it is, but not in example 7).

3.3 A PRAGMATIC THEORY OF EXPLANATION

Van Fraassen (1980) proposes the following analysis.[32] An explanation is an answer to a "why-question," that is a question of the form "Why is. . .?." Such a question consists in three elements being given, a subject P_k, a contrastive class $X = \{P_1, \ldots, P_k, \ldots\}$ and a relevance relation R. In this way van Fraassen proposes identifying a why-question Q with a triplet $\langle P_k, X, R \rangle$. The subject P_k is a proposition, the proposition the question is about (for example, that the tower measures 30 m high). It is one of the members of the contrastive class X, which includes other propositions which, intuitively, could have been true and which contrast with P_k (so in X we will also find the propositions that the tower measures 31 m or 29 m high, etc.). The relevance relation R is a relation between the propositions and the couples formed by a proposition and a contrastive class. A proposition A is relevant relative to Q if A is in relation R with the couple. Intuitively, A is in the relation R with the couple $\langle X, P_k \rangle$ if A does in fact constitute the sort of answer expected, in the given context, to the question regarding why, among all the propositions in X, it is P_k which is true. R could be, for example, the relation held between the motives of an action and the couple made up of the consequences of that action and the set of consequences of the other actions which were possible, or else R could be the relation held between an event and the couple made up of the causal consequences of that event and the set of consequences of the other events which were possible.

It is now possible to define what a direct answer to a why-question, $Q = \langle P_k, X, R \rangle$, is. It is an assertion of form:

(*) P_k rather than (the rest of) X because A.

An answer of this form asserts that P_k is true, that the other propositions in X are false, that A is true, and, finally, that A is a reason that P_k, that is that A is relevant relative to Q. Some questions may consider causal relations to be relevant, such as in counter-example 1, others may consider psychological motives to be relevant, such as in example 7. The whole weight of van Fraassen's analysis thus rests on the relation R, which is conceived as a contextual parameter that makes up an integral part of the question posed. This parameter corresponds to the extra-scientific dimension of explanation, since it is not science's job to say whether the question posed is such that the relevant reasons are, for example, causes or motives. But, as promised, this

[32] Van Fraassen owes much to the work of Bromberger (notably Bromberger, 1966) concerning the link between why-questions and explanation and to the logical analysis of questions initiated by Belnap (Belnap and Steel, 1976).

extra-scientific dimension does not equal some metaphysical escape route, but rather placing the asking aspect of explanation into our practices.

The above is only the first part of the theory of explanation van Fraassen proposes. It analyzes what an explanation is, but doesn't say what a good explanation is or under what conditions an explanation, as an answer to some specific question, is better than another. Van Fraassen suggests accounting for this in a probabilistic way, continuing on from works which re-read and criticized the IS model of explanation (notably, Salmon, 1971, in which the counter-example 6 to the IS model is presented, and Cartwright, 1979). The idea is that a good explanation is such that the reason *A* is *statistically relevant* relative to the subject P_k. We will say only a few words here about the possible analyses of the notion of statistical relevance.

First, remark that a large part of the problems to be resolved are found in the continuation of the mayor's syphilis example. This example indeed shows that it is not sufficient to say, in an IS model way, that *A* is statistically relevant relative to *B* if $p(B|A)$ is high. As a first attempt, we could say that there is statistical relevance when the probability is heightened by conditionalization.[33] Even if $p(B|A)$ is not very high, what matters is that $p(B|A) > p(B)$: all other things equal, it is more probable that one be struck by paresis if suffering from syphilis than it is probable, no other conditions considered, to suffer from paresis.[34].Nevertheless, this new proposition, or other analogous ones (for example, we could ask that *A* favors *B*, in the sense that $p(B|A) > p(B|{\sim}A)$), meets other counter-examples along its way. For example, taking the contraceptive pill (C) favors the onset of thrombosis (D).[35] But we don't necessarily get that $p(D|C) > p(D|{\sim}C)$. Indeed, pregnancy itself is accompanied with hormonal modifications associated with an increased risk of thrombosis. If the effect in question is sufficiently significant and if pregnancies are statistically sufficiently frequent, the fact that taking the pill greatly decreases the chances of pregnancy could result in $p(D|C) < p(D|{\sim}C)$. The whole idea is thus to propose a sufficiently refined probabilistic analysis of what to favor means, so as not to clash with counter-examples of this sort.

Second, it must be noted that Salmon initially proposed the statistical relevance model not as part of a pragmatic theory but to provide an analysis of causation acceptable to an empiricist, and thus to make possible a causal theory of explanation just as capable of evading Duhem's reproaches as the aporia of the DN and IS models. Salmon (1980) presents a set of problems which lead him to abandon the project in favor of a more "direct" version (with no probabilistic reduction of the concept of causality) of the causal theory of explanation which we will discuss in the following section. The probabilistic analysis of causality later enjoyed a revival via bayesian network theory

---

[33] To conditionalize *B* upon *A* is to go from $p(B)$, the probability of *B*, to $p(B|A)$, the probability of *B* given *A*.

[34] This idea plays a crucial role in Bayesian confirmation theory. See chapter 3 of this volume.

[35] The progestogen contained in contraceptive pills causes modifications to the vascular system and leads to coagulation, which "explains" the link between taking the pill and thrombosis.

which aims to provide a general framework for defining and modelling relations of statistical dependence.[36]

To conclude this discussion, let us come back to van Fraassen's pragmatic theory of explanation. Independently of the possibility of obtaining a satisfactory probabilistic analysis of what a good explanation is, does the pragmatic model provide a good analysis of what an explanation is, good or bad? This model may raise more questions than it answers. By making explanations depend on a contextual parameter, the relevance relation, van Fraassen's solution runs the risk of being charged with a certain form of relativism, in that what counts as an explanation is inescapably relative to a certain choice of relevance relation. In particular, if any relation can be a relevance relation, then anything and everything can count as an explanation (Kitcher and Salmon, 1987). And so it seems that this model too would have to be completed by a theory about what counts as a relevance relation in a why-question. But that comes down to asking that the theory be completed by a theory of what all types of possible explanation are, and by a theory of each of these types, and thus, in particular, by a theory of what causal explanation is, of what an explanation based on motives is, etc. Van Fraassen's analysis, even if it correctly identifies a form of relativity to the concept of explanation, a relativity which is crucial in explaining the contrast between examples 1 and 7, would only push back the problem of providing a more "substantial" theory of what an explanation is by one notch, since the problem would just appear again at the characterization of relevance relations level.

Furthermore, it is possible to contest the analysis proposed for the example of the knight's tower. The example is supposed to show that explanation is not, in absolute terms, directional, and that there is thus no reason to give priority to a particular type of explanations which are directional, such as causal explanations. But an objection to van Fraassen (Salmon, 1984a) would be to say that it is the knight's desire that the tower have a shadow of a certain length which enables us to explain that it is of a certain height, and that desire regarding the length of a shadow is not the same thing as the length of that shadow. In a detailed explanation of this shady story, the height of the tower would indeed play a causal role in explaining the length of the shadow. The knight had a tower of such a height built because a tower of such a height would produce a shadow of such a length. The desire explains the height of the tower which explains the length of the shadow, but the length of the shadow never explains the height of the tower. If this analysis is correct, example 7 does not give us valid reason to revise the moral of the story of counter-example 1, namely that explanations are directed. Indeed, example 7 would only illusorily escape from the fundamental asymmetry of explanation.

All these reasons together make it seem necessary to go beyond a purely pragmatic theory by proposing "substantial" theories of explanation to deal with the counter-examples brought against the DN model.

[36] For a presentation of probabilistic analyses of causality, see Hitchcock (2008); and on bayesian networks, see the seminal work of Pearl (2000).

4. Two Theories of Explanation for Going beyond the DN Model

4.1 CAUSAL THEORIES OF EXPLANATION

By "*causal theory of explanation*," we can mean any theory according to which "to explain an event[37] is to provide some information about its causal history" (Lewis, 1986, p. 217). Various versions of causal theory have been defended, notably by Salmon (1984a, 1994), Lewis (1986), Woodward (1989) and Strevens (2009). They differ first of all by the concept of causality brought into play.[38] Salmon analyzes causality in terms of causal processes, characterized as physical processes capable of transmitting marks. Lewis defines it in counterfactual terms, and Woodward in interventionist terms. As for Strevens, he leans on a "minimal" notion of causality and tries to account for the role of causal influence relations in explanation in as neutral a manner as possible, relative to the various analyses of causation. Here we will essentially present Salmon's version (1984a), known as the causal-mechanical (CM) model of explanation.[39] An example of a causal process in Salmon's sense is a billiard ball in motion. The process is made up of the billiard ball and its successive positions in space-time. This process is capable of transmitting a mark. If a certain modification in the structure of the process occurs (for example if the ball is marked with chalk when it is struck by the cue), this modification persists in all subsequent states unless something acts on it (the chalk mark is transmitted to the positions the ball occupies in space-time following its interaction with the cue). A causal interaction is an encounter in space-time between two causal processes which modifies the structure of each of them. To explain an event E is to show how E fits into a causal nexus, to say which causal processes and causal interactions lead to and constitute E. For example, if E is the collision of two billiard balls, explaining E consists of describing it as the interaction of two causal processes (the two balls in motion) and of describing these processes themselves, tracing back to the initial impetus transmitted to one of the two balls by the player's use of the cue, etc.

In the sciences, the correct identification of causal processes can mostly be done only at the level of unobservable entities—think, for example, of the explanation of an electrical phenomenon by the motion of free electrons. In that measure, Salmon's conception is inseparable from scientific realism.[40] In Salmon's terms, which are not far removed from those used by Duhem to cast explanation back to the side of

---

[37] This characterization would need completing to cover causal explanations, not of events but of laws.

[38] See chapter 3, "Causality," of this volume for a detailed presentation of Salmon, Lewis, and Woodward's conceptions. The current importance of the causality question in these debates even brought Cartwright to remark that "we no longer talk about explanation; its place has been taken by causation" (2006, p. 230).

[39] This causal model is also *mechanical*, in that the causal influences are conceived to propagate by contact and at a finite speed. This is a trait specific to Salmon's theory, not necessarily shared by all causal theories of explanation.

[40] *Scientific* realism is the hypothesis that science provides, or at least aims to provide, an exact description of the world. From the realist point of view, the theoretical entities posited by science, such as atoms or electrons, must be interpreted as being entities that really exist, and not as simply convenient

metaphysics, "To explain is to expose the internal workings, to lay bare the hidden mechanisms, to open the black boxes nature presents to us" (Salmon, 1989, p. 134). By means of this realist hypothesis, causal theories account for the link between explanation and understanding. To understand is to understand what really happens, and science's explanations enable us to understand things in so far as they reveal the hidden mechanisms at work in the production of phenomena.

The two principal arguments in favor of causal theories are, according to Salmon (1978), the asymmetries of explanation and the need for non-causal regularities to be explained. Causality is asymmetrical and temporally oriented: if A causes B, then A precedes B in time[41] and B does not cause A.[42] The impetus the billiard player relays to the white ball causes its collision with the black ball, but the collision with the black ball could not have caused the impetus. If to explain an event B is to explain it by means of one of its causes A, then the explanation inherits the properties of causality. If A explains B, then A precedes B and B does not explain A. In this way, counterexamples 1 and 2 disappear of themselves. The building's shadow cannot explain its height because it cannot cause that height. The drop in the barometer cannot explain the onset of the storm because it cannot cause that onset. Conversely, it can be said that the building's height explains the specific length of the shadow it casts because the height provides information about the causal history which produces the shadow. Likewise, the presence of cold, dry air at high altitude and of hotter, more humid air at a lower altitude explains the storm's breaking since these properties of air masses provide information about the causal history which produces the storm.

The second reason involves the unsatisfactory nature of non-causal regularities, such as, for example, the regularity described by the ideal gas law. According to Salmon's wording, "Non-causal regularities, instead of having explanatory force that enables them to provide understanding of events in the world, cry out to be explained" (1978, p. 687). One of the arguments put forward by Hempel in favor of the DN model's not being an essentially causal model was the existence of non-causal laws such as the ideal gas law. But such a law seems not to be at "the end of the explanatory chain." Imagine we explain, deductively, the pressure P exerted by a certain gas from the volume V of the gas, the quantity of matter n, and the temperature T, using the equation of ideal gases PV = nRT where R is the universal constant of ideal gases. What we have then is indeed a DN model explanation, but that explanation seems incomplete in so far

---

fictions destined to enable adequate descriptions of observable phenomena. This hypothesis and its implications are examined in chapter 4 of this volume.

[41] It is nevertheless possible to uphold the existence of simultaneous causality, and perhaps even "retrograde causality." Here we are supposing a "standard" conception in which causes precede effects. See chapter 3, "Causality," in this volume. Also, asymmetry is a consequence of temporal priority.

[42] *Prima facie*, we can envisage mutual causes: depression leads to excessive consumption of alcohol and excessive consumption of alcohol leads to depression. The plausibility of mutual causality relations depends on the level of the causality analysis. In the confines of a theory of causal processes seen as particular entities, it seems reasonable to consider that a particular state of depression leads to a particular consumption of alcohol, which is liable to lead to an exacerbation of the depressive state.

as the law itself demands an explanation. We can ask why the equation of ideal gases merits its worth, and this boils down to asking which are the underlying mechanisms that assure this worth. Statistical mechanics enables us to explain the ideal gas law in so far as it enables us to obtain the equation as a consequence of the motion of the molecules which make up the gas, and of the colliding of these molecules among themselves and against the walls of the container. Applied to this example, Salmon's theory is that the ideal gas law is not of itself explanatory, as it does not have a causal foundation, while its derivation in statistical mechanics is explanatory as it concerns the underlying causal mechanisms, mechanisms that are working at the microscopic level to produce the regularity observed at the macroscopic level. Thus, we have a veritable explanation only from the moment when, instead of calling on a non-causal regularity, we manage to reassess the phenomena to be explained as a series of causal processes and causal interactions.[43]

Note that the plausibility of causal theories of explanation depends first and foremost on the plausibility of the analysis of causation they provide, a question which goes beyond the subject matter of this chapter. The two arguments we have just given are thus, above all, incitements to the development of a theory of causality compatible, if not with Humean prerequisites, then at least with scientific method.

The causal theory of explanation must nevertheless face numerous objections. First, the generality of the model can be contested by disputing the claim that every explanation is a causal explanation. Certain principles of physics are considered as having explanatory worth without their necessarily being subject to a causal interpretation. Take the example of an application of the Pauli exclusion principle,[44] an example first proposed by Railton (1978). A star collapses on itself under the pressure of its own gravitational attraction. The collapse ceases because, if it continued, the Pauli exclusion principle would be violated. As Lewis puts it, "There was nothing to keep it out of a more collapsed state. Rather, there was just no such state for it to get into" (1986, p. 222). This example seems to constitute a counterexample to the CM model in particular and to causal theories of explanation in general, in so far as the evocation of the Pauli principle does not bring to light any causal mechanism which would explain the collapse ceasing.[45] It would be fitting to complete the theory of explanation by making room for other kinds of explanation beside the causal explanations—this is Railton's position (1980, pp. 736–739, cited by Salmon, 1989, p. 164) when he speaks of structural explanation in regards

[43] As we shall see, the analysis proposed for the example of derivation of the law of perfect gases, Salmon's favorite example, is problematic.

[44] The Pauli exclusion principle says that two fermions cannot simultaneously occupy the same quantum state. Fermions constitute a large family of elementary particles, among which we find electrons and the quarks that form neutrons and protons.

[45] The non-causal character of the explanation at play has been disputed. Skow (2013) argues that philosophers have been taking science wrong and that the actual explanation physicists accept is causal, involving outward-directed pressure due to the internal energy of the electron gas.

to explanations of this sort. Lewis (1986), on the contrary, considers that the objection is void, in so far as the Pauli principle provides negative information about the star's causal history, namely that the ceasing of the collapse has no cause. To the extent that negative information is information like any other, Lewis reckons that there is no problem in admitting that evoking the Pauli principle comes down to applying a causal explanation. But to say (negatively) that there is no cause, is to (positively) characterize the possible states of the system. It is this structural characterization which is explanatory. So Lewis's response may not provide exemption from a complementary theory of structural explanations. All the more so that structural explanations, in Railton's sense, are in no way exceptional in physics. Explanations which rely on principles of conservation, notably, can also be placed in this same group. Using Galileo's principle of relativity and the law which says that two bodies having opposite momentum will stick together after a perfectly inelastic impact, we can derive the law giving the momentum of the two bodies after a perfectly inelastic impact from their momentums before impact. But Galileo's principle of relativity, which asserts that the laws of mechanics are the same in all inertial frames, cannot be interpreted as a causal principle, and it is not clear how we could transpose Lewis's solution which consisted of reinterpreting the positive characterization of possible states as negative information regarding the absence of cause.

Second, the explanatory relevance of particular causal processes can be brought into question, in particular in the case of complex systems. A gas is just such a complex system, and from this point of view it is instructive to closely re-examine Salmon's example concerning the causal explanation of the ideal gas law, as Woodward (1989) suggests. In practice it is impossible to calculate the trajectories and causal interactions of each and every gas molecule, and the derivation of the law in statistical mechanics does not proceed along those lines. Here is a broad outline of how to derive the law taken from a manual of elementary physics (Giancoli, 2005, pp. 367–371). We begin by making certain hypotheses which characterize what exactly we call an ideal gas. In particular we suppose that the gas is composed of a very large number of molecules moving in random directions at various speeds, that the interactions between molecules are limited to collisions, that the collisions between molecules themselves and also with the wall of the container are perfectly elastic, etc. Then let us imagine that the gas is contained in a plane parallel container of length l. By simple application of the laws of mechanics, we first calculate the average force exerted by one molecule on one wall of the container of area A (the container's volume being thus l.A). The force exerted by one molecule over the wall is intermittent but in the presence of a large number of molecules, the total force can be assumed to be constant. The force exerted by one molecule is given by

$$F = \frac{\Delta(m.v)}{\Delta t} = \frac{2m.v_x}{2l / v_x} = \frac{m.v_x^2}{l}$$

where $\Delta(m.v)$ is the change in momentum, Δt the time between two collisions and v_x the horizontal velocity of the molecule moving toward the wall. The force exerted by all n molecules is then

$$F = \frac{m}{l}\left(v_{x_1}^2 + v_{x_2}^2 + \cdots + v_{x_n}^2\right)$$

Setting $\overline{v_x^2} = \frac{1}{n}(v_{x_1}^2 + v_{x_2}^2 + \cdots + v_{x_n}^2)$ as the average horizontal velocity, we get $F = \frac{m}{l} n.\overline{v_x^2}$. Furthermore setting $\overline{v^2} = \overline{v_x^2} + \overline{v_y^2} + \overline{v_z^2}$ as the average velocity of the molecules and assuming that velocities are the same along the three axes, we get the total force exerted on the wall by gas molecules as

$$F = \frac{m}{l} n.\left(\overline{v^2}/3\right)$$

Dividing both sides of the equation by the area A, replacing F/A by the pressure P, and then multiplying both sides by V = l.A eventually yields

$$PV = n.\left(m\overline{v^2}/3\right)$$

Provided that the absolute temperature is directly proportional to the average translational kinetic energy of the molecules in the gas (the multiplying constant being the ideal gas constant R), we arrive at the equation in Boyle's law, PV = nRT. This standard derivation is incontestably explanatory. But the derivation does not consist in itemizing a set of causal series. Nowhere is it a question of retracing singular molecular trajectories and jotting down the collisions. Rather, the whole derivation relies on the possibility of leaving aside these details (a possibility resulting from the hypothesis of an ideal gas). If a causal explanation consists in tracking, in accordance with Salmon's terms, the causal processes and interactions, then this derivation is not causal. Nevertheless, the criticism of the causal interpretation of deriving the ideal gas law from kinetic theory perhaps goes too far. We could always respond in Lewis's fashion that this derivation does indeed consist in giving causal information. Simply, this information does not involve singular causal processes but, for example, average values characterizing the causal interactions between molecules and the wall. The explanation is causal even if there are no singular causalities. This response to the initial objection in turn encounters a problem. Why is this general information about the causal processes at work explanatory? More explanatory in fact than, say, a full description of all the trajectories of all the gas molecules. Intuitively, part of the explanatory virtue of these general considerations resides in the fact that they rely on a theory, the kinetic theory of gases, which unifies the theory of gases and mechanics, and also in the fact that this theory accomplishes the identification of temperature with kinetic energy. But if this is what is explanatory, the causal theory of explanation doesn't tell us why. And so we come to addressing a reproach to causal theory analogous to the one already addressed to the pragmatic theory. It under-determines the choice of causal history traits which are to be considered as explanatory. Unificationist theory, presented in the following subsection, takes on this precise blind spot of causal theories.

The second objection to causal theories is even more onerous than the first in that it concerns the type of examples causal theories advance as being typical examples of explanation. Thus, to counter the objection, it does not seem possible to resort to the strategy of completing the causal theory of explanation, seen as a theory of causal explanations, by taking other kinds of explanation into account. As we have seen, the second objection initially concerns general causal explanations in their application to complex systems. Hitchcock (1995) maintains that analogous problems arise even in the case of simple system explanations which rely on bringing specific causal interactions and processes to light, since Salmon's theory is incapable of accounting for the distinction between the properties of those causal processes which are explanatory and those which are not, relative to a given event. Moreover, Batterman (2002) identifies and analyses a class of scientific explanations for which the detail of causal processes is essentially irrelevant. These explanations are formed from a deduction based on the study of the asymptotic behavior of the system considered, when either the number of elements in the system or the time-scale used in the study approach infinity. The explanation does not involve tracking a causal history but rather the identification of structural properties possessed by those systems which, at the limit, guarantee the stability correspondent with phenomenon to be explained. Taking up Batterman's analyses, Imbert (2008) has proposed revisiting the DN model by integrating a requirement of explanation relevance, according to which, "good explanations deduce nothing too much." This requirement is destined to fill the gap left not only by the initial DN model but also, as we have just seen, by the causal-mechanical model. We shan't pursue Imbert's proposition further here, though it is clear that what we expect from a theory of explanation is that it enable us to identify the conditions which distinguish a good explanation (maximum relevance) from an inferior one (reliant on more superfluous details).

4.2 UNIFICATIONIST THEORIES OF EXPLANATION

Other versions of causal theory to Salmon's seek to answer at least some of the objections we have related, but we are now going to turn to another style of explanation theory, the unificationist theories. By "unificationist theory of explanation" is to be understood any theory according to which a scientific explanation is an explanation by merit of the fact that it provides a unifying manner of accounting for a set of phenomena.[46] To unify is, on first consideration, to enable "the comprehending of a maximum of facts and regularities in terms of a minimum of theoretical concepts and assumptions" (Feigl, 1970, p. 12). It is then a question of precisely defining this balancing act between a minimum of inputs and a maximum of outputs. An initial version of the unificationist theory was proposed by

[46] Woodward (2009) includes an excellent presentation of the unificationist theory.

Friedman (1974).[47] The standard formulation is due to Kitcher (1989). In both cases what we have are theories of explanation of an entirely different style to Salmon's. They do not attempt to resolve the objections the DN model encounters by placing themselves on terrain largely foreign to empiricist philosophy, behind the lines of a realist interpretation of causation. Rather, they aim to deepen an intuition that already played an important role in the DN model,[48] namely that a dimension of generality is essential to explanation.

The appeal of the unificationist approach results first of all from its close harmony with the development of science, or at least with a certain interpretation of the development of science.[49] Indeed, scientists endeavor to account for an ever-growing diversity of phenomena using an ever decreasing number of principles. Galileo's law of falling bodies describes the motion of free falling bodies close to the earth's surface. Kepler's laws describe the motion of the planets around the sun. Newton's laws of motion and the universal law of gravitation enable equally well the derivation of the law of falling bodies as they do Kepler's laws. They make up a small set of principles allowing us to account for a vast assembly of phenomena concerning the motion of earthly as well as celestial bodies. Typically, Newtonian mechanics represents a progression in that it unifies what was previously separate. It provides an explanation of the regularities expressed by Galileo's and Kepler's laws, as well as their corresponding phenomena. Far from being an isolated case, Newtonian unification illustrates a serious trend in science. Thus, contemporary physics is directed toward research for the famous "Grand Unifying Theory" which would gather up three of the four fundamental forces (electromagnetic force, weak interaction and strong interaction) into just one, just as Maxwell's electromagnetic theory, in its time, unified the theories of electrical and magnetic forces.[50] To put it bluntly, unificationist theories of explanation account for the fact that science seeks *simultaneously* to explain and to unify by postulating that to explain *just is* to unify.

The slogan has its limits. In the last paragraph we understood unification to mean inter-theoretical unification. But the explanation of new and as of yet unexplained phenomena, for example, does not involve inter-theoretical unification. And so it behooves the disciples of explanation as unification to define exactly what is meant by unification. The central concept of Kitcher's theory is that of an argument pattern. An argument pattern is a certain model of argumentation used by a theory. Let's look at an example Kitcher (1981) gives himself. Presented is an argument pattern used within Newtonian mechanics to account for a system made up of a single body in motion:

---

[47] Kitcher (1976) opposes a set of technical difficulties to Friedman's formulation.

[48] Kitcher even presents the unificationist model as an "officious" model present since the very outset behind the "official" DN model (1981, p. 507).

[49] On unity in science and the problems in reducing one science to another, see the last chapter of this volume, devoted entirely to these questions.

[50] For a presentation of the electromagnetic theory from a unification perspective of this sort, see Morrison (1992).

(1) The force exerted on α is β.
(2) The acceleration of α is γ.
(3) Force = Mass × Acceleration
(4) (Mass of α) × (γ) = β
(5) $\delta = \theta$

We speak of a pattern because the argument contains schematic letters which, to obtain an argument, must be replaced in due form by expressions. For this reason, the argument pattern must include filling instructions indicating how the patterns should be instantiated. Say that α must be instantiated by an expression which names the object studied, that β is an algebraic expression denoting a function of spatio-temporal coordinates, γ a function which gives the acceleration of the body. δ must be replaced by an expression which expresses the position of α and θ is a function of time, in such a way as the instantiation of (5) specifies the body's different positions all along the motion considered. Finally, the last ingredient of the pattern, on top of the series of schematic sentences and filling instructions, is what Kitcher calls a classification. A classification, for each schematic statement of the argument, is an indication of its inferential status (i.e., is this an assumption or does it follow on from other statements?) accompanied by a list of instructions indicating the reasoning to be carried out in obtaining the statement in question when it is not an assumption. Hence, the classification would tell us that (4) must be deduced from (1), (2), and (3) by substitution of identicals, while (5) is extracted, in a more complex fashion, from (4) using the methods of functional analysis.

An argument is explanatory if it instantiates an explanatory argument pattern. The fact of an argument pattern's being explanatory is defined holistically by its membership to the best possible basis of argument patterns for the systematization of the set K of all statements which we accept. Such a basis is a set of patterns whose instantiations are arguments acceptable to anyone who accepts K and which enable the derivation of all the statements in K from a proper subset of K. A basis is all the better, that is to say that its explanatory power is all the stronger, where it contains only a small number of different argument patterns, where these argument patterns are homogeneous and where the patterns are stringent.[51] Kitcher's definition remains imprecise in so far as it does not provide systematic means for comparing any pair of sets of argument patterns with a view to deciding which is the better basis relative to a set K of beliefs. Nevertheless, it does constitute a first step in the precise formulation of a unificationist theory of explanation. Moreover, Kitcher's general strategy is clear. What makes a certain argument explanatory is not some isolated property of that argument. An argument is explanatory because it is associated with an optimal manner of systematizing our beliefs, that is to say because it instantiates an argument pattern

[51] An argument pattern is all the more stringent where (simplifying) the arguments that instantiate it have a similar logical structure and employ similar vocabulary.

which, completed with the help of other patterns, provides a basis representing the best unification of our beliefs possible.

The unificationist approach casts an interesting light on the link between explanation and understanding:

> Science advances our understanding of nature by showing us how to derive descriptions of many phenomena, using the same patterns of derivation again and again, and, in demonstrating this, it teaches us how to reduce the number of types of facts we have to accept as ultimate (or brute). (Kitcher, 1989, p. 432)

To understand is to not have to accept without understanding. Scientific theories enable us to reduce the quota of primitive facts which, for lack of a better alternative, we can but accept as such. Thanks to the unifications science has delivered, many things which we would otherwise have to quite simply accept can now be deduced from a small number of primitive facts and general arguments. Without Newton, Kepler's and Galileo's laws must be accepted in and of themselves. We understand the world better since Newton, because now we only have to accept the principles of Newtonian mechanics to account for everything that was beforehand accounted for by the combination of Kepler's and Galileo's laws. In this way a scientific explanation contributes to our understanding of the world that surrounds us precisely to the extent that it fits into a systematic process of reducing primitive facts.

As we have said, the unificationist theory sits into the continuation of the DN model. Consequently, a crucial test for the unificationist theory will be its ability to deal with the counter-examples brought against the DN model itself. Kitcher's strategy for resolving the asymmetry problems is to advance that, as it happens, unification produces asymmetry. Let's look at this by re-examining counter-example 1. It seems economical for our set of argument patterns to include a pattern for the derivation of shadow lengths from the height of the bodies which cast them. Let's call this the "height based" pattern. In accepting the height based pattern, it is no longer necessary to accept both the facts concerning heights and the facts concerning lengths of shadows as being primitive. It is enough to accept the facts concerning heights since the facts concerning lengths of shadows can be derived from them. But why would it not be just as economical for our set of explanatory patterns to include an argument pattern, let's call it the "shadow-based" pattern, covering derivations of body heights from the lengths of the shadows they cast? By accepting the shadow-based pattern, we would appear to reduce the number of primitive facts to be accepted just as much as by accepting the height based pattern, so that neither would be less explanatory than the other. It is here that the interaction of these patterns with other argument patterns which we hold for explanatory comes into play. Let's consider another argument pattern, the "history-based" pattern,[52] which enables us to derive some thing's dimensions by calling on the history of its origins and its development.

[52] Kitcher speaks of the "origin and development pattern of length explanation."

This very general pattern is applicable in equal measure to any sort of thing, be this organisms, technical objects or buildings. When the thing under consideration is a skyscraper or a tower, the history in question is that of its construction and later modification. The shadow-based pattern and the history-based pattern are in competition since they enable the derivation of the same types of facts, that is facts about some thing's dimensions. Must our explanatory resources include both of these patterns or just one, and, if this be the case, which one? As the two patterns are in competition, it is more economical to keep only one, if this is possible without deriving fewer facts. The "shadow-based" pattern is, in many situations, useless. Many things do not cast shadows, notably of course if they are not illuminated, or, if they do cast a shadow, some of their dimensions cannot be obtained using that shadow. On the other hand, all objects, and thus all objects which have a shadow, have a history, so that the history-based pattern is applicable. Thus, out of all the patterns of argumentation allowing the systematization of our beliefs, the most economical decision would be to keep only the history- based pattern and the height based pattern, but not the shadow-based pattern. The derivation of the height of the Empire State Building using the length of its shadow is not explanatory because the argument pattern it instantiates does not belong to the best possible systematization of our beliefs.[53] Kitcher's answer here relies on the holistic nature of the unificationist theory. In the particular cases behind these counter-examples, the history-based and shadow-based patterns are equivalent. Were it only a question of the Empire State Building, its height and the length of its shadow at a certain time of day, the systematization obtained by instantiating the shadow-based pattern is neither better nor worse than the systematization obtained by instantiating the history-based pattern. However, when we broaden the scope of the facts being considered, we realize that the history-based pattern accomplishes a unification superior to that possible using the shadow-based pattern. Regarding the skyscraper, neither pattern displays any intrinsic superiority. The superiority appears when we assess the systematizing value of both patterns on a larger scale.

This analysis of the Empire State Building counter-example illustrates well the fact that the model Kitcher proposes refines Hempel's model. With the DN model, any derivation of an *explanandum* E using laws accepted by some scientific theory T is an explanation of E in T. But not all acceptable derivations are equivalent, and it is for this reason that Kitcher makes argument patterns, rather than nomological statements, the fundamental elements in his analysis of explanations. What is important is not simply the possibility of deriving E, but the way in which E is derived. And by considering the explanatory power of the sets of argument patterns available globally, Kitcher accounts for what makes one way of deriving E better, or more explanatory, than another. So Kitcher's model is keener than Hempel's in the sense that it comes down to proposing keener individuation criteria for scientific theories. According to the DN

[53] If not all things actually have shadows, they all at least *potentially* have shadows, so we could derive something's dimensions from its disposition to cast shadows. For a detailed discussion of these complications see Kitcher (1989, p. 485).

model, two theories which enable the derivation of the same nomological statements are equivalent in terms of their explanatory power (and probably, for Hempel, equivalent in any terms). According to the unificationist model, two theories which enable the derivation of the same nomological statements can still diverge in regards to the patterns of argument used, and can thus differ in regards to their explanatory power (on the other hand, if two theories use the same patterns, they produce the same nomological statements).

The above can be attacked on at least two grounds. First, the adequacy of the solution proposed for the asymmetry problem can be questioned. Second, the very motivations of the unificationist theory are to be considered with caution.

Regarding the asymmetry problem, the problem comes from the fact that the unificationist theory, like the DN model and unlike causal theory, is not intrinsically directional. As we have seen, the unificationist solution to the counter-examples relies on an increase of the set of facts considered—Barnes (1992) refers to this as a "widening strategy." To overcome the absence of intrinsic asymmetry in the unificationist conception, it is crucial that the widening strategy be permanently available, and, beyond this, that there be reason to think that it really is the global unification properties of the patterns instantiated by particular explanations which are responsible for the asymmetry of explanation. Kitcher explicitly defends this last point:

> But the crucial point is that the 'because' of causation is always derivative from the 'because' of explanation. In learning to talk about causes [. . .] we are absorbing earlier generations' views of the structure of nature, where those views arise from their attempts to achieve a unified account of the phenomena. (1989, 477)

In other words, Kitcher can easily accept that the asymmetry in particular explanations is typically derived from the asymmetry of causality, because he considers that our judgements of causality themselves are based on the tried and tested explanatory force of the underlying argument patterns. We think things go in one direction (that A explains B, and not the other way round, because A causes B) because the derivation of B from A is part of a reasoning pattern which has shown itself to be fruitful (unifying and adequate). Coming back to our example, we do in fact make the judgement that the height of the building participates in producing a shadow of a certain length, and not vice versa; this causal judgement is certainly responsible for our preferring the explanation of the shadow's length by the building's height rather than the other way round. But that still doesn't mean that an analysis of causation should be substituted for a theory of explanation. On the contrary, if we believe Kitcher, perception of causal asymmetry is ultimately founded in the explanatory success of a certain argument pattern, in this instance the history-based pattern.

Kitcher's considerations here are extremely speculative. What reasons do we have to think that our judgements on causation derive from considerations on unification? When and how are comparisons regarding the unifying power of rival systematizations for our beliefs carried out?[54] Even independently of these criticisms, Barnes (1992) maintains that the widening strategy does not enable us to deal with all the counter-examples (and thus, *a fortiori*, that judgements on causation cannot be reduced to sedimentations of unifying explanations). Consider the case of a closed system whose laws are symmetrical from a temporal point of view, the example of the solar system in Newtonian mechanics for instance.[55] The Newtonian argument pattern is used to derive a system's later states from prior states. A reverse pattern, retrodictive, can be used to derive a system's prior states from its later states. By assumption, if the system's laws are reversible, then the degrees of unification presented by both patterns are identical. So it is not possible in unificationist terms to account for our refusal of the retrodictive pattern as being an explanatory pattern. Worse still, in the case of open systems (subject to interference from the outside), a retrodictive pattern can prove more fertile than a time respecting pattern. If the system is open, the future cannot be predicted from the past because outside intervention is always a possibility. But a present state of such a system can nevertheless enable certain inferences regarding its past states to be made. In particular, in applying the principle of entropy, if the system considered locally presents a small degree of entropy, this system state must have been caused by interaction with an outside element. For example, if the open system considered is the sand on a beach, footprints in the sand must have been produced by a rambler rather than by some internal evolution of the system itself (Grünbaum, 1963). So the problem of asymmetry does not seem to have left us.[56]

We should take the time now to step back a bit and assess the unificationist theory in light of the initial motivations proposed by its defenders. One of the promises of the unificationist theory was to shed light on the relationship between explanation and understanding. According to causal theory, to explain a fact F is to provide information about other facts, facts concerning F's causal history. Causal theory does not explain why it is causal facts which are explanatory. As we have seen, unificationist theory sheds light on the relationship between explanation and understanding in so far as unification equals better understanding. That unification is one of the facets of understanding is quite clear. That it be the only one, a lot less so. It could be that the most unified systematization, and thus, according to unificationist theory, the one which must serve as the basis for explanations, will not be the one which brings

54 See Woodward (2003) for a development of this criticism.

55 This example was already used in discussing retrodiction. A defender of the DN model could maintain that the future can be used to explain the past. Kitcher, on the contrary, commits himself, within the unificationist framework, to eliminating these conflicts with our intuitions. Barnes's objection is that, contrary to the case of the shadow, here the unificationist theory fares no better than the DN model.

56 It is nevertheless not certain that the widening strategy has had its final word. After all, Barnes's objection relies on the choice of certain system classes. Jones (1995a) maintains that by widening the classes considered it is possible to respond to Barnes's counter-examples in the same way we responded to counter-example 1.

the best understanding. Hence Humphreys (1993) compares two axiomatizations of propositional logic, one consisting in a single, quite complicated, axiom, and a more usual one based on axioms corresponding to the elementary inferences associated with each of the logical connectives. If we believe Kitcher, the first axiomatization would have to lead to a better understanding of propositional logic, because it has a higher degree of unification. However, it actually seems like the best understanding is reached via the second axiomatization, which is more natural and enables us to understand "where axioms comes from" (each axiom expresses part of the meaning of one logical connective). Now, as Kim (1994) remarks, it seems futile to try to account, by means of a logical comparison, for what makes a systematization more natural than another, or for which systematization makes us understand things the best.[57] The promise to account for the relationship between explanation and understanding would thus not be kept.

Another promise of unificationist theory was to be faithful to the general movement of science, it being understood that this movement is to endeavor to cover more and more phenomena with the help of ever fewer laws. The underlying image, at least according to a realist interpretation, is that of a world governed by a small number of fundamental laws which science would progressively manage to discover by the formulation of ever more general laws. This image of science, like this image of the world, has been contested. Maybe the world is just a complex overlapping of multiple, heterogeneous realities, maybe science only isolates small islands of regularities, in such a way that the ideas behind the unification theory would be irrelevant. Several philosophers of science (Dupré, 1993; Cartwright, 1999) have defended this slightly iconoclastic vision of things. Without a more precise formulation and without more arguments to prop it up, this objection has limited impact. It has the merit of bringing to light that the starting point of the unificationist theory can be seen as an incomplete point of view, if not to say an unwarranted presupposition, regarding science.

5. Questions for a Theory of Explanation

Having reached the end of this survey, what are the prospects and challenges that have come into view for a theory of explanation? Neither the causal theory, at least not in Salmon's mechanistic version, nor the unificationist theory seem fully satisfactory as they stand. At the same time, these two theories take on aspects of scientific explanation which are both complementary and important. Causal theory takes on the ontological side—it tells us what kind of relationship must be present between those facts enlisted under the *explanans* and that other fact which is the *explanandum*. The relationship in question is that the facts enlisted under the title *explanans* must be the cause of the fact

[57] The adjective "natural" is entirely vague, naming the problem rather than its solution. The problem, as a result, goes nowhere.

to be explained. The unificationist theory takes on the epistemological side:[58] it tells us how much more we know once we have an explanation, the epistemic gain that is brought about. The gain in question lies in the unification of our understanding of nature.

From this position we can envisage either defending the theory of complementarity between the two approaches, or else developing hybrids by picking out the best of what the two theories have to offer. In regards to the first option, Salmon (1989) concludes his forty year voyage of debates on scientific explanation by hinting at the possibility of peaceful coexistence. Here is the example Salmon uses. A young boy sitting in a plane and awaiting take-off is holding a helium-balloon. What happens to it at take-off? The balloon moves forward. Why? The movement can be explained in either a causal and mechanistic manner, describing what happens to the air molecules located in the cabin, or else in a unificationist manner, appealing only to the principle of equivalence between gravitational fields and acceleration set down by Einstein (see Salmon, 1989, pp.183–184, for a less succinct presentation of these explanations). If these two explanations are to be considered as equally valid, then we mustn't completely separate the causal and unificationist theories of explanation, on the contrary we must give account for their juncture. In particular, in order to respond to the objections leveled at causal theory at the end of section 4.1, this would involve accounting for how considerations of unification grant an understanding of precisely which causal information, among the available myriad, it is pertinent to retain.

Regarding the second option, various existing proposals can be seen as hybrid approaches. Kim (1994) openly seeks to provide a synthesis of this kind. Woodward's invariance theory (2003), constituting the most elaborated and most discussed rival proposition to the precedent ones around today, can also be interpreted in this way. It puts itself forward as a version of causal theory but nevertheless explicitly satisfies a demand for generality. Woodward's idea is as follows. To explain why A happens under certain circumstances B, it is not sufficient to deduce A from B, it must also be possible to say what would have happened in place of A had circumstances been (slightly) different from B. Explaining why Heckle is black is not a matter of invoking the general statement that all crows are black, as this statement doesn't provide us with any systematic relationship between variations governing a bird's belonging to a particular species and variations in the color of its plumage. On the other hand, consider the explanation of price fixation in a monopolistic market, an explanation that was presented in a different context in section 2.1 and one that is discussed by Woodward himself to illustrate this very point. In the case of some particular monopoly, the fixed price is explained as being the price at which the curves of marginal revenue and marginal cost intersect on the average revenue curve, for the monopoly in question. But we

[58] This distinction between an ontological and an epistemological side is inspired by Kim (1994), who distinguishes realist theories (focusing on the ontological side) and internalist theories (focusing on the epistemological side) of explanation. Salmon (1984b) proposes a tripartite division tying in epistemic, modal, and ontic theories.

can also tell what would have happened had things been slightly different, for example if economies of scale had been somewhat larger (altering the marginal cost curve and, hence, its point of intersection with the marginal revenue curve). This additional information is crucial in that it allows us to control the phenomena under consideration. Knowing why prices are set as they are is, among other things, knowing how to arrange for prices to be set differently (for example by means of technological innovation increasing economies of scale). One of the attractive aspects of Woodward's theory is thus the connection it proposes between the explanatory function of science and other uses of science, such as control or manipulation of phenomena.[59] We won't go into further detail on this manipulationist conception, but it seemed nevertheless worthwhile to introduce it as just one attempt to add a constraint of generality, in this instance a constraint of invariance into a causal approach to explanation.[60]

The "classic" debates, as we have presented them, are formed around the discussion of a certain number of counter-examples to the DN model. As a result of this, other important issues undoubtedly end up being left to one side. In particular the question arises of the degree of generality of a particular explanation theory and of the integration of different styles of explanation specific to the various disciplines. How about explanations in mathematics, for instance? The existence of an infinite number of prime numbers is an elementary fact of arithmetic. How would a mathematician explain this fact? A tempting answer is to say that the explanation is given in the proof of the theorem. For example, it is shown that given any prime number n, it is possible to find a prime number strictly greater than it and contained between $n + 1$ and $n! + 1$. To what extent is this explanation comparable to explanations in the empirical sciences? Can the explanatory character of a mathematical proof be accounted for within the framework of the theories of explanation we have presented? Kitcher argues that the unification model journeys quite naturally to the world of mathematics, in so far as proofs are based on axioms whose very purpose is to unify one or several domains of mathematical objects. It seems more difficult to make sense of causal theory within this context, even though it might be noted from the previous example that proofs tell us how prime numbers can be produced. The question does

[59] *Manipulation* is a precise technical concept Woodward uses in his definition of causation. The idea is that X is a direct cause of Y relative to a set P of parameters if it is possible to modify Y by an intervention on X that does not change the parameter values in P (Woodward, 2003, p. 59).

[60] Woodward unequivocally presents his manipulationist theory of causal explanation as a rival theory to those advanced by Salmon and by Kitcher. To be precise, he doesn't present it as a theory that accomplishes a synthesis between the two. Regarding Kitcher's views in particular, Woodward insists that, "Kitcher's account is thus fundamentally different in motivation from the manipulationist account" (2003, p. 360). Nevertheless, we do not see it as being unfaithful toward the manipulationist theory to insist on the reintroduction of a generality constraint into causal theory. This in contrast to the leading role Salmon gives to the description of particular causal histories and in affinity with the increased status accorded to the importance of mobilizing principles not limited to the specific case under consideration by the unification approach.

not only arise a propos the formal sciences. As is often the case with the focus in general philosophy of science, physics takes pride of place. But in the other sciences we find models of explanation that do not readily lend themselves to being brought into line with some such theory of explanation, initially conceived with physics in mind (see, for example, Sober (1983) on the indirect causal character of explanations in terms of balance). Another blind spot of theories of explanation concerns the relationship between explanation and understanding. As we have seen when discussing the views of Friedman (1974), Kim (1994), and Imbert (2008), some criticize either the DN model or causal theories by appealing to the necessary relationship between explanation and understanding. It's not a question of psychologizing the notion of explanation—having the feeling to understand is not all it takes to be in possession of a good explanation (Trout, 2002). However, a good explanation allows us to understand the phenomenon explained, and a good theory of explanation should account for this. In the absence of in-depth knowledge about what understanding is, appeals to the concept of understanding in the analysis of scientific explanations remain, as Kim (1994) acknowledges, of limited efficacy.

Were we, in concluding, to hazard a little forward glancing, it would be to say that a fully adequate theory of scientific explanation will need to gain ground on these two fronts—integrating a more detailed analysis of the styles of explanation present within the various disciplines or subdisciplines, while also becoming integrated into a grander theory on the very nature of understanding.

References

Armstrong, D. (1983). *What Is a Law of Nature?* Cambridge: Cambridge University Press.

Barnes, E. (1992). "Unification and the Problem of Asymmetry." *Philosophy of Science*, 59, 558–571.

Batterman, R. (2002). *The Devil in the Details, Asymptotic Reasoning in Explanation, Reduction, and Emergence.* Oxford: Oxford University Press.

Belnap, N. D., and Steel, J. B. (1976). *The Logic of Questions and Answers.* New Haven: Yale University Press.

Bromberger, S. (1966). "Why-Questions," *in* Colodny, R. G. (ed.) *Mind and Cosmos.* Pittsburg: University of Pittsburgh Press, pp. 86–111.

Carnap, R. (1950). *Logical Foundations of Probability.* Chicago: University of Chicago Press.

Carnap, R. (1966). *Philosophical Foundations of Physics.* New York: Basic Books.

Carroll, J. W. (2012). "Laws of Nature." *The Stanford Encyclopedia of Philosophy* (Spring 2012 Edition), Edward N. Zalta (ed.), http://plato.stanford.edu/archives/spr2012/entries/laws-of-nature/

Cartwright, N. (1979). "Causal Laws and Effective Strategies," *Nous*, 8, 419–437.

Cartwright, N. (1999). *The Dappled World.* Cambridge: Cambridge University Press.

Cartwright, N. (2006). "From Causation to Explanation and Back," *in* Leiter, B. (ed.), *The Future of Philosophy.* Oxford: Oxford Clarendon Press, pp. 230–245.

Dretske, F. (1977). "Laws of Nature." *Philosophy of Science*, 44, 248–268.

Duhem, P. (1908). *Sauver les phénomènes. Essai sur la notion de théorie physique de Platon à Galilée*. Paris: A. Hermann (Vrin, 2005).

Dupré, J. (1993). *The Disorder of Things: Metaphysical Foundations of the Disunity of Science*. Cambridge, MA: Harvard University Press.

Earman, J. (1984). "Laws of Nature: The Empiricist Challenge," *in* Bogdan, R. (ed.), *D. M. Armstrong*. Dordrecht: Reidel, pp. 191–223.

Eberle, R., Kaplan, D., and Montague, R. (1961). "Hempel and Oppenheim on Explanation." *Philosophy of Science*, 28, 418–428.

Feigl, H. (1970). "The 'Orthodox' View of Theories: Remarks in Defense as Well as Critique," *in* Radner, M., and Winokur, S. (eds.), *Theories and Methods of Physics and Psychology*, Minnesota Studies in the Philosophy of Science, vol. IV. Minneapolis: University of Minnesota Press, pp. 3–16.

Feigl, H., and Maxwell, G. (eds.) (1962). *Scientific Explanation, Space, and Time*, Minnesota Studies in the Philosophy of Science, vol. III. Minneapolis: University of Minnesota Press.

Friedman, M. (1974). "Explanation and Scientific Understanding." *Journal of Philosophy*, 71, 5–19.

Giancoli, D. (2005). *Physics, Principles with Applications*, 6th edition. Upper Saddle River: Pearson.

Grünbaum, A. (1963). *Philosophical Problems of Space and Time*. New York: Knopf.

Hempel, C. G. (1962). "Deductive-nomological vs statistical explanation," *in* Feigl H., and Maxwell, G. (eds.) (1962), pp. 98–169.

Hempel, C. G. (1965a). *Aspects of Scientific Explanation and Other Essays in the Philosophy of Science*. New York: Free Press.

Hempel, C.G. (1965b) "Aspects of Scientific Explanation," *in* Hempel (1965a), pp. 331–396.

Hempel, C. G., and Oppenheim, P. (1948), "Studies in the Logic of Explanation." *Philosophy of Science*, 15, 135–175.

Hitchcock, Ch. (1995). "Discussion: Salmon on Explanatory Relevance." *Philosophy of Science*, 62, 304–320.

Hitchcock, Ch. (2008). "Probabilistic Causation." *The Stanford Encyclopedia of Philosophy* (Fall 2008 Edition), Edward N. Zalta (ed.), http://plato.stanford.edu/archives/fall2008/entries/causation-probabilistic/

Humphreys, P. (1993). "Greater Unification Equals Greater Understanding?" *Analysis*, 53, 183–188.

Imbert, C. (2008). "L'opacité intrinsèque de la Nature." PhD thesis, Université Paris I.

Jones, T. (1995a). "How the Unification Theory Escapes Asymmetry Problems." *Erkenntnis*, 43, 229–240.

Jones, T. (1995b). "Reductionism and the Unification Theory of Explanation." *Philosophy of Science*, 62, 21–30.

Kaplan, D. (1961). "Explanation Revisited." *Philosophy of Science*, 28, 429–436.

Kim, J. (1963). "On the Logical Conditions of Deductive Explanation." *Philosophy of Science*, 30, 286–291.

Kim, J. (1994). "Explanatory Knowledge and Metaphysical Dependence." *Philosophical Issues*, 5, 51–65.

Kitcher, Ph. (1976). "Explanation, Conjunction and Unification." *Journal of Philosophy*, 73, 207–212.
Kitcher, Ph. (1981). "Explanatory Unification." *Philosophy of Science*, 48, 507–531.
Kitcher, Ph. (1989). "Explanatory Unification and the Causal Structure of the World," *in* Kitcher, Ph., and Salmon, W. (1989), pp. 410–505.
Kitcher, Ph., and Salmon, W. (1987). "Van Fraassen on Explanation." *Journal of Philosophy*, 84, 315–330.
Kitcher, Ph., and Salmon, W. (eds.), (1989). *Scientific Explanation*, Minnesota Studies in the Philosophy of Science, vol. XIII. Minneapolis: University of Minnesota Press.
Kyburg, H. (1965). "Comment." *Philosophy of Science*, 32, 147–151.
Lange, M. (1993). "Natural Laws and the Problem of Provisos." *Erkenntnis*, 38, 233–248.
Lewis, D. (1973). *Counterfactuals*. Cambridge, MA: Harvard University Press.
Lewis, D. (1986). "Causal Explanation," *in Philosophical Papers*, vol. II. Oxford: Oxford University Press, 214–241.
Morrison, M. (1992). "A Study in Theory Unification: The Case of Maxwell's Electromagnetic Theory." *Studies in History and Philosophy of Science*, 23, 103–145.
Pearl, J. (2000). *Causality: Models, Reasoning, Inference*. Cambridge: Cambridge University Press.
Railton, P. (1978). "A Deductive-Nomological Model of Probalistic Explanation." *Philosophy of Science*, 45(2), 206–226.
Railton, P. (1980). "Explaining Explanation: A Realist Account of Scientific Explanation and Understanding." PhD dissertation, Princeton University.
Salmon, W. (1971). "Statistical Explanation" *in* Salmon, W. (ed.) (1971), *Statistical Explanation and Statistical Relevance*. Pittsburgh: University of Pittsburgh Press, pp. 29–87.
Salmon, W. (1978). "Why Ask 'Why?' An Inquiry Concerning Scientific Explanation." *Proceedings and Addresses of the American Philosophical Association*, 51(6), 683–705.
Salmon, W. (1980). *Space, Time and Motion; A Philosophical Introduction*. Minneapolis: University of Minnesota Press.
Salmon, W. (1984a). *Scientific Explanation and the Causal Structure of the World*. Princeton, NJ: Princeton University Press.
Salmon, W. (1984b). "Scientific Explanation: Three Basic Conceptions." *PSA: Proceedings of the Biennial Meeting of the Philosophy of Science Association*, vol. 1984, pp. 293–305.
Salmon, W. (1989). "Four Decades of Scientific Explanation," *in* Kitcher, Ph., and Salmon, W. (1989), pp. 3–219.
Salmon, W. (1994), "Causality without Counterfactuals." *Philosophy of Science*, 61, 297–312.
Scriven, M. (1959). "Explanation and Prediction in Evolutionary Theory." *Science*, 130, 477–482.
Scriven, M. (1962). "Explanations, Predictions and Laws," *in* Feigl H., and Maxwell, G. (1962), pp. 170–230.
Skow, B. (2013). "Are There Non-Causal Explanations (of Particular Events)?" *British Journal for Philosophy of Science*, online first, doi: 10.1093/bjps/axs047.
Sober, E. (1983). "Equilibrium Explanation." *Philosophical Studies*, 43, 201–210.
Strevens, M. (2009). *Depth*. Cambridge, MA: Harvard University Press.
Trout, J. D. (2002). "Scientific Explanation and the Sense of Understanding." *Philosophy of Science*, 69, 212–233.
van Fraassen, B. (1980). *The Scientific Image*. Oxford: Oxford University Press.
van Fraassen, B. (1989). *Laws and Symmetry*. Oxford: Oxford University Press.

Xavier de Aguiar, T. R. (2005). "As simetrias do modelo hempeliano de explicação." *Kriterion*, 46, 138–152.

Woodward, J. (1989). "The Causal Mechanical Model of Explanation," *in* Kitcher, Ph., and Salmon, W., pp. 357–383.

Woodward, J. (2003). *Making Things Happen: A Theory of Causal Explanation*. Oxford: Oxford University Press.

Woodward, J. (2009). "Scientific Explanation." *The Stanford Encyclopedia of Philosophy* (Fall 2009 Edition), Edward N. Zalta (ed.), http://plato.stanford.edu/archives/fall2009/entries/scientific-explanation/

2

CONFIRMATION AND INDUCTION

Mikaël Cozic (Paris-Est Créteil, Institut Universitaire de France & IHPST)

1. Introduction

In the empirical sciences, hypotheses and theories are, in principle at least, compared to empirical data. They are assessed on the basis of such a comparison. Data may speak either in favor of or against a hypothesis. And they may support one hypothesis more than another. For instance, it is widely believed that the advance of Mercury's perihelion speaks for General Relativity theory and against the Newtonian theory. Or, again, that the fossil record supports the theory of evolution. These intuitive notions guide scientists in the development and assessment of their work. Philosophy of science approaches them with the general concept of confirmation. Before entering into confirmation theories, we will succinctly characterize this concept (and its close relatives) and introduce the way contemporary philosophy of science deals with it.[1]

[1] I am grateful to Anouk Barberousse, Isabelle Drouet, Paul Égré, David Miller, Philippe Mongin, Jan Sprenger, and Hervé and Denis Zwirn for their comments on the first versions of this chapter. I am also grateful to Christopher Robertson, who helped me to translate the original version of the chapter. Finally, I'd like to thank Brian Hill for his comments on a talk I gave on the problem of old evidence, Bernard Walliser for our numerous conversations on probabilities and inductive reasoning, and participants and organizers of the seminars "Probabilité, Décision, Incertitude" (Institut d'Histoire et de Philosophie des Sciences et des Techniques, Paris) and "Probabilismes" (Centre Cournot, Paris). Part of this work was done at the Institut d'Etudes Cognitives (Paris, ENS Ulm) with the support of the ANR-10-LABX-0087 IEC and ANR-10-IDEX-0001-02 PSL* grants.

1.1 CONFIRMATION AND CONFIRMATION THEORIES

The philosophical analysis of confirmation is developed in a strongly idealized framework. The sentences or propositions by which empirical data (the evidence) are reported are canonically denoted by "*E*", which is sometimes called an evidential statement. A hypothesis or a theory (we set aside the issue of distinguishing between these notions) is denoted by "*H*". The main question is to determine when the evidence *E* supports (or does not support) the hypothesis *H*. More specifically, four cardinal concepts (confirmation, disconfirmation, verification, and refutation) can be introduced in this context. When *E* speaks for (or against) *H*, we will say that *H* is confirmed (or, respectively, disconfirmed) by *E*. The expressions "speak for" and "speak against" are very vague. The concept of confirmation includes at least the idea that the data support or strengthen our confidence in the truth of *H*.[2] This feature of confirmation makes it distinct from Popper's notion of corroboration (to be discussed). Verification and refutation (or falsification) may be seen as limiting cases of confirmation and disconfirmation: a hypothesis is verified by empirical data when they confirm it maximally, that is, when they establish its truth. On the contrary, a hypothesis is refuted (or falsified) by empirical data when they disconfirm it maximally, that is, when they establish its falsity.

Some have questioned the legitimacy and significance of the concept of confirmation. (We will come back to this issue shortly.) But if we assume, on the other hand, that such a concept guides scientific reasoning, it is clear that its use relies on principles that are largely tacit. An analogy can be drawn with mathematical reasoning: when mathematicians establish their results, they are rarely very explicit on the logical principles on which they rely. One could argue that it is precisely the task of (deductive) logic to make explicit, codify, and analyze the principles of mathematical reasoning. By the same token, one can view the study of confirmation by the philosopher of science as consisting (partly at least) in making explicit, codifying, and analyzing the principles of reasoning that involve confirmation.[3] In the same vein, Hempel—one of the pioneers of contemporary confirmation theory—vindicates as a condition of material adequacy the requirement that a theory of confirmation "has to provide a reasonably close approximation to that conception of confirmation which is implicit in scientific procedure and methodological discussion" (1945b, p. 107).

1.2 CONFIRMATION AND DEDUCTION

Modern mathematical logic has codified deductive reasoning with undeniable success: it has rigorously characterized the intuitive idea according to which a sentence A is a logical consequence of a set of premises $\wp$ iff it is impossible for the premises to

---

[2] At this stage, we do not wish to take a stand on the issue of knowing whether the concept of confirmation is "subjective." Consequently, we do not exclude that the "strengthening of confidence" is objectively grounded.

[3] The analogy is made, for instance, by Carnap in "Inductive Logic and Science." Specifically, he claims that the goal of inductive logic is not to propose "new ways of thinking," but to make explicit "old ways."

be true and the conclusion false. Propositional and first-order logics give well-known examples of such a characterization. The relation of logical consequence plays an important role in the conceptual and formal analysis of the relation of confirmation. First of all, the verification of H by E is the case where E logically implies H. Similarly, the refutation of H by E is the case where E logically implies $\neg H$. Assume that an economic forecaster puts forward the hypothesis H = "the growth of the GDP in Europe will be at least 1.5 % in 2020." In principle, by the end of 2020, one should be in a position to collect a set of data that will verify or falsify H. Verification and refutation are very special cases, however. Confirmation and disconfirmation cannot be so easily characterized in terms of logical consequence: E may confirm (resp. disconfirm) a hypothesis H without logically implying it (or, resp., its negation). Indeed, this is the typical situation: if H is a universal sentence like "All P are Q" and if E is a particular sentence like "a is P and Q" (a so-called positive instance), then E is generally believed to confirm H whereas it does not imply it. On the one hand, Popper argued forcefully that most scientific hypotheses are not verifiable because of their universal form: a finite set of empirical data cannot imply any of them. On the other hand, one attributes to Duhem and Quine the claim that isolated scientific hypotheses are often not refutable by empirical data because they don't have observable implications without auxiliary assumptions (this is the so-called Duhem-Quine problem). As a consequence, the core of theories of confirmation lies in what happens beyond the limiting cases of verification and refutation: if E does not imply H (resp. $\neg H$), in which conditions does E confirm (resp. disconfirm) H?

1.3 DEDUCTION AND INDUCTION

One often distinguishes deductive from inductive reasoning, which is illustrated by some typical forms.[4] The most famous is perhaps

(1) Generalization or enumerative induction by which one infers a universal sentence like[5]

$$\text{All } P \text{ are } Q$$

from a set of positive instances of this sentence

$$a_1 \text{ is } P \text{ and } Q$$

$$a_2 \text{ is } P \text{ and } Q$$

$$\ldots,$$

$$a_n \text{ is } P \text{ and } Q$$

[4] See Vickers (2013).

[5] On enumerative induction, see Norton (2005, 2010) who notably stresses its importance in ordinary life and science: "The actual inductive practice of science has always used enumerative induction, and this is

Inductive reasoning is sometimes reduced to enumerative induction.[6] This is not satisfactory. In full generality, inductive reasoning does not necessarily go from particular premises to a universal conclusion. For instance, another form of inductive reasoning is the

(2) Singular inference by which one infers

$$b \text{ is } Q$$

from

$$a_1 \text{ is } P \text{ and } Q,$$

$$a_2 \text{ is } P \text{ and } Q,$$

$$\ldots,$$

$$a_n \text{ is } P \text{ and } Q, \text{ and}$$

$$b \text{ is } P$$

The central feature of an inductive reasoning is that it is ampliative: its premises do not imply its conclusion. There is "more" in the conclusion than in the premises. Induction in the narrow sense corresponds to forms of induction reasoning like (1) and (2). Induction in the broad sense covers any ampliative reasoning. There are numerous families of ampliative reasoning. Statistical reasoning provides lots of examples:

(3) Direct inference consists in inferring a proposition on a sample from a proposition on the whole population:

$$\frac{\text{80\% of patients have recovered after treatment T}}{\text{80\% of patients from hospital } H_1 \text{ have recovered after treatment T}}$$

(4) Predictive inference consists in inferring a proposition on a sample from a proposition on another sample:

$$\frac{\text{80\% of patients from hospital } H_1 \text{ have recovered after treatment T}}{\text{80\% of patients from hospital } H_2 \text{ have recovered after treatment T}}$$

(5) Inverse inference consists in inferring a proposition on the whole population from a proposition on a sample:

---

not likely to change. For example, we believe all electrons have a charge of -1.6×10^{-19} Coulombs simply because all electrons measured so far carry this charge" (2005).

6 See Mill (1843, Book III, Chap.II, § 1): "Induction is the process by which we conclude that what is true of certain individuals of a class is true of the whole class, or that what is true at certain times will be true in similar circumstances at all times." Note that, according to Mill, the conclusions of true inductions are "general propositions," that is, "one in which the predicate is affirmed or denied of an unlimited number of individuals."

$$\frac{\text{80\% of patients from hospital } H_1 \text{ have recovered after treatment T}}{\text{80\% of patients have recovered after treatment T}}$$

Deductive logic is coarse-grained: from its point of view, all ampliative inferences belong to the class of non-valid inferences. Yet intuitively, some ampliative inferences are better than others. Sometimes, the premises of an ampliative inference support strongly its conclusion, and sometimes not. Consider for instance (2) the singular inference. Specifically, compare the strengths of the inference which infers "*b* is *Q*" from "*b* is *P*" and two positive instances of the form "a_i is *P* and *Q*" with the inference which infers the same conclusion from "*b* is *P*" and a vast number of positive instances. Prima facie, the latter justifies its conclusion much more strongly than the former. Let us have a look at (2′) which infers

$$b \text{ is } \neg Q$$

from

$$a_1 \text{ is } P \text{ and } Q,$$
$$a_2 \text{ is } P \text{ and } Q,$$
$$\ldots,$$
$$a_n \text{ is } P \text{ and } Q, \text{ and}$$
$$b \text{ is } P$$

From the point of view of deductive logic, (2) and (2′) are on an equal footing: neither one nor the other are valid inference schemata, since the truth of their premises does not entail the truth of their conclusion. And yet, one would sooner rely on (2) than on (2′).

1.4 INDUCTION AND CONFIRMATION

Induction (in its broad sense) and confirmation are obviously very close concepts. This is manifest when one compares the premise *P* and the conclusion *C* of an inductive inference with the empirical data *E* and the hypothesis *H* of a confirmation relation. (i) In general, *P* does not logically imply *C*. (ii) *P* (resp. *E*) usually gives some amount of confidence in the truth of *C* (resp. *H*). (iii) This confidence in *C* (or in *H*) is a matter of degree.[7] According to Carnap (1950/1962), the problem of induction is essentially the same as the one raised by the confirmation relation. Nonetheless, there are, prima facie, some differences between the two that are worth noting. First, in principle, ampliative inferences are not restricted to premises reporting empirical data nor to conclusions expressing hypotheses or theories. Hence, the notion of confirmation implies some

[7] Let us stress that not all confirmation theories try to take the gradual character of confirmation into account. More on this to follow.

domain restrictions with respect to the more general concept of inductive or ampliative inference. Second, when one says that empirical data *E* confirms a hypothesis H, it is not clear whether one intends that the inference from *E* to *H* is a good ampliative reasoning (or that the inductive strength of the inference from *E* to *H* is high). It is indeed possible for one to intend that E strengthens our confidence in *H*.[8] This second point is not necessarily decisive, since it may be the case that the same kind of ambiguity is present both in the concept of confirmation and in the concept of ampliative inference. But whatever are our pre-theoretical concepts, the core (i)–(iii) common to both ampliative inference and confirmation is crucial from the philosophical point of view: it assumes the existence of a notion of inductive support. Theories of inductive reasoning and of confirmation both attempt to capture this notion of inductive support.

1.5 POPPER AGAINST INDUCTION AND CONFIRMATION

Before discussing theories of confirmation, it is worth remarking that the notion of confirmation—and more precisely the assumption that there exists something like inductive support—is not unanimously accepted by contemporary philosophers of science. For instance, Popper vigorously rejected it:

> The best we can say of a hypothesis is that up to now it has been able to show its worth, and that it has been more successful than other hypotheses although, in principle, it can never be justified, verified, or even shown to be probable. The appraisal of the hypothesis relies solely upon deductive consequences (predictions) which may be drawn from the hypothesis. There is no need to mention induction. (Popper, 1959, p. 317)

In Popper's view, scientific reasoning is a matter of deduction: one has first to (deductively) infer the observational consequences of a hypothesis, and then to compare these consequences with empirical data. If there is a divergence between observational consequences and empirical data, the hypothesis *H* is refuted or falsified. Up to this point, logic alone is sufficient. What happens if *H* is not refuted by empirical data? The fundamental assumption of confirmation theorists is that, from the epistemological point of view, something important may happen. For instance, *H* may be confirmed and our confidence in the truth of *H* may be strengthened. Not so for Popper: if *H* survives one or several tests, then it is "corroborated." (Popper uses this expression to mark his rejection of the concept of confirmation.)[9] The degree of corroboration of *H*

---

[8] More on this distinction later.

[9] See the footnote before section 79: "I introduced the terms 'corroboration' ('Bewährung') and especially 'degree of corroboration' ('Grad der Bewährung', 'Bewährungsgrad') in my book because I wanted a neutral term to describe the degree to which a hypothesis has stood up to severe tests, and thus 'proved its mettle'. By 'neutral', I mean a term not prejudging the issue whether, by standing up to tests, the hypothesis becomes 'more probable' in the sense of the probability calculus" (Popper, 1959, pp. 248–249).

is assumed to measure the degree to which H has survived severe tests, but not our confidence in the truth of H.

It is important to distinguish Popperian deductivism from the hypothetico-deductive theory of confirmation (HDTC), to be presented. Both views rely exclusively on logical concepts to explain empirical reasoning. Specifically, they take into account the logical consequences of the hypotheses. But the Popperian view is based only on pure deductive reasoning, whereas the HDTC builds a non-deductive concept of confirmation on the basis of logical concepts (HD-confirmation). A second point to stress is how radical Popper's anti-inductivism is. From his point of view, the fact that the hypothesis H survives numerous empirical tests does not justify an increased confidence in the truth of H. In that case, the present chapter should probably end right now, since the basic working assumption of confirmation theories just is that empirical data may increase (or decrease) our confidence in the truth of a hypothesis without implying it (or its negation). But Popper's conception meets serious objections. Let us content ourselves with mentioning one of the most famous, from W. Salmon (1981). Assume that an agent is in a choice situation, where he or she has to make some decision on the basis of what is predicted by alternative hypotheses. The corroboration of these hypotheses depends only on their past performances. Were this not the case, the concept of corroboration would have an inductive dimension, which is excluded by construction. But if this is so, it is not easy to see how corroboration could provide a rational foundation for choosing one of these hypotheses and for making predictions on its basis. Assume more specifically that H_1 is strongly corroborated whereas H_2, although not refuted, is not so. It is hard to see what, in Popper's view, rationally constrains the agent to relying on H_1 rather than on H_2, since he or she is not supposed to be more confident in the truth of H_1. By contrast, one of the strengths of the Bayesian confirmation theory (see below) is that it is fully integrated in a theory of rational action (so-called Bayesian decision theory).

1.6 CONTENT

This chapter is devoted to attempts that have been made since World War II to build a theory of confirmation. We will present, illustrate, and discuss the main rivaling theories. Section 2 will deal with some famous paradoxes of confirmation and will expose the two main qualitative theories of confirmation: the instantialist and hypothetico-deductive theories. Section 3 will lay down the foundations for Bayesianism, on which is based the currently dominant theory of confirmation: the Bayesian theory of confirmation. Section 4 deals with this theory specifically. The last section tackles the issues of the justification and objectivity of confirmation and inductive reasoning, with a focus on the Bayesian point of view. This survey is by no means exhaustive, but as a matter of fact, most of the theories of confirmation can be tied to one of these three theories. This is stressed by Norton (2005) whose claim encompasses the entire history of theories of inductive reasoning. He calls the three families of views "inductive generalization," "hypothetical induction," and "probabilistic induction." Among the theories of confirmation and inductive reasoning not developed in this survey for question of space figure notably Glymour's (1980) "bootstrapping" theory (but see below); "likelihoodism" (Edwards, 1972; Royall, 1997); and Mayo's (1996) approach, which puts learning from error and severe testing at the center of scientific reasoning.[10]

2. Instantialism and Hypothetico-Deductivism

Our study will begin with two basic confirmation theories: the instantialist (Hempel) and the hypothetico-deductivist. These two theories are qualitative: they do not build a measure of confirmation but a criterion by which one can decide, in principle, whether some empirical evidence E confirms a hypothesis H. Before presenting in some detail these theories, we will see that, even in this simple framework, confirmation theories meet with very serious challenges. Two examples will be given: the paradox of the ravens and Hempel's triviality result, which are both analyzed in Hempel's seminal contribution (1945a, b).

2.1 THE PARADOX OF THE RAVENS

It is not at all trivial to build a satisfactory theory of confirmation. One of the most spectacular expressions of problems thrown up by the project is the famous paradox of the ravens which shows how difficult it can be to make some intuitive properties of the confirmation relation compatible. Assume that the hypothesis *H* being studied is the following universal conditional sentence ("UC-sentence" for short):

"All ravens are black,"

which is symbolized in first-order logic by

$$\forall x(\mathrm{R}x \rightarrow \mathrm{B}x)$$

It seems quite natural to accept the following principle: if one observes an entity which has both properties (expressed by predicates) R and B, then this observation confirms the hypothesis H. The observation of a black raven confirms the hypothesis that all ravens are black. More formally, this means that a sentence like $(\mathrm{R}a \wedge \mathrm{B}a)$ confirms $H = \forall x(\mathrm{R}x \rightarrow \mathrm{B}x)$. $(\mathrm{R}a \wedge \mathrm{B}a)$ is called a positive instance of H and the principle itself is called Nicod's Criterion. Nicod's Criterion quite directly echoes enumerative induction (see 1.3.).

Another constraint on the relation of confirmation is the Equivalence Condition. According to this condition, if the evidence *E* confirms (resp. disconfirms) a hypothesis *H*, then it confirms (resp. disconfirms) any sentence H' that is logically equivalent to

---

[10] The insistence on severe testing is of course reminiscent of Popper's conception. In a nutshell, Mayo's view is that "data x in test T provide good evidence for inferring *H* (just) to the extent that hypothesis *H* has passed a severe test T with x." H has passed a severe test T with x if x agrees with *H* and test T would have "produced a result that fits H less well than x does, if *H* were false or incorrect" (Mayo, 2005).

H. The Equivalence Condition is normatively very attractive: to reject it would mean, as observed by Hempel (1945a), that the confirmation relation depends on the way in which the hypothesis is expressed. But the joint acceptance of Nicod's Criterion and the Equivalence Condition leads to paradoxical conclusions. The sentence "All ravens are black" is actually logically equivalent to "All non-black things are non-ravens" $(\forall x(\neg Bx \rightarrow \neg Rx))$. Hence, in virtue of the Equivalence Condition, E confirms "All ravens are black" iff it confirms "All non-black things are non-ravens." This in turn implies, given Nicod's Criterion, that any positive instance of H′, that is, any sentence of the form "a is a non-black non-raven" (which would be implied, for example, by the fact that a is a white sock), confirms "All ravens are black." This is, to say the least, counter-intuitive. As Goodman put it, "the prospect of being able to investigate ornithological theories without going out in the rain is so attractive that we know there must be a catch in it."[11]

2.2 HEMPEL'S TRIVIALITY RESULT

Another paradox is discussed in Hempel's seminal paper. Hempel identifies three properties of the confirmation relation as being so plausible that they should be considered as "conditions of adequacy" for a theory of confirmation:

- (C1) Supraclassicality Condition: if E implies H, then E confirms H.[12]
- (C2) Special Consequence Condition: if E confirms H and if H implies H', then E confirms H'.
- (C3) Consistency Condition: unless E is inconsistent, if E confirms H and H', then H and H' are not mutually inconsistent.

It turns out that the Special Consequence Condition trivializes the confirmation relation when it is taken together with another seemingly attractive property, that is, the

- (C4) Converse Consequence Condition: if E confirms H and H' implies H, then E confirms H'.[13]

The proof is very simple. Let E be any evidential statement and let H be any hypothesis. By the Supraclassicality Condition, since E implies itself, E confirms itself. $(E \wedge H)$ logically implies E therefore, by the Converse Consequence Condition,

[11] Goodman (1955), p. 70.

[12] It is called the 'Entailment Condition' by Hempel. The term 'supraclassicality' is used in the context of confirmation theories by Zwirn & Zwirn (1996).

[13] Let H′ be Newton's theory of gravitation and H Kepler's first law according to which the orbit of every planet of the Solar System is an ellipse with the Sun at one focus. Then let us assume for the sake of simplicity that H′ implies H. The Converse Consequence Condition means that any observation that confirms Kepler's first law confirms also Newton's theory of gravitation.

E confirms $(E \wedge H)$. But $(E \wedge H)$ implies H, hence by the Special Consequence Condition, E confirms H. Hence, any piece of evidence confirms any hypothesis! As we will see, Hempel's own reaction consists of rejecting the Converse Consequence Condition. This move presupposes the acceptance of the Supraclassicality and Special Consequence Conditions. LeMorvan (1999) and Moretti (2003), however, have strengthened Hempel's case by showing that a similar triviality result follows from the Converse Consequence Condition by assuming (almost) exclusively the Supraclassicality Condition, which is widely accepted. Note that another triviality results from the Special Consequence Condition (C2), together with the

- (C5) Conversion Condition:[14] if H implies E, then E confirms H

Since E is implied by $(E \wedge H)$, by (C5) E confirms $(E \wedge H)$. But $(E \wedge H)$ implies H and therefore by the Special Consequence Condition, E confirms H.[15]

2.3 HEMPELIAN INSTANTIALISM

Instantialist theories of confirmation (ITC) attach utmost importance to Nicod's Criterion, that is, to the idea that a sentence like "All Rs are Bs" is confirmed by its positive instances. Hempel's theory is basically a sophisticated form of instantialism. (More generally, it can be seen as a way of extending the notion of enumerative induction.)[16] It relies upon the notion of development. Let I be a finite set of individuals (in the logical sense of the term). Intuitively, the development of a hypothesis H with respect to such a set I is what H would assert in a world populated exclusively by I's members. For example, if $I = \{a,b\}$ and $H = \forall x \mathrm{P}x$, then the development of H is $(\mathrm{P}a \wedge \mathrm{P}b)$. By the same token, if $H' = \exists x \mathrm{P}x$, then the development of H' is $(\mathrm{P}a \vee \mathrm{P}b)$. Hempel's definition of confirmation is the following:

- E *H-confirms directly* H iff E logically implies the development of H with respect to the individuals mentioned in E.
- E *H-confirms* H iff H is logically implied by a set of sentences each of which is directly H-confirmed by E.

Some comments on this characterization are in order. (i) First, Hempel's theory has much wider scope than elementary instantialism, which only applies to hypotheses of the form "All Rs are Bs." (ii) The precise content of direct H-confirmation has to be stressed. The evidence E delineates a domain of individuals to which the hypothesis H is provisionally restricted (the development of H with respect to the set of individuals mentioned in E is the result of this restriction). Direct H-confirmation

[14] The name comes from Zwirn & Zwirn (1996).

[15] I am indebted to H. and D. Zwirn for having pointed out this very simple proof to me.

[16] This is stressed by Norton (2010, 2011).

not only requires that E be compatible with the development of H, but also that E imply this development—which is much stronger. Let us consider a simple example: If $H = \forall x Px$, then $E_1 = Pa$ directly H-confirms H since the domain delineated by E_1 is $\{a\}$. Now, the development of H with respect to $\{a\}$ is Pa, which is obviously implied by E_1. In contrast, the evidence $E_2 = (Pa \wedge Qb)$, where Q is any predicate distinct from P, does not directly H-confirm H since the domain delineated by E_2 is $\{a,b\}$ and E_2 does not imply the corresponding development of H, that is $(Pa \wedge Pb)$. One could find this judgment rather counter-intuitive. A supporter of the Hempelian theory could reply that there is something defective in this observation report: since the predicates are supposed to express observable properties, in principle if someone is in a position to observe b, he or she should be in a position to determine whether b has also property P (and mutatis mutandis for a and Q). (iii) The concept of H-confirmation gives the theory the means to extend itself significantly, so that it becomes a very liberalized version of instantialism. Consider $E_3 = (Pa \wedge Pb)$ and $H = \forall x Px$. E_3 directly H-confirms H, but not $H' = Pc$. However, H' is implied by H, therefore H' is H-confirmed (indirectly) by E_3. Intuitively, the fact that a and b have the property expressed by P gives us some confidence in the fact that the entity c (which is still unobserved) will also have the property expressed by P. Consequently, the concept of (indirect) H-confirmation gives us the means to account for forms of inductive inference like the singular inference which we have described in section 1.

What about the two paradoxes: the paradox of the ravens and Hempel's triviality result? Let us start with the latter. Hempel's theory satisfies conditions (C1)–(C3). Since it is not trivial (there exist some data which do not H-confirm some hypotheses), we have to conclude that H-confirmation violates the Converse Consequence Condition (C4).[17] As for the paradox of the ravens, let us consider again the evidence that an indoor ornithologist can "cheaply" obtain: "a is a white sock," which we will express as $E = (\neg Ra \wedge \neg Ba)$. Given that the hypothesis is $H = \forall x(Rx \to Bx)$ and that the domain delineated by E is $\{a\}$, H's development is $(Ra \to Ba)$. $(\neg Ra \wedge \neg Ba)$ logically implies $(Ra \to Ba)$. Therefore, "a is a white sock" directly H-confirms "All ravens are black": we are lead to the very same counter-intuitive result as the one induced by the joint acceptance of Nicod's Criterion and the Equivalence Condition. Hempel was fully aware of this consequence of his theory, and acknowledged that it is counter-intuitive. But according to him, our confirmational intuitions are at fault: we suffer from a "psychological illusion" that has to be dispelled. Hempel claims that two biases make it counter-intuitive that "a is a white sock" confirms "All ravens are black." The first one has to do with the interpretation of universal conditional sentences like "All P are Q." According to Hempel, we have the impression that such a sentence only concerns things which are P (and therefore that only a piece of evidence about a thing which is P could be relevant), whereas it actually says something about the whole domain of interpretation.

[17] This illustrates the distinction between instantialist and hypothetico-deductive theories of confirmation since (C4) is a straightforward consequence of the latter.

The second bias is more specific to confirmation. When we consider the confirmational power of $(\neg Ra \wedge \neg Ba)$ (with respect to $\forall x(Rx \rightarrow Bx)$), we have a tendency to instead consider the confirmational power of $\neg Ba$ about an entity a that is already known not to be R. In other words, we have a tendency to view the situation against an incorrect background of beliefs and information. If *a* is known to be non-raven, to learn that *a* is non-black is not informative for someone who is interested in the truth of "All ravens are black": whether or not *a* is black, the hypothesis will be satisfied. To judge whether $(\neg Ra \wedge \neg Ba)$ confirms $\forall x(Rx \rightarrow Bx)$, we have to assume that we neither know whether *a* is R nor whether it is B. It is Hempel's contention that, given these correct background assumptions, it becomes natural to judge that *E* actually confirms *H*. (Intuitively, before learning *E*, we were not sure that the relevant properties of *a* would be compatible with *H*, but after having observed these, we know that they are.)[18]

2.4 PROBLEMS FOR THE HEMPELIAN THEORY

The Hempelian theory is one of the most elegant and convincing ways to do justice to the instantialist intuition underlying Nicod's Criterion. However, it faces serious troubles of its own.[19]

A first issue concerns conditions (C1)–(C3), which Hempel sees as conditions of adequacy for any theory of confirmation. Carnap has pointed out that Hempel seems to confuse two distinct concepts of confirmation (Carnap, 1962, § 87).[20] This distinction is of general interest, and it is worthwhile to introduce it carefully. According to the absolute concept of confirmation, *E* confirms *H* if *E* gives good reason to think that *H* is true. By contrast, according to the incremental concept, *E* confirms *H* if it increases our degree of confidence in the truth of *H*.[21] Hempel's theory is certainly not a theory of absolute confirmation, since a positive instance of the sentence $\forall x(Rx \rightarrow Bx)$ directly H-confirms it. But if we turn to the incremental concept of confirmation, the Special Consequence Condition is not completely convincing: it seems possible that *E* increases our confidence in *H*, that *H* implies *H'*, and that, nonetheless, *E* does not increase our confidence in *H'*. Assume for instance that $E = Pa$, $H = (Pa \wedge Qb)$ and $H' = Qb$. In this case, learning *E* increases our confidence in *H*, *H'* is implied by *H*, but it is not clear why *E* should increase our confidence in *H'*. By contrast, the Special Consequence Condition is more convincing for the absolute concept of confirmation: if *E* gives good reason to believe that *H* is true, then it gives good reason to believe of any consequence *H'* of *H* that it is true. Recently, on the basis of a theoretical study of the relationship

---

[18] Hempel's ideas about the paradox of the ravens are without doubt thought-provoking. But it is not certain that they are really compatible with his own theory of confirmation. For a detailed analysis of this issue see Fitelson & Hawthorne (2006).

[19] Our discussion is inspired by Earman (1992, chap.3).

[20] For the rest of this chapter, however, I do not follow Carnap closely. For a more detailed discussion, see Huber (2007, 4.d; 2008). See also Salmon (1975).

[21] This is not Carnap's terminology. He uses confirmation as 'firmness' for the former concept and confirmation as 'increase in firmness' for the latter.

between abstract properties similar to Hempel's conditions of adequacy, Zwirn and Zwirn (1996) have distinguished two sets of such properties, one (which includes the Special Consequence Condition) corresponding to the absolute concept confirmation and the other to the incremental one.[22]

A second issue has to do with more specific performances of Hempel's theory. First, there are some counter-intuitive consequences. For example, as stressed by Earman (1992), the set of observations Ra_ia_j for $i = 1,2, \ldots, 10^9$ and $j = 1,2, \ldots 10^9 - 1$ does not H-confirm the hypothesis $\forall x \forall y Rxy$ since it does not imply its development for the appropriate individuals ($Ra_{10^9}a_{10^9}$ is "missing"). More importantly, the status of theoretical terms is unclear. When Hempel criticizes the hypothetico-deductive conception of confirmation,[23] he does right to draw attention to the fact that these terms play a crucial role in modern science. But it is hard to see how his theory can account for the fact that empirical data have confirmational power with respect to hypotheses where theoretical terms occur. In the technical exposition of his theory (Hempel, 1943), Hempel does not face this difficulty, since he considers a language containing only predicates that express observable properties and relations. Note, however, that in his *Theory and Evidence* (1980), C. Glymour proposes a "bootstrapping" theory of confirmation, which can be seen as a way to improve instantialism on this dimension.[24] In a nutshell, the idea is that evidence can be related to an (instance of *a*) hypothesis *H* featuring theoretical concepts with the help of a theory *T*.

A third difficulty is related to the general type of confirmation theory to which Hempelian instantialism belongs, that is, a purely syntactic theory of confirmation. These theories have to overcome an issue made salient by the "the new riddle of induction" or the "grue paradox" (Goodman, 1946, 1955). Let us consider the two following hypotheses:

$$H_1: \text{“All emeralds are green”}\ \forall x(\mathrm{E}x \rightarrow \mathrm{G}x)$$

$$H_2: \text{“All emeralds are grue”}\ \forall x(\mathrm{E}x \rightarrow \mathrm{GRUE}x)$$

By definition, something is "grue" iff either (a) it has been examined before *t* and is green, or (b) it did not get examined before *t* and is blue. It follows from this definition that if *a* has been observed before *t*, it is green iff it is grue. Assume that *a* has been examined before *t* and that it is a green emerald. The evidence can therefore be reported as $E = (\mathrm{E}a \wedge \mathrm{G}a \wedge \mathrm{GRUE}a)$. Hence *E* H-confirms (directly) both H_1 and H_2 (see Fitelson, 2008, for a detailed reconstruction). This conclusion is obviously counter-intuitive. First, it is hard to convince oneself that *E* confirms H_2 (the "grue hypothesis"). Moreover, the two hypotheses make incompatible predictions for the emeralds examined at *t* or afterward: they will be green according to H_1 but blue

[22] Actually, they identify a third set of properties that corresponds rather to an hybrid concept of confirmation.

[23] Hempel (1945b), sec. 7.

[24] For reasons of space, we will not expose and discuss Glymour's account of confirmation. The interested reader is referred to Christensen (1983, 1990), Glymour (1983), and Norton (2010, sec. 7) for discussions of this sophisticated theory.

according to H_2. The Hempelian theory seems to overgenerate. Goodman sees his "new riddle of induction" as an argument against "syntactical" theories of confirmation, that is, against the theories that base the confirmation relation on the logical form of the involved sentences. Indeed, the logical forms of H_1 and H_2 are symmetrical with respect to E whereas their confirmational behaviors are intuitively very distinct. Goodman draws the conclusion that a theory of confirmation based only on the logical form "misses" something essential in its target.[25] Goodman calls "projectible" a hypothesis which can be confirmed by its positive instances. His main negative claim is that the logical form alone cannot determine the projectibility of a hypothesis.

2.5 HYPOTHETICO-DEDUCTIVE THEORIES OF CONFIRMATION

Hempel (1945b, sec. 7) takes care to distinguish his theory from the hypothetico-deductive theories (HDTC).[26] HDTC is basically an elaboration of the Conversion Condition (C5) and can be described as follows. Let H be a hypothesis and K a set of background beliefs.[27] Assume that H and K (deductively) imply some observational consequence E. In this case E HD-confirms H (relative to background beliefs K):

- E HD-confirms H relative to K iff $(H \wedge K)$ logically implies E

Let us consider the following example. Ohm's Law states that the potential difference applied to an ohmic conductor (V) is equal to the product of the current that flows through it (I) and its resistance (R):

$$V = R \cdot I$$

Assume that we know, for a given conductor, its resistance R and the potential difference applied to it, V. In this situation, we can predict the current that flows through it. If this predicted value fits the measured value, Ohm's Law will be HD-confirmed by this observation, relative to known values R and V. Our description is of course considerably simplified: background beliefs should in principle be much richer. It should also include auxiliary assumptions, like the hypothesis that the ammeter with which the electric current has been measured is reliable. The preceding definition of HD-confirmation can be found wanting. If the background beliefs K already implies E, E will necessarily HD-confirm H. A straightforward improvement is the following:

- E HD-confirms H relative to K iff (i) $(H \wedge K)$ logically implies E and (ii) K does not imply E

---

[25] "Confirmation of a hypothesis by an instance depends rather heavily upon features of the hypothesis other than its syntactical form" (Goodman, 1955, pp. 72–73).

[26] For a recent overview, see Sprenger (2011).

[27] The next example will show the role of the background beliefs.

Generally speaking, according to HDTC, the confirmation relation is a kind of converse of the relation of logical consequence. One of the attractive features of the HDTC is that it seems to largely agree with the methodological practice of empirical sciences. It echoes directly the view of theory appraisal according to which (i) to empirically assess a theory, one needs first to draw its "predictions," and (ii) if these predictions turn out to be true, then our confidence in the theory is increased. Here is how Huygens distinguishes his method from the way the Geometers proceed in the Preface of his Treatise on Light (1690):

> . . . whereas the Geometers prove their Propositions by fixed and incontestable Principles, here the Principles are verified by the conclusions to be drawn from them; the nature of these things not allowing of this being done otherwise. It is always possible to attain thereby to a degree of probability which very often is scarcely less than complete proof.[28]

HDTC satisfies the Converse Consequence Condition (C4): if E HD-confirms H and H' implies H, then E HD-confirms H' since H' implies E by transitivity of the consequence relation (to simplify, we leave aside the background beliefs). Hence, the HD-confirmation relation is preserved by the logical strengthening of the hypothesis. By contrast, the Supraclassicality Condition (C1) is violated by HDTC. This is easily shown by considering $E = P(a)$ and $H = \exists x P(x)$. HDTC is thus unable to account for our intuition that, in this case, there exists a confirmation relation between E and H. It faces other serious problems. Improvements of the elementary version we just exposed have been regularly proposed since Hempel (1945a, 1945b), but there is currently no stable version.[29] (i) The first issue is the problem of irrelevant conjunction:[30] if E HD-confirms H, then for any hypothesis H', E confirms the conjunction of H and H'. This property follows directly from the monotonicity of the relation of logical consequence. (ii) The second difficulty is the dual side of the former. It is the problem of irrelevant disjuncti on: if E HD-confirms H, then for any evidence E', the disjunction of E and E' confirms H. Tentative solutions to these two "tacking paradoxes" are discussed by Sprenger (2011). (iii) The third issue is the problem of alternative hypotheses: often when E confirms some hypothesis H, it also confirms other hypotheses that are incompatible with H. Let us think, for instance, of the case where E reports the observation of two variables x and

[28] Huygens (1690), Eng. trans., p. vi. This passage is quoted by Maher (2004). The next phrases are also very interesting: "To wit, when things which have been demonstrated by the Principles that have been assumed correspond perfectly to the phenomena which experiment has brought under observation; especially when there are a great number of them, and further, principally, when one can imagine and foresee new phenomena which ought to follow from the hypotheses which one employs, and when one finds that therein the fact corresponds to our prevision. But if all these proofs of probability are met with in that which I propose to discuss, as it seems to me that they are, this ought to be a very strong confirmation of the success of my inquiry . . . "

[29] For some recent attempts, see Schurz (1991), Gemes (1998, 2005), and Sprenger (2013).

[30] It is also called the "problem of selective confirmation" by Gemes (1998).

y, and where the hypothesis states a relation between these variables. For any finite set of pairs (x, y), there exists an infinity of functions capable of inducing this set of pairs.

Let us mention one last drawback of HDTC, which is also a problem for any instantialist theory à la Hempel: it is not able to deal with hypotheses where "objective" probabilities (propensity, chance, or relative frequency) occur. Consider for instance H = "there is a 50% chance that a nucleus of radium 224 will decay during an interval of 3.5 days" and suppose that we can determine whether a radium nucleus decayed or not during some time interval. Yet H does not imply anything about a radium nucleus that could be verified or refuted by this observation.

3. Bayesianism

Up to now, we have only considered qualitative theories of confirmation, which intend to determine whether (but not to which degree) some piece of evidence confirms a hypothesis. By contrast the Bayesian theory of confirmation (BTC) deals both with the qualitative and quantitative concepts of confirmation. The BTC is based on Bayesian epistemology or Bayesianism. We will begin by introducing Bayesianism and its formal framework—the theory of probability.

3.1 DEGREES OF BELIEF AND PROBABILITY THEORY

Bayesianism is a multifaceted set of ideas, but contemporary Bayesian epistemology boils down essentially to the following three tenets:

- (B1) Gradualism: an adequate epistemology must consider degrees of belief (or credences) and not only "full" (or "categorical" or "binary") beliefs. The epistemic stance of agents toward propositions is a matter of degree. These degrees reflect how confident they are that these propositions are true.
- (B2) Probabilism: the degrees of belief of a rational agent can be represented by a probability distribution.
- (B3) Conditionalization: in the light of new evidence, a rational agent updates his or her degrees of belief by relying on conditionalization.

The remainder of this subsection will be devoted to tenets (B1) and (B2), the next one to (B3). Neither instantialism nor hypothetico-deductivism take into account degrees of belief. Bayesianism's first tenet is that these degrees of belief do matter crucially. As such, it is a vague claim. But probabilism (B2) helps to make it more precise. It states that the credences of a rational agent obey the theory of probability. Assume that the degree to which the agent believes that H is true is denoted by $P(H)$. Then (B2) states that $P(\cdot)$ is a probability distribution, namely, that the following axioms are satisfied by P:

(A1) $P(H) \geq 0$ for any H
(A2) $P(H) = 1$ if H is a logical truth
(A3) $P(H_1 \vee H_2) = P(H_1) + P(H_2)$ if H_1 and H_2 are logically incompatible

Axiom (A1) (resp. A2) expresses the fact that the minimal (resp. maximal) degree of belief is 0 (resp. 1). To any sentence is attributed a degree of belief between 0 and 1. (A3) is often seen as the crucial property of probability distributions and is called additivity.[31] Other properties follow straightforwardly from (A1) to (A3):

$P(H) = 1 - P(\neg H)$
$P(H) = 0$ if H is a logical falsehood
If H_1 and H_2 are logically equivalent, then $P(H_1) = P(H_2)$
$P(H_1) = P(H_1 \wedge H_2) + P(H_1 \wedge \neg H_2)$

Bayesians agree that if an agent is rational, then the agent's degrees of belief obey (A1)–(A3). Radical Bayesianism also holds the converse: as far as degrees of belief are concerned, an agent is rational as soon as he or she obeys (A1)–(A3). In other words, in this domain, there is no other norm of rationality than the one expressed by the theory of probability. In particular, if a rational agent is informed of objective probabilities, the agent is not forced to bring his or her degrees of belief into line with these probabilities. Another important feature of Bayesianism in the present context is that it supposes that scientists assign probabilities to hypotheses (and theories). This assumption plays a crucial role in the Bayesian analysis of confirmation, but it should be stressed, first, that it is often criticized, and second, that it is not an assumption all application of probability to inductive reasoning makes.[32] For instance, classical statistics does not rely on it.

3.2 CONDITIONALIZATION AND BAYES'S THEOREM

(B2) can be seen as the fundamental static (or "synchronic") Bayesian claim. By contrast, (B3) is a dynamic (or "diachronic") thesis that deals with belief revision. (B3) states that a rational agent has to revise his or her credences according to conditionalization. Upon learning that E is the case, his or her belief shifts from the initial (or prior) probability $P(H)$ to the new (or posterior) probability, $P(H \mid E)$ which is defined as the conditional probability:

- $(H \mid E) =_{\text{def}} P(E \wedge H) / P(E)$ when $P(E) > 0$

The verification that $P(. \mid E)$ satisfies (A1)–(A3) is left to the reader. It is worth pointing out two features of conditionalization which have been much discussed.

[31] Mathematicians usually rely on a slightly distinct axiomatization of probability. First, probabilities are assigned not to sentences but to sets (so-called "events"). Second, denumerable (and not finite) addivity is assumed. See Skyrms (1966) or Hacking (2001) for introductions to probability suited to philosophers.

[32] On both points, see notably Mayo (1996).

First, conditionalization is partial: if the evidence E has a null prior probability, then conditionalization does not constrain the posterior degrees of belief. Second, conditionalization applies to evidence viewed as certain. Philosophical objections can be raised against this assumption: are we ever certain that E is the case? R. Jeffrey, one of the main figures of contemporary Bayesianism, has proposed a generalization of conditionalization known today as the Jeffrey Rule. This rule deals with cases where an agent revises his or her beliefs on the basis of evidence whose probability is not necessarily maximal. Assume that after some observation, the probability of E changes from $P(E)$ (the prior) to $P^*(E)$. What should be the new probability distribution $P^*(\cdot)$? The Jeffrey Rule states that for any proposition H, $P^*(H) = P(H|E) \cdot P^*(E) + P(H|\neg E)P^*(\neg E)$. It is easy to check that in the limiting case where $P^*(E) = 1$, it boils down to conditionalization. In the remainder of the chapter, however, we will keep the usual idealization according to which people revise their beliefs upon receiving evidence that they consider as certain.

Bayes' Theorem is an obvious consequence of the definition of conditional probability:[33]

- (BT1) $P(H|E) = \left[P(E|H) \cdot P(H)\right] / P(E)$ when $P(H), P(E) > 0$

In the context of confirmation theory, Bayes' Theorem indicates how to determine the probability of H given the evidence E on the basis of prior probabilities (of E and H) and of the probability of E given H $P(E|H)$. $P(E|H)$ is sometimes referred to as the likelihood of H on E. It can be viewed as the degree to which the hypothesis H predicts E. It is easy to see that if H logically implies E, then $P(E|H)$ is maximal and that if H implies $\neg E$, then $P(E|H)$ is null. Often it would be quixotic to assume that $P(E)$ is directly known, though it can be computed if one knows $P(E(\neg H))$:[34]

- (BT2) $P(H|E) = \left[P(E|H)P(H)\right] / \left[P(E|H)P(H) + P(E|\neg H)P(\neg H)\right]$ when $P(H)$, $P(\neg H), P(E) > 0$

This second form of Bayes' Theorem can be generalized to the case where one considers n mutually exclusive and collectively exhaustive hypotheses $H_1, \ldots, H_n$. In this case, for any H_i $(0 \leq i \leq n)$,

- (BT3) $P(H_i|E) = [P(E|H_i).P(H_i)] / \Sigma_j [P(E|H_j).P(H_j)]$ where $P(H_j)$, $P(E) > 0$

3.3 JUSTIFYING BAYESIANISM

Why should the degrees of belief of a rational agent obey the theory of probability (B2)? At almost the same time, though independently, De Finetti (1937) and Ramsey

[33] See Hacking (2001), chap. 15 and Joyce (2007). Hacking (2001) includes several examples and exercises.

[34] In some areas (typically in medical statistics), $P(E|H)$ and $P(E|\neg H)$ are called the true positive rate (or sensitivity) and the false positive rate, respectively: if E is a positive answer of a test whose function is to determine whether H (e.g., a pregnancy test), $P(E|H)$ is the probability of a positive answer when H is true, and $P(E|\neg H)$ the probability of a negative answer when H is false.

(1926) constructed an argument known today as the Dutch Book Argument, the aim of which is to show that if an agent bets on the basis of non-probabilistic degrees of belief, "a book [can] be made against him by a cunning bettor" (Ramsey). More precisely, such an agent should be willing to accept a set of bets (a so-called Dutch Book) such that, whatever happens, he or she is sure to lose money. It can be shown that an agent has probabilistic degrees of belief iff he or she is invulnerable to a Dutch Book.

Assume for instance that (i) Paul believes to degree 0.4 that H is true and to degree 0.7 that H is false, and that (ii) Paul's degrees of belief are reflected in his betting odds. This means that Paul is willing to pay 0.4 m dollars for a bet that pays m dollars if H is the case, and nothing otherwise. Mary (the bettor) can offer Paul two bets such that a net loss is certain. Assume for instance that m = 10 dollars and that Mary offers

- Bet n°1 on H (price: 0.4×10 dollars), and
- Bet n°2 on $\neg H$ (price: 0.7×10 dollars)

If H is true, then Paul will obtain $10 - (0.4 \times 10 + 0.7 \times 10) = -1$ dollar. If H is false, then Paul will also lose 1 dollar. In both cases, Paul loses money. A similar argument, the Diachronic Dutch Book, has been proposed by David Lewis to justify conditionalization (Teller, 1973; Lewis, 1999, chap. 23 "Why Conditionalize?").

The justification of probabilism is a disputed issue. The Dutch Book Argument belongs to the family of pragmatic arguments in favor of probabilism. These pragmatic arguments are intended to show that non-probabilistic degrees of belief induce irrationality in action (or in preferences over options).[35] However, some object that pragmatic arguments reduce degrees of belief to their role in action and neglect their epistemic dimension. This led Joyce (1998) to put forward a purely epistemic argument in favor of probabilism. Joyce axiomatically characterizes a set of accuracy measures for degrees of belief and shows that for any such measure, if the degrees of belief of an agent are not probabilistic, there exists a probability distribution which is more accurate in every possible state of the world.

Before turning to Bayesian confirmation theory (BCT), it is worth noting that all Bayesian confirmation theorists do not attach the same importance to the justification of probabilism. It is obviously important for some of them (e.g., Howson & Urbach, 1989), but others leave it largely aside and focus instead on the ability of BCT to account for the scientific practice of confirmation.

4. Bayesian Confirmation Theory

4.1 DISTINCT CONCEPTS OF CONFIRMATION IN BCT

The distinction between absolute and incremental confirmation can easily be made precise in a probabilistic framework. On the one hand, E absolutely confirms H iff the

[35] The other main pragmatic argument is based on an axiomatic for preferences. The basic results are provided by contemporary decision theory, in particular by Savage (1954/1972).

probability of H given E is above some threshold k (typically $k \geq \frac{1}{2}$): $P(H|E) > k$ (the Threshold Criterion). For a detailed presentation of this absolute probabilistic concept, see Crupi (2014, § 3.1–3.2). On the other hand, E incrementally confirms H if the probability of H given E is higher than the probability of H:

- E B-confirms H iff $P(H|E) > P(H)$.
- E B-disconfirms H iff $P(H|E) < P(H)$
- E is neutral toward H iff $P(H|E) = P(H)$

The concept of B-confirmation is also sometimes called the "positive relevance" concept of confirmation. Assuming conditionalization, E B-confirms H from an agent's point of view if his or her confidence in H would be increased upon learning that E. Current BCT focuses on the incremental rather than on the absolute (probabilistic) concept of confirmation. One argument in favor of this priority is that, in some cases of conflict, our intuitions are more in line with the incremental concept (see Salmon, 1975). Suppose, for instance, that upon learning E, Paul's degrees of belief in H decreased—that is, $P(H|E) < P(H)$—but that Paul's degree of belief in H given E was above the threshold k. In this case, there is absolute but not incremental confirmation. But we are reluctant to say E confirms H. It's unclear, however, whether or not our intuitions are conclusive. Consider the following counterexample (Achinstein, 1978, 2001). Assume that Paul is a good swimmer and was in fine shape on Wednesday morning. The fact that he was swimming on Wednesday (E) is likely to increase the probability that he drowned on Wednesday (H). But we hesitate to say that E confirms H. And one potential reason for this hesitation could be that we do not consider H to be probable enough given E. For a discussion of this and other putative counterexamples, see notably Kronz (1992) and Maher (1996). Another reason to opt for the incremental concept is that, in a probabilistic framework, there are some intrinsic difficulties with the absolute one. It is well known that it may be the case that $P(H_1|E) > k$, $P(H_2|E) > k$ and $P(H_1 \wedge H_2 \mid E) < k$. In other words, the absolute concept of confirmation as modeled by the Threshold Criterion violates the

- (C6) Composition Condition: if E confirms H_1 and confirms H_2, then E confirms $H_1 \wedge H_2$

Absolute confirmation is arguably connected to acceptance in the sense that if E (absolutely) confirms H, then (under appropriate epistemic conditions) E is a reason to accept that H is the case.[36] Given this connection, the Composition Condition is very plausible as a condition of adequacy for absolute confirmation.[37] It turns out that these difficulties do not concern only the Threshold Criterion: Zwirn & Zwirn (1996, Thm. 7) have shown that any confirmation relation that satisfies a set of minimal

[36] Cf. Achinstein's (1978) "principle of reasonable belief" and Zwirn & Zwirn (1996).

[37] The same line of argument could be given for the Special Consequence Condition.

requirements, the Composition Condition, and that is not reducible to deduction, cannot be represented in a probabilistic framework.[38]

Before turning to the analyses that can be made on the basis of the concept of B-confirmation, two comments are in order. First, Bayesians often stress the importance of background beliefs *K*. Accordingly, they use a more fine-grained notion of confirmation:

- *E* B*-confirms *H* relative to *K* iff $P(H \mid E \wedge K) > P(H \mid K)$

To keep things simple, we will nonetheless rely on B-confirmation. Second, even if BCT is based on a quantitative concept of partial belief, the very concept of B-confirmation itself is a qualitative concept of confirmation. B-confirmation is silent on the degree to which *E* confirms *H*. However, one of the attractive features of BCT is that it allows one to also develop a quantitative concept, viz. a measure of confirmation. Such a measure is generically noted $c(H,E)$. A natural candidate for $c(H,E)$ is the difference between the prior and posterior probabilities of *H*:

$$d(H,E) = P(H \mid E) - P(H)$$

The measure $d(.,.)$ is positive (resp. negative) if *E* B-confirms (resp. B-disconfirms) *H*. It will be the default measure in the rest of this chapter, but there exist alternative measures in the literature (see Fitelson, 2001), to which we will come back further on.[39]

4.2 SOME BAYESIAN ANALYSES

In this subsection, we will see BCT "at work" by presenting a sample of Bayesian analyses. This sample by no means exhausts BCT's applications.[40] However, it should suffice to show why BCT is currently the dominant theory of confirmation.

---

[38] For the precise formulation of this result, I refer the reader to the original article. The notion of "representability" of a confirmation relation by a probabilistic concept is akin to the axiom of "formality" mentioned earlier. Note, however, that a confirmation relation satisfying these Conditions can be elaborated in other formal frameworks (notably the possibilistic framework of Dubois and Prades). I would like to thank H. and D. Zwirn for having brought these results to my attention.

[39] Note that one could also develop a quantitative measure of the absolute concept of confirmation. The most straightforward one is of course $c(H,E) = P(H \mid E)$. In Crupi (2014), it is shown how to differentiate axiomatically the quantitative measures that induce an absolute concept from those that induce an incremental one. (In what follows, I simplify the formulation of the results.) Both families of measures obey the axiom of formality according to which $c(H,E)$ depends only on $P(H \wedge E)$, $P(E)$, and $P(H)$, and the axiom of 'final probability' according to which $c(H, E_1) \leq c(H, E_2)$ iff $P(H \mid E_1) \leq P(H \mid E_2)$. One obtains measures of absolute confirmation by adding the axiom of 'logical equivalence' according to which if H_1 and H_2 are logically equivalent given *E*, $c(H_1, E) = c(H_2, E)$. By contrast, one obtains measures of incremental confirmation by adding the axiom of 'tautological evidence' according to which, if **T** denotes a tautology, $c(H_1, \mathbf{T}) = c(H_2, \mathbf{T})$.

[40] For instance, Bayesians have attempted to reconstruct the notion of "ad hoc hypothesis" or the idea that the variety of evidence has a special confirmational strength—on this last topic, see Horwich (1982, pp. 118 and sq) and Earman (1992, chap. 5, sec. 3).

- Bayes' Theorem and hypothetico-deductive theories.

BCT has the ability to account for lots of confirmational intuitions. From BCT and Bayes' Theorem (BT1), it follows immediately that

(1) All other things being equal,[41] the more *E* is probable given *H*,[42] the more *H* will be confirmed by *E*.
(2) All other things being equal, the less *E* is probable, the more the hypothesis *H* will be confirmed by *E* ("surprise principle," Joyce).
(3) *E* B-confirms *H* iff $P(E|H) > P(E|\neg H)$

Property (1) implies welcome relations between logical consequence and confirmation. (a) If *E* is logically incompatible with *H*, then $P(E|H) = 0$ and therefore the degree of disconfirmation of *H* by *E* is maximal (relative to the prior probability of *H*). (b) If *E* is logically implied by *H*, then $P(E|H) = 1$ and therefore the degree of confirmation of *H* by *E* is maximal (relative to the prior probabilities of *H* and *E*). BCT preserves a central feature of hypothetico-deductive theories: if *E* is logically implied by *H*, then *H* is confirmed by *E*. This follows from the fact that if *H* implies *E*, then $P(H \wedge E) = P(H)$. Hence, $P(H|E) = P(H)|P(E) > P(H)$ if $0 < P(H), P(E) < 1$ (Huygens's Rule, Jeffrey 1992).[43] In other words, if *E* and *H* are initially neither certainly false nor certainly true, *H* is necessarily B-confirmed by *E*. BCT therefore justifies the fundamental intuition underlying HDTC and an important part of actual scientific practice. Note that, in general, *E* B-confirms *H* iff $P(E|H) > P(E)$.

BCT is able to overcome some of the difficulties that HDTC faces. One of these is the problem of non-relevant conjunction: if *E* HD-confirms *H*, then necessarily *E* HD-confirms $(H \wedge H')$. BCT partially inherits this problem when *H* logically implies *E*: if $0 < P(H), P(H'), P(E) < 1$, then *E* B-confirms *H* and also $(H \wedge H')$. Nonetheless, this does not hold in full generality (as in HDTC): it is not true that if *E* B-confirms *H*, then for any *H'*, *E* B-confirms $(H \wedge H')$. (Unlike the notion of logical consequence, probabilistic dependence is not monotonic.) Moreover, when *H* logically implies *E*, if one relies on the difference measure, then the degree of confirmation conferred to *H* by *E* is higher than what *E* conferred to $(H \wedge H')$ (Earman, 1992, pp. 63–65).[44]

We are now in a position to compare more systematically the three main theories of confirmation studied so far from the point of view of their general properties.

Property (2) states that, all other things being equal, unexpected evidence has a strong confirmational power. The ceteris paribus clause is important: a hypothesis is (luckily) not confirmed by any improbable evidence. But if two data *E* and *E'* are

---

[41] The ceteris paribus clause is indispensable: if the two other variables $P(H)$ et $P(E)$ are not fixed, the claim is false.

[42] Or to use the terminology introduced previously: the more *H* "predicts" *E*.

[43] We will come back later to the case $P(E) = 1$.

[44] See Fitelson (2002) for a recent discussion of the Bayesian treatment of the problem of non-relevant conjunction.

TABLE 1.

Comparison of some properties of H-, HD-, and B-confirmation.

	H-confirmation	HD-confirmation	B-confirmation
Supraclassicality	Yes	No	Yes
Special Consequence	Yes	No	No
Converse Consequence	No	Yes	No
Conversion	No	Yes	Yes

predicted to the same degree by H, then H is more confirmed by the one which has the lowest initial probability. Bayesians see property (2) as a virtue of BCT.[45] Let us consider the following example (which is not supposed to be medically accurate). Even if scarlatina is invariably accompanied by a high fever and a rash, Paul's rash is arguably better evidence for scarlatina since it is a rarer symptom than a high fever. It is worth noting that, unlike property (1), property (2) is specific to BCT (by comparison with HDTC). Lastly, property (3) states that E B-confirms H exactly when H predicts "more" E than its negation does: it is more probable that E is true if H is true than if it is not.

- The ravens paradox[46]

Nicod's Criterion states that a positive instance $(Ra \wedge Ba)$ confirms its associate UC-sentence $\forall x(Rx \rightarrow Bx)$. If one accepts the Equivalence Condition, it implies that a positive instance $(\neg Ra \wedge \neg Ba)$ of "All non-black things are non-raven" confirms "All ravens are black." It is easy to check that the Equivalence Condition is necessarily satisfied by BCT. The questions of interest are therefore the following:

(Q1) Does BCT satisfy Nicod's Criterion?
(Q2) Are there situations where data like $(\neg Ra \wedge \neg Ba)$ B-confirm $\forall x(Rx \rightarrow Bx)$?
(Q3) Are there differences in the degrees to which $(Ra \wedge Ba)$ and $(\neg Ra \wedge \neg Ba)$ confirm $\forall x(Rx \rightarrow Bx)$?

As for Nicod's Criterion (Q1), it has been shown that it is not universally satisfied in BCT. Indeed, there are situations where Nicod's Criterion is not intuitive. Consider the sentence

"All foxes are outside Paris"

and assume that a fox has been observed outside Paris, but very close to the boundaries of the city. We have here a positive instance, but does it bring any support to its associate universal sentence? This does not seem to be the case. Foxes can move

45 E.g., Howson & Urbach (1989, pp. 86–88).

46 See Horwich (1982, pp. 54 and sq.), Earman (1992, pp. 69–73), Vranas (2004), Fitelson & Hawthorne (2006), Fitelson (2006a).

(at least, we can assume that this one is free to move) and if we have observed one fox very close to the boundaries of Paris, it does not seem unlikely that another one is in Paris, or that this one has been or will be in Paris. In short, this positive instance appears to decrease our confidence in its associate universal sentence.[47] Hempel (1967) disagrees, and claims that this kind of counter-example does not succeed in putting Nicod's Criterion into question. His objection is that it relies on background beliefs (in our example, about the geography, the foxes, and so on) whereas Nicod's Criterion should be understood as follows: if one relies on evidence $E = (Ra \wedge Ba)$ and nothing else, then E necessarily confirms its associate universal sentence. The idea of assuming nothing but E (or, equivalently, of assuming a degenerate background set of beliefs K containing only logico-mathematical truths) is problematic from a Bayesian point of view since it boils down to excluding individual differences, which are explicitly allowed by most versions of Bayesianism. By contrast, from the point of view of logical probability (e.g., Carnap, 1950/1962, more on this later), this idea is much more natural. Nevertheless, Maher (2004, sec. 3.8.) recently showed in a probabilistic neo-Carnapian framework that Nicod's Criterion was not valid for the incremental concept of confirmation either.

Let us turn now to the counter-intuitive conclusion of the paradox of the ravens: is it possible that evidence $(\neg Ra \wedge \neg Ba)$ confirms $\forall x(Rx \rightarrow Bx)$ (Q2)? BCT gives a positive answer to this question. It also may capture the intuitive idea that $(Ra \wedge Ba)$ confirms more the sentence $\forall x(Rx \rightarrow Bx)$ than $(\neg Ra \wedge \neg Ba)$ does—an idea first proposed by Hosiasson-Lindenbaum (1940) in one of the earliest contemporary study of confirmation—and that $(\neg Ra \wedge \neg Ba)$ brings very weak support to $\forall x(Rx \rightarrow Bx)$ (see Vranas, 2004; and Fitelson, 2006a) (Q3). To deliver such a result, one needs some assumptions: (i) the probability that a is a raven is very low compared to the probability that it is black and (ii) the probability that a is a raven or is non-black is independent of the probability of $\forall x(Rx \rightarrow Bx)$. Under these assumptions, one can show that

- $P(\forall x(Rx \rightarrow Bx) | (\neg Ra \wedge \neg Ba)) > P(\forall x(Rx \rightarrow Bx))$

  [i.e. $(\neg Ra \wedge \neg Ba)$ B-confirms $\forall x(Rx \rightarrow Bx)$]

- $c((\neg Ra \wedge \neg Ba), \forall x(Rx \rightarrow Bx)) = \varepsilon$ for a "small" ε

  [i.e. $(\neg Ra \wedge \neg Ba)$ to a small degree confirms $\forall x(Rx \rightarrow Bx)$]

- $P(\forall x(Rx \rightarrow Bx) | (Ra \wedge Ba)) > P(\forall x(Rx \rightarrow Bx) | (\neg Ra \wedge \neg Ba))$

  [i.e. $(Ra \wedge Ba)$ confirms more $\forall x(Rx \rightarrow Bx)$ than $(\neg Ra \wedge \neg Ba)$ does]

---

[47] Another counter-example is given by Good (1967): let us suppose that we know our world can be described by one of the two following hypotheses. According to the first hypothesis, there are 100 black ravens, no non-black ravens, and 1,000,000 other birds. According to the second, there are 1000 black ravens, 1 white raven, and 1,000,000 other birds. A positive instance of "All ravens are black" could increase our confidence in the second hypothesis and therefore B-disconfirm the UC-sentence.

Even if the independence assumption is disputed (see Vranas, 2004), the previous analysis shows the benefit that BCT can draw from the richness of the probabilistic framework, which allows one to discriminate between the confirmational abilities of $(Ra \wedge Ba)$ and $(\neg Ra \wedge \neg Ba)$.

- The Duhem-Quine problem

Dorling (1979) and Howson & Urbach (1989) put forward a Bayesian analysis of the Duhem-Quine problem.[48] The problem can be exposed as follows. Very often, a hypothesis H does not have empirical consequences by itself, but only in conjunction with some set of auxiliary hypotheses. Let A be their conjunction. Assume now that the empirical evidence is incompatible with $(H \wedge A)$: $(H \wedge A)$ implies $\neg E$ and E is the case. This means that $(H \wedge A)$—but not H alone—is refuted. One of the epistemic issues raised by this situation is to determine which proposition is likely to be false. As Duhem puts it,

> the only thing the experiment teaches us is that, among all the propositions used to predict the phenomenon and to verify that it has not been produced, there is at least one error; but where the error lies is just what the experiment does not tell us. (Duhem 1906/1914, Part II, Chap. VI, § II)

In such situations, even if there is a logical underdetermination, we often discriminate between the propositions involved in $(H \wedge A)$. Some are more disconfirmed than others by the observation that E is true. (Note that it is not necessarily the case that all these propositions are disconfirmed by E. In some situations, it may even be plausible that the probability of some of them is increased by E). BCT is able to describe these distinctions. Howson & Urbach (1989) give an example from chemistry. They consider the hypothesis H according to which the atomic weight of an element is a whole-number multiple of the atomic weight of hydrogen (Prout, 1815). In this case, the auxiliary assumptions A consist mainly in assuming the accuracy of the measuring technique. The measurements then available were not consistent with what H (together with A) predicts. According to Howson and Urbach, even if chemists initially believed strongly in H (let us say, to degree 0.9) and rather strongly in A (e.g., 0.6), they could justifiably revise their beliefs upon learning the results of the measurements in such a way that (i) their confidence in H was still very strong (e.g., 0.878), but (ii) their confidence in the reliability of the measurement technique decreased dramatically (e.g., 0.073). It is therefore a case of "light" B-disconfirmation of H and "massive" B-disconfirmation of A. A similar example has also been proposed by Dorling (1979). In general, if H is the hypothesis under examination, A the conjunction of auxiliary assumptions, and if $(H \wedge A)$ implies the

[48] See Earman (1992, pp. 83 and sq.).

negation of E, then several confirmational possibilities are open according to BCT. It can be the case that

- H (but not A) is B-disconfirmed by E (and conversely)
- H and A are B-disconfirmed[49]
- Neither H nor A are B-disconfirmed[50]

4.3 PROBLEMS FOR THE BCT

BCT's justifications and (perhaps more importantly) applications explain why it is currently the most popular confirmation theory. However, it faces serious difficulties. We now turn to two of them: the Popper-Miller objection and the so-called problem of old evidence.[51]

- The Popper-Miller objection

In a paper published in 1983, K. Popper and D. Miller presented an argument whose aim was to establish the impossibility of inductive logic. This argument directly questions the notion of increase in probability, which is at the core of BCT. Assume that H implies E, $0 < P(E) < 1$ and $P(H|E) \neq 1$. It can be shown that $P(H \vee \neg E) > P(H \vee \neg E | E)$.[52] In other words, the disjunction $(H \vee \neg E)$ is B-disconfirmed by E, even if H is B-confirmed by E. Why is this result problematic for BCT? Note that H is logically equivalent to $(H \vee E) \wedge (H \vee \neg E)$. It turns out that each of these two conjuncts is the weakest proposition that is strong enough to imply H in the presence of the other conjunct. Given that the first conjunct is implied by E, these properties lead Popper and Miller to see the second one as the content of H that "goes beyond E." If one follows this interpretation, the previous result means that, even if H is B-confirmed by E, the content of H that goes beyond E is necessarily B-disconfirmed by E. Popper and Miller conclude from what precedes that the idea that the increase in probability represents the inductive support conferred by E to H is an "illusion" and that "all probabilistic support is purely deductive." Another way to expose the objection begins by remarking that, in virtue of the initial assumptions,

$$d(H, E) = d(H \vee \neg E, E) + d(H \vee E, E)$$

In other words, the quantity of support conferred by E to H can be additively decomposed into (i) the support conferred to $(H \vee \neg E)$ and (ii) that conferred to $(H \vee E)$. Popper-Miller's result implies (under the assumptions that $0 < P(E) < 1$ and $P(H|E) \neq 1$)

[49] Hajek & Joyce (2008).

[50] See Salmon (1973) quoted by Earman (1992, p. 83).

[51] Other problems for BCT are discussed by Earman (1992, chap.4) and Norton (2011).

[52] Actually, it can be shown that $P(H \vee \neg E) - P(H \vee \neg E | E) = P(H \vee \neg E) - P(H|E) = P(\neg H \vee E) \cdot P(\neg E) > 0$

that $d((H \vee \neg E), E)$ is strictly negative. Gillies (1986) rephrases their objection by relying on this additive decomposition and assuming that $d(H \vee E, E)$ measures deductive support and $d(H \vee \neg E, E)$ inductive support.[53]

The Popper-Miller objection looks devastating for BCT and, more generally, for any probabilistic theory of confirmation based on the increase of probability. However, Bayesians have offered several replies. (i) A first issue concerns the interpretation of $(H \vee \neg E)$ as the content of H that goes beyond E. This view is rejected by supporters of BCT (Jeffrey, 1984; Howson & Urbach, 1989, 265).[54] Popper and Miller address this issue in Popper & Miller (1987, § 3). (ii) These supporters also argue that it is fallacious to infer the anti-inductivist conclusion from the additive decomposition of $d(H,E)$. Just because $d(H,E)$ can be decomposed into two functions, which admittedly cannot measure confirmation in isolation, this does not mean that it cannot measure it.[55] There exist other decompositions of $d(H,E)$. For instance,

$$d(H,E) = d(H \wedge \neg E, E) + d(H \wedge E, E)$$

from which one could draw the opposite conclusion.[56] It seems therefore problematic to ground the BCT's construal of inductive support on the first decomposition of $d(H,E)$. (iii) Finally, Eells (1988) points out that even if one accepts the main part of the Popper-Miller objection, it does not follow from the fact that E B-confirms only the content of H which it logically implies (i.e., $H \vee E$) that the probabilistic confirmation relation is purely deductive. It is easy to set up a pair of examples $[(H_1, E_1), (H_2, E_2)]$ such that, even if $d(H_1 \vee \neg E_1, E_1) = d(H_2 \vee \neg E_2, E_2) < 0$, E_1 B-confirms H_1 whereas E_2 B-disconfirms H_2. Consequently, the support conferred by E_i to H_i varies noticeably from one case to the other. According to Eells, this variation shows that, even if it is only the content of H which is logically implied by E that is supported, this support displays an essentially non-deductive dimension.[57]

- The problem of old evidence

Another concern for BCT, even more discussed than the Popper-Miller objection, is the problem of old evidence. This was formulated by C. Glymour in a set of arguments against BCT (1980, pp. 85 and sq.). It can be exposed as follows. During the second half of the Nineteenth Century, astronomical observations showed that the advance of Mercury's perihelion (574 arc seconds per century) diverged from what could be

[53] Popper and Miller do not state their case in this way.

[54] See also Zwirn & Zwirn (1993).

[55] This rejoinder is made in a debate with Gillies (Chihara & Gillies, 1988). See Earman (1992), p. 95 and Horwson & Urbach (1989), p. 264.

[56] See Dunn & Hellman (1986).

[57] One potential reply to this objection, and which is endorsed by D. Miller (personal communication), consists in denying that a relation depending essentially on a probabilistic measure is ipso facto inductive.

predicted on the basis of Newtonian theory.[58] Let E be the report of these observations and H the general theory of relativity (GTR). Assume moreover that H implies E. And consider the situation in 1915, when Einstein formulated GTR. Einstein knew the advance of Mercury's perihelion, hence from his point of view $P_{1915}(E)=1$. These observations were considered by Einstein (and by physicists in general) to be first-class empirical evidence for GTR. It is therefore quite natural to expect of a theory of confirmation that, when applied to this episode, it assigns a strong confirmational power to E. But it is a straightforward consequence from probability theory that $P_{1915}(H|E)=P_{1915}(H)$. It follows that E does not B-confirm H. The least we can say is that there is a large discrepancy between our evidential intuitions and B-confirmation.

The problem of old evidence is of course not restricted to the confirmation of GTR by the advance of Mercury's perihelion. If evidence E is known, E can neither B-confirm nor B-disconfirm any hypothesis H. This is the qualitative version of the problem. The quantitative version lies in the fact that when the probability of E goes to 1, $d(H,E)$ goes to 0.[59] The latter version shows that the problem of old evidence and the "surprise principle" (see 4.2.) are really two faces of the same coin. The problem lies at the very core of BCT, so that most contemporary confirmation theorists would agree with Maher (1996) that the problem of old evidence shows that BCT in its simplest form is inadequate. Therefore, it must be revised. Before presenting the revisions that have been proposed by Bayesians, it must be pointed out that there are at least two distinct problems of old evidence. On the one hand, there is the "increment problem"[60]: how could known data E increase our confidence in a hypothesis H? How, for instance, could Einstein's confidence in GRT have been increased by the consideration of the known astronomical data on the advance of Mercury's perihelion? On the other hand, there is the "survival problem":[61] how could the confirmational power of evidence E survive its being learned? In BCT, after an evidential statement has been learned, it can neither confirm nor disconfirm a hypothesis.

A first approach to the problem of old evidence consists in developing a "(logically) de-idealized Bayesianism." The basic idea is that in an epistemic situation like the one faced by Einstein in 1915, what increases his confidence in GRT is that, at some point, he becomes aware of the fact that GRT predicts the advance of Mercury's perihelion. In others words, Einstein would have learned some logico-mathematical knowledge. Standard Bayesianism assumes that agents are logically omniscient, i.e. that they believe all logical truths and all logical consequences of their beliefs. It has therefore to

[58] Newtonian theory can "only" account for 531 arc seconds per century (by applying a perturbative approach to the two-body system Sun-Mercury).

[59] Under the assumption that H implies E, if $P(E)=1-\varepsilon$ then $d(H,E)\leq \varepsilon/(1-\varepsilon)$.

[60] The increment problem roughly corresponds to the "historical problem of old evidence" (Garber, 1983), the "problem of new old evidence" (Eells, 1990), the "diachronic problem of old evidence" (Christensen, 1999), and the "problem of new hypothesis" or the "problem of logical learning" (Joyce, 1999).

[61] The survival problem corresponds roughly to the "ahistorical problem of old evidence" (Garber, 1983), the "synchronic problem of old evidence" (Christensen, 1999), and the "problem of evidential relevance" (Joyce, 1999).

be revised in such a way that it becomes capable of accounting for logical ignorance and logical learning (Garber, 1983, Jeffrey 1983). This de-idealized Bayesianism can at best solve the increment problem: once the agent has learned that E is a consequence of H, nothing is left for confirming H.

The survival problem motivates another kind of approach, which may be called "historicized Bayesianism." Assume that we are at time t and that E is known at t (hence $P_t(H|E) = P_t(H)$ where $P_t(\cdot)$ denotes the agent's epistemic state at t). E no longer has any confirmational power. One could be tempted into reasoning this way: if E has unfortunately lost its confirmational power upon being learned, why not go back through the agent's epistemic history to the time $t' < t$ at which he or she learned E. In standard BCT, confirmational judgments supervene on the agent's actual degrees of belief—if two agents have the same degrees of belief at t, then they have the same judgments. Historicized Bayesianism enlarges the basis on which these judgments supervene since it includes the whole epistemic history of the agent. But a shortcoming of this approach is precisely that it renders these confirmational judgments too dependent on the accidents of the agent's epistemic history (Christensen, 1999; see also Maher, 1996). Consider the following example.[62] Paul is wandering in the wood and discovers some stag droppings at t_1 (E_1), which strongly B-confirms the hypothesis that there is a stag in the wood. At t_2, he discovers some stag antlers (E_2), but given that the probability of H has been increased at t_1 by the discovery of the droppings, E_2 very weakly B-confirms H. Yet at the present time $t(>t_2>t_1)$, Paul could have the impression that E_1 and E_2 confirm H equally well. This is not the verdict of historicized Bayesianism, according to which E_1 has a stronger confirmational power than E_2. This is all the more counter-intuitive since, if chance had had it that Paul discover the antlers before the droppings, E_1 would have had a much weaker confirmational power than E_2.

A possible reaction to this issue consists in shifting from historicized Bayesianism to what may be called "counterfactual Bayesianism." According to this view, the confirmation conferred by E to H is determined by the increase in probability relative to the closest probability distribution where the agent does not know that E. This strategy boils down to considering the following question: if the agent did not initially know that E and then learned that E, would the probability of H be increased? Counterfactual BCT has famously been defended by Howson (1984, 1991) but has not been unanimously adopted. On the one hand, it is not clear that this approach can overcome all the difficulties faced by historicized Bayesianism. It all depends on the way one views the idea of the closest probability distribution where the agent does not know that E. In the scenario of the stag in the wood, if E_1 has probability 1 in the counterfactual probability distribution used for assessing E_1's confirmational power, then this power may be very weak. (At least, there will be some kind of symmetry: if E_2 also has probability 1 in the counterfactual probability distribution used for assessing E_1's confirmational power, then this power may also be very weak.) On

[62] Christensen (1999, pp. 444–445).

the other hand, counterfactual seems to be at best a solution to the survival problem. The problem of the increment, by contrast, is raised by the need to account for actual (not counterfactual) increase in probability.

The last approach we will sketch out has been put forward recently by Christensen (1999) and Joyce (1999). It relies on a measure of confirmation which is distinct from $d(.,.)$ and which can be motivated by the following remark: E B-confirms H iff $P(H|E)>P(H|\neg E)$, when $P(E)<1$. $s(H,E)=P(H|E)-P(H|\neg E)$ can thus be taken as another measure of confirmation. Prima facie, it is surprising to see the problem of old evidence tackled with $s(.,.)$ since it is undefined for $P(E)=1$. However, if one addresses the quantitative version (Christensen, 1999) or revises the Bayesian framework by allowing conditionalization on events with null probability (Joyce, 1999), then it turns out that $s(.,.)$ has some attractive features. Indeed, $s(.,.)$ can account for the fact that evidence E supports (even strongly supports) a hypothesis H though its prior probability be very close to 1. In general, $s(.,.)$ is able to neutralize the role of E's prior probability in confirmation.[63] For distinct reasons, neither Christensen nor Joyce view $s(.,.)$ as being the only appropriate measure for BCT. But even restricted in this way, their idea is not accepted by all Bayesians (see Earman, 1992, who discusses a similar approach, and Eells & Fitelson, 2000).

To sum up, there is currently no received solution to the problem of old evidence, which is still a major worry for BCT.

5. Bayesianism, Objectivity, and the Problem of Induction

We saw that an attractive theory of confirmation can be based on Bayesianism (§ 4). We saw also that the concepts of induction and confirmation are very closely related (§ 1). This naturally raises the issue of knowing whether Bayesianism can "solve" the famous problems raised by induction. This section will deal with this issue, which has been recently discussed (see Howson, 2000, and Strevens, 2004).[64]

5.1 THE PROBLEMS OF INDUCTION

Since the famous developments that D. Hume devoted to it (in the *Treatise of Human Nature*, 1739 and in *An Enquiry Concerning Human Understanding*, 1748), the problem of induction is viewed as one of the most fundamental problems in epistemology and general philosophy of science. From the Humean formulation to Goodman's "new

[63] This idea can be made precise as follows: $s(H,E)$ is invariant under learning a new probability for E according to Jeffrey's rule.

[64] This section is not intended to be a general overview of the problem of induction and will deal neither with classical treatments of the problem (e.g., Kant 1781/1787; Mill 1843, Book III) nor with contemporary ones which are alien to BCT. For such an overview, see Earman & Salmon (1992, Part II), Vickers (2014).

riddle of induction," the problem of induction has known many variations. It is worthwhile providing some preliminary clarification.

The problem of induction is often viewed as a problem of justification: what can be said in justification of the confidence we have in propositions that "go beyond" the empirical information that is available to us? The problem stems from the fact that empirical information does not conclusively demonstrate the truth or falsity of the propositions that go beyond it. Specifically, our empirical information does not logically establish the truth of the propositions we accept, nor the falsity of the propositions we reject. For instance, even if all the measurements we knew of were consistent with Ohm's Law, we would not have any logical guarantee that the Law is true. Notice that the case which is typically considered in the discussion of the induction problem is where one infers a universal sentence (often a UC-sentence) from a (finite) set of particular observations. However, we saw in Section 1 that inductive reasoning in the broad sense (or ampliative reasoning) goes beyond generalization (a.k.a. enumerative induction). An appropriate formulation of the problem of justification would therefore be as follows: how do we justify the "good" ampliative inferences on which we rely in ordinary life and in scientific reasoning? This version of the problem of induction, which corresponds closely to its traditional form, will be called the problem of the justification of induction-as-inference. Let $IND(P,C)$ be an inductive inference which infers the conclusion C from a set of premises P. For instance,

> P = "Up to now, any person who has jumped from the Eiffel Tower without a parachute died" and
> C = "The next person who jumps from the Eiffel Tower without a parachute will die."

Here, as in the general case, P does not imply C, therefore it is logically possible that P be true but not C. Is there a justification for the fact that we rely on $IND(.,.)$ to go from P to C? Hume famously argued against the existence of such a justification.[65] One way to state his argument is as follows. By assumption, the simple inference from P to C is not deductively secured. One may nonetheless claim that there is an implicit deductive inference based on P and a supplementary hypothesis. A candidate would be the hypothesis that Nature is (temporally) uniform, that is,

> U = If it has always been the case up to t that if x has property P then x has property Q, then it will be true of the next x observed after t that if it has property P then it will have property Q[66]

---

[65] We do not aim at exegetical rigor. We follow the common understanding of Hume, which assumes that a true justification must be deductive. This assumption is discussed and criticized in Stroud (1977, chap. 3).

[66] The role and status of this kind of principle of uniformity have been discussed since Hume (1739, I, III, VI).

Let us admit that *P* and *U* together imply *C*. Have we succeeded in justifying our use of *IND*(.,.)? Only if the supplementary hypothesis *U* is itself justified. But how can we justify *U*? It is neither a logical nor an analytic truth. Hence, if it is to be justified, it will be empirically. What could empirically support *U*? Maybe something like

> *P′* = It has always been true in the past that, when it has always been the case up to *t* that if *x* has property P then *x* has property Q, then it has also been true of the next *x* observed after t that if it had property P then it had property Q.[67]

P′ makes us arguably confident in *U*. How can this confidence be justified? *P′* does not deductively imply *U*. But if it is *IND*(.,.) which makes us infer *U* from *P′*, then our approach seems to be circular.[68] Hence, the argument goes, there is no sound justification for induction. This concludes our presentation of Hume's argument for inductive skepticism.

In what precedes, we have tacitly assumed that the inductive method *IND*(.,.), like the relation of logical consequence, is a matter of yes or no. This has been questioned by contemporary philosophers, and most notably by Carnap who, in his *Foundations* (1950/1962), sees inductive method rather as something which, given a set of premises *P* and a proposition *C*, determines the degree of support conferred by *P* to *C*. In this graded view of inductive method, *C* is no longer a "conclusion" in the sense that it would necessarily be reasonable for someone who accepts *P* to accept it. Carnap (1950/1962), §44 makes a similar point in terms of inference:

> The term 'inference' in its customary use implies a transition from given sentences to new sentences already possessed. However, only deductive inference is inference in this sense. If an observer X has written down a list of sentences stating facts which he knows, then he may add to the list any other sentence which he finds to be [logically implied] by sentences of his list. If, on the other hand, he finds that his knowledge confirms another sentence to a certain degree, he must not simply add this other sentence. The result of his inductive examination cannot be formulated by the sentence alone; the value found for the degree of confirmation is an essential part of the result.

---

[67] See also Mill (1843, book iii, chap. III) and Strawson (1952), pp. 251 and sq. on the "supreme premise of inductions."

[68] D. Hume (1748): "We have said that all concerning existence are founded on the relation of cause and effect; that our knowledge of that relation is derived entirely from experience; and that all our experimental conclusions proceed upon the supposition, that the future will be conformable to the past. To endeavour, therefore, the proof of this last supposition by probable arguments, or arguments regarding existence, must be evidently going in a circle, and taking that for granted which is the very point in question" (sec. IV, p. 26).

Let us accordingly assume that the inductive method *IND* is no longer a relation between sets of premises and conclusions but is a function which assigns to *P* and *C* the degree of support conferred by *P* to *C*. This change of view doesn't mean that the justification of induction problem disappears. In the same way that we asked why it is reasonable to accept *C* on the basis of *P*, we now ask why we should assign a degree of support or confidence of *r* to *C* on the basis of *P*. We may call this second, graded, version of the problem the problem of the justification of induction-as-support. (Note that both versions of the problem can be stated in a comparative way, irrespective of one's peculiar view of inductive method. Suppose that *IND* and *IND′* are two distinct inductive methods: the former is consistent with our inductive intuitions whereas the latter diverges strongly from them. What can justify our preference for *IND* over *IND′*?) The second version of the problem seems no easier to solve than the first one.[69]

In Section 2, when we discussed Hempelian instantialism, we introduced Goodman's grue paradox. One of the lessons one may draw from it is that a theory of confirmation which is based only on the logical form of the sentences is doomed to failure because there are intuitive evidential distinctions that it will not be able to account for. For instance, it will not be able to distinguish between the evidential bearing of our experience of emeralds on the "green" hypothesis and on the "grue" hypothesis. This raises a problem of induction which differs prima facie from the problems of its justification: it is the problem of the construction of an inductive method which "accords well with common sense and scientific practice" (Skyrms, 1966, p. 19). The distinction between the problem(s) of justification and the problem of construction is widely accepted. The latter is linked to the "new riddle of induction" that Goodman displays through the grue paradox. However, Goodman holds a conception of justification which blurs the distinction between the two problems. According to him, the justification of inductive reasoning proceeds in a way which is analogous to the justification of deductive reasoning. Both proceed via a back-and-forth between (potential) rules of reasoning and reasoning practice. A deductive rule is justified in so far as it accords with deductive practice, and our deductive practice is correct in so far as it obeys deductive rules. This idea, which is currently celebrated under the name of "reflective equilibrium," has to be understood dynamically: inferential practices and rules enter into a process of mutual adjustment up to the point where they reach a steady state. In the case of induction, this means that "predictions are justified if they confirm to valid canons of induction; and the canons are valid if they accurately codify accepted inductive practice" (1955, p. 64).

[69] The point is notably made by Goodman (1955, p. 62). Skyrms (1966, chap.2) provides an excellent reconstruction of the problem of the justification of induction-as-support. A survey of contemporary attempts to solve the problem is also given in the same chapter and, more recently, by Earman & Salmon (1992).

5.2 WHEN HUME MEETS BAYES

We are now ready to tackle the relation between Bayesianism and the problem of induction. On first glance, one may think that Bayesianism is able to solve it. (i) Bayesianism provides a framework and a criterion for characterizing the fact that evidence *E* supports the hypothesis *H* (and maybe that E supports H_1 more than H_2). (ii) BCT can account for lots of confirmational intuitions and practices. Hence, it can be considered as a plausible solution to the problem of the construction of an inductive method. (iii) Bayesianism can be justified, most famously by pragmatic arguments like the Dutch Book Argument. This suggests that BCT can also be viewed as a solution to the problem of the justification of induction.

Let us assume that evidence *E* B-confirms the hypothesis *H*. This means that, for a given individual, let's say Paul, his degree of belief in *H* is lesser than his degree of belief in *H* given *E*. It is however possible that for another agent, let's say Jean, who has distinct degrees of belief, *H* is B-disconfirmed by *E* without one of them being "wrong" from a Bayesian point of view. Paul and Jean just don't have the same degrees of belief. This situation displays a much discussed feature of BCT: its subjectivity. In some cases, the difference between Paul and Jean can be traced back to the fact that they received distinct information. But standard BCT allows Paul and Jean to have distinct degrees of beliefs even in the case where they have the same information. To put it in another way, standard BCT imposes very few constraints on the agents' priors. This feature is hard to reconcile with the expectations underlying the problems of induction. What we want to describe (and to justify) is an inductive method, something that tells us to which degree *E* supports *H*. (In the same way that a "deductive method" tells us what follows from what.) But in BCT, the answer to such a question depends on subjective elements, the individual's priors on *E* and *H*. This difficulty is hotly debated, and it is not easy to summarize these debates. Furthermore, it involves other fundamental issues like the interpretation of probability. In the remainder of the section, we will content ourselves with pointing out some salient elements of the discussion.

A straightforward reaction to the problem of subjectivity is to look for some "objective" priors. In its most extreme version, the idea is that if two individuals share the same evidence and background knowledge, they should rely on the same posterior probability. This view is often referred to as the "logical" view of probability and probabilistic confirmation, and is most prominently associated with the work of R. Carnap (1945, 1947, 1950/1962, 1952). The Carnapian project[70] is close to the ideas developed by Keynes in his *Treatise on Probability* (1921).[71] It aims at building an "inductive logic" which studies a relation of "partial implication" and is thus a generalization of deductive logic. Ideally, this inductive logic would deliver theorems like

[70] For a concise overview of the Carnapian program, see Hájek (2012, § 3.2). For a more detailed presentation, see Zabell (2011). For an introduction to inductive logic more generally, see Fitelson (2006b).

[71] On Keynes, see Gillies (2000), chap. 3.

'Evidence E confirms the hypothesis H to degree r' (*)

in the same way that deductive logic delivers theorems like

'Premise P implies consequence C' (**)

As Carnap puts it, "both statements [(*) – (**)] express a purely logical relation between two sentences" (1950/1962, § 10). The implementation of this project consists notably in imposing constraints on the set of possible probability functions in order to single out a (family of) logical probability functions. These constraints are typically axioms of symmetry (or of invariance), and they "may be regarded as representing the valid core of the old principle of indifference (or principle of insufficient reason)" (Carnap, 1962). A number of objections have been raised against Carnap's program (see e.g., Putnam 1963), so that it is largely abandoned today, despite still being defended by some philosophers (e.g., Maher, 1996, 2010). Among these objections figure the claims that the constraints envisioned are both too weak and too strong. On the one hand, these axioms are compatible with an infinite parametric family of probability functions (Carnap, 1952, 1963) among which the choice seems to be arbitrary. On the other hand, it is disputable that these axioms are truly logical constraints.

Another way of looking for objective constraints on priors consists in bringing them into line with chance (or "physical" probabilities). The second form of Bayes' Theorem (BT2) shows that if the likelihoods $P(E|H)$ and $P(E|\neg H)$ and the prior probability $P(H)$ are given, it is sufficient to determine $P(H|E)$. It turns out that, in a wide range of cases, likelihoods can be based on objective grounds. First, when H implies E or $\neg E$, the likelihood is trivially fixed (1 or 0) and is the same for every agent. Second, there exists a vast array of favorable cases: when the hypothesis H is statistical, that is, when it involves chance (or physical probability), or when H is connected to empirical data through auxiliary statistical assumptions. For instance, if

H = "the chance that a carbon 14 nucleus decays within 5370 years is one-half,"
A = "*a* is a carbon 14 nucleus," and
E = "*a* will decay within the next 5370 years"

then the probability of E given H (and the background assumption A) can be seen as being (objectively) one-half.[72] From a philosophical point of view, it is important to stress that mere Bayesianism does not require one to align his or her degrees of beliefs with (known) chances. This principle of alignment has been discussed for several decades by such names as the "principle of direct inference," "Miller's principle" or the "Principal principle" (Lewis, 1986). Assume that the hypothesis H states that the chance of E being true is r—for short, $Ch(E)=r$. In this case, a simple version of the principle is that

$$P(E|Ch(E)=r)=r$$ [73]

[72] The example is inspired by Hawthorne (2004/2012), which devotes special attention to likelihoods.

[73] One of the contributions of Lewis (1986) consists in making explicit the validity domain of the principle, i.e. in determining classes of situations where it seems reasonable to obey $P(E|Ch(E)=r)=r$.

Some wish to restrict BCT to cases where the likelihoods can be fixed by such an alignment with chance (Strevens, 2006). Hawthorne (2011) claims that even when likelihoods cannot be inferred in this way, members of a scientific community cannot strongly disagree about them. His argument to this effect is that the likelihood $P(E|H)$ expresses the probabilistic content of H. If Paul and Jean diverge strongly on $P(E|H)$, it is no longer clear that they are considering the same hypothesis H. In any case, the likelihoods are in general not sufficient for determining the probability of H given E. The prior probability of H is also needed. And it is not easy to see why Paul and Jean should have the same values for $P(H)$.

The problem of objectivity reappears in the way BCT addresses some philosophical issues and puzzles. Consider for instance the case[74] where E is a set of data which are implied by two rival hypotheses H_1 and H_2. In this case, the (ratio between) prior probabilities directly determine(s) the (ratio between) posteriors since $P(H_1|E)/P(H_2|E) = P(H_1)/P(H_2)$. Hence, empirical evidence cannot help to choose between the two hypotheses. Let H_1 be "All emeralds are green," H_2 be "All emeralds are grue," and E be all our past observations on emeralds. The preceding remark implies that the only means by which BCT can account for our inductive preference for H_1 over H_2 is to assume a prior preference for H_1 over H_2. One can therefore dispute that BCT truly explains our inductive behavior: it seems rather to describe it by assuming prior bias for "green" and against "grue." The problem of objectivity is also present in BCT's analysis of the Duhem-Quine problem, as stressed by Earman (1992, pp. 83–86). Indeed, BCT is capable of accounting for many sensible reactions to empirical refutation. But it may be the case that, given their respective priors, Paul should blame H (the target hypothesis) rather than A (the auxiliary assumptions), whereas the reverse is true of Jean. Yet it seems that a real solution to the Duhem-Quine problem should prescribe a uniform attitude to both Paul and Jean.

In response to these worries, Bayesians often put forward a class of results to the effect that, when each individual updates his or her degrees of belief by conditionalization, individual probabilities converge toward true hypotheses (Savage, 1954; Blackwell & Dubins, 1962; Gaifman & Snir, 1982; Schervish & Seidenfeld, 1990). This implies in turn that these probabilities converge to common values. These results hold under more or less restrictive assumptions—for instance, the assumption that the priors assign zero probability to the same propositions. The interpretation of these results, however, is not straightforward. For instance, the convergence typically holds "almost certainly" in the technical sense, that is, it is not secured in possible worlds to which a zero probability is assigned. Consequently, the characteristics of priors are still crucial, as is stressed by Earman (1992, chap.6, sec. 3–5) and Howson (2000, p. 210). Furthermore, another disputed issue is to know whether these long-term results have a decisive impact on the question of the justification of instantaneous confirmational judgments.

[74] Horwich (1982, p. 35).

C. Howson, one of the main supporters of BCT, claims in a recent book that Hume's argument for inductive skepticism is correct but does not preclude the existence of a logic of inductive inference (Howson, 2000)—this logic being nothing but BCT. As we saw, BCT depends on priors. These priors encode the agents' inductive commitments[75] but do not justify them (here lies the truth of Hume's argument). Howson claims therefore that "Inductive reasoning is justified to the extent that it is sound, given appropriate premises. These consist of initial assignments of positive probability that cannot themselves be justified in any absolute sense." (p. 238) To put it in terms which are closer to those used up to now: BCT is not an inductive method, but allows us to implement our inductive commitments coherently.

6. Conclusion

The concept of confirmation is at the heart of scientific reasoning. Together with the related concept of induction, it raises formidable philosophical problems. In this landscape, Bayesian confirmation theory is a rare species: it offers a set of flexible answers which are based on a general theory of rational belief and rational belief revision. However, BCT faces difficulties both from the point of view of the construction of an adequate theory of inductive reasoning (see, e.g., the problem of old evidence) and from the point of view of the problem of the justification of induction. If BCT is currently dominant despite these difficulties, it is partly due to a lack of convincing alternatives. This state of the art points to (at least) two research avenues, the first being motivated by the failures of BCT, the second by its successes. The first is the exploration of new alternative frameworks and the improvement of our understanding of the theoretical possibilities.[76] The second is the application of BCT to various episodes in the history of science. Bayesians have already made such applications (see, in particular, Howson & Urbach, 1989). But in view of its achievements, BCT definitely deserves to be more intensively applied.

References

Achinstein, P. (1978) "Concepts of Evidence," *Mind*, 87(345), 22–45.

Achinstein, P. (2001) *The Book of Evidence*, Oxford: Oxford University Press.

Blackwell, D. & Dubins, L. (1962) "Merging of Opinions with Increasing Information," *Annals of Mathematical Statistics*, 33, 882–887.

[75] This apt expression is due to Strevens (2004), which discusses Howson's view on the relation between BCT and the problem of induction. Strevens claims that the principle of alignment, although not inherent to Bayesianism, is a major source of inductive commitment for BCT. The idea that inductive commitments are rather described in than derived from (or justified by) the Bayesian framework is also endorsed by Norton (2011).

[76] Zwirn & Zwirn (1996) is an important step in this direction: the authors provide an axiomatic classification of theories of qualitative confirmation based on principles of adequacy à la Hempel.

Carnap, R. (1945) "On Inductive Logic," *Philosophy of Science*, 12, 72–97.

Carnap, R. (1947) "On the Application of Inductive Logic," *Philosophy and Phenomenological Research*, 8, 133–147.

Carnap, R. (1950/1962) *Logical* Foundations of Probability, Chicago: University of Chicago Press.

Carnap, R. (1952) *The Continuum of Inductive Methods*, Chicago: University of Chicago Press.

Carnap, R. (1962) "The Aim of Inductive Logic" in Nagel, E., Suppes, P., & Tarski, A. (eds.), *Logic, Methodology and Philosophy of Science*, Stanford: Stanford University Press, pp. 303–318.

Carnap, R. (1963) "Intellectual Autobiography" and "Replies and Systematic Expositions" in Schlipp, P. (ed.), *The Philosophy of Rudolf Carnap*, The Library of Living Philosophers, vol. XI, LaSalle: Open Court, pp. 3–83; 859–1012.

Chihara, C.S. & Gillies, D.A. (1988) "An Interchange on the Popper-Miller Argument," *Philosophical Studies*, 54, 1–8.

Christensen, D. (1983) "Glymour on Evidential Relevance," *Philosophy of Science*, 50, 471–481.

Christensen, D. (1990) "The Irrelevance of Bootstrapping," *Philosophy of Science*, 57, 644–662.

Christensen, D. (1999) "Measuring Confirmation," *Journal of Philosophy*, 96, 437–461.

Crupi, V. "Confirmation" (2014) *The Stanford Encyclopedia of Philosophy* (Spring 2014 Edition), E.N. Zalta (ed.), URL = http://plato.stanford.edu/archives/spr2014/entries/confirmation/.

De Finetti, B. (1937) "La prevision: ses lois logiques, ses sources subjectives," *Annales de l'Institut Poincaré*, 7, 1–68.

Dorling, J. (1979) "Bayesian Personalism, the Methodology of Scientific Research Programs, and Duhem's Problem," *Studies in the History and Philosophy of Science*, 10, 177–187.

Duhem, P. (1906/1914) *La théorie physique, son objet, sa structure*, Paris: Chevalier; Eng. trans. Wiener, Ph. *The Aim and Structure of Physical Theory*, Princeton, NJ: Princeton University Press, 1954.

Dunn, J.M. & Hellman, G. (1986) "Dualling: A Critique of an Argument of Popper and Miller," British Journal for the Philosophy of Science, 37(2), 220–223.

Earman, J. (1992) *Bayes or Bust? A Critical Examination of Bayesian Confirmation Theory*, Cambridge, MA: MIT Press.

Earman, J. & Salmon, W.C. (1992) "The Confirmation of Scientific Hypotheses" in Salmon, M. et al., (eds.), *Introduction to the Philosophy of Science*, Indianapolis & Cambridge: Hackett Publishers, pp. 42–103.

Edwards, A.W.F. (1972) *Likelihood*, London: Cambridge University Press.

Eells, E. (1988) "On the Alleged Impossibility of Inductive Probability," *British Journal for the Philosophy of Science*, 39, 111–116.

Eells, E. (1990) "Bayesian Problems of Old Evidence" in Wade Savage C. ed., *Minnesota Studies in the Philosophy of Science*, vol. 14, Minneapolis: University of Minnesota Press, pp. 205–223.

Eells, E. & Fitelson, B. (2000) "Measuring Confirmation and Evidence," The Journal of Philosophy, 97 (12), 663–672.

Fitelson, B. (2001) "Studies in Bayesian Confirmation Theory," PhD Dissertation, University of Wisconsin, Madison.

Fitelson, B. (2002) "Putting the Irrelevance Back into the Problem of Irrelevant Conjunction," *Philosophy of Science*, 69(4), 611–622.

Fitelson, B. (2006a) "The Paradox of Confirmation," *Philosophy Compass*, 1(1), 95–113.

Fitelson, B. (2006b) "Inductive Logic" in Sarkar, S. & Pfeifer, J. (eds.), *The Philosophy of Science. An Encyclopedia*, Oxford: Routledge, pp. 384–394.

Fitelson, B. (2008), "Goodman's 'New Riddle,'" *Journal of Philosophical Logic*, 37(6), 613–643.

Fitelson, B. & Hawthorne, J. (2006) "How Bayesian Confirmation Theory Handles the Paradox of the Ravens" in Eells, E. & Fetzer, J. (eds.), *Probability in Science*, Chicago: Open Court, pp. 247–275.

Gaifman, H. & Snir, M. (1982) "Probabilities over Rich Languages, Testing and Randomness," *The Journal of Symbolic Logic*, 47(3), 495–548.

Garber, D. (1983) "Old Evidence and Logical Omniscience in Bayesian Confirmation Theory" in Earman, J. (ed.), *Minnesota Studies in the Philosophy of Science*, vol. 10, Minneapolis: University of Minnesota Press, pp. 99–131.

Gemes, K. (1998) "Hypothetico-Deductivism: The Current State of Play; the Criterion of Empirical Significance: Endgame," *Erkenntnis*, 49, 1–20.

Gemes, K. (2005) "Hypothetico-Deductivism: Incomplete but Not Hopeless," *Erkenntnis*, 63, 139–147.

Gillies, D. (1986) "In Defense of the Popper-Miller Argument," *Philosophy of Science*, 53(1), 110–113.

Gillies, D. (2000) *Philosophical Theories of Probability*, London: Routledge.

Glymour, C. (1980) *Theory and Evidence*, Princeton, NJ: Princeton University Press.

Glymour, C. (1983) "Revisions of Bootstrap Testing," *Philosophy of Science*; 50, 626–629.

Good, I. J. (1967) "The White Shoe Is a Red Herring," *British Journal for the Philosophy of Science*, 12, 63–64.

Goodman, N. (1946) "A Query on Confirmation," *Journal of Philosophy*, 43(14), 383–385.

Goodman, N. (1955) *Fact, Fiction and Forecast*, Cambridge, MA: Harvard University Press.

Hacking, I. (2001) *An Introduction to Probability and Inductive Logic*, Cambridge: Cambridge University Press.

Hájek, A. (2012) "Interpretations of Probability," *The Stanford Encyclopedia of Philosophy* (Winter 2012 Edition), E.N. Zalta (ed.), URL = http://plato.stanford.edu/archives/win2012/entries/probability-interpret/.

Hájek, A. & Joyce, J. (2008) "Confirmation," in Psillos, S. & Curd, M. (eds.), *Routledge Companion to the Philosophy of Science*, London: Routledge, pp. 115–128.

Hawthorne, J. (2004/2012) "Inductive Logic," *The Stanford Encyclopedia of Philosophy* (Winter 2012 Edition), E.N. Zalta (ed.), URL = http://plato.stanford.edu/archives/sum2007/entries/logic-inductive/.

Hawthorne, J. (2011) "Confirmation Theory" in P.S. Bandyopadhyay & M. Forster (eds.), *Philosophy of Statistics, Handbook of the Philosophy of Science*, vol. 7, Amsterdam: North Holland, pp. 333–389.

Hempel, C. G. (1943) "A Purely Syntactical Definition of Confirmation," *The Journal of Symbolic Logic*, 8(4), 122–143.

Hempel, C. G. (1945a) "Studies in the Logic of Confirmation (I)," *Mind*, 54(213), 12–26

Hempel, C. G. (1945b) "Studies in the Logic of Confirmation (II)," *Mind*, 54(214), 97–121.

Hempel, C. G. (1967) "The White Shoe: No Red Herring," *British Journal for the Philosophy of Science*, 18, 239–240.

Horwich, P. (1982) *Probability and Evidence*, Cambridge: Cambridge University Press.

Hosiasson-Lindenbaum, J. (1940) "On Confirmation," *The Journal of Symbolic Logic*, 5(4), 133–148.

Howson, C. (1984) "Bayesianism and Support by Novel Facts," *British Journal for the Philosophy of Science*, 35(3), 245–251.

Howson, C. (1991) "The 'Old Evidence' Problem," *British Journal for the Philosophy of Science*, 42(4), 547–555.

Howson, C. (2000) *Hume's Problem. Induction and the Justification of Belief*, Oxford: Clarendon Press.

Howson, C. & Urbach, P. (1989) *Scientific Reasoning: The Bayesian Approach*, La Salle: Open Court.

Huber, F. (2007) "Confirmation and Induction" *in* Fieser, J. & Dowden, B. (eds.), *Internet Encyclopedia of Philosophy*, available at http://www.iep.utm.edu/conf-ind/.

Huber, F. (2008) "Hempel's Logic of Confirmation," *Philosophical Studies*, 139, 181–189.

Hume, D. (1739) *Treatise of Human Nature*, Oxford: Clarendon Press, 1960.

Hume, D. (1748) *An Enquiry Concerning Human Understanding*, Oxford Worlds Classics, Oxford: Oxford University Press, 2007.

Huygens, C. (1690) *Traité de la Lumière*, Leyden: Van der Aa. [Treatise on Light, trans. S.P. Thompson, New York: Dover, 1962].

Jeffrey, R. (1983) "Bayesianisms with a Human Face" in Earman, J. (ed.), *Minnesota Studies in the Philosophy of Science*, vol. 10, Minneapolis: University of Minnesota Press, pp. 133–154.

Jeffrey, R. (1984) "The Impossibility of Inductive Probability," *Nature*, 310, 433.

Jeffrey, R. (1992) *Probability and the Art of Judgment*, Cambridge: Cambridge University Press.

Joyce, J. (1998) "A Nonpragmatic Vindication of Probabilism," *Philosophy of Science*, 65(4), 575–603.

Joyce, J. (1999) *Foundations of Causal Decision Theory*, Cambridge: Cambridge University Press.

Joyce, J. (2007) "Bayes' Theorem" *The Stanford Encyclopedia of Philosophy* (Summer 2007 Edition), E.N. Zalta (ed.), URL = http://plato.stanford.edu/archives/sum2007/entries/bayes-theorem/.

Kant, I. (1781/1787) *Critique of Pure Reason*, trans./ed. P. Guyer and A. W. Wood, Cambridge: Cambridge University Press, 1997.

Keynes, J. M. (1921) *A Treatise on Probability*, London: Macmillan and Co.

Kronz, F. (1992) "Carnap and Achinstein on Evidence," *Philosophical Studies*, 67(2), 151–167.

LeMorvan, P. (1999) "The Converse Consequence Condition and Hempelian Qualitative Confirmation," *Philosophy of Science*, 66, 448–454.

Lewis, D. K. (1986) "A Subjectivist Guide to Objective Chance" *in Philosophical Papers*, vol. 2, Oxford: Oxford University Press, pp. 82–132.

Lewis, D. K. (1999) *Papers in Metaphysics and Epistemology*, Cambridge: Cambridge University Press.

Maher, P. (1996) "Subjective and Objective Confirmation," *Philosophy of Science*, 63(2), 149–174.

Maher, P. (2004) "Probability Captures the Logic of Scientific Confirmation" in Hitchcock, C. (ed.), *Contemporary Debates in the Philosophy of Science*, Oxford: Basil Blackwell, pp. 69–93.

Maher, P. (2010) "Explication of Inductive Probability," *Journal of Philosophical Logic*, 39, 593–316.

Mayo, D. G. (1996) *Error and the Growth of Experimental Knowledge*, Chicago: University of Chicago Press.

Mayo, D. G. (2005) "Evidence as Passing Severe Tests: Highly Probable versus Highly Probed Hypotheses," in Achinstein, P. (ed.), *Scientific Evidence. Philosophical Theories and Applications*, Baltimore, MD: John Hopkins University Press, pp. 95–127.

Mill, J-S. (1843) *A System of Logic Ratiocinative and Inductive, Being a Connected View of the Principles of Evidence and the Methods of Scientific Investigation in The Collected Works of John Stuart Mill*, vols. 7–8, Toronto: University of Toronto Press, London: Routledge and Kegan Paul, 1974.

Moretti, L. (2003) "Why the Converse Consequence Condition Cannot Be Accepted," *Analysis*, 63(4), 297–300.

Norton, J. D. (2005) "A Little Survey of Induction" in Achinstein, P. (ed.), *Scientific Evidence: Philosophical Theories and Applications*, Baltimore, MD: John Hopkins University Press, pp. 9–34.

Norton, J. D. (2010) "A Survey of Induction Generalization," Mimeo, available at http://www.pitt.edu/~jdnorton/homepage/research/ind_survey.html.

Norton, J. D. (2011) "Challenges to Bayesian Confirmation Theory" in P.S. Bandyopadhyay & M. Forster (eds.), *Philosophy of Statistics, Handbook of the Philosophy of Science*, vol. 7, Amsterdam: North Holland, pp. 391–439.

Popper, K. R. (1959/1968) *The Logic of Scientific Discovery*, London: Hutchinson.

Popper, K. R. & Miller, D. (1983) "A Proof of the Impossibility of Inductive Probability," *Nature*, 302, 687–688.

Popper, K. R. & Miller, D. (1987) "Why Probabilistic Support Is not Inductive," *Philosophical Transactions of the Royal Society of London*, Series A, 321(1562), 569–591.

Putnam, H. (1963) "'Degree of Confirmation' and Inductive Logic" in Schlipp, P. (ed.), *The Philosophy of Rudolf Carnap*, The Library of Living Philosophers, vol. 11, LaSalle: Open Court, pp. 760–782.

Quine, W. V. O. (1969) *Ontological Relativity and Other Essays*, New York: Columbia University Press.

Ramsey, F. P. (1931) "Truth and Probability," in Braithwaite, R. (ed.), *Foundations of Mathematics and Other Logical Essays*, London: Routledge & Kegan Paul, pp. 156–198.

Royall, R.M. (1997) *Statistical Evidence: A Likelihood Paradigm*, London: Chapman & Hall.

Salmon, W. (1975) "Confirmation and Relevance," in Maxwell, G. & Anderson, R. M. (eds.), *Minnesota Studies in Philosophy of Science*, vol. 6: *Induction, Probability, and Confirmation*, Minneapolis: University of Minnesota Press, pp. 3–36.

Salmon, W. (1981) "Rational Prediction," *British Journal for the Philosophy of Science*, 32, 115–125.

Savage, L. (1954/1972) *The Foundations of Statistics*, New York: Dover.

Schervish, M. J. & Seidenfeld, T. (1990) "An Approach to Consensus and Certainty with Increasing Evidence," *Journal of Statistical Planning and Inference*, 25, 401–414.

Skyrms, B. (1966) *Choice and Chance*, Belmont, CA: Dickinson.

Sober, E. (1994) "No Model, No Inference: A Bayesian Primer on Grue Problem" in Stalker, D. (ed.), *Grue! The New Riddle of Induction*, Chicago: Open Court, pp. 245–240.

Sprenger, J. (2011) "Hypothetico-deductive Confirmation," *Philosophy Compass*, 6, 497–508.

Sprenger, J. (2013) "A Synthesis of Hempelian and Hypothetico-Deductive Confirmation," *Erkenntnis*, 78, 727–738.

Strevens, M. (2004) "Bayesian Confirmation Theory: Inductive Logic, or Mere Inductive Framework?" *Synthese*, 141, 365–379.

Strevens, M. (2006) *Notes on Bayesian Confirmation Theory*, Mimeo. New York University.

Stroud, B. (1977) *Hume*, Routledge & Kegan Paul: London.

Talbott, W. (2006) "Bayesian Epistemology" *The Stanford Encyclopedia of Philosophy* (Fall 2006 Edition), E.N. Zalta (ed.), URL = http://plato.stanford.edu/archives/fall2006/entries/epistemology-bayesian/.

Teller, P. (1973) "Conditionalization and Observation," *Synthese*, 26(2), 218–258.

Vickers, J. (2013) "The Problem of Induction" *The Stanford Encyclopedia of Philosophy* (Spring 2013 Edition), E.N. Zalta (ed.), URL = http://plato.stanford.edu/archives/spr2013/entries/induction-problem/.

Vranas, P. (2004) "Hempel's Raven Paradox: A Lacuna in the Standard Bayesian Solution," *British Journal for the Philosophy of Science*, 55, 545–560.

Von Wright, G. H. (1965) *The Logical Problem of Induction*, Oxford: Basil Blackwell.

Zabell, S. L. (2011) "Carnap and the Logic of Inductive Inference" in Hartmann, S. & Woods, J. (eds.), *Handbook of the History of Logic*, vol. 10: Inductive Logic, Amsterdam: North Holland, pp. 265–310.

Zwirn, D. & Zwirn, H. P. (1993) "Logique inductive et soutien probabiliste," *Dialogue*, 32, 293–307.

Zwirn, D. & Zwirn, H. P. (1996) "Metaconfirmation," Theory and Decision, 41, 195–228.

3

CAUSALITY

Max Kistler (IHPST–Université Paris 1 Panthéon Sorbonne & CNRS)

IN 1912, BERTRAND RUSSELL recommended that philosophers eliminate causation from their stock of concepts. His argument relied on the premise that advanced sciences do not contain any concept corresponding to the intuitive notion of causation. However, Russell also argues that the notion of causation cannot possibly be reduced in purely scientific terms either. Now, if there is a conflict between an intuition of common sense and science, the naturalist attitude consists in resolving the conflict by following science instead of intuition. Thus, concludes Russell, philosophers should stop speaking of "causes." The debate launched by Russell's article continues to this day. On the one hand, many philosophers argue along lines similar to Russell's that the notion of causation has no equivalent in fundamental physics. One way to understand why causation plays such an important role in common sense without having any equivalent in physics is to interpret is as belonging to "folk science" (Norton, 2003). However, the debate concerning the presence of causation in fundamental physics continues.[1] It is for example argued that the distinction between timelike and spacelike distances in special relativity expresses a causal distinction: a distinction between distances that can be bridged by signals, which can be interpreted as causal processes, and distances that cannot be so bridged. On the other hand, there is now much less confidence that it is possible to generalize from physics to all other sciences. To the extent that nothing guarantees the effective reduction of all sciences

[1] See e.g. the debate between Frisch (2009a, 2009b) and Norton (2009).

to fundamental physics, causation might well be and remain a legitimate and even indispensable concept in other sciences even if it is not in physics.

The plan of this chapter is as follows. In the first section we will analyze Russell's reasons for holding that there can be no analysis of the concept of causation that is compatible with 20th century physics. We will see that the debate between "eliminativists" following Russell and philosophers holding that the concept of causation is as central to science as it is to common sense is structured by two distinctions: between microscopic and macroscopic entities and between concrete events and their measurable properties. It turns out that the debate on the legitimacy of the concept of causation is linked to the debate on the existence of laws of nature outside fundamental physics, laws that allow for exceptions, often called *ceteris paribus* laws. We will see that, even if it were correct that causation plays no role in the theoretical content of fundamental physics, it may be argued that the concept of causation is nevertheless legitimate and useful in many contexts. It does seem to be central not only for common sense, for example, in the context of our planning actions in light of their consequences, but also for all other sciences outside fundamental physics, such as biology and neuroscience, as well as for many projects involving the analysis of philosophical concepts in naturalistic terms. Thus, causation plays a central role in philosophical theories of intentionality, perception, knowledge and action.

After having thus justified the project of a philosophical analysis of the concept of causation, we shall examine the most important approaches that have been put forward and developed: in terms of counterfactual conditionals, in terms of probability raising, in terms of manipulability, and in terms of processes.

1. The central idea of the counterfactual analysis of causation is that, for any two events *c* and *e* that have actually occurred, *c* causes *e* if and only if it is true that: if *c* had not occurred, *e* would not have occurred.[2]
2. The central idea of the probabilistic analysis is that factor C exercises a causal influence on factor E if and only if an event of type C raises the probability of an event of type E.
3. The central idea of the manipulability analysis is that there is a causal relation between two variables C and E if and only if interventions modifying the value of C modify the value of E.
4. Finally, the central idea of the process analysis is that an event *c* causes another event *e* if and only if there is a physical process of transmission between *c* and *e*, for example, of a quantity of energy.

One difficulty one faces in comparing these approaches stems from the fact that they conceive of the terms of the causal relation in different ways: for some analyses, causes

[2] Lower-case variables represent concrete particular events; upper-case variables represent properties of objects or events.

and effects are singular events, whereas for others it is rather properties of events or "factors," which can be instantiated by numerous events.

Moreover, to understand the complex debate between the advocates of these theories, it is important to be conscious of the aims they pursue and of the criteria they use to judge their success. One can conceive of the task of a philosophical analysis of the concept of causation in at least two ways. Its aim can be taken to be (1) pure a priori analysis of the concept of causation as it is used by subjects, independently of the features of the actual world, as it is described by contemporary science, or (2) a partly empirical and partly conceptual enquiry on the "real essence" of causation, as it is in the actual world. According to this second interpretation of what it means to understand causation, causation is a natural kind of relation analogous to natural kinds of substances, such as water, gold or, for common sense, tigers. Common sense presupposes that such kinds of substances or animals possess a real essence that can be discovered by empirical science. In an analogous way, causation might have a real essence specific to our actual world. However, rather than beginning with these methodological reflections, we will take them up after having presented the debate on the counterfactual analysis: it is easier to think about the "metaphilosophical" question of the aim, method and criteria of adequacy of an analysis after having studied a sample of the debate.

1. Russell and the Elimination of the Concept of Causation

Russell's arguments are mainly directed against what is now often called "generic causation."[3] Singular causal judgments, such as "the fact that I have rubbed this match (i.e. the match that I see before my eyes) is the cause of the fact that it has lit," differ from generic causal judgments, such as: "in general, rubbing matches causes them to light." In Hume's conception,[4] the truth of a singular causal proposition depends on the truth of a generic causal proposition. The truth of the proposition that the singular event *c* causes the singular event *e* presupposes the truth of the generic proposition that events of the same type as *c* are followed by events of the same type as *e*. In other words, there can be no causation between *singular* events without an appropriate regularity at the level of *types* of events. We will see later that this thesis has been challenged, so as to dissociate singular causation from generic causation. If the existence of singular causal relations does not presuppose the existence of generic causal relations they instantiate, singular causation is no target for Russell's arguments. However, only a minority of contemporary analyses take singular causation to be independent of nomological relations at the level of types of events, factors or properties. To the extent that philosophical analyses of causation aim at explaining and justifying

[3] For a recent reevaluation of Russell's arguments against the possibility of constructing a concept of causality compatible with contemporary science, see Price and Corry (2007); Spurrett and Ross (2007).

[4] Many contemporary approaches to causation are deeply influenced by David Hume's (1739–1740, 1777) conception of causation.

the use of causal concepts in science, the generic concept remains the most relevant: it is generally taken for granted that the fact that this match has lit at time t can only be explained in terms of general propositions that apply to rubbings of matches at any place and at any time. Such an explanation might mention the general proposition that the energy produced in the form of heat by sufficiently strong rubbing triggers the chemical reaction of exothermic oxidation of any sample of phosphorus sesquisulfide (P_4S_3), which happens to be the substance that covers the head of ordinary matches.

1.1 THE PRINCIPLE OF CAUSALITY AND THE REPETITION OF EVENTS

Russell tries to establish the vacuity of the traditional "principle of causality" according to which "the same causes always have the same effects," or more precisely: "Given any event e_1, there is an event e_2 and a time-interval τ such that, whenever e_1 occurs, e_2 follows after an interval τ" (Russell, 1912/1992, p. 195). This is a "meta-law," stating that there are laws of succession involving types of events. Russell argues against the existence of such laws of succession—and thus against the principle of causality—in advanced sciences, by noting first that there can be recurring types of events only if these types are conceived (1) vaguely and (2) narrowly; and secondly that vaguely conceived events cannot be the target of scientific explanations whereas generalizations bearing on narrowly conceived events are not strictly true.

1. Events that recur are conceived *vaguely*: to use Russell's own example, events such as throwing of bricks—followed by breaking of windows—recur only if they are conceived in a way that abstracts away from microscopic details. There are no two throwings that resemble each other exactly in all microscopic details. The problem is that scientific explanation in its mature form requires one to be able to *deduce* the *explanandum* from the description of the situation playing the role of the *explanans*, together with statements of the laws of nature (see chapter 1 of this volume). Now, such a deduction is possible only if first the *explanans* contains a quantitative description of the cause, that is, is conceived "precisely" (Russell, 1912/1992, p. 200), second the laws of nature are also quantitatively precise, and third the *explanandum* is a quantitatively precise description of the effect. However, to the extent that events are conceived in this quantitatively precise manner—which is what makes their scientific explanation possible—they do not recur. To the extent that the antecedent of a universal conditional applies only to one event, the truth of the conditional is almost trivial: it is true if and only if its consequent is true in the unique situation in which its antecedent is true. Such a statement cannot be used to explain other events, which is a major function of laws. There cannot be strict laws containing quantitatively precise predicates that can be used for the explanation and prediction of new situations; there is room for strict regularities only in common sense and "in the infancy of a science" (p. 201).

2. Events that recur are *narrowly* conceived. There can only be recurring events if they are conceived locally, that is, as the content of a well-delimited region of space-time. There are many rubbings of matches of the same type only to the extent that the circumstances are not included in the rubbing events. However, to the extent

that one abstracts away from the person who rubs, the weather and other contextual factors, the regular lighting of matches when rubbed has exceptions: there may be factors present in the surroundings of the first event (the rubbing) that prevent the second event (the lighting) from occurring; in other words, the regularity exists only insofar as "all other things are equal," or *ceteris paribus*. The dialectic is similar to the case of vagueness: it is possible that a narrowly conceived event recurs, but insofar as the circumstances of the events are not taken into account, the regularity with which event *c* is followed by event *e* is not exceptionless because factors in the circumstances may interfere and prevent *e* from occurring even though *c* has occurred.

Generalizations bearing on narrowly conceived events cannot be used in scientific explanations because that requires strictly true universal propositions. "The sequence [. . .] is no more than probable, whereas the relation of cause and effect was supposed to be necessary" (Russell, 1912/1992, p. 201).[5] On the other hand, to the extent that the possibility of interference by factors present in the spatiotemporal vicinity of the antecedent event is diminished by including the surroundings of the events, the probability of their recurrence diminishes. "As soon as we include the environment, the probability of repetition is diminished, until at last, when the whole environment is included, the probability of repetition becomes almost *nil*." (p. 197)

Note that the first argument against the principle of causation questions only the existence of successions of *macroscopic* events conceived with *common sense* concepts: microscopic events, such as the interaction of an electron and a photon or the radioactive decomposition of a uranium-238 nucleus, recur even if they are precisely conceived. However, the second argument questions the strict recurrence of both microscopic and macroscopic events: if one considers a set of localized cause-events that are of strictly the same type but does not take their surroundings into consideration, such cause-events are not necessarily followed be the same effect-events, because these effects can be influenced by events occurring in the neighborhood of the cause-events.

Thus, Russell's conclusion also covers microscopic events: "As soon as the antecedents have been given sufficiently fully to enable the consequent to be calculated with some exactitude, the antecedents have become so complicated that it is very unlikely they will ever recur." (Russell, 1912/1992, p. 198). In sum, there are no macroscopic events that are both precisely conceived and recur; microscopic events may recur even when they are precisely conceived; however, the succession of microscopic events only recurs to the extent that the events are conceived locally, without taking their surroundings into consideration. Thus, the principle of causality "same cause, same effect" is according to Russell, "utterly otiose" (p. 198), to the extent that what would allow for repetition ("same cause"), that is, conceiving of macroscopic events vaguely, or including spatiotemporal surroundings for microscopic events, either makes them inappropriate for being used in the exact sciences (for the former) or prevents them from recurring (for the latter).

5 Russell does not consider probabilistic causation because he takes necessitation to be a defining condition of causality.

1.2 THE FUNCTIONAL LAWS OF MATURE SCIENCE

Russell's second argument against the possibility of finding scientific legitimacy for the notion of cause consists in showing that the laws that are used in the explanations of mature sciences cannot be interpreted as causal laws. The laws used in mathematical physics, for example, in "gravitational astronomy" (Russell, 1912/1992, p. 193), have the form of functions: in a system of masses subject only to the force of gravitational attraction, it is possible to represent the configuration of the system at a given moment as a function of that moment and of the configuration and speeds at some other moment (or as a function of the configurations at two other moments).[6] Although it is true that such a function "determines" the configuration of the system, this does not justify the idea that this determination is *causal*. Russell has two reasons for holding that "in the motions of mutually gravitating bodies, there is nothing that can be called a cause, and nothing that can be called an effect" (Russell, 1912/1992, p. 202).

The first is that this determination is purely logical and indifferent to the direction of time: Newton's laws, together with the law of gravitational attraction, make it possible to calculate the configuration of a system of masses at some time in the past as a function of its configuration at some future time, in exactly the same way in which they make it possible to calculate the characteristics of the system at some future time on the basis of its characteristics at some moment in the past. Given that the traditional concept of causation requires that the cause precedes the effect, this functional determination cannot be interpreted as being causal.[7].

The second reason concerns the terms of the relations: causality relates particular, or concrete events, whereas functional equations relate values of measurable quantities. In other words functional equations relate *properties* of concrete events rather than events themselves. The equation expressing the law of gravitation—or law of universal attraction—indicates the value of the force of gravitational attraction between two massive bodies as a function of their masses and distance. The equation expressing Newton's first law says that the numerical value of the product of the acceleration of a massive object and its mass equals the numerical value of the total force acting on the object. These laws hold for all massive objects, however diverse they may be in other respects. Although the problem of induction is one obstacle to the knowledge of a law, there is another problem concerning our knowledge of functional laws such as the two just mentioned: It is practically impossible to test a hypothesis bearing on a law expressing a constant proportion of the values of certain magnitudes because these magnitudes are not instantiated in isolation, but by concrete events which also depend on other properties.

---

6 The configuration of a system is the set of the positions and speeds of each of its components.

7 This traditional assumption has been challenged by the elaboration of the concept of *backward causation*, which is intended to apply in particular to certain processes in particle physics. Cf. Dowe (1996). Simultaneous causation raises its own problems.

There are two reasons why a law such as the law of gravitation cannot be tested directly. (1) The first is that there is no system of two masses that is not also subject to the attraction of other masses, in general, at a greater distance. (2) The second is that massive objects also have other properties that can give rise to other forces. Russell concludes that the quantitatively exact laws of mature sciences are not causal because the referents of their terms are not—as causes and effects would have to be—directly accessible to experience. "In all science we have to distinguish two sorts of laws: first, those that are empirically verifiable but probably only approximate; secondly, those that are not verifiable, but may be exact" (Russell, 1912/1992, p. 203). The first type of laws corresponds to the "causal laws" of common sense and of sciences at the beginning of their development, whereas the laws of mature sciences belong to the second type: they cannot be interpreted as causal since their terms do not refer to concrete events.

1.3 CETERIS PARIBUS LAWS

The problem raised by Russell has been the object of a rich literature on so-called *ceteris paribus* laws.[8] It has been noted that the interpretation of many quantitative laws presents us with a dilemma.

Either

1. One supposes that laws bear on concrete objects or events that are directly accessible to experience. If so then it turns out that these laws have exceptions or, in other words, hold only *ceteris paribus*;

Or

2. One supposes that laws bear neither on particular objects nor on particular events. Then it becomes hard to understand how it is nevertheless possible that such laws are being used to produce scientific explanations and predictions.

Hempel gives the following example. For every bar magnet *b*, "if *b* is broken into two shorter bars and these are suspended, by long thin threads, close to each other at the same distance from the ground, they will orient themselves so as to fall into a straight line" (Hempel, 1988, p. 148). This generalization is not true without exception of the movement of concrete bar magnets: in certain circumstances, like when a strong air current blows in the direction perpendicular to the orientation of the magnet or when there is a strong external magnetic field, the two halves of the magnet do not align. Similarly, if one takes the law of gravitational attraction to bear on concrete massive

[8] See e.g. the special issue of *Erkenntnis* (2002) 57(3).

objects, so that it determines the net force acting on them (which in turn determines their acceleration) as a function of their masses and their distances, the law has numerous exceptions:[9] an object with mass m_1 that is at distance d of a second object with mass m_2, is in general not subject to a net force $\frac{Gm_1m_2}{d^2}$ in direction of this second object (nor accelerated with $\frac{Gm_2}{d^2}$ in its direction).

However, it is not necessary to conclude from this, with Cartwright (1983), that the laws "lie."[10] Several strategies are available for reinterpreting functional equations and other nomological statements in such a way that they turn out true, despite the fact that the evolution of concrete objects and events often does not (strictly speaking) match with these equations and statements. One strategy consists in taking laws to bear only on systems that are in *ideal* situations, which means in particular that they are isolated.[11] For certain laws, such as the law of gravitational attraction, this has the consequence that the laws bear on no real system (because no real system is ideal in the sense of being isolated from external gravitational influences). Moreover, even if there were isolated systems this strategy faces the difficulty of explaining how a law that is true only of idealized situations can nevertheless be used for the prediction and explanation of facts concerning real systems.

Another strategy consists in taking laws to bear on abstract models rather than on real systems. Smith (2002) proposes to solve the problem of interpreting *ceteris paribus* laws by distinguishing between fundamental laws and equations of movement. Fundamental laws do not directly apply to real concrete systems. The law of universal gravitation determines the force with which two masses attract each other. However, this law cannot be used to *directly* calculate the movement of real objects, to the extent that no real object is exclusively subject to the gravitational attraction due to its interaction with a single other object. Every real object is attracted by many other massive objects, over and above being in general subject to other forces. Smith presents the law of universal gravitation as featuring in an algorithm or "recipe" for constructing a model. The last step of the algorithm leads to an equation of movement that is specific for a concrete system. In this sense, it does not have, according to Smith, the generality required for a law. Smith's fundamental laws correspond to the laws of which Russell says that they are not verifiable but can be exact. Among these fundamental laws, there are in particular the laws determining the different forces that are exerted on an object as a function of its properties and the other objects represented in a model A that contains a partial specification of the properties of a concrete system C under consideration. If C does not evolve as predicted by model A, this indicates simply that A represents C only incompletely. In this case, it may be necessary to improve

---

[9] Cartwright (1983), pp. 57–58; Hempel (1988), p. 150; Pietroski and Rey (1995, p. 86); Smith (2002).

[10] The title of Cartwright's book says, ambiguously, "How the Laws of Physics Lie," which could also mean "How the laws of physics stand." However, in her introduction, Cartwright explains that this is not the intended interpretation: "laws in physics [. . .] must be judged false" (Cartwright 1983, p. 12).

[11] Silverberg (1996); Hüttemann (1998).

A by including in it additional objects, properties and interactions. The equations of movement that are calculated (on the basis of model A) in order to represent the evolution of sets of concrete systems C correspond to the laws, of which Russell says that they are "empirically verifiable but probably only approximate (Russell, 1912/1992, p. 203), because nothing prevents a certain concrete system C to be subject to the influence of factors not represented in A.

In a similar spirit, Cummins (2000) has suggested distinguishing between "general laws of nature," whose domain of application is unlimited, and "*in situ* laws," which apply only to systems of a particular type, such as planetary systems or living beings, by virtue of the constitution and organization of these systems. If such a system, which Cartwright (1999) calls a "nomological machine," evolves according to a (system) law, its evolution can be seen as a causal process. In contrast with general laws of nature, system laws are not strict. Exceptions result from influences that perturb the evolution of the system from outside.[12] These perturbations can be the objects of causal judgments. According to Menzies (2004), every causal statement presupposes a model (constituted by a natural kind and a law applying to that kind). A factor is judged to be a cause if it makes a difference to the evolution of the system, relative to the background of the normal evolution of the model.[13] In one of Menzies' examples, a person who has been smoking for years develops cancer. Intuitively, the fact that the person is born and the fact that she has lungs are not causes of her cancer although both are necessary conditions. Menzies explains this intuition by suggesting that the identification of a cause normally constitutes the response to a "contrastive why-question" (Menzies, 2004, p. 148), of the form: "why did the man get lung cancer rather than not?" (Menzies, 2004, p. 149). The real history is compared with a fictive (or "counterfactual") history, in which the person does not develop any cancer. The facts of being born and of having lungs are not causes because they also feature in the fictive history.

Russell's analysis shows that laws having the form of quantitatively precise functional dependencies as they are used in mathematical physics cannot be interpreted as directly expressing regularities among observable events; more particularly, they cannot be interpreted as generalizations expressing the succession of causes and effects. This raises the general problem of understanding the relation between laws or models as they are used in the advanced sciences and their use for the prediction and explanation of real concrete systems. As the contemporary debate on *ceteris paribus* laws shows, this difficulty is not specific to the scientific justification of *causal* judgments. The same difficulty arises, for example, in the context of the determination of the spatial conformation of a macromolecule, on the basis of its components and the laws governing their interactions by virtue of their properties. Here, the

[12] Cf. Kistler (2006).

[13] Menzies's idea that a cause is a factor that "makes a difference" relatively to a background makes use of Mill's (1843) analysis of the distinction between causes and conditions, and of Mackie's (1974) conception of the background as the "causal field." Similar ideas can be found in Lewis's (2000) analysis of causation in terms of influence, and in Hitchcock's (1996a; 1996b) and Woodward's (2003, 2004) work.

notion of causality does not come into play because the dependence at issue of the macroproperty on the microproperties is simultaneous dependence between different properties of the same object. While the difficulty of understanding the application of models to real systems raises an important challenge to philosophy of science, it is not specific to the justification of causal judgments. The same can be said of the problem of induction: As Russell notes, it poses a principled obstacle to the knowledge of causal generalizations. However, the problem of induction is a general problem that arises just as well in the context of the knowledge of non-causal generalizations.

2. The Reduction of Causation to Deductive-Nomological Explanation

The most specific challenge raised by Russell's arguments is the justification of the characteristic features of causality, first and foremost its asymmetry, that is, an event *c* cannot be both the cause of a second event *e* and its effect. Russell argues that no asymmetry of this sort exists at the level of the functional laws of physics. However, this does not show that there are no asymmetric relations in reality; it only shows that the scientific explanation of the source of this asymmetry must be found somewhere other than these functional laws.

The fact that the notion of cause does not appear in fundamental physics does not make the project of a philosophical analysis of this notion illegitimate. The laws of fundamental physics and causal judgments do not apply to the same objects: the values of the variables that figure in the former are determinate quantities that characterize certain *properties* of substances or events, whereas the terms of causal relations are concrete events. Given that causal judgments regularly occur not only in the judgments of common sense but also in many philosophical projects and in judgments bearing on the experimental testing of scientific theories,[14] the project of a naturalistic analysis of causation has been very actively pursued during the 20th century, beginning with Russell himself.[15]

The so-called deductive-nomological (DN) analysis of causation has been dominant during the first half of the 20th century. It can be seen as a contemporary version of the traditional reduction of causality to regularities and laws of nature. However, this reductive analysis of causation in the tradition of 20th century logical empiricism takes a form that distinguishes it from its philosophical predecessors. Instead of beginning, like Hume, with the analysis of the *idea* of causality that arises from the *experience* of the regular repetition of certain successions of events, and instead of

[14] Cf. Putnam (1984).

[15] In 1914, Russell explains that "there is, however, a somewhat rough and loose use of the word 'cause' which may be preserved. The approximate uniformities which lead to its pre-scientific employment may turn out to be true in all but very rare and exceptional circumstances, perhaps in all circumstances that actually occur. In such cases, it is convenient to be able to speak of the antecedent event as the 'cause' and the subsequent event as the 'effect'" (Russell 1914/1993, p. 223). Russell (1948/1992, p. 471ff.) presents a more elaborate theory of causation.

suggesting, like Galileo, Newton, and many others, to substitute the notion of law for the notion of cause, the DN analysis aims at analyzing first of all causal *explanation*, as it is accomplished in the sciences (see chapter 1 of this volume). According to this analysis, it is equivalent to say that C causes E and to say that C figures as a premise in a DN explanation of E: the effect E is the *explanandum*—what is to be explained—and occupies the role of the conclusion of the argument, and the cause is the content of one of the premises that together constitute the *explanans*—that which explains. Here is how Carnap justifies his analysis of causation in terms of DN explanation: "What is meant when it is said that event B is caused by event A? It is that there are certain laws in nature from which event B can be logically deduced when they are combined with the full description of event A" (Carnap 1966/1995, p. 194).[16] It is essential for a scientific explanation that the link between the premise designating the cause and the conclusion designating the effect be provided by one or several laws of nature. If E were a *logical* consequence *of C alone*, their link would be logical or conceptual, which would be incompatible with the generally accepted Humean thesis that causation is a contingent relation. In retrospect, the attempt to reduce causation to deducibility with the help of laws appears as an attempt to *eliminate* causality and to replace it by mere laws. Such an analysis may well keep the word "causality" but the DN analysis deprives the word of its content: to say that C figures in a *causal* explanation of E means nothing more than to say that C figures in a *scientific* explanation of E. If all scientific explanations are causal, the concept of causation loses its discriminative content.

The main reason why the DN analysis has widely been abandoned is that it has become clear that some scientific explanations are *not* causal:[17] there is a specific difference between non-causal and causal explanations that the DN analysis denies. Many physical explanations using functional dependences do not intuitively correspond to causal relations: when the thermal conductivity of a copper wire is deduced from its electric conductivity or vice versa (according to the Wiedemann-Franz law, which says that the values of these two properties of metals are proportional), none of them appears to be the cause of the other. In the same way, when the temperature of a sample of gas that can be considered to be "ideal" (in the sense of falling in the domain of validity of the ideal gas law according to which the product of pressure P and volume V of a sample of ideal gas equals the product of the volume V it occupies, the number n of moles contained in the sample and the universal gas constant R: $pV = \mathrm{n}\mathrm{R}T$) is deduced from its pressure, given the volume it occupies, it seems intuitively clear that the pressure of the gas is not the cause of its temperature. Pressure and volume characterize the same individual sample at the same time; their correlation can be explained

[16] Popper also identifies causal explanation with scientific explanation, in the framework of the DN model: "To give a *causal explanation* of an event means to deduce a statement which describes it, using as premises of the deduction one or more *universal laws*, together with certain singular statements, the *initial conditions*" (Popper 1935/2002, p. 38; italics are Popper's).

[17] I cannot develop here the reasons that have led to abandoning the classical conception of logical empiricism, i.e. the assimilation of causation to scientific explanation in the form of a deductive-nomological argument. See chapter 5 of Barberousse, Kistler, Ludwig (2000).

by processes at the level of the molecules composing the gas. The ideal gas law being symmetrical, DN explanations that can be constructed on its basis cannot be causal without contradicting the asymmetry of causation. If the fact that $P(x,t)$ (the pressure of sample x of gas at time t) is proportional to $T(x,t)$ sufficed to establish that $P(x,t)$ causes $T(x,t)$, $T(x,t)$ would cause $P(x,t)$ for the same reason.

3. The Analysis in Terms of Counterfactual Conditionals

Given the number and the diversity of the counterexamples that have been found against the analysis of causation in terms of DN explanation, many philosophers have found it judicious to abandon that analysis. In a passage that marks a turning point in philosophical thinking on causality, David Lewis writes in 1973: "I have no proof that regularity analyses are beyond repair, nor any space to review the repairs that have been tried. Suffice it to say that the prospects look dark. I think it is time to give up and try something else. A promising alternative is not far to seek" (Lewis 1973/1980, p. 160). The basic alternative idea Lewis has in mind can be found in Hume's *Enquiries Concerning Human Understanding*. After his famous definition of causation in terms of succession, Hume offers a second definition: a cause is "an object, followed by another, [. . .] where, if the first object had not been, the second never had existed" (Hume 1777, p. 76).[18] This second definition contains the leading idea of what is now known as the counterfactual analysis of causation: the proposition "*c* causes *e*" means that "if *c* had not occurred, *e* would not have occurred either." The latter proposition is often represented by the expression "C □→ E."[19] This analysis is intended to be a priori, in the sense that its aim is not to discover the physical nature of real causal processes, but rather something that is implicitly known by every competent speaker of English (or any other language containing a synonym of the word "cause"), namely the meaning of the concept expressed by the predicate "causes." In the tradition of logical empiricism, the use of counterfactuals was considered methodologically suspect. Indeed, determining the truth value of a counterfactual proposition requires evaluating possibilities, which are not observable.[20] However, the elaboration of a formalism in which modal and counterfactual propositions can be interpreted in terms of possible worlds has given new life to the project of an analysis of causation in counterfactual terms. The strength of the counterfactual approach rests on the initial plausibility of the idea that a cause "makes a

[18] Hume does not develop this new idea, nor does he comment on the fact that it is not equivalent to the analysis of causation in terms of regularity.

[19] In Lewis's terminology, upper case C represents the proposition that the event named by the corresponding lower case letter *c* has occurred. Except when quoting Lewis, I stick to the usual convention of using lower case letters like *c* and *e* for events and upper-case letters for predicates and propositions.

[20] J. St. Mill (1843) analyzes the counterfactual "if A occurred, then B would have occurred" in terms of the possibility to deduce B from A together with a set of auxiliary propositions S, which must necessarily contain laws of nature. Thus understood, the counterfactual analysis is equivalent to the DN analysis.

difference," an idea that can be expressed in a quite straightforward way by a counterfactual conditional.[21]

David Lewis's contribution to the counterfactual analysis of causality has determined the orientation of all subsequent research in this framework. Lewis proposes to conceive of the semantic evaluation of counterfactuals in terms of the similarity of possible worlds. The terms of causal relations and of counterfactuals are events, where "event" is understood "in the everyday sense" (1986b, p. 161) of a particular happening at a determinate place and time.

The strategy adopted by Lewis for determining the truth conditions of counterfactuals consists in comparing different possible worlds with respect to their global similarity with respect to the actual world, where "actual" is understood in the modal sense. It starts with the thesis according to which the counterfactual proposition expressed by "if C were the case, E would be the case" is true in a world *w* if and only if (1) C is not true in any possible world or (2) if some world in which both C and E are true is closer to *w* than all possible worlds in which C is true but E false. When one asks whether *c* causes *e*, one presupposes that *c* has occurred, and that C is therefore true in the world *w*. On the basis of this presupposition, the second clause determines the truth value of the counterfactual.

Lewis's analysis of the causal relation in counterfactual terms is indirect; it uses causal dependence as an intermediate concept. If *c* and *e* are two distinct actual events,[22] *e* depends causally on *c* if and only if it is true that "if *c* had not occurred, *e* would not have occurred." Causation is then defined by the existence of a set of intermediate events constituting a chain reaching from the cause *c* to the effect *e: c* is a cause of *e* if and only if there is a finite chain of intermediate events e_1, e_2, ... e_k, between *c* and *e*, such that the second link of the chain depends causally on the first, and in general if, for every *n*, the *n*th link depends causally on the preceding (n–1)th link. The events *c* and *e* must be distinct in the sense that the space-time region in which *c* occurs must not overlap the region in which *e* occurs. With this restriction, the analysis avoids the problem of wrongly classifying non-causal dependence relations as causal: it is clear that the truth of the counterfactual "if John had not said 'hello', he would not have said 'hello' loudly" does not reveal the existence of any causal relation.[23]

The counterfactual analysis can account for both deterministic and indeterministic causality. In a world in which there are indeterministic laws, *e* depends causally on *c* (where *c* and *e* are distinct events occurring in the actual world) if and only if, if *c* had not occurred, the probability of the occurrence of *e* had been much less than it actually was (Lewis 1986c, p. 176).

---

[21] Mackie (1974, chap. 2) has enriched the counterfactual analysis by the distinction between the background "causal field" and the salient factor that appears intuitively to be the cause insofar as it "makes a difference" with respect to the background.

[22] In the general case where *c* and *e* are possible events, it must be true both that "if *c* had not occurred, *e* would not have occurred" and "if *c* had occurred, *e* would have occurred."

[23] Cf. Kim (1973); Lewis (1986a).

Several objections have been raised against Lewis's analysis of causation. Two sorts of counter-examples have been found: "false positives" seem to show that counterfactual dependence is not sufficient for the existence of a causal relation, whereas "false negatives" seem to show that it is not necessary either. We will look at some of these counterexamples and the lessons to be drawn from them. However, rather than taking these criticisms as refutations, advocates of the counterfactual analysis regard these problems as indications of a need for improvement.

A first difficulty for the counterfactual analysis stems from the existence of so-called *backtracking* counterfactuals, according to which a past event depends counterfactually on a present or future event. Take a wave on the ocean. It seems correct to say: "if a given wave summit had not been at x at time t, it would not have been at x-dx at time t-dt," where "x-dx" represents the location of the wave summit at a moment t-dt preceding t. Such backtracking counterfactuals seem to be true in conditions in which some event *c* is a sufficient condition for some later event *e*, in the sense that, once *c* had happened, nothing could have intervened to prevent *e* from happening. In such a situation, it seems true that, if *e* had not occurred, *c* would not have occurred either. Take a situation in which a bomb explodes at instant t after having been triggering by a detonator, and suppose that the triggering is sufficient for the explosion, in the sense that the explosion could not have been prevented once the triggering had occurred. It seems correct to say: if the bomb had not exploded, its detonator would not have been triggered. Now, if there are true backtracking conditionals, counterfactual dependence is not sufficient for (nor, a fortiori, equivalent to) causal dependence, because the future event cannot be the cause of the past event,[24] although the past event depends counterfactually on the future event. The wave summit at (x, t) does not cause the wave summit at (x-dx, t-dt), although the wave summit at (x-dx, t-dt) seems to depend counterfactually on the wave summit at (x, t); similarly, the triggering of the detonator depends counterfactually on the explosion of the bomb but the explosion of the bomb does not cause the triggering of the detonator. In other words, the counterfactual analysis seems to predict wrongly that effects sometimes cause their own causes.

Lewis solves this problem by arguing that the use of backward counterfactuals does not correspond to our "standard" (Lewis 1979/1986, p. 35) strategy of judging the similarity among possible worlds.[25] The justification of this thesis depends on a contingent but real asymmetry of our actual world. According to Lewis (1979/1986, p. 49), a set of conditions is a "determinant" of a given event if these conditions, together with the laws of nature, are sufficient for the occurrence of the event. Among the determinants of an event, there are its causes as well as the traces it leaves behind. The asymmetry of

[24] I put the possibility of backward causation to one side here. It remains controversial whether and how backward causation might be conceived and whether such a concept can be applied to certain physical processes. Cf. Faye (2010).

[25] Given that counterfactuals are in general vague and given that that their evaluation depends on the context, Lewis (1979/1980, p. 32–35) acknowledges that there are particular contexts, in which we take backward counterfactuals to be true. However, he argues that these particular contexts should be excluded from the evaluation of those counterfactuals that must be used for the analysis of causal dependence.

the actual world is grounded on the fact that events have in general few determinants preceding it (its causes) but a large number of determinants following it (its traces). Lewis calls this fact the "asymmetry of overdetermination" (Lewis 1979/1986, p. 49): ordinary events have in general only one cause. It is a contingent fact characteristic of the actual world that events are only exceptionally overdetermined by many causes. If one considers the waves that propagate from a perturbation localized at a point on the surface of a lake, there is only one common cause of numerous perturbations on the surface of the water, whereas the event at the origin of the wave has numerous traces: the origin of the wave is overdetermined by the traces in its future, whereas these traces are not overdetermined by the point-like cause in the past.

Here is how Lewis justifies his thesis that backward counterfactuals are not relevant for the analysis of the meaning and truth value of causal statements. To judge whether *e* depends counterfactually on *c*, it is necessary, according to the counterfactual analysis, to evaluate the counterfactual "if *c* had not occurred, *e* would not have occurred." This requires considering possible worlds in which *c* does not occur. Such worlds differ from the actual world, for in the actual world, both *c* and *e* occur. Among those possible worlds in which *c* does not occur, those that determine the truth value of the counterfactual by determining the truth value of the consequent *e*, are the worlds that are closest to the actual world. Lewis gives several weighted criteria for determining whether a world is "closer" to the actual world. The first two criteria in order of decreasing importance are

1. Avoiding "big, widespread, diverse violations" (Lewis 1979/1986, p. 47) of the laws of the actual world
2. Maximizing the spatiotemporal region in which there is perfect match with respect to particular facts of the actual world.[26,27]

Recall that the relevant possible worlds all differ from the actual world by the fact that *c* does not occur in them. In the framework of events that are determined according to deterministic laws, this divergence is accompanied either by a vast divergence of states of affairs with respect to the causal histories leading respectively to *c* (in the actual world) and to non-*c* (in the possible worlds under consideration), or by a violation of the laws, that is, by the fact that the possible worlds under consideration do not

[26] The technical sense of the expressions "fact" and "state of affairs" as they are used in contemporary philosophy has its origin in Wittgenstein's *Tractatus* (1921). According to an important interpretation, a fact ("Tatsache" in German) is what makes true a descriptive statement: the satisfaction of a predicate by an object. The concept of a "state of affairs" ("Sachverhalt" in German) is more general in the sense that it also applies to what is possible, what could be the case. If it is possible that object *a* satisfies predicate P, then "*a* is P" expresses a "state of affairs." If *a* is actually P, "*a* is P" also expresses a fact.

[27] Lewis mentions avoiding small divergence with respect to laws or facts as separate criteria: "(3) It is of the third importance to avoid even small, localized, simple violations of law. (4) It is of little or no importance to secure approximate similarity of particular fact, even in matters that concern us greatly" (Lewis 1979/1986, p. 48).

perfectly obey the laws of the actual world. Lewis argues that the analysis of our practice of making and evaluating counterfactuals shows that we consider to be closest to the actual world those worlds that resemble the actual world perfectly for their entire history up to the time of *c*, and differ from the actual world by virtue of a localized violation of the laws of nature at a moment just before the time of *c*. We judge such worlds to be closer to the actual world than worlds that do not contain any such "miracles," but differ from the actual world by a great number of facts concerning a large part of their history.

At this point, the "asymmetry of overdetermination" comes into play to guarantee that counterfactuals are evaluated according to the "standard" interpretation, that is, in such a way that the future depends counterfactually on the past but not vice versa. Given the asymmetry of overdetermination, the worlds in which the miracle takes place in the *past* of *c* are closer to the actual world than worlds in which the miracle takes place in the *future* of *c*.[28] A miracle that would be sufficient to make a non-*c* world "reconverge" toward the actual world so as to resemble the actual world perfectly for the *future* of *c*, would have to be much more extended than the miracle required to prevent *c* in a world that resembles perfectly the actual world with respect to the *past* of c. From this reasoning, Lewis concludes that the relevant possible worlds always contain a miracle occurring at a moment immediately preceding the antecedent of the counterfactual. This "standard" choice of the relative importance of the criteria of similarity between possible worlds, taken to be implicit in our practice of evaluating counterfactual propositions, together with the contingent asymmetry of the actual world, guarantees that all backtracking counterfactuals are false. Consider a "backward" counterfactual of the form "if *e* had not occurred, *c* would not have occurred" where *c* and *e* are events that occur in the actual world and where *e* occurs *later* than *c*. The possible worlds that are relevant for its evaluation are those in which the antecedent non-*e* is true by virtue of a "tiny miracle" that occurs *immediately before* the occurrence of *e* in the actual world. Thus, the miracle occurs *after* the occurrence of *c*; therefore, *c* occurs in the closest possible world in which the antecedent of the counterfactual is true; therefore, the consequent of the backtracking counterfactual is false, and thus the counterfactual itself is false as well.

The argument that establishes that backward counterfactuals are systematically false also provides a solution to what Lewis (1986b, p. 170) calls "the problem of epiphenomena": consider an event *c* that causes two effects, *e* and *f*, but where *e* does not cause *f* nor does *f* cause *e*. Lewis's analysis seems to predict wrongly that *e* causes *f* because there seems to be a chain of counterfactual dependences between *e* and *f*: if *c* is necessary in the circumstances for *f* then *f* depends counterfactually on *c*; and if *c* is sufficient for *e* then *c* seems to depend counterfactually on *e*: if *e* had not occurred, *c* would not have occurred. Now if Lewis's argument is correct to the effect that our

[28] That is, the past with respect to the moment at which *c* occurs in the actual world. An event *e* in world w_1 appears as a miracle with respect to world w_2 if the circumstances in which *e* occurs (in w_1) are not in conformity with the laws of w_2. Then *e* is a miracle in w_1 relative to w_2.

criteria for evaluating counterfactuals guarantee, in the context of the asymmetry of overdetermination, that backward counterfactuals are always false, then the latter counterfactual is false, and there is not after all any chain of counterfactual dependence between the epiphenomena *e* and *f*.

Several objections have been raised against this reasoning. Horwich (1987) notes that the asymmetry of overdetermination is known only by science and a posteriori. Insofar as it is not an aspect of reality that is known a priori by all competent speakers, a conceptual analysis of the concept of causation cannot make use of it.[29] Several authors have questioned the scientific correction of Lewis's (and Popper's 1956) thesis according to which events have typically few determinants preceding them but many determinants following them, or, in other words, few causes and many traces. Concerning the deterministic and symmetric laws of classical mechanics, this difference is in fact illusory. Elga (2000) has shown that, even for counterfactuals whose antecedent expresses an irreversible event in the thermodynamic sense (of an increase in entropy), Lewis is wrong to say that worlds in which the antecedent is true by virtue of a miracle that occurs immediately *before* the antecedent are closer to the actual worlds than worlds in which the miracle occurs *after* the antecedent. Elga illustrates this point with a situation in which Gretta smashes, in the actual world w_1, an egg in her pan at 8 o'clock. Consider the closest worlds in which Gretta does not smash any egg at 8 o'clock. According to Lewis, it needs only a tiny miracle, for example, in a process taking place in Gretta's brain just before 8 o'clock, say at 7:59, that guarantees that she does not smash any egg. Such a world w_2 containing a miracle at 7:59 resembles the actual world perfectly with respect to all facts in the whole of history right up to 7:59, and diverges from it only after the time of the miracle. However, Elga shows that there is a world w_3 that shares, contrary to the actual world, the whole set of facts pertaining to the future beginning from a moment just after 8, say from 8:05, so that there is in w_3, after 8:05, a smashed egg just as in the actual world w_1. These are worlds in which Gretta smashes no egg but in which the miracle that guarantees the convergence with respect to the actual world is not larger than the miracle that occurs in world w_2. Elga has us consider a process that corresponds to the process taking place in the actual world from 8 to 8:05 but which evolves in the opposite direction, like when one watches a film in the wrong direction. The egg that has been smashed in the pan "*uncooks*" beginning at 8:05 and returns in the eggshell. This process is in agreement with the laws of physics although it is very improbable because it depends in an extremely sensitive manner on its initial conditions: if one produces an tiny change in the positions and speeds of the molecules at 8:05, a more banal process will take place, in which the egg remains in the pan and starts cooling down. Thus it suffices to have a tiny miracle at 8:05, to guarantee that the entire past changes, including Gretta's act of smashing an egg at 8 o'clock. With such a small miracle at 8h05, the whole past in w_3 is different from what it is in the actual world, and does in particular not contain Gretta's

[29] Lewis answers this objection in (1979/1986, p. 66).

smashing any egg at 8 o'clock. However, worlds w_1 and w_3 resemble each other perfectly for all times after 8:05. Thus, w_3, in which the miracle that ensures that the smashing does not occur happens *after* 8 o'clock (the time of the smashing in w_1), does not differ more from w_1 than world w_2, in which the miracle occurs *before* 8 o'clock.

We have seen that Lewis defines causation indirectly, using the notion of causal dependence as an intermediary between counterfactual dependence and causation: *c* is a cause of *e* if and only if there is a finite chain of intermediate events e_1, e_2, ... e_k, between *c* and *e*, such that the second link of the chain depends causally on the first, and in general if, for every n, the *n*th link depends causally on the preceding (*n*-1)th link. Causal dependence is then, as we have seen, reduced to counterfactual dependence.

This analysis solves two difficulties: first it guarantees the transitivity of the causal relation, and second it allows justifying the intuition that a "pre-empted" cause is only a potential rather than an actual cause.

1. Counterfactual dependence is in general not transitive: it is easy to find examples where it is true that A □→ B and that B □→ C, but false that A □→ C. The reason is that the evaluation of a counterfactual depends on the background circumstances of the antecedent, and that the antecedents in a series of counterfactuals do not in general share their backgrounds. When the causal relation is reduced to a chain of events in which each link depends counterfactually on the preceding link (instead of reducing it directly to causal dependence), the first and the last link of a causal chain are guaranteed to be linked as cause and effect, whereas the last link does in general not counterfactually depend on the first. However, this aspect of Lewis's analysis has also given rise to an objection. Several authors claim that there are counter-examples to the transitivity of causation. In particular, such counter-examples concern judgments in which an absence, or a particular aspect of an event, play the role of cause or effect, or judgments in which the causal link is grounded on a double prevention.[30] In an example offered by Ehring (1987), someone puts potassium salts in the fireplace, which brings about a change of the color of the flame from orange to purple. Later, the flame lights a piece of wood next to the fireplace. There is a causal chain between the act of putting potassium salt in the fireplace and the lighting of the piece of wood. However, it seems false to say that the first event causes the last.[31] The transitivity of causation can be defended against certain counter-examples by showing that the appearance of the existence of a causal chain is due to too coarse a conception of the terms of the relevant causal relations. If the terms of the causal relations are not concrete events, but *facts* bearing on these events, there does not appear

[30] See Bennett (1987); Hall (2004a).

[31] Other examples of this kind can be found in McDermott (1995), Hall (2000/2004b), and Paul (2004).

to be any chain linking the act of throwing salt in the fire to the lighting of the piece of wood: the salt is causally responsible for the fact that the flame changes color; however, the cause of the lighting is not the fact that the flame changes its color but rather the fact that its gives off heat.[32] It can also be defended by denying that there are causal relations with "negative" terms, such as absences or omissions: such relations correspond often to non-causal explanations, which can give an illusory impression of causality. Such statements describe a situation lacking any causal process, which is implicitly contrasted with a background situation in which there is such a causal process.[33] If this is correct, explanatory chains containing double prevention do not in general indicate the existence of a causal chain. To use an example that Hitchcock (2001) attributes to Ned Hall,[34] a hiker sees a rock falling, which causes him to duck so as to avoid being hit by the rock. The fact that he has not been touched might seem to be a cause of the pursuit of the trek. This is a case of *double prevention*, in the sense that the ducking prevents the rock from preventing the pursuit of the hiker's trek. It seems wrong to say that the falling of the rock caused the pursuit of the trek, although there seems to be a causal chain from the first event to the last. However, it can be denied that it is a causal chain, thereby defending the transitivity of causation, by denying that the negative fact of *not* being touched by the rock can be either an effect or a cause.

2. The second problem that the introduction of a chain of intermediate events solves arises in the context of situations of "preemption" and cases involving "redundant causation." Such situations are frequent, for instance in biology. For example, sometimes it is said that evolution brings about both a main mechanism important for an organism's survival, and a backup mechanism, something that takes over in case of failure of the main mechanism.[35] Other examples involve human actions. One of the paradigm cases of preemption in the literature involves two snipers, S_1 and S_2, who aim at the same victim at the same time. S_1 decides to fire (event *a*); this decision causes her shot, which causes the death of the victim (event *c*). S_2 who sees S_1 shoot does not shoot and thus does not cause *c*; S_2's determination to fire (event *b*) is not followed by S_2's firing: the process is interrupted by S_2's seeing S_1 fire. This situation shows that counterfactual dependence is not necessary for causation: *a* causes *c* although *c*, the victim's death, does not counterfactually

[32] See Kistler (2001). Paul (2004) offers a similar analysis, in which she argues that causation links aspects of events, rather than events themselves.

[33] Cf. Kistler (1999/2006); Hall (2004a); Kistler (2006).

[34] According to Hitchcock (2001, p. 276), this example features in an unpublished version of Hall (2004a).

[35] The main mechanism for the orientation of honey bee workers relies on the perception of the location of the sun, but backup mechanisms are available for situations in which the sun is not directly visible: one relies on the perception of patterns (of the ultraviolet component) of polarized light, another on the perception of landmarks (Winston 1991, pp. 163–164).

depend on *a*. If *a* had not happened, S_2 would have fired. The event *b*, corresponding to S_2's determination to fire, would have caused S_2's firing, which would have caused *c*; in short, *c* would have happened even without *a*.

The requirement of the existence of a chain of intermediate events solves this difficulty: for event *a*, the positions of the bullet on its trajectory from *a* to *c* constitute such a chain. By contrast, given that S_2 does not fire, there are, for all times following S_2's noticing that S_1 has fired, no intermediate events between *b* and *c* on which the death of the victim depends counterfactually and which depend on *b*. Lewis's analysis yields the intuitively correct result that *b* is no cause of the death of the victim. This type of situation is called "early preemption," because, insofar as the potential causal chain between *b* and *c* is interrupted early, that is, a sufficiently long time before *c*, there exists a chain of events between *a* and *c* to which no parallel chain between *b* and *c* corresponds.

However, this solution is ineffective in cases of what has been called "late preemption," in which there is a continuous chain of events between events *b* and *c*, but where *b* still does not cause *c*. Hall (2004a, p. 235), for instance, considers the situation in which two children (Suzy and Billy) throw rocks at a bottle. Suzy throws her rock a little earlier than Billy, so that her rock smashes the bottle (event *c*). However, Billy's rock follows closely behind Suzy's rock, so that there is not only a chain of events between Suzy's throw and *c*, but also between Billy's throw and *c*. Nevertheless, to the extent that Suzy's rock reaches the bottle a moment before Billy's, Suzy's but not Billy's rock is the cause of *c*.

In "Postscripts to 'Causation,'" Lewis (1986c) introduces the concept of "quasi-dependence" to solve the problem of late preemption. In cases of late preemption, in spite of the presence of the preempted event *b*, and in spite of the fact that there is an entire parallel chain from *b* to *c*, the "preempting" event *a* causes *c*. The reason why the presence of the redundant cause *b* does not deprive *a* of being efficacious in causing *c* is the fact that causality is an *intrinsic* quality of the process localized between *a* and *c*. According to Lewis, each event in the chain between *a* and *c* is *quasi-dependent* on its predecessor because the process intrinsically resembles—that is, if only the events localized on the chain linking *a* to *c* are taken into account—processes whose elements are fully counterfactually (and therefore causally) dependent on their predecessors. Event *a* (Suzy's throw) is the cause of *c* because *a* intrinsically resembles possible throws that Suzy executes in the absence of any of Billy's throws. Event *c* is quasi-dependent on Suzy's throw because *c*'s counterpart in such possible situations (where Suzy throws but Billy doesn't) is counterfactually dependent on the counterpart of Suzy's throw.

However, there are even more problematic cases of preemption that involve a chain of intermediate events that makes the effect *c* "quasi-dependent" on an earlier preempted event *b* (which is *not* a cause of *c*). Schaffer (2000b) calls this sort of situation "trumping preemption": a major and a sergeant shout orders at a corporal. Both shout "Charge!" at the same time, and the corporal decides to charge. Given that a soldier obeys the orders of the higher-ranking soldier, the cause of the corporal's decision is the major's order, not the sergeant's. However, the corporal's decision is quasi-dependent both on the sergeant's and on the major's order. The chain reaching from one of the orders to the corporal's decision is intrinsically similar to a chain that, in the absence of the second order, guarantees counterfactual dependence along the links of the chain and therefore the existence of a causal relation. Quasi-dependence is therefore not, after all, sufficient for causation.

This difficulty has led Lewis (2000) to devise a new version of his counterfactual account, in terms of "influence." Lewis suggests that *the fact* that the occurrence of *e* is counterfactually dependent on the occurrence of *c* is not by itself sufficient for *c* being a cause of *e*; there is the further requirement that *the way* in which *e* occurs and *the moment* at which *e* occurs also depend counterfactually on the manner and the moment in which *c* occurs. Lewis's new analysis employs the notion of the alteration of an event. An alteration of an actual event *e* is a possible event that differs slightly from *e*, either by its properties or by the moment at which it occurs. If an event *c* influences another event *e*, there is "a pattern of counterfactual dependence of whether, when and how on whether, when and how" (Lewis 2000/2004, p. 91). More precisely: "Where C and E are distinct actual events, let us say that C *influences* E iff there is a substantial range C_1, C_2, . . . of different not-too-distant alterations of C (including the actual alteration of C) and there is a range E_1, E_2, . . . of alterations of E, at least some of which differ, such that if C_1 had occurred, E_1 would have occurred, and if C_2 had occurred, E_2 would have occurred, and so on" (Lewis, 2000/2004, p. 91; emphasis Lewis's).[36] Just as in his original analysis, the fact that *c* causes *e* is reduced to the existence of a chain of intermediate events in which each link influences the following link.

Another objection against the counterfactual analysis concerns the fact that it does not respect the common sense distinction between causes and background conditions. Now one might consider rejecting this distinction (as did Mill) since the distinction only reflects the interests of human observers; but "philosophically speaking," background conditions are causes in the same sense as salient factors that common sense recognizes as causes. However, to the extent that the aim of the counterfactual analysis is not the nature of causation as it is in reality, but the structure of our naïve concept of causation, it seems essential that the analysis respects this distinction. To accomplish this, one can hypothesize that ordinary causal statements like "c causes e" in fact contain implicit comparisons to a "normal" background situation. This can be made explicit in a paraphrase of a form such as "*c* rather than *c** has caused *e* rather than *e**." The correct counterfactual analysis would then be: "if *c** had occurred rather than *c*, *e** would have occurred rather than *e*."[37] This idea is closely related to the intuition that a cause makes a difference with respect to its effects: one compares, though often implicitly, the situation as it is when the cause is present to the situation, as it

[36] I have kept Lewis's notation, where the upper case letter "C" represents "the proposition that *c* exists (or occurs)" (1986b, p. 159), where lower case "*c*" represents a particular event.

[37] Cf. Hitchcock (1996a, 1996b); Maslen (2004); Schaffer (2005).

would have been if the cause had been absent. If the effect is present in a situation in which the cause is present but absent where the cause is absent, one has good reason to think that the cause is responsible of this difference. To use Achinstein's (1975) example, the cause of Socrates's death is his drinking hemlock, because this is the factor that makes the crucial difference with respect to his death. Many other characteristics of the situation, such as the fact that Socrates's drinking hemlock occurred *at dusk*, are not causes of his death. The time at which the drinking occurred made no difference to the hemlock's fatal effect.

4. Methodology

The successive modifications of the counterfactual analysis are motivated by the attempt to avoid two sorts of counter-examples. "False positives" for a proposed analysis are situations featuring two events that the analysis presents as being related as cause and effect, where intuitively they are not so related. "False negatives" are on the contrary situations in which an event *c* is intuitively the cause of another event *e*, but where the analysis yields the result that it is not. These are the two possible forms of mismatch between a given analysis and intuition. The research on improving the counterfactual analysis is driven by the presupposition that the main criterion of adequacy of a philosophical analysis of the concept of causation is agreement with common sense intuitions. However, this choice of the criterion of adequacy is controversial. The diversity of extant analyses of the concept of causation can be explained at least in part by the existence of different ways of conceiving the aim and method of such an analysis. A major disagreement opposes a priori and a posteriori analyses.

1. Advocates of the counterfactual analysis want to provide a "conceptual analysis" of a concept mastered by everyone (at least everyone within the language community of speakers of some natural language containing causal vocabulary). Just like other common sense concepts, people use causal concepts to reason about possible or counterfactual situations in addition to reasoning about actual situations. For example, causal concepts are also used to reason about the consequences of science fiction novels, where facts and even laws of nature may differ widely from the actual world. If the aim of the philosophical analysis of causation is an analysis of this common sense concept, the analysis must be such that it applies to all possible worlds to which the concept of causation applies. Moreover, insofar as the common sense concept of causation is not informed by scientific knowledge about the physical nature of the causal processes of the actual world, scientific knowledge appears irrelevant to the philosophical analysis of the concept. Therefore a conceptual analysis can be conducted in a purely a priori manner. The adequate method consists in carefully spelling out "from the armchair" one's spontaneous intuitions on a certain number of fictitious

situations. And although these situations can reflect real world scenarios, such as children throwing rocks at bottles or soldiers shouting orders, the a priori analysis of our naïve concept of causation can just as well make use of intuitions concerning unreal or even physically impossible situations, such as situations in which magicians cast spells. In a situation conceived by Schaffer (2004, p. 59), Merlin casts a spell that transforms a prince into a frog. Magical causal interactions of this sort are not constrained by physical laws and can act at spatial and temporal distance without any causal intermediaries.

2. A theory can start with the analysis of the common sense concept, but then make corrections in order to obtain better coherence and systematicity without thereby abandoning the framework of a priori constraints. It is, for example, intuitively correct to judge both that an ice cube (more precisely the melting of the ice cube) in a glass of water causes the water to cool down, and that the cooling of the water (more precisely the fact that the water gives off heat) causes the melting of the ice cube. Taken together, the set of these two judgments violates the asymmetry of causation, which is, as we have seen, a central component of the concept of causation. It can be concluded that at least part of the naïve intuitions on this situation must be incorrect. However, there does not seem to be any reason to take one to be incorrect rather than the other.
3. There is an alternative way of conceiving of the aim of the philosophical analysis of causation. Causation can be taken to be a concept of a "natural kind" of relation whose real essence must be discovered a posteriori. This is the way in which process theories of causation conceive of their task. From such a perspective, the causal relation whose "real essence" one tries to discover does not exist in all possible worlds. In this framework, one may look for a scientific reason for following one intuitive judgment rather than the other in the case of the two judgments that together violate the asymmetry of causation. The judgment that the cooling of the water causes the melting of the ice cube corresponds to the physical transference of heat, whereas there is no physical process corresponding to the other judgment.[38]

From the point of view of the project of conceptual analysis, an approach that takes into account physical constraints on possible causal interactions seems to "suffer from a lack of ambition" (Collins et al. 2004, p. 14). For a priori approaches, the analysis of the concept of causation must apply in all possible worlds to which the concept of causation applies, and in particular in "worlds with laws very different from our own" (Collins et al. 2004, p. 14). Limiting one's reflection to those causal processes that are

[38] One may of course describe the process of diffusion of heat in a negative way. Instead of saying that the water transfers heat onto the ice cubes, one can say that the presence of a colder object diminishes the heat contained in the water.

possible in the actual world given its laws, appears as "not merely unfortunate but deeply misguided" (Collins et al. 2004, p. 14) from the point of view of advocates of conceptual analysis who aim at finding an "account that has a hope of proving to be not merely true, but necessarily so" (Collins et al. 2004, p. 14).

Defenders of the idea that the causal relation is a natural kind of relation whose nature needs to be discovered on the basis of both conceptual and empirical constraints, can reply that we have here two different though related projects. The difference between the research on the naïve concept of causation and the research on what the essence of causation is in the actual world is analogous to the difference between the psychological research on "naïve physics," or "folk physics" and research in physics, or between psychological research on "folk biology" and biological research. Naïve physical concepts and naïve convictions on the properties and the evolution of physical objects determine only very partially the concepts and theories of scientific physics. In an analogous way, our a priori convictions on the nature of causation might only partially constrain the theory of causation as a natural relation existing in the actual world. The nature of such a natural relation must at least in part be discovered by empirical research.

One may try to reconcile the project of a priori conceptual analysis with the project of discovering the nature of causation as a natural kind of process (as it is in the actual world) in the framework of what has been called the "Canberra plan."[39] It proceeds in two steps, the first of which belongs to conceptual analysis: one discovers the constraints that a real relation must satisfy so as to be a candidate for being the causal relation. Transitivity and asymmetry are among these conceptual constraints. In a second step, which is empirical, one discovers which actual relations or processes satisfy the constraints discovered in the first step. The idea is to apply to the concept of causation a general strategy for reducing common sense concepts to scientific concepts, which is known as functional reduction (Jackson 1998, Kim 1998). In the first conceptual step, one shows, for example, that the concept of water is a functional concept that applies to a substance insofar as it satisfies a certain number of functional conditions: it is liquid at temperatures between 10°C and 30°C, it is transparent but refracts light with a characteristic refraction index, it freezes at 0°C and boils at 100°C under atmospheric air pressure at sea level etc. In the second step, it is empirically discovered that substances that satisfy these conditions in the actual world are mostly composed of H_2O molecules.

5. Causation as a Process

As we have seen, an important motivation of the counterfactual analysis has been the discovery of various sorts of "false positives" for the deductive-nomological analysis.

[39] This expression has been introduced by O'Leary-Hawthorne and Price (1996) by reference to the Australian National University at Canberra, in the context of the analysis of the concepts of truth, reference, and belief. Lewis (2000/2004, p. 76) applies it to the analysis of the concept of causation.

Some facts can, on the background of laws of nature, play the role of premises and conclusions of deductive arguments, without being linked as causes and effects. However, certain situations that refute the deductive-nomological analysis are also false positives that refute the counterfactual analysis. In certain background conditions, given two effects e_1 and e_2 of a common cause *c*, e_1 can serve as a premise in an argument whose conclusion describes e_2, and vice versa. Now, in appropriate circumstances, e_1 and e_2 can also be counterfactually dependent on each other. This parallel is certainly no coincidence: nomological dependence (which is according to the DN analysis a crucial part of what makes causal propositions true) creates counterfactual dependence. This is the case both when the nomological dependence goes together with causation and when it does not. For this reason, counterfactual dependence seems to be too weak to guarantee causation. We have already considered the debate about Lewis's suggestion that the counterfactual dependence between e_1 and e_2 is not sufficient for causation because it is grounded on causal dependences between e_1 and the common cause *c* and between *c* and e_2, and because the second counterfactual dependence is backward. This solution does not apply to cases of counterfactual dependence between aspects of an event or situation: given a sample *g* of gas (which approximately satisfies the conditions for being an "ideal" gas) and the ideal gas law $pV = nRT$ (where *p* represents pressure, *V* Volume, *T* temperature, *n* the number of moles of gas, and R the universal gas constant), if *g* had not been at temperature *T* (supposing its volume to be held fixed), it would not have had pressure *p*. If the kinetic energy of the molecules contained in *g* had not been *E*, the temperature of *g* would not have been $T = 2E/3k_B$ (where k_B represents Boltzmann's constant). It is one of the central conceptual constraints on the causation relation that its terms must occupy distinct spatiotemporal regions. "C and E must be distinct events—and distinct not only in the sense of nonidentity but also in the sense of nonoverlap and nonimplication" (Lewis 2000, p. 78). Pressure and temperature of the same sample of gas at the same moment cannot be linked as cause and effect because there is no spatiotemporal distance between these instances of properties. The same is true of the relation between the temperature of the sample of gas and the mean kinetic energy of its molecules. These examples of dependence between different properties of a given system at a time show that for such properties, counterfactual dependence is not sufficient for causation.

This problem (as well as the problem that counterfactual dependence is not necessary for causation either, as preemption scenarios seem to show) can be avoided by analyzing causation in terms of a local process that stretches between two events that are localized in space and time. There are several versions of such process accounts of causation. One of its historical sources is Russell's (1948/1992) analysis of causation in terms of "causal lines," which is inspired by the physical notion of a world line. The concept of a world line can be obtained from the spatiotemporal trajectory of an object. In a three-dimensional representation of the position of the Earth in space, its trajectory around the Sun appears as an ellipse. In a four-dimensional representation, in which the temporal dimension is represented as a fourth dimension alongside the three spatial dimensions—following at this point the unification of the spatial and temporal dimensions required in physics by the theory of relativity—the Earth's trajectory appears as its world line, which is an open curve in 4-dimensional space-time.

A causal line is a world line that satisfies an additional condition: along the line there are qualities or structures that are either constant or change in a continuous and smooth manner: "Throughout a given causal line, there may be constancy of quality, constancy of structure, or gradual change in either, but not sudden change of any considerable magnitude." (Russell 1948/1992, p. 477) This condition is supposed to guarantee that causation grounds our acquisition of knowledge. For Russell, as for Hume, the only way in which we can justify beliefs whose subject matter goes beyond what is immediately given to our senses consists in relying on causation. The perception of a table provides knowledge of the table, and not only of the sensory impressions from the table. This is so because these sense impressions are linked by a causal chain to the table, or more precisely to events of interaction between light and the surface of the table. Russell defines the notion of a causal line with respect to the possibility of justifying our inferences to what happens at some distance from ourselves: "A 'causal line', as I wish to define the term, is a temporal series of events so related that, given some of them, something can be inferred about the others whatever may be happening elsewhere" (Russell 1948/1992, p. 477). Any inference of this sort is inductive, and therefore fallible. In this context, Russell notes that an inference to an effect from a given cause is more reliable than a "backward" inference from an effect to a cause. The reason is that events of the same type can have different causes. Now, the inferences that provide us with knowledge of the world external to our sense organs belong to this second and more fragile sort of inferences.

Russell defines causal lines as world lines whose qualitative continuity can serve as inductive justification to enhance our knowledge beyond our perceptions. The fact that causal lines are defined by an epistemic requirement makes them inadequate as a basis for a metaphysical account of causation because this would make the existence of causal processes and relations dependent on human inferences. The fallibility of inferences grounded on the continuity of causal lines shows that such a causal line can only be a fallible indicator of the existence of a real causal process; however, being a causal line is neither necessary nor sufficient for being a real causal process. It is not sufficient because the continuity of structure or quality can also characterize "pseudo-processes" (Salmon 1984). Pseudo-processes are world lines that give human observers the illusory impression of a causal process. Their qualitative continuity qualifies them as Russellian causal lines, even though they are not real causal processes. Take Salmon's (1984, p. 141–142) spot of light cast on the inner wall of a hollow cylinder by a projector rotating at its center. The world line characterized by the series of places on the wall at the times at which the light spot appears on them is a causal line without being a causal process. The trajectory of the spot of light along the inner wall of the cylinder can exhibit perfect qualitative continuity. However, it is no causal process because spots of light at successive moments do not exercise any causal influence on one another: the light spot that appears at x at t does not cause the spot that appears at the immediately following place and time; rather, each spot is the end point of a causal process originating in the projector. Being a causal line is not necessary for being a causal process either because continuity of structure is not necessary: some causal processes are characterized by large and fast qualitative changes, for example, when several particles of different types follow each other in a "cascade" of radioactive decomposition.

Taking his inspiration from Russell's causal lines and Reichenbach's (1956) concept of a mark, which is defined as a local modification of structure, Salmon (1984) has suggested analyzing the concept of causal process as a process that (1) has structure or qualities that are either permanent or only changing continuously and (2) is capable of transmitting a mark. The light spot gliding along the wall of the cylinder is not a causal process because, if one modifies its color by inserting a red filter between the projector and the wall at one point, this modification will not propagate to the subsequent evolution of the spot.

This analysis in terms of continuity of structure and mark transmission raises several difficulties:[40] causal processes that are characterized by large and fast qualitative changes are counterexamples to the requirement of continuity of structure. Insofar as a world line is subject to changes that are fast relative to the scale of human observation, so that its observation does not give to an ordinary human observer the impression of qualitative constancy or of continuous change, it is neither a Russellian causal line nor a causal process as defined by Salmon. Salmon begins with the Russellian concept of a causal line, which requires the existence of a structure that is preserved along the line, and adds the additional requirement of mark transmission. "A given process, whether it be causal or pseudo, has a certain degree of uniformity—we may say, somewhat loosely, that it exhibits a certain structure. The difference between a causal process and a pseudo-process, I am suggesting, is that the causal process transmits its own structure, whereas the pseudo-process does not" (Salmon 1984, p. 144). A world line that is subject to fast and important qualitative changes, relative to the scale of what it observable by an ordinary human, does not even satisfy the conditions that Salmon imposes on processes: "processes can be identified as space-time paths that exhibit continuity and some degree of constancy of character" (Salmon, 1994, p. 298; repr. *in* Salmon, 1998, p. 249). A fortiori, it cannot be a causal process. On the other hand, there seem to be pseudo-processes capable of transmitting marks. Kitcher (1989, p. 463) mentions derivative marks: when a passenger in a car holds a flag out of the window, the shadow cast by the car as it passes along a wall bears the mark of the flag. Moreover, the analysis of the notions of mark and of causal interaction seems to be circular: A mark is a modification of structure introduced into a process by a causal interaction, but an interaction is causal if it leads to the introduction of a mark.

A tradition going back to the 19th century[41] identifies causal processes with processes of transmission of energy, momentum (Aronson 1971, Fair 1979), or more generally, of a quantity of a conserved quantity (Salmon 1994; Kistler 1998; 1999/2006). This

[40] These difficulties have led Salmon (1994) to abandon it.

[41] See Krajewski (1982).

approach is motivated by a "mechanist" intuition, according to which causal influence propagates only by contact and with finite speed. This intuition manifests itself when one considers certain situations that are problematic for theories analyzing causation in terms of nomological regularity or counterfactual dependence. Thunderstorms follow regularly upon sudden falls of barometer readings. They also depend counterfactually on them: if the barometer had not fallen, there would not have been a thunderstorm. However, the reason for which the barometer reading is nevertheless not a cause of the thunderstorm is that the barometer does not take part in the mechanism of the genesis of the thunderstorm. Some authors deny the possibility that a quantity of energy can be transferred in the strict sense: the reason is that particular quantities of energy lack the individuality required to give sense to the idea that it remains the same quantity across time (Dieks 1986). For this reason, the most elaborate version of the process theory in terms of conserved quantities (Dowe 1992a; 2000) does not make use of the concept of transmission, but uses instead Russell's concept of the "continuous manifestation" of a conserved quantity. By the continuous manifestation of a property by a world line, Dowe means that this property characterizes all points on the line, which does not require any form of transmission. This makes his account vulnerable to the objection that certain pseudo-processes manifest conserved quantities, without thereby being causal.[42] We have already considered the light spot gliding over the internal wall of a hollow cylinder. The trajectory of this spot constitutes a perfectly homogeneous world line: in the conditions stipulated by this thought experiment, the light spot contains, or manifests, exactly the same energy at each instant; each instant is qualitatively perfectly similar to each other. Nevertheless, the world line constituted by the trajectory of the light spot is not a causal process. The causal process responsible for the light spot is the process of propagation of light from the projector to the wall.

Theories that analyze causation in terms of transmission or continuous manifestation of conserved quantities avoid the problems, mentioned previously, of the relation between two effects of a common cause and of redundant or preempted processes. The fact that two events are effects of a common cause does not entail that there is a causal relation between those events, since no process of transference may relate them. Moreover, the fact that a process P_1 is accompanied by a second redundant (preempted) process P_2 does not prevent P_1 from transmitting conserved quantities. Consider again two snipers shooting at the same victim from which they are separated by the same distance. Imagine that sniper S_1 shoots a tiny moment earlier than sniper S_2, so that the bullet shot by S_1 kills the victim. In this case, S_2's shot (event *b*) does not cause the victim's death (event *c*). Neither the probabilistic nor the counterfactual analysis can account for the intuition that what the makes S_1's shot (event *a*) the cause of the victim's death must be some feature that is localized at the process linking *a* to *c*.[43] Both the probabilistic and the counterfactual analysis make the existence of

[42] See Salmon (1994, p. 308); Kistler (1998, 1999/2006).

[43] The probabilistic analysis will be presented in the next section.

a causal relation between *a* and *c* depend on factors that are *not* localized between *a* and *c*. If sniper S_1's shot takes place in a situation in which sniper S_2 also shoots, there is no counterfactual dependence between *a* and *c*: given S_2's shot, it is not true that, had S_1 not shot, the victim would not have died. One of our intuitions seems to indicate that the existence of a causal relation between *a* and *c* can only depend on processes situated between *a* and *c*, and that it cannot depend on events and processes that do not interfere with the processes between *a* and *c*.[44] On the other hand, the analysis according to which causation is grounded on a process of transmission takes into account this intuition of locality, according to which the existence of a causal relation between *a* and *c* only depends on processes between *a* and *c*. If *a* transmits something, say an amount of energy, to *c*, *a* is a cause of *c*, whether or not other events such as *b*, also have a causal impact on *c*.

However, transference theory encounters several important problems.

1. We have already mentioned the objection that the transmission analysis suffers from a lack of ambition, because its target is causation as it is in the actual world, rather than the general concept that applies to all possible worlds. However, this is only an objection to the extent that one shares the presupposition that conceptual analysis is the only legitimate or at least the only sufficiently ambitious aim of philosophical theories of causation.
2. Transference analyses can also be suspected of a lack of ambition of another sort: they seem to apply only to physical causal processes. Therefore the transference analysis seems inadequate for ordinary causal judgments involving non-physical properties, arguably for example psychological properties. To illustrate: the fact that the doorbell rings wakes Peter up. The noise of the doorbell seems to be the cause of his waking up, but it does not seem to be relevant to consider the underlying causal process from the point of view of energy transmission.[45] Indeed the application of the analysis to causal judgments of common sense presupposes that all causes and effects are physical. In reply, there are several ways of articulating the content of ordinary causal judgments with transference theory. The causal judgment that the doorbell wakes Peter up does not directly make reference to energy transmission. The dependence of his awakening on the propagation of sound waves, their transduction in nerve signals and the transmission of the latter to Peter's auditory cortex is the object of several "special" sciences, such as acoustics, psychophysics, physiology and neurophysiology. In a

[44] Lewis's (1986c) notion of quasi-dependence makes whether *c* causes *e* depend on possible worlds in which there is a process between *c** and *e** that is intrinsically similar to the process between *c* and *e* and where *e** depends indirectly (through a chain of dependence) counterfactually on *c*.* However, whether *c** causes *e** in those possible worlds is not only a matter of the intrinsic characteristics of the local process between *c** and *e*.*

[45] See Collins et al. (2004), p. 14.

physicalist framework, it is supposed that all these facts supervene on the set of physical facts.[46] If this is correct, the process of the doorbell waking Peter up may supervene on a physical process of transmission. The relevant properties of which the causal judgment states the causal dependence may even be specific forms of conserved quantities. The picture that emerges from this possibility has two parts: two conditions together make true the judgment that the fact that *c* (the activation of the doorbell at time t) is F (makes a specific sound) is causally responsible for the fact that *e* (Peter at the moment immediately following t) is G (wakes up). It is made true by 1) a process of transmission from cause *c* to effect *e* and 2) a law of nature expressing the dependence of G on F (Kistler 1999/2006). To judge that the doorbell wakes Peter up there must be an "in situ" law according to which, in ordinary, nonexceptional circumstances, doorbells wake sleeping people up, or at least raise the probability of their waking up. A different approach consists in articulating the condition of transmission with a counterfactual condition: according to Menzies (2004), the two facts that (1) the cause "makes a difference" to the effect and that (2) there is a process from cause to effect are both necessary and together sufficient for the existence of a causal relation. Transmission guarantees the existence of a process between *c* and *e* (Menzies's condition 2). The fact that *c* is F makes a difference with respect to the fact that *e* is G, to the extent that, if *c* had not been F (if the doorbell had made no sound), *e* would not have been G (Peter would not have wakened) (Menzies' condition 1).

3. The ordinary concept of transmission being causal, the transference approach seems condemned to circularity. However, circularity can be avoided by redefining the concept of transmission. Given two distinct spatio-temporal regions x and y, a quantity *A* is transmitted between x and y if and only if *A* is present both at x and at y.
4. If transmission is construed in this way, causality is not asymmetric. However, it can be argued that the asymmetry of causation is a physical characteristic of causality as it is in the actual world, rather than flowing from a conceptual constraint. Our region of the universe contains a plethora of irreversible processes that are all oriented in the same direction, as is guaranteed by the second law of thermodynamics. Such a physical ground

[46] Roughly, a first set of properties (or predicates) M is said to "supervene" on a second set P if and only if it is impossible that two objects differ with respect to a property of set M, without differing with respect to any property of set P. Physicalism is the doctrine according to which the set of mental properties supervenes on the set of physical properties. The truth of physicalism implies that a person cannot change mentally without changing physically and that there cannot exist a copy (or "clone") of a person p that differs from p mentally without differing from p physically. Several concepts of supervenience have been elaborated. One important difference between them concerns the interpretation of the concept of necessity (or impossibility) that is used in their definition. Cf. Kim (1990) and the introduction to Savellos and Yalcin (1995).

of the asymmetry of causation can also ground the direction of time (Reichenbach 1956; Lewis 1979/1986; Hausman 1998; Savitt 2006).

5. Transmission processes are everywhere. Events that are spatiotemporally sufficiently close to each other are, for example, often linked by transmissions of photons. Therefore, transmission theory seems condemned to lead to an inflation of true causal judgments. A first reply to this objection is that those plethoric causal judgments are true but lack communicational relevance. A second reply is that the relevant causal processes can be chosen on perfectly objective grounds, on the basis of the properties of the effect that is indicated in the *explanandum* of the causal explanation one is looking for. If one asks for the cause of Peter's waking up, the relevant causal process is at the physiological and psychological level and leads to the instantiation of the physiological and psychological properties constitutive of waking up.
6. It has been argued (Curiel 2000; Lam 2005) that the theory of general relativity does not guarantee global energy conservation, so that energy cannot be transmitted. In reply, it may be said that local conservation of energy is sufficient to guarantee the existence of local transmission and local causation, even if it turns out that the applicability of the concept of causation to large scale cosmological events and processes is more restricted than common sense would have expected.
7. Transmission theory seems to be refuted by a much less technical problem: there are many true causal propositions both in common sense and in science where negative facts play the role of causes or effects. Important types of propositions of this sort involve omission or prevention. If I kill a plant by *omitting* to water it, it seems that I have caused its death without having transmitted anything to it.[47] If on the contrary I *prevent* the plant's death by watering it, the event of the plant's death does not take place and cannot therefore be the object of any transmission. Schaffer (2000a) argues that there are many common sense causal propositions bearing on situations in which no transmission seems to be involved. Striking cases are propositions expressing double prevention, in which something or someone prevents the prevention of an event. Schaffer (2006) offers the example of the terrorist who prevents the sentinel in the control tower of the airport from preventing a collision of two airplanes.

Causal propositions in which the cause and/or the effect is/are a negative fact(s) are incompatible with three intuitive properties of causation noted by Hall (2000/2004b): a causal process is local (in the sense that the cause is linked to the effect by an intermediate series of events), intrinsic (it does not depend on what happens or is the case

[47] The example is Beebee's (2004). More precisely, I do not transmit anything relevant to the plant, although there are no doubt innumerable irrelevant processes linking me to it, such as transmission of photons.

elsewhere), and transitive. If *a* can cause *b* by omission, prevention, or double prevention, then certain causal relations obey neither to locality nor to intrinsicality nor to transitivity. Three (incompatible) consequences can be drawn from this.

1. Omissions are not instances of causality although they appear to us as such, for example, because we tend to conflate causal and non-causal explanation or because we conflate moral responsibility with causality (Dowe 2000; Armstrong 2004; Beebee 2004; Kistler 2006).
2. Propositions involving omission and prevention can be truly causal, which means that locality, intrinsicality and transitivity are not after all necessary conditions for causation (Schaffer 2000a).
3. There are two concepts of causation or two aspects of the concept of causality: One corresponds to counterfactual dependence (or to probability raising or to nomological dependence), the other corresponds to the existence of a transmission process. According to Hall (2000), these two concepts of causality are even independent of each other.

6. The Probabilistic Analysis

There are two strategies for discovering laws in general and causal laws in particular on the basis of data bearing on complex situations. The first uses statistical correlations expressed in conditional probabilities that can be found in the data; the second uses controlled experiments. Each of these methods can be used to construct an analysis of causation: the former has inspired the probabilistic analysis of causation that will be discussed presently; in the next section, we will examine the analysis of causation in terms of intervention or manipulation.

In the complex situations explored by such sciences as economics, sociology, epidemiology or meteorology, laws and causal relations do not manifest themselves as exceptionless regularities: not all smokers get lung cancer. In macroeconomics, the so-called Phillips curve represents the dependence between the rate of inflation and the unemployment rate; it implies that the higher the unemployment rate is, the slower is the raise of salaries, and that if on the contrary unemployment is decreasing, salaries and indirectly inflation tend to rise; however, it turns out that that a high unemployment rate can coexist, for quite long periods, with strong inflation.

In the perspective of improving the analysis of causation in terms of regularity, the probabilistic analysis is built on the idea of associating causation with the influence of one factor on a second factor, where this influence need not be universal but must only be statistically significant. The fundamental hypothesis is that factor A has a causal influence on factor B if and only if the probability of B given A is greater than the probability of B given the absence of A.

(PR, Probability raising) A is a cause of B if and only if $P(B \mid A) > P(B \mid \text{non-}A)$

There are two sorts of motivations for switching from an analysis of causation in terms of universal regularities to an analysis in terms of probability raising. The first reason is that lawful and causal influences are, in complex situations, often masked by other influences and therefore do not manifest themselves in the pure form of a universal regularity, as it happens in the examples just mentioned. The second reason is the hypothesis that there are intrinsically statistical laws, in the sense that, even in a situation in which nothing interferes, some causes only raise the probability of their effects without necessitating them. It is controversial whether there are any laws of this kind outside of quantum physics, but the capacity of the probabilistic analysis to take laws of this kind into account gives it an advantage over analyses of causation in terms of universal regularities.

Two remarks before we consider the development of the fundamental hypothesis (PR). The first is that the probabilistic analysis assimilates ontology to epistemology: the causal relation is identified with what allows us to discover causal influences in complex situations, that is, the inequality of conditional probabilities. The second is that the probabilistic analysis does not apply—at least not directly—to causal relations and processes between particular events, but only to relations of causal influence between "factors," properties or types of events. The formalism that is a central part of this approach presupposes that the terms of the causal relation can be subjected to the operations of propositional logic, such as negation and conjunction. This requires construing the terms of the causal relation as facts (Vendler 1967a, 1967b; Bennett 1988; Mellor 1995) or types of facts rather than as particular events (Davidson 1967).

Condition (PR) is faced with two difficulties that it shares with the DN and the counterfactual account.

1. Probability raising is symmetrical: if A and B are statistically positively correlated, so that P(A | B) > P(A | non-B), it is also true that P(B | A) > P(B | non-A).
2. The effects of common causes are generally statistically correlated although one effect is no cause of the other. If smoking (F) raises both the probability of lung cancer (C) and the probability of heart attack (I), C and I are *ceteris paribus* also positively correlated with each other. One of the reasons of the success of the probabilistic analysis is that this second problem can quite straightforwardly be solved with the condition of the absence of a "screening factor."[48] If A and B are statistically positively correlated, a third factor C is called a "screening factor" with respect to A and B if the positive correlation between A and B disappears if the probabilities are calculated conditionally on the presence or absence of C. Formally, in such a situation we have P(B | A) > P(B | non-A), but P(B | A & C) = P(B | non-A & C) and P(B | A & non-C) = P(B | non-A & non-C).

[48] This concept has been introduced by Reichenbach (1956).

The concept of a screening factor can then be used to complete the probabilistic analysis. Factor A, instantiated at instant t, is cause of factor B, instantiated at the same time or later, if and only if two conditions are satisfied:

1. P(B | A) > P(B | non-A)
2. There is no factor C, instantiated at t or earlier, which screens off the correlation between A and B.

This condition solves the problem that positive statistical correlation is in general not *sufficient* for causation, as shown by the correlation between effects of common causes. However, there are also situations in which such a positive correlation is not *necessary* for causation. There are situations in which the presence of factor A, which is a cause of factor B, nevertheless *diminishes* the probability of B. If smokers (M) practice more sport (S) than non-smokers, making M positively correlated with S, it is possible that the beneficial effect of S, which diminishes the risk of cardio-vascular illness (CV), overcompensate for the negative effect of M, which enhances the risk of CV. In such situations a factor M may diminish the probability of its effect CV:

P(CV | M) < P(CV | non-M)

There is a solution to this problem, different versions of which have been proposed by Cartwright (1979, p. 423) and Skyrms (1980). In Cartwright's version, A causes B if and only if the probability of B is higher in the presence of A than in its absence, in all sets that are homogeneous with respect to all causes of B that are not effects of A.

A causes B if and only if $P(B \mid A \& C_i) > P(B \mid \text{non-}A \& C_i)$ for all C_i, where C_i are causes of B that are not caused by A.

A "test situation" is characterized by holding fixed the set of factors that cause B but are not caused by A. Insofar as a test situation excludes all indirect causal influence from A on B, it provides a means for evaluating by purely statistical means whether A causes B. This strategy may, for example, justify the intuitive judgment that M causes CV: in a test situation, the conditional probability of CV given M is evaluated within a set of persons who all have the same level of sports practice (S). In such a situation, the probability of CV given M is greater than given not-M.

However, the proposal to analyze the causal influence from A on B in terms of the raising of probability in test situations changes the nature of the project of probabilistic analysis. First, in the form proposed by Cartwright and Skyrms, the analysis cannot any more serve as a basis for the reduction of the concept of causality: indeed, the *analysans* essentially contains the concept of cause. In order to determine whether A causes B, it is already required to know all other causes of B, or more precisely all factors that cause B independently of A.

Second, the requirement of measuring conditional probabilities in sets that are homogeneous with respect to all factors that can influence the probability of B but are not correlated with A is incompatible with one of the major motivations of the probabilistic approach: its aim was to provide a method for detecting causal influences in situations where correlation is imperfect, because the presence of interfering factors prevents the universal correlation of cause and effect. However, insofar as intrinsically indeterministic laws are not taken into account, in a situation in which all causes of B that are independent of A are held fixed, if A causes B, $P(B \mid A) = 1$. Indeed, probabilities lower than 1 measure the net effect of unknown factors that are independent of A and influence B negatively or positively.

We have already mentioned another important problem for the probabilistic analysis: statistical correlation is symmetrical, so that if the probability of B is larger in the presence of A than in its absence, the probability of A is also larger in the presence of B than in its absence. There are several proposals for what should be required in addition to probability raising, in order to distinguish cause and effect. One possibility is to simply stipulate that the factor that is instantiated earlier in time is the cause, and the factor instantiated later, the effect. However, this idea does not fit well with a theory first of all devised for causal relations between general factors, rather than between particular instances of these factors. Moreover, such a stipulation precludes the possibility of so-called backward causation, that is, causal processes evolving in the direction opposite to the direction of time. Finally, it makes it impossible to reduce the direction of time itself to the direction of causation. A traditional approach to explaining the origin of the asymmetry of time consists in making the hypothesis that it derives from the asymmetry of causation: the fact that instant t_2 is later than instant t_1 is grounded on the fact that an event occurring at t_1 may cause an event occurring at t_2, but that the opposite is not possible.[49] However, the probabilistic analysis can be defended against this objection if the direction of time can be grounded on something other than the direction of causation. According to one hypothesis, the asymmetries of causation and time both derive from the asymmetry of some fundamental physical processes. These are often taken to be thermodynamically irreversible processes, characterizing the evolution of systems whose entropy rises. Other processes that have been suggested as possibly grounding the asymmetry of causation are intrinsically asymmetric microphysical processes, such as the disintegration of K-mesons, or "kaons."[50]

It has also been suggested that the difference between cause and effect might be an effect of the perspective of an observer or human agent, in the sense that, independently of the perspective of the agent, at the level of the objective dependence among factors in the world, causation is symmetric.[51]

The most influential proposal to account for the asymmetry of causation in terms of probabilistic conditions is due to Reichenbach (1956) who has suggested using common causes in order to determine the direction of causation (and time). If A and B are positively correlated and if C is a screening factor, such that the correlation between A and B

[49] This would require some refinement to take account of special relativity.

[50] These decomposition processes "violate" the symmetry with respect to temporal inversion ("T"). Cf. Dowe (1992b, p. 189).

[51] Fair (1979), Price (1992); Menzies and Price (1993); Price (2007).

disappears both in the presence and in the absence of C, and such that the presence of C raises both the probability of A and of B, the triplet ACB is called a "conjunctive fork." If the factor C is instantiated in the past of A and B, and if there is no factor D satisfying the same conditions as C but instantiated in the future of A and B, ACB constitute an open fork in the direction of future (and C is a common cause of the two effects A and B); if the only factor D that satisfies these conditions is instantiated in the future with respect to A and B, ADB constitute an open fork directed toward the past; if finally there is both a factor C in the past and a factor D in the future that satisfy the indicated conditions, ACBD constitute a closed fork. Reichenbach's hypothesis is that the direction from cause to effect (which is also the direction of time) is the direction in which open forks dominate.

Finally, there are numerous attempts to improve the analysis of the notion of causation by a synthesis of conceptual elements of different approaches. One such analysis does so in terms of probabilistic counterfactuals. This theory, suggested by D. Lewis (1986c) and elaborated by Noordhof (1999, 2004), analyzes the causal relation between particular events in the following way: "For any actual distinct events, e_1 and e_2, e_1 causes e_2 iff there are events $x_1, \ldots, x_n$ such that x_1 probabilistically depends on $e_1, \ldots,$ e_2 probabilistically depends on x_n" (Noordhof 1999, p. 97). Probabilistic dependence is then analyzed in terms of a counterfactual condition on the chances of the corresponding types of events:[52] "e_2 *probabilistically-depends* on a distinct event e_1 iff it is true that: if e_1 were to occur, the chance of e_2's occurring would be at least x, and if e_1 were not to occur, the chance of e_2's occurring would be at most y, where x is much greater than y" (Noordhof 1999, p. 97).

7. Manipulability and Structural Equations

One of the most fruitful recent developments in this field is the philosophical analysis of models that have been elaborated in artificial intelligence. The relevant models represent research strategies for analyzing causal structures that are employed in sciences like economics that study causal influences in complex systems. This approach makes use of statistical analysis of conditional probabilities, and in some versions at least (Pearl 2000) analyzes causation in terms of counterfactuals involving experimental interventions or manipulations.[53] As with the probabilistic approach, the analysis of causation in terms of interventions or manipulations is grounded on an analysis of the logic implicit in scientific research on causes. In the social sciences like sociology, economics, and also psychology, the analysis of conditional probabilities is used to

[52] Chances are single-case probabilities, "as opposed to finite or limiting frequencies" (Lewis 1986c, pp. 177–178).

[53] Another version has been worked out by Spirtes, Glymour, and Scheines (2000). Woodward (2003) has elaborated a philosophical analysis on causation on the basis of the works of Spirtes, Glymour, and Scheines (2000) and Pearl (2000). Keil (2000, 2005) has offered an original analysis of causation in terms of manipulation that makes no use of the technical apparatus of structural equations and directed graphs.

extract information on causal influences among different factors. However, in experimental sciences, interventions are a crucial additional method for discovering causal influences. The experimenter manipulates a given "cause" variable under conditions in which other variables are under control, to observe subsequent variation in "effect" variables, which indicates causal influence. Causal graphs and structural equations are formal tools that have been developed to build models of causal structures on the basis of information obtained in this way. The philosophical analysis of such models of the logical form of the scientific research for causes has led to a complete renewal of older philosophical theories of causation in terms of "manipulation" or "intervention."

According to one traditional analysis of causation not yet mentioned so far, a cause C of an effect E is an action that would give a human agent a means to obtain E if she decided to make C happen.[54] However, in this form, such an account suffers from two major defects, circularity and anthropocentrism. The latter is implicit in the thesis that an event can be a cause only if its occurrence can be the result of the decision of a human agent. Von Wright (1971) has argued that although the fact that the human capacity to intervene in events in the experimental sciences is indispensable for the analysis of our *knowledge* of causal relations, we should not conclude from this that human action is essential to the *metaphysics* of causation. It will be shown how recent manipulationist (or interventionist) accounts reply to the objection of anthropocentrism. As for circularity, it seems impossible to build a non-circular analysis of causation that is grounded on the notion of intervention, insofar as an intervention is a causal process. For this reason, recent manipulability theories of causation such as Woodward's (2003) do not aim at a reductionist analysis of the notion of causation, but only at analyzing the logic of causal reasoning in the context of experimental interventions.

Here are some key ideas that structure the approach to causation in terms of interventions, using the formal tools of structural equations and causal graphs. The causal structure of a complex system is represented by a model built from a set of variables V and a set of structural equations that express functional relations among these variables. Let us use Menzies's (2008) analysis of a toy situation often used in the philosophical literature: two kids throw rocks at a bottle to smash it. We have already encountered this situation as an example of preemption: Billy's throw does not smash the bottle although it would have had Sally not thrown her rock an instant earlier, so that it smashed the bottle before Billy's rock could. To represent the relevant actual and possible causal influences in this situation, the following variables can be used. In this case, all variables have only two values ("1" in case the event described by the variable occurs, "0" in case it doesn't), but the formalism can also be used with variables with more than two and also continuous values.

- $BT = 1$ if Billy throws a rock, otherwise $BT = 0$
- $ST = 1$ if Sally throws a rock, otherwise $ST = 0$

[54] Cf. Gasking (1955); Menzies and Price (1993).

- *BH* = 1 if Billy's rock hits the bottle, otherwise *BH* = 0
- *SH* = 1 if Sally's rock hits the bottle, otherwise *SH* = 0
- *BS* = 1 if the bottle shatters, otherwise *BS* = 0

Each variable is associated to a structural equation. A variable is called "*exogenous*" if its value is determined by factors external to the causal system whose model is being built. In the example, *BT* and *ST* are exogenous variables, insofar as their values are not determined by the values of other variables within the model. Therefore, the structural equations for this variables, *BT* = 1 and *ST* = 1, do not contain any other variables, but simply stipulate their values. By contrast, the value of an *endogenous* variable is a function of other variables within the system. The equation for the endogenous variable *SH* is *SH* = *ST*, which means that the value of SH is determined by the value of ST: if Sally throws a rock, the rock reaches the bottle (*ST* = 1 and *SH* = 1) and if she doesn't, the rock doesn't reach the bottle (*ST* = 0 and *SH* = 0). The preemption of the process beginning with Billy's throwing his rock is expressed by the equation for *BH: BH* = *BT* and non-*SH*. Billy's rock reaches the bottle only if (1) Billy throws the rock and if (2) the rock thrown by Sally does not reach it. The variable representing the smashing of the bottle is also endogenous: *BS* = *SH* or *BH*. The bottle gets smashed either if Sally's rock reaches it or if Billy's rock reaches it.

The content of a set of structural equations can also be represented in a structural graph. Figure 1 shows a graph representing the structural equations defining our situation: each variable corresponds to a node in the graph. An arrow going from variable *X* to variable *Y* represents the fact that the value of *Y* depends on the value of *X*; in this case, *X* is called a "parent" of *Y*. A *directed path* from *X* to *Y* is a set of arrows leading from *X* to *Y*. Each arrow and each structural equation represents a set of counterfactual conditionals. Once a model is constructed, it can be used to determine the truth-value of new counterfactuals that do not simply correspond to one arrow. Say we want to know what would have happened if the rock thrown by Sally had not reached the bottle. To find this out, one sets the variable corresponding to the antecedent of the counterfactual to the value it would have if the antecedent were true. In this case, one sets *SH* = 0. This represents an "atomic intervention" (Pearl 2000, p. 70). It is equivalent to what Lewis calls a "miracle." One does not take into consideration the past that might have led to the truth of the antecedent. Rather, the value of the antecedent

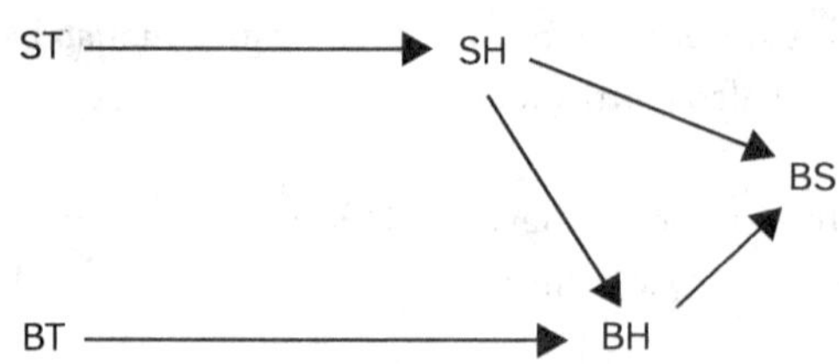

FIGURE 1 Structural graph representing causal influences
Source: Menzies 2008.

(here, *SH*) is set while the values of all variables corresponding to the past of the antecedent keep the values they have in actuality. In the graphical representation, this means that all arrows leading to the variable *SH* are erased, which is equivalent to transforming *SH* into an exogenous variable. In the manipulationist interpretation of this formalism, this corresponds to a localized experimental intervention on variable *SH*, which comes from outside the system and is direct in the sense that it is not obtained indirectly by intervening on factors that influence *SH* within the system. As with Lewis's concept of a miracle, this guarantees that no "backtracking" counterfactual can be true. When the value of variable *X* is modified, the variables situated in the past of *X* are left untouched. In the standard representation, these are the variables figuring at the left of *X*. The values that the variables to the right of *X* take in a situation in which *X* takes the stipulated value can then be determined on the basis of the equations corresponding to the arrows starting at *X*.

Pearl (2000, p. 70) defines the causal effect of *X* on *Y*, written "P(*y*/do(*x*))," as the probability distribution of the different values *y* of *Y*, given that an intervention ("do") has fixed *x* as the value of variable *X*. This has the consequence that all factors different from *X* that also influence *Y* are included in *X*'s impact on *Y*. To avoid this result, Woodward (2003) imposes additional constraints on interventions I appropriate to determining whether *X* causes *Y*. (1) *I* must be the only cause of *X*, in the sense that all other influences on *X* must be cut. (2) *I* must not cause *Y* through any paths that do not go through *X*. The administration of a placebo pill in the following situation does not fulfill this condition. *I* is the ingestion of the pill; *X* is the action of the pill on the body after its ingestion; *Y* is recovery. By the definition of a placebo, if *I* is efficacious in changing the value of *Y*, its influence does not flow through *X*, that is, changes in the body brought about by the absorption of the pill. In such a situation, the fact that *I* influences *Y* does not mean that *X* causes *Y*. (3) *I* must not be correlated with any cause that influences *Y* through any path that does not go through *X*. If, in order to find out whether the indication *X* of a barometer causes the thunderstorm Y, my interventions *I* on *X* depend on (my knowledge of) air pressure, then *Y* may vary as a function of the values that *I* imposes on *X*, whereas X does of course not cause *Y*. (4) The values of all possible causes of *Y* that are not situated on a path from *I* through *X* to *Y* must be held fixed.

In this context, Woodward (2008) defines the causal effect of *X* on *Y* by the difference of the values of *Y* that corresponds to the difference between two values *x* and *x** of the variable *X*, on which one intervenes via I.

(CE) ("causal effect") $Y_{\mathrm{do}(x),\,\mathrm{Bi}} - Y_{\mathrm{do}(x^*),\,\mathrm{Bi}}$

where "$Y_{\mathrm{do}(x),\,\mathrm{Bi}}$" represents the value of the variable *Y* given that an intervention has set variable *X* to value *x*, in circumstances B_i.

If the relation between *X* and *Y* is deterministic, *X* is a cause of *Y* if and only if there are pairs of values *x* and *x** ($x^* \neq x$), such that (CE) differs from zero; if the relation is indeterministic, *X* is a cause of *Y* if and only if there are pairs of values *x* and *x**, such that there are values of *Y* whose probability is different for the two values of *X*.

The structural equations model shares with the counterfactual analysis the idea that causation must be defined in models in which the past corresponds to actuality but the putative cause has a counterfactual value; however, it avoids at least some of the counterexamples to the counterfactual analysis. Thus, it yields the intuitively correct result in the preemption case considered earlier.

In Figure 1, *BH* is the only intermediate variable between *ST* and *BS* that is not on the path *ST*—*SH*—*BS*. Thus, in order to judge whether *ST* causes *BS*, *BH* must be held fixed at its actual value $BH = 0$. If one considers the counterfactual situation in which the value of the putative cause *ST* is changed so as to become $ST = 0$, the value of *BS* determined by the equations also differs from its actual value, to become $BS = 0$. This means that *ST* causes *BS*.

Lewis's analysis fails to yield the correct result in this case because *BS* does not counterfactually depend on its cause *ST*, because $BS = 1$ even if $ST = 0$. The interventionist analysis avoids this difficulty by "freezing" on their actual values all variables that are not on the path connecting the putative cause to its putative effect.

Here is an interpretation of this formal difference. In the structural equations model, the antecedent can be taken to represent, not a fact in a different possible world, but a situation resulting from an experimental intervention. In Lewis's analysis, the evaluation of a counterfactual requires holding fixed all events in the past of the putative cause (described by the antecedent of the counterfactual), whereas the structural equations model requires holding fixed the values of all variables that are not situated on the path between the putative cause and effect. This difference has formal consequences: Lewis's analysis makes causation transitive, whereas it isn't necessarily transitive in the structural equations model.[55]

The structural equations model provides the means of distinguishing different causal notions that can all be expressed by the common sense word "cause." The fact that it allows defining different causal notions shows the fecundity of this approach, although it cannot provide a non-circular analysis of causation. A variable *X* can influence another variable *Y* in two independent ways in such a way that these influences cancel each other out. Starting the engine of a car *X* raises the temperature of the engine *Y*,[56] but *X* also causes the onset of the ventilation system *Z*, which lowers the temperature of the engine. It is possible that the positive direct influence from *X* on *Y* is exactly compensated by the negative influence of *X* on *Y* via *Z*, such that *X* has zero net influence on *Y*. In such a case, it seems both intuitively correct to say that starting the engine raises the temperature of the engine and that it does not. However, this involves no paradox insofar as the two judgments contain different notions of causation that are expressed by the same common sense term.[57] The former is correct if "raises" is taken to express the concept of being a *contributing* cause, the latter is correct if "raises" is taken to express the concept of being a *total* cause.

---

[55] Cf. Hitchcock (2001).

[56] Hesslow (1976) gives a structurally similar example.

[57] Cf. Woodward (2003, p. 50 sq.).

In the situation sketched, *X* is not a "total cause" of *Y*, as defined by condition (CE). However, *X* is a "contributing cause":

> (CC) *X* is a contributing cause of *Y* if and only if the value of *Y* changes as a consequence of a change of the value of *X*, where the values of all variables different from *X* and Y are held fixed, and in particular those that lie on paths between *X* and *Y*.

Indeed, if we hold *Z* in our example fixed, we find that an intervention on *X* modifies the value of Y, so that starting the engine is a contributing cause of the rise of temperature of the engine, although applying condition (CE) shows that it is not a total cause of the rise of temperature: if *Z* is not held fixed, starting the engine does not make the temperature rise.

Older versions of the manipulability theory make the judgment "*X* causes *Y*" depend on the possibility of acting on *X*. This seems to make it impossible to apply the concept of causation to events that are in principle outside the sphere of influence of human interventions. However, eruptions of volcanoes and explosions of supernovae seem to be causes although no possible human action could ever bring them about or modify them. This problem is solved in recent versions of the interventionist analysis, in which the notion of intervention is defined without any reference to human action. Analyses of causation in terms of structural equations and directed graphs avoid anthropocentrism because the intervention that sets the value of the putative cause is no longer required to be the result of a human action. Natural events entirely independent of all intentional actions can satisfy the formal conditions on an intervention modifying the value of the putative cause. Such a "natural experiment" provides just as good a basis for judging causal influence as intentional interventions by human experimenters. Neuropsychology is one important field of research where hypotheses on the causal influence of the activation of specific brain regions are evaluated by such "natural experiments": the hypothesis that the activation of brain region *X* causally influences the activation of brain region Y is confirmed by the observation that a modification of *X* due to accident or illness is systematically followed by a modification of *Y*.

However, there seem to be causal relations on which even interventions as defined by these new theories seem to be impossible. To judge whether the gravitational attraction of the moon causes the tides, one must examine the consequences of an intervention on the position or the mass of the moon. It can be doubted whether "interventions" on the moon are physically possible: such an intervention would require modifying the position or the mass of the moon by some means that does not also *directly* influence the tides.

8. Conclusion

Philosophical research on causation has developed into a rich and complex field. Since the once dominant deductive-nomological analysis has been abandoned, several alternative approaches based on very different premises have been developed. Each can claim a certain extent of success insofar as it can account for intuitions or alleged facts about causation that provide counterexamples against rival accounts. But each also has its own counter-examples. The confusion that threatens can be greatly diminished by realizing that different approaches do often not share their goals. Most traditional philosophical analyses pursue the aim of a priori conceptual analysis, whereas others, such as analyses in terms of manipulability or process theories take as their criterion of success fit with the logic of scientific research about causal relations or with the structure of reality as described by present-day physics. Although these aims may seem incompatible, there are also efforts to construct a synthetic theory that can preserve what is correct from several seemingly incompatible theories. According to one hypothesis of this sort, different theories are applicable to different domains of phenomena and of scientific research. The probabilistic analysis may provide an adequate analysis of causal judgments in economy and other social sciences, whereas theories in terms of transmission processes and conserved quantities may be adequate for physical causation. The counterfactual conception may seem most adequate to account for common sense causal judgments. Such a "regionalist" conception is not the only form of pluralism or relativism, according to which there is more than one concept of causation.[58] Something may be a cause in the sense of one of these concepts, without being a cause in the sense of others. In the counterfactual sense, the rock thrown by Sally is not the cause of the bottle's breaking because the breaking does not depend on Sally's action. It would have broken anyway in the context of the backup cause constituted by Billy's well-aimed throw. However, in the sense of causation as a physical process, Sally's throwing her rock does cause the bottle's breaking. More ambitious syntheses aim at constructing a unified theory that can account for all situations, making use of conceptual ingredients taken from different (and incompatible) theories. Examples are the probabilistic counterfactual analysis (Noordhof 1999), and the theory according to which causation can be analyzed in terms of the raising of the probability of a process (Schaffer 2001). The conception of functional reduction provides another framework for a synthetic account. According to this approach, causation is a concept whose conditions of application are in part a priori and in part a posteriori. It applies to causation a two-stage model of reduction devised by Armstrong (1968) and Lewis (1972) to solve the mind-body problem. The first step of pure a priori conceptual analysis aims at making explicit the "functional profile" of a given concept: these are the constraints that determine the set of objects to which the concept applies. To use one of the paradigmatic examples from the mind-body problem, pain is a state of a subject A that is caused by damage to the body of A and causes characteristic mental states and behavior, such as the desire that the pain ceases and actions aiming at interrupting or diminishing what causes the damage. This first conceptual step of the analysis is independent of empirical research and aims at the a priori conditions of application of the

[58] Hitchcock (2007) provides a classification of types of pluralism about causation.

concept. The second step aims at discovering those natural objects, states, or processes that possess, in the actual world, the functional profile found in the first step. For cognitive concepts such as pain, it is conceivable that one finds different natural states or processes occupying a given functional role in different cognitive systems, for example, animals of different species. If this is correct, there would be a general concept of pain although the concept applies to different types of states or processes in animals of different kinds.

Applying this strategy to the analysis of causation, it is conceivable that different sorts of natural relations or processes play the role of causation in different fields. In this way one is led to a pluralist conception, in which it would be coherent to judge, for example, that probability raising occupies the conceptual role of causation in epidemiology and economy, that counterfactual dependence occupies it in the explanation of human actions, that the existence of a mechanism plays the role of causation in biology, whereas the existence of a transmission process plays the role in physics. There would be both a general concept of causation corresponding to a priori conceptual constraints, such as spatiotemporal distinctness of cause and effect and asymmetry, and "regional" concepts of causation, specific to different domains of inquiry.[59]

References

Achinstein, P. (1975) "Causation, Transparency, and Emphasis," *Canadian Journal of Philosophy* 5, 1–23.

Armstrong, D. M. (1968) *A Materialist Theory of Mind*, revised ed., 1993, London: Blackwell.

Armstrong, D. M. (2004) *Truth and Truthmakers*, Cambridge: Cambridge University Press.

Aronson J. J. (1971) "The Legacy of Hume's Analysis of Causation," *Studies in the History and Philosophy of Science* 2, 135–165.

Barberousse, A., Kistler, M., and Ludwig, P. (2000) *La philosophie des sciences au XXe siècle*, Paris: Flammarion.

Beebee, H. (2004) *Causing and Nothingness*, *in* J. Collins, N. Hall, and L. Paul (eds.), *Causation and Counterfactuals*. Cambridge, MA: MIT Press, 291–308.

Bennett, J. (1987) "Event Causation: The Counterfactual Analysis," *Philosophical Perspectives*, 1, 367–368, repr. *in* Sosa and Tooley (1993), 217–233.

Bennett, J. (1988) *Events and Their Names*, Indianapolis and Cambridge: Hackett.

Carnap, R. (1966/1995) *An Introduction to the Philosophy of Science*, New York: Dover. Original edition, *Philosophical Foundations of Physics: An Introduction to the Philosophy of Science*, New York: Basic Books, 1966.

Cartwright, N. (1979) "Causal Laws and Effective Strategies," *Noûs*, 13, 419–427, repr. *in How the Laws of Nature Lie*, Oxford: Clarendon Press, 1983.

Cartwright, N. (1983) *How the Laws of Physics Lie*, Oxford: Clarendon Press.

Cartwright, N. (1999) *The Dappled World, A Study of the Boundaries of Science*, Cambridge: Cambridge University Press.

[59] I thank Andrew McFarland for helpful comments and for improving the language of this chapter.

Collins, J., Hall, N., and Paul, L.A. eds. (2004) *Causation and Counterfactuals*, Cambridge, MA: MIT Press.

Cummins, R. (2000) "How Does It Work? vs. What Are the Laws? Two Conceptions of Psychological Explanation," *in* F. Keil and R. Wilson (eds.), *Explanation and Cognition*, Cambridge, MA: MIT Press, 117–145.

Curiel, E. (2000) "The Constraints General Relativity Places on Physicalist Accounts of Causality," *Theoria* (San Sebastian), 15, 33–58.

Davidson, D. (1967) "Causal Relations," *in* Davidson D. (1980), *Essays on Actions and Events*, Oxford: Clarendon Press 1980.

Dieks D. (1986) "Physics and the Direction of Causation," *Erkenntnis* 25, 85–110.

Dowe, Ph. (1992a) "Wesley Salmon's Process Theory of Causality and the Conserved Quantity Theory," *Philosophy of Science*, 59, 195–216.

Dowe, Ph. (1992b) "Process Causality and Asymmetry," *Erkenntnis*, 37, 179–196.

Dowe, Ph. (1996) "Backwards Causation and the Direction of Causal Processes," *Mind*, 105, 1–22.

Dowe, Ph. (2000) *Physical Causation*, Cambridge: Cambridge University Press.

Ehring, D. (1987), "Causal Relata," *Synthese* 73, 319–328.

Elga, A. (2000) "Statistical Mechanics and the Asymmetry of Counterfactual Dependence," *Philosophy of Science*, Supp. 68, 313–324.

Fair D. (1979) "Causation and the Flow of Energy," *Erkenntnis*, 14, 219–250.

Faye J. (2010) "Backward Causation," *Stanford Encyclopedia of Philosophy*, http://plato.stanford.edu/entries/causation-backwards/.

Frisch, M. (2009a) "'The Most Sacred Tenet?' Causal Reasoning on Physics," *British Journal for the Philosophy of Science*, 60, 459–474.

Frisch, M. (2009b), "Causality and Dispersion: A Reply to John Norton," *British Journal for the Philosophy of Science*, 60, 487–495.

Gasking, D. (1955) "Causation and Recipes," *Mind*, 64, 479–487.

Hall, N. (2004a) "Two Concepts of Causation," *in* Collins et al. (eds.) (2004), 225–276.

Hall, N. (2000/2004b) "Causation and the Price of Transitivity," *in* Collins et al. (eds.) (2004), 181–204.

Hausman, D. (1998) *Causal Asymmetries*, Cambridge: Cambridge University Press.

Hempel, C. G. (1988) "Provisos: A Problem Concerning the Inferential Function of Scientific Theories," *Erkenntnis*, 28, 147–164.

Hesslow, G. (1976) "Two Notes on the Probabilistic Approach to Causality," *Philosophy of Science*, 43, 290–292.

Hitchcock, Ch. (1996a) "The Role of Contrast in Causal and Explanatory Claims," *Synthese*, 107, 395–419.

Hitchcock, Ch. (1996b) "Farewell to Binary Causation," *Canadian Journal of Philosophy*, 26, 335–364.

Hitchcock, Ch. (2001) "The Intransitivity of Causation Revealed in Equations and Graphs," *Journal of Philosophy*, 98, 273–299.

Hitchcock, Ch. (2007) "How to Be a Causal Pluralist," *in* P. Machamer and G. Wolters (eds.), *Thinking about Causes*, Pittsburgh: University of Pittsburgh Press, 200–221.

Horwich P. (1987) *Asymmetries in Time*, Cambridge, MA: MIT Press.

Hume, D. (1739–1740) *Treatise of Human Nature*, Selby-Bigge, L. A., and Nidditch, P. H. (eds.), Oxford: Clarendon Press, 1978.

Hume, D. (1777) *Enquiries Concerning Human Understanding and Concerning the Principles of Morals*, L. A. Selby-Bigge and P. H. Nidditch (eds.), Oxford: Clarendon Press, 1975.

Hüttemann, A. (1998) "Laws and Dispositions," *Philosophy of Science*, 65, 121–135.

Jackson, F. (1998) *From Metaphysics to Ethics: A Defense of Conceptual Analysis*, Oxford: Clarendon Press.

Keil, G. (2000) *Handeln und Verursachen*, Frankfurt a. M.: Vittorio Klostermann.

Keil, G. (2005) "How the *Ceteris Paribus* Laws of Physics Lie," *in* Jan Faye et al. (eds.), *Nature's Principles*, Dordrecht, Kluwer, 167–200.

Kim, J. (1973) "Causes and Counterfactuals," *Journal of Philosophy* 70, 570–572.

Kim, J. (1990) "Concepts of Supervenience," repr. *in* J. Kim, *Supervenience and Mind*, Cambridge: Cambridge University Press, 1993, 53–78.

Kim, J. (1998) *Mind in a Physical World*, Cambridge, MA: MIT Press.

Kistler, M. (1998) "Reducing Causality to Transmission," *Erkenntnis*, 48, 1–24.

Kistler, M. (1999/2006) *Causation and Laws of Nature*, Londres: Routledge, 2006. Published in French as *Causalité et lois de la nature*, Paris: Vrin, 1999.

Kistler, M. (2001) "Causation as Transference and Responsibility," *in* W. Spohn, M. Ledwig, and M. Esfeld (eds.), *Current Issues in Causation*, Paderborn: Mentis, 115–133.

Kistler, M. (2006) "La causalité comme transfert et dépendance nomique," *Philosophie*, 89, 53–77.

Kitcher P. (1989), "Explanatory Unification and the Causal Structure of the World," *in* P. Kitcher and W. C. Salmon (eds.), *Minnesota Studies in the Philosophy of Science, Vol. XIII: Scientific Explanation*, Minneapolis: University of Minnesota Press, 410–505.

Krajewski, W. (1982) "Four Conceptions of Causation," *in* W. Krajewski (ed.), *Polish Essays in the Philosophy of the Natural Sciences*, Dordrecht: Reidel, 1982, 223–235.

Lam, V. (2005) "Causation and Space-Time," *History and Philosophy of the Life Sciences*, 27, 465–478.

Lewis, D. (1972) "Psychophysical and Theoretical Identifications," *Australasian Journal of Philosophy* 50, 249–258, repr. *in* D. Chalmers (ed.), *Philosophy of Mind: Classical and Contemporary Readings*, New York: Oxford University Press, 2002, 88–94.

Lewis, D. (1979/1986) "Counterfactual Dependence and Time's Arrow," with Postscripts, *in Philosophical Papers, vol. 2*, New York: Oxford University Press, 32–66.

Lewis, D. (1986a) "Events," *in Philosophical Papers, vol. 2*, New York: Oxford University Press, 241–269.

Lewis, D. (1986b) "Causation," *in Philosophical Papers, vol. 2*, New York: Oxford University Press, 159–172.

Lewis, D. (1986c) "Postscripts to "Causation," *in Philosophical Papers, vol. 2*, New York: Oxford University Press, 172–213.

Lewis, D. (2000) "Causation as Influence," *in* Collins et al. (eds.) (2004), 75–106.

Mackie J. L. (1974) *The Cement of the Universe*, Oxford: Clarendon Press.

Maslen, C. (2004) "Causes, Contrasts, and the Nontransitivity of Causation," *in* Collins et al. (eds.) (2004), 341–357.

McDermott, M. (1995) "Redundant Causation," *British Journal for the Philosophy of Science*, 46, 523–544.

Mellor, D.H. (1995) *The Facts of Causation*, London: Routledge.

Menzies, P. (2004) "Difference-making in Context," *in* Collins et al. (eds.) (2004), 139–180.

Menzies, P. (2008) "Counterfactual Theories of Causation," *in Stanford Encyclopedia of Philosophy*, http://plato.stanford.edu/entries/causation-counterfactual/

Menzies, P., and Price, H. (1993) "Causation as a Secondary Quality," *British Journal for Philosophy of Science*, 44, 187–203.

Mill, J.S. (1843), *A System of Logic, Ratiocinative and Inductive*, London: Parker, repr. of the 1891 edition, Honolulu: University Press of the Pacific, 2002.

Noordhof, P. (1999) "Probabilistic Causation, Preemption and Counterfactuals," *Mind*, 108, 95–125.

Noordhof, P. (2004) "Prospects for a Counterfactual Theory of Causation," *in* Ph. Dowe and P. Noordhof (eds.), *Cause and Chance*, New York: Routledge, 188–201.

Norton, J. (2003) "Causation as Folk Science," *Philosopher's Imprint* 3, *repr. in* H. Price, and R. Corry (eds.), *Causation, Physics, and the Constitution of Reality: Russell's Republic Revisited*, Oxford: Clarendon Press, 2007, 11–44.

Norton, J. (2009) "Is There an Independent Principle of Causality in Physics?," *British Journal for the Philosophy of Science*, 60, 475–486.

O'Leary, J., and Price, H. (1996) "How to Stand Up for Non-Cognitivists," *Australasian Journal of Philosophy*, 74, 275–292.

Paul, L.A (2004) "Aspect Causation," *in* Collins et al. (eds.) (2004), 205–224.

Pearl, J. (2000) *Causality. Models, Reasoning, and Inference*, Cambridge: Cambridge University Press.

Pietroski, P., and Rey G. (1995) "When Other Things Aren't Equal: Saving *Ceteris Paribus* Laws from Vacuity," *British Journal for the Philosophy of Science*, 46, 81–110.

Popper, K. R. (1935/2002) *The Logic of Scientific Discovery*, London and New York: Routledge. Original edition: *Logik der Forschung*, Vienna: Julius Springer, 1935.

Popper, K. R. (1956), "The Arrow of Time," *Nature*, 177, 538.

Price, H. (1992) "Agency and Causal Asymmetry," *Mind*, 101, 501–520.

Price, H. (2007) "Perspectival Causation," *in* H. Price and R. Corry (eds.) (2007), 250–292.

Price, H., and Corry R. (eds.) (2007) *Causation, Physics, and the Constitution of Reality: Russell's Republic Revisited*, Oxford: Clarendon Press.

Putnam, H. (1984) "Is the Causal Structure of the Physical Itself Something Physical?" repr. *in* H. Putnam (1990), *Realism with a Human Face*, ed. J. Conant, Cambridge, MA: Harvard University Press, 80–95.

Reichenbach, H. (1956), *The Direction of Time*, Berkeley: University of California Press, 1991.

Russell, B. (1912) "On the Notion of Cause," *Proceedings of the Aristotelian Society*, 13 (1912–1913) and *Scientia (Bologna)*, 13 (1913), repr. *in Mysticism and Logic* (1917), repr. London: Routledge, 2004, and in J. G. Slater (1992) (éd.) *The Collected Papers of Bertrand Russell, vol. 6: Logical and Philosophical Papers 1909–13*, London and New York: Routledge, 193–210.

Russell, B. (1914/1993) *Our Knowledge of the External World*, London: Routledge.

Russell, B. (1948/1992) *Human Knowledge, Its Scopes and Limits*, London: Routledge,

Salmon, W. (1984) *Scientific Explanation and the Causal Structure of the World*, Princeton, NJ: Princeton University Press.

Salmon, W. (1994) "Causality without Counterfactuals," *Philosophy of Science*, 61, 297–312.

Salmon, W. (1998), *Causality and Explanation*, New York, Oxford: Oxford University Press.

Savellos, E.E., and Ü.D. Yalçin (eds.) (1995) *Supervenience: New Essays*, Cambridge: Cambridge University Press.

Savitt, St. (ed.) (2006) "The Arrows of Time," *Studies in History and Philosophy of Modern Physics*, 37(no. 3), 393–576.

Schaffer, J. (2000a) "Causation by Disconnection," *Philosophy of Science*, 67, 285–300.

Schaffer, J. (2000b) "Trumping Preemption," *Journal of Philosophy* 97, 165–181, repr. *in* Collins et al. (eds.) (2004), 59–73.

Schaffer, J. (2001) "Causes as Probability-Raisers of Processes," *Journal of Philosophy*, 98, 75–92.

Schaffer, J. (2005) "Contrastive Causation," *Philosophical Review*, 114, 297–328.

Schaffer, J. (2006) "Le trou noir de la causalité," *Philosophie*, 89, 40–52.

Silverberg, A. (1996) "Psychological Laws and Non-Monotonic Logic," *Erkenntnis*, 44, 199–224.

Skyrms, B. (1980) *Causal Necessity*, New Haven and London: Yale University Press.

Smith, Sh. (2002) "Violated Laws, *Ceteris Paribus* Clauses and Capacities," *Synthese*, 130, 235–264.

Sosa, E., and Tooley, M. (eds.) (1993) *Causation*, Oxford: Oxford University Press.

Spirtes, P., Cl. Glymour, and R. Scheines (2000) *Causation, Prediction and Search*, 2nd ed., Cambridge, MA: MIT Press.

Spurrett, D., and D. Ross (2007) "Notions of Cause: Russell's Thesis Revisited," *British Journal for the Philosophy of Science*, 58, 45–76.

Vendler, Z. (1967a) "Causal Relations," *Journal of Philosophy*, 64, 704–713.

Vendler, Z. (1967b) "Facts and Events," *in Linguistics and Philosophy*, Ithaca, NY: Cornell University Press, 12–146.

von Wright, G.H. (1971) *Explanation and Understanding*, Ithaca, NY: Cornell University Press.

Winston, M.L. (1991) *The Biology of the Honey Bee*, Cambridge, MA: Harvard University Press.

Wittgenstein, L. (1921) *Tractatus logico-philosophicus*, trans. C. K. Ogden and Bertrand Russell, London: Routledge, 2001.

Woodward, J. (2003) *Making Things Happen: A Theory of Causal Explanation*, Oxford: Oxford University Press.

Woodward, J. (2004) "Counterfactuals and Causal Explanation," *International Studies in the Philosophy of Science*, 18, 41–72.

Woodward, J. (2008) "Causation and Manipulability," *Stanford Encyclopedia of Philosophy*, http://plato.stanford.edu/entries/causation-mani/.

4

METAPHYSICS OF SCIENCE AS NATURALIZED METAPHYSICS

Michael Esfeld (University of Lausanne, Department of Philosophy)

1. What Is Metaphysics of Science?

Metaphysics of science is a metaphysics because it puts forward ontological claims (that is, claims about what there is in the world) by basing itself on science—instead of conceptual analysis, common sense, or intuitions. By "metaphysics," one does in this context not mean a theory that claims to refer to a domain of being beyond the empirical realm, but in the Aristotelian sense, a theory that seeks to achieve a general and fundamental understanding of the empirical world itself (see Aristotle, Metaphysics, book 4). Today's metaphysics of science is considered as being part and parcel of analytic philosophy broadly conceived, which, since its metaphysical turn, no longer focuses on the analysis of language. It is instead a systematic and argumentative enterprise that seeks to achieve a comprehensive view of the world and our place in it—in short, pursuing what philosophy has been since its beginnings in Plato and Aristotle. The trait that distinguishes metaphysics of science from standard analytic philosophy is its being anchored in science: one bases oneself on science in doing metaphysics. It is therefore common today to use the term naturalized metaphysics.

Indeed, standard analytic metaphysics seeks to find out truth about the constitution of the world mainly based on conceptual analysis linked with common sense realism and intuitions. This enterprise is not hostile to science: a global supervenience thesis to the effect that everything that there is in the world supervenes on a basis that is investigated by fundamental physics is common ground. But standard analytic metaphysics does not show any particular interest in research about what this basis may be like given our current fundamental physical theories. Jackson's (1998) plea for conceptual analysis is a good illustration of this situation.

Naturalized metaphysics is opposed to the methodology employed by this type of metaphysics. The term "naturalized metaphysics" characterizes not so much the ontological stance of naturalism—a stance shared by most adherents to conceptual analysis –, but the method that seeks to find out truth about the constitution of the world by means of a close examination of our current fundamental physical theories. In particular, any metaphysical claim is to be motivated and justified by the content of our best scientific theories—by contrast to conceptual analysis, common sense, or intuitions. The book by Ladyman and Ross (2007) is the most forceful articulation of this type of metaphysics.[1]

This opposition in methodology does not necessarily lead to an opposition in the content of the metaphysics thus obtained, given in particular the nearly universal acceptance of the mentioned global supervenience thesis. However, naturalized metaphysics often comes with a good deal of polemic against analytic metaphysics: it is suggested that the best candidate for truth about the constitution of the world that we can currently achieve can be read from our fundamental physical theories and that what thus can be extracted from physics contradicts much of what is commonly accepted in standard metaphysics.

This chapter investigates how metaphysics of science qua naturalized metaphysics can work when taking fundamental physics as a guideline. To start with, I will consider the ontology of Newtonian mechanics (section 2), followed by an enquiry into the special theory of relativity and its alleged philosophical consequences for the metaphysics of time (section 3). I will then go into the options for an ontology of quantum physics (section 4) and examine how these options depend on the stance that one takes with respect to laws of nature and modality (section 5). By contrast to what Ladyman and Ross (2007) suggest, it will become increasingly clear during this investigation that there is no one-way road from physics to metaphysics, but that any ontology of physics has to bring in both the physical theory in question and considerations from standard metaphysics.[2] In a nutshell, there neither is a neo-positivist way of deducing metaphysics from physics, nor a neo-rationalist realm of investigation for metaphysics that is independent of physics. What we need is a metaphysics of science or a naturalized metaphysics that is a natural philosophy as practiced in the 17th and 18th century, when physics and metaphysics were treated as forming a seamless whole.

---

1 See furthermore the papers in Ross, Ladyman, and Kincaid (2013) for discussion, as well as Ney (2012).

2 See also the balanced positions of Callender (2011), as well as Chakravartty (2013), who examines to what extent naturalized metaphysics has to go into the topics of standard metaphysics that are usually considered as being far from science. On the other end of the spectrum, see Monton (2011), whose argumentation, however, is based on the claim that our current fundamental physical theories are false.

2. Newton's Natural Philosophy

Classical mechanics proposes an ontology of matter in motion: the fundamental physical domain consists in moving particles, with the laws of nature accounting for the way in which the particles move.[3] Thus, Newton famously writes at the end of the "Opticks" (1704):

> . . . it seems probable to me, that God in the Beginning form'd Matter in solid, massy, hard, impenetrable, moveable Particles . . . the Changes of corporeal Things are to be placed only in the various Separations and new Associations and motions of these permanent Particles. (Question 31, p. 400 in the edition Newton 1952)

Newton's natural philosophy (philosophia naturalis) can be considered as seeking to reply to three questions. The first is this one: What are the physical objects? Newton's answer is that matter consists in particles that are distributed in a background space, a particle being a material object that is so small that it is localized at a point in space, thus being indivisible. Hence, some points of space are occupied—where a particle is localized—whereas others are empty.

If one adopts a sparse view of physical properties, there is no reason to make use of the notion of properties as far as this basic characterization of matter is concerned. Matter is primitive stuff, and it is a primitive fact that some points of space are occupied whereas others are not. There is a good reason for conceiving matter in terms of particles, that is, in terms of points of space being occupied or empty. If one considered matter to be a continuous stuff distributed all over space (that is, gunk), then one would have to maintain that there is more stuff at some points of space and less stuff at others in order to be able to accommodate variation. But it could not be a primitive fact that there is more stuff at some points of space and less at others; a property of the stuff would be needed to account for that difference. However, as we will see shortly, all the properties that classical mechanics attributes to matter concern its temporal development, not simply the fact that there is matter. The view of matter consisting in particles can easily take into account the fact that there is more matter in some regions of space than in others: in some regions of space, more points are occupied than in others.

In Newtonian mechanics, the distribution of matter in a background space develops in a background time. That is to say, as time passes, there is change in which points of space are occupied and which are empty. That change is such that the particles persist in the sense of enduring, each moving on a continuous trajectory. An alternative view would be to admit just single events, with no continuous sequences of events. But again, taking matter to have a continuous existence—instead of events popping in a

[3] See Maudlin (2012, chs. 1–2) for an excellent recent examination.

discontinuous way in and out of existence as time passes—seems to be the simpler view. Consequently, each particle has an identity in time by which it distinguishes itself from all the other particles. The particles can therefore with good reason be regarded as substances.

The fact that there is change implies that Newton has to answer a second question: What are the laws of the temporal development of the physical objects? More precisely: What are the properties of the physical objects so that certain laws describe their behavior? Consequently, the need for a commitment to properties arises in Newton's philosophia naturalis when it comes to an account of the temporal development of the physical objects. Change in position as time passes means that the particles have the property of velocity, which is the first temporal derivative of position. That is to say, over and above having an initial position, the particles have an initial velocity, and this initial velocity makes them move in a certain manner. The property of velocity of each particle is conserved, as long as it is the only property that is taken into consideration. Velocity thereby gives rises to Newton's first law, which says that given an initial velocity, particles move on a straight line with constant velocity (inertial motion).

However, it is an empirical fact that there is not only change in the points in space that particles occupy as time passes, but also change in their state of motion, that is, change in velocity. That is why it is necessary to attribute more properties to the particles than just an initial velocity. Newton does so in taking the particles to be equipped with mass. In virtue of possessing mass, particles accelerate in the sense that they attract each other (gravitational mass) as well as resist to acceleration (inertial mass), acceleration being the change of velocity in time and thus the second temporal derivative of position. Newton's second law describes how properties change the state of motion of particles by accelerating them. In doing so, Newton introduces the notion of forces. Thus, in virtue of possessing mass, particles exert a force of attraction on each other, namely the force of gravitation. However, there is no need to subscribe to an ontological commitment to forces over and above a commitment to properties of the particles such as their mass. Given the masses of the particles at a time t and their positions and velocities at t, the acceleration of the particles at t is determined (modulo the gravitational constant). Forces are a device to calculate the consequences that the presence of properties such as mass has for the change of the state of motion of the particles, but no addition to being.[4] The same goes for other properties that account for the change of the state of motion of particles in classical physics, such as their charge, giving rise to acceleration due to electromagnetic interaction: there is charge determining that interaction, but no force that acts in nature over and above there being charged particles.

Finally, Newton's natural philosophy has to answer a third question: How do the physical objects and their properties explain the observable phenomena? As the

---

[4] See e.g. Jammer (1957, pp. 243–245). As regards the contemporary discussion about the ontological status of Newtonian forces, see notably Bigelow, Ellis, and Pargetter (1988), Wilson (2007), and Massin (2009).

quotation from Newton's "Opticks" (1704) shows, Newton answers this question by maintaining that (a) all macrophysical objects are composed of microphysical particles and that (b) all differences in macrophysical objects can be traced back to the position (configuration) and the change of position (motion) of the microphysical particles. That is to say, the properties that account for the temporal development of the position of the microphysical particles (that is, their initial velocity and their mass, as well as their charge) thereby also account for all the variations in the macrophysical objects.

Newton's theory is a paradigmatic example of natural philosophy in that physics and metaphysics come together in this theory in an inseparable manner. Newton's theory is not a naturalized metaphysics in the sense of being a positivist metaphysics: the assumption that there are particles and that properties of the particles have to be admitted that change the state of motion of the particles by accelerating them cannot be derived from any observation. It is an ontological postulate. But Newton's theory is not a rationalist metaphysics either: there is no a priori justification of the commitment to particles and properties that accelerate them. Making these assumptions yields a theory that is both physical-mathematical and metaphysical in one, being a universal physical theory that has the ambition to provide for a complete ontology of nature, and whose justification consists in its success in predicting and explaining the observable phenomena.

3. Relativity Physics, Quantum Non-Locality, and the Metaphysics of Time

There is, however, a stumbling block in Newton's theory, namely the assumption that particles interact instantaneously across empty space. Maxwell's field theory of electromagnetism developed in the 19th century provides the means to overcome this stumbling block: in virtue of being charged, particles create a field, and their interaction is transmitted by the field and thus retarded. Hence, instead of action at a distance, there is local action: interactions propagate from a space-time point to its neighboring points. They are thereby transmitted with a finite velocity. In fact, the velocity of light is the upper limit for the propagation of effects.

In the special theory of relativity, Einstein (1905) draws the consequences of the field solution to the problem of action at a distance in Newtonian mechanics. This theory is built on the following two principles:

1. All inertial reference frames are equivalent for the description of physical phenomena.
2. The velocity of light is a constant, being independent of the state of motion of its source and thus the same in all inertial reference frames.

Principle (1) is taken over from pre-relativistic physics, going back to Galilei. Principle (2) implements the field solution to the problem of action at a distance in Newtonian mechanics. It implies that the Galilean transformations are no longer applicable when switching from one inertial reference frame to another one. They have to be replaced with the Lorentz transformations. The latter unify space and time in the following sense: only the four-dimensional, spatio-temporal distance between any two events occurring at space-time points is an invariant. This is the reason for the claim that following the special theory of relativity, space and time are not separate entities, but are unified in a four-dimensional space-time.

In order to draw metaphysical conclusions from these two principles, one does not have to assume that the special theory of relativity is true. Indeed, strictly speaking, it is false, since the general theory of relativity no longer treats space-time as a non-dynamical background. Nonetheless, these two principles carry over from the special to the general theory of relativity. One can therefore presume that these two principles put a constraint on any future theory of space-time, whatever the further content of such a theory may be.

These two principles suggest certain consequences for the metaphysics of time. They entail that there is no objective simultaneity, because any two events that are simultaneous in one inertial reference frame are not simultaneous in other inertial reference frames, and all inertial reference frames are equivalent; in other words, there is no unique foliation of space-time into spatial hypersurfaces that are ordered in time. Consequently, any metaphysics of time that is based on the tenses—the past, the present, the future—being objective features of the world and that ties existence to tense is incompatible with these two features.

In particular, presentism is refuted by these two principles: presentism is the view that only what is present exists. What is past no longer exists, and what is future does not exist as yet. Presentism, thus construed, takes for granted that there is a unique foliation of four-dimensional space-time into three-dimensional spatial hypersurfaces that are ordered in time.[5] It is the view that these hypersurfaces come into and go out of existence such that always only one such hypersurface exists—the present one. Monton (2006, p. 264) characterizes this view as "Heraclitean presentism," because its central tenet is the reality of change in the sense of events coming into being and going out of being. Presentism thus is opposed to eternalism according to which everything that there is in nature simply exists. In the context of the special and the general theory of relativity, the latter position is known as the view of the block universe: everything that there is in space-time simply exists. Consequently, there is no temporal becoming in the sense of something coming into being as time passes.

There are strategies available to avoid drawing the metaphysical conclusion of the block universe view from relativity physics. The most prominent strategy that one can try is solipsism: if one assumes that only the space-time point at which one is situated—my "here" and "now"—exists, then no contradiction with the mentioned two principles arises (see Stein 1968 for setting out that option and e.g. Harrington 2008 for endorsing it). But solipsism certainly is not a serious metaphysical stance based

[5] But see Fine (2005, ch. 8, § 10, pp. 298–307) for a view that relativizes existence to inertial frames.

on science. Wüthrich (2013, sections 3–6) examines the various strategies envisaged in the literature to avoid the conclusion of the block universe view and convincingly argues that all these strategies are desperate. In short, taking the metaphysical position of presentism to be refuted by relativity physics and regarding eternalism in the form of the block universe view as vindicated by relativity physics is a straightforward metaphysical conclusion from space-time physics, if anything ever is a straightforward metaphysical conclusion from a scientific theory.

However, metaphysics of science is not concerned with metaphysical conclusions that one may draw from one particular scientific theory. Any metaphysics of science, whatever methodology it pursues, seeks to develop a coherent and complete vision of nature on the basis of our mature scientific theories. Thus, in order to build metaphysical conclusions on a particular scientific theory, two conditions have to be met: (a) the principles of the scientific theory in question on which the metaphysical conclusions at issue are based have to be such that we have reason to believe that these principles put a constraint on any future successor theory of the scientific theory in question. (b) The scientific theory in question either has to be itself a complete fundamental physical theory or there have to be no other contemporary mature scientific theories that challenge its principles. As argued earlier, condition (a) is satisfied in this case. However, condition (b) is not fulfilled: quantum physics is a mature science that has at least the same scientific standing as relativity physics. Quantum physics challenges the conjunction of the two principles on which relativity physics is based.

A popular way of setting out that challenge invokes what is known as the collapse of the wave-function in a quantum measurement process. However, whether such a collapse really occurs as a process in nature is a controversial issue, as we will see in the next section. The collapse view of measurement does not meet condition (a). There is no challenge to the block universe metaphysics stemming from wave-function collapse in quantum mechanics (see Callender 2008). Nonetheless, it is true that quantum physics calls the conjunction of the two principles on which relativity physics is based into question.

John Bell, in one of his last papers entitled "La nouvelle cuisine" (1990, reprinted in Bell 2004, ch. 24), formulates a principle of local causality: "The direct causes (and effects) of events are nearby, and even the indirect causes (and effects) are no further away than permitted by the velocity of light" (quoted from Bell 2004, p. 239). No particular notion of causation is implied here (see Bell 2004, p. 240). The idea is that whatever events whose occurrence contributes to determining the probabilities for a given event to happen at a certain space-time point are located in the past light-cone of that event. This is one way of formulating the principle of local action that is implemented in classical field theories and that overcomes Newtonian action at a distance. Since relativity physics endorses this principle, it can waive the commitment to a unique temporal order of events and thus the commitment to an objective simultaneity: whatever contributes to determining a given event is situated in its past light cone; consequently, there is no need to settle for a unique temporal order of events that are situated outside each other's light cones.

Consider the thought experiment of Einstein, Podolsky, and Rosen (EPR) (1935) in the version of Bohm (1951, pp. 611–622). Two elementary quantum particles are prepared in an entangled spin state at the source of the experiment (such as two systems of spin 1/2 in the singlet state). Later, when they are far apart in space so that there is no interaction any more between them, Alice chooses the spin parameter to measure in her wing of the experiment and obtains an outcome, and Bob does the same in his wing of the experiment. Alice's setting of her apparatus is separated by a spacelike interval from Bob's setting of his apparatus. The following figure illustrates this situation:

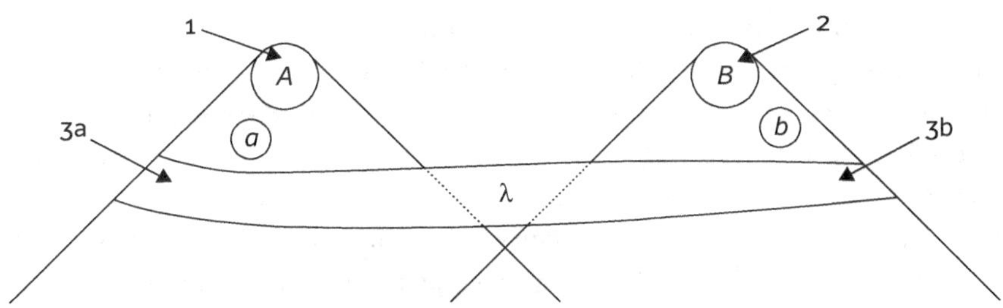

FIGURE 1 The situation that Bell considers in the proof of his theorem.
Figure taken from Seevinck (2010, appendix) with permission of the author.

In this figure, a stands for Alice's measurement setting, A for Alice's outcome, b stands for Bob's measurement setting, B for Bob's outcome, and λ ranges over whatever in the past may influence the behavior of the measured quantum systems according to the theory under consideration (which may be standard quantum mechanics, or a theory that admits additional, so-called hidden variables).

Bell's principle of local causality—or locality for short—can then be formulated in the following manner:

$$
\begin{aligned}
P_{a,b}(A \mid B, \lambda) &= P_a(A \mid \lambda) \\
P_{a,b}(B \mid A, \lambda) &= P_b(B \mid \lambda)
\end{aligned}
\tag{1}
$$

That is to say: the probabilities for Alice's outcome depend only on her measurement setting and λ. Adding Bob's setting and outcome does not change the probabilities for Alice's outcome. The same goes for Bob. The theorem that Bell proved in 1964 (reprinted in Bell 2004, ch. 2) establishes that quantum mechanics violates (1). That is to say, for some measurement settings, even if the probabilities for Alice's outcome A depend only on her setting a and the past state λ, it is then necessarily so that the probabilities for Bob's outcome B depend not only on his setting b and the past state λ, but also on Alice's setting a and outcome A, although b and B are separated by a spacelike interval from a and A. Moreover, any theory that reproduces the well-confirmed experimental predictions of quantum mechanics has to violate (1). This conclusion applies not only to quantum mechanics, but also to quantum field theory.[6] One can therefore say that

[6] See Bell (2004, ch. 24). See Hofer-Szabó and Vescernyés (2013), as well as Lazarovici (2014) for the current discussion.

Bell's theorem puts a constraint on any—present or future—physical theory that is to match the experimentally confirmed predictions of quantum mechanics.

The proof of Bell's theorem does not depend on the truth of quantum mechanics or quantum field theory. The theorem then establishes that any theory that complies with the predictions for macrophysical measurement outcomes of quantum mechanics or quantum field theory—whatever its content may be—cannot satisfy the locality principle (1). One can limit the point at issue of Bell's theorem to correlations between space-like separated macrophysical measurement outcomes, such as the directions in which pointers point. In other words, even if one abstains from hypotheses about the microphysical constitution of such macrophysical events, one still gets Bell's theorem.

Nonetheless, the proof of Bell's theorem requires more than the locality principle (1): it requires also that the measurement settings a and b are independent of the past state λ. Failure of such independence can arise in two different ways: either the measurement settings exert some influence on λ, or λ somehow influences the measurement settings. It is obvious from Figure 1 that the first option involves influences travelling backward in time. Indeed, any attempt to save the locality principle (1) by relying on backward causation retroactively correlates the measurement settings a and b with the past state λ: the settings a and b influence the outcomes A and B, which in turn retroactively influence λ.[7] The second option contradicts the presupposition that the measurement settings can be freely chosen by an experimental physicist or a random generator.

However, the assumption of such an independence is not specific for Bell's theorem, but applies to any experimental evidence: if the behavior of the measured system that produces the measurement outcome were correlated with the parameter that is measured on the system, then no conclusions about the constitution of nature would be possible on the basis of experimental evidence. Furthermore, this assumption does not imply any sort of indeterminism. A physical theory with a completely deterministic dynamics can satisfy this assumption—as does for instance Bohmian mechanics in the quantum case.[8] It is therefore a well-grounded conclusion to maintain that quantum physics refutes the locality principle (1).[9]

If the probabilities for what happens in a given space-time region are influenced by what happens in regions that are separated by a spacelike interval from that region, then the mentioned two principles on which the special and the general theory of relativity are built are challenged. It would, however, be unwarranted to conclude that there are signals travelling with a velocity that is much higher than the velocity

---

[7] See Price (1996, ch. 8 and 9) for a prominent such attempt. See the papers in *Studies in History and Philosophy of Modern Physics* 38 (2008), pp. 705–784, for discussion.

[8] See the exchange on this issue between Bell, Shimony, Horne, and Clauser in Bell et al. (1985). The so-called free-will theorem by Conway and Kochen (2006, 2009) does not show anything standing up to scrutiny that is not already given by Bell's theorem. See notably Tumulka (2009), Goldstein et al. (2010), and Wüthrich (2011).

[9] See Maudlin (2011, chs. 1–6), Norsen (2009), Seevinck (2010), and Seevinck and Uffink (2011) for the current state of the discussion on Bell's theorem.

of light. There is no precisely formulated version of quantum theory that includes superluminal signals, although one can contemplate models of quantum non-locality that are built on the idea of superluminal signals, as notably Chang and Cartwright (1993, section III) do. If events that occur in a space-time region that is separated from a given space-time region by a spacelike interval contribute to determining what happens in the latter region, this suggests that there is a unique or objective temporal order between these events. In other words, the principle that is challenged is the one of the equivalence of all inertial reference frames (special relativity) so that there is no unique foliation of space-time into three-dimensional, spatial hypersurfaces that are ordered in time (general relativity).

Even if one takes the EPR-correlations between space-like separated events to require a unique temporal order of these events, one does not have to contradict any of the experimental evidence for the special and the general theory of relativity. Quantum non-locality then implies that there is more structure of space-time than is admitted by relativity physics, but this additional structure is not accessible by experience—otherwise, one could use quantum non-locality for superluminal communication. However, this is not possible. The reason is, in brief, that one cannot control the measurement outcomes A and B (and if there are additional, so-called hidden variables in λ, one cannot have full access to these variables). The conflict between quantum physics (quantum mechanics, quantum field theory) and relativity physics (special relativity, general relativity) does not arise on the operational level, but only on the ontological level (cf. Albert 2000). Even if one regards quantum non-locality as evidence for there being a unique foliation of space-time into spatial hypersurfaces that are ordered in time, one is by no means committed to going back to endorse an ether that serves as the privileged inertial frame. On the contrary, one may contemplate the idea that the distribution of mass in the universe fixes the objective foliation of space-time, or that the universal wave-function does so (see Dürr et al. 2013b).

Introducing on the basis of quantum non-locality the assumption that there is a unique foliation of space-time rules out the inference from relativity physics to eternalism in the form of the block universe metaphysics. However, this assumption does not as such contain an argument that favors presentism over eternalism. One can even raise doubts whether this assumption is compatible with the presentism that draws its support from common sense, given notably that this unique foliation of space-time is not empirically accessible (see Callender 2008). In any case, in metaphysics of science, arguments based on common sense (or on intuitions about time, or on an a priori analysis of the concept of time) are not admissible. If one sets out to make a case for presentism, one has to develop positive arguments for this metaphysics of time based on science. The lesson of the tension between relativity physics and quantum non-locality as established by Bell's theorem is not that physics favors presentism over eternalism, or that physics is neutral with respect to this metaphysical debate, but only the following methodological one: building metaphysical conclusions on a physical theory requires spelling out how this theory can accommodate all the available evidence in its domain, and assessing this evidence involves both physics and metaphysics in an inseparable manner. I will come back to this conclusion in section 5, showing there that the stance that one takes with respect to this tension depends on one's views about laws of nature and objective modality.

4. The Problem of the Referent of Quantum Physics

Let us now turn to quantum mechanics. In this case, there is no straightforward answer to the question of what this theory tells us about the world, supposing that it is true or approximately true. Instead of being a theory like Newtonian mechanics in which the physics itself implements a certain ontology, we have to engage in the business of interpreting quantum mechanics, its interpretation involving to settle for a specific mathematical formulation of quantum mechanics that then enables an answer to the question of what the theory tells us about the world. Again, physics and metaphysics are inseparable, since metaphysical considerations determine the choice of the mathematical formulation of the physical theory.

The following, easily accessible thought experiment suggested by Einstein at the Solvay conference in Brussels in 1927 illustrates this situation (my presentation is based on de Broglie's version of the thought experiment in de Broglie 1964, pp. 28-29, and on Norsen 2005): consider a box which is prepared in such a way that there is a single elementary quantum particle in it. The box is split in two halves that are sent in opposite directions, say from Brussels to Paris and Tokyo. Suppose that Alice in Tokyo opens the box she receives and finds it to be empty. If Alice's box is empty, it then is a fact that there is a particle in the box that Bob receives in Paris.

The textbook quantum formalism represents the particle in the box by means of a wave-function. When the box is split and the two halves are sent to Paris and to Tokyo, the wave-function represents the particle in terms of a superposition of its being in the box that travels to Paris and its being in the box that travels to Tokyo. The operational meaning of this representation is that there is a 50% chance of finding the particle in the box that travels to Paris and a 50% chance of finding the particle in the box that travels to Tokyo. When Alice in Tokyo opens the box she receives and finds it to be empty, this representation changes such that the wave-function represents the particle to be located in the box that travels to Paris. That sudden change is known as the collapse of the wave-function.

The problem with this formalism is the following one: if one takes the collapse of the wave-function upon measurement to represent a process that occurs in nature, one is committed to what Einstein called "spooky action at a distance"—the local operation of opening the box in Tokyo creates the fact that there is a particle in the box in Paris. If, by contrast, one takes the collapse of the wave-function upon measurement to represent an updating of information of the observer, such that before opening the box the observer does not know where the particle is, one is committed to the view that textbook quantum mechanics is an incomplete physical theory—the particle then was all the time in the box travelling to Paris, and the formalism of textbook quantum mechanics is unable to represent its trajectory. Bell's theorem then shows that one cannot complete quantum mechanics in terms of a local dynamics—that is, a dynamics complying with the locality principle (1) (although in this particular case of one particle in a box, a local account is possible and provided by de Broglie's and Bohm's quantum theory). But Bell's theorem does not settle the issue of whether or not the representation in terms of the wave-function is a complete representation of quantum objects and whether or not wave-function collapse indicates a process that occurs in nature.

Indeed, the problem of understanding quantum mechanics goes deeper than answering the question of what wave-function collapse stands for. Generally speaking, in quantum mechanics, the phase space of classical mechanics is replaced with a configuration space each point of which represents a possible configuration of particles in three-dimensional space.[10] Thus, if there are N particles, the configuration space has 3N dimensions. On this configuration space a quantum state of the particles is defined, which can be expressed in terms of a wave-function that is a field in configuration space and that develops in time according to the Schrödinger equation. This has the consequence that whenever one considers the states of two or more quantum systems, the occurrence of entangled states of these systems is generic, and their states will in general remain entangled, unless wave-function collapse occurs.

This formalism runs into the following problem of understanding: on the one hand, it seems to be committed to particles, since the dimension of the configuration space is defined by the number of particles considered. On the other hand, the law (i.e., the Schrödinger equation) is not a differential equation that is about the temporal development of a particle configuration in three-dimensional space (i.e., the development of particle positions and thus particle trajectories), but about the temporal development of a wave-function in configuration space. The fundamental problem of understanding this formalism therefore is that there is an underdetermination of what its referent is: Is it objects in ordinary space? Or is it a wave-function in configuration space? By way of consequence, metaphysics has to come in to settle the very issue of what the formalism of quantum mechanics is about.

It is usually taken for granted that the natural world consists in matter distributed in three-dimensional space or four-dimensional space-time and that the task of physics is to develop an account of matter and its temporal development (plus an account of space and time themselves). However, in a famous paper about realism in quantum mechanics, Albert (1996) claims the contrary:

> . . . it has been essential (. . .) to the project of quantum-mechanical realism (in whatever particular form it takes . . .) to learn to think of wave functions as physical objects in and of themselves. And of course the space those sorts of

[10] See North (2013, section 2) for an argument as to why one should regard configuration space, and not Hilbert space, as the fundamental state space of quantum mechanics.

> objects live in, and (therefore) the space we live in, the space in which any realistic understanding of quantum mechanics is necessarily going to depict the history of the world as playing itself out (. . .) is configuration space. And whatever impression we have to the contrary (whatever impression we have, say, of living in a three-dimensional space, or in a four-dimensional space-time) is somehow flatly illusory. (Albert 1996, p. 277, emphasis in the original; see also Albert 2013 as well as Ney 2011 and North 2013)

This stance is known as wave-function realism. It is motivated by attributing a literal meaning to the fact that the Schrödinger equation is about the temporal development of a wave-function in a high-dimensional space: this stance takes that wave-function to be the object of quantum mechanics. Hence, the wave-function is an ordinary physical object, as particles or fields are ordinary physical objects in three-dimensional space in classical mechanics. The drawback of this move is that it cannot attribute a literal meaning to the fact that the dimension of the space in which the wave-function exists is defined in terms of the number of particles in three-dimensional space. Indeed, when one adopts this stance, the term "configuration space" becomes obsolete: there is no given configuration of anything that points of this space represent. The physical reality is the wave-function—to be precise, the wave-function of the universe—existing as a field in configuration space.

If one endorses this stance, the obvious task then is to develop an account of our experience of objects localized in three-dimensional space and moving in that space. In the meantime, Albert (2013) has withdrawn his claim from 1996 that doing so implies regarding our impression of living in a three-dimensional space as "somehow flatly illusory" and announces a forthcoming account of the objects of common sense in functional terms, so that from wave-function realism one can derive common sense realism instead of having to reject the latter. However, such an account has not been accomplished as yet, neither in Albert's version of wave-function realism, nor in the contemporary versions of what is known as Everettian quantum mechanics.[11]

---

[11] Wallace (2012) is the most detailed contemporary version of Everettian quantum mechanics. He rejects wave-function realism, maintaining, in brief, that the quantum state (which can be represented by the wave-function of the universe) is a state instantiated in four-dimensional space-time, developing in such a way that there are *many* four-dimensional space-times existing in parallel ("multiverse," "branches of the universe"; see in particular chs. 2 and 8). Nonetheless, the problem of developing an account of the experience of objects as well as ourselves being localized in *one* four-dimensional space-time arises in this theory in the same manner as in the one of Albert. Furthermore, Wallace's position amounts to what is known as *super-substantivalism* (Sklar 1974, pp. 221–224). If one maintains that physical properties or states are properties or states of space-time itself instead of being instantiated by objects localized in space-time, thus avoiding a commitment to what is known as a primitive ontology of objects in space-time, one has to elaborate on a theory of how physical properties can be properties of space-time itself—otherwise, one only performs what Sklar (1974, pp. 166–167, 222–223) describes as a linguistic trick, namely changing language in attributing the properties that are usually ascribed to objects in space-time to space-time itself.

Even if such an account were set out in detail, wave-function realism would not be established as simply following from the formalism of quantum mechanics. It would still require metaphysical argument to justify the conclusion that the very high dimensional space on which the wave-function of the universe is defined is the realm of physical reality, instead of three-dimensional space or four-dimensional space-time. The reason is that it is impossible to satisfy both elements, namely, that (a) the state space of quantum mechanics is a configuration space whose dimension is determined by the number of particles existing in three-dimensional space and whose points represent possible given configurations of those particles, and that (b) the fundamental law of quantum mechanics is a dynamical equation that is about the temporal development of the wave-function in that space. In other words, one cannot have both (a) a configuration space that represents a physical reality outside that space and (b) a fundamental law that is about the temporal development of an object inside that space.

Wave-function realism endorses (b) and abandons (a). The other option is to endorse (a) and to abandon (b). If one takes quantum mechanics to be about a physical reality in three-dimensional space or four-dimensional space-time, one is committed to what is known as a primitive ontology of quantum mechanics (this term goes back to Dürr, Goldstein and Zanghì 2013a, ch. 2, end of section 2, originally published 1992). That ontology is primitive at least in the sense that it cannot be inferred from the formalism of quantum mechanics, but that it has to be put in as the referent of that formalism. Consequently, the fundamental law then has to be a law that describes the temporal development of the elements of the primitive ontology in three-dimensional space or four-dimensional space-time, and that law can obviously not be the Schrödinger equation (see Allori et al. 2008 for the structure of primitive ontology theories).

The de Broglie-Bohm quantum theory, going back to de Broglie (1928) and Bohm (1952) and known today as Bohmian mechanics (see the papers in Dürr, Goldstein and Zanghì 2013a) is the oldest and most widely known primitive ontology theory of quantum mechanics. Bohmian mechanics endorses particles as the primitive ontology, adding the position of the particles as additional, so-called hidden variable to the formalism of textbook quantum mechanics (so that the wave-function is not a complete representation of physical reality, since it does not represent the actual particle positions). Consequently, there is at any time one actual configuration of particles in three-dimensional space, and the particles move on continuous trajectories in physical space. The fundamental law of Bohmian mechanics is the guiding equation, describing the temporal development of the position of the particles in three-dimensional space. The wave-function, as it figures in the guiding equation, has the job to determine the velocity of the particles at any time t given their position at t. The Schrödinger equation then comes in as an additional law, describing how the wave-function itself develops in time.

Strictly speaking, only the universal wave-function—that is, the wave-function of the configuration of all the particles in the universe—fulfills this job: strictly speaking, the velocity of any particle at t depends on the position of all the particles in the universe at t via the universal wave-function. That is the way in which Bohmian mechanics takes into account the non-locality established by Bell's theorem. Nonetheless, in many situations, the position of distant particles is de facto irrelevant for the trajectory of a given particle (as in the case of the particle in Einstein's boxes). Since this theory is committed to particles moving on continuous trajectories, there is no need for wave-function collapse as a process in nature to account for measurement outcomes: these consist simply in certain particle configurations, developing according to the guiding equation. If one assumes that the initial particle configuration of the universe is typical in a precise mathematical sense, it is possible to derive in Bohmian mechanics Born's rule for the calculation of probabilities for measurement outcomes via what is known as the quantum equilibrium hypothesis (see Dürr, Goldstein and Zanghì 2013a, ch. 2, originally published 1992).

Instead of subscribing to the Bohmian guiding equation, one can also go for a modification of the Schrödinger equation such that this equation includes conditions under which the wave-function localizes spontaneously in configuration space, thus enabling it to represent objects that are localized in three-dimensional space. The most precise proposal in that respect is the one going back to Ghirardi, Rimini, and Weber (GRW; 1986). However, the GRW law still is about the temporal development of the wave-function in configuration space, by contrast to a differential equation that is about the temporal development of objects in three-dimensional space. One therefore still has to put in a primitive ontology as the referent of the GRW formalism. There are two proposals in that respect developed in the literature.

The one proposal is committed to gunk in the sense of a matter density field in three-dimensional space: the temporal development of the wave-function represents the temporal development of the matter density in space-time, with the spontaneous localization of the wave-function in configuration space (its collapse) representing the spontaneous contraction of gunk in certain locations so that measurement outcomes and, in general, well-localized macroscopic objects are accounted for (see Ghirardi, Grassi, and Benatti 1995). Again, the dynamics is non-local, since the spontaneous contraction of gunk can occur all over space, independently of spatial distances. Thus, in the mentioned case of a particle in a box, when the box is split in two halves, the matter density of the particle stretches over both the half-boxes and spontaneously localizes in one of them upon measurement.

The other proposal is committed to single events, known as flashes, occurring at space-time points: whenever there is a spontaneous localization of the wave-function in configuration space, this collapse of the wave-function represents an event occurring in space-time, in the sense of a flash appearing centered on a space-time point. More precisely, the dynamics being non-local again, the collapse of the wave-function represents the spontaneous occurrence of spacelike separated, but nonetheless correlated flashes. The flash-events are all there is in space-time. Hence, there is no continuous distribution of matter in physical space, namely no trajectories or world-liness of particles, and no field—such as a matter density field—either. There only is a sparse distribution of single events in space-time (see Bell 2004, ch. 22, originally published 1987, and Tumulka 2006).

In any case, one needs a principle or a law that establishes the link between the primitive ontology and the GRW equation such that the GRW equation can fulfill the function of describing the temporal development of the elements of the primitive ontology in three-dimensional space (cf. Monton 2004). In a coherent formulation of the theory, that principle or law has to stand as the fundamental one, if the theory is to be about the temporal development of matter in three-dimensional space.

Against this background, consider the following three metaphysical claims that are often put forward as following directly from the formalism of quantum mechanics:

1. Quantum mechanics refutes the standard metaphysical view of objects, namely that objects are individuals, possessing an identity that distinguishes each object from all the other ones. The whole debate about the status of quantum particles takes for granted that the standard view is refuted and that the point at issue following quantum mechanics only is whether a notion of weak discernibility can bestow some sort of individuality on quantum particles (Saunders 2006) or whether even this is not possible (Ladyman and Bigaj 2010).
2. Since the formalism of quantum mechanics does not specify any particular objects in space-time as its referent and since interpreting quantum physics as being about objects as traditionally conceived leads to an underdetermination between an ontology of individuals and an ontology of non-individuals, quantum mechanics grounds a metaphysics of structures, known as ontic structural realism—by contrast to the object-based metaphysics that is taken for granted in mainstream analytic philosophy. This is the central claim of the naturalized metaphysics argued for by Ladyman and Ross (2007).
3. Quantum physics has implications for the metaphysics of modality: in particular, it refutes Lewis's thesis of Humean supervenience. This claim is widespread in the literature on the metaphysics of quantum physics since the seminal paper of Teller (1986). Maudlin (2007, ch. 2, pp. 51–64) turns it into a forceful attack on Humeanism in general based on quantum physics.

However, as regards the first two claims, they fail to consider the issue of what exactly the formalism of quantum mechanics represents. There is no point in seeking to draw metaphysical consequences from a formalism that contains operators which are introduced in order to allow for the calculation of probabilities of measurement outcomes—doing so would amount to what is known as naïve realism about operators (see Daumer et al. 1996. And there is no point in proposing a metaphysics of structures without considering how these structures are instantiated in the physical realm (see Esfeld 2013). Concerning the third claim, I will show in the next section that it goes through only if one presupposes an anti-Humean metaphysics of laws of nature. If, by contrast, one endorses Humeanism about laws,

Lewis's thesis of Humean supervenience can be literally true even in the light of quantum physics.

The considerations in this section seek to establish that one cannot read off a metaphysics from the formalism of quantum mechanics, because the formalism as such does not specify its referent. The first task for naturalized metaphysics in this area therefore is to chart out the options for providing a referent for quantum mechanics. By way of consequence, what one proposes as the mathematical formulation of a fully developed quantum theory depends on metaphysical considerations. Moreover, all the known options for specifying the referent of the formalism of quantum mechanics are committed to objects, and these options cover all the traditional kinds of objects—namely particles, gunk (a matter field), and single events.

If one takes the universal wave-function to be the referent of quantum physics, then the object to which quantum physics is committed is a field, albeit a field in an extremely high dimensional space by contrast to a field in four-dimensional space-time. But the field then has definite numerical values at the points of that high-dimensional space so that these values can be regarded as intrinsic properties occurring at the points of that space. Furthermore, the dynamical law for the temporal development of this field in that space is local (as long as it is given by a linear dynamical equation such as the Schrödinger equation). Consequently, if one takes the universal wave-function to be the referent of quantum physics, one obtains a traditional field ontology with a local dynamics as in classical field theory. There hence is in this case no motivation for basing a metaphysics of structures on quantum physics (see Albert 1996, p. 283, n. 7).

If, by contrast, one goes for a primitive ontology of matter distributed in ordinary space-time as the referent of the formalism of quantum physics, then there is continuity in the space in which the physical reality plays itself out, namely four-dimensional space-time, and there is continuity in objects from classical to quantum mechanics—particles, a matter field, single events being the options for a primitive ontology of quantum mechanics as outlined earlier. However, there then is change in the mathematical structure of the theory from classical to quantum mechanics, since the law for the temporal development of these objects then has to be a non-local one, in order to meet the conditions set by Bell's theorem. One can then regard this non-local law as being grounded in a modal structure that takes all the physical objects as its relata and that determines their temporal development (in a deterministic or probabilistic manner; see Esfeld 2013). Nonetheless, in any case, these physical objects come with their own identity conditions: particles and single events are absolutely discernible due to their position in physical space, and a matter field is one continuous object distributed all over physical space. In sum, there is no point in seeking to draw metaphysical conclusions about objects directly from the formalism of quantum mechanics. One first has to settle what one takes to be the referent of that formalism, and doing so brings in metaphysical considerations. Once this has been done, there then is no longer an issue about the discernibility or the identity conditions of quantum objects.

5. Physics and the Metaphysics of Modality

Let us assume that the referent of the formalism of quantum mechanics is the distribution of matter in ordinary space-time. It then seems obvious that entanglement shows that quantum mechanics refutes Humean metaphysics, in particular David Lewis's thesis of Humean supervenience:

> It is the doctrine that all there is to the world is a vast mosaic of local matters of particular fact, just one little thing and then another. (. . .) We have geometry: a system of external relations of spatio-temporal distance between points. . . . And at those points we have local qualities: perfectly natural intrinsic properties which need nothing bigger than a point at which to be instantiated. For short: we have an arrangement of qualities. And that is all. . . . All else supervenes on that. (Lewis 1986, pp. ix–x)

It seems evident that Lewis's view of the supervenience basis, consisting exclusively in the distribution of intrinsic physical properties located at space-time points, is contradicted by the fact of there being relations of quantum entanglement (see Teller 1986 and for a recent statement of this view Humphreys 2013, pp. 56–57). Notably Maudlin (2007, ch. 2, pp. 51–64) argues that quantum entanglement (the non-separability of quantum states) refutes not only Lewis's conception of the supervenience basis, but thereby also the Humean rejection of objective modality. The formalism of quantum physics thus has a direct bearing not only on what there is in the fundamental physical domain of the actual world, but also on which philosophical views of modality—and thereby of laws of nature—are admissible.

Lewis's reason for proposing a metaphysics of fundamental physics that recognizes only intrinsic properties located at space-time points is free combinatorialism: one can hold any local quality occurring at a space-time point fixed and vary all the other local qualities, the result always is a possible world. Any property occurrences can be combined with any other property occurrences, no instantiation of a property poses any restrictions on what the world has to be like beyond the space-time point at which the property in question is instantiated. Hence, there are no necessary connections in the world.

Philosophers with a favorable attitude toward Humeanism reacted to the challenge from quantum physics by trying to adapt Humeanism so that quantum entanglement is taken into account. The most important suggestion in this respect is to admit irreducible relations of entanglement over and above the spatio-temporal relations to the ontological ground floor of Humeanism (Darby 2012) and to envisage developing a Humean version of ontic structural realism on the basis of including such relations (Lyre 2010). However, recognizing irreducible relations of quantum entanglement considerably restricts free combinatorialism and arguably implies a commitment to some sort of objective modality, since these relations tie the temporal development of—in the last resort all—quantum systems together, whatever their spatial distance may be. Thus, considering the EPR experiment with the same parameter measured in both wings of the experiment, if in one wing the measured quantum system behaves in such a way that the measurement outcome is spin up, then it is necessarily so that in the other wing the measured quantum system behaves in such a way that the measurement outcome is spin down. If one does not want to talk about two systems in this respect, one can also formulate this point by saying that if in one wing of the experiment the pointer of the measuring apparatus indicates the outcome spin up, then it is necessarily so that in the other wing of the experiment the pointer of the measuring apparatus indicates the outcome spin down. Consequently, in any case, if there are relations of quantum entanglement in the supervenience basis, these relations pose a constraint on what can and what cannot happen elsewhere in space-time.

Furthermore, one can adapt Humeanism to quantum physics by adopting wave-function realism and admitting the very high-dimensional configuration space of the universe instead of four-dimensional space-time as the realm of physical reality (Loewer 1996). In this case, as mentioned at the end of the last section, everything is local in that space. Nonetheless, it is a considerable change of Humean metaphysics, which is inspired by common sense realism, to switch to configuration space as the stage of the Humean supervenience basis.

However, in recent years, it has become clear that no such adaptation is necessary. Humeanism is not refuted by quantum physics. More precisely, Lewis's thesis of Humean supervenience can be literally true even in the light of the empirical evidence for quantum entanglement. The background that enables Humeanism to stand firm is the development of primitive ontology theories of quantum physics, as outlined in the preceding section. To recap, the primitive ontology consists in the distribution of matter in three-dimensional space or four-dimensional space-time; that distribution is the referent of the formalism of quantum physics. Furthermore, a law is admitted as that what fixes (in a probabilistic or a deterministic manner) the temporal development of the distribution of matter in physical space, given an initial configuration of matter. That's all. In particular, according to the primitive ontology theories, the quantum mechanical wave-function is part and parcel of the law instead of being a physical entity on a par with the primitive ontology.

Since the primitive ontology is in any case constituted by local matters of particular fact—"local beables" to use Bell's famous term (Bell 2004, ch. 7)—the only move that the Humean has to make is this: instead of admitting the law as an entity that exists in addition to and independently of the primitive ontology, governing or guiding the temporal development of the primitive ontology, the Humean has to regard the law as supervening on the distribution of matter throughout the whole of space-time, that is, the entire mosaic of "local beables" or local matters of particular fact. This move has been made with respect to Bohm's quantum theory in recent literature (see Callender unpublished; Esfeld et al. 2014, section 3; Miller 2014). It is obvious that it can be extended also to the GRW matter density ontology and the GRW flash ontology (see Callender 2015 and Esfeld 2014). It is no objection to this move that the quantum mechanical wave-function does not supervene on the configuration of matter in space at any given time, since the Humean claims only that it supervenes on the entire distribution of matter in the whole of space-time. In a nutshell, on the Humean view, the universal wave-function is fixed only at the end of the world. If the entire distribution of matter in space-time were still to leave room for different universal wave-functions, that difference would not make any empirical difference and could therefore be dismissed by the Humean as a mathematical surplus structure.

Indeed, already Bell himself recognized this position as a coherent stance in the paper in which he introduced the notion of "local beables" (1975; "beable" is Bell's neologism for what exists by contrast to "observable," that is, what can be observed):

> One of the apparent non-localities of quantum mechanics is the instantaneous, over all space, 'collapse of the wave function' on 'measurement'. But this does not bother us if we do not grant beable status to the wave function. We can regard it simply as a convenient but inessential mathematical device for formulating correlations between experimental procedures and experimental results, i.e., between one set of beables and another. (Quoted from Bell 2004, p. 53)

Bell makes two points in this quotation: (1) It is not mandatory to grant beable status to the wave-function. If one admits "local beables," one has an ontology of the physical world. Not granting beable status to the wave-function does not, however, commit one to an instrumentalist attitude to the wave-function, as Bell suggests here. Humeanism is distinct from instrumentalism (Miller 2014, section 5, stresses this point). The Humean only has to maintain that the primitive ontology is the full ontology, with everything else supervening on it. That is why Humeanism is also not touched by recent claims about experimental evidence in favor of the reality of the wave-function (Pusey, Barrett, and Rudolph 2012; Colbeck and Renner 2012): these claims challenge only the view that the wave-function represents nothing but the information about probabilities for measurement outcomes that is available for an observer. However, on Humeanism, the universal wave-function is not relative to observers: it is an objective matter of fact in supervening on the entire distribution of the "local beables." Since any experimental evidence consists in "local beables," the Humean is in the position to accommodate whatever experimental evidence there may be.

(2) Given that it is the wave-function which is entangled and which correlates "local beables" whatever their spatial or spatio-temporal distance is, if one does not grant beable status to the wave-function, there is no reason to admit non-supervenient relations of entanglement (or of dependence or of influence) among the "local beables" over and above their occurrence at space-time points. In being entangled, the wave-function establishes such correlations, but these are no addition to what there is over and above the occurrence of the "local beables" at space-time points, since the universal wave-function and its temporal development supervene on the entire mosaic of these "local beables." Hence, even if one regards ordinary space-time as the realm of physical reality also in quantum physics, if one adopts Humeanism, there is again no motivation to go for ontic structural realism based on quantum physics.

By way of consequence, if one takes the primitive ontology to be the full ontology, there is no problem with Lorentz-invariance, since the primitive ontology consists entirely in local matters of particular fact. However, as soon as one takes the entanglement of the wave-function in configuration space to refer to dependency relations among some such local matters of particular fact over and above their simple occurrence at space-time points, then Bell's theorem implies that there is no Lorentz-invariant theory of these dependency relations or influences possible: there then is a fact for any given flash-event, particle position or matter density value at a space-time point of whether or not the occurrence of that flash, particle position or matter density value depends on where other flash-events, particle positions or matter density values occur at spacelike separated locations. This conclusion is not called into question by the work of Tumulka (2006 and 2009) who has shown for the GRW flash ontology that the GRW law is Lorentz-invariant and that there is no problem with Lorentz-invariance as long as one considers only probabilities for entire distributions of flashes in space-time;[12] this important result does not touch on the fact that for the occurrence of each new flash, it remains a meaningful question to ask whether or not the occurrence of that flash depends on or is influenced by where at a spacelike separated location other flashes occur, and there is no Lorentz-invariant answer to that question available (see Esfeld and Gisin 2014). Consequently, the metaphysical issue of Humeanism vs. anti-Humeanism about laws of nature and objective modality has direct implications for the issue of whether or not a Lorentz-invariant quantum ontology of matter distributed in space-time is available. For the Humean, it is no problem to obtain such an ontology, whereas for the anti-Humean, such an ontology is not available.

Since the Humean does not grant beable status to the wave-function, there is nothing that determines the temporal development of an initial configuration of "local beables." The particle positions simply happen to develop in such a way that there are, as far as Bohmian quantum mechanics is concerned, continuous particle trajectories; the matter density values just happen to develop in such a way that the matter density takes a certain shape making true the GRW law, and the flash-events just happen to occur in such a manner that they make true a law of the GRW type. There is nothing that drives, guides or forces them to do so. This is simply what the general Humean attitude toward laws and objective modality implies. One may have reservations about that attitude. But there is nothing in quantum physics that obliges one to abandon it. In brief, it is "anti-Humeanism in, anti-Humeanism out," or "Humeanism in, Humeanism out." If one assumes that the wave-function is some sort of a real entity over and above the primitive ontology, then quantum physics comes out anti-Humean. If, by contrast, one bases oneself on the empiricist idea that the primitive ontology is the full ontology, then one obtains a Humean ontology of quantum physics.

[12] Bedingham et al. (2014) seek to achieve a similar result for the GRW matter density field ontology.

This is not to say that physics has no bearing on the metaphysics of modality. One important argument in this respect is that Humeanism cannot explain why the regularities of physics—such as the law of gravitation—always turn out to be well-confirmed (and thus, for instance, why a human being cannot fly into the air without technical aid). On Humeanism, there is no constraint at all on which local matters of particular fact can and which ones cannot occur in the future of any given local matter of particular fact; the laws of nature supervene only on the mosaic of the local matters of particular fact in the whole of space-time. Hence, what the laws of nature are depends on what there will happen in the future of any given local matter of particular fact, instead of that future depending on the laws of nature. Dispositional essentialism, the most prominent contemporary form of anti-Humeanism about laws of nature, by contrast, provides for such an explanation: to take up the example of the law of gravitation, if it is essential for the property of mass to exercise a causal role as described by the law of gravitation (whatever the correct law of gravitation may be), then this is the reason why the regularities of physics concerning gravitation always turn out to be well-confirmed (and thus why it would be futile for a human being to try to fly into the air without technical aid) (see notably Bird 2007). The occurrence of mass then poses a constraint on what there can be in the future of any given such occurrence. By the same token, turning to quantum physics, dispositional essentialism (or modal ontic structural realism in this case) can and Humeanism cannot explain why the outcomes of an EPR type experiment always turn out to be correlated (namely because there is a dispositional property or modal structure of entanglement instantiated by the configuration of objects in physical space).[13] However, an argument of this type is a general argument from physics against Humeanism, and not a refutation of Humeanism by a particular physical theory.

Against this background, let us reconsider classical mechanics and draw some general conclusions. As pointed out in section 1, the commitment to properties of the primitive matter—that is, the particles—arises in Newton's theory, because there is something that plays a causal role in the temporal development of the trajectories of the particles. Thus, in virtue of possessing mass, the particles accelerate each other. Referring to the property of gravitational mass instantiated by the particles provides for a causal explanation of the acceleration of the particles independently of whether or not a medium is indicated by means of which the influence that particles exert on each other's state of motion is transmitted and independently of whether or not time passes between the presence of gravitational mass (the cause) and the acceleration of the particles (the effect).

In this vein, Blondeau and Ghins (2012) argue that the "general form of a causal law is an equation that exhibits the following mathematical form:

$$E = \frac{\partial_x}{\partial_t} = C_1 + \ldots C_n \tag{2}$$

[13] See Dorato and Esfeld (2010), as well as Esfeld et al. (2014).

E refers to the effect, whereas the causes C_i can, but need not, be functions of time. The above general form reads: C_1, C_2, . . . are the causes of the infinitesimal variation of the property x of a system, that is, of the effect E" (p. 384). The decisive point is that any law fitting into this form is asymmetric in that what appears on the right side induces a certain temporal development of the quantity on the left side, but not vice versa, without any time having to pass between the presence of the causes C_1 . . . C_n and the effect E, that is, the manner in which x develops in time. Thus, on Newton's law of gravitation, the presence of gravitational mass induces a change in the velocity of the particles without any time passing between the presence of mass and the acceleration of the particles.

For the Humean, the causal role that the properties referred to by "C_1 + . . . C_n" is a contingent one: it varies from one possible world to another, supervening on the distribution of the local matters of particular fact in the world in question as a whole. Thus, it is contingent that the property we refer to as "mass" exercises the role expressed in Newton's law of gravitation in the actual world (assuming, for the sake of the argument, that Newton's law is the correct law of gravitation for the actual world). The anti-Humean, by contrast, does not regard that causal role as contingent. For the dispositional essentialist, properties are dispositions whose essence it is to exercise a certain causal role, such that whenever the property in question is instantiated in a possible world, it exercises the same role in any world. The law expresses that role.

Hence, as far as classical mechanics is concerned, the Humean and the anti-Humean can both agree that the matter distributed in space-time (i.e., the particles) instantiates certain properties such as mass, or charge. Their dispute concerns the issue of whether or not the causal role that these properties exercise according to the laws of classical mechanics (fulfilling the scheme indicated earlier) is contingent or necessary (essential) to them. However, when it comes to quantum physics, the Humean can no longer recognize such properties: whereas mass and charge can be considered as intrinsic properties of particles "which need nothing bigger than a point at which to be instantiated" (Lewis 1986, p. x), there are no such intrinsic properties as far as the features that are specific for quantum physics are concerned. If one seeks for properties in quantum physics that fill in the scheme provided by formula (2), these can only be relations or structures of entanglement, relating all the objects in physical space (be it particles, a matter density field, or flashes in the sense of single events). Thus, the dispositional essentialist can regard the quantum mechanical wave-function as referring to a dispositional property or modal structure instantiated by the configuration of matter as a whole and determining (in a deterministic or probabilistic manner) the temporal development of the configuration of matter.[14] But the Humean cannot admit relations or structures of entanglement on pain of destroying Humean supervenience.

Consequently, quantum physics does after all have a repercussion for Lewis's Humean ontology: in the light of quantum physics, one can no longer maintain that

---

[14] See Dorato and Esfeld (2010) for dispositionalism about the GRW ontologies. For dispositionalism about Bohmian mechanics, see Belot (2012, pp. 77–80) and Esfeld et al. (2014, sections 4–5).

the mosaic of local matters of particular fact consists in "local qualities: perfectly natural intrinsic properties which need nothing bigger than a point at which to be instantiated" (Lewis 1986, p. x). There are no such qualities or intrinsic properties in quantum physics. Quantum entanglement rules out that such local, intrinsic properties could do any work as far as the features that are specific for quantum physics are concerned. Nonetheless, quantum entanglement notwithstanding, Lewis's thesis of Humean supervenience can be literally true in quantum physics, as the primitive ontology theories show. The only adaptation that is necessary to obtain this result is to consider the mosaic of local matters of particular fact as being constituted by primitive stuff distributed in space-time. That stuff can be particles or flashes, with a particle or a flash being at a space-time point signifying that there is stuff located at the point instead of the point being empty, or that stuff can be gunk in the sense of a continuous matter density field.

The option to maintain that the matter distributed in space-time and constituting the Humean supervenience basis consists in primitive stuff instead of local qualities (intrinsic properties) is not limited to quantum physics, but also available for classical mechanics. Also with respect to classical mechanics, one can subscribe to the view that mass and charge, like the quantum mechanical wave-function, are only variables that appear in the best system, that is, the system that achieves the best balance between being simple and being informative in describing the distribution of local matters of particular fact—such as particle positions—throughout the whole of space-time (see Hall unpublished §5.2). In other words, there are not mass and charge instantiated as intrinsic properties occurring at space-time points over and above particle positions signifying that a space-time point is occupied by stuff instead of being empty, as there is no wave-function instantiated as a relation or structure in space-time over and above the elements of whatever may be the primitive ontology of quantum physics. Adopting this stance removes the stock objections against Humeanism from quidditism and humility: if there were intrinsic properties instantiated at space-time points that exercise a causal role contingently, their essence would be a pure quality (a quiddity) to which we could moreover have no epistemic access (humility; see Lewis 2009).[15] Again, it is evident that the relationship between physics and metaphysics goes in both directions, with the physics here shaping Humean metaphysics in such a way that a central metaphysical objection against Humeanism no longer applies.

---

[15] There also is a sort of humility implied by the primitive ontology theories of quantum mechanics: in order for these theories to make the right empirical predictions and to rule out exploiting quantum non-locality for superluminal signaling, they have to limit the epistemic access that we can have to the elements of the primitive ontology in the sense that we cannot know the exact initial conditions (that is, the exact initial particle configuration in Bohmian mechanics, the exact initial matter density distribution in the GRW matter density ontology, or the exact initial configuration of flashes in the GRW flash ontology). However, this is not an ignorance of the types of properties or entities that there are in the actual world, but only an ignorance of initial conditions, albeit a principled one.

In conclusion, this chapter has sought to show the following:

(a) How physics and metaphysics match in Newton's philosophia naturalis
(b) How even what seems to be a clear-cut case of metaphysical conclusions following directly from the formalism of a physical theory (presentism being ruled out by special relativity) is called into question when one takes the whole of contemporary fundamental physics into account
(c) How specifying what the very referent of the formalism of quantum mechanics is draws on metaphysical considerations
(d) How the stance that one takes in the metaphysics of laws and modality shapes the options that are available for an ontology of quantum physics

In sum, far from separating physics from metaphysics, the physics of the 20th century calls for natural philosophy in the sense of an enterprise that regards physics and metaphysics as forming a seamless whole in the enquiry into the constitution of the world, at least as much as the physics of the 17th and the 18th century did.

References

Albert, David Z. (1996): "Elementary quantum metaphysics." In: J. T. Cushing, A. Fine, and S. Goldstein (eds.): *Bohmian mechanics and quantum theory: An appraisal*. Dordrecht: Kluwer, pp. 277–284.

Albert, David Z. (2000): "Special relativity as an open question." In: H.-P. Breuer and F. Petruccione (eds.): *Relativistic quantum measurement and decoherence*. Berlin: Springer, pp. 1–13.

Albert, David Z. (2013): "Wave function realism." In: D. Albert and A. Ney (eds.): *The wave function: essays on the metaphysics of quantum mechanics*. Oxford: Oxford University Press, pp. 52–57.

Allori, Valia, Goldstein, Sheldon, Tumulka, Roderich, and Zanghì, Nino (2008): "On the common structure of Bohmian mechanics and the Ghirardi-Rimini-Weber theory." *British Journal for the Philosophy of Science* 59, pp. 353–389.

Bedingham, Daniel, Dürr, Detlef, Ghirardi, Gian Carlo, Goldstein, Sheldon, Tumulka, Roderich, and Zanghì, Nino (2014): "Matter density and relativistic models of wave function collapse." *Journal of Statistical Physics* 154, pp. 623–631.

Bell, John S. (2004): *Speakable and unspeakable in quantum mechanics*. Cambridge: Cambridge University Press. Second edition. First edition 1987.

Bell, John S., Shimony, Abner, Horne, Michael A., and Clauser, John F. (1985): "An exchange on local beables." *Dialectica* 39, pp. 85–110.

Belot, Gordon (2012): "Quantum states for primitive ontologists. A case study." *European Journal for Philosophy of Science* 2, pp. 67–83.

Bigelow, John, Ellis, Brian, and Pargetter, Robert (1988): "Forces." *Philosophy of Science* 55, pp. 614–630.

Bird, Alexander (2007): *Nature's metaphysics. Laws and properties*. Oxford: Oxford University Press.

Blondeau, Julien, and Ghins, Michel (2012): "Is there an intrinsic criterion for causal lawlike statements?" *International Studies in the Philosophy of Science* 26, pp. 381–401.

Bohm, David (1951): *Quantum theory*. Englewood Cliffs: Prentice-Hall.

Bohm, David (1952): "A suggested interpretation of the quantum theory in terms of 'hidden' variables." *Physical Review* 85, pp. 166–193.

Callender, Craig (2008): "Finding 'real' time in quantum mechanics." In: W. L. Craig and Q. Smith (eds.): *Einstein, relativity, and absolute simultaneity*. London: Routledge, pp. 50–72.

Callender, Craig (2011): "Philosophy of science and metaphysics." In: S. French and J. Saatsi (eds.): *The continuum companion to the philosophy of science*. London: Continuum, pp. 33–54.

Callender, Craig (unpublished): "Discussion: the redundancy argument against Bohm's theory." Manuscript. http://philosophyfaculty.ucsd.edu/faculty/ccallender/publications.shtml

Callender, Craig (2015): "One world, one beable." *Synthese* 192, pp. 3153–3177.

Chakravartty, Anjan (2013): "On the prospects of naturalized metaphysics." In: D. Ross, J. Ladyman, and H. Kincaid (eds.): *Scientific metaphysics*. Oxford: Oxford University Press, pp. 27–50.

Chang, Hasok, and Cartwright, Nancy (1993): "Causality and realism in the EPR experiment." *Erkenntnis* 38, pp. 169–190.

Colbeck, Roger, and Renner, Renato (2012): "Is a system's wave function in one-to-one correspondence with its elements of reality?" *Physical Review Letters* 108, 150402.

Conway, John H., and Kochen, Simon (2006): "The free will theorem." *Foundations of Physics* 36, pp. 1441–1473.

Conway, John H., and Kochen, Simon (2009): "The strong free will theorem." *Notices of the American Mathematical Society* 56, pp. 226–232.

Darby, George (2012): "Relational holism and Humean supervenience." *British Journal for the Philosophy of Science* 63, pp. 773–788.

Daumer, Martin, Dürr, Detlef, Goldstein, Sheldon, and Zanghì, Nino (1996): "Naive realism about operators." *Erkenntnis* 45, pp. 379–397.

de Broglie, Louis (1928): "La nouvelle dynamique des quanta." In: *Electrons et photons. Rapports et discussions du cinquième Conseil de physique tenu à Bruxelles du 24 au 29 octobre 1927 sous les auspices de l'Institut international de physique Solvay*. Paris: Gauthier-Villars, pp. 105–132. English translation in G. Bacciagaluppi and A. Valentini (2009): *Quantum theory at the crossroads. Reconsidering the 1927 Solvay conference*. Cambridge: Cambridge University Press, pp. 341–371.

de Broglie, Louis (1964): *The current interpretation of wave mechanics. A critical study*. Amsterdam: Elsevier.

Dorato, Mauro and Esfeld, Michael (2010): "GRW as an ontology of dispositions." *Studies in History and Philosophy of Modern Physics* 41, pp. 41–49.

Dürr, Detlef, Goldstein, Sheldon, and Zanghì, Nino (2013a): *Quantum physics without quantum philosophy*. Berlin: Springer.

Dürr, Detlef, Goldstein, Sheldon, Norsen, Travis, Struyve, Ward, and Zanghì, Nino (2013b): "Can Bohmian mechanics be made relativistic?" *Proceedings of the Royal Society* A 470, p. 2162

Einstein, Albert (1905): "Zur Elektrodynamik bewegter Körper." *Annalen der Physik* 17, pp. 891–921.

Einstein, Albert, Podolsky, Boris, and Rosen, Nathan (1935): "Can quantum-mechanical description of physical reality be considered complete?" *Physical Review* 47, pp. 777–780.

Esfeld, Michael (2013): "Ontic structural realism and the interpretation of quantum mechanics." *European Journal for Philosophy of Science* 3, pp. 19–32.

Esfeld, Michael (2014): "Quantum Humeanism, or: physicalism without properties." *Philosophical Quarterly* 64, pp. 453–470.

Esfeld, Michael, and Gisin, Nicolas (2014): "The GRW flash theory: a relativistic quantum ontology of matter in space-time?" *Philosophy of Science* 81, pp. 248–264.

Esfeld, Michael, Lazarovici, Dustin, Hubert, Mario, and Dürr, Detlef (2014): "The ontology of Bohmian mechanics." *British Journal for the Philosophy of Science* 65, pp. 773–796.

Fine, Kit (2005): "Tense and reality." In: K. Fine (ed.): *Modality and tense: philosophical papers*. Oxford: Oxford University Press, pp. 261–320.

Ghirardi, Gian Carlo, Grassi, Renata, and Benatti, Fabio (1995): "Describing the macroscopic world: Closing the circle within the dynamical reduction program." *Foundations of Physics* 25, pp. 5–38.

Ghirardi, Gian Carlo, Rimini, Alberto, and Weber, Tullio (1986): "Unified dynamics for microscopic and macroscopic systems." *Physical Review D* 34, pp. 470–491.

Goldstein, Sheldon, Tausk, Daniel V., Tumulka, Roderich, and Zanghì, Nino (2010): "What does the free will theorem actually prove?" *Notices of the American Mathematical Society* 57 (11), pp. 1451–1453.

Hall, Ned (unpublished): "Humean reductionism about laws of nature." http://philpapers.org/rec/HALHRA

Harrington, James (2008): "Special relativity and the future: a defense of the point present." *Studies in History and Philosophy of Modern Physics* 39, pp. 82–101.

Hofer-Szabó, Gábor, and Vecsernyés, Péter (2013): "Bell inequality and common causal explanation in algebraic quantum field theory." *Studies in History and Philosophy of Modern Physics* 44, pp. 404–416.

Humphreys, Paul (2013): "Scientific ontology and speculative ontology." In: D. Ross, J. Ladyman, and H. Kincaid (eds.): *Scientific metaphysics*. Oxford: Oxford University Press, pp. 51–78.

Jackson, Frank (1998): *From metaphysics to ethics. A defence of conceptual analysis*. Oxford: Oxford University Press.

Jammer, Max (1957): *Concepts of force: a study in the foundations of dynamics*. Cambridge (Massachusetts): Harvard University Press.

Ladyman, James, and Bigaj, Tomasz F. (2010): "The principle of the identity of indiscernibles and quantum mechanics." *Philosophy of Science* 77, pp. 117–136.

Ladyman, James, and Ross, Don (2007): *Every thing must go. Metaphysics naturalized*. Oxford: Oxford University Press.

Lazarovici, Dustin (2014): "Lost in translation. A comment on 'Noncommutative causality in algebraic quantum field theory.'" In: M. C. Galavotti et al. (eds.): *New directions in the philosophy of science. The philosophy of science in a European perspective*. Volume 5. Cham: Springer, pp. 555–560.

Lewis, David (1986): *Philosophical papers*. Volume 2. Oxford: Oxford University Press.

Lewis, David (2009): "Ramseyan humility." In: D. Braddon-Mitchell and R. Nola (eds.): *Conceptual analysis and philosophical naturalism*. Cambridge, MA: MIT Press, pp. 203–222.

Loewer, Barry (1996): "Humean supervenience." *Philosophical Topics* 24, pp. 101–127.

Lyre, Holger (2010): "Humean perspectives on structural realism." In: F. Stadler (ed.): *The present situation in the philosophy of science*. Dordrecht: Springer, pp. 381–397.

Massin, Olivier (2009): "The metaphysics of forces." *Dialectica* 63, pp. 555–589.
Maudlin, Tim (2007): *The metaphysics within physics*. Oxford: Oxford University Press.
Maudlin, Tim (2011): *Quantum non-locality and relativity*. Third edition. Chichester: Wiley-Blackwell.
Maudlin, Tim (2012): *Philosophy of physics*. Volume 1. *The arena: Space and time*. Princeton, NJ: Princeton University Press.
Miller, Elizabeth (2014): "Quantum entanglement, Bohmian mechanics, and Humean supervenience." *Australasian Journal of Philosophy* 92, pp. 567–583.
Monton, Bradley (2004): "The problem of ontology for spontaneous collapse theories." *Studies in History and Philosophy of Modern Physics* 35, pp. 407–421.
Monton, Bradley (2006): "Presentism and quantum gravity." In: D. Dieks (ed.): *The ontology of spacetime*. Amsterdam: Elsevier, pp. 263–280.
Monton, Bradley (2011): "Prolegomena to any future physics-based metaphysics." In: J. Kvanvig (ed.): *Oxford Studies in philosophy of religion*. Volume III. Oxford: Oxford University Press, pp. 142–165.
Newton, Isaac (1952): *Opticks or a treatise of the reflections, refractions, inflections and colours of light*. Edited by I. B. Cohen. New York: Dover.
Ney, Alyssa (2011): "The status of our ordinary three dimensions in a quantum universe." *Noûs* 44, pp. 1–36.
Ney, Alyssa (2012): "Neo-positivist metaphysics." *Philosophical Studies* 160, pp. 53–78.
Norsen, Travis (2005): "Einstein's boxes." *American Journal of Physics* 73, pp. 164–176.
Norsen, Travis (2009): "Local causality and completeness: Bell vs. Jarrett." *Foundations of Physics* 39, pp. 273–294.
North, Jill (2013): "The structure of a quantum world." In: D. Albert and A. Ney (eds.): *The wave function: essays on the metaphysics of quantum mechanics*. Oxford: Oxford University Press, pp. 184–202.
Price, Huw (1996): *Time's arrow and Archimedes' point. New directions for the physics of time*. Oxford: Oxford University Press.
Pusey, Matthew F., Barrett, Jonathan, and Rudolph, Terry (2012): "On the reality of the quantum state." *Nature Physics* 8, pp. 475–478.
Ross, Don, Ladyman, James, and Kincaid, Harold (eds.) (2013): *Scientific metaphysics*. Oxford: Oxford University Press.
Saunders, Simon (2006): "Are quantum particles objects?" *Analysis* 66, pp. 52–63.
Seevinck, Michiel P. (2010): "Can quantum theory and special relativity peacefully coexist?" Invited white paper for Quantum Physics and the Nature of Reality, John Polkinghorne 80th Birthday Conference. St Anne's College, Oxford, September 26–29, 2010. http://arxiv.org/abs/1010.3714.
Seevinck, Michiel P., and Uffink, Jos (2011): "Not throwing out the baby with the bathwater: Bell's condition of local causality mathematically 'sharp and clean.'" In: D. Dieks, W. Gonzalez, S. Hartmann, T. Uebel, and M. Weber (eds.): *Explanation, prediction and confirmation*. New trends and old ones reconsidered. Dordrecht: Springer, pp. 425–450.
Sklar, Lawrence (1974): *Space, time and spacetime*. Berkeley: University of California Press.
Stein, Howard (1968): "On Einstein-Minkowski space-time." *Journal of Philosophy* 65, pp. 5–23.
Teller, Paul (1986): "Relational holism and quantum mechanics." *British Journal for the Philosophy of Science* 37, pp. 71–81.
Tumulka, Roderich (2006): "A relativistic version of the Ghirardi-Rimini-Weber model." *Journal of Statistical Physics* 125, pp. 825–844.

Tumulka, Roderich (2009): "The point processes of the GRW theory of wave function collapse." *Reviews in Mathematical Physics* 21, pp. 155–227.

Wallace, David (2012): *The emergent multiverse. Quantum theory according to the Everett interpretation*. Oxford: Oxford University Press.

Wilson, Jessica (2007): "Newtonian forces." *British Journal for the Philosophy of Science* 58, pp. 173–205.

Wüthrich, Christian (2011): "Can the world be shown to be indeterministic after all?" In: C. Beisbart and S. Hartmann (eds.): *Probabilities in physics*. Oxford: Oxford University Press, pp. 365–390.

Wüthrich, Christian (2013): "The fate of presentism in modern physics." In: R. Ciunti, K. Miller, and G. Torrengo (eds.): *New papers on the present—focus on presentism*. München: Philosophia, pp. 91–131.

5

THEORIES AND MODELS

Marion Vorms (Birbeck College, London)

THE NEWTONIAN THEORY of gravitation, the Darwinian theory of evolution, Einstein's theory of general relativity, these are all cases of remarkable achievements in scientific inquiry. There is no doubt that one of the major tasks of scientists is to produce theories, conceived of as explanatory systems for external world phenomena. As a consequence, the study of the nature and development of scientific knowledge requires a description and analysis of the scientific theories that are the repositories of such knowledge. It is far from clear though how one should define, and study, theories. In fact, throughout the 20th century philosophers of science have dealt with this question without reaching any consensus. Some of them would even deny that the notion is of any relevance for the analysis of scientific knowledge and would suggest that one adopt other units of analysis, such as models or paradigms.[1]

This chapter is dedicated to a critical examination of the notion of theory as a unit of analysis for the study of scientific knowledge. The main issues at stake concern the content of theories—namely what they tell us about the world—rather than their justification (see chapter 2 on confirmation). Although the notion of theory is also used in mathematics—and its use in the empirical sciences derives in some way from its use in mathematics—this chapter will concentrate on the study of theories

[1] The author wishes to thank Anouk Barberousse and Jean Gayon for having inspired, supported, and supervised the doctoral research from which this chapter is one product, as well as Mikaël Cozic and Paul Humphreys for helpful comments and suggestions, and Henri Galinon for useful discussions on Carnap's view of theoretical terms. This project has received funding from the European Union's Horizon 2020 research and innovation programme under the Marie Skłodowska-Curie grant agreement No. 660187.

in the empirical sciences, namely theories dealing with phenomena in the external world. Indeed, one of the main problems arising from the study of theories is how their concepts and hypotheses relate to the observable phenomena.

Most of the following analysis will consist in presenting and criticizing two major proposals made by philosophers of science from the analytic tradition for defining theories. The first, known as the "received view," is commonly attributed to the logical empiricists, although there is no one unique conception of theories uncontroversially held by all of them. The second, called the "semantic view of theories," is often presented as a criticism of the former (which was called "syntactic" by contrast) and overtook it as the new orthodoxy in the 1960s. Beyond their differences, these two approaches have in common to aim at *formal reconstructions of theories*. One goal of the present chapter is to clarify what, precisely, that means. By presenting the main tenets of the two aforementioned formal approaches to theories, it aims at highlighting their internal limits by questioning both their shared assumptions and the relevance of the very notion of theory underlying the project of formal reconstruction. By contrast, alternative ways of construing scientific theorizing—if not scientific theories—will be sketched out. It is important to keep in mind that the analyses of the present chapter are themselves underlain by a certain view of theorizing which, in turn, drives the interpretation of the theses that are presented and criticized. This view will be made explicit in the course of the chapter. In a nutshell, it consists in an agent-centered approach to theorizing, according to which the study of the very content of theories must account for the way they are used and understood in practice.

Section 1 introduces the main issues arising in the study of theory content. Leaning on the case of classical mechanics, which serves as a recurrent example throughout this chapter, it clarifies one shared assumption of the formal approaches, namely that the content of a theory is independent from the way it is formulated, understood, and used by scientists. Section 2 presents the received view, first as it was expressed by Rudolf Carnap (1956, 1966), and then as Ernest Nagel (1961) cast it. It is argued that some tensions and internal contradictions in Nagel's views are symptomatic of the ineluctable failure of the logical empiricist view of theories. Section 3 sketches out the main tenets of the semantic view of theories, mostly relying on Patrick Suppes's (1957, 1960, 1962, 1967) works. Although the semantic view seems to overcome some of the obstacles encountered by the logical empiricists, and also to better achieve what could be presented as their common goal, it is argued that the very project of formal theory reconstruction is doomed to partial failure. Section 4 briefly presents some recent developments and proposals from critics of the semantic view, whose intention is to better capture scientific theorizing.

1. What Is the Content of a Scientific Theory?

One of the central tasks of philosophy of science is to clarify the content of existing scientific theories. What does this mean, and why is it worth the effort? After some introductory remarks on theories, the example of classical mechanics will be presented as a means to highlight the specific issues that arise when trying to define what the content of a theory is. Then, the common goal of formal approaches to theories will be presented as one particular way of implementing such a clarification project. The section will conclude with the presentation of some formal tools necessary for understanding the rest of the chapter.

1.1 PRELIMINARY REMARKS ON THEORIES

1.1.1 Theories as Representational and Inferential Tools

The term "theory" commonly refers to the form under which scientists express the knowledge resulting from their observations and experimentations in a given domain of phenomena. It is rather uncontroversial that the major function of theories is to allow for the prediction and explanation of the empirical phenomena. In order to do so, theories express general hypotheses about the phenomena they describe—in contrast to merely observational reports. For instance, Newtonian mechanics states that force is the product of mass by acceleration ($\boldsymbol{F} = m\boldsymbol{a}$); Mendelian genetics says that, during the formation of germ cells, pairs of genes segregate independently; Ricardo's labor theory of value says that the exchange value of a good is determined by the labor needed to produce it. These theoretical hypotheses (often called "laws" or "principles") offer more than a mere description of the phenomena—however rich and informative such description may be. They establish relations between concepts, such as force, mass, or gene, which do not always refer to observable things or processes. This is where the predictive and explanatory power of theories lies: by using these concepts to represent a given state of affairs (by relating them to the observable phenomena) one can predict what will happen or explain what has happened (or what happens in general) by virtue of the relationship the theory establishes between these concepts.

Such predictions and explanations result from the inferences the theoretical hypotheses enable one to draw. For instance, in order to predict or explain the behavior of a given physical system, one would represent it by means of the concepts of mass and force, thus obtaining a set of equations (drawn from the hypothesis that $\boldsymbol{F} = m\boldsymbol{a}$) that would enable calculations leading to the desired prediction or explanation. Similarly, the Mendelian theory of heredity, by introducing the concept of gene and describing the behavior of hypothetical entities this concept refers to (by means of Mendel's probabilistic laws), enables the prediction and explanation of the distribution of observable characteristics in individuals of successive generations.

Hence, it is by representing the phenomena in a certain way—by means of hypotheses using certain concepts—that a theory allows for predictions and explanations to be made. A theory is both a tool of *representation* and a tool of *inference* or computation. This would probably not be contested by any philosopher and can be taken as a basic assumption for any study of theories. But the rest of the chapter will show that, however uncontroversial, this basic characterization is not unproblematically compatible with other assumptions commonly made by philosophers of science, particularly in the logical empiricist tradition, which was center stage for most of the 20th century.

1.2 CLARIFYING THE CONTENT OF THEORIES

The most mature scientific theories (particularly in the physical sciences) often take the form of a deductive system of hypotheses which are supposed to encapsulate all that theories have to say about the world—their *content*. Most often, this content is expressed by means of statements in natural language, extended by scientific concepts (which sometimes, like with the concept of force, belong to ordinary language, but adopt a meaning differing from their ordinary one). Moreover, expressing the content of a theory often requires the use of specific formalisms, such as differential equations in physics.

Because theoretical terms such as "mass" or "force" do not have a straightforward empirical meaning (unlike ordinary language terms such as "stone" or "table"), the content of theories featuring such terms calls for thorough analysis. The function of theories is to help us understand the empirical world by predicting and explaining the observable phenomena. Hence, one needs to ensure that they do in fact have empirical meaning, by contrast with metaphysical speculations. For instance, in order to warrant the scientificity of mechanics, one needs to be sure that the concept of force has stronger explanatory power than opium's *virtu dormitiva*. In order to do so, one needs to clarify the empirical meaning of this concept—its link to the empirical phenomena. Similarly, as a way to understand what Mendelian genetics tells us about heredity, one has to ensure that the concept of gene is not meaningless and that it allows for the formulation of fruitful hypotheses about the transmission of characteristics from generation to generation. Clarifying the content of a theory by inquiring into its empirical meaning may also help in determining its limits and shedding light on its relations with other theories. For instance, clarifying the concept of gene is a way to analyze the relation between classical and molecular genetics.

Theoretical concepts, particularly in the case of systematic theories like classical mechanics, are often (partially) defined by their relations with other theoretical concepts. In order to grasp their meaning, one also needs to clarify the logical structure of the theory—the deductive links between its various hypotheses and the way these hypotheses relate to the phenomena. This is the reason why, since the end of the 19th century, philosophers of science—who in many cases were also scientists[2]—have assigned themselves the task of clarifying the logical structure and the empirical (or physical) meaning of the existing theories. This conceptual clarification remains one of the unanimously acknowledged specific tasks of philosophy of science:

[2] See (Hertz, 1894; Mach, 1883; Poincaré, 1902, 1905; Boltzmann, 1897; von Helmholtz, 1847; Kirchhoff, 1877). Their reflections took place in the context of the "crisis of physics" (Poincaré, 1905) and of the explanatory model of classical mechanics that was to give birth to relativity theories and quantum physics.

> The role of philosophy of science is to clarify conceptual problems and to make explicit the foundational assumptions of each scientific discipline. The clarification of conceptual problems or the building of an explicit logical foundation are tasks that are neither intensely empirical nor mathematical in character. They may be regarded as proper philosophical tasks directly relevant to science. (Suppes, 1968, p. 653)

The formal approaches to theories, to which this chapter is dedicated, embody a particular way—which itself was diversely implemented—of handling this clarification project. Before saying more about it, let us now have a look at the case of classical mechanics, as a way to raise certain issues and better clarify the goal of formal approaches.

1.2 WHAT IS THE CONTENT OF CLASSICAL MECHANICS?

Classical mechanics is one of the most famous examples of a scientific theory and has long been the gold standard of what a theory ought to be. A first way to identify its content is to define the domain of phenomena to which it applies and within which it is valid: classical mechanics deals with the motion of macroscopic bodies, for small velocities (in comparison with the speed of light). Yet such a definition is not sufficient for identifying the content of a theory. There can be various contradictory or incompatible theories with the same intended domain. One also wants to know *what the theory says* of these phenomena, that is, the hypotheses by means of which it enables us to predict and explain these phenomena. In the case of classical mechanics, these hypotheses seem to be well identified: they consist of Newton's three laws, and particularly the famous fundamental principle of mechanics (Newton's second law, $\boldsymbol{F} = m\boldsymbol{a}$), from which the equations of motion can be drawn. The content of classical mechanics could thus be defined as the set of all the deductive consequences of Newton's principles. The systematic presentation of these principles seems to raise no particular problem, as Newton had already exposed them in axiomatic form in his 1687 *Principia*.

However, there exist different formulations of classical mechanics, among which the Newtonian, or vectorial one, and the analytical, or variational ones (which themselves divide into Lagrangian formulation, Hamiltonian formulation, and Hamilton-Jacobi theory).[3] These formulations do not merely correspond to successive historical presentations of the Newtonian principles, each of which would be more powerful and complete than the preceding one (the latter thus falling into disuse). In fact, all the aforementioned formulations are still taught and used today in various scientific

[3] These names correspond to the scientists who introduced crucial mathematical novelties into the theory. However, today's Lagrangian formulation, for instance, does not correspond to Lagrange's (1788) historical presentation. Neither does the Newtonian formulation correspond to Newton's (1687) *Principia*. These issues are not central to the core analysis of the chapter but were studied in some depth in Vorms (2009).

TABLE 1.

The Formulations of Classical Mechanics: Main Differences

Newtonian (Vectorial) Formulation	Lagrangian (Analytical, Variational) Formulation
Cartesian coordinates $\mathrm{F} = m\mathrm{a} \qquad \mathrm{a} = \frac{d^2\mathrm{r}}{dt}$	Generalized coordinates (qi) $L = T - V \qquad \frac{d}{dt}\left(\frac{\partial L}{\partial \dot{q}i}\right) - \frac{\partial L}{\partial qi} = 0$
Concept of force Newton's second law	Concept of energy Hamilton principle (least action)
Differential calculus	Variational calculus

contexts. As a way to understand why this is so, and how much this matters for an analysis of the content of classical mechanics, let us have a quick look at the main differences between the Newtonian and the Lagrangian formulations, which are schematically summarized in Table 1.[4]

First of all, the two formulations do not represent the configuration of physical bodies in the same coordinate systems. Newtonian formulation uses Cartesian (and sometimes polar) coordinates, and represents physical bodies as discrete systems of particles. Analytical formulations use generalized coordinates (q_i), which take into account the constraints of the system.[5] Representing the state of a system in the Newtonian formulation consists in identifying the positions and velocities of its various particles, along with the forces being applied on them, at which point one can write down the equations of motion using Newton's second law.[6] In the Lagrangian framework, it is unnecessary to write down equations for all applied forces, rather it is a matter of representing the initial conditions of the system in generalized coordinates by taking the degrees of freedom of the system into account, thus enabling the Lagrangian equations of motion to be written down. Whereas the Newtonian equations feature force as a core concept, the Lagrangian equations are expressed in terms of energy, which is a scalar quantity (rather than a vectorial quantity, such as force is).[7]

[4] For extended presentations of the different formulations of classical mechanics, see Lanczos (1970) or Goldstein (1950/2002).

[5] A constrained system is a system whose different components do not move freely, independently from each other, or independently from any external constraints (e.g., a ball rolling on a given surface, or two masses kept at a constant distance from each other). In many textbooks, one motivation for introducing the analytical formulations of classical mechanics is that the Newtonian formulation cannot practically handle some typical constrained systems, such as a pendulum consisting of two masses connected by a rigid stick of fixed length.

[6] Velocity is the first derivative of the position function **r**, of which acceleration **a** is the second derivative. Bold characters stand for vectorial quantities.

[7] *L*, the Lagrangian of the system, is defined as the difference between the kinetic and the potential energy (*T* and *V*) of the system.

Moreover, the fundamental principle of the analytical formulations is not a vectorial, differential equation, but rather a variational principle (Hamilton's principle of least action). In brief, the formulations differ by their fundamental principles, their core concepts, and the mathematical form of their equations.

These differences have important practical consequences, as applying one or the other formulation does not amount to the same operation. In order to predict and explain the behavior of a given system, the inferential procedures by means of which one can access the consequences of the principles are not the same. Moreover, the different formulations are not practically applicable to the same problems: the Newtonian formulation cannot be applied to some constrained systems, whereas, on the other hand, the Lagrangian formulation cannot be used to solve problems involving friction. Considered through the conceptual architecture of classical mechanics and its relations with other physical theories, there are many consequences of the Hamiltonian formulation that cannot be drawn from the Newtonian one—unless the problem in question can be reformulated in Hamiltonian terms. Consequently, if one knows only Newtonian mechanics and has no notion of Hamiltonian formalism, then one cannot, in practice, access some of the consequences that are, in principle, deducible from the Newtonian principles, when subjected to a Hamiltonian reformulation.[8] Finally, it can also be argued *prima facie* that the formulations also differ greatly from the point of view of the understanding they offer us: the Newtonian formulation represents motion as an instantaneous, local phenomenon, stating a relation of proportionality between force and acceleration, whereas the Lagrangian formulation expresses a condition (according to which a given integral, called A, has to be stationary) regarding the motion taken globally. The former speaks of forces, whereas the latter is about energy. In some sense—to be analyzed further—they do not seem to tell us the same thing about the phenomena.

Despite all these differences, the formulations of classical mechanics are almost uncontroversially considered as expressions of one and the same theory.[9] By contrast with particular theories whose intended domains overlap despite their not being entirely identical, such as Newtonian mechanics and Einstein's general relativity, contradictory predictions cannot be drawn from the different formulations of classical mechanics. Indeed, their principles are inter-deducible. As a result, the set of all their deductive consequences—their deductive closure—is necessarily one and the same set. The logical equivalence of the principles guarantees the equivalence of the whole set of empirical consequences even though, in practice, some consequences cannot be reached by way of such and such formulation.

---

[8] This is where consideration of the problem from a diachronic point of view—by studying the successive mathematical reformulations and conceptual reorganizations of classical mechanics—is enlightening: Were the consequences of Hamiltonian mechanics already there in the Newtonian theory, even before variational principles were introduced?

[9] "Uncontroversally" except by some philosophers who have thoroughly considered the issue and taken classical mechanics as problematic rather than as the paradigmatic, ready-made example of a clearly identified scientific theory (see e.g., North, 2009, Balzer et al., 1987).

The importance of the practical differences highlighted earlier may be diversely evaluated. In a related manner, whether the practical application of theories should feature in an account of the content of theories is itself a debatable question. However, it is worth emphasizing that, whatever the case may be, it might not be as easy as first seems to isolate the objective content of a theory (even one as canonical as classical mechanics) under the form of a small number of principles together with their deductive consequences. This is precisely what formal approaches to theories aim to do (at least in the logical empiricist tradition). Formal reconstruction is an attempt at extracting and presenting the objective content of theories, beyond differences in their formulation.

1.3 THE MOTIVATION FOR FORMAL RECONSTRUCTION

One of the leading ideas of formal approaches to theories, in both the syntactic and the semantic traditions, is that some aspects of the actual formulations of theories are inessential to their content. According to this view, what a theory says depends neither on how it says it—the language in which it is expressed—nor on how agents understand and reason with it. Consider classical mechanics again. True, representing the phenomena by means of the concept of force does not provide us with the same understanding and does not allow us to form the same mental images as representing them in terms of energy. But according to formal approaches, this should be treated as a merely psychological effect of the language chosen to express the principles of mechanics. Concluding that the two formulations do not have the same content would amount to ignoring the distinction between metaphysical and proper scientific explanations.[10] The mathematical formalisms, as well as the terms, used to express the principles of mechanics are merely the concrete "wrapping" of the objective theoretical content. The task of the philosopher is to see beyond the psychological, inessential effects of the formulations and capture this hard theoretical core.

Although the legitimacy of the distinction between a difference in formulation and a difference in content might appear as rather uncontroversial to anyone seeking for objectivity in science (nobody would claim that the content of a theory changes when it is translated from English into French, for example), the formulations of classical mechanics do provide us with an example highlighting the fact that this distinction is not always so easy to draw. Clarifying this distinction is precisely one of the goal of formal reconstruction. Formal approaches aim at making explicit the content of theories by means of formal tools, borrowed from logic and mathematics. When

[10] The distinction between the goals of physics and of metaphysics, as formulated by Pierre Duhem (1914), also underlies philosophy of science in the analytic tradition. Duhem considered that the goal of physical theory is merely to represent empirical laws, explanation being strictly the prerogative of metaphysics. The logical empiricists would reintroduce the notion of explanation, taking care to deprive it of any metaphysical dimension, through Hempel's famous "deductive-nomological" model of explanation (see Hempel and Oppenheim, 1948; Hempel 1965a, and chapter 1 in this volume).

various formulations of a theory exist, formal reconstruction should thus be able to show that their content is identical—or, failing this, to conclude that they do not express the exact same theory. Besides this shared assumption—whose limits the present chapter aims at highlighting—, the two approaches studied here diverge on many aspects, one of which is the formal tools chosen to implement the reconstruction project. In order to understand this, as well as the rest of the chapter, a few technical definitions are necessary.

1.4 FORMALIZATION AND AXIOMATIZATION

As will be seen, formal reconstruction of scientific theories is often referred to as "formalization," or "axiomatization." Understanding the meaning of these two terms is a necessary prerequisite to any analysis of the formal approaches to theories and the differences between them.

1.4.1 The Formal Toolbox

A *formal language* is a language whose vocabulary and rules of construction and transformation (of expressions) are explicitly defined. More precisely, a formal language consists of

- A syntax, itself consisting of
  - A vocabulary (a set of symbols)
  - Rules of construction defining well-formed expressions
  - Rules of transformation indicating the authorized inferences
  - A semantics consisting of the set of interpretation rules of the well-formed expressions, which allows for the formation of meaningful statements[11]

The **formalization** of a proposition or of an argument, in the strictest sense of the notion, consists in presenting it under a form, which distinguishes its syntax from its semantics. Consider the following argument: "If Mary comes, Peter comes. Mary comes. Hence Peter comes." By means of the formal implication symbol "→" and the logical consequence symbol "⊢," the argument can be formalized as follows:

$$Pa \rightarrow Pb, Pa \vdash Pb$$

As such, it is purely syntactic. Knowledge of the manipulation rules of the symbol "→" is enough to establish that this has the form of a valid argument. But we need the semantics to interpret the statements that make up the argument; to know that the predicate symbol Px means "x comes" and that the constants a and b refer to Mary and

[11] Sometimes "formal language" refers only to the syntax, semantics being the interpretation of the formal language.

Peter respectively. An interpretation for a formal language fixes a domain of objects and assigns, within this domain, referents to the non-logical terms of its vocabulary and a truth-value to the statements made thereof.

A **model** for a statement is an interpretation in which this statement is true. It is commonly said that such an interpretation "satisfies" this statement. A statement is true, by definition, of its models; it is true of the world (true *simpliciter*) if the world is a model of it—if it is realized in the world. The model of a statement, or of a set of statements, can be a set of abstract elements such as a mathematical structure, but it can also be a set of concrete objects satisfying the statement. For instance, the uninterpreted (purely syntactic) statement "*x*"*y* (*Rxy* ⟷ *Ryx*) can be given as a model the set of all male human beings, the predicate *Rxy* meaning "*x* is *y*'s brother" (the statement thus says that for any man, if he is the brother of another man, then the latter is also the brother of the former).

A **formal theory** (or **calculus**) is a set of statements expressed in a formal language. Most often, one uses this expression to refer to the pure syntactic skeleton, namely the uninterpreted axioms. Here is a simple geometric example, borrowed from Van Fraassen (1980, p. 42). Consider theory *T* consisting of the following five statements (axioms):

A1 For any two lines, there is at most one point that lies on both.
A2 For any two points, there is exactly one line that lies on both.
A3 On every line there lie at least two points.
A4 There are only finitely many points.
A5 On any line there lie infinitely many points.

Let us ignore the usual meaning of "point" and "line" and consider them as uninterpreted symbols. As such, one can claim that they are not totally devoid of meaning, since the relations assigned to them by the theory already provide, so to

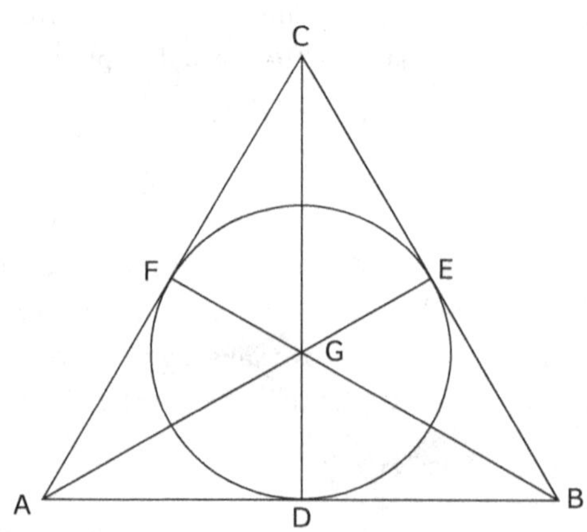

FIGURE 1 The Seven Point Geometry (from Van Fraassen, 1980, p. 42)

speak, an implicit definition. But as of yet they have no referents. These relations only constrain their interpretation (the referents one could assign to them). Even before interpreting *T* by assigning referents to its terms, one can identify logical relations between these statements, in virtue of which, for instance, one can claim that theory *T* is consistent (that there is no logical inconsistency among its axioms). Then one can interpret it, that is, assign a reference domain to it, within which it can be either true or false. Figure 1 represents a geometrical structure (called the "Seven Points space") that satisfies *T*, and hence is a model of *T*. In this figure, seven elements are called "points" (A, B, C, D, E, F, and G), thus satisfying A5. It can be easily demonstrated that the other statements are likewise satisfied.

One calls **axiomatization** the presentation of a theory under the form of a deductive system organized so that its whole content is comprised in a determinate set of statements that are the axioms of the theory. The logical consequences of the axioms are the theorems of the theory. The whole set of the deductive consequences of the theory is called its "deductive closure."

Strictly speaking, *axiomatization and formalization are mutually independent*. One can present a theory under an axiomatic, hypothetico-deductive form without distinguishing its syntax from its semantics.[12] And one can formalize statements without presenting them in a systematic way. In the case of an axiomatization in formal language, the formal theory, namely the syntactic, uninterpreted structure of the theory (in the earlier example, the *Ai*, before they are assigned a domain of discourse) is also called "formal system" or "formal calculus."

A theory can be axiomatized in different languages. Languages differ from each other in their logical and non-logical vocabularies. A formalization *stricto sensu* is executed in a language whose non-logical vocabulary is entirely uninterpreted; it is strictly syntactic. The axiomatizations of mathematical theories are often couched in a language that supposes that some mathematical notions are already defined, such as the notions of set theory, matrix algebra, or other branches of mathematics. In such cases, the axiomatization comes without *stricto sensu* formalization, since the syntax is not strictly speaking uninterpreted (as it has a prior mathematical interpretation).

1.4.2 Formalization, Axiomatization, and the Formal Reconstruction of Theories

The formal approaches to scientific theories, to which this chapter is mostly dedicated, are inspired from the axiomatic method in formal logic and in mathematics; this they extend to formal reconstructions of *empirical* theories. Since the latter, by contrast with mathematical theories, are supposed to have referents in the empirical world, one particular problem that arises is how to relate the logical or mathematical structure to empirical phenomena.

---

[12] The first axiomatizations, such as Euclid's, were not formalized, strictly speaking. But the modern axiomatizations in logic and mathematics imply a prior formalization. Hilbert's (1899) axiomatization of geometry, by contrast with Euclid's, does not presuppose any knowledge of the meaning of "point" or "line." Rather, it gives an implicit definition of these terms by describing their relations.

One of the major differences between the two formal approaches examined here lies in how they tackle this problem. For logical empiricists, the empirical interpretation of theories is provided by correspondence rules, which directly relate the uninterpreted axioms of the calculus to statements describing the empirical phenomena. From this perspective, the whole semantic content of theories *is*—reduces to—their empirical content. Advocates of the semantic view, on the other hand, dissociate the semantic interpretation of the theory from its empirical application. This reveals in the difference between the formal tools they choose for axiomatizing theories: formalization[13], *stricto sensu*, for the former; set theory and other mathematical languages (but especially set theory) for the latter. If the precise definitions given for "formalization" and "axiomatization" were to be respected, then one should speak of formalization only for the syntactic conception, and of axiomatization for both.[14] In this chapter, we will stick to this rule, keeping "formalization" to refer to *stricto sensu* formalization; "axiomatization," and more generally "formal reconstruction," will be used to refer both to the logical empiricists' strictly formalized and to the semantic conception's axiomatic methods. However, it is worth noting that "formalization" and "axiomatization" are often used in an interchangeable way by proponents of formal reconstructions of theories.[15] What matters is that the type of axiomatization/formalization in play be specified at all times (*stricto sensu* or admitting of mathematical objects). The distinction between the two is at the core of the comparison this chapter proposes of the two aforementioned approaches.

2. The Logical Empiricist View of Theories, and Its Limits

This section is dedicated to the conception of theories that was put forward by the logical empiricists, members and heirs of the Vienna Circle, particularly Rudolf Carnap, Carl Hempel, and Ernest Nagel. Although this conception is generally considered as having occupied the role of orthodoxy until the 1960s—Hilary Putnam (1962) named it "the received view"—there exists no one standard version of it. In the last decades of the 20th century, advocates of the semantic view reconstructed it as a body of doctrine they labeled the "syntactic view" (see in particular Suppe 1977b), but it cannot be, as

[13] Advocates of the semantic theory often attribute to logical empiricists, and particularly to Carnap, the project of formalizing theories in first-order logic. First-order languages admit in their logical vocabulary only first-order existential and universal quantifiers (which means that they admit only one type of variable, namely individual variables), as well as the usual verifunctional connectors. It seems that this attribution is historically inaccurate. See note 41.

[14] For an analysis of the distinction between formalization, axiomatization, symbolization, and modelization, see the first 18 pages of Mongin (2003) on axiomatization in economics.

[15] In fact, Suppe (1977b, p. 113) uses them in the opposite way: "Axiomatization consists in the establishment of an axiomatic calculus, and thus consists in an essentially syntactical formalization. Formalization encompasses both the syntactical techniques of axiomatization and the semantic techniques of model theory." Similarly, Suppes (1968) speaks of "formalization" in the general sense of formal reconstruction.

such, attributed to any philosopher in particular. Most of its main tenets were hotly debated and controversial among the logical empiricists themselves, who, moreover, had their own continuously evolving conceptions. Nevertheless, their claims, as well as their debates, can be viewed as arising from the same primary inspiration, which consists in reducing the meaning of any statement or set of statements to its empirical content, accepting as nonempirical elements only logical rules. From this derives a normative criterion of scientificity for theories: a scientific theory should in principle be formalizable in a way that would isolate the logical skeleton of its principles (its syntax), and make explicit the links between such a skeleton and the empirical world (its semantics). What might be lost in the formalization process is not to be considered as part of the content of the theory.

This chapter will first deal with Carnap's late views (1956, 1966), which are the most systematic expression of the logical empiricist picture of theories, as well as the most developed effort which stands by the original, founding principles of logical empiricism. Then Nagel's (1961) version will be presented, with an emphasis placed on its internal tensions. Under the interpretation proposed in this chapter, Nagel's views embody the very limits of the logical empiricist formalist program.

2.1. SCIENTIFIC THEORIES ACCORDING TO RUDOLF CARNAP

The formalist program in the philosophy of science originates in the works of the Vienna Circle's members, who aimed at denouncing the pretensions of metaphysics by showing that its discourses were deprived of any meaning. Their objective was to implement this project by means of a critical analysis of the language of science (see Carnap et al. 1929; Carnap 1932). Such analysis was to lead to the definition of a demarcation criterion between meaningful and meaningless statements (see Carnap 1934a,b), the same criterion also serving to distinguish between science and metaphysics (or, more generally, discourses that illegitimately take on scientific form). In a nutshell, the demand for a criterion of "cognitive significance" for the statements of science corresponds to the idea that any piece of knowledge is either logico-mathematical or else empirical, that is, grounded in sense experience. What distinguishes empirical science statements from metaphysical pseudo-statements is that the former, unlike the latter, express matters of fact, which are true or false depending on the state of the world. Metaphysical statements might have an expressive, emotional significance, but they do not convey any factual knowledge: they do not say anything about the world; hence they have no truth-value.[16] Only statements with an empirical content are meaningful, and their meaning in fact reduces to this empirical content (the only

[16] The logical empiricist's analysis of language relies on the distinction between synthetic statements, which express matters of fact and are true or false depending on the state of the world, and analytic statements, which are true or false depending on their logical form and on the meaning of the terms they feature, independently of the state of the world (typically, logical contradictions, tautologies, and definitions are all analytical statements).

nonempirical elements being logical rules). As a consequence, the meaning of any term has to be reducible, at least in principle, to a finite set of observational statements.

This strictly reductionist conception of meaning was initially aimed at establishing a criterion of significance for any statement about the empirical world—scientificity being equated to empirical meaningfulness. However, the difficulties it raised led Carnap to evolve toward a more liberal, but also more complex, view of meaning in which the units of cognitive significance are neither terms nor statements but rather theories as wholes. This is how he came to formulate a conception of scientific theories and the rules for their formal reconstruction.

2.1.1 The Distinction between Theoretical and Observational Terms

Carnap distinguishes, within the language of scientific theories, between an observational language (L_O) and a theoretical language (L_T). L_O and L_T have in common their logical vocabulary, the terms of which do not have referents.[17] Besides this logical vocabulary, statements in L_O only use terms referring to observable entities, properties, relations and processes.[18] The whole set of these terms constitutes the observational vocabulary V_O. Statements in L_O can be used to describe either a particular state of affairs or a singular event, as well as to express empirical generalizations or phenomenal laws, in which case they take a universal logical form. For instance, a statement such as "all incandescent bodies are hot" is an observational statement, although it expresses a generalization. Statements in L_T are theoretical principles or laws. It is in principle from them that one can deduce and explain phenomenal laws and particular empirical descriptions. Terms in L_T belong to the theoretical vocabulary V_T and are called "theoretical terms." They do not refer to observable entities or processes. Famous examples of theoretical terms are "force," "energy," "field," "electron," and, in science of days past, "ether" and "phlogiston."

Note that what distinguishes the theoretical from the observational is not the (universal) form of theoretical statements (statements in L_O may be the result of a generalization process), but rather the nature of the (non-logical) terms they feature, which seem to make reference to non-observable entities or processes. The meaning of theoretical terms is thus a core issue for the definition of theories.

---

[17] See Suppe's (1977b) reconstruction of the syntactic conception of theories for a finer-grained description of the different languages and sublanguages of theories according to Carnap. He distinguishes in particular between a strict observational language, with no quantifiers and no modalities, and an extended observational language.

[18] See Carnap (1966, chap. 23), for a definition of the observable. There was a significant shift, in Carnap's works, from the problem of defining observational, or "protocol" sentences, to the problem of defining theoretical terms. In the end, "observational" equals "extra-theoretical," the problematic issue being the definition of the theoretical. In the context of the formal reconstruction of theories, a term is observational—nontheoretical—when it can be understood independently of the theory at stake. And a term is theoretical when it cannot be given a fully observational definition. Criticisms of the distinction between L_O and L_T can be found in Putnam (1962) and Achinstein (1965, 1968).

Although the justification of the statements in L_O may raise a problem (the problem of inductive inference posed by any empirical generalization, see chapter 2 on confirmation), their meaning is not specifically problematic: L_O is fully empirically interpreted. But V_T cannot satisfy the reductionist demand according to which every statement in L_T should be deducible from a finite set of statements in L_O, and hence that every term in V_T be explicitly defined by means of a finite chain of statements in L_O. Indeed, if every term in a theory were reducible to observational statements, then what this theory said would be nothing more than a mere observational report; it would just be "an abstract, artificial device for bringing order into the large mass of experiences in somewhat the same way that a system of accounting makes it possible to keep orderly records of a firm's financial dealings" (Carnap, 1966, chap. 26, p. 248). On this point, Carnap seems to reject a radically positivistic, instrumentalist position such as Pierre Duhem's, for whom the aim of a theory is merely to classify experimental laws.[19] Such an accounting system would not deserve to be called a "theory." Indeed, we expect of a scientific theory that it state an explanation in a deeper sense than that in which an empirical generalization such as "all incandescent bodies are hot" furnishes an explanation of the heat coming from such or such a particular incandescent body. Moreover, one expects that it allow for the derivation of novel empirical laws from the theoretical laws; hence the latter cannot be mere summaries of a conjunct of empirical laws. In other words, the reductionist demand on theoretical terms amounts to depriving theories of their predictive and explanatory power.

The logical positivist faces a hard challenge, which sounds like a dilemma:[20] warranting the cognitive significance of the terms of the theory, while authorizing the theory to give "more" than a mere observational report. Theoretical terms are indispensable for the predictive and explanatory power of theories. But if one accepts terms whose meaning cannot be exhausted by a set of observational sentences, how can one account for the cognitive significance of the theory, and hence for its empirical meaningfulness? "How can the right of a scientist to speak of theoretical concepts be justified, without at the same time justifying the right of a philosopher to use metaphysical terms?" (Carnap 1966, chap. 26, p. 248).

---

[19] For Duhem, a scientific theory is "an abstract system whose aim is to summarize and classify logically a group of experimental laws without claiming to explain these laws" (Duhem, 1914, p. 7). From this perspective, theoretical laws, namely those hypotheses featuring concepts with no straightforward empirical referent (theoretical concepts, such as force or energy), do not give anything more than the set of propositions about empirical observations ("experimental laws"), of which they are nothing but concise and logically ordered presentations. Theoretical concepts, in this view, are nothing but convenient symbols with no meaning at all.

[20] See Hempel (1958, pp. 49–50), who presents the "theoretician's dilemma" as arising from the consideration that "if the terms and the general principles of a scientific theory serve their purpose, i.e., if they establish definite connections among observable phenomena, then they can be dispensed with since any chain of laws and interpretative statements establishing such a connection should then be replaceable by a law which directly links observational antecedents to observational consequents." Hence the dilemma: "If the terms and principles of a theory serve their purpose they are unnecessary, as just pointed out, and if they don't serve their purpose they are surely unnecessary."

Already in "Testability and Meaning" (1936–1937), Carnap had conceded that theoretical terms cannot be explicitly defined by means of observational terms. He progressively renounced on the model of translation altogether, for the model of a "partial interpretation," presented in subsection 2.1.2. About "Testability and Meaning," he writes in 1956:

> At the time of that paper, I still believed that all scientific terms could be introduced as disposition terms on the basis of observation terms either by explicit definitions or by so-called reductive sentences, which constitute a kind of conditional definition [. . .]. Today I think, in agreement with most empiricists, that the connection between the observation terms and the terms of theoretical science is much more indirect and weak than it was conceived either in my earlier formulations or in those of operationism. Therefore a criterion of significance for L_T must be very weak. (Carnap, 1956, 53)

As we will now see, Carnap's proposal (1956, 1966) is that a theory, taken as a whole, acquires its meaning through correspondence rules, which provide it with a partial interpretation. Theoretical terms and statements themselves both draw their meaning from the meaning of the theory.

2.1.2 Theories as Formal Calculi: Correspondence Rules and Partial Interpretation

In his search for a more liberal criterion of cognitive significance, Carnap was progressively led to formulate a conception of scientific theories—and not only of scientific terms and statements—which corresponded to the most standard version of the received view. As we will see, it is the scientific theory as a whole that has meaning; theoretical terms and statements do not have any meaning per se.

Following the model of axiomatizations in mathematical logic, Carnap (1956, 1966) proposes to conceive of theories as logical calculi, together with interpretation rules. Theories thus consist essentially of two sorts of postulates: theoretical principles or "*T*-postulates," which are the theory's axioms (the whole set of them is called *T*), and correspondence rules or "*C*-postulates" or "*C*-rules" (the whole set of them is called *C*). *T*-postulates only contain terms from V_T, whereas *C*-postulates are mixed statements, each of them containing at least one theoretical term and one observational term in a non-dispensable way. Their conjunct *TC* is a *partially interpreted* system of axioms. What does that mean?

The theoretical principles *T* furnish an implicit definition of the terms they feature. For instance, $\boldsymbol{F} = m\boldsymbol{a}$ furnishes an implicit definition of the terms "force" and "mass" in mechanics. Strictly speaking, the terms so defined do not have any meaning yet, nor does the statement in which they appear. What is called here "implicit definition" is in fact the imposition of a constraint on what one can call "force" or "mass" (just like point and line in modem axiomatizations of geometry, such as Hilbert's 1899). This constraint is established by the expression of a relation that has to be satisfied, but no particular cue is given about how to empirically interpret these terms, which do not have any extra-theoretical meaning; they are purely blind symbols. That some terms from ordinary language are used to name theoretical concepts (e.g., "force" or "work") should not mislead us: the mental representations associated with the term "force" are not part of its meaning.[21] *T* is a purely formal—syntactical—theory, or a form of theory, with no semantic interpretation. Statements in *T* are not strictly speaking assertions: "insofar as the basic theoretical terms are only implicitly defined by the postulates of the theory, the postulates assert nothing, since they are statement-forms rather than statements" (Nagel, 1961, p. 91).

It is the correspondence rules *C* that provide the empirical interpretation of the formal calculus *T*, thus ensuring that the theory as a whole has a cognitive significance. They generally present themselves as specifications of experimental procedures for applying the theory to the phenomena. These rules are mixed statements containing terms from V_T and terms from V_O, which relate statements in L_T to statements in L_O. For instance, the statements "If there is an electromagnetic oscillation of a specified frequency, then there is a visible greenish-blue color of a certain hue" and "The temperature (measured by a thermometer and, therefore, an observable in the wider sense explained earlier) of a gas is proportional to the mean kinetic energy of its molecules" (Carnap, 1966, chap. 24, p. 233) are correspondence rules. Hence, correspondence rules draw a bridge between theoretical laws and empirical laws, without which theories would be totally useless. Thanks to correspondence rules, one can draw empirical predictions from the theory.

The whole empirical meaning of the theory is given by the correspondence rules. Each correspondence rule uses various theoretical terms, which in turn feature in various correspondence rules. Hence, the unit of cognitive significance is not the term or the statement but the theory itself, which, as a whole, has a set of empirical consequences reached by means of the correspondence rules.

So conceived, the empirical interpretation of the theory is doomed to be incomplete (as the expression "partial interpretation" suggests). This incompleteness allows for the addition of novel correspondence rules that in turn allow for the derivation of new empirical laws from theoretical laws.

> A postulate system in physics cannot have, as mathematical theories have, a splendid isolation from the world. Its axiomatic terms—"electron," "field," and so on—must be interpreted by correspondence rules that connect the terms with observable phenomena. This interpretation is necessarily incomplete. Because it is always incomplete, the system is left open to make it possible to add new rules of correspondence. Indeed, this is what continually happens in the history of physics. I am not thinking now of a revolution in physics, in which an

[21] Hempel (1970), rejecting the idea of a clear-cut distinction between theoretical and observational vocabularies, rejects outright the project of formalization of theories and insists on the importance of fuzzy concepts and of the use of natural language in the formulation of theories.

> entirely new theory is developed, but of less radical changes that modify existing theories. [. . .]
>
> There is always the possibility of adding new rules, thereby increasing the amount of interpretation specified for the theoretical terms; but no matter how much this is increased, the interpretation is never final. In a mathematical system, it is otherwise. There a logical interpretation of an axiomatic term is complete. Here we find another reason for reluctance in speaking of theoretical terms as "defined" by correspondence rules. It tends to blur the important distinction between the nature of an axiom system in pure mathematics and one in theoretical physics. (Carnap, 1966, pp. 237–238)

The incompleteness of the interpretation provided by correspondence rules is thus characteristic of empirical theories and distinguishes them from mathematical theories (which are only theoretical), on the one hand, and from observational reports (which are only empirical), on the other. But the problem of the surplus meaning of *T* (the part of its meaning that is not provided by *C*) remains.

2.1.3 The Problem of Theoretical Terms: The Ramsey-Carnap Solution

If *TC* provides only a partial interpretation of the terms in V_T, what does the rest of their meaning consist of? This "surplus meaning" is indispensable for the explanatory role of theories and for the possibility of deriving new empirical laws, but it poses a major challenge to the logical empiricist. How are we to warrant the cognitive significance of the theory if part of its meaning, which one may call its proper theoretical meaning, is not captured by its empirical consequences? The thesis of partial interpretation, and the holistic conception of the meaning of theories which comes with it, have led some logical empiricists, like Hempel (1950, 1951, 1965b) to give up the idea of a criterion of cognitive significance for theories relying on a clear-cut distinction between the empirical part of their meaning and the surplus meaning that partial interpretation does not define.

But Carnap refuses such a move. Against what he calls a "skeptical position" (1956, p. 39), he adopts the following strategy: he claims that any question concerning the surplus meaning of these terms is in fact a disguised linguistic question. This relies on the idea that the distinction between the empirical content of a theory (what it does tell us about the world) on the one hand, and the (psychological) effects of the language one has chosen to express this content (e.g., a language containing the term "electron") on the other hand, corresponds to the distinction between the synthetic and the analytical part of the theory.[22]

Carnap's formalization of such a distinction consists in a logical trick, known as the "Ramsey-Carnap" method. This relies on a proposition made by Ramsey (1929),

[22] This is why Carnap claims that, "a sharp analytic-synthetic distinction is of supreme importance for the philosophy of science" (Carnap, 1966, chap. 27, p. 257).

which consists in eliminating all theoretical terms featuring in the statement of the postulates of a theory *TC* by means of a simple logical manipulation: having couched *TC* under the form of a statement corresponding to the conjunct of all statements it contains (*T* and *C*), V_T terms are substituted with variables, and to the resulting open formula is added an existential quantifier for each of these variables. The statement so obtained (the existential closure of the formula), called the "Ramsey statement of the theory" (*RTC*), is logically and empirically equivalent to *TC*. Indeed, it has strictly the same observational content. But instead of theoretical terms, it features bound variables.

RTC expresses the synthetic part of *TC*: everything *TC* says about the world is contained in *RTC*. Now it remains only to account for the analytical part of *TC*. Carnap proposes to conceive of *TC* as equivalent to the conjunct of two statements: F_T, which expresses the whole factual content of the theory (it corresponds to *RTC*), and A_T, which is devoid of any factual content and plays the role of the meaning postulates for all terms in V_T.[23] The postulate A_T (which is not an assertion) of a theory *TC* is $RTC \rightarrow TC$. This statement is empirically empty; indeed, the whole factual content of *TC* is already contained in *RTC*. The postulate A_T exhausts what the theory says about the terms in V_T: it says that, if *RTC* is true (if it is empirically realized), then one must understand the terms in V_T so as to make *TC* true. *TC* is the logical consequence of the conjunction of *RTC* and A_T (equivalent to $RTC \rightarrow TC$).

This "solution" enables Carnap to claim that the debate between realism and instrumentalism reduces to the pragmatic question of a choice of language. Indeed, the non-empirical part of the theory (A_T) is not an assertion. If the electricity theory seems to say something more about electrons than what is contained in *RTC*, it is a mere effect of the language used to formulate it. The question "Do electrons exist?" is thus a disguised linguistic question. Either it is internal to the linguistic framework, in which case the answer is trivially positive, or else it is external to this framework, and so is a metaphysical question with no cognitive significance.

Surplus meaning is thus harmless. But claiming that *RTC* and *TC* say the same thing about the world does not amount to claiming that they say the same thing *tout court*:

> Ramsey merely meant to make clear that it was *possible* to formulate any theory in a language that did not require theoretical terms but that said the same thing as the conventional language.
>
> When we say it "says the same thing," we mean this only so far as all observable consequences are concerned. It does not, of course, say *exactly* the same thing. The former language presupposes that theoretical terms, such as "electron" and "mass," point to something that is somehow *more* than what is supplied by the context of the theory itself. Some writers have called this the "surplus meaning"

[23] Meaning postulates are linguistic conventions given or registered from use in a linguistic community. Together with logical rules, they constitute the analytical part of a language. See Carnap (1952, 1955).

> of a term. When this surplus meaning is taken into account, the two languages are certainly not equivalent. The Ramsey sentence represents the full *observational content* of a theory. It was Ramsey's great insight that this observational content is all that is needed for the theory to function as theory, that is, to explain known facts and predict new ones. (Carnap, 1966, chap. 26, p. 254)

This may sound like an implicit admission of failure, immediately followed by the most explicit statement of the fundamental assumption of Carnap's formalist project. He somehow solves the problem by a *coup de force*: there might be a surplus meaning, he admits, but it has no role in the predictive and explanatory function of the theory. The whole predictive and explanatory function of the theory relies on its observational content: "the Ramsey sentence has precisely the same *explanatory and predictive power* as the original system of postulates" (Carnap, 1966, chap. 26, p. 252). Hence, if one is interested in the theory insofar as it allows for prediction and explanation, as one should be, then one can reduce its content to its empirical element. Rather than a solution to the problem, Carnap makes explicit the fundamental assumption that he must accept. This, however, is not tenable.

Although Ernest Nagel (1961) presents his view of theories as in agreement with Carnap's (as does Carnap with Nagel's in 1966), his conception precisely relies on the rejection of the assumption according to which the predictive and explanatory function of a theory is entirely reducible to the expression of its observational content. The following presentation of Nagel's views will focus on their differences with Carnap's. It will be argued that Nagel, by hopelessly trying to hold together the fundamental requirements of logical empiricism and the (implicit) rejection of Carnap's aforementioned assumption, highlights the internal limits of the formalist program.

2.2 ERNEST NAGEL AND THE LIMITS OF THE LOGICAL EMPIRICIST PROGRAM

Nagel's view of theories is expressed in chapters 5 and 6 of his 1961 book entitled *The Structure of Science*. At first sight, it differs from Carnap's view on one significant point, namely the addition of one component, called "model," into the formal picture of theories. We will see that this change opens a real breach in the logical empiricist construal of theories.

2.2.1 The Three Components of Theories

Nagel (1961) describes theories as essentially consisting of three components (instead of two, as in Carnap's view):

> (1) an abstract calculus that is the logical skeleton of the explanatory system, and that "implicitly defines" the basic notions of the system; (2) a set of rules that in effect assign an empirical content to the abstract calculus by relating it to the concrete materials of observation and experiment; and (3) an interpretation or

model for the abstract calculus, which supplies some flesh for the skeletal structure in terms of more or less familiar conceptual or visualizable materials. (Nagel, 1961, 90)

The first two components jointly correspond to *TC* in Carnap's picture. The third one is specifically aimed at furnishing an interpretation to the skeleton: this suggests that correspondence rules do not by themselves suffice to turn the uninterpreted set of formal calculus axioms into meaningful statements.

What does the third component added by Nagel correspond to? The term "model" is highly equivocal, and Nagel's use of it is not devoid of ambiguity. *Prima facie*, Nagel seems to have in mind the logical notion of model (see e.g., Nagel, 1961, p. 96): models are structures that provide an interpretation to the formal theory they satisfy (see subsection 1.4). But Nagel's words about models tend to also emphasize the representational, intentional function of models: anything presenting some relevant analogy with the system under study can be used as a representation of this system. A typical example of such analogical (also called "iconic")[24] models is the miniature model of a plane; the structural features it shares with the real plane enable one to draw inferences concerning the real plane by reasoning with, and manipulating the model. A famous example in science is the so-called billiard-ball model in the kinetic theory of gases, which represents the motion of molecules as analog to the observable, well-known motion of macroscopic bodies. This model prompts us to imagine molecules of gas *as* billiard balls, thus facilitating both our understanding and use of the theory's equations. As this example shows, analogical models are often used to represent unobservable (e.g., microscopic) and poorly known phenomena under the traits of familiar objects and phenomena. Nagel distinguishes these "substantial" analogies from "formal" ones. Typically, formal analogies are those prompted by the use, in certain scientific domains, of a formalism (for instance, a type of equation) belonging to another domain. The two domains thus have in common a structure of abstract relations. Some principles of relativity theory, for instance, are formulated by analogy with the fundamental principle of Newtonian dynamics: the equations used to express them have the same mathematical form. "The example illustrates how the mathematical formalism of one theory can serve as a model for the construction of another theory with a more inclusive scope of application than the original one" (Nagel, 1961, p. 111).[25]

The question of whether and how the equivocal character of the notion of model can be dispelled is a subtle one, having to do with other tensions within Nagel's views. For the moment, let us just admit that models enable one to form a mental representation

[24] The notion of iconic model was already discussed by Campbell (1920). In the 1960s, various philosophers (Hesse, 1966: Black, 1962) emphasized the importance of analogies and metaphors in scientific theories and put the spotlight on the notion of models. But the specificity of Nagel's account consists in trying to integrate this notion into the logical empiricist view of formal reconstruction.

[25] The difference between these two types of analogy is less marked than it might seem. Formal and substantial analogies often come together.

of what theories say, by embodying their logical structure in a cognitively tractable way. This is the sense in which they provide *TC* with an interpretation: they make it intelligible to us.

2.2.2 Models, Explanation, and Understanding

That analogical models have an important heuristic, and probably pedagogical value, is rather uncontroversial. But Nagel also claims that models play a fundamental role in explanation. Coming from a logical empiricist, this claim is somewhat puzzling. It is hard to see how it can make sense within the standard picture of explanation, as expressed in Hempel's deductive-nomological account (Hempel and Oppenheim, 1948; Hempel 1965a; see chap. 1 of this volume), which Nagel otherwise contributes to developing.[26] In this account, explanations have the exact same form as predictions, and consist in deductive arguments from nomological premises (together with statements about initial conditions) to their empirical consequences. Explanation in this sense seems to be the job of correspondence rules, which ensure the relation between theoretical principles and the empirical phenomena.

Nagel thus seems to have in mind a different notion of explanation, one that is closely tied to the psychological and pragmatic notion of understanding that Hempel pushes outside the scope of his logical account of explanation.[27] Nagel's words suggest that, if explanation reduces to the drawing of observational statements from the first two components of the theory, then it amounts to a pure, blind, calculus that does not yield any understanding. Indeed, such operation is not, by itself, intelligible: it is not a proper piece of reasoning, as it is not a cognitive activity involving the manipulation of mental representations. Without the model, the theory may allow for the drawing of predictions, but this is clearly separated from the representational dimension of theorizing.

Is this a merely terminological issue regarding the word "explanation"? After all, Nagel may well be in agreement with the orthodox logical empiricist picture of theories and of explanation, while in parallel highlighting the psychological dimension of theorizing and understanding—for which he uses the term "explanation" in a non-orthodox way. This interpretation would, however, miss the point. What Nagel suggests is that inferences leading to prediction or explanation *cannot* be drawn in the absence of a model. The pure

---

[26] The subtitle of his *Structure of Science* is *Problems in the Logic of Scientific Explanation*. The book is presented as a development of the logical empiricist conception of explanation.

[27] In Hempel's account, the "feeling of understanding" that may arise from a good explanation is a psychological and pragmatic phenomenon. "Very broadly speaking, to explain something to a person is to make it plain and intelligible to him, to make him understand it. Thus construed, the word 'explanation' and its cognates are pragmatic terms: their use requires reference to the persons involved in the process of explaining. [. . .] Explanation in this pragmatic sense is thus a relative notion: something can be significantly said to constitute an explanation in this sense only for this or that individual" (Hempel, 1965a, pp. 425–426). This is clearly not what the formal account of explanation seeks to capture. Just as, in Carnap's views, mental images should not be considered as part of the cognitive content of theories, this psychological phenomenon is not relevant for a logical analysis of explanation in Hempel's views.

logical skeleton, together with correspondence rules, is not the object of any reasoning process. In order to be reasoned with, it has to be mentally represented, which means that it has to be presented under a certain form that is mentally tractable. And this particular form is what he calls a model. Models need not be visualizable images relying on a familiar analogy: the use of a given formalism—for example, second-order differential equations—is already a certain presentation of the syntactic skeleton, which is not, taken alone, accessible to the scientist.[28] Hence, theories, when expressed and reasoned with, are already equipped with a model. Explanation, in Nagel's view, relies on the actual implementation of the deduction in a scientist's mind, which requires a model—an interpretation of the theory that a cognitive agent can handle.

True, one could still consider that models are a necessary cognitive interface between theories and scientists, but that they do not have to enter into the characterization of theories—as they concern only the pragmatic and psychological dimension of theorizing. Emphasizing the psychological indispensability of models, as such, is not really controversial. Carnap himself acknowledges that scientists need models to reason and to develop theories.[29] Precisely, formalization is aimed at distinguishing

---

[28] In his famous considerations on two types of minds (the "abstract" ones, and the "ample" or "imaginative" ones) and the kind of theoretical construct that corresponds to each of them (the abstract theory for the former and the mechanical models for the latter), Duhem explicitly considered that the use of symbolic algebra belongs to the second category. His distinction is not between mathematical language on the one hand and visualizable images on the other: equations, just like diagrams, belong to the concrete "wrapping" of the abstract theories. Imaginative minds need them, but abstract minds can do without them. The ability to manipulate algebraic symbols, which Duhem calls "calculus," is in fact an imaginative one, rather than a purely intellectual (logical) one (which he calls "reasoning"): "mathematicians have created procedures which substitute for [the] purely abstract and deductive method another method in which the imaginative faculty plays a greater part than the power of reasoning. Instead of studying directly the abstract notions with which they are concerned, [. . .] they submit the numbers furnished by measurement to manipulations performed according to the fixed rules of algebra; instead of deducing, they *calculate*. Now this manipulation of algebraic symbols (which we may call calculus, in the largest meaning of the word) presupposes, on the part of the creator as well as of the one who uses it, much less power to abstract and much less skill in arranging one's thoughts in order than aptitude for expressing diverse and complicated combinations. These may be formed with certain visible and traceable signs in order to see off-hand the transformations permitting one to pass from one combination to another" (Duhem 1914, p. 63).

[29] In a section entitled "Understanding in Physics," Carnap (1939, 67–68) clearly states that models should not be part of the formal reconstruction of theories: "The possibility and even necessity of abandoning the search for an [intuitive] understanding [. . .] was not realized for a long time. When abstract, nonintuitive formulas, as, e.g., Maxwell's equations of electromagnetism, were proposed as new axioms, physicists endeavored to make them 'intuitive' by constructing a 'model,' i.e., a way of representing electromagnetic micro-processes by an analogy to known macro-processes, e.g., movements of visible things. Many attempts have been made in this direction, but without satisfactory results. It is important to realize that the discovery of a model has no more than an aesthetic or didactic or at best a heuristic value, but is not at all essential for a successful application of the physical theory. The demand for an intuitive understanding of the axioms was less and less fulfilled when the development led to the general theory of relativity and then to quantum mechanics, involving the wave function. Many people, including physicists, have a feeling of regret and disappointment about this. Some, especially philosophers, go so far as even to contend that these modern theories, since they are not intuitively understandable, are not at all theories about nature but 'mere formalistic constructions,' 'mere calculi.' But this is a fundamental misunderstanding of the function of a physical theory. It is true a theory must not

the logical core from such psychologically indispensable wrapping, which should not be considered as part of the theory.[30] However, that is not what Nagel says; as we have seen; in his picture, *TC* alone is not a proper theory. Why does Nagel insist on counting models among the essential components of theories?

2.2.3 Semantics Splitting

Let us start by turning the question around: if the interpretation of the syntactic skeleton is provided by models, what function do correspondence rules serve? The answer is that they warrant the empirical grip of the theory. Indeed, models, even though they provide an interpretation to the axioms of the theory and hence make them intelligible—or sensible —, do not establish any link between axioms and experience. The model of a theory, which gives flesh to its content, is not intended as a true representation of the empirical world. Models ensure the intelligibility of the theory; correspondence rules warrant its empirical character.

Hence, Nagel creates a gap between the meaning of theories—what they say—and their empirical consequences. We thus end up with two different things: on the one hand, what the theory says *tout court* and, on the other hand, what can be drawn from it about the empirical world. This is clearly at odds with Carnap's program, which consisted, in the spirit of the verificationist theory of meaning, in defining the cognitive significance of a statement by its empirical truth conditions. Nagel thus operates a splitting in the semantics of the theories:[31] theories receive their empirical interpretation from correspondence rules, but this interpretation does not make them meaningful. This splitting between the empirical consequences of a theory and its semantic content enables Nagel to take a subtle position regarding the debate between realism and instrumentalism, something he calls the question of the "cognitive status of theories."

2.2.4 The Cognitive Status of Theories

Do theories aim at being true statements about the world (realism), or merely at facilitating predictions with no pretension to truth (instrumentalism)? Let us first consider *TC* (the first two components of theories). As it more consists of a set of recipes for drawing empirical predictions than of a set of assertions, *TC* does not say

be a 'mere calculus' but possess an interpretation, on the basis of which it can be applied to facts of nature. But it is sufficient, as we have seen, to make this interpretation explicit for elementary terms; the interpretation of the other terms is then indirectly determined by the formulas of the calculus, either definitions or laws, connecting them with the elementary terms."

[30] One reason not to consider models as proper parts of theories is that, so construed, they are context-dependent. A familiar type of equation, such as the equations of classical mechanics and because of the scientific centrality of the theory featuring them, becomes a standard of intelligibility. And standards of intelligibility are doomed to change.

[31] This splitting can be described as Nagel's way to fill the gap left by the incompleteness of the interpretation provided by correspondence rules in Carnap's picture.

anything. Indeed, as the interpretation is not complete, correspondence rules *C* do not make the statements in *T say* anything (neither do they provide theoretical terms with any meaning). Correspondence rules are not *translation* rules, but rather *transformation* rules: they are inferential "tickets" allowing one to draw empirical statements by adequately manipulating the syntactic skeleton. It is in fact inaccurate to claim that empirical statements can be *deductively* drawn from the syntactic skeleton by means of the correspondence rules:

> [. . .] a theory functions as a 'leading principle' or 'inference ticket' *in accordance with which* conclusions about observable facts may be drawn from factual premises, not as premises *from which* such conclusions are obtained. (Nagel, 1961, pp. 129–130)

A deduction is a reasoning process, which includes premises that are meaningful statements. The only meaningful statements here are statements about the empirical world ("factual premises"), but statements in *T* are not premises—they are not proper assertions; transformation rules cannot feed them with any content.

As a consequence, for those who consider that theories reduce to *TC*, instrumentalism is the only option: theories so construed do not say anything (not even, as a descriptivist would argue, about the empirical world).[32] But Nagel does not hold such a view, as he considers models as essential components of theories. Through models, theories do indeed say something. However, they are not to be taken as true representations of the world: modeling molecules of gas *as* billiard balls does not imply believing that molecules of gas are billiard balls. Models give flesh to the theory, by embodying what they say, but they themselves do not say anything about the empirical world. Hence:

- *TC* alone does not say anything at all.
- Models do not say anything about the world (hence have no truth-value).

The distinction between *TC* and models (just as the distinction between L_T and L_O), however, does not correspond to any tangible difference between elements of real theories: in practice, *TC* is not isolated from its model(s), as theories always come expressed in natural language. As a consequence, their principles look like statements, and scientists may tend to take them as premises. This is what Nagel suggests when he says that, "theories are usually presented and used as premises,

[32] Nagel distinguishes between three positions regarding the cognitive status of theories: realism, instrumentalism, and descriptivism. Whereas the latter considers that theories contain proper assertions, restricting these to the empirical statements that can be drawn from them, instrumentalism is the view that theories do not say anything: they are pure instruments, and they do not contain any assertion (even empirical). Nagel's argument, here, is that descriptivism amounts to instrumentalism, because of the incompleteness of the interpretation.

rather than as leading principles" and that "some of the most eminent scientists [. . .] have conducted their investigations on the assumption that a theory is a *projected map* of some domain of nature, rather than a set of *principles of mapping*" (1961, p. 139). In fact, the "contextual" distinction between premises and inference tickets (p. 138)—or between projected maps and projection rules—corresponds to the distinction between the logician's formal reconstruction of theories, and their actual formulation(s) by scientists at work.[33] True, one can still hold, in a Carnapian spirit, that the psychology of scientists does not have to be taken into account in an analysis of theories; saying that, in practice, theories are always presented under a certain form, as such, would not harm the Carnapian view very much, as it is not intended to account for scientific practice and then scientists' psychology. But what Nagel teaches us is that the aforementioned "contextual" distinction is not a distinction between the proper content of theories and scientists' psychological, subjective, understanding of it. Indeed, prediction and explanation themselves require that the theory be formulated in a cognitively tractable way—be equipped with a semantics. Formalization, insofar as it aims at dissociating the logical structure of theories from their semantics, prevents us from actually retrieving their content. In other words, contrary to what Carnap suggested, the Ramsey sentence is not all that matters for prediction and explanation. Prediction and explanation are obtained by reasoning on already interpreted statements—theoretical language is nothing but a part of natural language, which only some specialists master, but which does not have a separate semantic status.[34] It is not only "in practice" that theories are to be interpreted as expressed in (i.e., an extension of) natural language; if their predictive and explanatory status is taken seriously, then one must acknowledge that they must be cognitively represented. A theory broken down by separating its syntax from its semantics is simply not a scientific theory anymore, as it cannot fulfill its predictive and explanatory function.

2.2.5 Theories as Mental Representations

In section 1, we suggested that it is rather uncontroversial to say that theories are tools of representation and inference. The analysis suggests that the logical empiricist formalist program does not do justice to this. Indeed, Carnap acknowledges that theories aim at providing explanations and predictions, but he seems to neglect that, in order to obtain them, one needs to draw inferences on a cognitively tractable representation. The inferential function of theories is closely tied to their representational

[33] Like in Carnap's view, the debate between instrumentalism and realism appears as a pragmatic issue from the Nagelian perspective too, but in a slightly different sense. Here the pragmatic difference is not between different languages (Should I choose a language containing the term "electron"?), but between the use of theories' statements as "inference tickets" or as premises: Should I take expressions containing the term "electron" as pure calculation tools (in which this term is a blind symbol), or as statements saying something about the world, and in particular about electrons?

[34] See Schaffner's (1969) criticism of the notion of correspondence rules.

dimension: it is by representing the phenomena a certain way—by using a certain language and formalism—that a theory enables one to draw inferences.[35]

As a consequence, the very project of formalizing a theory like classical mechanics as a way to capture its content, going beyond its various formulations, may appear as misguided. Indeed, the various formulations of mechanics offer different representational and inferential tools: despite their being logically equivalent, they represent the phenomena in different ways, thus facilitating different types of inferential processes. In Nagel's terms, the various formulations of classical mechanics can be considered as so many models of the same syntactic core. Interestingly though, in his chapter on mechanical explanations (Nagel, 1961, chap. 7), Nagel treats the various formulations of mechanics as providing one unique type of explanation, which seems incoherent with his views on models and explanation. Our diagnostic is that such incoherence, like the tension mentioned earlier between the function of correspondence rules and of models and what we called the "splitting" of the semantics of theories, are the result, in Nagel's thought, of his attachment to the empiricist project of a syntactic formalization of theories, which is in fact incompatible with his views on theories as cognitive objects, or mental representations.[36] However, the untenability of his position highlights the very limits of the logical empiricist project.

2.3 CONCLUSION: WHAT IS THE "SYNTACTIC CONCEPTION" OF THEORIES?

The logical empiricists' view of theories, presented in this section, was a posteriori labeled "syntactic conception" by some advocates of the semantic view. The reason for this should now be clear: for the logical empiricists, the core of the theory is its syntactic skeleton, and the project of formalization consists in distinguishing this syntactical core from its empirical interpretation. Nagel himself, whose view of the semantics of theories is ambiguous, starts by acknowledging that the first component of theories is its syntactic skeleton. Now, the logical empiricists' formalization project is more the description of the ideal canonical form a scientific theory should in principle be able to take, rather than a plan of actual formalization of existing theories.

Even though it does not capture any unified body of doctrine, the somewhat caricatural picture portrayed by the later generations is not totally fabricated. Beyond their variety, the authors defending the views grouped under the umbrella name "syntactic conception" share a construal of theories as a network of principles and laws whose

[35] As Heinrich Hertz famously claimed, in our efforts to "draw inferences as to the future from the past, we [. . .] form for ourselves images or symbols of external objects; and the form which we give them is such that the necessary consequents of the images in thought are always the images of the necessary consequents in nature of the things pictured" (Hertz, 1894, p. 1). And he acknowledged that these images, being "produced by our mind," are "necessarily affected by the characteristics of its mode of portrayal" (1894, p. 2).

[36] That Nagel's works embody such a major break with the logical empiricists' dogma was already acknowledged by Alexander Rosenberg, who sees Nagel's book (rather than Quine's works) as the "locus classicus of philosophical naturalism" (Rosenberg, 2000, p. 7).

link to the observable world has to be accounted for in linguistic terms. Nagel's untenable position is the result of his attachment to this image of theories as syntactic networks to be related to the empirical realm by correspondence rules, together with his acknowledgement that scientific theories necessarily include some undetachable interpretation.

As we will see in section 3, the semantic conception rejects outright this linguistic image of theories. Acknowledging that one cannot separate the syntactic skeleton of theories from their interpretation, its proponents go to formal semantics and set theory to borrow the tools required for their project of axiomatization of theories, the content of which is no longer identified with their empirical consequences. Theories do not acquire their meaning through correspondence rules that furnish, *post hoc*, uninterpreted axioms with an empirical interpretation. Rather, a theory always comes immediately equipped with an interpretation, which consists of a set of mathematical structures. Thus, giving up the model of linguistic translation in favor of the model of interpretation (in the set theory sense), the semantic conception nevertheless remains a formal approach: its goal is to account for the content of theories in a way that allows one to dismiss those aspects of theories that are related to their actual linguistic formulation(s) as well as to the way they are used and understood by agents. According to advocates of the semantic view, the failure of logical empiricism is not due to this latter goal in itself, but rather to the way logical empiricism implemented it (its linguistic, syntactic aspect).

3. The Semantic View of Theories

Before being explicitly put forward as a new theory of theories, the semantic conception originated in the works of the logician Evert Beth (1940) and was then developed in the 1950s by Patrick Suppes (1957, 1960, 1962, 1967). Suppes is, in many respects, the father of the semantic approach, and one of its most important advocates. Most of this section is devoted to his views. Beside him, the "big names" of the semantic conception are Frederick Suppe (1971, 1977b, 1989), Bas van Fraassen (1980, 1987, 1989, 1991), and Ronald Giere (1979, 1988, 2006). Another school, more neatly distinguished from Suppes's legacy, but nevertheless pertaining to it, is the so-called structuralist school of Wolfgang Balzer (1985), Ulises Moulines (1975), Joseph Sneed (1975, 1976), and Wolfgang Stegmüller (1976).[37] Finally, some philosophers and scientists have proposed actual axiomatizations of existing scientific theories, in particular in physics (Hughes 1989, van Fraassen 1991) and in biology (Beckner 1959, Beatty 1982, Lloyd 1988, Thompson 1989, 2007).

The common ground for all proponents of the semantic view is that the linguistic aspect of the logical empiricists' formalist enterprise misled them into questions of philosophy of language that are of no relevance to the philosophy of science. Instead,

[37] See Balzer et al. (1987).

the "central dogma" of the semantic conception is that theories should be conceived as sets (or families) of models, rather than as sets of linguistic statements. A theory is thus an extra-linguistic entity, corresponding to the set of models satisfying its various formulations. Formal semantics and set theory are thus taken as a superior formal framework to study the content of theories. As will be seen, this choice of a different set of formal tools reveals a totally different conception of the content of theories.

Moreover, the goal of formal reconstruction is not the same here as in the logical empiricists' works. Rather than the definition of a criterion of scientificity, the semanticists' program aims at exploring the structure of existing scientific theories through axiomatization, as a way to clarify their content and solve conceptual problems. By contrast, the ideal formalization of the syntactic conception, besides being practically unrealizable, would not—if realized—be able to shed light on important aspects of theories and their relations to other theories, something the semantic approach contributes to highlighting.

3.1 THEORIES AS FAMILIES OF MODELS

The semantic conception relies on the rejection of the following two assumptions: 1. that one can (at least in principle) isolate the content of a theory by reconstructing its syntactic skeleton, and 2. that a theory acquires its meaning through the establishment of a linguistic relation between such skeleton and empirical statements.

Nagel's untenable hybrid image of the semantics of theories somehow announced the ineluctable divorce between empirical application and semantic content. The semantic conception goes one step further: it pronounces such a divorce and just gets rid of the idea of correspondence rules. The content of theories is now entirely determined by the sets of models described by the theory. "Models," here, are to be understood in their logico-mathematical sense. Hence, according to the semantic conception, the content of a theory is given by its *mathematical theory*, namely the set of models that satisfy its various formulations.

Already in the 1950s, Suppes (1957) was undertaking a program of axiomatization of various theories (particularly the classical mechanics of particles), inspired from set-theoretical axiomatizations of mathematical theories.[38] The procedure consists in defining a predicate in terms of notions of set theory—a "set-theoretic predicate" (see Suppes, 1957, 249–253). Suppes's hypothesis is that this method of axiomatization, originally designed for (and up until then exclusively applied to) mathematical theories, can be used to reconstruct the content of *empirical* theories. The resulting formal reconstruction takes the form of a set of axioms that, together, define a predicate. Any entity that satisfies this predicate—of which this predicate is

[38] This was along the line of both Hilbert's (1899) works on the foundations of mathematics and the Bourbaki group's in the 1940s and 1950s.

true—is a model of the theory so expressed. Let us consider the case of Newtonian particle mechanics, as axiomatized by Suppes (1957, 294):

DEFINITION 1. A system $\beta = \langle P, T, s, m, f, g \rangle$ is a system of particle mechanics if and only if the following seven axioms are satisfied:

KINEMATICAL AXIOMS

AXIOM P1. The set P is finite and non-empty.

AXIOM P2. The set T is an interval of real numbers.

AXIOM P3. For p in P, s_p is twice differentiable on T.

DYNAMICAL AXIOMS

AXIOM P4. For p in P, m(p) is a positive real number.

AXIOM P5. For p and q in P and t in T,

$$f(p, q, t) = - f(q, p, t).$$

AXIOM P6. For p and q in P and t in T,

$$s(p, t) \times f(p, q, t) = - s(q, t) \times f(q, p, t).$$

AXIOM P7. For p in P and t in T,

$$m(p)D^2s_p(t) = \Sigma_{q \in P} f(p, q, t) + g(p, t).$$

The seven axioms provide the definition of a system of particle mechanics: any structure, either abstract or concrete, which satisfies the kinematical and dynamical description of system β is a system of particle mechanics. The content of this theory consists of the whole set of these systems (of the structures satisfying axioms P1–7). These are models in the logical sense. It is worth mentioning a terminological point. The "mathematical model" of a theory, in Suppes's terms, is the class of all the structures satisfying the axioms of this theory, namely the class of all its logical models.

Although linguistically expressed, these axioms are not of the same kind as the theoretical postulates (*T*) in Carnap's formalization. Axioms, here, are not *uninterpreted statements* expressing laws—or rather forms of law—to which one has to assign empirical referents.[39] Suppes's axiomatization does not rely on a syntactic formalization. The axioms used to define the set-theoretic predicate make use of all the required mathematical apparatus, without the need to give a *stricto sensu* formalization of this apparatus (cf. subsection 1.4.1). Suppes's method, as with modern axiomatizations of geometry, does not consist in relating syntactic statements to their semantics, the latter being conceived of on the model of empirical verifiability. Rather, it consists in choosing a set of elements and mathematical objects (relations, functions, operations on this set) and then imposing on them the conditions that are expressed by the

[39] The relevance of the classical notion of law is questioned by the advocates of the semantic view. Giere entitles his 1988 book *Science without laws*, and van Fraassen (1989) proposes to replace it by the notion of symmetry. Furthermore, Lloyd (1988), following Beatty (1982), shows that the notion of law underlying the hypothetico-deductive image of science conveyed by logical empiricism is particularly inadequate for studying evolutionary biology.

axioms.[40] Consider axiom P7: it corresponds to the expression of Newton's second law. A purely syntactic formulation of this law would consist in a statement establishing a relation between two variables *F* and *a*, without any interpretation for these terms (representing them as vectors is already providing them with an interpretation, if not physical, at least mathematical). In such a syntactic formalization, the empirical interpretation of these terms would be provided by correspondence rules that would specify the operations enabling us to assign values to these variables when studying the behavior of an empirical system. Here, however, axiom P7 describes a mathematical object, namely a class of structures, defined in given domain, and satisfying certain conditions.

The models so defined are not models of an uninterpreted syntactic structure but rather mathematical objects described in a language containing the mathematical notions required to directly present the structure of our most sophisticated physical theories.

For Suppes, whereas the logicist project of studying the foundations of mathematics is of prime importance for the philosophy of mathematics and logic, the project of formalizing (*stricto sensu*) our empirical theories is both hopeless and vain. It is hopeless because some physical theories have a structure as complex as theories in pure mathematics, which are not formalizable in first-order logic.[41] And it is vain because it consists in deliberately depriving oneself of the tools needed to explore the structure of theories—and such exploration, rather than the study of the language of theories, should be the goal of axiomatization. Such a laborious project appears totally unnecessary when one gives up the (mistaken) idea that the content of theories reduces to their empirical meaning.

The semantic view thus substitutes the exploration of the mathematical structure of theories for the logical analysis of their language, as a means to study their content. *Prima facie*, this move might just appear as an improvement of the logical empiricists' program based on the choice of more appropriate tools. However, it amounts to rejecting their program altogether, as consisting in grounding the content of theories on their empirical interpretation.

So now, with semantic interpretation and empirical application clearly separated, how are theories related to the empirical world, and how do they acquire an empirical content? What makes a theory, so construed, an *empirical* theory? The answer is that it

[40] "In a modern presentation of geometry we find not the axioms of Euclidian geometry, but the definition of a Euclidian space. Similarly Suppes and his collaborators sought to reformulate the foundations of Newtonian mechanics, by replacing Newton's axioms with the definition of a Newtonian mechanical system" (van Fraassen, 1987 p. 109).

[41] As mentioned in note 13, Carnap's project did not imply a formalization in first-order logic. He explicitly admits mathematical elements within the non-descriptive apparatus of the formal language (Carnap, 1956, p. 43). However historically inaccurate, Suppes's characterization of the syntactic conception reveals that his rejection of this view does not merely correspond to the choice of a different kind of formal tool (set theory *versus* first-order logic), but rather to a thoroughly different conception of the axiomatic enterprise itself.

is the models themselves — rather than their linguistic description—that are related to the empirical phenomena.

3.2 THE EMPIRICAL APPLICATION OF THEORIES: LOGICAL MODELS AND PHYSICAL MODELS

The semantic conception also rejects, as an abusive and fruitless idealization, the notion of correspondence rules. A quick glance at scientific practice teaches us that theories are applied through the use of models (ranging from idealized systems such as the simple pendulum, to concrete, three-dimensional representations such as the double helix model of the DNA). However, as Suppes himself acknowledges, the models defined by the set-theoretic predicate, as described, are "highly abstract, non-linguistic entities, often quite remote in their conception from empirical observations" (Suppes 1967, p. 57). They seem to have little to do with concrete, or even idealized systems used in scientific practice. How can such abstract entities relate to the observable phenomena?

This is where one of the strongest assumptions of the semantic conception comes in. According to Suppes, the models used by scientists to represent phenomena in their day-to-day practice can also be construed as logical models[42] (Suppes, 1960, pp. 12–13). Theorizing can be adequately described by means of the formal tools of model theory: it consists in establishing a certain relationship between structures at different levels of abstraction.[43] At each level, a theory describes what a possible realization could be: at the top of the hierarchy, the fundamental theory describes the most abstract models; then comes a theory of experiment that describes the possible empirical realizations of the theory; and, finally, a statistical theory defines how models of data (Suppes, 1960, 1962, 1957) can be built from concrete experience. In what is usually called "the application of theory to experience," the elements being compared are not the abstract theory on the one side and the empirical data on the other, but rather, on the one side the models of the experiment and on the other the models of data, the latter constituting concrete realizations of the experiment models (Suppes, 1967, p. 62; 1962, p. 253) and a "highly schematized version of the experience" (Suppes, 1960, p. 300). This deeply questions the naive conception of the confrontation between theory and experience:

> One of the besetting sins of philosophers of science is to overly simplify the structure of science. Philosophers who write about the representation of scientific theories as logical calculi then go on to say that a theory is given empirical meaning by providing interpretations or coordinating definitions for some

[42] More precisely, the logical notion of model refers to a particular structure, whereas the physical notion of model, such as Bohr's atom, refers to a class of isomorphic models (cf. van Fraassen, 1980, p. 44).

[43] Advocates of the semantic view diverge on this: van Fraassen and Suppe would speak of isomorphism, whereas Giere speaks of similarity (or resemblance).

> of the primitive or defined terms of the calculus. What I have attempted to argue is that a whole hierarchy of models stands between the model of the basic theory and the complete experimental experience. Moreover, for each level of the hierarchy there is a theory in its own right. Theory at one level is given empirical meaning by making formal connections with theory at a lower level. Statistical or logical investigation of the relations between theories at these different levels can proceed in a purely formal, set-theoretical manner. The more explicit the analysis the less place there is for non-formal considerations. Once the empirical data are put in canonical form [at the level of models of data], every question of systematic evaluation that arises is a formal one (Suppes, 1962, p. 260–261).

3.3 THE COGNITIVE STATUS OF THEORIES: WHAT DO SCIENTIFIC THEORIES SAY?

Given the separation operated by the semantic view between the semantic interpretation of the theory and its empirical application, it seems legitimate to raise the question of the cognitive status theories have. Theories as such do not tell us anything *about the world*: they are mathematical constructs that can be used to represent real world phenomena. But what is the status of this latter representational relationship: is it to be construed in terms of truth-value? Or are the relationships between models at different levels of abstraction of a different type? Are the advocates of the semantic view instrumentalists or realists?

At the level of the semantic interpretation of the theory, if one can speak of truth then it is in the sense of truth relative to a model. As Ronald Giere puts it:

> The relationship between some (suitably interpreted) equations and their corresponding model may be described as one of characterization, or even definition. We may even appropriately speak here of "truth." The interpreted equations are *true of* the corresponding model. But truth here has no *epistemological* significance. The equations truly describe the model because the model is defined as something that exactly satisfies the equations. (1988, p. 79)

However, as Bas van Fraassen (1987, p. 106) notes, a scientific theory "must be the sort of thing that we can accept or reject, and believe or disbelieve [. . .]. To put it more generally, a theory is an object for epistemic or at least doxastic attitudes." In fact, when a scientist uses a theory, asserting that there is a certain isomorphism or similarity relationship between the models at different levels of the hierarchy, she is making a *theoretical hypothesis*. This hypothesis can take various forms. A realist would consider that some of the models described by the axioms of the theory do truly represent some portions of the empirical world. An instrumentalist would say that they represent these portions in such a way that they allow predictions to be made, without implying that these theoretical structures actually correspond to the underlying structure of the world. On this issue, the advocates of the semantic conception have different opinions.[44]

However, there is another, subtler issue concerning the theoretical hypothesis made by the user of a theory: Is it, or is it not, part of the theory? It is certainly not part of the axioms that describe the model. Hence, if by "theory" one refers to the axiomatic definition of a class of models only, then this hypothesis is extra-theoretical. However, the advocates of the semantic view seem to hesitate on this point. Indeed, if one considers that the hypothesis does not belong to the theory itself, one ends up with a picture of theories that does not match the intuitive notion at all: a theory (even in empirical science) is a mere tool that does not say anything, and that does not even bear on a particular domain of phenomena—it is, in fact, a mathematical theory that may be used to represent the empirical world. This may sound odd. And that may be the reason why van Fraassen sometimes suggests that the theoretical hypothesis enters into the characterization of the theory, for instance when he endorses Giere's formula stating that "a theory consists of (a) the *theoretical definition*, which defines a certain class of systems; (b) a *theoretical hypothesis*, which asserts that certain (sort of) real systems belong to that class" (van Fraassen 1987, p. 109).[45]

This, however, is far from a trivial issue, as Giere, who advocates for the integration of the theoretical hypothesis, himself acknowledges:

> A compromise is to say that a theory includes both statements defining the population of models and hypotheses claiming a good fit between various of the models and some important types of real systems. The price we pay for trying to have our cake and eat it is that a theory turns out to be a rather heterogeneous type of thing. It includes both definitions and empirical hypotheses. But that may be a small price for capturing the diverse intuitions of what a theory is.
>
> My only objection to this compromise is that it puts too much emphasis on matters linguistic. It focuses attention on the statements that define the population of models rather than on the models themselves. I would prefer to substitute the models for the definitions. Newton's laws and the force laws would remain, though only implicitly, and not in linguistic garb. They would be embodied in the models. (Giere 1988, p. 85)

In other words, integrating the hypothesis into the theory—thus drawing a heterogeneous image of theories—amounts to acknowledging that pragmatic aspects of

---

[44] Giere (1988) and Suppe (1989) advocate realist positions ("constructive realism" for the former, "quasi-realism" for the latter). Van Fraassen's "constructive empiricism" is a form of instrumentalism that distinguishes between a general theoretical structure and empirical substructures; only the latter are supposed to directly represent the empirical phenomena (see van Fraassen, 1980, p. 64).

[45] Van Fraassen elsewhere advocates an instrumentalist view of theories, which he sometimes presents as a normative conception of theories and sometimes as a description of scientists' epistemic attitudes (see van Fraassen, 1980, chap. 2 and 3). This hesitation can be related to the one presented here, regarding the status of the theoretical hypothesis.

the use of theories enter into the characterization of theories themselves. What Giere suggests by emphasizing the linguistic aspects is that studying the nature of the theoretical hypothesis implies paying attention to the way agents use and understand theories in practice and, therefore, paying attention to their actual formulations. Hence, integrating the theoretical hypothesis into the content of a theory amounts to assigning limits to its formal reconstruction: these pragmatic aspects cannot themselves be formalized.[46]

However, this does not challenge the semantic view, as such. Van Fraassen does explicitly acknowledge the pragmatic aspects of theory acceptance (1980, p. 4). As for Suppes (1967, p. 66), he calls for the development of cognitive approaches to the use of language, as a way to solve the debate between realism and instrumentalism: to him, this is not a question that can be tackled with formal tools—at least not until the use of language can be precisely modeled. The goal of formal reconstruction is thus to describe relations between models, without any assumption about the epistemic commitment that comes with the establishment of these relations. On one side, there is the formal reconstruction of the content of theories and their empirical application, and, on the other side, a (nonformal) analysis of scientific practice and the agents' epistemic attitudes. However, it is difficult, indeed practically impossible, to establish a clear-cut distinction between the formal characterization of theories and the way they are used by agents, as we shall now see.

3.4 WHAT IS THE PURPOSE OF THE SEMANTIC VIEW, AND DOES IT ACHIEVE IT?

It is now time to review the program of the semantic view, to take stock of its achievement, and to assess whether it is in fact more successful than that proposed by the logical empiricists, as it claims to be. In order to do so, we must be quite clear about what its purpose is.[47] One first thing to note, which may dispel some hasty criticisms, is that its goal (at least in Suppes's version) is not to give a definition of theories, in the sense of a list of necessary and sufficient criteria for the identification of given bodies of knowledge as theories.

> It does not seem to me important to give precise definitions of the form: *X* is a scientific theory if, and only if, so-and-so. What is important is to recognize that

[46] Note that Giere elsewhere advocates a pragmatic and cognitive account of theorizing, which he tries to integrate within the semantic view by construing the notion of model both in its logical sense and in the sense it has in the cognitive study of reasoning. Hence, he does not deny the importance of language or of representations in general. His rejection of the "linguistic" here is rather to be understood as a rejection of the syntactic view. But it is far from clear that including the theoretical hypothesis within the theory implies a return to a linguistic view such as the one put forward by logical empiricists. See subsection 3.4.2 of this chapter on the confusion between formulation and formalization.

[47] What is proposed here is a certain understanding of the semantic view. Although it stands as a body of doctrine in a clearer way than the syntactic one does, there are still important divergences between authors.

> the existence of a hierarchy of theories arising from the methodology of experimentation for testing the fundamental theory is an essential ingredient of any sophisticated scientific discipline. (Suppes, 1967, pp. 63–64)

The goal of the semantic view is not to force theories into a predefined mold. Quite the opposite; it aims at designing tools to explore the content of different types of bodies of knowledge, commonly acknowledged as theories.[48]

A point on which the semantic view has received much criticism during the last decades (Frigg, 2002, 2006; Suárez, 1999, 2003) is that it does not account for the representational use of theories. Formal tools allow for the identification of structural relationships between models (e.g., isomorphism, embedding) in virtue of which one model can be said to represent another one. But here, "representation" refers to a structural relationship, which has nothing to do with the intentional aspects of representation. Formal reconstruction accounts neither for how phenomena are structured into models of data nor for how agents use these models and these relations to draw inferences about the phenomena they study. Although this is certainly true, it should be viewed as an external criticism: this highlights the limitations of the semantic view's scope. But neither Suppes nor van Fraassen claim to give a formal account of these pragmatic aspects of theorizing.

Now, once its purpose is clearly delineated, the question still remains of whether the semantic view is successful or not. The rest of this section will tackle this very issue. In conclusion, we will formulate arguments for an internal criticism of both the semantic view and the formal reconstruction project in general.

3.4.1 The Virtues of Formal Reconstruction

Let us come back to the initial motivation and goals of formal reconstruction, as stated in subsection 1.3. We have seen that axiomatization is a way to present the objective content of a theory, beyond the perspective effects prompted by its actual formulations. According to Suppes (1968), this has the following virtues. First of all, by making totally explicit the concepts used in various theories (Suppes, 1968, p. 654), the axiomatization enterprise opens the path for a standardization of scientific language. In turn, this should make inter-disciplinary communication easier, as well as facilitating scientific teaching; this also opens a promising route toward the unity of science (Suppes, 1968, p. 654). Moreover, by isolating the self-contained assumptions of a theory, one avoids ad hoc verbalizations (Suppes, 1968, p. 655). Last but not least, Suppes states that a good axiomatization, by isolating the minimal hypotheses that are indispensable for the expression of a theory, enables one to reach an otherwise inaccessible level of generality and objectivity (Suppes, 1968, p. 656).

As a program of actual axiomatization of existing theories, the semantic conception has undeniably proved fruitful. It provides tools for studying inter-theoretical

48 See (Suppe, 1977, pp. 62–66), and also Rapoport's (1958) taxonomy.

relations by establishing representation theorems: Suppes (2002) has applied this method to various scientific domains (probability theory, space-time theories, classical and quantum mechanics, language theories). In physics, and more precisely quantum mechanics, Hughes's (1989) and van Fraassen's (1991) works show that the semantic conception is able to shed light on some problems internal to scientific theories. In biology, various axiomatizations have been proposed for evolutionary theories and population genetics (e.g., Beatty, 1982; Lloyd, 1988; Thompson, 1989, 2007). They have contributed to clarifying concepts and to examining the status of different principles, also shedding light on the kind of explanation and confirmation at play in this domain, in which the notion of law is considered highly problematic. As a research program, the semantic conception of theories, through its concrete implementation, shows one of its alleged superiorities over the logical empiricist conception: it offers powerful tools for studying existing scientific theories rather than describing the ideal form that a theory should in principle be able to take, but that none takes in fact.

Beside this undeniable success, however, there are some reasons to question whether the semantic view really provides us with the tools to identify the content of theories beyond their actual formulations. Whether or not this is Suppes's goal is not clear—and there are reasons to think it is not. In fact, Suppes's view of axiomatization does not really distinguish a good axiomatization from a good formulation[49]—and there would be scientific reasons to adopt axiomatizations as actual formulations.[50] However, a closer look at the claims of some other advocates of the semantic view, together with a reconsideration of the example of classical mechanics, ends up raising a few challenging issues for the semantic view.

3.4.2 Formulation and Formalization: A Confusion

One of the often advanced superiorities of the semantic conception is that it enables the dismissal of unessential aspects of theory formulations. If various formulations are really expressions of the same theory, axiomatization should be able to show that, even though they seem to say different things, they are in fact descriptions of the same mathematical structures—they are mathematically equivalent.

Although this may sound like a common goal of the two formal approaches studied here, advocates of the semantic view often charge the logical empiricists with a serious confusion between theories and their formulations; moreover, this confusion is supposed to prevent them from accounting for the identity of, for example, classical mechanics, beyond the variety of its formulations. By defining theories as sets of

[49] Indeed, once the project of formalization *stricto sensu* is given up, a particular axiomatization of a theory is not essentially different from a formulation thereof.

[50] This is a notable difference from the logical empiricists' formalization project: Carnap explicitly states that it would be absurd for a scientist to use the Ramsey formulation of theories. On the relation between formal reconstructions and formulations, see the debate between Suppes and Kuhn during the discussion of Kuhn's (1969) paper at the symposium on the structure of scientific theories that gave birth to Suppe's book (1974/1977a, pp. 511–153).

statements, so the argument goes, one is condemned to the claim that different sets of statements must be different theories.[51] However, this accusation is ungrounded and relies on a harmful ambiguity of the term "formulation," together with a misinterpretation of the nature of formalization as conceived by the logical empiricists.

True, logical empiricists construe theories as linguistic entities. But this only means that they propose the reconstruction of their content by means of uninterpreted statements directly related to the empirical world and which acquire a meaning through this relation. These statement forms, before being interpreted, are not at all similar to statements in natural language. Hence, Suppe's charge against the syntactic conception, according to which it identifies theories with formulations of theories, seems to rely on a confusion between natural language (even enhanced with theoretical terms and mathematical formalisms) and formal language. More precisely, this seems to be a confusion between the formulation of a theory and its logical formalization.

Hence, there is no contradiction, for the logical empiricists, in stating that the various linguistic formulations of classical mechanics are expressions of the same logical skeleton—which does not mean that one can easily come up with such a skeleton. In fact, as highlighted in section 1, the common assumption of the two formal approaches is that it is possible, at least in principle, to define what a theory says in such a way that this content, being independent from the agents' understanding of it, is unaffected by the diversity of its formulations. This means that it should be possible to distinguish between what has to do with the content of the theory, on the one hand, and what is a mere effect of its formulation(s), on the other. We have seen, in section 2, that logical empiricists fail in this project. Now, the question is still open as to whether the axiomatic method put forward by the semantic view is or is not able to deal with the identity issue of classical mechanics.

3.4.3 Formal Reconstruction and the Identity of Classical Mechanics

How does the semantic view handle the case of classical mechanics? Does it succeed in showing the structural equivalence of the models satisfying its different formulations? Or does it end up with different axiomatizations, showing that, in the end, the formulations of mechanics are different theories? After all, it would be perfectly acceptable, given the aim of formal reconstruction, to show that the differences between the formulations of classical mechanics are theoretical differences. Whatever the answer, a successful formal reconstruction should come up with a clear answer to the issue of the identity of classical mechanics.

---

[51] "To say that something is a linguistic entity is to imply that changes in its linguistic features, including the formulation of its axiom system, produce a new entity. Thus on the Received View, change in the formulation of a theory is a change in theory." (Suppe, 1989, pp. 3–4) Later, about classical mechanics, he writes: "it is the same theory regardless which formulation is employed" (p. 82). "The mistake, I think, was to confuse a theory with the formulation of a theory in a particular language" (van Fraassen, 1987, p. 109).

Without repeating the whole argument here,[52] let us just state that a closer look into how advocates of the semantic view have handled the case of classical mechanics reveals that, in the end, letting some pragmatic aspects of their use into the picture of theories is unavoidable. Let us consider Suppes's (1957) axiomatization. It clearly appears as an axiomatization of classical mechanics *under its Newtonian formulation*,[53] and as such it does not say anything about its equivalence with other formulations. As the axiomatization language is not essential (this is one of the semantic view's slogans), one chooses as primitives and as mathematical tools the one that is most convenient to one's given goal. In a reconstruction of Newtonian mechanics, it is natural to use the mathematical tools through which this theory is actually expressed. Suppes does not aim at isolating and individuating theories by giving one and only one formal reconstruction of them, but rather at studying the structural relations between different theories—or different formulations of theories. Hence, if the different formulations of classical mechanics are equivalent, this does not necessarily imply that one should end up with one and only one axiomatization. Rather, it implies that one should be able to show the equivalence of the structures described by these various formulations. One might start by separately axiomatizing the formulations and then coming up with mathematical arguments showing their structural identity.

However, a closer look reveals that, in the end, the answer to the question of the equivalence/difference of the models described by the different formulations itself depends on pragmatic factors. Balzer, Moulines, and Sneed (1987) have tried to answer the question of whether axiomatization enables one to show the equivalence of the Newtonian and the Lagrangian formulations. After proposing an axiomatization of each of them (pp. 103–108 for the Newtonian one and pp. 149–155 for the Lagrangian one), they show that: 1. their empirical equivalence can be shown if one specifies their intended domain, which amounts to trivializing the equivalence (pp. 292–295), and that; 2. their complete equivalence is far from being trivial (p. 303). In fact, their three attempts at showing the equivalence fail, from which they conclude that equivalence cannot be shown. For the moment, one could consider that these failed attempts only prove that the issue of equivalence in the formulations of classical mechanics is not a trivial one. However, there are good reasons to conclude that formal tools are not enough, by themselves, to show either the equivalence, or the difference of the formulations. In fact, extra-axiomatic—pragmatic—factors necessarily come into the picture, as we shall now see.

We have seen in section 1 that the different formulations of mechanics describe the dynamics of physical systems in different coordinate systems. As Jill North (2009) has shown, this results in differences in the geometrical structure of the space in which the

[52] It is given in Vorms (2011a).

[53] Indeed, the variables correspond to the concepts of mass and force, and the fundamental axioms are expressed in terms of second-order time derivative.

state of the systems is represented.[54] These differences do not, and cannot, imply any empirical differences: the dynamical constraints imposed by the equations describe the same set of possible trajectories. However, the structure of the underlying geometrical spaces imposes different constraints on the geometrically possible movements. Hence, in order to settle the question of the identity of classical mechanics by means of formal reconstruction, the latter should provide us with a criterion to distinguish between those structures that are actually part of the content of classical mechanics and those that are only an effect of the formulation chosen to describe them. In the present case, one could propose to distinguish between the structure of the geometrical space in which the equations describe the systems' trajectories, on the one hand, and the models intended to have an empirical counterpart, on the other. This would amount to saying that the formulations of mechanics have the same content, although they describe this content at different levels of generality. In this view, the geometrical structure of the space in which trajectories are described is not part of the content of the theory—the theory does not say anything about geometrical structure.

One can certainly defend such a view. However, the formal tools alone are not enough to warrant doing so. This view depends on what one considers to be the object of classical mechanics. Let us recall what Suppes (1968) emphasizes as one virtue of axiomatization: axiomatization enables one to identify the fundamental hypotheses of a theory and, in some cases, to show the equivalence between formulations. But depending on what one considers to be the object of classical mechanics, one will consider that the geometrical structure assigned to the space state either is, or is not, part of the fundamental hypotheses. Arguably, Newton's second law, insofar as it describes the dynamics of systems in a vectorial space, says that the geometrical structure of the world is vectorial. Moreover, if one conceives of mechanics not only as the science of motion in space and time, but also as a theory of space and time as constraints to motion, then, certainly, the different formulations do not rely on the same fundamental hypotheses. In this view, axiomatization enables us to highlight that some substructures of the formulations are identical, but that their theoretical "superstructure" is different. On the other hand, if one considers that classical mechanics only consists in a description of the dynamical evolution of physical bodies in space and time, then axiomatization enables one to show the equivalence of the formulations beyond their apparent, but inessential, differences.

Here is what the analysis has shown: formal reconstruction, by itself, does not suffice to prove either that the formulations of classical mechanics express the same content or that they imply structural differences. It can be *used to show* either of these claims, depending on what one takes to be the object of the theory. In fact, depending on this choice, the same axiomatization will not be made. True, axiomatization is nothing more than a certain way of presenting structures, but depending on what one

[54] Hamiltonian formulation assigns a symplectic structure to space, whereas the Lagrangian one assigns it a metric structure, and the Newtonian one a vectorial structure.

considers to be essential aspects of it, one will be able to show the equivalence or not.[55] In a nutshell, axiomatization is not enough to determine the matter: it is itself partially determined by the epistemic commitments and choices of the agents who build it and draw conclusions from it. Depending on the generality level at which one chooses to axiomatize mechanics—whether one chooses to separately axiomatize the three formulations first and then to compare them, or to axiomatize the Hamiltonian one first, in order to then show that it describes the same structures at a higher level of generality, one may conclude either that the fundamental hypotheses of the formulations are equivalent, or else the opposite.

To summarize, axiomatization alone is not sufficient to distinguish between a genuine structural (theoretical) difference and a pure formulation difference. Again, this distinction depends on the way agents use and understand the theory. Depending on what one considers to be the object of mechanics, and on one's epistemic attitude toward it, the difference between the formulations of classical mechanics can be treated either as a mere formulation difference or else as a genuine theoretical difference. In other words, the boundary between what is due to formulation and what should be isolated by formal reconstruction is determined by parameters which formal tools cannot account for. This is not to deny that axiomatization in the semantic style is a powerful tool for exploring and clarifying the content of theories. Rather, it highlights the internal limits of the formalist project by taking one of its favorite examples—the issue of the identity of classical mechanics beyond its formulations. Although they can certainly shed light on the structural relations between the formulations, formal tools cannot, by themselves, settle the issue, which is, in the end, a pragmatic one.

4. Toward a Pragmatic and Cognitive Approach to Theorizing

Our analysis of both the syntactic and the semantic approaches was intended to highlight the internal limits of any project of formal theory reconstruction. In a nutshell, our main criticism consisted in showing that one cannot capture the content of a theory by entirely abstracting away from the pragmatic and cognitive aspects of theorizing. Whether they rely on this kind of argument or on other criticisms, there have been various proposals, during the last decades, in favor of a *practical turn* in the philosophy of science.[56] A distinctive feature of most of these proposals is that they reject the relevance of theories (whether conceived of as sets of statements or as families of models—in the logical sense) as central units of analysis. The motto is that, in practice, scientists construct, manipulate, and reason with models, rather than apply theories to real world phenomena in a hypothetico-deductive way. As a consequence, analysis of scientific knowledge and activity should concentrate on the models

55 Consider again the choice by Balzer et al. (1987) to separately axiomatize Newtonian and Lagrangian mechanics, as, they claim, their basic concepts and fundamental law are different (p. 149).

56 In 2006, advocates of this new perspective created the Society for Philosophy of Science in Practice.

used by scientists, in learning as well as in research: models, rather than theories, are the representational devices that allow for the prediction and explanation of the empirical phenomena. Models, here, are not construed (only) as mathematical abstract structures, but also as more concrete devices which are partially independent from theories.[57] As such, their elaboration and use requires skills that cannot be reduced to the mere implementation of logical rules, but rather include invention, imagination, and rules of thumb. Thus, insofar as it is centered on modeling practices, scientific theorizing cannot be fully captured by formal reconstruction. The advocates of this new perspective on science can be described as being the heirs of Thomas Kuhn, insofar as Kuhn (1962/1970) emphasized that scientific knowledge is composed as much of knowing-how as of knowing-that;[58] indeed, in parallel to his thoughts about incommensurability, in which paradigms are viewed as global, encompassing conceptual entities (see chapter 6 of this volume), Kuhn also insisted on the local and concrete dimension of the representing and experimenting devices at the core of scientific training and practice.[59]

Without blurring the boundaries between epistemological and socio-historical approaches to science (see chapter 7 of this volume), this new perspective contributes to developing a fruitful dialogue between the two disciplines. Indeed, for philosophers aiming at accounting for the actual practice of science, case studies are doomed to play a central role, rather than a purely illustrative or anecdotal one. Moreover, attention to public, visual representations, and to the concrete, material aspects of scientific practice, is a traditional topic within the social studies of science (e.g., Latour and Woolgar, 1979; Lynch and Woolgar, 1990). Another notable aspect of this practical turn is the growing interest paid to cognitive science. Adopting the agents' point of view, and focusing on their situated understanding and representational practices, some philosophers of science and cognitive scientists try to clarify the cognitive underpinnings of model-based reasoning as a way to shed new light on the development of theoretical knowledge.[60]

Giving justice to the variety of this growing and heterogeneous field of research would go too far beyond the scope of this chapter. In the following, we will first focus

[57] The topic of models as autonomous, mediating instruments was initially developed by Cartwright (1999) and by Morgan and Morrison (1999), who initiated a new perspective on classical topics such as theory confirmation, explanation, measurement, etc.

[58] This pragmatic turn in the philosophy of science—which consists in studying what scientists *do*, rather than (only) the abstract structures that are supposed to represent the phenomena—also shows in the vocabulary used to describe scientific knowledge: rather than theories, one would speak of "theorizing." For instance, Ian Hacking's (1983) book is entitled *Representing and Intervening*, thus suggesting that Hacking is more interested in the act of representing than in representation as a relation between two entities.

[59] In fact, these two aspects of Kuhn's thought (a global, holistic view of paradigms on the one hand, and a local, fine-grained analysis of scientific practice on the other) are rather incompatible.

[60] In particular, some philosophers of science (Nersessian, 1984, 1992a, 1992b, 1999, 2002a, 2002b, 2007, 2008; Giere, 1988, 1992, 2006; Magnani, Nersessian, & Thagard, 1999; Magnani & Nersessian, 2002) find in the cognitive theory of mental models (Johnson-Laird, 1983; Gentner & Stevens, 1983) a fruitful hypothesis to explain the use of models in both scientific learning and theory development.

on one particular view of models as representational devices, put forward as a criticism of the semantic view and which has been quite influential. We will then suggest that this view is still centered on too abstract a conception of representation and theorizing, and, finally, we will sketch out new directions for the study of theorizing that are now under exploration.

4.1 MODELS AND MODELING: A NEW ORTHODOXY

Whether they insist on external representational devices or on internal (mental) representations, the various proposals belonging to the pragmatic and cognitive turn take *models* as central units of analysis. The term "model" is far from being univocal though, its referents ranging from concrete, three-dimensional objects to mental representations, including abstract mathematical structures, fictional or imaginary entities, equations, diagrams, etc. All these devices can be qualified as "representations," though in rather different senses. Our goal here is not to force all these uses of the term into one unique mold, which would be both difficult and pointless. Rather, paying heed to this polysemy, we will focus on one particular type of entity referred to by the term "models," namely idealized systems such as the simple pendulum in classical mechanics, perfectly isolated populations in population genetics, or perfectly rational agents in economics. The central role of these kinds of models in scientific practice is at the core of one important type of criticism that has been brought against formal approaches. In the course of the presentation we will find other uses of the term, whose links with the one studied here should appear more clearly.

In subsection 3.2, we have seen that one of the strongest assumptions of the semantic view is that the models used in scientific practice, such as the simple pendulum, can be construed as logical models. This enables advocates of the semantic view to describe the empirical application of theories as the establishment of a structural relation between theoretical models and physical models, which themselves are isomorphic (or structurally analogous in some way) to the model of the phenomena. In so doing, they claim to be closer to scientific practice than the logical empiricists. This assumption has been much criticized (Downes, 1992; Morrison, 1999; Cartwright, 1999; Suárez, 1999; Frigg, 2002, 2006, 2010; and Godfrey-Smith 2006); on various grounds, these critics argue that this analysis is still much too abstract and remote from actual scientific practice, and that scientific models cannot be accounted for in purely structural terms. As suggested in subsection 3.4, many of these criticisms can be viewed as emphasizing the external limits of formal approaches, without showing them to be internally flawed.[61] In any case, this does not make their positive claims

[61] In fact, Ronald Giere (1988, 2006) advocates the semantic view while proposing a conception of the use of models that is very close to the one presented in this section, also drawing on some results in cognitive science to account for the psychological processes underlying such a use. We will not present his view in this section; as far as the questions at stake here are concerned, it is closer to Frigg's or Godfrey-Smith's views.

less relevant for the study of scientific practice. Let us now focus on one of them, which has been put forward in various forms (e.g., Suárez, 1999; Frigg, 2002, 2006, 2010; and Godfrey-Smith, 2006).

One core idea is that the establishment of a symmetrical, structural relationship between theoretical models (i.e., the mathematical structures described by the theory) and physical models (such as the simple pendulum) is not enough to account for the intentional aspects of representation, which involves (at least) three entities: an agent, a *representatum*, and a *representans*. Hence, more attention should be paid, so the argument goes, to the details of the process by which agents *use* physical models to *represent* the external world phenomena.

Consider the equations of a theory like Newtonian mechanics. They do not apply, as such, to the empirical phenomena. Rather, they describe a mathematical structure (which satisfies them). Representing the behavior of a grandfather clock by means of an equation of motion requires a series of operations involving idealization, abstraction, and approximation procedures (e.g., ignoring frictions and air resistance, but also non relevant properties such as the color and materials of the clock).[62] Whatever its level of precision, an equation of motion does not describe, strictly speaking, the behavior of a grandfather clock, but rather of an "idealized version" (Frigg, 2010) of it, such as the simple pendulum.

An important part of scientific theorizing, according to these critics, consists in designing these kinds of idealized versions of empirical systems, which are commonly called "models."[63] This essentially consists of two operations (Godfrey-Smith, 2006; Frigg, 2010): (1) presenting a hypothetical idealized system by means of a description—such as a linguistic statement ("imagine a point mass suspended in the void to a mass-less thread") or an image like Figure 2, which enables one to write down the equation describing its behavior; (1) stating that this system *represents* the portion of the empirical world under study (e.g., the motion of the pendulum in some actual grandfather clock). In what sense is this account different from the semantic view, according to which theoretical models can be considered as representing models of the phenomena? *Pace* the semantic view, Frigg (2010) and Godfrey-Smith (2006) argue that idealized systems like the simple pendulum do not reduce to models in the logical sense, that is, to mathematical structures. According to them, the simple pendulum, or the isolated population, are not structures, but rather "imagined concrete things": "An imaginary population is something that, if it was real, would be a flesh-and-blood population, not a mathematical object" (Godfrey-Smith, 2006, pp. 734–735). Certainly, many features of a flesh-and-blood population, or of a real, wooden clock, are irrelevant. "When asked to imagine an evolving population, we will usually be told what the mating system is, but not the number of toes that the organisms have" (Godfrey-Smith, 2006, p. 735). Model systems are schematic, and idealized: "what is important

[62] For a detailed analysis and comparison of idealization and abstraction, see Thomson-Jones (2005).

[63] These considerations do not only apply to physics. See Godfrey-Smith (2006) for examples in population genetics and evolutionary theory.

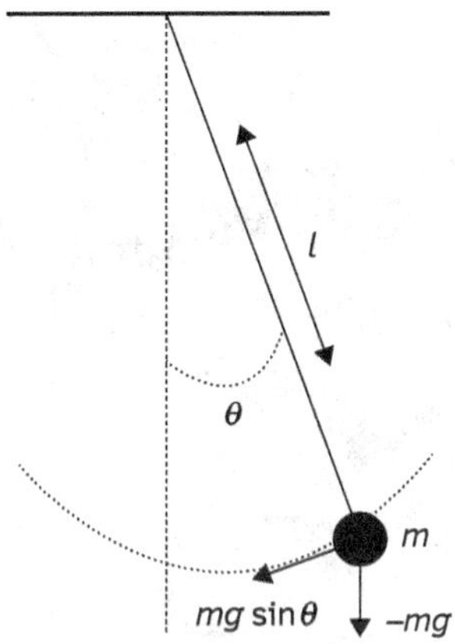

FIGURE 2 The simple pendulum.

is usually not a single imagined system but a *collection* of them" (p. 735). Nevertheless, they are not abstract like a mathematical object, since they share features with real objects that mathematical structures do not have. When one imagines an object like the simple pendulum, one represents it under the traits of a concrete object, which is only underdetermined by the mathematical structure it instantiates. And this is crucial to its representational role: it is in virtue of the concrete aspects of the simple pendulum that one is able to represent a grandfather clock *as* a simple pendulum. Or, in other words, it is in virtue of these concrete aspects of the imaginary pendulum that agents succeed in representing a real pendulum (the bob of a grandfather clock) as an object consisting of a point mass suspended to a mass-less thread, whose behavior can then be described by means of the simple pendulum equation. In Kuhnian terms, it is thanks to this imaginary entity that agents understand the meaning of the equations, and that they are able to see concrete empirical states of affairs *as* Newtonian states of affairs, and thus to predict and explain them.

Hence, in the conception of scientific representation put forward by Godfrey-Smith (2006), as well as by Frigg (2010), an imaginary entity, which they propose to construe as a fictional entity, plays the central role.[64] Various representational relations are at play, as appears in Figure 3. The equations of motion *describe* a mathematical structure ("model structure" in Frigg's picture). This model description is necessarily presented in some particular form (mathematical formalism, words, diagram, etc.).

[64] Godfrey-Smith (2006, p. 735) claims that "although these imagined entities are puzzling, [. . .] they might be treated as similar to something that we are all familiar with, the imagined objects of literary fiction." Frigg (2010) develops this conception of models as fictions by drawing from Kendall Walton (1990). This conception of scientific representation as analogous to fiction was already found in more ancient works in the philosophy of science, such as Vaihinger's (1911). More recently, various philosophers of science have found in philosophical analysis of literary fiction some fruitful tools to account for scientific representation (see Cartwright, 1983; Fine, 1993; Elgin, 1996; Barberousse & Ludwig, 2009), as well as the collection of essays dedicated to this topic edited by Suárez (2009). Giere (2009), however, argues against this conception and prefers to construe models as "abstract" entities.

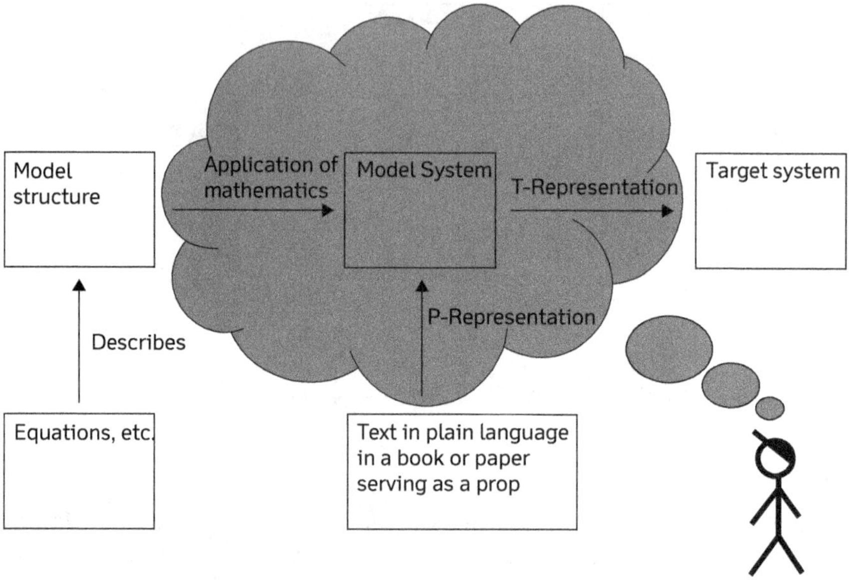

FIGURE 3 Scientific representation according to Frigg (2010)
Source: Frigg (2010).

Its representational relation with the model structure is quite simple: it specifies, or defines a structure of which it is true. As noted in subsection 3.3, truth, here, has no epistemological significance: the model structure is trivially true of its definition.[65] Thus, one can establish a mathematical relation between this structure and the imaginary system (the "model system"). The imaginary system, itself, is represented by means of a linguistic description (or by a diagram), just like Julien Sorel is described by means of statements in *Le Rouge et le Noir*. This relation is what Frigg calls "P-representation." Finally, there is another representational relation between the (fictional) model system and the (real) target system, which Frigg calls "T-representation." Hence, according to both Frigg (2010) and Godfrey-Smith (2006), the mathematical structures specified by the theory cannot be said to represent the real world phenomena, because the picture is much more complicated and because the central representational role is played by an entity that is neither a pure mathematical structure nor a real-world object.

We will now argue that the complicated picture arising from this analysis—implying that one clarifies both the ontological status of fiction and its representational role—is based on still too abstract a conception of representation. This is not to deny that idealizations such as the simple pendulum should be treated as fictions; rather we contest that the representational dimension of theorizing lies in the relation between such fictional entities and real systems.

[65] See Thomson-Jones (2006) for a critical appraisal of the semantic view, based on a detailed analysis of the different ways in which something can play the role of a truth-maker.

4.2 FOCUSING ON REPRESENTATIONAL FORMS

Theorizing certainly relies on idealization, and models such as the simple pendulum do play an important role in theory understanding. As Nagel taught us, imaginary entities such as perfectly elastic billiard balls give flesh to the theory, enabling us to represent what it says. Moreover, as Kuhn (1969) famously emphasized, scientific training relies on the resolution of simple problems involving such fictional models. This is in fact how one becomes capable of solving further, more complex problems: the use of a theory to study real systems requires that one represents these systems in an idealized way—*as* simple pendulums, perfectly rational agents, or perfectly isolated populations. However, when one focuses on the details of the use of models in prediction and explanation, it becomes clear that this does not consist in using the simple pendulum *as such* to represent real systems.

Consider the resolution of a problem in classical mechanics. First, one needs to identify the forces applied to the system (while approximating and abstracting away from irrelevant features) so as to write down the equations of motion under the appropriate form. As Kuhn (1969) taught us, one acquires this skill by solving increasingly complex problems (*exemplars*). Then, once the initial conditions are known, one can proceed to the solution of the equation. Whether applied to an imaginary case (like in textbook exercises) or to a real case, the representations that one manipulates are equations whose particular form is crucial for the inferences one can draw. As we have seen in section 1, representing a real world situation "by means of the simple pendulum" does not consist in the same representational and inferential operations if one uses a Newtonian rather than a Lagrangian equation. This is not specific to the case of classical mechanics: one never reasons with a fictional entity *in abstracto*, but rather with a particular, concretely formatted representation or description (even though it is only mentally represented). The simple pendulum, which may be a useful fictional entity for understanding classical mechanics and for learning how to identify the relevant parameters in a real pendulum, is not, in itself, a representation. One needs to represent it under a certain form—Newtonian, Lagrangian, or Hamiltonian equations. And representing real systems by means of the pendulum means nothing more than using this particular form to represent them.[66]

Things become clearer when one considers another type of theoretical representation, such as the double helix model of DNA. Consider the concrete, material model Watson and Crick constructed and with the aid of which they reasoned. One can certainly claim that this metallic construction is the concrete representation of an idealization. Indeed, just like the equation of the pendulum, it does not strictly speaking

---

[66] "Students in physics regularly report that they have read through a chapter of their text, understood it perfectly, but nonetheless had difficulty solving the problems at the chapter's end. Almost inevitably their difficulty is in setting up the appropriate equations, in relating the words and examples given in the text to the particular problems they are asked to solve. Ordinarily, also, those difficulties dissolve in the same way. The student discovers a way to see his problem as like a problem he has already encountered. Once that likeness or analogy has been seen, only manipulative difficulties remain" (Kuhn, 1969, p. 470).

describe a particular DNA molecule, but rather expresses a theoretical hypothesis according to which DNA molecules can be represented under a double helicoid form. If one refers to Frigg's diagram, the metallic construction Watson and Crick worked on plays the same role as the statements ("text in plain language") that describe the imaginary system, though it is not itself an imaginary system.

However, it is both onerous and pointless to state that the role of this concrete, material model is to enable us to access an imaginary entity, which in turn would itself serve as a representation of real molecules. True, the material model is used to draw inferences about the structure of DNA molecules in general, abstracting away from the specifics of particular, real molecules. It certainly expresses a general theoretical hypothesis. But in order to draw inferences about real molecules, one would use the material model as a representation of these molecules (which implies being able to select the relevant features and abstract away from the dissimilarities—for example, DNA molecules are not metallic) rather than as a description of an abstract entity that would itself represent the molecules.[67] It is the concrete, real model that allows one to draw inferences. One could certainly consider that all the material, three-dimensional models of DNA molecules belong to a class of equivalent representations. But this class does not correspond to the idealized models that Frigg, Godfrey-Smith, and Giere have in mind; indeed, they are equivalent for scientific reasoning only insofar as they facilitate the same inferences.[68] The simple pendulum, as we have seen, can be described under various formulations that are not equivalent from this inferential point of view. As such, it is still too abstract to be reasoned with.

Hence, in order to get a clearer view of scientific representation, we ultimately suggest that one should pay more attention to the concrete representational devices that are used, and to their particular format (see Vorms, 2011b). Rather than abstract models, local, concrete artifacts (Knuuttila, 2011) seem to be appropriate units of analysis for shedding light on the articulation between the representational and the inferential dimensions of theorizing. Some philosophers of science have already started exploring such a route, in diverse ways. In the philosophy of biology and of chemistry, the importance of diagrams and images is now rather commonly

[67] Quite surprisingly, Ronald Giere (2006, pp. 105–106), after emphasizing the importance of the concrete manipulation of various types of representations (diagrams, equations, drawings, etc.), states that "the expert is using the external representations in order to *reconstruct* aspects of the abstract model relevant to the problem at hand. [. . .] Watson and Crick's physical model of DNA, for example, also served the purpose of specifying some features of an abstract model of DNA, such as the pitch of the helix and the allowable base pairs. Other features of the physical model, such as being made partly of sheet metal, have no counterpart in the abstract model." It seems more relevant to note that they have no counterpart in the real molecules. Moreover, if one sticks to Giere's conception of representation as relying on resemblance relations, it is hard to see how an abstract model can resemble a real-world object at all. If one wants to speak of resemblance, it should rather be between the concrete representation and the real system.

[68] Of course, all representations of the simple pendulum are equivalent in the same sense as the different formulations of classical mechanics are equivalent—a sense quite remote from the inferences agents actually draw.

acknowledged (Griesemer 1991a, 1991b; Perini, 2005; Sheredos et al., 2013; Wimsatt, 1990; Griesemer and Wimsatt, 1989; Woody, 2004), and philosophers of economics have emphasized the relevance of attention to the materiality of representation (Morgan and Knuuttila, 2012; Morgan, 2012). Another growing field that is worth mentioning in conclusion is the study of computational science: as Paul Humphreys (2004) notably argued, since computer simulations have become central to entire fields of research, the computational aspects of the "templates" that are implemented by machines become at least as important as the representational content of the theories they are drawn from. In fact, as he suggests, focusing on the "syntax"—here, "syntax" refers to the particular form of a given representation, rather than on its logical, syntactic skeleton, which the formalist enterprise aims at extracting—of the particular devices that are used may lead to a reorganization of the disciplinary landscape. In such a new cartography, sciences would also be grouped according to the formats they use, rather than (only) according to the domain of phenomena they stand for.

References

Achinstein, P. (1965), "The Problem of Theoretical Terms," *American Philosophical Quarterly*, 2, pp. 193–203.

Achinstein, P. (1968), *Concepts of Science: A Philosophical Analysis*, Baltimore: Johns Hopkins University Press.

Balzer, W. (1985), "On a New Definition of Theorecity," *Dialectica*, 39, pp. 127–145.

Balzer, W, Moulines, C.U., and Sneed, J. (1987), *An Architectonic for Science. The Structuralist Program*, Dordrecht: D. Reidel Publishing Company.

Barberousse, A., and Ludwig, P. (2009), "Models as Fictions," *in* Suárez, M. (ed.), *Fictions in Science. Philosophical Essays on Modelling and Idealisation London*, Routledge, pp. 56–73.

Beatty, J. (1982), "What's Wrong with the Received View of Evolutionary Biology?" *Proceedings of the 1980 Biennial Meetings of the Philosophy of Science Association*, vol. 2, East Lansing, MI: Philosophy of Science Association, pp. 397–426.

Beckner, M. (1959), *The Biological Way of Thought*, New York: Columbia University Press.

Beth, E. (1940), "Toward an Up-to-Date Philosophy of the Natural Sciences," *Methodos*, 1, pp. 178–185.

Black, M. (1962), *Models and Metaphor*, Ithaca, NY: Cornell University Press.

Bolztmann, L. (1897), *Vorlesungen über die Principe der Mechanik*, Leipzig: Johann Ambrosius Barth.

Campbell, N.R. (1920), *Physics: The Elements*, Cambridge: Cambridge University Press.

Carnap, R. (1932), "Die Physicalische Sprache als Universalsprache der Wissenschaftssenschaft," *Erkenntnis*, 2, pp. 432–465.

Carnap, R. (1934a), *Die Aufgabe der Wissenschaftslogik*, Vienna: Gerold, Einheitswissenschaft.

Carnap, R. (1934b), *Logische Syntax der Sprache*, Vienna: Julius Springer, Schriften zur wissenschaftlichen Weltauffassung.

Carnap, R. (1936/1937), "Testability and Meaning," *Philosophy of Science*, 3, pp. 419–471; 4, pp. 1–40.

Carnap, R. (1939), "Theories as Partially Interpreted Formal Systems," *in Foundations of Logic and Mathematics, International Encyclopedia of Unified Science*, 1(3), Chicago: University of Chicago Press.

Carnap, R. (1952), "Meaning Postulates," *Philosophical Studies*, 3(5): 65–73.

Carnap, R. (1955), "Meaning and Synonymy in Natural Languages," *Philosophical Studies*, 6(3): 33–47.

Carnap, R. (1956), "The Methodological Character of Theoretical Concepts," *in Minnesota Studies in the Philosophy of Science. The Foundations of Science and Concepts of Psychology and Psychoanalysis*, vol. 1, Minneapolis: University of Minnesota Press, pp. 38–76.

Carnap, R. (1966), *Philosophical Foundation of Physics*, New York: Basic Books.

Carnap, R., Hahn, H., and Neurath, O. (1929), "Wissenschaftliche Weltauffassung. Der Wiener Kreis," *in Veröffentlichung des Vereines Ernst Mach Heft I*, Vienna: Artur Wolf Verlag.

Cartwright, N. (1983), *How the Laws of Physics Lie*, Oxford University Press.

Cartwright, N. (1999), *The Dappled World: A Study of the Boundaries of Science*, Cambridge: Cambridge University Press.

Downes, S. (1992), "The Importance of Models in Theorizing: A Deflationary Semantic View," *in* Hull, D., et al. (eds.) *Proceedings of the Philosophy of Science Association*, vol. 1, East Lansing, MI: Philosophy of Science Association, pp. 142–153.

Duhem, P. (1914), *La Théorie Physique, son objet, sa structure*, 2nd ed., Paris: Chevalier et Rivière. English translation, Wiener, P. (1954), *The Aim and Structure of Physical Theory*, Princeton, NJ: Princeton University Press.

Elgin, C. Z. (1996), *Considered Judgment*, Princeton, NJ: Princeton University Press.

Fine, A. (1993), "Fictionalism," *Midwest Studies in Philosophy*, 18, pp. 1–18.

Frigg, R. (2002), "Models and Representation: Why Structures Are Not Enough," *Measurement in Physics and Economics Project Discussion Paper Series*, DP MEAS 25/02, London School of Economics.

Frigg, R. (2006), "Scientific Representation and the Semantic View of Theories," *Theoria*, 55, pp. 49–65.

Frigg, R. (2010). "Models and Fiction," *Synthese*, 172(2), pp. 251–268.

Gentner, D., and Stevens, A. (1983), *Mental Models*, Mahwah, NJ: Lawrence Erlbaum Associates.

Giere, R. N. (1979), *Understanding Scientific Reasoning*, New York: Holt, Rinehart and Winston.

Giere, R. N. (1988), *Explaining Science. A Cognitive Approach*, Chicago: University of Chicago Press

Giere, R. N., ed. (1992), *Cognitive Models of Science*, Minnesota Studies in the Philosophy of Science, Vol. XV, Minneapolis: University of Minnesota Press.

Giere, R. N. (2006), *Scientific Perspectivism*, Chicago: University of Chicago Press.

Giere, R. N. (2009), "Why Scientific Models Should Not be Regarded as Works of Fiction," *in* Suárez, M. (ed.), *Fictions in Science. Philosophical Essays on Modelling and Idealisation*, London: Routledge, pp. 248–258.

Godfrey-Smith, P. (2006), "The Strategy of Model-Based Science," *Biology and Philosophy*, 21, pp. 725–740.

Goldstein, H. (1950/2002), *Classical Mechanics*, Reading, MA: Addison-Wesley, 3rd ed.

Griesemer, J.R. (1991a), "Material Models in Biology," *in* Fine, A., Forbes, M., and Wessels, L. (eds.), *PSA 1990: Proceedings of the Biennial Meeting of the Philosophy of Science Association*, vol. 2, East Lansing, MI: Philosophy of Science Association, pp. 79–93.

Griesemer, J. R. (1991b), "Must Scientific Diagrams Be Eliminable? The Case of Path Analysis," *Biology and Philosophy*, 6, pp. 155–180.

Griesemer, J. R., and Wimsatt, W. C. (1989), "Picturing Weismannism: A Case Study of Conceptual Evolution," *in* Ruse, M. (ed.), *What the Philosophy of Biology Is. Essays for David Hull*, Dordrecht: Kluwer Academic Publishers, pp. 75–137.

Hacking, I. (1983), *Representing and Intervening*, Cambridge: Cambridge University Press.

Hempel, C. G. (1950), "Problems and Changes in the Empiricist Criterion of Meaning," *Revue Internationale de Philosophie*, 41(11), pp. 41–63.

Hempel, C. G. (1951), "The Concept of Cognitive Significance: A Reconsideration," *Proceedings of the American Academy of Arts and Sciences*, 80, pp.61–77.

Hempel, C. G. (1958), "The Theoretician's Dilemma," *in* Feigl, H., Scriven, M., and Maxwell, G. (eds.), *Concepts, Theories, and the Mind-Body Problem. Minnesota Studies in the Philosophy of Science*, vol. 2, Minneapolis: University of Minnesota Press. Reprint. *in* Hempel, 1965b, pp. 173–226.

Hempel, C. G. (1965a), "Aspects of Scientific Explanation," *in* Hempel (1965b), pp. 331–496.

Hempel, C. G. (1965b), *Aspects of Scientific Explanation and Other Essays in the Philosophy of Science*, New York: Free Press.

Hempel, C. G. (1970), "On the 'Standard Conception' of Scientific Theories," *in* Radner, M., and Winokur, S. (eds.), *Analyses of Theories and Methods of Physics and Psychology*. Minnesota Studies in the Philosophy of Science, vol. 4, Minneapolis: University of Minnesota Press, pp. 142–163.

Hempel, C. G., and Oppenheim, P. (1948), "Studies in the Logic of Explanation," *Philosophy of Science*, 15, 135–175.

Hertz, H. (1894), *Die Prinzipien der Mechanik. Gesammelte Werke*, vol. 3, Leipzig: Barth.

Hesse, M. (1966), *Models and Analogies in Science*, Notre Dame, IN: Notre Dame University Press.

Hilbert, D. (1899), *Grundlagen der Geometrie*, Leipzig: Teubner.

Hughes, R. (1989), *The Structure and Interpretation of Quantum Mechanics*, Cambridge, MA: Harvard University Press

Humphreys, P. (2004), *Extending Ourselves. Computational Science, Empiricism, and Scientific Method*, Oxford: Oxford University Press.

Johnson-Laird, P. (1983), *Mental Models*, Cambridge: Cambridge University Press.

Kirchhoff, G. (1877), *Vorlesungen über mathematische Physik: Mechanik*, Leipzig: B. G. Teubner.

Knuuttila, T. (2011), "Modelling and Representing: An Artefactual Approach to Model-Based Representation," *Studies in History and Philosophy of Science Part A*, 42(2), pp. 262–271.

Kuhn, T. S. (1962/1970), *The Structure of Scientific Revolutions*, Chicago: University of Chicago Press (1970, 2nd edition, with postscript).

Kuhn, T. S. (1969), "Second Thoughts on Paradigms," *in* Suppe, F. (ed.), 1974/1977a, pp. 459–482.

Lagrange, J.-L. (1788), *Mécanique analytique*.

Lanczos, C. (1970), *The Variational Principles of Mechanics*, 4th ed., New York: Dover.

Latour, B., and Woolgar, S. (1979), *Laboratory Life: The Construction of Scientific Facts*, Princeton, NJ: Princeton University Press.

Lloyd, E.A. (1988), *The Structure and Confirmation of Evolutionary Theory*, New York: Greenwood Press.

Lynch, M.E., and Woolgar, S. (1990), *Representation in Scientific Practice*, Cambridge, MA: MIT Press.

Mach, E. (1883), *Die Mechanik in Ihrer Entwicklung Historisch-Kritisch Dargestellt*, Chicago: Open Court Publishing Company. Trans. T. J. McCormack, 1893.

Magnani, L., Nersessian, N.J., and Thagard, P., eds. (1999), *Model- Based Reasoning in Scientific Discovery*, Berlin: Springer.

Magnani, L., and Nersessian, N. J., eds. (2002), *Model-Based Reasoning. Science, Technology, Values*, Berlin: Springer.

Mongin, P. (2003), "L'axiomatisation et les théories économiques," *Revue économique*, 54(1), pp. 99–138.

Moulines, C.-U. (1975), "A Logical Reconstruction of Simple Equilibrium Thermodynamics," *Erkenntnis*, 9.

Morgan, M.S. (2012), *The World in the Model: How Economists Work and Think*, Cambridge: Cambridge University Press.

Morgan, M.S., and Knuuttila, T. (2012), "Models and Modelling In Economics," *in* Mäki, U. (ed.), *Philosophy of Economics. Handbook of the Philosophy of Science*, 13, Amsterdam: Elsevier, pp. 49–87.

Morgan, M.S., and Morrison, M., eds. (1999), *Models as Mediators: Perspectives on Natural and Social Science*, Cambridge: Cambridge University Press.

Morrison, M. (1999), "Models as Autonomous Agents," *in* Morgan and Morrison (1999), pp. 38–65.

Nagel, E. (1961), *The Structure of Science: Problems in the Logic of Scientific Explanation*, New York: Harcourt, Brace and World, Inc.

Nersessian, N. J. (1984), *Faraday to Einstein: Constructing Meaning in Scientific Theories*, Dordrecht: Kluwer.

Nersessian, N. J. (1992a), "How Do Scientists Think? Capturing the Dynamics of Conceptual Change in Science," *in* Giere, R. N. (ed.), *Cognitive Models of Science*, Minneapolis, University of Minnesota Press, pp. 3–45.

Nersessian, N. J. (1992b), "In the Theoretician's Laboratory: Thought Experimenting as Mental Modeling," *Proceedings of the Biennial Meeting of the Philosophy of Science Association*, 2, pp. 291–301.

Nersessian, N. J. (1999), "Model-Based Reasoning in Conceptual Change," *in* Magnani, L., Nersessian, N. J., and Thagard, P. (eds.), *Model-Based Reasoning in Scientific Discovery*, New York: Kluwer Academic/Plenum Publishers, pp. 5–22.

Nersessian, N. J. (2002a), "Maxwell and 'the Method of Physical Analogy': Model-Based Reasoning, Generic Abstraction, and Conceptual Change," *in* Malament, D., *Essays in the History and Philosophy of Science and Mathematics*, Lasalle, IL: Open Court, pp. 129–166.

Nersessian, N. J. (2002b), "The Cognitive Basis of Model-Based Reasoning in Science," *in* Carruthers, P., Stich, S., and Siegal, M. (eds.), *The Cognitive Basis of Science*, Cambridge: Cambridge University Press, pp. 133–153.

Nersessian, N. J. (2008), *Creating Scientific Concepts*, Cambridge, MA: MIT Press.

Nersessian, N. J. (2007), "Mental Modeling in Conceptual Change," *in* Vosniadou, S. (ed.), *International Handbook of Conceptual Change*, London: Routledge, 2008, pp. 391–416.

Newton, I. (1687), *Philosophiae Naturalis Principia Mathematica.*

North, J. (2009), "The 'Structure' of Physics: A Case Study," *Journal of Philosophy*, 106, pp. 57–88.

Perini, L. (2005), "Explanation in Two Dimensions: Diagrams and Biological Models," *Biology & Philosophy*, pp. 257–269.

Poincaré, H. (1902), *La Science et l'Hypothèse.*

Poincaré, H. (1905), *La Valeur de la Science.*

Putnam, H. (1962), "What Theories Are Not," *in* Nagel, E., Suppes, P., and Tarski, A. (eds.), *Logic, Methodology and Philosophy of Science: Proceedings of the 1960 International Congress*, Palo Alto, CA: Stanford University Press, pp. 240–251.

Ramsey, F. (1929), "Theories," *in* Braithwaite, R. B., ed. (1950), *The Foundations of Mathematics and Other Logical Essays*, Patterson, NJ: Littlefield and Adams, pp. 212–236.

Rapoport, A. (1958), "Various Meanings of Theory," *American Political Science Review*, 52: 927–988.

Rosenberg, A. (2000), *Darwinism in Philosophy, Social Science and Policy*, Cambridge: Cambridge University Press.

Schaffner, K.F. (1969), "Correspondence Rules," *Philosophy of Science*, 36(3), pp. 280–290.

Sheredos, B., Burston, D.C., Abrahamsen, A., and Bechtel, W. (2013), "Why Do Biologists Use So Many Diagrams?," *Philosophy of Science*, 80, pp. 931–944.

Sneed, J. (1975), *The Logical Structure of Mathematics*, London: Routlege and Kegan Paul.

Sneed, J. (1976), "Philosophical Problems in the Empirical Science of Science: A Formal Approach," *Erkenntnis*, 10, pp. 114–146.

Stegmüller, W. (1976), *The Structure and Dynamics of Theories*, New York: Springer-Verlag.

Suárez, M. (1999), "Theories, Models, and Representation," *in* Magnani, L., Nersessian, N., and Thagard, P. (eds.), *Model-Based Reasoning in Scientific Discovery*, London: Kluwer Academic/ Plenum Publishers, pp. 75–83.

Suárez, M. (2003), "Scientific Representation: Against Similarity and Isomorphism," *International Studies in the Philosophy of Science*, 17(3), pp. 225–244.

Suárez, M., ed. (2009), *Fictions in Science: Philosophical Essays on Modelling and Idealisation*, London: Routledge.

Suppe, F. (1971), "On Partial Interpretation," *The Journal of Philosophy*, 68(3), pp. 57–76.

Suppe, F. (1974/1977a), *The Structure of Scientific Theories*, Champaign: University of Illinois Press.

Suppe, F. (1977b), "Introduction," *in* Suppe, 1974/1977a, pp. 3–241.

Suppe, F. (1989), *The Semantic View of Theories and Scientific Realism*, Champaign: University of Illinois Press.

Suppes, P. (1957), *Introduction to Logic*, Princeton, NJ: Van Nostrand.

Suppes, P. (1959), "Axioms for Relativistic Kinematics with or without Parity," *in* Henkin, L., Suppes, P., and Tarski, A. (eds.), *The Axiomatic Method with Special Reference to Geometry and Physics*, pp. 191–307.

Suppes, P. (1960), "A Comparison of the Meaning and Use of Models in Mathematics and the Empirical Sciences," *Synthese*, 12(2–3), pp. 287–301.

Suppes, P. (1962), "Models of Data," *in* Nagel, E., Suppes, P., and Tarski, A. (eds.), *Logic, Methodology and Philosophy of Science: Proceedings of the 1960 International Congress*, Palo Alto CA: Stanford University Press, pp. 252–261.

Suppes, P. (1967), "What Is a Scientific Theory?," *in* Morgenbesser, S., (ed.), *Philosophy of Science Today*, New York: Basic Books, pp. 55–67.

Suppes, P. (1968), "The Desirability of Formalization in Science," *The Journal of Philosophy*, 65(20), pp. 651–664.

Suppes, P. (2002), *Representation and Invariance of Scientific Structures*, Stanford, CA: CSLI Publications.

Thompson, P. (1989), *The Structure of Biological Theories*, Ithaca: State University of New York.

Thompson, P. (2007), "Formalisations of Evolutionary Biology," *in* Matthen, M. and Stephens, C. (eds.), *Handbook of the Philosophy of Science. Philosophy of Biology*, vol. 2., Amsterdam: Elsevier, pp. 485–523.

Thomson-Jones, M. (2005), "Idealization and Abstraction: A Framework," *in* Jones, M. and Cartwright, N. (eds.) *Correcting the Model: Idealization and Abstraction in the Sciences*, pp. 173–217.

Thomson-Jones, M. (2006), "Models and the Semantic View," *Philosophy of Science*, 73(5), pp. 524–535.

Vaihinger, H. (1911), *The Philosophy of "as If,"* English translation, 1924, London: Kegan Paul.

van Fraassen, B. C. (1980), *The Scientific Image*, Oxford: Oxford University Press.

van Fraassen, B. C. (1987), "The Semantic Approach to Scientific Theories," *in* Nersessian, N. J. (ed.), *The Process of Science*, Dordrecht: Martinus Nihoff, pp. 105–124.

van Fraassen, B. C. (1989), *Laws and Symmetry*, Oxford: Clarendon Press.

van Fraassen, B. C. (1991), *Quantum Mechanics: An Empiricist View*, Oxford: Oxford University Press.

von Helmholtz, H. L. (1847), *Über die Erhaltung der Kraft*, Berlin.

Vorms, M. (2009), *Théories, modes d'emploi: une perspective cognitive sur l'activité théorique dans les sciences empiriques*, PhD dissertation, University Paris 1. https://tel.archives-ouvertes.fr/tel-00462403/document.

Vorms, M. (2011a), *Qu'est-ce qu'une théorie scientifique?* Paris: Vuibert.

Vorms, M. (2011b), "Representing with Imaginary Models: Formats Matter," *Studies in History and Philosophy of Science*, 42(2), pp. 287–295.

Walton, K.L. (1990), *Mimesis as Make-Believe: On the Foundations of the Representational Arts*, Cambridge, MA: Harvard University Press.

Wimsatt, W. (1990), "Taming the Dimensions-Visualizations in Science," *Proceedings of the Biennial Meeting of the Philosophy of Science Association, 1990*, vol. 2: Symposia and Invited Papers, pp. 111–135.

Woody, A. I. (2004), "More Telltale Signs: What Attention to Representation Reveals about Scientific Explanation," *PSA 2002 Proceedings, Philosophy of Science Association*, 71, pp. 80–793.

6

SCIENTIFIC CHANGE

Anouk Barberousse (Sorbonne Université) and Marion Vorms (Birbeck College, London)

1. INTRODUCTION

Many philosophers of science have highlighted the importance of a systematic study of scientific change. For them, the dynamic of science is governed by *theoretical* change, theories following on one from the other. As Laudan et al. (1986) point out, it is the existence of scientific theories, as well as the power of prediction and control they carry, which is the principal explanation behind the position science holds in our culture. Scientific knowledge resides in its theories, and scientific change is the history of the passage from one theory to another; most historians and sociologists of science, on the other hand, do not share these positions.

Even if the topic of scientific change is central to philosophy of science, no consensus is forthcoming in the direction of one particular approach. The first sign of discord is with the abundant vocabulary used in describing the different phases of the evolution of science and with its lack of coherency: "paradigms" are not the same as "research programs," which in turn are not the same as "research traditions." The word "theory" also takes on different meanings in the mouths of different scholars of scientific change. Today's philosophers of science do however agree on a certain number of arguments, laid out by Laudan et al. (1986), which have emerged since the middle of the 20th century. Philosophers' interest in the subject of scientific change and the history of science indeed results from criticism, in the 1960s, of the logical and nonhistorical approach to scientific theories put forward by the successors to the Vienna

Circle. Here is a list of the arguments which seem to be established today and which we will come back to along the course of this chapter:[1]

- The most important units for understanding scientific change are relatively stable conceptual structures, larger in scale than theories, whose internal structure plays only a minor role.
- These conceptual structures are rarely abandoned when faced with purely empirical difficulties.
- Empirical data is not sufficient for determining the choice of one theory among others which cover the same phenomena.
- The *potential* successes of sets of theories are just as important as their proven successes when it comes to theoretical choice.
- No observation is neutral with regard to the theories within which it plays a role.

The questions about which there is no such agreement are the following:

- Do the conceptual structures which stay the most stable throughout the history of science change gradually or rapidly?
- What are the relationships between one of these conceptual structures and its successor?
- What quantity of empirical content is conserved?
- What are the causes of change?
- Are the methodological values associated with these conceptual units liable to change radically?

In studying these questions, philosophy of science is in competition with other approaches which are also focused on them. What is its specific legitimacy? Can it force a method onto history of science? These questions provide the background to the four themes which form the framework of this chapter. The first is that of the continuity or discontinuity of scientific change. Here different versions of the incommensurability thesis between the phases of scientific development will be presented and discussed. This first section will be the longest by far, as it will contain presentations of distinctions and arguments which are also at the heart of the subsequent sections. It will be followed by a section analyzing the stakes of the incommensurability debate for the notion of scientific progress, a notion which sparked many debates during the 20th century. In the third section, we will examine the various explanations which have been proposed for the evolution patterns of science presented in the preceding parts. Some of these explanations suggest that scientific change is necessary or rational. Whether it is so or not will be the subject of the fourth section.

[1] We will take from Laudan et al. (1986) only those positions, which will be elaborated on in the rest of this chapter.

2. Is Scientific Change Continuous?

A naive representation of the history of science would say that it consists in the accumulation of knowledge gained by scientists over time: each generation of researchers leans on the results of their predecessors and, bit by bit, builds up the edifice of scientific knowledge. One first distinction allows us to go beyond that representation, namely the distinction between positive knowledge (the set of all observed facts consolidated into bodies of knowledge which are universally recognized and shared) and theories. The quantity of facts observed within the domain of those phenomena liable to receive scientific explanation certainly does seem to increase with the passing centuries: the assertion that researchers find new facts would be difficult to contest. In contrast, the hypothesis that new theories are developed out of previous ones is more problematic. The result of this is that a large part of the discussions on scientific change, and particularly those involving its continuous (or otherwise) nature, revolve around the distinction between observed facts and theories, whether this be with a mind to enriching or rejecting that distinction.

The body of current scientific knowledge does, admittedly, conserve whole parts of previous theories, and some of our mathematical knowledge, for example, has not changed at all since Antiquity. In the field of empirical science, it sometimes seems like new theories are presented as generalizations of previous ones whose approximations they correct: for example, Einstein's general theory of relativity provides a theoretical framework which allows for the deduction and explanation of Newton's laws. However, during certain episodes of the history of science, an overturning can occur such that it seems as though the positive knowledge inherited from previous generations is reorganized within new theoretical systems which explain and describe it by means of entirely new principles and concepts: these are the "scientific revolutions" which Kant (1787), then Koyré (1957, 1961, 1966), and then Kuhn (1962) spoke of. One of the most famous examples of such an overturning is called quite simply the Scientific Revolution; this episode stretches from the end of the 15th century to the beginning of the 18th, and marks the birth of modern science.

In this section, we will examine the question of whether scientific change is continuous or not. It includes the following questions. In what manner does the passage from one theory, or set of theories, to another take place? What relationship exists between the successive phases of the history of science? We will see that one of the most heatedly debated questions is: can a theory be compared to the one which preceded it? Before answering these questions, an understanding much be reached, as was already suggested in the introduction to this chapter, regarding the unit of analysis for scientific activity. Historians and philosophers have suggested that theories are too small as units and that larger units must be found to study.

2.1 SCIENTIFIC CHANGE ACCORDING TO LOGICAL EMPIRICISM

As has been mentioned, the classical approach to scientific change describes it as the increase of a knowledge set by means of a methodical, empirical investigation and the rational examination of our beliefs about the world. According to this view, the scientific method, developed in the 17th century, allows for the discovery of new facts which are then organized into a rational system which describes and explains them. Each generation holds on to what is true from previous generations' beliefs, corrects what is false and adds new truths. Such a vision thus relies on the premise that a rational method exists which enables us, if followed correctly, to get closer to the truth. Another underlying premise is that of a clear distinction between observed facts and the laws which allow for their organization into an explanatory system. New facts spawn new explanations, which are then either confirmed or contradicted by the facts.

This vision relies on an image of science one could call "positivist"; it was denounced by Kitcher (1993; see chap. 7 of this volume) as belonging to "legend." In the twentieth century, a (sophisticated) version of this positivist image was developed by the successors to the Vienna Circle, among whom Carnap and Hempel. As we shall see, the logical formalization of scientific theories they undertake is often accompanied, in practice, by the omission of the genuinely dynamic and historical aspect of science. It is this double characteristic which so-called historicist approaches criticize, the most famous of these being Thomas S. Kuhn's, which will be presented shortly.

The logical empiricist approach deliberately ignores the way in which, historically speaking, scientific theories are developed. To borrow Hans Reichenbach's (1938) distinction, they look at the "context of justification" and leave the "context of discovery" to the psychologists, sociologists and historians. The context of discovery, according to this distinction, is the set of social, historical and psychological events leading to the formulation of a scientific result. The context of justification, which the philosopher of science has the task of laying out, refers to the context within which a theory is formulated and founded rationally, independently of the contingent circumstances which led to its elaboration. For logical empiricists, history of science has only an illustrative role, since the logical reconstructions they put together are a long way from the theories actually employed by scientists.

In their view, a scientific theory, if it wishes to be more than just vain metaphysics, must contain no term which, ultimately, cannot be defined using terms which refer to observable entities or processes. It is this criterion which guarantees what they call "cognitive significance" (as opposed to expressive or affective significance which poetry, for example, explores and exploits) and, consequently, truth-value and scientific legitimacy. The verificationist theory of meaning, which characterized the first logical positivism movement, advances that, as a criterion for the cognitive significance of a synthetic statement, it must be deducible from a finite set of statements which employ only observational terms. This amounts to saying that a statement of fact's cognitive meaning is entirely dictated by its conditions of truth and falsity. Historicist critics of logical empiricism have taken this theory of meaning as their target, but have seldom commented on later developments.

Before moving onto the historicist positions, let us present the notion of *inter-theoretical reduction*, one of the favored tools of logical empiricists for describing the relationships between scientific theories. On the one hand, this approach is particularly revealing of the presuppositions contained within their conception of scientific change; on the other hand, it has focused a large number of subsequent criticisms. For logical empiricists, a scientific theory can either contradict its predecessor, in which case it purely and simply replaces it, or else it can "reduce" it. Before anything else, let us underline the major presupposition of this conception, that theories can always be *compared* with each other.

The concept of inter-theory reduction (see Nagel, 1961, chap. 11) allows us to study the relationships between successive theories or between theories dealing with different phenomenal domains (physics and biology, for example). So this concept is relevant both to the question of scientific change and to the question of the unity of science (cf. chap. 8 of this volume). To say that a theory T_1 reduces another one T_2, is to say that all the phenomena explained and predicted by T_2 are predictable and explainable with T_1, in accordance with the deductive-nomological model of explanation presented in chapter 1 of this volume. In other words, for T_2 to be reduced by T_1, the former must be a logical consequence of the latter. The explanatory force of T_2 must be contained within T_1. Nagel distinguishes two types of reduction, homogeneous reductions (where the two theories include the same set of concepts) and non-homogeneous reductions.

The first kind is unproblematic from the perspective of logical empiricism: all the concepts of T_2 are either present in T_1, or else are clearly and thoroughly definable using the concepts of T_1. The example Nagel uses is the explanation of Galileo's and Kepler's laws by Newton. The Newtonian theory does indeed allow for the deduction of both Galileo's law of falling bodies, excepting for one caveat (Galileo's law does not involve the distance separating the centers of mass of the earth and the free-falling body), and also of Kepler's laws of areas and periods, thus contributing to the unification of two scientific domains which had until then been distinct: the study of the motion of terrestrial bodies, and the study of the motion of celestial bodies. According to Galileo's law, $x = 1/2gt^2$, where x is the distance traveled by the falling body, g is a constant, and t is time. No concept contained in this law is missing from the Newtonian theory of motion which, on the contrary, contains new concepts, such as those of mass and force. Hence, Galileo's law is a logical consequence of Newton's universal law of gravity $\boldsymbol{F} = -G\,Mm/\boldsymbol{r}^2$, where $\boldsymbol{F}$ is the gravitational force exerted between two bodies of masses M and m, G is a constant, and $\boldsymbol{r}$ is the distance between the two bodies. As we will see, the case of homogeneous reduction, seen as unproblematic by Nagel, was subject to extremely cutting criticism at the hands of Kuhn. In the case of non-homogeneous reduction, the reduced theory contains concepts which do not belong to the reducing theory, and so means must be found to satisfactorily connect them to the concepts of the reducing theory. This type of relationship between theories poses particular problem when it comes to the quest for unity of the sciences (see chap. 8 of this volume), but not especially when it comes to the study of scientific change.

It is plain here that what characterizes this approach to the relationships between theories is that it is fundamentally ahistorical. Indeed, the relationships between successive theories concerning the same domain of phenomena, which are at the heart of scientific change studies, receive the same treatment as the (not necessarily temporal) relationships between theories which don't share the same domain of phenomena, and which may be contemporaneous.

2.2 HISTORICIST CRITICISMS

As we have seen, the logical empiricists are the modern representatives of the cumulative conception of scientific development. Since the 1960s, a number of historians and philosophers of science have proposed approaches that are radically opposed to the logical empiricist view. These criticisms have often been called "historicist" because of their insistence on the essentially dynamic—and thus historic—nature of scientific activity. They are characterized by a certain number of common points, among which is the incommensurability thesis. According to this thesis, if the fundamental hypotheses of a scientific domain change throughout its history, then theories of this domain do not only say different things about the phenomena, but also do not speak about the same thing.

In this section, we will begin by presenting Kuhn's criticisms of logical empiricism, before looking at his own conception of scientific change and the incommensurability of paradigms. Following this, we will present Feyerabend's principal theses on incommensurability, and then finish by proposing some criticisms of the incommensurability thesis.

2.2.1 Kuhn's Criticisms of Logical Empiricism

The Structure of Scientific Revolutions (Kuhn, 1962) caused significant waves in the world of history and philosophy of science, spreading out to the intellectual world as a whole. The book is the source of concepts which have become common currency in philosophy and history of science, the most famous being "paradigm" and "scientific revolution." This latter concept is accorded a crucial role in his analysis of scientific change.

Kuhn's approach opposes, point by point, the conception of science put forward by logical empiricism. This opposition rests on a fundamental disagreement regarding the relationship of human knowledge to its past and the essentially dynamic nature of this knowledge. For Kuhn, the content of a science, along with the reasoning and research methods which characterize it, are all tightly related to its historical development. He does not content himself to opposing another conception of scientific change to the one advanced by logical positivism; his intention is to take the historicity of science seriously. This simple change in perspective is a challenge for philosophy of science itself, as we shall see further on in this section.

In the introduction to his collection *Scientific Revolutions*, Ian Hacking (1981) lists nine aspects regarding which the image of science advanced by Kuhn differs from that held by positivist philosophers.[2] The explicit presuppositions that Kuhn opposes point by point are as follows:

- *Realism*. Science is an endeavor towards discovering the world, supposed to be unique. Statements which are true relative to the world are so independently

[2] These aspects do not exclusively concern scientific change; on the contrary, the frontal opposition stems from taking the history of science into consideration and, consequently, from the realization that scientific change, as such, is a problem for philosophy of science, science being an essentially, and not incidentally, dynamic phenomenon.

of what scientists think; and for each aspect of the world there exists one single best description.

- *Demarcation*. A solid distinction exists between scientific theories and other kinds of belief.
- The scientific endeavor is *cumulative*. Though false starts are common, science, as best it can, is built on what is already known.
- *Theory/observation distinction*. There is a definite contrast between observational reports and theoretical statements.
- *Foundations*. Observation and experience assure the foundations and justifications of hypotheses and theories.
- Theories have a *deductive structure* and theory testing takes place by the deduction of observational findings using theoretical postulates.
- Scientific concepts are *precise*, and the terms used in science have a fixed meaning.
- There is a *context of justification* distinct from the *context of discovery*.
- *The unity of science*. There should exist only one science. The less fundamental sciences are *reducible* to the more fundamental. Sociology is reducible to psychology, psychology to biology, biology to chemistry, chemistry to physics.

Later in this section, we will examine some of Kuhn's arguments against these logical empiricist positions, starting with the last one.

2.2.2 Normal Science and Paradigms

According to Kuhn, far from moving towards the ideal of one unified science, the development of the sciences consists rather in the life and death of successive sciences whose respective zeniths are marked by their periods of "normal science." As soon as a specific science has been individuated, it goes through the following characteristic sequence: *normal science—crisis—revolution—new normal science*. This sequence describes the life and death of successive sciences which, contrary to the positivist ideal of a unified science, succeed each other without any possibility for generalization, reduction or unification. Let us look in more detail at how Kuhn conceives of this succession, starting with the definition he gives for the concept of normal science.

Normal science is "everyday" science, as it is practiced in research laboratories; the science which is taught in the manuals and which attracts public and private financial support. It is much more the result of an accomplishment than some set of eternal questions and practices.

Periods of normal science are characterized by the fact that the principal activity of researchers is the resolving of "puzzles" by which they try to both expand the resolution techniques which have already proven their worth and eliminate the problems remaining within an established body of knowledge. In doing so, they bring about minor modifications to the theories in place by increasing their field of application, and they develop technologies derived from these theories.

Kuhn calls this phase of scientific activity "normal" because it is during this phase that the norms regarding the questions to be solved, the methods for responding to them, and the standards of rationality and scientific legitimacy all come together. The manuals are the vehicles of these norms, providing the exemplars that every student must know by heart in order to belong to the community in question. They also present history of science as a heroic narrative (with "*whiggish*" or conservative features) and thus contribute to the creation of a "mummified" image of science, all the while being indispensable to the development of normal science. Hence, a major characteristic of normal science is being conservative. If every new hypothesis that popped into the head of dreaming students were to be taken seriously, no scientific progress would occur: "normal science does not aim at novelties of fact or theory and, when successful, finds none" (Kuhn, 1962, 52). Normal science isn't ever charged with verifying, let alone falsifying, the central hypotheses of theories. This model, therefore, is in direct opposition to the logical empiricist position according to which scientific activity is a matter of confronting theory and experience.

The theoretical notion Kuhn created to better explore the normal science phase of scientific development is the concept of *paradigm*. As has been often stated (Masterman, 1970), the term "paradigm" is extremely polysemic. In 1969, and again in 1970, Kuhn revisited the concept and gave two principal meanings: in the first, a paradigm is a set of values common to a scientific community, a set of methods, of standards, of generalizations; in the second, a paradigm is an accepted way of resolving problems (and is, thus, one aspect of the first meaning). Paradigm, in the first sense, is also known as "disciplinary matrix" and comprises four types of elements:

- "Symbolic generalizations," which are "those expressions, deployed without question or dissent by group members [. . .]. They are the formal or the readily formalizable components of the disciplinary matrix" (Kuhn, 1962, 182); an example of this is Newton's second law, expressed by $\boldsymbol{F} = m\boldsymbol{a}$.
- "Models," which Kuhn quite vaguely defines as what assure scientists' membership to some school of thought, providing them with "analogies" and sometimes an "ontology." For example, the mechanist model of nature is a model of natural phenomena intelligibility.
- Values, which "are more widely shared among different communities than either symbolic generalizations or models" (Kuhn, 1962/1970, 184), like the quantitative character and accuracy of predictions, and the simplicity, coherence and plausibility of theories.
- "Exemplars," which are "the concrete problem-solutions that students encounter from the start of their scientific education, whether in laboratories, on examinations, or at the ends of chapters in science texts" (Kuhn, 1962/1969, 254–255). Exemplars are paradigms in the second sense.

So, within the paradigm framework, scientific activity consists of resolving "puzzles" and attaining greater precision on a greater variety of situations. The "puzzles" scientists try to resolve are problems which the community consider to be scientific. These problems must occur inside the paradigm in order to be formulated within its system of concepts. That is why,

> paradigms provide scientists not only with a map but also with some of the directions essential for map-making. In learning a paradigm the scientist acquires theory, methods, and standards together, usually in an inextricable mixture. Therefore, when paradigms change, there are usually significant shifts in the criteria determining the legitimacy both of problems and of proposed solutions. (Kuhn, 1962, 109)

2.2.3 Crises and Revolutions

Seen through Kuhn's conception of things, scientific change is essentially a shifting of paradigms. At certain moments in history, anomalies will arise in some branch of science that nothing seems capable of accounting for. This is a crisis. Only a complete rethinking of the theoretical and experimental tools, a "revolution," can lead to the elimination of these anomalies.

Crisis periods are characterized by the multiplication of such anomalies, which appear as so many cracks in the paradigm. These anomalies are problems that are impossible to solve definitively within the framework of the current paradigm. So, when the first anomalies appear, efforts are made to integrate them into the paradigm by adding ad hoc hypotheses, hypotheses whose only purpose is to explain the anomalies themselves, without their addition being otherwise justified. With time these anomalies become more and more urgent. This explains why crisis periods can be recognized by the multiplication of competing theories, resembling a pre-scientific stage.

In Kuhn's view, the shift from one paradigm to another during a revolution does not take place because the new paradigm provides better answers to the problems of the old paradigm, nor because experimental proofs are found which support the new paradigm's theories, nor even because the metaphysical framework supplied is more fitting. The revolution takes place because new theoretical efforts present a new way of looking at things and, in this way, create their own new, challenging problems. One of Kuhn's fundamental theses is that a theory can only be abandoned once another valid theory is ready to take its place. During a revolution, it is not uncommon for the old problems to be covered up or forgotten about, especially if the revolution spans a generation change. So there is neither reduction nor generalization. That is why

> a new theory, however special its range of application, is seldom or never just an increment to what is already known. Its assimilation requires the reconstruction of prior theory and re-evaluation of prior fact, an intrinsically revolutionary process that is seldom completed by a single man and never overnight. No wonder historians have difficulty in dating precisely this extended process that their vocabulary impels them to view as an isolated event. (Kuhn, 1962, 7)

2.2.4 Paradigm Incommensurability according to Kuhn

The conception of scientific change developed by Kuhn has an important consequence which is that two successive paradigms are "incommensurable," meaning that no outside standard can be used as a basis for comparing them. Given the amplitude this concept's influence has attained, we will present its implications in detail in this section.

According to Kuhn, the elaboration of a new paradigm relies on a thorough and complete redefinition of the corresponding scientific discipline. This implies a total transformation of the criteria for distinguishing a genuinely scientific solution from a purely metaphysical speculation, a juggling around of words, or a mathematical trick: the concepts of problem and explanation deemed admissible change radically. Indeed, Kuhn compares a paradigm shift to a change in our world view. Like Hanson (1958) before him, he sees an analogy with *Gestalt psychology*, which studies the psychological processes of perception, and the events that occur during a paradigm shift. In the same way that we can alternately see a rabbit or a duck in the same image, during a paradigm change, our world view transforms in such a way that we can stop seeing falling stones and see pendulums instead.[3] However, we can never see both at the same time, and neither can we place ourselves at some higher vantage point from where we could compare both world views. The world is always viewed in the framework of a given paradigm and no transcendent criteria exist for comparatively judging between paradigms.

The consequences of this thesis are many and far-reaching. It is clear that, for Kuhn, it is paradigms that determine which questions and answers are the right ones: with the arrival of a new paradigm, the old answers lose their pertinence and even become unintelligible. Further still, Kuhn asserts that "paradigm changes do cause scientists to see the world of their research engagement differently. In so far as their only recourse to that world is through what they see and do, we may want to say that after a revolution scientists are responding to a different world." (Kuhn 1962, 111) Further on, Kuhn clarifies the meaning of this assertion:

> Though the world doesn't change with a change of paradigm, the scientist afterwards works in a different world. [. . .] What occurs during a scientific revolution is not fully reducible to a reinterpretation of individual and stable data. In the first place, the data are not unequivocally stable. A pendulum is not a falling stone, nor is oxygen dephlogisticated air. Consequently, the data that scientists collect from these diverse objects are [. . .] themselves different. (Kuhn, 1962, 121)

What we see here is the extreme radical nature of the Kuhnian view: even data, despite being long considered as particularly stable elements of scientific activity, change in

[3] While Aristotle saw motion as belonging to each individual body, Galileo saw the motion of pendulums and of bodies in a vacuum as belonging to one and the same theoretical setting.

meaning during a revolution. It is not easily understood exactly what Kuhn means by this. Furthermore, this argument has been subject to innumerable interpretations. For the purpose of this chapter, we will follow the lead of many authors, including Shapere (1966), and focus on the principal presupposition which seems the most fruitful in interpreting Kuhn's thesis, namely, that it is the very scientific *terms* themselves, as much the theoretical terms as those used in representing data, whose meaning changes during a revolution. Kuhn provides a particularly striking example of such a change in meaning:

> the Copernicans who denied its traditional title 'planet' to the sun were not only learning what 'planet' meant or what the sun was. Instead, they were changing the meaning of 'planet' so that it could continue to make useful distinctions in a world where all celestial bodies, not just the sun, were seen differently from the way they had been seen before. (Kuhn, 1962, 128–129)

Later in this chapter we will see that the case for scientific terms changing their meaning sparked important debates which make up the conceptual framework of current discussion on scientific change.

2.2.5 Incompatibility of Successive Theories

Kuhn's approach leads him to consider cases of so-called homogeneous reduction as both highly problematic and emblematic of the positivists' failure to account for scientific development. The typical example is Einstein's explanation of Newton's theory, which he does by making it one particular case of his new theory (Kuhn, 1962/1969, 98–106). Indeed, it can be shown that, in the case of macroscopic phenomena involving velocities considerably lower than that of light, Newton's laws constitute an extremely precise approximation of Einstein's theory. So Einstein's theory encompasses an understanding of why Newton's theory is true for this group of phenomena.

For Kuhn, such a vision of things is both logically erroneous and historically improbable. On the second point, Kuhn's argument consists of showing that the positivist theory of reduction renders scientific change genuinely inconceivable. Here is the somewhat caricaturist view Kuhn presents of the logical empiricist criterion for the cognitive significance of theoretical statements: by demanding that only those statements which can be entirely reduced to a set of statements describing observable phenomena be considered as scientific, logical positivists are led to "restrict the range and meaning of an accepted theory so that it could not possibly conflict with any later theory that made predictions about some of the same natural phenomena" (Kuhn, 1962, 98).

Kuhn's interpretation of the positivist argument (Kuhn, 1962, 99–100) can be roughly reconstructed as follows: Newton's theory is only false if applied to very large velocities; in its capacity as a genuinely scientific theory, it does not claim to be applicable to these cases since it has not been tested for this application; consequently it is true:

> In so far as Newtonian theory was ever a truly scientific theory supported by valid evidence, it still is. Only extravagant claims for the theory - claims that were never properly part of science - can have been shown by Einstein to be wrong. (Kuhn, 1962, 99)

In defining scientific legitimacy by empirical verifiability, positivists are brought to restricting what theories say to what has already been actually verified. In this way they are, by definition, protected from error. This renders impossible the invalidation of "any theory which has ever been successfully applied to any range of phenomena at all" (1962, 100).

Besides this inability to conceive even the possibility of theoretical change, the reductionist position, says Kuhn, also suffers from a "logical lacuna." Contrary to what is affirmed by the theory of homogeneous reduction, Newton's law cannot be deduced from Einstein's theory, not even as an approximation. Indeed, their very terms, especially the term "mass," have different meanings for the two different theories. Thus, that a law with the same symbolic expression as Newton's law can be deduced from the Einsteinian theory, this in no way permits claiming it to actually be Newton's law, because the symbol *m*, for example, does not have the same referent in both contexts, the concept of mass not having the same definition for Newton as it does for Einstein.

Through a posteriori reconstruction, history can often give us the impression that there is compatibility between a new theory and the theory preceding it; but this compatibility is the result of progressive assimilation and is a historical illusion. Consequently, Kuhn concludes that the two theories are fundamentally incompatible, "in the sense illustrated by the relation of Copernican to Ptolemaic astronomy: Einstein's theory can be accepted only with the recognition that Newton's was wrong" (Kuhn 1962, 98).

This second part of the argument against the reduction theory, which opens a logical lacuna in all cases of reduction, including those said to be "homogeneous," is advanced further, more systematically and more radically by Feyerabend from 1962 onward.

2.2.6 Feyerabend's Incommensurability Theses

Feyerabend's theory of incommensurability, more closely than Kuhn's, appears as a criticism of Nagel's notion of reduction. One of the points where Kuhn and Feyerabend are in deep disagreement is the importance of their historicist discourse: Feyerabend's intention is to produce a normative discourse, which turns out to be both anarchist and pluralist. Kuhn, on the other hand, claims to describe what actually happens in the history of science, so his conclusion is genuinely conservative. However, it is not this aspect which will occupy our attention in this section, but rather the conceptions Feyerabend developed on the subject of scientific terms and their meanings. The reason for this is the influence these had on the debates of the 1960s and 1970s, the repercussions of which are still felt today.

Feyerabend goes as far as claiming that no term whatsoever, neither observational nor theoretical, is shared by two theories. His primary presupposition is that meanings depend on *theoretical context*, to be understood in the broadest sense possible, including the set of all the beliefs held by scientists active at the moment in question. Indeed, his notion of theory is broader—and vaguer—than that of the logical empiricists, because in his view "scientific theories are ways of looking at the world; and their adoption affects our general beliefs and expectations, and thereby also our experiences and our conception of reality" (Feyerabend, 1962, 29). Another fundamental argument of Feyerabend's is that the theory accepted is presupposed by the sort of language employed, so that any change in belief or theory implies a change in meaning of all the theory's terms. Feyerabend therefore defends a form of radical semantic holism, which explains his argumentation toward the incomparability of terms from different theories.

In his 1965 article, Feyerabend brings to light two principles that, in his view, are the cornerstones of the logical empiricist theory of explanation (see chap. 1 of this volume) and of the Nagelian conception of reduction which accompanied it (Feyerabend, 1965, 163):

1. The consistency condition: the only admissible theories for a domain are either those containing already used theories in this domain, or theories that are logically compatible with the latter.
2. The meaning invariance condition, much debated after its appearance (see the special edition of the journal *Philosophy of Science* dedicated to this subject: no. 38(4), 1971, as well as Martin, 1971, 1972): meanings should be invariant relative to scientific progress.

Feyerabend then attacks both of these conditions by seeking to show (i) that scientific theories cannot be logically compatible with each other, and (ii) that the meaning of each term we use depends on the theoretical context in which it appears. Words mean nothing in isolation; they draw their meaning from the theoretical system they belong to. This dependence on theory also stretches to observational terms. Indeed, the meaning of all scientific terms, even observational ones, depends on the theory within which they are used.

Feyerabend therefore claims that the meanings of theoretical terms do not depend (as was claimed by the logical empiricist tradition) on the fact that they be interpreted with the help of a previously and independently incorporated observational language: each theory specifies its own language of observation. The influence of Quine's criticism of the "dogmas of empiricism" is very clear: Quine (1951) showed that the two pillars upholding the logical positivist philosophy, the reductionism of theoretical terms and the distinction between analytical statements and synthetic statements, are in fact simply two sides of the same dogma. In Quine's work, this criticism leads to a rejection of the distinction between theoretical and observational statements: all statements making up our knowledge, when taken altogether, form our conceptual scheme, characterized by the interdependence of these statements. Those we call "observation sentences" are simply found closer to the edges of the scheme, and are thus more readily abandoned and modified. Kuhn and Feyerabend's incommensurability theses and their accompanying semantic holism are an application of Quinian theses (which are rather theses on language and knowledge in general) to the scientific domain.

More generally, Feyerabend's position implies an inversion in the relationships between theory and observation. Whereas theories have a meaning independent from observation, observational statements draw their meaning from the theories featuring them (see Feyerabend, 1965, 213).

Like Kuhn, Feyerabend takes on the traditional empiricist conception according to which, first, a theory must be tested against objective facts (independent of the theory) and, second, a theory is chosen over another one because it gives a better account of the facts—facts which remain the same for both theories. According to Feyerabend, philosophical arguments concerning the fundamental points of theories "are invariably circular. They show what is implied in taking for granted a certain point of view, and do not provide the slightest foothold for a possible criticism" (Feyerabend, 1965, 151). A notable consequence of this thesis is that, in order for criticism to be possible, alternative theories must first be developed (in an anarchic way): "We must choose a point outside the system or the language defended in order to be able to get an idea of what a criticism would look like" (1965, 151).

2.2.7 Criticisms of Incommensurability

Three types of criticism leveled against Kuhn's and Feyerabend's incommensurability theses deserve to be retained. The first type deals with Kuhn's attempt to show that the rigidity of the empiricist conception of theories, which demands that, ultimately, each term be strictly definable using terms referring to observable phenomena, renders scientific change, theory testability, and error all impossible. Indeed, if everything theories can say must already have been observed, then theories are neither useful nor falsifiable.[4] We can however reproach Kuhn for his ignorance of the efforts made by logical empiricists to allow for the very possibility of a theory saying more than what is merely observed. In fact this is precisely the aim of debates on the meaning of theoretical terms which have occupied philosophers of science for several decades. The cognitive significance criterion sought by Carnap has gradually been liberalized to leave place, among other things, to the possibility of applying a theory to domains of phenomena for which it was not initially conceived, and to the deduction, from theoretical laws, of new empirical laws (Carnap, 1966, chap. 25). However, Kuhn's argument does have the merit of showing that, if philosophers tighten up too

[4] The logical empiricists themselves saw a genuine problem in this. According to the "theoretician's dilemma," the expression coming from Hempel (1958), theoretical terms, if they be reducible to observational terms, are useless, in the sense that the meaning of the theory will go no further than the set of the observational statements. But if theoretical terms are not entirely definable in observational terms, then understanding how empirical predictions can be deduced is impossible.

much over theoretical language and the meaning of the terms used in theories, this will make them less attentive to what is actually happening: in practice, scientists do overstep the strict limits of the observed and do make errors and modify their theories.

Second, as shown by Shapere (1964, 1966), Kuhn's position also comes down, in an opposing manner, to denying all continuity in the movement from one paradigm to another, or at least to making unintelligible the fact that two theories in two different paradigms can, in one way or another, speak about the same thing. In other words, Kuhn doesn't give us the tools for understanding how two successive theories on celestial motion have more in common with each other than, for example, an astronomical theory and a biological theory. Indeed, according to Kuhn, "the physical referents of these Einsteinian concepts [space, time, and mass] are by no means identical with those of the Newtonian concepts which bear the same name" (Kuhn, 1962, 102). He further asserts that "in the passage to the limit, it is not only the forms of the laws that have changed. Simultaneously we have had to alter the fundamental structural elements of which the universe to which they apply is composed" (Kuhn, 1962, 102). However, these supposedly radical changes do not at all prevent these same men, depending on the context and their aims, from switching from one paradigm to another. Scientific practice itself goes against Kuhn's claims; much more, the very possibility of this discussion presupposes some continuity.

Third, the theory of meaning adopted by Kuhn and Feyerabend, and which is at the foundation of their incommensurability theses, underwent a series of systematic criticisms at the hands of Shapere (1966). Rejecting the logical empiricists' distinction between theoretical and observational languages, on the one hand, and between meaningful and meaningless statements, on the other hand, the advocates of historicism claim that the meaning of all scientific terms, both observational and theoretical, is determined by the theory or paradigm underlying them (see Shapere, 1966, 50).

However, the principal difficulty with Feyerabend's radical position and, to a lesser degree, Kuhn's as well, is that it offers no criteria for judging if something counts as a change in meaning or a change of theory.

> We are given no way of deciding either what counts as a part of the 'meaning' of a term or what counts as a 'change of meaning' of a term. Correspondingly, we are given no way of deciding what counts as a part of a 'theory' or what counts as a 'change of theory'. (Shapere, 1966, 55)

However, since change in meaning and theory or paradigm change are interdependent, we find ourselves stuck in a circle. The concepts of "paradigm" and of "theory" can be employed, depending on the case, on very different levels. Sometimes they become so broad and general that it is no longer possible to say what is or is not to be included (see Shapere, 1966, 66).

Added to this lack of criteria for identifying change in theory and meaning is a rigid conception of the very concept of meaning and difference in meaning itself. Kuhn and Feyerabend consider the change in meaning of a term to be an all-or-nothing affair and don't at all think of the possibility of similarity in meaning.

> Two expressions of sets of expressions must either have precisely the same meaning or else must be utterly and completely different. If theories are not meaning-invariant over the history of their development and incorporation into wider or deeper theories, then those successive theories (paradigms) cannot *really* be compared at all, despite apparent similarities which must therefore be dismissed as irrelevant and superficial. (Shapere 1966, 68)

The absence of a criterion of identity for meaning, added to this radical conception of meaning change, renders the tools of analysis Kuhn and Feyerabend proposed ineffective for the study of scientific change, even though this is precisely why they were created, to remedy the positivists' excessive logical rigor. Wishing to underline the importance of taking the dynamic aspect of science into account, they end up advancing a conception which, through being excessively radical, also misses its target.

> If the concept of the history of science as a process of "development-by-accumulation" is incorrect, the only alternative is that it must be a completely noncumulative process of replacement. There is never any middle ground and, therefore, it should be no surprise that the rejection of the positivistic principles of meaning invariance and of development-by-accumulation leave us in a relativistic bind, for that is the only other possibility left open by this concept of difference of meaning. But this relativism, and the doctrines which eventuate in it, is not the result of an investigation of actual science and its history; rather, it is the purely logical consequence of a narrow preconception about what "meaning" is. (Shapere, 1966, 68)

Shapere suggests that the perspective opened by Kuhn and Feyerabend's historicist criticisms, a perspective on studying scientific practice, science as it is done, and the accounting for its essentially dynamic aspect, is in turn shut down at their own hands, and for a reason similar to the one they had attacked the positivists for: paying too much attention to logical and linguistic aspects when the philosopher of science's eye should be trained on the actual history.

In the end, they themselves turn historical examples, like the changeover from Newtonian to Einsteinian mechanics, into "anecdotes" intended to illustrate preconceived theories. Moreover, their conception reveals itself to be incapable of accounting for phenomena that are common in the history of science, like the existence of different versions—successive or simultaneous—of one theory, as there are, for example, for classical mechanics (Shapere, 1964). The notion of paradigm or theory change radicalizes the difference between successive theories—considered to be incommensurable—and is deaf to intra-theory changes which are, however, characteristic of normal scientific activity.

In conclusion, let us remark that Kuhn has responded to the numerous criticisms of the incommensurability thesis that place the debate on the border of philosophy of language and philosophy of science, quite far from Kuhn's initial intention in *The*

Structure of Scientific Revolutions. He admits that a distinction must be made between incommensurability and incomparability, and he proposes the notion of "partial incommensurability" (Kuhn, 1983). This allows us to account for the fact that we are able to understand theories that are indeed incompatible with ours and that we can, to a certain measure, compare them. The historian of science is thus presented as an interpreter rather than a translator: her role is not to translate past theories into the language of contemporary science (an impossible task) but to learn to speak, for example, the language of phlogiston chemistry so as to understand what experiments led Presley to write what he wrote.

3. How Is Scientific Progress Defined?

In the traditional view of science portrayed above, the notion of scientific progress is meaningful and even inherent to the very idea of science, since it is progress which distinguishes science from other human activities like art or religion. For thinkers influenced by the Enlightenment movement, there exist clear standards for evaluating scientific advances; that scientific progress exists is taken to be self-evident.

It is, nevertheless, possible to go beyond this supposed self-evidence and remark that the notion of progress does deserve to be analyzed and not simply presumed to be a natural part of scientific activity. It is particularly dependent on the goals that science is set: the search for truth, for precision, for error avoidance, or for theoretical explanation and unification, simplicity of descriptions, and so forth. In assigning one or the other of these goals to scientific activity, an appropriately fitting notion of progress will have to be created. Which criteria are to be used in evaluating this progress will also have to be indicated, to avoid begging the question. Such a normative approach hangs general scientific advancement on researchers' individual goals.

According to another approach, known as "naturalist," the notion of progress must be *defined* using scientific developments: there is no independent notion of it. Here we see an important dividing line, among the very people who accept that the notion of scientific progress is meaningful, between the disciples of a normative approach and those of a "naturalist" approach.

However, not all researchers are willing to assert that scientific development always constitutes progress. Indeed, such a conception of science is smashed by Kuhn and Feyerabend's work, as could have been guessed in light of what we have already seen. For Kuhn, for example, scientific development is to be compared to biological development rather than any progression voluntarily directed towards some conscious goal.[5] It is, of course, a unidirectional and irreversible process (Kuhn, 1962/1969, 206), but one which manages neither to give a match for what is "really there" (1962/1969, 206), nor to attain or even give more and more precise approximations of the truth. Thus,

[5] Other philosophers, albeit without sharing Kuhn's views on other ideas, have also advanced an evolutionist conception of scientific development, Popper (1972) and Toulmin (1961), for example.

"later scientific theories are better than earlier ones for solving puzzles" (1962/1969, 206) and making predictions: the evolution of science is subject to no other standard than the solving of puzzles. Likewise, for Feyerabend, the traditional notion of scientific progress is obsolete, since the evolution of knowledge occurs by complete replacement rather than successive subsumption. Any innovative researcher will start over from the beginning (Feyerabend, 1965, 199).

The richest debates on scientific progress have taken place within the normative approach of it. In this section, we concisely present them. Discussion of Kuhn and Feyerabend's conceptions regarding the more general question of the motor behind scientific change will be reserved for the next section.

The primary motivation to adopting a normative approach to scientific progress is that scientists themselves generally have an opinion regarding the nature of the normative criteria to be used in evaluating the choices being made, that have been made or, indeed, that could have been made by scientific communities. They consider the question of choices being good or bad to be a meaningful one; consequently, asking about the criteria behind these choices is legitimate, as long as it is understood that these criteria cannot be limited exclusively to research activity or the abilities it requires. These criteria must be relative to the goals of science and to the results obtained. Indeed, there is no necessary link between research quality and scientific progress, as Niiniluoto (2007) has highlighted.

A proponent of scientific realism will assign scientific research the goal of pursuing the truth. It is, however, a delicate matter to formulate a theory of scientific progress which defines it in relation to this goal, since there is no method that easily enables us to decide if, when, or to what extent this goal has been reached. Furthermore, as highlighted by Isaac Levi (1967), the goals of research are many and they cannot be reduced to the search for the truth, not even for a proponent of scientific realism. Levi proposes defining these goals as a weighted combination of different epistemic utilities, which are sometimes in conflict with each other. These different normative theories of scientific progress can be thought of as depending on as many different ways of thinking about these epistemic utilities. So we can consider, like Levi himself or like Popper (1934, 1963), that the goal of scientific research is a certain combination (as of yet imprecise) of truth and of informational content (since the discovery of new tautologies does not constitute progress in any meaningful kind of way), or else, like Hempel, that its goal is rather explanatory and predictive power. Exactitude, coherence, the scope of the phenomena accounted for, simplicity and fruitfulness are other goals of scientific activity often mentioned (Kitcher, 1993). The empirical success of theories, their applicability to numerous phenomena and precise predictive ability, all remain, in any case, a minimum criterion which, as we shall see presently, is nevertheless not entirely sheltered from difficulty.

Let us now give some examples of the difficulties encountered when trying to precisely define some criteria for scientific progress, starting with the most immediately testable of them: a theory's empirical success. One first idea for defining a theory's empirical success is to associate it with the number of true empirical statements implied by it, as well as with the few empirical counter-examples which could be opposed to it. In this way, moving from theory T_1 to theory T_2 constitutes progress if T_2 has more true observational statements as consequences and if fewer empirical counter-examples can be opposed to it. The structuralist philosophers of science (Balzer et al., 1987), as well as Lakatos and Musgrave, adopt such a definition. It is the subject of several criticisms. First of all, it presupposes the possibility of isolating observational statements from theoretical ones, a hypothesis which is at the root of many debates (see, among others, Carnap, 1956 and Maxwell, 1962). Also, it supposes that enumerating the observational statements consequent to a theory is an easy feat, yet to do this, a further criterion of relevance is needed and this would be difficult to define. Further, besides the observational consequences of theories, there is another source of empirical success or failure which it doesn't take into account: conceptual evolutions. Laudan (1977, 1981), in this regard, has proposed defining empirical success as depending on the number of empirical problems it resolves *and* on the number of conceptual problems these solutions give rise to.

An alternative proposition, also put forward by Laudan (1977), is to adopt a theory's capacity for effectively resolving problems as a criterion for scientific progress. The difficulty with this proposition is in finding a framework which would allow for the identification and enumeration of the problems in question, as Rescher (1984) has pointed out. A radical version of this proposition boils down to superposing scientific progress onto technological progress (Rescher, 1977): indeed, it is easier to identify technological problems than strictly scientific ones.

The criterion which has raised the most discussion is the verisimilitude criterion. Intuitively, it does seem satisfactory and relatively easy to define scientific progress as a path towards the truth. So, a theory T_2 will be deemed closer to the truth than theory T_1 if it has more true consequences and fewer false ones (Popper 1963, 1972). However, such a definition does not allow for the comparison of two false theories—yet such a comparison could be sought after, for example, to compare Newtonian mechanics and the phlogiston theory. We know that these two theories are false (if Newtonian mechanics is considered as a *general* theory of motion and not as a theory of exclusively low velocity motions), but it would be good to be able to say that the phlogiston theory is more false than Newtonian mechanics. Applying the criterion Popper defined, this is impossible. Basing himself on the approach developed by Tichy (1974), Niiniluoto (1987) defines a notion of verisimilitude based on the distance between the (partial) answers a problem receives and its true answer, which is the target set when the problem is posed. Niiniluoto introduces two parameters, one indicating the benefit there is in giving an answer close to the target, and the other about the benefit of answers which exclude not only false statements but also statements located far from the target. Thus, verisimilitude theories provide a simple way of deciding to what extent a scientific theory has fulfilled its objective.

4. What Is the Driving Force Behind Scientific Change?

We have just seen that debates about the nature or structure of scientific change have been, and still remain, lively. This is also the case with debates about its causes. When considering scientific development since the origins of modern science in the 17th century, scientific activity appears intrinsically to produce new results, new hypotheses and new discussions. At times, attempts are made to characterize these advances as the results of a set of rational precepts gathered together under the title of "scientific method." Even if it has become more and more evident throughout the course of the 20th century that there cannot be one exclusive scientific method, the search for a unique explanatory pattern of scientific change has held on to its appeal among philosophers of science.

4.1 POPPER AND THE FALSIFIABILITY OF THEORIES

One of the first to wholeheartedly set off on such a quest was Popper. *The Logic of Scientific Discovery* does propose an explanatory pattern for scientific development, and gives arguments against other such explanatory propositions, in particular inductivism, in its various forms. Given the sheer scale of the reactions Popper's ideas have sparked, it is worth recalling them here in brief.

Against all attempts at a logical inductive formalization enabling the confirmation of theories to be defined and measured using observable data, Popper advances a strictly deductive conception of the scientific method. His criticism of inductivism is accompanied by a challenge to Carnap's verificationist criterion of cognitive significance. Indeed, since no universal statement can be verified or even confirmed on the basis of one or several instances of it (cf. chap. 2 of this volume), science in no way consists in seeking confirmation for theoretical laws using empirical data (which would make them more and more probable), but in the search for the most informative hypotheses possible, namely, the most falsifiable ones, those most likely to meet with counter-examples.

A theory's falsifiability is measured by its degree of improbability, given relevant available knowledge; the more falsifiable, but still not falsified, a theory is (that is, the better it stands up to testing), the more it is corroborated by experience. But its corroboration cannot be measured on a probability degree model. For Popper, a theoretical hypothesis is never probable. There is no induction from experience that enables any level of probability to be established. Rather, the pattern goes like this: face to face with experience, the scientist proposes a theory (through an inventive process which obeys no rational method); she then compares the deductive consequences with the experiment which, if it seems to fit the theory, corroborates it and, if not, falsifies it. The more possibilities for falsification a theory throws up the more informative and innovative it is and the more its corroboration will count as progress. So the driving force behind scientific research, according to Popper, is the search for falsification.

Furthermore, Popper puts forward the falsifiability of theories as a criterion for their scientific legitimacy; this allows him to rule out "false sciences" such as psychoanalysis, which are not falsifiable because they rely on the successive integration of ad hoc hypotheses. The criterion of demarcation between science and nonscience that Popper proposes clearly distinguishes itself from the logical empiricist criterion, centered on cognitive significance.

Popper's battle against inductivism in all its forms is generally considered to have been settled by a defeat. As chapter 2 of this volume shows, the current dominant theory of confirmation is the Bayesian theory, one of Popper's targets. Popper's other adversaries, who side themselves with either Kuhn's theses or with slight variants on them, are also of the opinion that falsificationism cannot be viewed as a valid description of the causal mechanisms of scientific change.

4.2 KUHN AND "THE ESSENTIAL TENSION"

According to Kuhn, in periods of normal science, the goal of researchers is by no means to falsify available theories, but rather to better and better corroborate various elements of the paradigm through new applications, as well as through the development of new mathematical apparatus for expressing them. In many respects, *The Structure of Scientific Revolutions* is a systematic attack on *The Logic of Scientific Discovery*—the title itself bears witness to this. As a result, for Kuhn, there is no such thing as falsifying experiences:

> Anomalous experiences may not cannot be identified with falsifying ones. Indeed, I doubt that the latter exist. [. . .] no theory ever solves all the puzzles with which it is confronted at a given time; nor are the solutions already found often perfect. On the contrary, it is just the incompleteness and imperfection of the existing data-theory fit that, at any time, define many of the puzzles that characterize normal science. If any and every failure to fit were ground for theory rejection, all theories ought to be rejected at all times. (Kuhn, 1962, 146)

Furthermore, Kuhn denies that theories are ever abandoned due to being falsified:

> once it has achieved the status of paradigm, a scientific theory is declared invalid only if an alternate candidate is available to take its place. No process yet disclosed by the historical study of scientific development at all resembles the methodological stereotype of falsification by direct comparison with nature. (Kuhn, 1962, 77)

This is clearly a refutation of Popper's main argument. However, are we obliged to buy into the description of periods of normal science Kuhn proposes here? Is normal science really a "pursuit not directed to novelties and tending at first to suppress them" (Kuhn, 1962, 64)? *The Structure of Scientific Revolutions* contains numerous examples to back up this description, but the task to be completed in making it an incontestable account of actual science remains a herculean one.

Kuhn's principal line of argument regarding the explanation of scientific change involves enlisting descriptions of actual scientific activity that clash with Popper's explanatory generalizations. On a more general level, Kuhn criticizes the very explanatory undertaking itself, regardless of whether it relies on a verificationist or a falsificationist schema.

Just like his adversaries who defend verificationism, Popper's thesis is based on the illusory idea of a genuine confrontation of theories and facts, something Kuhn denounces outright.

> To the historian, at least, it makes little sense to suggest that verification is establishing the agreement of fact with theory. All historically significant theories have agreed with the facts, but only more or less. There is no more precise answer to the question whether or how well an individual theory fits the facts. But questions much like that can be asked when theories are taken collectively or even in pairs. It makes a great deal of sense to ask which of two actual and competing theories fits the facts *better.* (Kuhn, 1962, p. 147)

In the next section we will come back to how Kuhn proposes describing in what way a theory—or a paradigm—beats another one by fitting the facts *better*. We will see that the deciding factors in the victory of one paradigm over another are essentially external ones. For now, let us note that Kuhn's criticism of the project to unlock an explanatory pattern of theory change is focused on the fact that it deliberately fluffs over what he calls the "built-in mechanism" of normal science, which is just as responsible for a paradigm's internal progress, described as "mopping-up operations" the aim of which is to "force nature into [a] preformed and relatively inflexible box" (see the final section), as it is for the appearance of anomalies which give rise to the invention of new theories.

> By focusing attention upon a small range of relatively esoteric problems, the paradigm forces scientists to investigate some part of nature in a detail and depth that would be otherwise unimaginable. And normal science contains a built-in mechanism that ensures the relaxation of the restrictions that bound research whenever the paradigm from which they derive ceases to function effectively. (Kuhn, 1962, 24)

So Kuhn describes the driving force of scientific change, for which he refuses to give an explanatory pattern, as an extreme form of conservatism that leads scientists to resist novelties that would threaten the paradigm for as long as possible, until the moment the paradigm implodes under the weight of the anomalies it helped to reveal. This tension between tradition and innovation is what Kuhn (1959) calls "the essential tension." This implosion gives rise to the appearance of theoretical novelties, an appearance which is placed, as also for Popper, beyond the scope of any rational explanation: for both authors, there is no controlled exercise of reasoning which can explain the appearance of new hypotheses. Novelty emerges, but cannot in any way be prompted by exercises of reasoning. It should be noted, on this point, that Kuhn, all the while insisting on the importance of change in science, renders the very possibility of change fundamentally incomprehensible.

He credits "tradition" with the ability to mysteriously spawn novelty: "The very fact that a significant scientific novelty so often emerges simultaneously from several laboratories is an index both to the strongly traditional nature of normal science and to the completeness with which that traditional pursuit prepares the way for its own change" (Kuhn, 1962, p. 65). This notion of tradition regroups a significant part of the irrational elements of scientific activity both in Kuhn's conception and also in Feyerabend's and all disciples of science studies who follow them in this regard, as we shall see in the following section.

Let's take a moment to look at the complex arrangement of positions we have here: the irrational nature of the appearance of theoretical novelty is a point that moves Kuhn closer to Popper, but also to Feyerabend with whom he also shares very similar views on incommensurability, and yet Feyerabend and Popper are united in vehemently opposing Kuhn's conservative description of normal science. For Popper and Feyerabend equally, the advancement of science relies on the daring and inventiveness that push scientists to proposing new hypotheses. As we have previously remarked, it is also important to distinguish between Popper and Feyerabend's normative project and Kuhn's descriptive ambition.

Finally, Kuhn's position leads him not only to criticizing the idea that the invention of new hypotheses is the result of a rational process (just like Popper), but also to questioning the very idea of the discovery of new facts. Indeed, he dedicates numerous pages to examining this classical category in the positivist vision of the history of science.

The example Kuhn uses is the discovery of oxygen: in 1774, Priestly isolated a gas which, in his theoretical system, he could not define as a distinct gas and which he believed to be "dephlogisticated" air; in 1777, Lavoisier recognized this substance as a distinct gas, oxygen. Kuhn remarks that the category of "discovery," insofar as it presupposes a clear distinction between facts and theory, is symptomatic of a history of science which poses the wrong questions:

> Was it Priestley or Lavoisier, if either, who first discovered oxygen? [. . .] Discovery is not the sort of process about which the question is appropriately asked. The fact that it is asked [. . .] is a symptom of something askew in the image of science that gives discovery so fundamental a role. (Kuhn, 1982, 54)

Then, further on:

> discovering a new sort of phenomenon is necessarily a complex event, one which involves recognizing both that something is and what it is. Note, for example, that if oxygen were dephlogisticated air for us, we should insist without hesitation that Priestley had discovered it, though we would still not know quite when. But if both observation and conceptualization, fact and assimilation to theory, are inseparably linked in discovery, then discovery is a process and must take time. (Kuhn, 1962, 55–56)

So the explanatory approach to scientific change is doomed from the moment it imagines change as some simple event that can be precisely dated and that results from confronting the theory with the observed facts.

4.3 NEW PHILOSOPHICAL APPROACHES

Among those philosophers of science who are opposed to Kuhn's line of research, and who continue to seek an explanation to scientific change, the question, "In what way do scientific theories evolve?" has been judged to be the best lead to follow in finding a convincing one. This question has given rise to several different approaches to the exploration of possible inter-theory relationships. These approaches can be called "internal" because their exclusive object of research is theory components, disregarding in this the people who use the theories. Thus, Balzer et al. (1987) proposed renovated formal tools for analyzing the notion of inter-theory reduction; the notion of inter-theory correspondence was also the center of several propositions. More recently, Kitcher (1993) developed an approach that takes scientific communities and their practices into account by focusing the analysis on how things pass from a state of competition between theories to a state of consensus (for yet another approach, see Mongin, 2009).

Several case-studies have developed in recent years (see Hartmann, 2002, for an overview) around the notion of correspondence proposed by Post (1971). This notion takes its inspiration from the correspondence principle formulated by Bohr in regards to quantum mechanics. Post's proposed principle consists of asserting that any new theory, in order to be acceptable, must be able to explain the well-confirmed elements of its predecessor. And this is effectively, according to him, what is seen throughout the history of science. The aim of this principle then is to take the historical development of science seriously all the while refuting Kuhn's theses of incommensurability and what are known as "Kuhn-losses," this latter thesis referring to the idea that, in the course of every scientific revolution, certain explanatory aspects of the abandoned paradigms are lost, because they have no corresponding aspect within the terms of the new paradigm. The correspondence principle claims to be applicable even in recognized cases of scientific revolution where the whole theoretical framework is overhauled. In these cases, says Post, low level descriptive structures are particularly stable and it is the most fundamental and least experimentally confirmed aspects which are modified; thus, the periodic table in chemistry stays in place despite quantum mechanics taking the place of the old theory of chemistry:

> The periodic system is the basis of inorganic chemistry. This pattern was not changed when all of chemistry was reduced to physics, nor do scientists ever expect to see an explanation in the realm of chemistry which destroys this pattern. The chemical atom is no longer strictly an atom, yet whatever revolutions may occur in fundamental physics, the ordering of chemical atoms will remain. (Post, 1971, p. 237)

5. Is Scientific Change Rational? Is It Necessary?

We finish this chapter with a twofold question that is more delicate to answer than it may seem. It could indeed be thought, if one were a proponent of scientific realism (see chap. 4 of this volume), that the scientific change which brought us from the Ptolemaic conception of the universe to super-string theory, for example, is not only perfectly rational but also necessary, since it was, to a large extent, steered by the world itself via our interactions with it. In this conception, nothing outside of the norms of rationality takes part in the development of science which, furthermore, could not have followed any other path.

In this section, we will begin by presenting a famous argument in favor of scientific realism which is based on an analysis of scientific development. Then we will discuss the possibilities that are respectively available to the anti-realists and the realists in regards to the twofold question of the rationality and necessity of scientific change. We will show that the argument in favor of realism, despite all appearances, does not enable us to confirm that the scientific realism position fits alongside the position that scientific change is at once rational and necessary.

5.1 THE NO-MIRACLE ARGUMENT FOR REALISM

Putnam (1975, p. 73) and then Boyd (1983) developed what Bas van Fraassen (1980, p. 39) baptized the "ultimate argument" for scientific realism, a meta-philosophical argument also known as the "no-miracle argument," due to Putnam's wording of it: "The positive argument for realism is that it is the only philosophy that doesn't make the success of science a miracle" (Putnam, 1975, p. 73).

The argument is presented as a two-fold abductive argument. First, if a scientific theory is approximately true then, typically, it meets with empirical success. Further, if the central terms of a scientific theory possess authentic referents then, generally, this theory meets with empirical success. Our theories do meet with empirical success. It can thus be concluded with probability that our theories are approximately true and that their terms possess authentic referents.

Second, if the prior theories of a mature science are approximately true, and if their central terms possess authentic referents, then the most recent theories preserve the old ones in the guise of limiting cases. Scientists, then, aim to preserve old theories as limiting cases of new theories and, generally speaking, they manage to do this. Therefore, in all probability, a mature science's old theories are approximately true and their central terms possess authentic referents.

Let us look at the presuppositions contained within this pro-realism argument. To start with, they imply that a convincing answer has been offered to Kuhn and Feyerabend's propositions on the meaning of scientific terms. As we have seen, both their theses consist of knocking holes in the very ideas of theory truth and the reference stability of theory terms. In order to account for the fact that successive theories on heat, electricity, celestial motion and so forth, do indeed deal with the *same* set of phenomena, and that there therefore exist terms which Shapere (1969) calls "transtheoretical," Putnam (1973a, 1973b) developed a theory of reference for theoretical terms inspired by Kripke's (1972) theory of proper nouns. The causal theory of reference asserts that the relationship between a term and its referent consists in the causal chain of relationships between the utterances of a term and the instances to which these utterances refer. This implies that the properties attributed to a term's referent through reference do not necessarily belong to it. This theory enables us to account for something that neither the positivists nor their historicist critics manage to describe: the retention of a term's reference beyond the changes in meaning brought about by changes in our beliefs and theories about what the term refers to. This allows us understand that there is a common referent for the term "water," for example, as it is employed by a scientist who knows its chemical composition is H_2O and also as it was employed by our ancestors who knew nothing at all of chemical composition.[6]

Furthermore, it is evident that Putnam's theory implies that the realist hypothesis is the only one which can explain science's empirical success. However, this is accepted as a presupposition before any analysis of the notion of empirical success itself. Moreover, the argument presupposes that the notions of truth and reference can play a causal explanatory role in epistemology (this indeed being one of the goals of the causal theory of reference).

The premise of the argument dealing with approximate truth can also, in turn, be criticized. This premise asserts that scientific theories (in mature sciences) are typically approximately true and that the newest theories are closer to the truth than the former ones. Yet what is the relationship between a theory's approximate truth and its empirical success? Even if it were false to say that the approximate truth of a theory implies its empirical success, it is possible that a theory's empirical success be a sign of its approximate truth and, therefore, that from a theory's empirical success one could legitimately conclude its being approximately true.

Such reasoning supposes that the theoretical terms of successful theories possess authentic referents, yet this is not the case: phlogiston chemistry, the caloric theory of heat, the physiologist theory of vitalism, and so forth, are so many examples of empirically successful theories which turned out to be non-referential, and that the "pessimist meta-induction" argument can use in opposing defenders of realism. According to this argument, particularly advanced by Laudan (1981), the history of science presents us with a succession of theories which met with definite empirical success and which are nevertheless considered to be false today. Some of these, classical mechanics for example, are still used and taught today, something which forces us to admit the clear distinction there is between the usefulness and the truth of a theory. From the proven falsity of all past theories, the pessimist can conclude to the general rule that all scientific theories are false and that our current theories will be falsified in turn by new

[6] One of the flaws in Putnam's theory is that, while it seems particularly satisfactory when it comes to accounting for the reference point of terms referring to natural elements, like water, it flounders when it comes to accounting for the way in which the terms of former theories, now considered to be non-referential, like, for example, "phlogiston," nevertheless enabled scientists to formulate truths (to the extent that these theories met with some amount of empirical success). The notion of "reference potential," proposed by Philip Kitcher (1978, 1982, 1993) in the wake of Putnam's realist theses, takes on this very problem.

theories, and so on; indeed, there is no reason to believe that our current theories will meet with a happier end than any of those they replaced.

Defenders of scientific realism can respond that the only theories concerned are mature theories. However, we have no criterion which would allow us to delimit mature theories from immature ones. It falls back onto the realists to show the link between the increase in precision of our characterization of the (unobservable) structure of phenomena and the improvement of our predictions, explanations and manipulations on the phenomenological (observable) level.

The above analysis supposes that every proponent of scientific realism would stand behind the claim that scientific change is necessary, and that, consequently, the borders between realists and anti-realists, on the one hand, and between proponents and opponents of the necessity of scientific change claim, on the other hand, become confused. But this analysis is incomplete. Not only does it pass over the fact that an anti-realist may well consider scientific change to be both rational and necessary, but it also leaves aside the possibility for a realist to remark that scientific change has largely been of an irrational nature.

5.2 ANTI-REALIST OPTIONS

What are the various positions open to the anti-realist? She may consider (and this is actually a common attitude) scientific procedures to be largely rational and, furthermore, refuse the Kuhnian distinction between periods of normal science and periods of revolution. In this case, she will transpose her diagnostic that the actions of individual scientists are rational onto the rationality of the global evolution of science. Furthermore, she may also consider that the scientific procedures adopted at a given time necessarily led to the results that the scientists of that time have obtained, without any alternative science being accessible. These two attitudes, however, are independent of each other. The anti-realist may consider scientific procedures to be rational, but also consider that they do not lead on to necessary results and that the history of science could have been very different.

Andrew Pickering is one of the anti-realists who have most strongly defended the contingency of scientific change thesis. According to him, nothing in the history of science that we have known was inevitable. Let us take the example he gives in *Constructing Quarks* (1984). According to him, particle physics, without any challenging to the rationality of its scientific procedures, could have declined the quark route and conserved the older models rather than postulating the vast array of sub-atomic particles that today populate the standard model, this being the accepted theory among all physicists for the representation of sub-atomic phenomena. According to Pickering, the alternative physics that he proposes (though, of course, without developing it) would have become just as successful in terms of empirical predictions as the physics we know today. It would, however, be radically different from it, to the point that, apart from the empirical predictions, there would be no common elements to be found between them.

As Hacking (1999, chap. 3) points out, the majority of physicists, and many molecular biologists as well, consider Pickering's position to be quite simply absurd. They insist that another path for physics in particular, and for science in general, is totally inconceivable. The force of Pickering's argument is to highlight that such an inconceivability only ever comes to light after the fact; furthermore, he claims that even now alternative directions are still accessible.

The argument for the contingency of the direction of science is at once provocative and rich. Let's look at some developments of it. It frequently occurs during the execution of experiments that the world "resists," that, for example, the measuring apparatus don't give the expected results or don't behave in the way that had been predicted. According to Pickering, amplifying Duhem's thesis on the experimental under-determination of theories (Duhem, 1914), scientists can react in several ways to this resistance, all of which boil down to *adapting*. They can revisit the fundamental theory which is supposed to govern the phenomena they are studying with a view to modifying it; they can revise their beliefs about the apparatus used in the experiment; they can also change the theoretical model describing the apparatus, which equates to changing the interpretation of the experiment's results; or, in the case of *big science*, they can even transform the "phenomenology" of the experiment, that is, its results, which, in particle physics, are not at all transparent and can only be obtained at the cost of an immense interpretative undertaking. When a team meets with "resistance," nothing, according to Pickering, predetermines how its members will "adapt" to it, nor what changes they will carry out in order to manage to recover some sort of relatively robust or reproducible harmony between the theories, the apparatus, the models of these apparatus, and the phenomenology of the experiments. So, for Pickering, it is not the phenomena alone which determine the way in which the "resistance" is overcome and which new equilibrium we move towards; it is rather the dialectic of the resistance and the adaptation.

Such a conception of scientific development, already defended by Duhem, is not particularly iconoclastic: the under-determination thesis claims that several theories can exist for one set of data which they represent symbolically in such a way as to facilitate their prediction, but certainly this thesis does not claim that the data is constructed by the theory. This is, however, the case for the consequences Pickering draws from the contingency of scientific development. Indeed, as Hacking (1999, chap. 3) points out, they come down to the possibility of a physics without Maxwell's equations, without the second law of thermodynamics, or even without the famous equation $E = mc^2$, a physics in which we would be totally lost, but whose predictive power would be just as successful as that of our actual physics.

To go that far, it would seem necessary to accept an anti-realist premise. However, a disciple of scientific realism could stand his ground despite accepting that, yes, from time to time the dialectic of the resistance and the adaptation introduces a certain contingency, even if this only relates to levels of financing, the construction time for large apparatus, the gathering together of a team, and so forth. But as soon as it has been admitted that not everything in scientific development is absolutely predetermined by the state of the physical world, then it is difficult to resist soritical style arguments, arguments of the form: a million grains of sand form a heap; if one grain is removed, what remains is still a heap, and so on; but when there are only five grains left, it is difficult to accept the conclusion that if one grain is removed we are still, nevertheless, before a heap of sand. Indeed, where should the limit to contingency be introduced in such a way as to tie in with Nobel prize winner (along with Abdus Salam and Steven Weinberg) Sheldon Glashow's intuition that "Any intelligent alien anywhere would have come upon the same logical system as we have to explain the structure of protons and the nature of supernovae" (1992, p. 28, quoted by Hacking, 1999, p. 75)? If we admit a part of contingency in the development of science, then it becomes difficult to fill the gap between that contingency and Glashow's realist convictions.

5.3 REALIST OPTIONS

Let us now look briefly at the possibilities open to the realist. There exists at least one confirmed proponent of scientific realism who does not believe scientific change to be rational. Karl Popper (1963), as seen in the previous section, denies there is any notion of inductive logic that acts as a guiding force in choosing between hypotheses and that enables us to determine their confirmation. While the schema for theory falsifiability may allow us to describe a theory's effective confrontation with experience as obeying some strictly deductive logic, nothing of the sort allows us to describe the invention of new hypotheses, contrary to what is suggested by the logic of scientific discovery that Hanson (1958) tries to formulate. Popper sees the progress of science as following a dialectic of conjectures and refutations, entirely distinct from the positivist image of the hypothesis obtained by induction and then confirmed by experience. The appearance of a new conjecture—as opposed to the appearance of a new hypothesis obtained by induction—is genuinely irrational, if by rationality we mean a set of rules that logic is supposed to realize.

Just like with Kuhn (see section 2), the appearance of a new conjecture is inexplicable by any logical, argumentative means. This does not prevent Popper from talking about rationality in scientific method: this rationality is that of the critical mind, of the search for falsifiability, of risk taking and of boldness.

5.4 THE IRRATIONALITY AT THE HEART OF SCIENCE

In finishing, we will present one of the strongest diagnostics of irrationality which has been made of scientific change, namely Kuhn's. As it will have been understood from the first section of this chapter, the passage from one paradigm to another is not guided by any rational procedure at all.

Beyond the problem of the appearance of novelty, which Kuhn, like Popper, places outside all possibility of rational explanation, the problem of choosing between competing paradigms arises: once novelty has emerged, one of the paradigms must indeed win out over the other. On this point, no norm of transcendent rationality can serve as a guide in comparing the competing theories: the choice "between competing paradigms proves to be a choice between incompatible modes of community life."(Kuhn, 1962, p. 94)

Since paradigms cannot be compared, the passage from one to another cannot be the fruit of some rational argumentation but rather the result of a process which Kuhn describes in places as an instantaneous event comparable to *Gestalt* switches, and in others as analogous to a religious conversion:

> normal science ultimately leads only to the recognition of anomalies and to crises. And these are terminated, not by deliberation and interpretation, but by a relatively sudden and unstructured event like the gestalt switch. [. . .] No ordinary sense of the term *interpretation* fits these flashes of intuition through which a new paradigm is born. Though such intuitions depend upon the experience, both anomalous and congruent, gained with the old paradigm, they are not logically or piecemeal linked to particular items of that experience as an interpretation would be. (Kuhn, 1962, pp. 122–123)

He borrows Max Planck's image, where the adoption of a new theory is a process that cannot take place on an individual level, but is rather more like the extinction of one species and the emergence of another:

> a new scientific truth does not triumph by convincing its opponents and making them see the light, but rather because its opponents eventually die and a new generation grows up that is familiar with it. (Max Planck, 1940, quoted by Kuhn, 1962, p. 151)

The triumph of one paradigm over another cannot occur through rational argumentation that would convince a group of scientists of their error by force of proof. Being irrational, its nature is more of a persuasive enterprise.

The first reason for this is that rationality is only defined within the bounds of a paradigm. Kuhn offers quite a sad image for this, which contrasts with the flamboyance of a new paradigm emerging that was conceived beyond all rational constraints, as indicated between the lines of the following quote:

> Mopping-up operations are what engage scientists throughout their careers. They constitute [. . .] normal science. Closely examined, whether historically or in the contemporary laboratory, that enterprise seems an attempt to force nature into the preformed and relatively inflexible box that the paradigm supplies. No part of the aim of normal science is to call forth new sorts of phenomena; indeed those that do not fit the box are often not seen at all. Nor do scientists normally aim to invent new theories, and they are often intolerant of those invented by others. Instead, normal-scientific work is directed towards the articulation of those phenomena and theories that the paradigm already supplies. (Kuhn, 1962, 24)

Kuhn goes even further in his criticism of the claim that scientific activity is rational. Though he may consider that the very notion of rationality is relative to a paradigm, he is still very far from affirming that all the decisions made by scientists during periods of normal science and ending in progress are rational. On the contrary, the notion of *tradition* is, for Kuhn, a determining factor in the analysis of scientific activity:

> Scientists work from models acquired through education and through subsequent exposure to the literature often without quite knowing or needing to know what characteristics have given these models the status of community paradigms. And because they do so, they need no full set of rules. The coherence displayed by the research tradition in which they participate may not imply even the existence of an underlying body of rules and assumptions that additional historical or philosophical investigation might uncover. (Kuhn, 1962, 46)

6. Conclusion

The subject of scientific change calls for numerous philosophical tools from various domains (notably, those from philosophy of language for tackling the problem of the meaning of theoretical terms) and is an invitation to rethinking the big questions of philosophy of science concerning, for example, the nature of scientific theories, the relationship between theory and experience, and scientific realism. It is not simply a question among others: the place given to scientific change in the study of science determines how other questions are posed and dealt with.

To emphasize scientific change is equally to contribute to challenging the delimitation of the various disciplines whose object is science, particularly philosophy of science and history of science. As we have seen, the prioritizing of the essentially dynamic nature of science by the historicist critics led them to redefine the *units of analysis* of scientific activity, thus opening up the philosophical approach to the historical aspects of science. In doing this, not only do they redefine the objects the philosopher of science should occupy herself with, they also claim to dictate a method to history of science itself.

Today we are witnessing a resurgence in empirical studies on the sciences, centered around case studies (see the corresponding discussion in chap. 7). In contrast with the global approach which characterizes the historicist criticisms of logical positivism, the aim of these studies is to get up close with scientists' actual practices by looking at more "contained" units than those set out in the very large concepts of paradigm or research program. An immediate task for philosophy of science is to define the methodological tools and principles of these approaches. One of the ways of doing this which is being developed today is to tap into the results of cognitive science; for example, some developmental psychologists collaborate with philosophers and historians of science in order to weave links between studies on the cognitive development of young children and studies on conceptual change in science (Carey, 1985; Gopnik, 1996; Spelke, 1991).

References

Balzer, W., Moulines, C. U., and Sneed, J. (1987) *An Architectonic for Science. The Structuralist Program*, Dordrecht: Reidel.

Boyd, R. (1983) "On the Current Status of the Issue of Scientific Realism," *Erkenntnis* 19, p. 45–90.

Carey, S. (1985) *Conceptual Change in Childhood*, Cambridge, MA: MIT Press.

Carnap, R. (1956) "The Methodological Character of Theoretical Concepts," in Feigl, H., and Scriven, M. (eds.), *The Foundations of Science and the Concepts of Science and Psychology*, Minnesota: University of Minneapolis Press, pp. 38–76.

Carnap, R. (1966) *Philosophical Foundations of Physics*, London: Blackwell.

Duhem, P. (1914) *La Théorie Physique, son objet, sa structure*, 2nd ed., Paris: Chevalier et Rivière. English Translation Phillip Wiener, *The Aim and Structure of Physical Theory*, Princeton, NJ: Princeton University Press, 1954.

Feyerabend, P. K. (1962) "Explanation, Reduction, and Empiricism," in Feigl, H. and Maxwell, G. (eds.), *Minnesota Studies in the Philosophy of Science: Scientific Explanation, Space, and Time*, Minneapolis: University of Minnesota Press, pp. 28–97.

Feyerabend, P. K. (1965) "Problems of empiricism," in Colodny, R. G. (ed.), *Beyond the Edge of Certainty*, Englewood Cliffs, NJ: Prentice-Hall, pp. 145–260.

Gopnik, A. (1996) "The Scientist as a Child," *Philosophy of Science*, 63(4), pp. 485–514.

Hacking, I. (ed.) (1981). *Scientific Revolutions*, Oxford: Oxford University Press.

Hacking, I. (1999) *The Social Construction of What?*, Harvard: Harvard University Press.

Hanson, N. R. (1958) *Patterns of Discovery: An Inquiry into the Conceptual Foundations of Science*, Cambridge: Cambridge University Press.

Hartmann, S. (2002) "On Correspondence," *Studies in History and Philosophy of Modern Physics*, 33B, pp. 79–94.

Hempel, C. G. (1958) "The Theoretician's Dilemma," in Feigl, H., Scriven, M., and Maxwell, G. (eds.), *Concepts, Theories, and the Mind-Body Problem, Minnesota Studies in the Philosophy of Science*, vol. 2, Minneapolis: University of Minnesota Press, pp. 37–98.

Kant, I. (1787) *Kritik der reinen Vernunft*, Riga: Johann Friedrich Hartknoch.

Kitcher, P. (1978) "Theories, Theorists, and Theoretical Change," *The Philosophical Review*, 87(4), pp. 519–547.

Kitcher, P. (1982) "Genes," *The British Journal for the Philosophy of Science*, 33(4), pp. 337–359.

Kitcher, P. (1993) *The Advancement of Science*, Oxford: Oxford University Press.

Koyré, A. (1957) *From the Closed World to the Infinite Universe*, Baltimore: John Hopkins Press.

Koyré, A. (1961) *La Révolution astronomique: Copernic, Kepler*, Borelli, Paris: Hermann.

Koyré, A. (1966) *Études d'histoire de la pensée scientifique*, Paris: Gallimard.

Kripke, S. (1972) "Naming and Necessity," in D. Davidson and G. Harman (eds.), *Semantics of Natural Language*. Dordrecht: Reidel, pp. 253–355.

Kuhn, T. S. (1959) "The Essential Tension: Tradition and Innovation in Scientific Research," in *The Third (1959) University of Utah Research Conference on the Identification of Scientific Talent*, C. Taylor, Salt Lake City: University of Utah Press, 162–174.

Kuhn, T. S. (1962) *The Structure of Scientific Revolutions*, Chicago: University of Chicago Press (1970, 2nd edition, with postscript).

Kuhn, T. S. (1983) "Commensurability, Comparability, Communicability," *PSA 1983: Proceedings of the 1983 Biennial Meeting of the Philosophy of Science Association*, ed. P. Asquith and T. Nickles, East Lansing MI: Philosophy of Science Association, pp. 669–688.

Lakatos, I., and Musgrave, A., eds. (1970) *Criticism and the Growth of Knowledge*, London: Cambridge University Press.

Laudan, L. (1977) *Progress and Its Problems*, Berkeley: University of California Press.

Laudan, L. (1981) "A Confutation of Convergent Realism," *Philosophy of Science*, 48, pp. 19–49.

Laudan, L., Donovan, A., Laudan, R., Barker, P., Brown, H., Leplin, J., Thagard, P., and Wykstra, S. (1986) "Scientific Change: Philosophical Models and Historical Research," *Synthese*, 69(2), pp. 141–223.

Levi, I. (1967) *Gambling with Truth*, New York: Knopf.

Martin, M. (1971) "Referential Variance and Scientific Objectivity," *British Journal for the Philosophy of Science*, 22, pp. 17–26.

Martin, M. (1972) "Ontological Variance and Scientific Objectivity," *British Journal for the Philosophy of Science*, 23, pp. 252–256.

Masterman, M. (1970) "The Nature of a Paradigm," in Lakatos, I., and Musgrave, A. (eds.), (1970) *Criticism and the Growth of Knowledge*, Cambridge: Cambridge University Press, pp. 59–89.

Maxwell, G. (1962) "The Ontological Status of Theoretical Entities," *Minnesota Studies in the Philosophy of Science*, vol. 3, Minneapolis: University of Minnesota Press, pp. 3–27.

Mongin, P. (2009) "Duhemian Themes in Expected Utility Theory," Brenner, A., and Gayon, J. (eds.), *French Studies in the Philosophy of Science*, Amsterdam: Springer, pp. 303–357.

Nagel, E. (1961) *The Structure of Science: Problems in the Logic of Scientific Explanation*, Chicago: Harcourt Brace.

Niiniluoto, I. (1987) *Truthlikeness*, Dordrecht: Reidel.

Niiniluoto, I. (2007) "Scientific Progress," in *The Stanford Encyclopedia of Philosophy*, Zalta, E.C. (ed.), URL = <http://plato.stanford.edu/entries/scientific-progress/>.

Pickering, A. (1984) *Constructing Quarks: A Sociological History of Particle Physics*, Chicago: University of Chicago Press.

Planck, M. (1940) *Scientific Autobiography and Other Papers*, trans. F. Gaynor, New York: Philosophical Library.

Popper, K. (1934) *Logik der Forschung*, Berlin: Springer.

Popper, K. (1963) *Conjectures and Refutations: The Growth of Scientific Knowledge*, London: Routledge.

Popper, K. (1972) *Objective Knowledge: An Evolutionary Approach*, Oxford: Clarendon Press.

Post, H. (1971) "Correspondence, Invariance, and Heuristics," *Studies in History and Philosophy of Science*, 2, pp. 213–255.

Putnam, H. (1973a) "Explanation and reference," in Pearce, G., and Maynard, P. (eds.), *Conceptual Change*, Dordrecht: Reidel.

Putnam, H. (1973b) "Meaning and Reference," *The Journal of Philosophy*, 70(19), pp. 699–711.

Putnam, H. (1975) "The Meaning of 'Meaning'," in *Language, Mind, and Knowledge: Minnesota Studies in the Philosophy of Science VII*, Minneapolis: University of Minnesota Press, pp. 131–193.

Quine, W. V. (1951) "Two Dogmas of Empiricism," *The Philosophical Review*, 60, pp. 20–43.

Reichenbach, H. (1938) *Experience and Prediction*, Chicago: University of Chicago Press.

Rescher, N. (1977) *Methodological Pragmatism: A Systems-Theoretic Approach to the Theory of Knowledge*, London: Blackwell.

Rescher, N. (1984) *The Limits of Science*. Berkeley/Los Angeles: University of California Press.

Shapere, D. (1964) "The Structure of Scientific Revolutions," *The Philosophical Review*, 73, pp. 383–394.

Shapere, D. (1966) "Meaning and Scientific Change," in Colodny, R. G. (ed.), *Mind and Cosmos: Essays in Contemporary Science and Philosophy*, Pittsburgh: University of Pittsburgh Press, pp. 41–85.

Shapere, D. (1969) "Notes Towards a Post-Positivistic Interpretation of Science," in Achinstein, P., and Barker, S. (eds.), *The Legacy of Logical Positivism*, Baltimore: Johns Hopkins Press, pp. 115–160.

Spelke, E. S. (1991) "Physical Knowledge in Infancy: Reflections on Piaget's Theory," in Carey, S. and Gelman, R. (eds.), *The Epigenesis of Mind: Essays on Biology and Cognition*, Philadelphia: Lawrence Erlbaum Associates, pp. 133–169.

Tichý, P. (1974) "On Popper's Definitions of Verisimilitude," *The British Journal for the Philosophy of Science*, 25, pp. 155–160.

Toulmin, S. (1961) *Foresight and Understanding: An Enquiry into the Aims of Science*. Bloomington: Indiana University Press.

van Fraassen, B. C. (1980) *The Scientific Image*, Oxford: Oxford University Press.

7

PHILOSOPHY OF SCIENCE AND SCIENCE STUDIES

Anouk Barberousse (Sorbonne Université)

1. Introduction: A Violent Conflict

Philosophy of science is not the only discipline to take science as its object. Both history of science and sociology of science share in this ambition. These three disciplines rarely come into conflict: indeed, the questions they pose about their object differ, they have developed diverging methods of investigation, and they advance perspectives that complement each other.

Another approach to science, close to the history and sociology of science but posing yet different questions, has been developing for the last 40 years or so. This is "social studies of science" or just "science studies." Science studies' mission is to renew the analysis of scientific activity by renouncing a certain number of suppositions commonly found in the other approaches. Specifically, it views science as just one human activity among many, without according it any special privilege relative to truth, objectivity, rationality, or the justification of the statements it produces. The modus operandi is to study "science as it is done" and not to develop a normative conception of science.

The voluntarily provocative results of science studies are rarely taken into account by philosophers of science; however, the intellectual and institutional stakes of such an enterprise are not so easily disregarded, as this chapter will set out to show. Indeed, its intention is to take stock of the relationships between philosophy of science and science studies.

Two observations form the root of this chapter:

1. Even though philosophy of science fits right into the vast domain of science studies, since its object of study is science, in actual fact it is absent from the field. In this way, the institutional separation between philosophers of science and specialists of science studies is total. Also dissimilar is the training they involve.
2. Moreover, science studies' point of origin is a philosophical preoccupation: Bloor and Collins, who were among the first proponents of a social study of science entirely distinct from the traditional philosophy of science of their time (the 1960s), wanted in this way to undertake what they called a "materialist" analysis of science. Their goal in suggesting such an analysis was to avoid calling on hypotheses generally accepted by philosophers of science and which they labeled as "metaphysical," such as, for example, the hypothesis of a correspondence between statements and facts, or that of a causal relationship between facts and the beliefs of an agent. In contrast to these "metaphysical" hypotheses, they suggested granting central place to the practical and body-connected aspects of scientific activity. Bruno Latour, for his part, admits to seeking another philosophical language for the analysis of science, another framework for understanding the world to the one proposed by academic philosophy. His ambition, huge as it is, is to shift the whole worldview which the West has placed on itself since the 17th century and to jointly redefine both science and society. More recently, in his *Introduction aux science studies* (2006), Dominique Pestre stated that his book "addresses epistemological questions" (p. 8).

This explicit claim to a philosophical approach is rarely taken seriously by philosophers of science, who are quick to point out the inconsistencies in their adversaries' positions. Thus, they point out the difficulty there is in entirely foregoing on any recourse to the concepts of truth, objectivity or justification in analyzing scientific activity. They also emphasize the fundamental difficulty there is in placing the social interactions between researchers and the interactions between researchers and the natural phenomena they study on the same explanatory level. As a result, dialog between philosophy of science and the aforementioned "science studies" amounts to little more than heated, if not violent, exchanges. At stake in these exchanges is the very legitimacy of scientific discourse, the specialists of science studies believing the legitimacy of philosophical discourse to be ill-gotten, and the philosophers of science reacting vehemently to that accusation.

Even accounting for the significance of both philosophical and institutional aspects in this debate, one cannot help but be struck by the virulence of the exchanges between philosophers of science and specialists of science studies. The insults fly, the most common of these being those of "positivism" on the one hand, and "relativism" on the other. These terms, it must be pointed out, have often lost all clear meaning. The specialists of science studies consider themselves as being at the forefront of a new way of understanding science which sweeps away the despotism of the philosophical categories of truth, of reality, or of rationality. If it was they who kicked off the offensive, the philosophers of science are not taking it sitting down, and the accusation of philosophical naivety is frequent, with Laudan going so far as to label science studies the "pseudo-science of science." Apart from Joseph Rouse (1987, 1993), Arthur Fine (1996) and Ian Hacking (1999), philosophers of science who make the effort (considerable as it is) to try and faithfully reconstruct the philosophical positions of their adversaries are few and far between. Recently, Paul Boghossian took the relativist bull by the horns, so to speak, by proposing a thorough critique of the "fear of knowledge" (2006). Science studies specialists do sometimes go into detailed presentations of their principal philosophical targets, such as Popper or Bachelard (although not for the same reason), but often they content themselves to flatly leveling their accusations in one go at "The Vienna Circle."

In this respect, it can be said that specialists in science studies sometimes seem to consider contemporary philosophy of science to be nothing other than the development of the legacy left by Bachelard, Popper and "The Vienna Circle." This is undoubtedly a large reason for the strong dismissive reaction philosophers of science have towards their body of work. Indeed, from within philosophy of science, it seems clear that, for one thing, Bachelard's and Popper's positions are at odds on many points, for another that Popper was primarily an *opponent* to the Vienna Circle, and finally that the Vienna Circle's legacy is far from being a coherent and unified whole (see, for example, Richardson and Uebel, 2007). Furthermore, the positions philosophers of science espouse are often more nuanced than those held by certain scientists, such as Weinberg, a fervent defender of a bold form of realism, but specialists of science studies often assimilate both sets. In short, philosophy of science has nothing of the monolithic whole often caricatured by specialists of science studies—in the same way that there exist profound divergences between Bloor, Latour, Pickering, and so forth.

Let us further point out that the so-called Science War is a prolongation of this violent quarrel and this mutual wanton ignorance. The "Science War" is the declared opposition, from within both parties, between "hard" sciences (physics and biology) and "soft" sciences (social sciences, human sciences, cultural studies). Philosophers of science, for the most part, are situated in the hard sciences "corner," while specialists of science studies have readily taken up the defense of the soft sciences. Of course, the origins of the "Science War" are many; however, the fact that specialists of science studies and philosophers of science have predominantly regrouped on opposing sides has certainly vulcanized the divisions. Prompted by Alan Sokal's hoax of 1996, which denounced publication procedures within the human sciences and the humanities, Sokal and Bricmont's 1997 book first appeared in France, followed by responses, including Jurdant and Savary (1998). While in the United States the "Science War" remained confined to the academic world, in France it was widely spoken of in the media and the entire intellectual milieu stuck its oar in.

This chapter is organized around three questions whose aim is to analyze the conflict of legitimacy between two rival discourses on science. As we have seen, many science studies specialists take up an openly philosophical starting point and yet laud methods of investigation which are radically opposed to the methods employed by philosophers of science. So, the first question will be (1) Which is the best method for studying science? The examination of this question will take up most of this chapter.

One of the major criticisms specialists of science studies make against philosophers of science concerns their ignorance of aspects of scientific activity which they themselves judge to be extremely important. The second section of this chapter will aim at showing how philosophers of science have recently become aware of some of these aspects. It will thus be dedicated to the question (2) How can the intrinsically collective nature of scientific activity be broached seriously? Indeed, the relative silence of philosophers of science on this question is one of the weapons of choice for science studies specialists. In this section will be presented both sides, the sociological and the philosophical, of social epistemology, as well as some possibilities for dialog.

Science studies places itself within the continuity of current reflections on other human activities, such as politics or culture. The comparisons within science studies between scientific activity and other cultural activities are many. It is those aspects common to the various human activities which are then highlighted. By contrast, philosophy of science maintains scarce relationship with correspondent philosophical disciplines such as philosophy of art, political philosophy, and philosophy of history. This asymmetry between philosophy of science and science studies is due to the theoretical choices adopted, but also the result of philosophy of science's history, the course of which is today undergoing a change in direction. It is for this reason that the third question to be tackled is: (3) What kind of relationships may or should philosophy of science foster with other disciplines like philosophy of art, political philosophy, or philosophy of history? This final section will take the form of a brief conclusion to the chapter.

This chapter does not claim to encompass the full diversity of the science studies field, which is far from being unified, so diverse are its objects of study, methods, and basic assumptions. Its aim is rather to show that, contrary to appearances, philosophy of science can benefit from opening a dialog with science studies, and vice-versa.

2. Which Is the Best Method for Studying Science?

Philosophy of science, following its institutionalization in Europe at the beginning of the 20th century and its development, from the 1930s onwards, in the U.S.A. (Moulines, 2006), became the object of many criticisms. By means of an introduction to this section, I will speak of just one of them, despite its retrospective character, according to which the philosophy of science of the first half of the 20th century contributed heavily to the propagation of what Philip Kitcher (1993, ch. 1) calls a "legendary" conception of science. According to that legend, science is guided by noble goals, whose affair is the search for truth and which are better and better achieved as time moves forward. These successes, crowning achievements of human reasoning, are explained by the intellectual qualities and exemplary morals of the scientists and by the use of THE scientific method which, since the 17th century, has led to the creation of objective criteria for assessing statements, thus avoiding bias, confusion, and superstitions. Many versions of this legend have been spun, philosophers of science having had, for example, diverging opinions on the nature of scientific method. Nevertheless, up until the 1950s it underpinned a large part of their work.

The necessity for a methodological overhaul was felt in philosophy of science not with the appearance of science studies as we know it, but rather with the publication, at the end of the 1950s, of works by Norwood Russell Hanson (1958) and Thomas Kuhn (1962), and also, later, Paul Feyerabend (1975), which unapologetically questioned the "legend" described by Kitcher, as well as the philosophical work it underpinned (see chap. 6). The schematic, if not to say simplistic, nature of the history of science which was channeled by the dominant philosophy of science (that is, the *received view* of philosophy of science) was denounced with force by these authors and by other historians or sociologists of science. The basis for this criticism was that the schematic conception of the actual history of the sciences, which was encouraging philosophers of science to work only from a limited number of examples described in a particularly poor fashion, was leading to erroneous inferences about the nature of scientific activity and its development (see chap. 6).

Hanson, Kuhn, and Feyerabend level several criticisms at philosophers of science, as well as at certain sociologists of science such as Merton (1973). The first is for considering science as a set of knowledge of a purely conceptual nature, sheltered from any social influence, and which comes to the world as a product of purely "knowing" minds. They also rebuke philosophers of science for not taking the institutional aspects of scientific activity into consideration (which is what Merton does). But the biggest criticism concerns not questioning one of their major assumptions—that the development of science constitutes a rational progression. Kuhn, like Feyerabend, denies that there are any logical norms governing this progression and asserts that scientific practice is, on the contrary, governed by local habits of thought. A richer and more varied vision of science was desperately needed, in their opinion, to replace the philosophers' schematic conception of things. They advance that such a vision can only be formulated on the basis of precise empirical case studies, historical or contemporary, judged to be particularly "interesting."

Of course, a philosopher will jump in with the question: "For who, or in what perspective, are these cases supposed to be interesting?" Certain specialists of science studies, and particularly Dominique Pestre (see, for example Pestre, 2006, and also the analysis presented in Keucheyan, 2008), have no problem with granting an absolute value to this property of being "interesting," pointing to Paul Veyne (for example, 2006) to back them up. From the philosophy of science perspective, a case will be judged all the more "interesting" if the ends for which it was chosen are known. In the case of science studies, these ends are determined by the methodological principles which will be presented in this section.

First, we will set out answers to the question concerning the best method for studying science in the form of a dilemma that philosophers of science have had to face since the beginning of the study of science. This will be followed by the exposition of the principal theses put forward by their proponents, concerning, on the one hand, the historicity of the concepts that play a major role in philosophy of science—such as the concepts of empirical proof, demonstration, or truth—and, on the other hand, questions of methodology. In finishing, we will discuss Rouse's claim (1987, 2002), according to which science studies could enable philosophy of science to exit sterile debates (about scientific realism, the nature of confirmation, of explanation, etc.) by the high road, so to speak, these debates having been, in his view and in Fine's (1996), a burden to it for almost 60 years.

2.1 A DILEMMA

Philosophers of science often consider history of science to be the empirical basis of their generalizations (but for a more fine-grained view, see Nickles, 1995). If they hold onto a simplistic conception of that empirical basis, then their inferences will most probably lead to flawed or biased conclusions. Does this mean, however, that philosophers of science must become historians to better ground their work? This option leads straight into a dilemma:

(i) either the philosopher of science acquires training as a historian of science and abandons her initial discipline
(ii) or she doesn't acquire this training and leaves herself open to accusations of excessive simplification

It seems to me that this dilemma must be taken seriously, since the two disciplines of history and philosophy of science are growing further and further apart, as much from the methodological angle as from the perspective of their objects. This observation of distancing belongs not only to historians of science (there was, for example, Robert Fox who answered negatively the question in the conference "History and Philosophy of Science: Towards a New Alliance?," Paris, October 2002) but also to those philosophers of science who are nevertheless tuned into history of science (see Ernan McMullin at the first Integrated History and Philosophy of Science conference, Pittsburgh, October 2007—it was in fact in reaction to this situation that this series of conferences was created).

However, it can happen that the same person happily occupy both functions, alternately philosopher and historian of science. In this case, the history of science practiced is often a so-called "internal" history, that is, one which finds its objects and modes of explanation within the field of scientific activity and, more generally, the field of thought. This practice of history follows Lakatos's intuition that scientific development is able to account for its own rationality because of the internal logic of scientific discoveries. It is neglected more and more by professional historians of science who, as we shall see, adopt a wider, different vision of their work, while philosophers of science remain attached to analyses of this sort.

This dilemma has two aspects. The first, and older, is about knowing how to connect the descriptive elements simultaneously to the normative elements belonging to philosophy of science and to research into well founded generalizations. Specifically, it is the norms of *justification* which must find their place within the descriptions and narrations. The richer a description is, that is, the better it gives account of "science as it is done" (what historians and sociologists accuse philosophers of forgetting) the better it can individuate the case described and, thus, the less it will allow for generalizations. How, then, should the philosopher of science connect search for generalizations to descriptions of detail, enabling her thus to respond to the demands of a more empirically satisfying conception of science?

The second aspect of the dilemma is more recent. It involves the repartition of disciplines across the various component elements of science studies, in the broad sense. Up until the 1960s, historians of science, for the most part, had a philosophical education and saw their historical work as an extension of that background. Dialog between historians and philosophers of science was aided by that shared pool of references and practices of reflection. Today, by contrast, historians of science claim a strong professional specificity: they consider themselves to be closer to historians than to philosophers and cast more of a critical eye on the history of science that their predecessors practiced. Dialog with philosophers of science is practically nonexistent, which has led to many historians of science choosing the first branch (*i*) of the dilemma and educating their students accordingly.

Science studies specialists now borrow their questions from general history and the social sciences. They also seek to satisfy the demands of general historiography, that is: not to anticipate on the future, and to analyze the attitudes of the protagonists in action, and independently of the outcome of the debate being studied. As we shall see further on, David Bloor, within the remit of the "Strong Program in the Sociology of Scientific Knowledge," proposed a set of methodological rules for science studies which brings them closer to history and to the social sciences. Their common order is to systematically refuse any explanations grounded on the fact that their interactions with nature justify scientists in nurturing certain beliefs. We must call on other explanatory factors, sociological factors for the most part, in order to avoid falling into the error of "judged history," that is, a retrospective history retold in light of knowledge which we have today. According to the proponents of the Strong Program, the description we today give to the interactions between, for example, Galileo and natural phenomena is biased by the understanding we have of these phenomena today, whence the rule of systematically ignoring these interactions. It is with a mind to satisfying these methodological orders, or others of the same type, that specialists of science studies choose themes which have never attracted the attention of more traditional historians of science, and this in order to show the rich potential of the epistemological questions they set themselves. We will briefly examine some of these new themes.

The "controversy studies" are a first example instantiating a certain reinvigoration when compared to the classic subjects in history and in philosophy of science (the most famous example of this is Rudwick, 1985). In a controversy study, one examines, as precisely as possible, the various steps of the debate without presupposing that the consensus arrived at in the end guides its development in any way, so as not to replicate the major flaw of historians of science of the "old school" and philosophers of science mentioned earlier—the flaw of "judged history." When the history of science is told in this latter manner, we are placed at the vantage point of modern scientists, with all they know about the domain in question. This vantage point allows us to judge, retrospectively, the errors of the protagonists of the past, something science studies specialists bar themselves from doing. It is the unfolding of the controversy itself, rather than its outcome, which they see as being worthy of study. Thus, all aspects of the original exchanges are taken into consideration, as much those involving findings and the arguments developed from them as those involving the competitive relationships between scientists, laboratories, or nations. There is no confining to what scientists of today consider to be important, nor even to what the protagonists of the controversy themselves identified as being "scientific." Specialists of science studies claim to refrain from all judgment in that respect.

The development of the methodology for controversy studies impelled science studies specialists to choose contemporary episodes where the debate was not yet settled, in such a way that it be impossible to commit the error of judged history (see, in particular, Collins, 2004, who recapitulates two decades of sociological research on controversies regarding the detection of gravitational waves). More generally, specialists in science studies criticize philosophers of science for sticking only to matured disciplines for their objects of investigation and they insist on the richness of studying budding disciplines, still under construction.

Another particularly revealing theme of the radical change advocated by specialists in science studies, regarding the previously commonplace presuppositions on the nature of scientific activity, is the theme of "consensus emergence." Previously it was considered that once an experimental result had been obtained or a theoretical hypothesis confirmed, then the scientific debates were closed. Historical studies have shown that, at least in some cases, this is not at all true, and that agreement between scientists came about slowly, along the course of discussions bringing in not only experimental results but also methodological values or cultural elements (see, for example, Warwick, 1992, 1993). It is more through cultural integration of practices and procedures, rather than exchanges of arguments, that consensus tends to emerge. The conclusion that science studies specialists draw from this is that the notions of explanation, of confirmation, or of invalidation of a theoretical hypothesis (see chaps. 1 and 2), in the way they are analyzed by philosophers of science, are far from capable of accounting for the workings of scientific communities. On the contrary, all controversies are to be solved one by one; arguments of various types are put forward, which have little to do, according to the authors of these historical or sociological studies, with the patterns described by philosophers of science. Historical contingency is often called to the rescue as an explanatory principle for the way in which controversies lead to consensus, in order to underline the fact that such emergence is not within the remit of the criteria discussed by philosophers of science. These studies show that the notion of *justification*, which is at the heart of philosophical analyses of science, possesses no universal legitimacy as far as the science studies specialists are concerned. In light of this observation, the question which immediately springs to mind is what, for them, distinguishes scientific activity from other human activities? The most frequently given answer is "nothing connected to rationality, but rather particularly sophisticated practices of social domination."

Just as controversy studies and research into the emergence of scientific consensus point to a major change in the way specialists in science studies view scientific activity, so too the research begun by Peter Galison (1987) on the way in which "experiments end" bears witness to a new conception of experimentation's place within science. What mattered for philosophers of science up until the 1990s was the *result* of experiments, and not the way in which these results were acquired, this being seen as non-problematic. The observed facts needed no analysis; they were seen as being delivered by means of an unequivocal process. Galison, on the other hand, showed the full richness of the experimental process. In particular, he insisted on the fact that "the" result of an experiment is not a legitimate concept. In contemporary physics, an experiment always yields numerous results and answers numerous questions; it is the reason why it is not at all clear at what moment an experiment should be stopped. According to Galison, the end of an experiment is decided by means of a negotiation and not, for example, because it has brought us to the explanation of a phenomenon, or to the confirmation of a theoretical hypothesis.

More generally, science studies specialists underline that the process of experimental work implies actions and choices whose complexity is that of all human acts and whose study sheds light on the results obtained. In this way, for each case, it must be carefully determined which elements convinced the experimenters of the fact that a certain phase of their work was complete (see Atten and Pestre, 2002). Atten and Pestre insist on the irreducible singularity of each case, referring to the collection led by Revel and Passeron (2005), whose objective was to resolve what is given as the central problem of the human sciences: "How, from singular configurations, does one arrive at generalizations?"

Let us finally mention one other set of phenomena which had been ignored by philosophers of science, what specialists in science studies call the "civilities of proof," that is, in their view, the social rules within which scientific practices and procedures garner meaning. Shapin and Schaffer (1985), in particular, have revealed the appearance in 17th century Britain of forms of sociability which guaranteed, socially speaking, the truth of facts reported in the observational accounts presented before the *Royal Society*. These forms of sociability, say Shapin and Schaffer, explain why the results pronounced by a learned or a noble man are accepted as legitimate and reliable by the audience, even if they did not witness the experiments themselves. Moreover, Shapin

(1984) has shown the importance of the "literary technologies" developed by Boyle, among others, to convince his readers.

The majority of the examples mentioned have been subject to scrupulous historical or sociological investigations which a philosopher of science would judge to be outside of her abilities. On the other hand, the methodological and epistemological presuppositions put to work in these investigations require the philosopher of science's critical analysis (such as the one conducted by Sargent, 1988). Indeed, science studies specialists either adopt philosophical positions that they assert with force, often in quite a confused way, as Fine (1996) has highlighted, or else they base themselves on implicit presuppositions which are no less philosophical. Once these ambiguities have been overcome, we rapidly become aware that the philosophical theories being defended, sometimes implicitly, by science studies specialists are worthy of individual discussion—be this only to avoid these theories monopolizing the discussion when science studies specialists take on the role of experts on the relationships between science and society. Conducting this discussion is a way of getting out of the dilemma: it allows us to show the relevance of philosophy of science while sticking as close as possible to recent historical and sociological analyses, revealing their hidden forces. Thus, we shall see, we are dealing with a traditional role of clarification augmented with a role of watchman over the coherency of the conclusions reached within science studies.

2.2 HISTORICALLY SITUATED CONCEPTS

In this subsection, we will analyze some of the positions openly defended by science studies specialists relative to the historically situated nature of the central concepts of philosophy of science. As we have seen, specialists in science studies vigorously criticize philosophers of science, pointing to their small regard for the empirical reality of "science as it is done," to use their favored expression, contrasting implicitly with "science as it should be done," the supposed object of study for philosophers of science. One of their principal arguments is that philosophers of science think of certain concepts as being a-temporal, like the concepts of empirical proof, of objectivity, of distinction between facts and opinions, of rationality, of pure science, and of explanation. Specialists in science studies, on the other hand, resoundingly claim to have shown the deeply historical character of these concepts (see in particular Shapin and Schaffer, 1985; Shapin, 1994; Daston and Galison, 2007; Atten and Pestre, 2002). Above all, science itself, they say, is a historical object: to think of science has being given once and for all would be historically incorrect.

Opposing the position, according to which results and demonstrations would have a universal scope, science studies specialists claim that proofs, be they empirical or formal, and to the extent that they are intended to convince, always have a contingent dimension. More precisely, in their view legitimizations are never transparent. It is for this reason that it is necessary to study the way in which proofs are "administered," and not be content to simply present the published, textual arguments. The relationship between the proof and the proven is, indeed, not provided in advance, since the reason for which we are convinced is always a *singular* combination of circumstances. Indeed, judging the relevance of empirical demonstrations and results is the very heart of scientific work. Thus it is essential to closely analyze precisely what is convincing at any given time for a group or an individual and to be aware of judgments *in situation*. As we can see, one immediate consequence of the argument that the ways in which justifications are administered are historically situated is that rational reconstruction methodology, and its variants developed by Lakatos (1978), Laudan (1977), or Toulmin (1961) (see also chap. 6), is null and void in history of science. In order to save this methodology, if it must be saved, it falls to philosophers of science to respond to the statements issued by the science studies specialists, in particular by seeking to lay open the universal core of the concept of confirmation (see chap. 2).

One avenue to follow in answering the science studies specialists is to deepen the analysis of presuppositions which underpin the statement saying that results and demonstrations can only truly convince in situ. For example, Pestre (2006) insists on the "variety of rationalities" genuinely put to work within scientific activity along its history. However, it is not easy to know what this term "rationality" refers to in this context, even more so in the plural. In general, philosophers of science call up only one concept of rationality, the definition of which may have several variants, but which does not have a meaning as broad as the one Pestre puts forward. One clarification is needed here, for if, ultimately, it turned out that all the protagonists were in agreement on a minimal conception of rationality, the science studies specialists' statements would lose plausibility.

The considerations alluded to here, as well as the example of Hertz's work on electromagnetic waves taken up by Atten and Pestre (2002), lead them to posing the following question: What is the legitimacy of a vantage point of 35,000 feet, that is, a vantage point far removed from the practices and presuppositions of the actors in the historical account? Such a vantage point is the high place philosophers of science are accused of arbitrarily projecting themselves. According to Pestre, it is a place where statements of knowledge would never have to be corrected and where legitimizations would be transparent. Pestre here denounces the fiction of a world to which only perfect spirits would have access, where everything would be unequivocal, in word and in act, and where communication would encounter no loss. It is true that philosophers of science make use of quite a few idealizations regarding agents' cognitive capacities. Thus they often suppose that agents are endowed with logical omniscience (they are capable of accessing the whole set of logical consequences of all their beliefs), or that their capacity for logical calculation is infinite (see, however, the discussions of this question in chaps. 2 and 16). However, are these idealizations, which allow for the study of the conditions under which certain goals, judged to be eminently scientific (such as the explanation of phenomena or phenomenological laws, or the confirmation of theoretical hypotheses), are reached, justified? Philosophers of science insist on their fruitfulness: they enable the updating of the norms of justification.

In the end, does the historicization of the concepts of proof or objectivity constitute a major problem for philosophers of science? Could a devastating methodological objection against the methods of philosophy of science really be formulated by simply updating the historically situated nature of these concepts? Two analyses of these questions are possible:

a. The historicization of the central concepts of philosophy of science simply demands that philosophical discourse about scientific proofs, confirmation, explanation, and objectivity remain within the correct historical era, and also specifies that it is not meant to be a-temporal, that is, that it shouldn't separate the idealized analyses (the conceptual core of hypotheses and practices) from the elements which rely on historical context. Besides which, there is no uniform position among philosophers on the question of the a-temporality of these concepts, which sends us back to debates in philosophy of language and logic, and in metaphysics about the nature of truth, of language, of the relationship between language and the world; debates which are far from resolved. To choose this option is to claim compatibility between a strong normative core and historical variants which can differ greatly from each other, so much so, it can be difficult to uncover an invariable conceptual core.
b. According to another analysis, the legitimacy of a-temporal analyses must be maintained at the cost of *minor* adjustments to account for historical contexts and their evolution. In this case there is an insistence on the strong normativity of the concept of rationality.

2.3 EMPIRICAL STUDIES, WHATEVER THE COST

Science studies specialists see themselves, for the most part, as being more empirical with respect to their object than philosophers of science. As we saw, they support the case-study methodology and advocate the benefit of "thick" descriptions inspired from the anthropological methods of Clifford Geertz (1973). A thick description of some behavior includes its context, in such a way as to make that behavior intelligible for someone who did not witness it. Furthermore, specialists of science studies want, insofar as is possible, to avoid using standard categorizations (and especially those introduced by philosophers of science) without first questioning them. It was Bruno Latour who, in 1987, went the furthest in this direction by refusing to take up the distinction between humans and non-humans in his conceptualization of agency: in his view, just as much account should be taken of what things "do" as what humans do.

However, any description, as faithful to the facts as it may claim to be, is always the result of theoretical choices, sometimes implicit, regarding, in particular, the categorization and choice of what is "interesting." Thus the slogan "Let's be empirical" doesn't say enough from a methodological point of view. Science studies specialists have tried to say more, as we shall see in the rest of this section.

Some science studies specialists have sought to formulate the theoretical framework which seemed to them the best suited to their descriptive quest. The first to work towards this were David Bloor and Barnes, the two principal advocates of the "Strong Program" in sociology of science. To present that undertaking in a nutshell, we can say that their main effort consisted of replacing philosophy of science with sociology, understood in a highly empiricist way (by contrast, see chap. 14). Sociology, according to Bloor and Barnes, is indeed an empirical discipline, purely descriptive, and which therefore does not fall victim to the normative and reductionist failings of philosophy of science.

To replace philosophy of science with sociology is to adopt a *bona fide* philosophical position, alternative to the position identified as dominant in philosophy in the 1960s and 1970s. Barnes's and Bloor's main targets are Popper and the members of the Vienna Circle, taken as a whole and, as we have already seen, in total ignorance of their internal debating. The criticisms science studies specialists level at philosophers of science are often no more than attacks on straw men; by contrast, the methodological principles by which disciples of the "Strong Program" wish to replace the theoretical presuppositions of their adversaries are clearly opposed to certain positions genuinely defended by Popper or by members of the Vienna Circle, as we shall see in the remainder of this section.

The driving force behind the Strong Program is its refusal of an unduly intellectualized conception of knowledge. On the contrary, its disciples draw up the hypothesis according to which the transmission, distribution, maintenance and changing of beliefs and practices are all open to exclusively sociological causal explanations. They insist on the importance of taking the instruments, the experimental techniques, the know-how, and the knowledge of the working body into account. In their view, scientific knowledge, far from being disembodied, is always tied in with physical places and domains of production and validation. But Bloor and Collins go further than demanding account for what is linked to instrumentation and the physical in scientific activity. They also reject the majority of the classically opposed couples which form the structure of traditional approaches, such as form-content, knowledge-context, logic of justification-contingency of discoveries.

2.3.1 The Four Principles of the Strong Program

According to the *principle of causality*, all statements produced by the actors in the history we tell must be brought into their context, that is, into the intellectual, social, and cultural framework which legitimizes them and within which they can be held as true. Or in other words, scientific statements should not be considered as eternally true and necessarily accepted as such, but like any other kind of statement, whose general meaning is highly dependent on the context. Indeed, it seems like only logical statements and mathematical statements, when they are expressed in formal languages, escape this contextual dependence.

If we stick only to acknowledging the contextual dependence of the meaning of statements, all that will be seen in the principle of causality is an obvious recommendation. However, what the principle of causality imposes goes well beyond that rather banal demand. Indeed, it obliges us to consider the intellectual, social and cultural context as being the veritable *cause* of the statements—far more so than the interactions between the scientist in question and the world, for example. For followers of the Strong Program, a causal explanation can only be sociological. This is why the statements studied must be causally "related" to the social and cultural environment of the actors who produce them: this is the only way of understanding why they and their interlocutors hold these statements as meaningful and truthful (when such is the case).

The principle of causality is, among the four principles of the Strong Program, undoubtedly the most surprising for philosophers of science, who generally consider that what makes us hold a statement as true is its *content* rather than the social and cultural environment in which it is enunciated. Besides which, it seems reasonable to suppose that the content itself, when dealing with scientific statements, depends simultaneously on what is observed and on the knowledge thus-far acquired, whatever the precise modalities of that dependence may be. Thus, philosophers of science generally consider that once the meaning of a statement has been determined in context, the epistemic attitude adopted in respect of it (whether we hold it as true, false, dubious, probable, etc.) depends more on what it is about than on the context in which it was enunciated.

Science studies specialists often put forward Forman's article (1971) on the profound influence which the author says "Weimar culture" wielded on the idea of causality developed by contemporary physicists and mathematicians. The article is seen as showing that social and cultural context can have a causal influence on the acceptance of scientific statements. In it, Forman studies the reception of statements from a revolutionary theory, quantum mechanics, and tries to show that Weimar culture, within which the absence of causal determination, individuality and visualization are important elements, favored acceptance of the new theory, itself indeterministic and whose interpretation at the time was greatly steeped in discussions about the visualizability of the trajectories of quantum objects. More precisely, according to Forman, German mathematicians and physicists, under the influence of their ambient culture, itself largely determined by Germany's defeat in World War I, had a tendency, from 1921 onwards, to reject causal conceptions. However, it doesn't really seem that with this long article Forman has shown anything other than a coincidence between a certain cultural climate and a certain interpretation of quantum mechanics. He establishes no causal relationship, in the strict sense, between the two—in any case, not in any strict enough sense to render causal attribution unproblematic.

The second principle of the Strong Program is the principle of *impartiality*, which dictates that the person studying an episode of the history of science account for it without taking sides in favor of the truth or falsity of the statements pronounced or written down by the protagonists of the episode, nor in favor of the rationality or irrationality of their attitudes either. She must therefore recount the episode in question as if she didn't know the outcome of the described debates; this consists, more precisely, in removing all traces of retrospective knowledge from her narration.

A counter-intuitive consequence of this principle is that the historian cannot relate true beliefs to the attitudes and actions of the agents by the unique merit of their being true. Yet this is what an apparently convincing analysis of the relationships between beliefs and actions invites us to do (see for example Ramsey, 1926, according to whom beliefs can be seen as guides to action: true beliefs are thus more reliable guides than false beliefs, and can enter into an explanation for the success of action). Moreover, the demand it imposes seems artificial, if not simply contrary to the norms of erudition: pretending to ignore an important element of a historical episode leads at the very least to misplaced convolutions, especially in those cases where the reader herself is aware of the outcome of the episode in question. To build up an intelligible narrative, it seems, on the contrary, necessary to use all available information.

We can see that the principle of impartiality is essentially methodological but that it is founded, just like the principle of causality, on an unusual conception of the relationships between beliefs and the world. That conception is difficult to characterize precisely; at best we can point out what it doesn't include. Thus, according to the proponents of the Strong Program, interactions between agents and the world seem to have but little influence on their epistemic attitudes, this constituting yet another reason for the skepticism of philosophers of science regarding this research program.

The *principle of symmetry* is an extension of the principle of impartiality and requires that the historian of science apply identical (symmetrical) presuppositions in the explanations of the beliefs of all protagonists in a debate, whatever the truth value or empirical adequacy of these beliefs may be. Here we rediscover the argument stating that the truth value of a belief should not be regarded as an *explanans* for an agent's possession of that belief, or for the actions she may undertake on its basis. Had we to apply this argument to the explanation of everyday actions, we would find this most difficult: we wouldn't be able to understand why trains fill up with passengers, for example, even though, by simply recalling that the explanation sought depends on the fact that the passengers believe their train departs at such a time, from such a station and that, furthermore, these beliefs are true, then the phenomenon is not at all puzzling.

The three principles presented so far—the principles of causality, impartiality and symmetry—open up two possible interpretations of the manner in which the proponents of the Strong Program envisage the relationships between belief and action.

(i) Either they support (implicitly) a completely heterodox conception of these relationships, according to which the truth of a belief plays no role in the success of the action it founds. Here, it is purely the social and intellectual context which is responsible for the success (or failure) of actions in so far as they are founded in beliefs. This relativist option is quite widely shared, this being a particularly direct wording of it: "As we come to recognize the conventional and artificial status of our forms of knowing, we put ourselves in a position to realize that it is ourselves and not reality that is responsible for what we know" (Shapin and Schaffer, 1985, p. 344). Besides the details of such a conception remaining to be laid out, the burden of proof of its superiority over the common conception rests with those who defend it.

(ii) Or else the disciples of the Strong Program adopt the common conception of the relationships between belief and action when dealing with everyday life, but propose a radically different conception when it comes to scientific activity. In this case, they must give convincing reasons for such a schism within the beliefs and practices of scientists as this seems quite implausible.

Finally, the *principle of reflexivity*, which has sparked heated debates, demands that the three principles that the social explanations sought to obey be universal. This constraint results from a large number of science studies specialists wanting to be scientists themselves by leaning on observations and avoiding the expression of any norm whatsoever.

After the Strong Program, another research program was launched, under the moniker EPOR: *Empirical Program of Relativism* (see Collins, 1981), whose goal was to precisely describe the "fabrication" of scientific statements. It is the proponents of EPOR who developed the methodology of controversy analysis.

All the principles of the Strong Program, as well as those of EPOR, have been subject to discussions within the science studies community. These principles are far from universally accepted by historians and sociologists of science who have followed the science studies turn; they do, however, share some larger methodological precepts like the refusal of explanations unmindful of actors' realities and, in contrast, the search for sociological explanations, within which care is taken to place scientific acts inside the social contexts which give them meaning. So the theoretical option taken by the disciples of science studies can be labeled as *particularist*: they consider no general explanation to be valid in history of science and only particular explanations to be acceptable. Philosophers of science, on the other hand, favor the search for general explanations in so far as they are, in their view, the most likely candidates for making scientific activity intelligible. In doing so, they are led to calling on idealizations and simplifications, which they are bound to justify, as with any scientific enterprise.

2.4 EXITING STERILE DEBATES BY THE HIGH ROAD?

It can be considered, as, for example, Jo Rouse (1987) has done, that the theoretical options of certain science studies specialists represent an exit door to the endless debates which have been the bread and butter of philosophy of science for almost sixty years. Philosophers agree neither on the problem of induction nor on the nature of empirical confirmation nor on the best position to adopt in respect of scientific theories: realist or anti-realist, and so forth. Science studies specialists refuse to take part in these debates and claim them to be inane, for the reason that the notions in question are eminently relative to the historical and social contexts in which they are used.

By exiting the debates which structure the philosophy of science field, we end up at those (no less inextricable) debates which structure the science studies field. One of these debates, which is rarely made plain (see, however, Pestre, 2006, 42), stems from the vigorous criticism leveled at "judged" history. The resulting demand for the investigator is to suspend all retrospective judgment. In describing a historical situation, one must proceed as if its scientific outcome were unknown. However, readers of these historical accounts' background knowledge is often reduced to what *current* science has to say about the situation in question. How then can a link be established, when conceiving a historical account, between the methodological requisites of science studies and the readers' expectations in terms of understanding? Aren't we obliged, in proceeding as a historian, to practice judged history, at least to a certain extent? Pestre writes: "I lean on the latest science to construct my argument—thus maybe demonstrating that it is impossible (for me) to not also be, in practice, a partisan of judged history" (2006, p. 42). But then, is not the very core of the science studies enterprise put at risk?

Another debate runs through science studies, that of the legitimacy of the principle of reflexivity and the recourse to current scientific norms that it imposes. Professing to perform the science of science is to voluntarily bow to current scientific norms—norms which are the very object of study, and whose mutually dependent links to the social context in which they appear we seek to reveal. Is this demand of radical reflexivity tenable? Some philosophers of science get out of this cycle with ease by recalling the philosophical, that is, non-scientific, nature of their work when they take scientific norms as their object. Science studies specialists rule out that solution straightaway. In doing so they risk sliding into a methodological cycle.

Bourdieu also supported the demand for reflexivity, particularly in his final work (2001), but from a completely different perspective, since his principal preoccupation was to protect science from economic, political and religious interests while acknowledging its historical and social nature. In this way he was trying to show the possibility of a rationalist approach to sociology of science, founded, among other things, on the concepts of *habitus* and of scientific capital.

In contrast, the majority of science studies specialists have a far more negative idea of science. As Fine (1996) highlights, many of them see themselves as a sort of romantic avant-garde to the anti-science crusade. Hence Pickering (1984, 413) affirms that "there is no obligation upon anyone framing a view of the world to take account of what 20th century science has to say [. . .]. World views are cultural products; there is no need to be intimidated by them." However, a recent sociological study (Keucheyan, 2008) has shown that practitioners of a mildly radical form of science studies, put into practice at the French *Centre de Sociologies de l' Innovation*, then directed by Bruno Latour and Michel Callon, did, on the contrary, display a certain reverence with respect to their fields of study—and thus also to the results produced by the scientists of these fields. (This reverence is even openly justified by the obligation to not place oneself in the position of superiority adopted by the expert, seen as contemptuous, and attributed to Bourdieu.) Rather, in Latour's texts, a solid continuity between ordinary and scientific knowledge is affirmed. This is another tension within the ideas adopted by specialists of science studies.

The biggest source of tension, however, remains one of the most famous slogans of social science studies, that is, constructivism, which often takes the form of an extreme reductionism, incompatible with other presuppositions of that approach, as Fine (1996) has shown. For the majority of the specialists of this field, scientific concepts are entirely reducible to the social interactions within which they are put into practice. Thus, using Wolgar's radical wording, "The argument is not just that social networks mediate between the object and observational work done by participants. Rather, the social network constitutes the object (or lack of it). [...] There is no object beyond discourse, [...] the organization of discourse is the object. Facts and objects in the world are inescapably textual constructions" (1988, pp. 65, 73). In a more general manner, according to this approach science is totally reducible to sets of social configurations.

It would be difficult to make this radical form of constructivism compatible with the particularism claimed by the majority of the specialists of science studies. On the one hand, they consider that one of the goals of science, as social institution, is to self-perpetuate. This general goal governs the analysis of other more particular themes such as interest, social influences, reward infrastructure, and training protocols, which all form a sort of network. Thus, within the constructivist approach, we have a visible means of evaluating the overall practical rationality of scientific activity, as Fine (1996) highlights. On the other hand, the explanatory presuppositions of that approach are particularist, as we saw earlier. So a conflict exists between the presupposition stating that scientific activity is governed by a *global* practical rationality and the proposition stating that the explanatory elements are irreducibly *particular* and localized.

We see here that science studies does not escape the threat of profound internal conflict. Thus nothing guarantees that choosing this path be a simple solution to the problems that badger philosophy of science.

3. How Can the Intrinsically Collective Nature of Scientific Activity Be Seriously Broached?

Kuhn to start with, and then his historian and sociologist successors, all had their hearts set on radically criticizing a long-established and widespread idealization in philosophy of science, that of the isolated scholar or the learned individual facing the world alone. In an extreme version, the isolated scholar replicates all the experiments and all the reasonings of his contemporaries in order to verify their validity. Even if that idealization makes possible the study of individual faculties of knowing and therefore casts light on certain undoubtedly important conditions of scientific activity, everyone knows that science "is not done" in this way. However, the isolated scholar is often used as a model for developing philosophical positions about confirmation or induction, in the sense that, when studying these questions, one always takes the viewpoint of an individual agent with her inferential capacities (see chap. 2). It is for the sake of convenience that we envision the work of an individual mind rather than seeking to represent the collective work, which we nevertheless know is determinant in modern science.

Science studies specialists reject the individualist model from the outset. Where Kuhn insisted on the importance of scientific *communities*, his successors wished to analyze the mechanisms giving structure to these communities, from a social perspective of course, but also from the perspective of the consequences of these continuous epistemic interactions on the elaboration of scientific results. When the actors in science are envisioned like this, as communities rather than as individual agents, with seems necessary when we study the various domains of the current "*big science*" (particle physics, molecular and genome biology, for example), then new questions come into view which had been largely neglected by the philosophers.

An important aspect of scientific activity, too long disregarded in philosophy of science, is that the vast majority of the knowledge that scientists acquire about their fields comes neither from experiments nor from reasoning that they have conducted themselves, but from the testimony of others, be this teachers or peers. For a long time, the specific epistemological questions that this mode of knowledge acquisition raises were not broached within philosophy of science. The empirical work of certain specialists of science studies may have much to teach us about this, for they show, for example, that learning to attune the confidence we place in others is just as essential as learning to be critical and skeptical, skills which are generally ranked ahead of learning trust management (see Pestre, 2006).

The data gathered by science studies specialists has come at just the right time for a project which may seem obvious but which has nonetheless made little inroads at time of writing: to build a bridge between theory of knowledge (a field in which the epistemology of testimony is well developed) and philosophy of science. Indeed, what is now being developed, within theory of knowledge, is a set (reasonably heterogeneous) of attempts to form a *social epistemology*, that is, a theory whose goal is to go beyond the idealization of the learned individual facing the world alone. As we will see, epistemology of testimony is the primary component of this type of approach.

A particularly ironic aspect of the current situation is the fact that the term "social epistemology" is also claimed by certain specialists of science studies who are trying to develop a purely descriptive approach to knowledge as an intrinsically social phenomenon. There is little contact between these two sides of social epistemology, apart from a few articles in the collection directed by A. Bouvier and B. Conein (2007). In this section some of the philosophical aspects of social epistemology will be presented, along with correspondent themes from the science studies, in order to point out that philosophy of science, if it takes due note of the knowledge acquired by social epistemology in the first sense, could be a real driving force in science studies, in the broad sense.

3.1 EPISTEMOLOGY OF TESTIMONY

Science studies specialists often put testimony and the question of its reliability at the center of their analyses. For them, the fact that scientific activity inextricably involves relying on others for the advancement of any knowledge-based enterprise indicates that the epistemic norms the philosophers are after are obsolete. However, certain philosophers of knowledge put their time into explaining how, and to what extent, learning by testimony is, in certain circumstances, just as rational as learning by the intermediary of perception or reasoning.

Among them, John Hardwig, in two articles where he reveals himself to be mindful of applications to philosophy of science (1985 and 1991), has analyzed the relationships between trust, necessary for all learning by testimony, and rationality. Hardwig firstly recalls that in a classical view of individual knowledge, the two sources authorized to justify belief, and thus capable of turning it into knowledge, are perception and reasoning. According to that view, when an individual learns something by testimony, we cannot, *strictly speaking*, say she knows it. Yet, within the scientific domain, as with everyday life, the notion of epistemic authority plays a major role—otherwise we would never learn from reading scientific journals. We constantly invest our trust in purveyors of information, in experts, in other words we confer a certain epistemic authority on them. In what way is this act of deference rational (as we must presume it is, unless we wish to brand the majority of our epistemic life as irrational)?

Hardwig (1985) analyzes the structure of this recourse to epistemic authority and shows that it can rightfully be taken as a source of justification for belief and for knowledge, that is to say, that we cannot just settle for blanket criticism of arguments of authority. To put it another way, when we place our trust in experts, we don't (always) leave with right opinions, but with actual knowledge. According to Hardwig, we have good reason to believe a proposition if we have good reason to believe that others have good reason to believe it. Consequently, rationality sometimes imposes that we *not* think for ourselves, a precept that Hardwig slams as being a romantic and completely unrealistic ideal. One of the possibly counter-intuitive effects of Hardwig's work is that the intellectual autonomy of the individual is undermined—this in turn prompts a reexamination of our concept of rationality, a conclusion which, on the surface, concords with the conclusions of the specialists of science studies, but which, above all, reveals an undoubtedly rich lead for understanding in what way trust, even if it is partially blind, plays such an important role in scientific activity.

3.2 COLLABORATIVE RELATIONSHIPS AND DISTRIBUTED KNOWLEDGE

As Hardwig, not to mention Thagard (1993, 1994, 1997, 2006) as well as Kitcher (1993, ch. 8), point out, acts of epistemic deference are one of the conditions for another huge phenomenon in contemporary scientific activity: collaboration and organization of cognitive work. Thagard (1997) highlights the predominance of collaboration in contemporary science, and lays out different types of collaboration: between employer and employee, teacher and apprentice, and between peers, of two sorts, intra-disciplinary peers or inter-disciplinary peers. He takes up the criteria proposed by Goldman (1992) for the evaluation of epistemic practices and analyzes in which conditions the different types of collaboration are productive for scientific practice.

Goldman's premise (1992; see also 1999, 2000, 2004) is that all research which puts epistemic collaboration to work has truth as its goal; Thagard, for his part, adopts a more neutral premise by supposing that the goal is rather the gaining of results which will be useful for moving forward. The first criteria of evaluation for an epistemic practice (collaborative practice, in this case), is its reliability, that is, the relationship between the number of reliable results and the total number of convictions created by that practice. Thagard shows that within the framework of scientific activity, collaboration is often far more reliable, using these criteria, than strictly individual work. The second criterion concerns the "strength" of an epistemic practice, that is, its capacity to help researchers achieve useful results. Assuming good organization, collaboration is also "stronger," in this sense, than individual research. The third criterion is productivity, or the capacity for the analyzed epistemic practice to lead many researchers to a large number of useful results. And the fourth is speed. It is clear that collaboration is, generally speaking, faster than individual research, and, in this sense, generally more productive too. Finally, the fifth criterion is efficiency: an epistemic practice is more efficient than another if it manages to limit, compared to other practices, the cognitive cost of gaining useful results. Again, collaboration is generally much more efficient in this sense than individual research, even if it is more likely to move the individual researchers towards a certain breaking up of their work. Overall, it is clear that how we define what counts as a useful result will play a determining role here. Indeed, if we have lowish standards, then we risk accepting a higher number of errors; if, on the contrary, our standards are too high, we risk pushing collaboration towards the break-up of cognitive work.

Granting that collaboration is, at least part of the time, necessary and epistemically productive, the question arises how to organize so that it is maximally epistemically productive. This question is addressed by Kitcher (1993, chap. 8). In particular, he attempts to give formal criteria for the conditions which must be brought together in order for a certain epistemic authority to be granted to a peer of some scientific community, thus defining how such an act of granting should be gaged. His analyses allow him to turn his attention towards a question posed several years earlier by Kuhn, the question of the balance between tradition and innovation within a community (Kuhn, 1977). Kitcher's method is of course founded on a certain number of strong idealizations about the cognitive capacities of individual agents; it is thus open to this very kind of criticism by adherents to more descriptive approaches. Nonetheless, it reveals a specifically philosophical way of analyzing a major fact in scientific activity, and thus opens a space . . . of collaboration with other approaches.

The *distributed* nature of current scientific knowledge is also part of the themes traditionally ignored by philosophers of science yet at the heart of science studies. A major characteristic of knowledge creation within the large research teams of big science is that no single person has a global epistemic mastery over the experiments being conducted, as Hardwig (1985), among others, underlines. Each individual has epistemic access to only a very limited part of the experiment and must therefore trust the other members of the team in order to ensure proper coordination of the work as well as the validity of the results obtained. Some have gone as far as identifying an epistemological paradox in this situation, since defining a collective subject of knowledge is no mean feat. However, some, such as Nelson (1993), imagine just such a collective subject, going up against the vast majority of current approaches in epistemology (see also chap. 14), yet offering an area of discussion with other perspectives too.

3.3 SITUATED KNOWLEDGE

Many specialists in science studies insist on the fact that scientific knowledge is always *situated*: situated in a social, historical, and geographical context. Thus, according to Pestre (2006), it is the practical ways of judging things, ways of assessing experimental acts in the heat of the moment, pitches, people, and so forth, that are at the heart of scientific activity, more so than the standards or norms governing these practices. As a result, one important task for the study of science as it is done is to give a detailed description of the cultural integration of the doings and sayings surrounding socially identified apparatus and professional bodies. Because they reject all the traditional philosophy of science presuppositions in one fell swoop, science studies specialists dismiss, as we have seen, the legitimacy of general notions like knowledge and justification whose value is supposed to be eternal and absolute.

Philosophers of science and epistemologists seek rather to define such notions in the most satisfying way possible. This frontal opposition is nevertheless softening with time, since philosophers and specialists of cognitive science are currently developing their own concept of situated cognition. Admittedly, the components of this concept are different to those making up the concept used in science studies. However, in the similarities between the two approaches, we can see the possibility of a fertile coming together of philosophy of science and science studies.

4. Concluding Remarks: The Relationships between Philosophy of Science and Its Neighbors

One of the good things in the confrontation between philosophy of science and science studies is that it allows for questioning about the relationships that philosophy of science and other philosophical disciplines keep. In the same way that the sciences do not advance independently of the economic, social and cultural evolutions which support them, so also it would be a mistake to think that philosophy of science advances more productively within its ivory tower than by benefiting from the progress of other philosophical domains.

Science studies, as we have seen, developed amidst an outright rejection of the philosophical approach to science which dominated history of science up until the 1960s. One of their central motivations was that history of science is a historical discipline among others and that it is not legitimate for it to be so long cut off from general history, and from social, political and cultural history. Some went so far as to say that history of science was a sub-discipline of cultural history.

This positioning was accompanied by a methodological overhaul, the consequence of which was that historians of science actively participated in epistemological debates which have animated general history since the 1980s. They were particularly receptive to the methodology of micro-history (see Revel, 1989). In the same way, sociologists of science were engaged in debates specific to sociology and the human sciences in general, as well as showing a keen interest for ethnomethodology (see Lynch, 1993). The discussions dealt with the delimitation of objects of investigation: science studies specialists were seeking to widen the scope of their case studies by not confining themselves to those areas judged the most interesting by scientists themselves. So they attempted to give science a different image to the one spontaneously adopted by the majority of scientists through granting more place to disciplines outside of "pure science," a category itself defined by scientists.

In the same way, science studies specialists sought to no longer take as writ that the goals of science were those dictated by the scientists and philosophers. Their aim is to show the wide variety of these goals, not by presupposing what they could be, but by "making them emerge" from their descriptions of the interactions between actors.

This method leads science studies specialists to compare their object, that is, science as it is done, with other possible objects of sociological analysis, such as art or other cultural practices. Philosophers of science, generally speaking, are not so keen on such comparisons, and few relationships exist between philosophy of science and philosophy of art or history, or political philosophy. Philosophy of science maintains a rich dialog with metaphysics (see chap. 4) and epistemology, but it has few links with philosophy of history, philosophy of law, or political philosophy.

If science studies are at least partially justified in analyzing scientific activity as being structured first and foremost by social and political relationships (and only secondly by epistemological problems), then putting a dialog in place between philosophy of science and political philosophy, if not moral philosophy, is indeed desirable, as Fuller (1998) and Rouse (1987), for example, point out by analyzing the political dimensions of cognitive authority. However, to resolve this question, what must first be determined is the validity of the results the science studies specialists claim to have obtained regarding the interpenetration of social structures and epistemological questions within scientific activity. Those philosophers of science who have delved into this question remain generally skeptical towards this aspect of science studies, by reason of the problems evoked in the first section of this chapter.

References

Atten, M., and Pestre, D. (2002) *Heinrich Hertz, L'administration de la preuve*, Paris: PUF, Collection Philosophies.

Barnes, B. (1977) *Interests and the Growth of Knowledge*, London: Routledge & Kegan Paul.

Barnes, B., and Bloor, D. (1982) "Relativism, Rationalism, and the Sociology of Knowledge," in Hollis, M. and Lukes, S. (eds.) *Rationality and Relativism*, Oxford: Blackwell, pp. 21–47.

Biagoli, M. (1999) *Science Studies Reader*, London: Routledge.

Bloor, D. (1976) *Knowledge and Social Imagery*, London: Routledge.

Bloor, D. (1981) "The Strengths of the Strong Programme," *Philosophy of the Social Sciences*, 11, pp. 199–213.

Boghossian, P. (2006) *Fear of Knowledge. Against Relativism and Constructivism*, Oxford: Oxford University Press.

Bourdieu, P. (2001) *Science de la science et réflexivité*, Paris: Éditions Raison d'Agir; English translation: *Science of Science and Reflexivity*, Chicago: University of Chicago Press, 2004.

Bouvier, A., and Conein, B. (eds.) (2007) *L'épistémologie sociale. Une théorie sociale de la connaissance*, Paris: Éditions de l'Ecole des Hautes Études en Sciences Sociales.

Coady, C.A.J. (1992) *Testimony: A Philosophical Study*, Oxford: Clarendon Press.

Collins, H. (1981) "Stages in the Empirical Programme of Relativism," *Social Studies of Science*, 11(1), pp. 3–10.

Collins, H. (1985) *Changing Order. Replication and Induction in Scientific Practice*, London: Sage.

Collins, H. (2004) *Gravity's Shadow: The Search for Gravitational Waves*, Chicago: University of Chicago Press.

Daston, L., and Galison, P. (2007) *Objectivity*, New York: Zone Books.

Feyerabend, P. (1975) *Against Method*, London: Verso.

Fine, A. (1996) "Science Made-Up: Constructivist Sociology of Scientific Knowledge," in Galison, P., and Stump, D. (eds.) *The Disunity of Science. Boundaries, Contexts, and Power*, Stanford: Stanford University Press, pp. 231–254.

Forman, P. (1971) "Weimar Culture, Causality and Quantum Theory, 1918–1927: Adaptation by German Physicists and Mathematicians to a Hostile Environment," *Historical Studies in the Physical Sciences*, 3, pp. 1–115.

Fuller, S. (1988/2002) *Social Epistemology*, Bloomington: Indiana University Press.

Galison, P. (1987) *How Experiments End*, Chicago: University of Chicago Press.

Geertz, C. (1973) "Thick Description: Toward an Interpretive Theory of Culture," in *The Interpretation of Cultures*, New York: Basic Books, pp. 3–30.

Goldman, A. (1992) *Liaisons: Philosophy Meets the Cognitive and Social Sciences*. Cambridge, MA: MIT Press.

Goldman, A. (1999) *Knowledge in a Social World*, Oxford: Oxford University Press.

Goldman, A. (2000) "Social Epistemology," in *Stanford Encyclopedia of Philosophy*, http://plato.stanford.edu/entries/epistemology-social/.

Goldman, A. (2004) *Pathways to Knowledge: Private and Public*, Oxford: Oxford University Press.

Hacking, I. (1999) *The Social Construction of What?*, Harvard: Harvard University Press.

Hanson, N. R. (1958) *Patterns of Discovery: An Inquiry into the Conceptual Foundations of Science*, Cambridge: Cambridge University Press.

Hardin, R. (2002) *Trust and Trustworthiness*, London: Russell Sage Foundation.

Hardwig, J. (1985) "Epistemic Dependence," *Journal of Philosophy*, 82(7), pp. 335–349.

Hardwig, J. (1991) "The Role of Trust in Knowledge," *Journal of Philosophy*, 87(12), pp. 693–708.

Jurdant, B. and Savary, N. (eds.) (1998) *Impostures scientifiques, les malentendus de l'affaire Sokal*, Paris: La Découverte.

Keucheyan, R. (2008) "L'imagination constructiviste. Une enquête au Centre de Sociologie de l'Innovation," *L'année sociologique*, 58(2), pp. 409–434.

Kitcher, P. (1993) *The Advancement of Science. Science without Legend, Objectivity without Illusion*, Oxford: Oxford University Press

Knorr-Cetina, K. (1981) *The Manufacture of Knowledge: An Essay on the Constructivist and Contextual Nature of Science*, Oxford: Pergamon Press

Kuhn, T. S. (1962) *The structure of scientific revolution*, Chicago: University of Chicago Press (2nd ed. 1970, with a postscript)

Kuhn, T. S. (1977) *The Essential Tension. Selected Studies in Scientific Tradition and Change*, Chicago: University of Chicago Press

Lakatos, I. (1976) *Proofs and Refutation. The Logic of Mathematical Discovery*, Cambridge: Cambridge University Press.

Lakatos, I. (1978) *The Methodology of Scientific Research Programmes: Philosophical Papers Volume 1*, Cambridge: Cambridge University Press.

Latour, B. (1984) *Microbes: guerre and paix*, Paris: La Découverte.

Latour, B. (1987) *Science in Action*, Cambridge: Harvard University Press.

Latour, B., and Woolgar, S. (1979) *Laboratory Life: The Social Construction of a Scientific Fact*, 1st ed. London: Sage.

Laudan, L. (1977) *Progress and Its Problems. Towards a Theory of Scientific Growth*, Berkeley: University of California Press.

Laudan, L. (1981) "The Pseudo-Science of Science," *Philosophy of the Social Sciences*, 11, pp. 173–198.

Licoppe, C. (1996) *La formation de la pratique scientifique*, Paris: La Découverte.

Lynch, M. (1993) *Scientific Practice and Ordinary Action: Ethnomethodology and Social Studies of Science*, New York: Cambridge University Press.

Merton, R. K. (1973) *The Sociology of Science: Theoretical and Empirical Investigations*, Chicago: University of Chicago Press.

Moulines, C. U. (2006) *La philosophie des sciences. L'invention d'une discipline*, Paris: Éditions Rue d'Ulm.

Nelson, L. H. (1993) "Epistemological Communities," in Alcoff, L., and Potter, E. (eds.) *Feminist Epistemologies*, New York: Routledge, pp. 121–159.

Nickles, T. (1995) "Philosophy of Science and History of Science," *Osiris* 2nd series, 10, pp. 138–163.

Pestre, D. (2006) *Introduction aux Sciences Studies*, Paris: La Découverte.

Pestre, D., and Cohen, Y. (eds.) (1998) "Histoire des techniques," *Annales. Histoire, Sciences sociales*, 4–5.

Pickering, A. (1984) *Constructing Quarks: A Sociological History of Particle Physics*, Chicago: University of Chicago Press.

Pickering, A. (1995) *The Mangle of Practice: Time, Agency, and Science*, Chicago: University of Chicago Press.

Popper, K. (1934) *Logik der Forschung*, Berlin: Springer, English translation *The Logic of Scientific Discovery*, London: Routlege, 1959.

Ramsey, F. P. (1926) "Truth and Probability," in Ramsey, F. P. (1931) *The Foundations of Mathematics and Other Logical Essays*, R. B. Braithwaite (ed), London: Paul Kegan, Ch. VII, pp. 156–198.

Revel, J. (1989) "L'histoire au raz du sol," in Levi, G. (ed.), *Le pouvoir au village*, 2nd ed., Paris: Gallimard.

Revel, J. and Passeron, J.-C. (eds.) (2005) *Penser par cas*, Paris: Editions de l'Ehess.

Richardson, A. and Uebel, T. (eds.) (2007) *The Cambridge Companion to Logical Empiricism*, Cambridge: Cambridge University Press.

Rouse, J. (1987) *Knowledge and Power: Toward a Political Philosophy of Science*, Ithaca, NY: Cornell University Press.

Rouse, J. (1993) "What Are Cultural Studies of Scientific Knowledge?" *Configurations*, 1.1, pp. 57–94.

Rouse, J. (2002) *How Scientific Practices Matter*, Chicago: University of Chicago Press.

Rudwick, M. (1985) *The Great Devonian Controversy*, Chicago: University of Chicago Press.

Sargent, R.-M. (1988) "Explaining the Success of Science," in Fine, A. and Leplin, J. (eds.) *PSA 1988, Volume 1*, E. Lansing, Michigan: Philosophy of Science Association, pp. 55–63.

Schmitt, F. F. (ed.) (1994) *Socializing Epistemology. The Social Dimensions of Knowledge*, Lanham: Rowman & Littlefield.

Shapin, S. (1982) "The Sociology of Science," *History of Science*, 20, pp. 157–211

Shapin, S. (1984) "Pump and Circumstance: Robert Boyle's Literary Technology," *Science Studies*, 14(4), pp. 481–520.

Shapin, S. (1994) *A Social History of Truth: Civility and Science in Seventeenth-Century England*, Chicago: University of Chicago Press.

Shapin, S., and Schaffer, S. (1985) *Leviathan and the Air Pump*, Princeton, NJ: Princeton University Press.

Sokal, A. (1996) "Transgressing the Boundaries: Towards a Transformative Hermeneutics of Gravity," *Social Text*, 46/47, pp. 217–252.

Sokal, A., and Bricmont, J. (1997) *Impostures intellectuelles*, Paris: Odile Jacob.

Thagard, P. (1993) "Societies of Minds: Science as Distributed Computing," Studies in History and Philosophy of Science, 24, pp. 49–67.

Thagard, P. (1994) "Mind, Society, and the Growth of Knowledge," *Philosophy of Science*, 61, pp. 629–645.

Thagard, P. (1997) "Collaborative Knowledge," *Noûs*, 31, pp. 242–261.

Thagard, P. (2006) "How to Collaborate: Procedural Knowledge in the Cooperative Development of Science," *Southern Journal of Philosophy*, 44, pp. 177–196

Toulmin, S. (1961) *Foresight and Understanding: An Enquiry into the Aims of Science*, Bloomington: Indiana University Press

Veyne, P. (2006) *Le quotidien and l'intéressant*, entretiens avec Catherine Darbo-Peschanski, Paris: Hachette Littératures

Warwick, A. (1992) "Cambridge Mathematics and Cavendish Physics, Cunningham, Campbell and Einstein's Relativity, 1905–1911, Part 1: The Uses of Theory," *Studies in the History and Philosophy of Science*, 23(4), pp. 625–656

Warwick, A. (1993) "Cambridge Mathematics and Cavendish Physics, Cunningham, Campbell and Einstein's Relativity, 1905–1911, Part 2: Comparing Traditions in Cambridge Physics," *Studies in the History and Philosophy of Science*, 24(1), pp. 1–25

Woolgar, S. (1988) *Science, the Very Idea*, London: Tavistock Publications

Zammito, J. H. (2003) *A Nice Derangement of Epistemes: Post-positivism in the Study of Science from Quine to Latour*, Chicago: The University of Chicago Press

8

REDUCTION AND EMERGENCE

Pascal Ludwig (Sorbonne Université, Sciences Normes Décision)

1. Introduction

The aim of scientific disciplines and theories is to explain phenomena which may at first glance seem quite disparate. Neuroscience studies chemical and electrical phenomena at the scale of neuronal connections and the networks these form within the brain. Psychology, on the other hand, tries to explain human behavior as a consequence of contentful mental causes: desires, intentions, beliefs, wishes, sensations, emotions, and so on. Today there is almost full consensus to support the existence of genuinely psychological explanations. But does it then follow that psychological phenomena possess a nature of their own, irreducible, distinct from the nature of the chemical and electrical phenomena studied by neuroscience? To believe so means taking on a form of ontological pluralism, of which the multifarious variants of mental dualism are the most striking illustrations. Otherwise, one may wish to maintain that the set of all scientific theories, psychology included, presents a unified image of the world. Such a debate concerns the relationships between science and ontology, and is closely tied to the question of physicalism. By "physicalism," we refer to the idea that all existent entities in the world are of a physical nature, and that all the properties these entities have are in turn either physical properties or properties which can be related, one way or another (to be clarified presently), to physical properties. Initial impressions will be of a certain arbitrariness in the characterization of physicalism. An entity or property "of a physical nature" is just an entity or property described by physical theories. Yet the borders of physics themselves are hazy. Shouldn't "physics" refer only to fundamental physics? Or maybe to all scientific domains studied in physics departments? We will come back to these questions. For the moment, let us make do with pointing out the difficulty in correctly defining the physicalist position.

The debate around physicalism and reduction is closely tied to a central question in philosophy, that of the unity of science. Scientific practice is organized into multiple disciplines: physics, biology, anthropology, economics, and so on. But does this disciplinary multitude match an actual ontological heterogeneity in the underlying phenomena, or is it merely the provisional effect of our limited human perspective of the world? Is it possible, in theory at least, to trace back all scientific disciplines and see them only as specialized, applied branches of theoretical physics? The unity of science position has a preponderant standing in the history of 20th-century philosophy of science, particularly within the history of logical positivism.[1] It can, however, be interpreted in two different ways, namely with a weak and a strong interpretation.[2] According to the weak interpretation, the unity of science results from the unity of its empirical basis. For the logical positivists, who endorsed a verificationist conception of meaning, observation was the only source of justification for meaningful statements that could communicate information about the world. In this chapter though, it is to a stronger interpretation that we give our attention: namely, the reductionist interpretation of the unity argument. For a reductionist philosopher, a logical relationship exists between the diverse scientific theories, a relationship which must allow, at least in theory, for them to all be traced back down to fundamental physics. From a metaphysical point of view, the reductionist considers the special sciences—by which, in keeping with (Fodor, 1974), we mean all disciplines which cannot be traced back to fundamental physics in any obvious way—to be nothing other than round-about ways of talking about physical phenomena. On the contrary, pluralist philosophers consider that genuinely autonomous levels of phenomena do exist, in parallel to the level of physical phenomena. According to them, the laws of the special sciences cannot be derived from those of fundamental physics.

I shall begin this chapter by showing that, if the principle of causal closure of the physical world is accepted, then ontological pluralism comes up against a decisive difficulty: causal overdetermination. Given that, according to the principle of causal closure, physical effects all possess a physical cause, non-physical causes lose all explanatory power and become epiphenomenal. If ontological pluralism is given up, and if the theories of the special sciences, such as psychology, are not in fact without value, then a reductionist explanation must be provided as to why these theories possess explanatory power. I will present the different reductionist strategies that seem conceivable today.

---

[1] Cf. Carnap (1966).

[2] Cf. Kistler (2007).

2. Emergentism, Ontological Pluralism, and Causal Overdetermination

The diversity of natural phenomena is not chaotic but in fact well-ordered: all phenomena that the special sciences aim to explain seem correlated to physico-chemical phenomena. Let us consider a macroscopic phenomenon, such as boiling a certain quantity of water. The terms "water" and "boiling" do not, of course, belong to the vocabulary of physics or of chemistry. Nonetheless, there is a definite correlation between the presence of water and the presence of molecules of H_2O, a correlation between the increase in heat of the water and the increase of the mean kinetic energy of these molecules, and, finally, a correlation between the water boiling and some activity in the H_2O molecules. To take another example, there is a correlation between instances of pain in a person's mind and the activity of certain fibers of the nervous system. How can these correlations be accounted for?

An initial suggestion is based on the concept of emergence. Emergence is conceived of by scientists and philosophers as a relationship between complex phenomena based on simpler phenomena where the complex phenomena ontologically depend on the simple phenomena but can nevertheless not be reduced to them. It is to Georges Henri Lewes (1875) that we owe the term "emergent," and his characterization of emergent phenomena is still relevant today:

> Thus, although each effect is the resultant of its components, the products of its factors, we cannot always trace the steps of the process, so as to see in the product the mode of operation of each factor. In this latter case, I propose to call the effect an emergent. (. . .) The emergent is unlike its components in so far as these are incommensurable, and it cannot be reduced either to their sum or their difference. But on the other hand, it is like its components, or, more strictly, it is these: nothing can be more like the coalescence of the components than the emergent which is their coalescence.[3]

Lewes, as well as Alexander, Morgan, and Broad, the three great names of emergentism in Great Britain in the 1920s, tried to find a middle route between dualism and reductionism.[4] It would be judicious, before considering abstract definitions, to begin with a few examples.

It is sometimes said that the liquidity and transparency of water are emergent on the molecules of oxygen and hydrogen found in structured collections of water molecules. Two things are meant by this. First off, that there is an ontological dependence between the macroscopic properties of liquidity and transparency and the properties of water molecules: the former simply could not exist without the latter, and for the former to have any occurrences, the same must also be the case for the latter. Still,

[3] Lewes (1875, pp. 368–369).

[4] Cf. Alexander (1927), (Morgan (1923), and Broad (1925). On the British emergentism movement and its fate, see also Andler, Fagot-Largeaut, and Saint-Sernin (2002, pp. 439–1048) and McLaughlin (1992).

liquidity and transparency cannot be thought of as properties of molecules, and reducing them to being properties of aggregates seems to be quite difficult.

Life is a second very important example often evoked in support of the idea of emergence (Bedeau and Humphreys, 2008, p. 2; Malaterre, 2008). Just consider the relationship between an organism and the collection of cells that make it up. In one sense, the cells ontologically make up the organism. Nevertheless, the characteristic properties of living creatures can be said to emerge from all their cells taken together, since there is no easy way to define them in exclusively cellular terms. Since emergentism is characterized in terms of its contrast to reductionism on the one side and dualism on the other, it is important to begin by clarifying these positions.

2.1 CLASSICAL REDUCTIONISM, DUALISM, AND EMERGENTISM

While the notion of emergence is certainly not defined clearly by the British emergentists, they do insist on the following aspect: there can be said to be emergence from one phenomenal level with respect to another when there is systematic dependence without reduction from one level to another. Thus, Alexander writes: "The higher quality emerges from the lower level of existence and has its roots therein, but it emerges therefrom, and it does not belong to that lower level, but constitutes its possessor a new order of existent with its special laws of behaviour."[5] But what is to be understood by "reduction"? Though it may be an obvious anachronism in discussing the British emergentists, we will nonetheless begin with a linguistic analysis of this concept, due to Ernest Nagel (Nagel, 1961). According to Nagel, for there to be reduction a certain logical relationship needs to exist between two theories, the reducing theory T_1 and the reduced theory T_2 (see chapter 3). Stating the aim of a scientific theory as providing explanations for a set of phenomena, a necessary condition for all inter-theoretical reductions is easy to formulate: all phenomena the theory to be reduced can explain must be equally explainable by the reducing theory. However, a phenomenon's explanation by a theory, according to the nomologico-deductive conception of explanation, takes the form of a deduction of the proposition that describes the phenomenon's occurrence on the basis of the theory's laws and a description of the initial conditions (see chapter 1). If all the laws of the reduced theory can be logically derived from the laws of the reducing theory, then the first theory is seen to be a particular case of the second, and it is thus clear that all phenomena explainable in terms of the first theory will be in terms of the second as well. So, for instance, the Galilean law of free falling bodies can be deduced from the Newtonian theory of gravitation and, for that reason, we can consider the former to have been reduced to the latter. We see exactly the same result with Kepler's theory of planetary motion. For example, that the motion of a planet around the sun, caused by the latter's force of attraction, will have the form

[5] Alexander (1920; 1927, pp. 46–47).

of an ellipse can be deduced from Newton's principles, in direct accordance with Kepler's theory.

Here we have emphasized the importance of deducing T_2 from T_1 when it comes to reducing the first theory on the basis of the second. But in order to speak of deducing one set of propositions from another set, it must first be confirmed that these propositions do in fact refer to the same entities, if not in an obvious manner then at least after analysis and some definition work. Thus, conceptual links must be established between the vocabularies of the theory to be reduced and the reducing theory. In certain cases, the establishment of such links poses no real problem. For instance, the planets spoken of in Kepler's theory can be easily described as bodies in motion on which certain forces act, such terms enabling us to apply the laws of Newtonian theory to them.[6]

In other cases, however, it is not self-evident that such inter-theory conceptual links can be found. In such cases, we must speak of heterogeneous reduction, because the vocabulary of the theory to be reduced is not present in that of the reducing theory. Consider the classic example of the relationships between thermodynamics—a theory to be reduced, whose aim is to explain certain macroscopic phenomena—and statistical mechanics—a reducing theory. There are certain concepts used in the theory to be reduced not used in the reducing theory. So the macroscopic concept of temperature features in the formulation of the laws of thermodynamics, in the Boyle-Mariotte law for example, but for the precise reason that it refers to a macroscopic property it is never openly encountered in statistical mechanics. But one only needs to study the treatises of statistical mechanics to see that a correspondence can be achieved: the temperature of a gas can be identified with the mean kinetic energy of the molecules that make it up. Each instance where the concept "temperature" is applied to a phenomenon it must also be possible to apply the concept "mean kinetic energy" to it. Such a systematic correspondence between two predicates is what Nagel calls a "bridge principle," a proposition with the following logical form:

(1) $\forall x(Px \leftrightarrow Qx)$

(2) For every set of molecules *x*, the temperature of *x* is *P* if and only if the mean kinetic energy of *x* is *Q*.

The Nagelian "bridge principle" notion is both ambiguous and problematic. Ambiguous because the exact nature of the connection that bridge principles are supposed to establish is not entirely clear. Is it a purely conceptual connection? Or simply a nomological connection? We will see a little further on that the interpretation of reductionism closely depends on how this question is answered. For the moment, let us just point out that it is precisely the existence of such "bridges," be they conceptual or nomological, that anti-reductionists deny.

---

[6] For an in-depth analysis of this example, see Kistler (2007).

Refusing to reduce theory T_2 to theory T_1 just is to maintain that there exists what philosophers of mind (since the work of Joseph Levine) call an "explanatory gap" between the two theories (Levine, 1983, 1993; Chalmers, 1996). In other words, it is to affirm that there are certain phenomena that theory T_2 can explain but which escape the explanatory power of theory T_1. In philosophy of mind, the most discussed example of an explanatory gap involves conscious experience. Frank Jackson's thought experiment is well known:

> Mary is a brilliant scientist who is, for whatever reason, forced to investigate the world from a black and white room via a black and white television monitor. She specialises in the neurophysiology of vision and acquires, let us suppose, all the physical information there is to obtain about what goes on when we see ripe tomatoes, or the sky, and use terms like "red," "blue," and so on. She discovers, for example, just which wave-length combinations from the sky stimulate the retina, and exactly how this produces via the central nervous system the contraction of the vocal chords and expulsion of air from the lungs that results in the uttering of the sentence "The sky is blue." (It can hardly be denied that it is in principle possible to obtain all this physical information from black and white television, otherwise the Open University would of necessity need to use colour television.)

What will happen when Mary is released from her black and white room or is given a color television monitor? Will she learn anything or not? It seems just obvious that she will learn something about the world and our visual experience of it. But then it is inescapable that her previous knowledge was incomplete. But she had all the physical information. Ergo there is more to have than that, and Physicalism is false (Jackson, 1982).[7]

Possessing all imaginable physical information about the sight of colors, according to Jackson, provides no understanding or explanation of what the conscious experience of them is like. If we consider all statements (potentially formulated in the first person) concerning the experience of colors as a theory (in a slightly broadened sense), we can say that there is an explanatory gap between the phenomena described by this theory and physics.

Let it be well understood that the existence of an explanatory gap between two sets of phenomena is a question of knowledge, thus an epistemological and not a metaphysical question. As Joseph Levine points out with respect to Jackson's thought experiment: "what is at issue is the ability to explain qualitative character itself; why it is like what it is like to see red or feel pain."[8] In this way, in the case of life, it was commonly said before the mid twentieth century that it was not possible to understand, explain or predict properties of living organisms on the basis of properties of physico-chemical entities.

---

[7] See also Ludlow, Nagasawa, and Stoljar (2004).

[8] Levine (1993, p. 128). See also Nagel (1974) for an alternative expression of the idea of an explanatory gap between subjective experience and the physical domain.

The presence of an explanatory gap being a phenomenon of epistemological nature does not prevent our wishing to account for it through metaphysical considerations. This being the case in the various forms dualism can have.[9] Consider the example of conscious mental states. For a dualist, it is because these states are of a non-physical nature that their existence can be neither predicted nor explained by physical theories. Dualists are thus ontological pluralists: they consider that there is not just one class of things in nature, but rather various classes of things. So Cartesian dualism considers that there are both things with extension—material bodies—and also essentially thinking entities—minds. A vitalist, on the other hand, thinks that living creatures have a non-physical nature and that their behavior can therefore not be predicted nor their properties explained by the laws of physics.

2.2 SUPERVENIENCE AND MINIMAL FORMS OF PHYSICALISM

Emergentism, in its most attractive forms, does not however go hand in hand with substance dualism. Indeed, the British emergentists, as we have seen, do not consider that living creatures, for example, belong to a different domain of reality than that of physical beings. Unlike the vitalists, they maintain that living creatures are material substances, that is, beings that can be decomposed into material parts. So we must try to understand how adhesion to a minimal form of physicalism, excluding substance dualism, and rejection of reductionism can co-exist.

We will speak of non-reductionist physicalism with regard to the following double metaphysical thesis. First, there is no substance in the world which cannot be decomposed into physical parts: there is nothing other than the entities described by fundamental physics and the aggregates formed out of these entities. However, the aggregation of these fundamental entities leads, once a certain level of complexity has been reached, to the emergence of totalities governed by laws of a different level to those of physics and which are impossible to deduce from the laws of physics. Minimal physicalism, unlike Cartesian dualism, seems fundamentally monist, since it recognizes only one kind of fundamental substances: physical substances. Nevertheless, from these base entities emerge levels of reality autonomous to the physical level, with each level possessing its own laws and thus its own principles of explanation.[10]

So the position of non-reductionist physicalism is subtle, because within it are affirmed both a systematic dependence of one set of properties—the emergent properties—with respect to another, as well as the irreducibility of the former to the latter. In order to precisely express the monist idea of a systematic dependence of emergent properties with respect to physical properties, it will be useful to call on the concept of supervenience. It will be said that a set of properties X (for example,

---

[9] Cf. Chalmers (1996, 2002) on the connection between the explanatory gap and contemporary versions of dualism.

[10] This idea has its origin in Putnam (1975a), but it is Fodor (1974) who gave it its most influential expression. See also Lycan (1987), Dupré (1993), Horgan, (1993), as well as Kim (1989) for a critical evaluation.

psychological or biological properties) supervene on a set of properties Y (for example, physical properties) when the following conditions are met:[11]

- First of all, two entities (or two states, or two events) cannot differ with respect to the properties belonging to X without differing with respect to the properties belonging to Y. This amounts to saying that for a non-reductionist physicalist it is not possible for two organisms to differ with respect to their biological properties (for example) without also differing with respect to their physical properties.
- Moreover, it is impossible for an entity to possess a property M belonging to X if it does not also possess a property P belonging to Y, which is called its "realization" or its "realizing property."
- Finally, the occurrence of a realizing physical property is necessarily a sufficient condition of the property it realizes. In other words, it is necessary that when an entity (or event, or state) possesses P it also possess M. Nevertheless, it must be pointed out that possession of the realizing property P is only a sufficient condition for possession of M, and not a necessary condition. So the property M can have an occurrence without the property P having one.

This definition is not content to just account for the idea of a systematic dependence between sets of properties. Indeed, a dualist could accept the idea that the occurrence of a mental, or biological property is nomologically tied to the occurrence of a physical property. Its aim is also to capture the idea of an existential dependence between the two sets of properties. If, for example, we accept the thesis of biological properties supervening on physical properties, it is not possible for an organism to possess a biological property without at the same time possessing a physical property that realizes it. Thus, some type of material structure must exist that can be characterized in the vocabulary of physics and that realizes the biological property of being a heart.

Note that the constraint of realizability of emergent properties does not get in the way of their autonomy, nor of the existence of laws at the emergent level that are irreducible to laws of physics. Starting in the 1970s, such a stratified conception of the world began to enjoy renewed interest, due to the attention given to disciplines other than physics, notably the special sciences (Lycan, 1987; Dupré, 1993; Horgan, 1993). The conditions to satisfy if one hopes to reduce a given special science, say, political economics, to physics are extremely sharp: one would have to manage to derive every one of the special science's laws from the laws of physics, but above all one would have to manage to establish connections, by means of bridge laws, between the vocabulary of economics and that of the material sciences. Fodor (1974) maintains that such a condition could never be satisfied. His argument leans entirely on his analysis of one

[11] The interpretation of non-reductionist, minimal physicalism in terms of supervenience is due to Jaegwon Kim. See the articles collected in Kim (1993, 1998).

example; Gresham's law, according to which, "in a monetary system with two types of money, the bad money drives out the good." A reductionist will have to attempt to find physical mechanisms likely to realize this law. For this, it is obviously necessary to be able to describe monetary exchanges using only the vocabulary of physics. Such a task is not insurmountable: after all, there must exist physical devices which realize monetary exchanges, and so it must be possible to describe these devices in the vocabulary of the material sciences. The problem, according to Fodor, lies elsewhere, in the infinite diversity of forms these realizations could have: "a physical description which covers all such events must be wildly disjunctive. Some monetary exchanges involve strings of wampum. Some involve dollar bills. And some involve signing one's name to a check." (Fodor, 1974, p. 103) Supposing that laws allowing physics to be linked to political economics do exist, these laws would possess the form (4) and not the form (3):

(3) $\forall x(Px \leftrightarrow Qx)$

(4) $\forall x\left(Px \leftrightarrow Q_1\, or \ldots or \ldots Q_n x\right)$

But as Fodor points out, bridge laws cannot tie predicates of the special science to indeterminate disjunctions of physical predicates. Indeed, an indeterminate disjunction of predicates can certainly not describe a natural property: in general, it is not possible to create a natural property simply by creating a new predicate by disjunction with the help of two old predicates denoting heterogeneous properties. For example, one cannot speak of the natural property of being a heart or of weighing 22 kg. Furthermore, Fodor rightly insists on the indeterminate character of the disjunction described by (4), and it is certainly not clear how a physical property could correspond with such an indeterminate disjunction.

The consequences Fodor draws from his argument are important: since no physical property can be designated by a "wildly disjunctive" predicate, so no physical property is identical to the properties that the vocabulary of the special science—the theory of monetary exchanges in this case—describes. Here anti-reductionism meets functionalism, introduced into philosophy of mind by Hilary Putnam in the 1960s (Putnam, 1967).

A predicate like "exchange of money" does seem to characterize a causal role: being an exchange of money, for a physical device or sequence of events, amounts to fulfilling a certain causal function which can be identified by its typical effects. The same analysis can be applied to numerous predicates of the special sciences. Thus, "being a heart" is being a physical structure capable of pumping blood. It is with respect to its causal role, or its function, that the heart is identified. But this role underdetermines the type of structures liable to realize it. Since what matters are the effects of the structure, it seems that a machine—an artificial heart—can be considered to possess the property of being a heart, presuming that it correctly fulfills its function.

So, from a metaphysical point of view, anti-reductionism places functional properties in the foreground, characterizing them as causal roles of physical properties. In order to rigorously define what a functional property is, let us take Jaegwon Kim's lead and begin with a set of basic structural properties.[12] In general, this involves physical properties, though this is not necessary: more generally, it involves the set with respect to which the causal roles are defined.

> P is a functional property with respect to the basic set B if and only if having the property P amounts to having a certain property Q of B, to be called the "realizing property" of P, satisfying a causal condition C with respect to the other properties of B.

In this definition, the condition C corresponds to the causal role with whose help the functional property is defined. It results that one and the same functional property can be realized by various material structures. Thus, a biological function, defined by its typical effects for the organism, and thus by a condition C, can be realized by diverse structural properties, depending on the species. Think, as an example, of the diverse ways of producing a wing bestowing flight on an organism that we find in birds and mammals. Besides which, it appears clearly that functional properties characterized in this way are second-order properties, since their nature is expressed by a causal condition imposed on other properties. It can also be observed that the structural properties with respect to which the second-order functional properties are defined are not necessarily physical properties. From this point of view, the supervenience thesis adds a physicalist component to functionalism, since it specifies that the supervenient properties must be realized by physical properties.

To conclude the discussion, we can propose a definition of an emergent property that makes use of the concept of supervenience[13]:

> Let P be a property of the entity E. P is emergent if and only if P supervenes on the properties of the parts of E.

According to this definition, a biological property of an organism, such as "having a heart," emerges because it supervenes on the physical properties of the parts of the organism. This definition of emergence has the merit of capturing the double intuition of emergentists: on the one hand, emergent properties depend ontologically on material properties since they are realized by properties of the parts of the entity possessing them; but on the other hand, they do not reduce to these material properties. We can therefore sum up the position of anti-reductionist physicalism with the following two theses (Kim, 2005a, pp. 33–35):

> T1: the properties of the special sciences supervene on physical properties
> T2: the properties of the special sciences are irreducible to physical properties

The image of the relationships between the special sciences and physics resulting from the conjunction of these two theses is appealing. It reflects both the intuition

[12] Kim (1998, p. 20).

[13] Cf. Van Cleeve (1990, p. 222).

that physics possesses a generality which grants it a special metaphysical standing (see chapter 3), as well as the emergentist conviction that the special sciences describe levels of reality possessing autonomous laws. We will see, however, that a very important argument (due to Jaegwon Kim) can be opposed to non-reductionist physicalism.

2.3 SUPERVENIENCE AND CAUSAL EXCLUSION

From a metaphysical point of view, it seems essential for the properties described by the predicates of the special sciences to possess genuine causal efficiency. The anti-reductionists' motivation comes from their conviction that these properties "genuinely matter" in the constitution of the world, and that they are thus destined to play an ineliminable role in our explanations, including when these explanations claim to describe causal relations. Consider the example of psychological properties and states, specifically conscious states. When an anti-reductionist affirms that pain, as a conscious experience, emerges from the base of all cerebral properties, she typically considers this state to possess causal powers, thus indicating the possibility of including it in causal explanations of behavior. For example, the anti-reductionist would maintain that an occurrence of intense pain causally explains the subject's withdrawing her hand from the burning surface. Therefore, we must consider thesis T3 to be just as important for anti-reductionists as T1 and T2:

> T3: the properties described by the special sciences possess causal efficiency.

Yet Jaegwon Kim maintains that T1, T2 and T3 are incompatible, in that they cannot all be true. This thesis is called *Kim's trilemma* (Kim, 2005, pp. 30–35).

The core of the problem is found in what can be called the causal overdetermination of events consisting in the instantiation of an emergent property. The following principle results:

> Principle of causal closure of the physical world: Every event consisting in the instantiation of one or more physical properties has a physical cause which is sufficient for causally explaining its occurrence.

The principle of causal closure is not one that can be justified a priori. Conceiving of possible worlds in which, for example, miracles occur implies no logical contradiction. In such worlds, there are physical effects that have no physical causes. However, the principle of causal closure does seem to be very well confirmed. In terms of the mind, for instance, positing the existence of cerebral effects with no cerebral cause seems barely plausible. No observation whatsoever could justify such a hypothesis, which would, moreover, be incompatible with the principle of conservation of energy.[14]

---

[14] The principle of physical causal closure is incompatible with Descartes's substance dualism, which supposes that there are physical events, in the pineal gland, which have strictly mental causes but no physical ones. Nonetheless, the idea is not in contradiction with the principle of conservation of motion

Taking the principle of closure as a given, let us suppose that supervenient causality is possible, that a given event E possessing an emergent property M, as a result of its possessing this property M, causes an event E′ giving rise to an emergent property M′. According to the supervenience thesis, there must be a physical property P′ which realizes M′.[15] In other words, it is in virtue of possessing the realizing property P′ that E, possesses the emergent property M′. The question that now arises is; why does the event E′ possess the property M′? The solution seems to impose itself naturally: because it possesses the physical property P′ which is the realizing property of M′. But the result is that the causality relationship between the instantiations of emergent properties becomes indirect: it is because its occurrence causes the occurrence of P′ that the occurrence of M causes that of M′. Already, this first result is significant: supervenience, that is to say, dependence between different levels of reality, precludes causal autonomy of levels. Thus, if a mental event—for example, a pain—is the cause of another mental event—the desire for the pain to cease—and if we accept the supervenience thesis, it follows that the pain event is the cause of its mental effect only in so far as it is also the cause of a physical effect. But the supervenience thesis carries a second implication: the occurrence of the emergent property M must itself depend on the occurrence of its physical realizing property. Moreover, thesis T2, the irreducibility of emergent properties to physical properties, implies that P is not identical to M. So we find ourselves in the following situation:

(i) It is in virtue of the occurrence of its realizing property P that M has an occurrence.
(ii) It is in virtue of the occurrence of M that P′ has an occurrence.
(iii) It is in virtue of the occurrence of P′ that M′ has an occurrence.

Let us add that (i) must not be read as expressing a causal connection. According to the supervenience thesis, the relationship between the occurrence of a realizing property and the occurrence of the emergent property it realizes just is not a causal relationship, but rather a relationship of determination. The sequence "occurrence of P—occurrence of M—occurrence of P′" is therefore not a causal sequence. The result is that when we think over the cause of the occurrence of P′, there seems to be only three possibilities: either the occurrence of P is the cause of P′; or else it is the occurrence of

---

as Descartes conceives of it: in his conception only the quantity of motion is conserved in a physical system, and not its direction. So at the pineal gland, the mind intervenes by acting on the direction of certain motions. What must furthermore be underlined is the idea that an action of the mind on the body is compatible with the principles of Newtonian physics, since this physics admits the existence of forces acting at a distance. If the action of gravitational force is accepted, there is nothing to preclude the existence of other types of force: chemical force, magnetic force, cohesive force . . . and, why not, mental force. It would be necessary to wait for the formulation of the grand principle of energy conservation within a system, as well as for the application of this principle (by Helmholtz in the 19th century) to the case of living systems, before the principle of physical causal closure imposed itself on the scientific community. On the history of this principle, see the appendix in Papineau (2002).

[15] The following elaboration is mostly taken from Kim's recent presentation of the case (Kim, 2005, pp. 39–45). See also Kim (1998).

M; or else it is the joint occurrence of M and P. However, according to Kim, we have to accept the following principle of causal exclusion:

> Principle of causal exclusion: No singular event can have more than one sufficient cause capable of explaining its occurrence at a given moment.

This principle obviously obliges us to choose between M and P as being the best candidate for "sufficient cause" of P′. This is where the principle of causal closure of the physical world appears: since P′ is a physical property, its occurrence at t must have a physical cause at t. Then the conclusion seems inescapable: it is the occurrence of P, the physical realizing property of M, that is the cause of the occurrence of P′, the realizing property of M′, thus excluding M which appears to play no causal role in the occurrence of P′. We see that the causal role of M in the causation of P′ is somewhat preempted (see chapter 3) by the occurrence of its realizing property P, and this because of our adhesion to the principle of physical causal closure: because in explaining the occurrence of the physical property P′ we have a choice between the emergent cause M and the physical cause P, this principle forces our hand towards choosing P. So it must be noted that the conclusion of Kim's argument is not justified a priori: it takes on the empirical nature of the principle of closure.

It can thus be seen that the joint affirmation of T1 and T2 implies the negation of T3: if supervenient and irreducible emergent properties exist then these properties are found to be devoid of causal powers. Therefore, supposing that emergent properties do exist, they must be epiphenomenal.

Given that the argument of causal exclusion has most often been put to use in discussions exclusively involving the question of mental causality, it is important to emphasize that it is in fact applicable on all natural levels where one could be tempted to suppose the existence of emergent properties. As Ned Block, for example, points out, once the premises T1 and T2 have been affirmed in conjunction the argument can be generalized (Block, 2003). We see then that Kim's argument gets to the very heart of the stratified conception of the world and of scientific explanation: if functional properties only play causal roles through the action of their structural realizing properties, can we really consider them as playing an important role in scientific explanations? The consequences of the argument are so disastrous for the stratified conception of the world that certain authors have not hesitated to consider them a *reductio ad absurdum* of the entire argument. As Ned Block writes:

> First, it is hard to believe that there is no mental causation, no physiological causation, no molecular causation, no atomic causation but only bottom level physical causation. Second, it is hard to believe that there is no causation at all if there is no bottom level of physics.[16]

Kim's argument seems valid; to consider it as a *reductio ad absurdum* nevertheless requires the rejection of one of its premises. So then, we must examine the following

[16] Cf. Block (2003, 138).

possibilities: (1) rejecting premise T3 which leads to the thesis that emergent properties do not have causal powers, and thus to epiphenomenalism, (2) rejecting the principle of causal closure of the physical world, which leads to forms of dualism other than epiphenomenalism, (3) rejecting premise T1, that is, the supervenience thesis, and (4) rejecting thesis T2 on the irreducibility of emergent properties, which leads to a reconsideration of reductionism. Before entering into the details of the discussion, let it be noted that Kim himself favors the final option.

2.4 VERSIONS OF DUALISM

Let us first consider the consequences of rejecting premise T3 along with the consequences of rejecting the principle of causal closure. If T3 is abandoned, then the idea that emergent properties possess causal efficiency is also abandoned, that is, the idea that their existence has a real transforming impact in the world, from the causal processes point of view. This position is known as "epiphenomenalism."[17] According to epiphenomenalism, which has mainly been discussed in philosophy of mind, mental states are caused by physical states but have no causal efficiency whatsoever of their own. A pain state, for example, is determined by a cerebral state, but it can itself cause nothing: in the functionalist perspective it would be said that it is its cerebral realizing property that preempts its causal power.

The advantage of epiphenomenalism is that it is compatible with the principle of causal closure of the physical world: mental properties exist, they are irreducible to cerebral properties, but their occurrences cannot cause anything. Physical effects must therefore have physical causes, in keeping with the principle of closure. Epiphenomenalism is also compatible with numerous forms of dualism: Descartes's substance dualism, property dualism, as well as with emergentism. An epiphenomenal emergentist considers irreducible emergent properties to exist which are genuinely novel with respect to the physical properties of the entity possessing them, but also considers that there is no emergent downward causality, in the following sense: although the occurrence of physical properties can cause the occurrence of emergent properties, the reverse is not true, since the latter are causally inert.

The main argument against epiphenomenalism comes from its apparent incompatibility with our naive way of conceiving of emergent properties. Consider the case of pain. According to our naive psychology, the occurrence of a pain experience—let's say, a burnt hand—causally explains the full range of appropriate behavior: withdrawal of the hand, desire for the pain to cease, avoidance of the situation where the burn happened, etc. But, such causal explanations are excluded if we accept epiphenomenalism: given that the occurrence of epiphenomenal properties cannot cause anything, it can of course not cause behaviors or actions specifically either. Worse still, if we consider

[17] A forceful defense of *epiphenomenalism* can be found in (Huxley, 1874), even though the term itself is not featured in this oft-cited article. See also Campbell (1970), Jackson (1982), Robinson (1988), as well as the discussion in Chalmers (2002).

that perceiving of the fact that an object possesses a property does rely on a causal relationship, which seems difficult to deny, then epiphenomenal properties cannot be perceived. In that case, making transparency an epiphenomenal emergent property would imply that transparency can never be perceived. Of course, one could always respond by calling on the existence of regularities tying the occurrence of epiphenomenal properties to the occurrence of physical properties and maintaining that it is thanks to these regularities that occurrences of epiphenomenal properties can be known. Except that the resulting vision of the world seems most complicated and, above all, it would render the role of emergent properties within scientific explanation entirely secondary.

A second dualist option consists of rejecting the principle of causal closure. After all, this principle, as we have seen, is not justified a priori, so it could logically turn out to be false. Such a rejection leads to interactionist dualism, which can in turn take different forms: Cartesian interactionist substance dualism, interactionist property dualism, or emergentism. We will predominantly look at the last position. What distinguishes it from epiphenomenalism is that it makes room for emergent downward causality: complex physical systems possess irreducible emergent properties and the possession of these properties can have effects not only on the emergent property level but also on the physical level. This clearly implies negating the principle of closure, since there are now physical events that are not caused by the instantiation of physical properties but rather by the instantiation of emergent properties.

The main objection to interactionist dualism then, of any variety, is in the fact that the principle of closure seems to be so well confirmed. However, at least two responses to this objection are possible. First of all, one can insist on the possibility that emergent downward causal relationships exist in certain domains but that they have not yet been discovered (Popper and Eccles, 1977). Speculating on the future of scientific developments is nonetheless highly risky. In a more ambitious way, one can also try to indicate certain fundamental physical phenomena on which an emergent downward causality could act. Generally the most talked about domain, especially when it comes to discussing the emergence of conscious phenomena, is quantum mechanics (Chalmers, 2002; Hodgson, 2002 for an overview presentation). There is an interpretation of quantum mechanics according to which an interaction with a (macroscopic) measuring device has an effect on the quantum process which cannot be explained at the scale of the process itself. It is called "collapse of the wave function." The evolution of a quantum process, left to itself, is given by Schrödinger equation taking the wave function as an argument. The wave function describes the state of the system, which may be an "entangled" (see chapter 3 and chapter 11). An entangled state is an inextricable combination of "pure" states, which are the only ones that can be observed. Within an entangled state, each pure state possesses a certain probability of being revealed during measurement. Indeed, during interaction with a measuring device, the state measured is always a pure state. It is as if the measuring device "chose" just one of the pure states among all of those building up the entangled state. So, according to this interpretation (although see Albert, 1992 for other interpretations) we have a downward causal action: that of the measuring device on the quantum processes, which the description by way of the wave function and Schrödinger equation cannot account for. A weakness of this response—apart from its supposing a specific interpretation of quantum mechanics which nothing indicates a priori to be the best one—resides in its only being valid in the specific case of quantum phenomena: emergent causality would thus still remain unexplained in all other domains of reality.

2.5 EMERGENCE WITHOUT SUPERVENIENCE

The second strategy to answer Kim's trilemma amounts to rejecting the supervenience thesis, the idea that emergent properties are supervening properties with respect to a basis of given structural properties. Is it nevertheless possible to find a coherent intermediary position between dualism on the one side and reductionism on the other, that does not accept the supervenience thesis? To succeed in this, one must succeed in defining the systematic dependence between emergent properties and the basis with respect to which they emerge in a novel way, and to do this without slipping into dualism.

Such an approach is certainly not a hopeless one. First, it must be pointed out that the interpretation in terms of supervenience is inapt for accounting for certain intuitions. Indeed, emergentism insists on the "many-layered" nature of reality: according to this position, there exist distinct levels of explanation and reality which correspond to different scales found in nature (Lycan, 1987). But as Jaegwon Kim himself points out, the structural properties and functional properties defined by causal conditions on these structural properties "are properties of the same entities and systems." If a perfectly rigorous vocabulary is adopted, then the second-order functional properties would of course have to be distinguished from their first-order realizing properties. But this distinction matches with neither a difference in level of reality nor with a difference of scale.

In order to better respect emergentist intuitions and avoid getting trapped in Kim's trilemma, it is advisable to abandon the functionalist idea that emergent properties should be realized by lower level structural properties capable of preempting their causal powers. Then it becomes a case of discovering how the properties in question emerge: what exactly is the dependency relationship they have with respect to lower level properties? In an important article (Humphreys, 1997a), Paul Humphreys makes the following suggestion. Let us suppose, for the sake of argument, that distinct levels $N_0 \ldots N_j$ in the properties of nature do exist. Perhaps it is fitting to conceive of the emergent properties of a given level N_i as being ontologically constituted by the fusion of properties of level N_{i-1}. An emergent property, according to Humphreys, must be conceived of as a new totality, yes created on the basis of lower level properties, but irreducible to these properties, which from a metaphysical point of view cease to exist in this fusion. More precisely, the fusion of several properties is understood as "a unified whole in the sense that its causal effects cannot be correctly represented in terms of the separate causal effects [of the base properties]."[18] Of course, it is no longer possible, within this theoretical framework, to speak of supervenience: an entity can very well possess the emergent property of level N_i resulting from the fusion of two properties

[18] Humphreys (1997a), in Bedau and Humphreys (2008 p. 117).

of level N_{i-1} without however possessing any property whatsoever of level N_{i-1}. The fusion, it must be pointed out, appears then as an ontological operation, not as a logical one. Recalling an old emergentist slogan, a property obtained through fusion is supposed to differ from the logical sum of its parts.

An entity can indeed possess the fusion of two properties P and Q without possessing these properties separately. The causal powers of properties resulting from fusion are thus genuinely new, just as the fundamental intuition of emergentism would have it, therefore providing an escape route from Kim's trilemma. And yet, Humphrey's theory is still definitely physicalist. The properties resulting from a fusion depend existentially on the properties that are fused, since the former could not have existed without the latter.

This conception of emergence as a fusion, proposed by Paul Humphreys, is without doubt appealing: it responds well to emergentists' principal motivations in wishing to understand both the autonomy of emergent properties by granting them new causal powers as well as their dependence with respect to underlying lower level properties.[19] It is however permitted to wonder whether in many cases it is not more plausible to interpret "emergent" properties as complex properties, wholes structured with the help of fundamental logical operations rather than the help of the metaphysical operator "fusion." In several authors, particularly David Armstrong and more recently Jaegwon Kim, we find the idea that numerous natural properties can be decomposed in this way. For example, the property of being a molecule of water can be defined as the complex property of being a whole composed of two hydrogen atoms and one oxygen atom bound together in a certain way. In such a whole the parts obviously do not disappear, no more than the parts or the properties that they instantiate. Armstrong speaks of "structural properties" in regard to complex properties whose instantiation by a whole depends on the instantiations of properties of certain of its parts (of a necessarily lower scale) and on the relationships that these parties hold with each other (Armstrong, 1978, chap. 18. See also Kim, 1998; and Kistler, 2005).

The fundamental question for the philosopher is to know whether it is indispensable in analyzing the most interesting cases of emergence to call on the notion of fusion such as Humphrey conceives of it, or whether the notion of structural property does suffice. Once again, quantum mechanics offers the favored ground of investigation. Indeed, quantum theory admits of states which seem to exactly match the notion of emergence Humphreys defined on the basis of the fusion operator. So the entangled states are described by the inseparable entanglement of several pure states. An entangled state cannot be described in the language of quantum mechanics as being logically "composed" of pure states. Moreover, the notion of complex wholes does not apply to this type of state either, their being of a specifically quantum nature. So, it seems, we cannot consider the property of being an entangled state as a structural property in Armstrong's sense, although the concept of fusion can be used.

[19] Cf. also Humphreys (1997b, c), O'Connor (1994), O'Connor and Wong (2005), and the introduction to the volume Bedau and Humphreys (2008).

Humphreys prudently highlights that metaphysical consequences cannot be directly drawn from the vocabulary of quantum mechanics: due to ongoing debates around the interpretation of this theory, epistemological precautions must be taken. However, it does in his view constitute an example of emergence to be taken into consideration.

3. Reductive Explanations

So it turns out that none of the anti-reductionist solutions to Kim's trilemma is completely satisfactory. And thus we find ourselves confronted with a problem well known in philosophy of mind: the existence of "explanatory gaps" which separate the domains we are tempted to regard as emergent from those domains they seem to emerge from. It may seem unusual to speak of "explanatory gaps" in such a context. However, certain philosophers, inspired by logical positivism, had already long ago observed that the notion of emergence may be relative to the state of scientific theories at a given moment. Thus, in the following text, Carl Hempel and Paul Oppenheim maintain that properties are emergent relative to a theory once the occurrences of these properties cannot be deduced from the principles of the theory:

> (. . .) the emergentist assertion that the phenomena of life are emergent may now be construed, roughly, as an elliptic formulation of the following statement: Certain specifiable biological phenomena cannot be explained, by means of contemporary physicochemical theories, on the basis of data concerning the physical and chemical characteristics of the atomic and molecular constituents of organisms. (Hempel and Oppenheim, 1948, 151; republished in Bedau and Humphreys, 65)

According to Hempel and Oppenheim, emergence is therefore not an absolute characteristic of certain families of properties but a relative characteristic: properties appear to us as emergent at a certain moment, relative to our best theories, while we cannot explain them, that is, while we cannot deduce their occurrences in the appropriate circumstances. The existence of an explanatory gap can then be explained as a mere gap in our knowledge at a given moment. If emergence is relative, and if it is above all an epistemic phenomenon, then nothing can be deduced, metaphysically, from our inability to explain occurrences of pertinent phenomena in a given theoretical framework, if not just that the framework is maybe not developed enough to allow for their explanation.

Let us consider the most discussed example, the emergence of consciousness, taking the case of pain experience as our starting point. On the one hand, we know that pain is rigidly correlated to the excitation of certain fibers; it is a supervenient property, or at least one that is systematically linked to cerebral properties. For this reason, we would want to provide a reductive explanation for the occurrence of pain. We would want, in other words, to understand the nature of this property within an entirely physicalist framework. But what exactly is a reductive explanation, and how would a reductive explanation manage to fill in an explanatory gap?

3.1 THE FAILURE OF CLASSICAL REDUCTIONISM

In the history of science, "explanatory gaps" have often been erased through reduction. Hence, the Newtonian theory of motion reduced the theory of celestial motion to dynamics by unifying the physics of sublunary and superlunary motions. Starting in the 1920s, Heitler and London were able to use quantum mechanics to deduce certain chemical properties of molecules from the physical properties of their atomic parts. This was an important event in the history of contemporary science as it demonstrated the possibility, through the application to chemistry of principles from quantum physics, of reductively explaining certain phenomena that could seem to be emergent. Moreover, a new discipline, quantum chemistry, would be born from these tentative applications of physics. Two points are to be underlined regarding this example. First of all, the notion of deduction does seem to play a central role in the reductive explanations of chemical phenomena. How and ever, the existence of reductive explanations does not guarantee the existence of a reduction, in the strict sense, of chemical theory based on the principles of quantum physics. Predictions are limited, at least in the beginning, to relatively simple cases which rely on quantum models of small molecules, like the hydrogen molecule.

For the moment, the only reduction model at our disposal is Ernest Nagel's. Let us recall that, according to Nagel, a theory T_2 can be said to be reduced to a theory T_1 if and only if the laws of T_2 can be logically deduced from those of T_1, fleshed out by a certain number of bridge principles. Extremely ambitious, since one speaks of reduction only when the principles of one theory can be derived from those of another, the Nagelian model of reduction raises insurmountable problems.[20]

The source of the difficulty is in the bridge principles. These principles have the status of empirical, contingent laws, justified by observation. Their importance to the Nagelian notion of reduction cannot be overestimated. It is indeed easy to see that once bridge principles have been discovered between T_1 and T_2, reduction is but a formality.[21] With the help of bridge principles, all the statements of T_2 can be translated into T_1. Once the translation has taken place, there are two possibilities. If all the laws of T_2 can be considered as theorems of T_1, then the reduction is over. But suppose that this is not the case, and that at least one law of T_2, translated into the vocabulary of T_1, cannot be deduced from the laws of T_1. The reduction still does not fail! Because the law is formulated in the vocabulary of T_1, and because it can be supposed that it is justified by observation, there is nothing to oppose its addition to the theory T_1. Certainly, in

[20] The history of Nagelian reductionism is complex. Following on from the famous work of historians and philosophers of science (Kuhn, 1962; Feyerabend, 1962), it has become clear that the reductions actually carried out throughout history were often accompanied by a modification of the theory reduced. Thus, Newtonian mechanics allows for the derivation of an *approximation* of Galileo's laws and not the exact version of these laws. In order to take this critique into account, Nagel's model was modified. See Schaffner (1967, 1992) and Bickle (1998). As we will mainly insist on the role of bridge principles, it must be highlighted that the reduction model defended in Schaffner (1967) interprets these principles differently from Nagel (1961). On these questions, see Bickle (1998) and Kistler, (2007).

[21] Cf. Kim (2005, p. 99).

this case we do not get a reduction of T_2 to T_1, but rather a reduction of T_2 to a theory T_1' which can be seen as a natural extension of T_1.

So the discovery of bridge principles is sufficient, in principle, for carrying out inter-theory reduction. But can it be said that they are sufficient for filling an explanatory gap? Let us again consider the case of pain. Suppose that empirical correlations between pain events and events that can be explained in a neurophysiological vocabulary (like the electrical activation of certain cerebral fibers) are discovered. Is the empirical establishment of such correlations sufficient for explaining pain neurophysiologically, for understanding its nature? This can certainly be doubted.

First, the establishment of a correlation does not, in itself, constitute a physicalist explanation. Even a dualist philosopher could admit the existence of bridge laws establishing correlations between brain states and pain experiences. More precisely, using bridge laws in the deduction of pain phenomena is begging the question for the physicalist. To fill the explanatory gap between two theories, one would manage to deduce the occurrence of one family of phenomena exclusively using the explanatory resources of the reducing theory. In the case of pain, the pain phenomena would therefore have to be exclusively derived with the aid of neurophysiological laws. But this is exactly what Nagelian reductionism doesn't manage to do, since its derivation depends entirely on bridge principles. By using bridge principles that mention purely psychological properties, the reductionist supposes an understanding of certain psycho-physical laws relating not only to physical properties but also to psychological ones; she also supposes that she can then use these laws in her reduction activity. This supposition is clearly not legitimate. We cannot, for example, suppose that we understand pain well enough to formulate bridge laws, if our aim is to provide a purely physicalist explanation of pain.

So Nagelian reductionism seems to be circular, because the derivations of the principles from the theories to be reduced contain, via the intermediary of the bridge principles, mention of the precise properties whose nature must be understood using a vocabulary limited to that of the reducing theory. Jaegwon Kim further proposes the following constraint, which any reducing explanation must satisfy

> NC: Non-circularity principle: The explanatory premises of a reductive explanation of a phenomenon involving property F (e. g. an explanation of why F is instantiated on this occasion) must not refer to F.[22]

The theory used in a reductive explanation of a phenomenon, in other words, must not mention any other property apart from those belonging to the ontology of the reducing theory.

The result is a challenge for the reductionist: how can the explanatory gap be filled in a given domain of phenomena without violating the principle of non-circularity? In other words, how does one manage to derive a particular theory T, without circularity, from the totality of the statements of physics and chemistry? At first glance, this challenge

[22] Cf. Kim (2005, p. 105).

may seem extremely difficult to meet. It is indeed indispensable, as we have seen, to establish connections between the theory to be reduced and the reducing theory if we hope to arrive at a reducing explanation for a family of phenomena. The function of bridge principles, in the Nagelian approach to reduction, is to put such connections in place. Yet we have just seen that this approach is circular. So the difficulty is as follows: to manage to connect the two theories involved all the while respecting the NC principle.

To meet the challenge, it is important to distinguish between two different ways for mentioning properties. We can speak of a "substantial mentioning" apropos of a statement which genuinely communicates information through use of a predicate describing a property P. Mentions of properties realized through Nagelian bridge principles are substantial in this exact sense, since these principles are contingent, empirically justified statements. However, there also exist non-substantial uses of predicates, in particular in statements expressing definitions. These only express the meaning of the predicates. The following statement, for example, says nothing substantial about bachelors:

(5) All bachelors are unmarried.

However legitimate the desire to avoid question-begging may be, it does seem reasonable to weaken the NC principle by reformulating it in the following way:

> NC: Principle of non-circularity: The premises of a reductive explanation of a phenomenon of type P involving the emergent property F must not mention F in a substantial way.

This new principle authorizes the formulation of statements connecting a theory to be reduced to a reducing theory, on condition that these statements communicate no empirical information about the emergent properties of the theory to be reduced. This amounts to saying that the connecting statements must consist of analyses of concepts, or at least that they must express necessary propositions and not contingent propositions as Nagel wished.

Two main approaches are in competition for the best way of accomplishing such a neo-reductionist program, corresponding to the two main contemporary versions of physicalism, and also to two types of necessary statements liable to perform the connection between physics and the theories to be reduced.

It is fitting, in presenting the debate which has recently developed between these two approaches, to set out from the following conditional statement which I shall call the Reducing Implication (RI):[23]

(RI): Necessarily (Propositions of physics P ⇒ Propositions of special science S)

This implication expresses the derivability of the propositions of the special science to be reduced using the full set of all propositions of physics. Here we consider physics to be

---

[23] Cf. Chalmers (1996, 2002) and Stoljar (2006).

the reducing theory, but of course all the discussions to follow can be generalized to cases where the reducing theory is a special science of some higher level than physics.

Let us begin by observing that the supervenience thesis implies the truth of (RI), something easily shown by reduction to the absurd. If one supposes that (RI) is false, then there is a possible world in which all the propositions of physics are true, but in which at least one proposition of the special science to be reduced is false. In other words, there is a possible world that is completely indiscernible from the real world from the perspective of physical facts, but which one would still be able to distinguish from the real world in the perspective of facts described by the special science S. But the existence of such a possibility is excluded by the supervenience thesis. And so (RI) must constitute a necessary implication. According to this statement, the propositions of the special science S must be true if one supposes the propositions of physics to be true.

The statement (RI) expresses the fundamental reductionist intuition according to which the truths of the special sciences are metaphysically implied by the truths of physics. Rejecting (RI), we just saw, is equivalent to rejecting the supervenience thesis, which is equivalent to the adoption of a strong emergentist position: either some version of dualism or else the non-reductionist emergentism we presented earlier. Before entering further into the interpretation of (RI), the motivations that could be invoked by its disciples should be recalled.

In favor of (RI), all the familiar physicalist arguments mentioned already can be put to work. However, one important difficulty is to be noted. (RI) makes reference, with no further precision, to "physics." Moreover, this is also the case with the physicalist position as we have informally described it up until now. But what is a "theory of physics" or a "truth of physics"? Without claiming to answer this delicate question in a fully satisfactory way, we will get along with stating that it is a theory which explains the behavior of paradigmatic objects that we consider to be "physical" objects. Daniel Stoljar, to whom we owe this conception of physicalism, illustrates it with the aid of an analogy to mechanisms.[24] We all have a more or less clear idea of what a paradigmatic machine is: there is easy agreement on the fact that elevators, planes, or computers are paradigmatic machines, while this is not the case with flowers, mushrooms, or cows. This preconception allows us to define mechanical truths as truths which it is necessary to mention in order to explain the essential nature of machines. In a similar way, we can start out from our common notion of what paradigmatic physical things are in order to characterize physical truths.

Against (RI), all the intuitions associated with the explanatory gap notion can be opposed. Several forms have been given to these intuitions, mainly in philosophy of mind. To simplify the discussion we will concentrate on the Knowledge Argument, which constitutes a particularly striking presentation of these intuitions and which we have already seen earlier in this chapter. Let us simply recall that, according to the conclusion of that argument, all physical truths about the vision of colors can be known without one's having to know what it is like to see red, and thus being ignorant of at

[24] Cf. Stoljar (2006, pp. 29–30).

least one psychological truth, which seems like a good reason for rejecting (RI). Let us point out that the argument can be generalized to any domain once the existence of an explanatory gap is suspected. A philosopher who considers the properties of living beings to be emergent will, for example, certainly maintain that all physico-chemical properties of an organism could be known without one's being able to derive all the relative biological truths of that organism.

Contemporary reductionists agree on accepting (RI), but they do differ markedly in their beliefs on how it can be justified. Two cases can be distinguished:

(i) either (RI) is a proposition that can be justified a priori
(ii) or (RI) is a true proposition which can only be justified a posteriori

We will examine these two options successively: each leads to its own variety of physicalism, which we will name "type A physicalism" and "type B physicalism," using the terminology introduced by David Chalmers.[25]

3.2 FUNCTIONALISM AND CONCEPTUAL ANALYSIS: TYPE A PHYSICALISM

The origins of type A physicalism can be traced back to David Lewis's work on the definition of theoretical terms and to David Armstrong's work on functionalism in philosophy of mind, though it is only very recently that this approach has been much developed.[26] According to this first variant of reductionism, reducing explanations rely on a functional physicalist analysis of the concepts making up certain fundamental propositions of the special sciences. Let us consider an example from David Chalmers in order to illustrate this idea; the example is sexual reproduction.[27] It seems as if a reductive explanation can be given for the biological phenomenon of reproduction. Indeed, there is reproduction when two organisms produce one (or many) other(s). This last statement is part of conceptual analysis and not empirical research. In fact, understanding the meaning of the concept of "sexual reproduction" is sufficient in order to be in a position of knowing that it is a process through which two organisms produce one (or many) other(s). This conceptual analysis enables the causal role—or function—of reproduction to be identified, this being the production of one (or many) organism(s) from other organisms. Above all it allows for the mechanism which realizes that function to be identified, since it can be supposed that there is a series of types of physical events allowing two organisms to produce one (or many) other(s). Two aspects are clearly brought out by this example. First of all, the functional analysis of reproduction allows the establishment of a necessary connection between a biological predicate, the "sexual reproduction" predicate, and the vocabulary of physics and chemistry. In doing this, it allows the proposal of a reductive explanation of reproduction. Second, this reduction

[25] Cf. Chalmers (2002).
[26] Cf. Armstrong (1964), Lewis (1970, 1980), Jackson (1998), Chalmers and Jackson (2001), Polger, (2002), Kim (1998, 2005a).
[27] Cf. Chalmers (1996, p. 44).

does not rely on the empirical discovery of a bridge law but rather on conceptual analysis: the proposition "there is reproduction if and only if two organisms produce one (or many) other(s)" is thus justified a priori through our mastering of the concept of reproduction, and not through observation. If a vitalist objector were to criticize our reductive explanation by advancing that we have not explained reproduction but merely the way a cellular process may lead to the production of a complex physical entity similar to some first complex physical entity, then, according to Chalmers, we should retort that it is the vitalist who has not understood the concept of reproduction. According to this concept, to reproduce, by definition, consists in nothing more for a complex physical entity than to produce another similar to it through a cellular process.

According to type A physicalism, a reductive explanation of a process or a phenomenon does not rely on the discovery of bridge laws, but on a physicalist analysis of the vocabulary of the special science; thus, it satisfies the (NC) principle of non-circularity. So we see that within this approach, the functional analysis of special science concepts plays a crucial role. As distinct from anti-reductionist functionalists, type A physicalists consider that these concepts denote physical properties rather than functional properties. This amounts to maintaining that there aren't genuinely any functional properties, only functional ways of characterizing first-order physical properties or, if we prefer, functional descriptions of these properties. The reductive explanation of the fact that an entity possesses a property P of a special science entails the following steps, the first of which is purely a priori, the second empirical:

(i) First of all, a functional a priori analysis of the concept that designates P must enable the identification of a causal role, or a function, corresponding to possession of P.
(ii) Second, our empirical knowledge of the world, and particularly of physics, enables us to determine which physical property (or structured set of physical properties) realizes this causal role

This conception of reduction can be illustrated with the example of contemporary molecular genetics. According to type A functionalists, the property of having gene X is not a functional property but rather a physico-chemical property described functionally. Thus, the blue-eye gene is a physico-chemical property that can be characterized with the help of the following causal role: transmitting (under certain conditions) the phenotypical property of having blue eyes from parents to children. Knowledge of this causal role then enables the identification of a chemical mechanism capable of realizing it, which will be localized in some strand or other of the DNA molecule.

It is certainly plausible to analyze the concept of the gene as a functional concept. In this light, the philosopher of science Lenny Moss, for example, writes, "The concept of the gene began not with an intention to put a name on some piece of matter but rather with the intention of referring to an unknown something, whatever that something might turn out to be, which was deemed to be responsible for the transmission of biological form from between generations."[28] We can, however, still ask whether all the fundamental concepts of the special sciences are open to this type of analysis. Many authors, Jaegwon Kim and David Chalmers in particular, consider that concepts dealing with conscious states, like concepts of color sensation, could never be analyzed functionally (Chalmers, 1996; Kim, 2005a). In their opinion, the consequence of this is that the explanatory gap which exists between conscious phenomena and the natural sciences will never be filled.

3.3 A POSTERIORI IMPLICATIONS? TYPE B PHYSICALISM

According to type B physicalists, conceptual analysis alone does not allow for the establishment of a bridge between physics and the special sciences. For a connection to be established between these domains, one must turn to theoretical identity statements (to employ the terms found in the work of S. Kripke and H. Putnam).[29] These identity statements are of the following sort:

(5) Water = H_2O
(6) Heat = mean kinetic energy of molecules
(7) Pain = stimulation of C-fibers

Type B physicalists maintain that these statements are not analytical but empirical: they can only be justified a posteriori. Unlike Nagel however, they do consider that these are necessary statements.[30] The property of being composed of H_2O molecules, for example, belongs to the nature, or the essence of the aqueous substance: there is no possible world in which water could exist without being identical to the substance composed of molecules of H_2O. In other words, this is a case of the famous "necessary *a posteriori*" introduced by Saul Kripke in *Naming and Necessity* (Kripke, 1980).

If the disciples of type B physicalism are correct, statements capable of establishing a connection between the theory to be reduced and the reducing theory are necessary but a posteriori: it is not enough to just carry out an analysis on the concepts of the special science in order to arrive at a reductive explanation. So, the conditional (RI) cannot be justified a priori, even though it is indeed necessary. Thus it resembles (8) more than (9):

(8) If Zorro committed the robbery, then Don Diego de la Vega committed the robbery.
(9) If a bachelor committed the robbery, then an unmarried person committed the robbery.

[28] Lenny Moss, What Genes Can't Do (2003, p. 2).
[29] Cf. Kripke (1980), Putnam (1975a).
[30] Cf. Loar (1990), Hill (1991), Block and Stalanaker (1999), and Papineau (2002). Type B physicalism finds its origin in the theory of identity of U. T. Place and S. S. Smart (cf. Place, 1956; Smart, 1959; and Feigl, 1967). Contrary to the eliminativism defended by Churchland (1985), this really is a reductionism, since the phenomena the reductive explanations deal with are considered to have genuine existence, even if their true nature may be misunderstood before the theoretical identity statements have been established. Type B physicalism is also anticipated in Schaffner (1967) insofar as it interprets "bridge principles" as identity statements and not as descriptions of natural laws. See also Enç (1983), as well as Bickle (1998, 1999).

The two statements (8) and (9) are necessary; but the truth of (8), unlike that of (9), cannot be justified by pure conceptual reflection. It must be discovered empirically that Zorro is none other than Don Diego in order to arrive at a position to know that (8) is necessarily true. In order to better understand this position, let us use a specific case of a reductive explanation; the explanation of the psychological pain phenomenon. Let us suppose neurophysiology shows that absorbing a certain medication M limits stimulation of C-fibers. The following reasoning constitutes a reductive explanation for the psychological efficiency of the medication:

(i) Conjunction of the neurophysiology propositions.
(ii) Absorption of M reduces stimulation of C-fibers.
(iii) Pain = stimulation of C-fibers.
(iv) Therefore, absorption of M reduces pain.

From an explanation point of view, the crucial steps are (i) and (ii). Indeed, if pain, metaphysically speaking, is nothing more than the stimulation of C-fibers, then explaining the fact that absorbing M reduces the stimulation of C-fibers amounts quite precisely to explaining the fact that absorption of M reduces pain. For this reason, we can maintain that, even though (iii) is a posteriori, the reductive explanation avoids any objection of circularity. In fact, the concept of pain is used in a perfectly non-substantial way in this reasoning: if pain really is nothing other than the stimulation of C-fibers, then speaking of pain or speaking of C-fiber stimulation amounts to speaking about precisely the same phenomenon. Type B physicalism does however come up against two serious difficulties.

The first concerns the justification of theoretical identity statements. These statements, as we have seen, are justified by a posteriori reasoning and not by conceptual analysis. But by what kind of reasoning exactly? Here there seems to be a consensus among type B physicalists to call on the notion of inference to the best explanation, even though the approaches differ considerably in their details. The point where they agree deals with the following central idea: theoretical identity statements can be rationally accepted, because they allow for explanations of certain phenomena that would not be so easily available were they to be rejected. The disagreements are about phenomena whose explanation would justify accepting identity statements.

According to (McLaughlin, 2001), these phenomena are the correlations between the occurrence of properties to be reduced and the occurrence of physical properties capable of reducing them. We know, for example, that possession of a conscious psychological property in a person is regularly associated with possession of a neurophysiological property in the brain. Of themselves, these correlations between states of mind and states of body do not constitute a direct reason for adopting a reductionist position, since their existence is perfectly compatible with the different forms of dualism discussed earlier. They do, however, constitute an indirect motivation for accepting identity statements between mental states and physical states, since these identities, according to McLaughlin, provide the "best explanation" of their existence.

The thesis according to which the identity of A and B constitutes the best explanation of the co-occurrence of A and B seems appealing: after all, is not the best explanation for the fact that Don Diego de la Vega is always to be found in the vicinity of places where Zorro has been simply that Don Diego is none other than Zorro himself? Certain philosophers, particularly Block and Stalnaker (1999), do however raise a problem: the very idea of correlation relies on the idea of a difference between the events which are correlated. Yet if Don Diego is none other than Zorro, then an event comprised of the instantiation of a property P by Don Diego will be absolutely identical to an event comprised of the instantiation of P by Zorro. So, Don Diego's entrance into a bank is not exactly correlated but rather identical to Zorro's entrance into that bank. According to this point of view, accepting an identity statement can in no way enable a correlation to be explained: rather it should be said that the question of how the correlation exists has been erased: it no longer arises.

According to Block and Stalnaker, this shows that it is an inference to the best explanation which allows for theoretical identity statements to be justified, but this inference is not based on an explanation of psycho-physical correlations (or whatever other kind of correlation between events described by the theory to be reduced and those described by the reducing theory). Rather it is based on the explanation of phenomena described by the science to be reduced. Block and Stalnaker write on this matter:

> Why do we suppose that heat = molecular kinetic energy? Consider the explanation given above of why heating water makes it boil. Suppose that heat = molecular kinetic energy, pressure = molecular momentum transfer and boiling = a certain kind of molecular motion (we are alluding to an empirical identity claim, not the a priori behavioral analysis considered earlier). Then we have an account of how heating water produces boiling. If we were to accept mere correlations instead of identities, we would only have an account of how something correlated with heating causes something correlated with boiling. Further, we may wish to know how it is that increasing the molecular kinetic energy of a packet of water causes boiling. Identities allow a transfer of explanatory and causal force not allowed by mere correlations. Assuming that heat = mke, that pressure = molecular momentum transfer, etc., allows us to explain facts that we could not otherwise explain. Thus, we are justified by the principle of inference to the best explanation in inferring that these identities are true. (1999, 23–24)

This text reveals clearly the main difficulty with type B physicalism. Block and Stalnaker recognize that identity statements, in and of themselves, are devoid of explanatory power: they merely allow the "transfer" of available explanations from the reducing science—in this case, statistical mechanics—to the theory to be reduced. It is thus strange to speak of "inference to the best explanation." We can admit that statistical mechanics provides the best explanation of the reason increasing mean kinetic energy causes a certain kind of molecular motion, that is to say, provided that we accept the theoretical identity statements, the best explanation of the reason why heat causes a packet of water to boil. But the identity statements, in and of themselves, play no role in the explanation. Accepting them only amounts to accepting that the phenomena described by statistical mechanics are in fact exactly the same as those described by thermodynamics, and that the explanations available for the former are also available for the latter.

This discussion leads us to the second big problem with type B physicalism, resulting from the first: it represents a reductionism which, when well understood, does not claim to fill in the explanatory gaps but rather to make them disappear, denying they exist in any genuine way. By proposing a neurophysiological explanation for pain phenomena, the type B physicalist does not claim to explain the phenomenal nature of pain with the help of neurophysiology, but rather denies that there is anything to explain beyond cerebral phenomena. It is of course questionable whether a philosopher ingrained with emergentist intuitions would be able to accept such a point of view.[31]

4. Conclusion: Physicalism and the Limits of Science

Come to the end of this discussion of contemporary reductionist positions, we see that none are fully protected from significant objection. Type A physicalism's merit is that it proposes to fill the explanatory gaps which, according to emergentist intuitions, exist between theories of the special sciences and theories of physics. However, it maintains that conceptual analysis must enable the establishment of bridges between reducing theories and theories liable to be reduced. Though it may be easy to see how these bridges can be established in certain cases—in the life sciences, for example—many philosophers doubt of this possibility in other areas. In this regard, the most discussed case is that of conscious experience: many authors doubt that a conceptual connection between what it is like to see red, for example, and neurophysiological theories could ever be established. On the other hand, we have seen that embracing type B physicalism amounts to dismissing the explanatory gap problem as being badly formulated rather than genuinely trying to answer it.

So it can be asked, in conclusion, whether certain physicalist explanations, although certainly existing in an absolute sense, are not liable to remain beyond the reach of our theorizing activity. This is a position that has been strongly defended, particularly by (McGinn, 1999), a propos of conscious experience. According to McGinn, there is fundamentally no doubt that subjective experiences are metaphysically nothing other than states of a physical entity; from this point of view, experience does not differ from phenomena like respiration or digestion. Nevertheless, this same philosopher maintains that the explanatory gap separating our best neurophysiological theories from the first person descriptions we can give of our experiences is destined to forever remain unfilled.

This position has sometimes been presented as a form of "mysterianism": certain natural phenomena will always escape our understanding due simply to the limits of

[31] The question to be answered is whether reductionism, in this extreme version, does not become confused with the eliminativism defended by Churchland (1985).

our mind. It is not however a position of irrationalism, and it can be described in a less pessimistic way than that given by McGinn. We could in fact maintain that all natural phenomena are physical phenomena—so that the reductive implication is necessary and true—and that it is thus possible, a priori, to derive all truths of the special sciences from the truths of physics, it is just that we do not have the adequate conceptual tools for this derivation project at our disposal at the moment.

In a recent publication (Stoljar, 2006), Daniel Stoljar names this position the "epistemic conception," as it amounts to attributing the existence of inter-domain explanatory gaps to the limitations of our conceptual framework. Contrary to McGinn, Stoljar does not consider that the epistemic gap caused by the inadequacy of our current concepts for understanding the nature of conscious experiences could in principle never be filled. He simply maintains that, for the moment, we cannot propose reductive explanations for these phenomena.

It seems important to point out the affinity that exists between type A physicalism and the epistemic conception defended by Stoljar. In both cases, it is admitted that an explanatory gap must, at least in theory, be capable of being filled. According to type A physicalists, the gap can be filled with the help of a conceptual analysis which already comes available with our current theories; with the epistemic conception, on the other hand, it is indeed possible that such an analysis could one day be produced, it is just that we do not yet dispose of the theoretical tools which would enable its formulation. For this reason, a disciple of the epistemic conception insists, without doubt correctly, on the explanatory limits of the scientific theories of any given moment of history.

References

Albert, D. Z. (1992), Quantum Mechanics and Experience, Cambridge, MA: Harvard University Press.

Alexander, S. (1920), *Space, Time and Deity*, London: Macmillan.

Andler, D., Fagot-Largeault, A., and Saint-Sernin, B. (2002), *Philosophie des sciences*, vol. 2, Paris: Gallimard.

Armstrong D. (1964), *A Materialist Theory of Mind*, New York: Humanities Press.

Armstrong D. (1978), *A Theory of Universals*, vol. 2, Cambridge: Cambridge University Press.

Atler, T., and Walter, S. (2007), *Phenomenal Concepts and Phenomenal Knowledge: New Essays on Consciousness and Physician*, New York: Oxford University Press.

Bedau, M., and Humphreys, P. (2008), *Contemporary Readings in the Philosophy of Emergence*, Cambridge, MA: MIT Press.

Bickle, J. (1998), *Psychoneural Reduction: The New Wave*, Cambridge, MA: MIT Press.

Bickle, J. (1999), "Multiple Realizability" in E. Zalta (ed.), *The Stanford Encyclopedia of Philosophy*, https://plato.stanford.edu/.

Block, N. (ed.) (1980a), *Readings in Philosophy of Psychology*, vol. 1, Cambridge, MA: Harvard University Press.

Block, N. (1980b), "Introduction: What Is Functionalism?" in N. Block (1980a), pp. 171–184.

Block, N. (2003), "Do Causal Powers Drain Away?" *Philosophy and Phenomenal Research*, 67, pp. 133–150.

Block, N., and Fodor, J. (1972), "What Psychological States Are Not," reprinted in Block (1980a), pp. 79–99.

Block, N., and Stalnaker, R. (1999), "Conceptual Analysis, Dualism, and the Explanatory Gap," *Philosophical Review*, 108, pp. 1–46.

Boyd, R., Gasper, P., and Trout, J. D. (eds.) (1991), *The Philosophy of Science*, Cambridge, MA: MIT Press.

Broad, C. D. (1925), *The Mind and Its Place in Nature*, London: Routledge and Kegan Paul.

Campbell, K. K. (1970), *Body and Mind*, New York: Doubleday.

Carnap, R. (1966), *Philosophical Foundations of Physics*, New York: Basic Books.

Chalmers, D. (1996), *The Conscious Mind*, Oxford: Oxford University Press.

Chalmers, D. and Jackson, F (2001), "Conceptual Analysis and Reductive Explanation," *The Philosophical Review*, 110, pp. 315–361.

Chalmers, D. (2002), "Consciousness and Its Place in Nature," in D. Chalmers (ed.) *Philosophy of Mind: Classical and Contemporary Readings*, New York: Oxford University Press, pp. 247–272.

Churchland, P. (1985), "Reduction, Qualia, and the Direct Introspection of Brain States," *Journal of Philosophy* 82, pp. 8–28.

Dupré, J. (1993), *The Disorder of Things*, Cambridge, MA: Harvard University Press.

Enç, B. (1983), "In Defense of Identity Theory," *Journal of Philosophy* 80, pp. 279–298.

Farrell, B. A. (1950), "Experience," *Mind*, 59, pp. 170–198.

Feigl, H. (1967), *The "Mental" and the "Physical": The Essay and a Postscript*, Minneapolis: University of Minnesota Press.

Feyerabend, P. K. (1962), "Explanation, Reduction and Empiricism," in H. Feigl and G. Maxwell (eds.), *Minnesota Studies in the Philosophy of Science*, vol. 3: *Scientific Explanation, Space, and Time*, Minneapolis: University of Minnesota Press, pp. 28–97.

Fodor, J. (1974), "Special Sciences, or The Disunity of Science as a Working Hypothesis," *Synthese* 28, pp. 77–115, repr. in R. Boyd, P. Gasper, and J. D. Trout (eds.) (1991).

Hempel, C., and Oppenheim, P. (1948), "Studies in the Logic of Explanation," *Philosophy of Science*, 15(2), pp. 135–175. Reprinted in M. Bedau and P. Humphreys (2008), pp. 61–67.

Hill, C. (1991), *Sensations*, Cambridge: Cambridge University Press.

Hodgson, D. (2002), "Quantum Physics, Consciousness and Free Will," in R. Kane (ed.), *The Oxford Handbook of Free Will*, New York: Oxford University Press, pp. 85–110.

Horgan, T. (1993), "Nonreductive Physicalism and the Explanatory Autonomy of Psychology," in S. Wagner and R. Warner (eds.), *Naturalism: A Critical Appraisal*, Notre Dame: University of Notre Dame Press, pp. 295–320.

Humphreys, P. (1997a), "How Properties Emerge," *Philosophy of Science* 64, pp. 1–17

Humphreys, P. (1997b), "Aspects of Emergence," *Philosophical Topics* 24, pp. 53–70

Humphreys, P. (1997c), "Emergence, Not Supervenience," *Philosophy of Science* 64, pp. S337–S345.

Huxley, T. (1874), "On the Hypothesis That Animals Are Automata, and Its History," *Fortnightly Review*, 95, pp. 555–580. Reprinted in his *Collected Essays*. London, 1893.

Jackson, F. (1982), "Epiphenomenal Qualia," *Philosophical Quarterly* 32, pp. 127–136.

Jackson, F. (1998), *From Metaphysics to Ethics: A Defense of Conceptual Analysis*, Oxford: Clarendon Press.

Kemeny, J. G., and Oppenheim, P. (1950), "On Reduction," *Philosophical Studies*, 7, pp. 6–19.

Kim, J. (1989), "The Myth of Nonreductive Physicalism," reprinted in Kim (1993), pp. 265–285.

Kim, J. (1992), "Multiple Realization and the Metaphysics of Reduction," reprinted in Kim (1993), pp. 309–336.

Kim, J. (1993), *Mind and Supervenience*, Cambridge: Cambridge University Press.

Kim, J. (1998), *Mind in a Physical World. An Essay on the Mind-Body Problem and Mental Causation*, Cambridge, MA: MIT Press.

Kim, J. (2005), *Physicalism, or Something Near Enough*, Princeton, NJ: Princeton University Press.

Kistler, M. (2000), "Réduction fonctionnelle et réduction logique," *Philosophiques* 27(1), pp. 27–28.

Kistler, M. (2005), "Réduction "rôle-occupant," réduction "micro-macro," et explication réductrice a priori," *Dialogue*, 44(2), pp. 225–248.

Kistler, M. (2007), "La réduction, l'émergence, l'unité de la science and les niveaux de réalité," *Matière Première* 2, 2007, pp. 67–97.

Kripke, S. (1980), *Naming and Necessity*, Cambridge, MA: Harvard University Press.

Kuhn, T. (1962), *The Structure of Scientific Revolutions*, Chicago: University of Chicago Press.

Levine, J. (1983), "Materialism and Qualia: the Explanatory Gap," *Pacific Philosophical Quarterly*, 64, pp. 354–361.

Levine, J. (1993), "On Leaving Out What It's Like," in M. Davies and G. W. Humphreys (eds.), *Consciousness*, Oxford: Blackwell, pp. 543–557.

Lewes, G. H. (1875), *Problems of Life and Mind*, vol. 2, London: Trübner & Co.

Lewis, D. (1970), "An Argument for the Identity Theory," *Journal of Philosophy*, 67, pp. 203–211.

Lewis, D. (1980), "Mad Pain and Martian Pain" in N. Block (ed.), *Readings in the Philosophy of Psychology*, vol. 1, Cambridge, MA: Harvard University Press, pp. 216–222.

Lewis, D. (1995), "Should a Materialist Believe in Qualia?" *Australasian Journal of Philosophy*, 73, pp. 140–144.

Loar, B. (1990), "Phenomenal States," in J. Tomberlin (ed.), *Philosophical Perspectives IV: Action Theory and the Philosophy of Mind*, Atascadero, CA: Ridgeview, pp. 81–108.

Ludlow, P., Nagasawa Y., and Stoljar D. (2004), *There's Something about Mary. Essays on Phenomenal Consciousness and Frank Jackson's Knowledge Argument*, Cambridge, MA: MIT Press.

Lycan, W. G. (1987), "The Continuity of Levels of Nature," reprinted in Lycan (ed.) (1990), pp. 77–96.

Lycan, W. G. (ed.) (1990), *Mind and Cognition*, Oxford, Blackwell.

Malaterre, C. (2008), "Les origines de la vie: émergence ou explication réductive?" Thèse de doctorat de l'université Paris I—Panthéon-Sorbonne.

McGinn, C. (1999), *The Mysterious Flame: Conscious Minds in a Material World*, New York, Basic Books.

McLaughlin, B. (1992), "The Rise and Fall of British Emergentism," in A. Beckermann, H. Flor and J. Kim (eds.), *Emergence or Reduction? Essays on the Prospects of Nonreductive Physicalism*, Berlin: De Gruyter, pp. 49–93.

McLaughlin, B. (1997), "Emergence and Supervenience," *Intellectica*, 25, pp. 25–43.

McLaughlin, B. (2001), "In Defense of New Wave Materialism: A Response to Horgan and Tienson," in C. Gillett and B. Loewer (eds.), *Physicalism and Its Discontents*, Cambridge: Cambridge University Press, pp. 319–30.

Morgan, C. L. (1923), *Emergent Evolution*, London: Williams and Norgate.

Mosse, L. (2003), *What Genes Can't Do*, Cambridge, MA: MIT.

Nagel, E. (1961), *The Structure of Science*, New York: Harcourt, Brace and World.

Nagel, T. (1974), "What Is It Like to Be a Bat?" *Philosophical Review*, 83, pp. 435–450.

O'Connor, T. (1994), "Emergent Properties," *American Philosophical Quaterly*, 31, pp. 91–104.

O'Connor, T., and Wong, H. Y. (2005), "The Metaphysics of Emergence," *Noûs* 39(4), pp. 658–678.

Oppenheim, P., and Putnam, H. (1958), "Unity of Science as a Working Hypothesis" in H. Feigl, M. Scriven, and G. Maxwell (eds.), *Minnesota Studies in the Philosophy of Science*, vol. 2, Minneapolis: University of Minnesota Press, pp. 3–36.

Papineau, D. (2002), *Thinking about Consciousness*, Oxford: Oxford University Press.

Place, U. T. (1956), "Is Consciousness a Brain Process?" *British Journal of Psychology*, 47, pp. 44–50.

Polger T. W. (2002), *Natural Minds*, Cambridge, MA: MIT Press.

Popper, K., and Eccles, J. (1977), *The Self and Its Brain: An Argument for Interactionism*, New York: Springer.

Putnam, H. (1967), "The Nature of Mental States," reprinted in Putnam (1975), pp. 429–440.

Putnam, H. (1975), *Mind, Language, and Reality: Philosophical Papers*, vol. 2, Cambridge: Cambridge University Press.

Richardson, R. (1979), "Functionalism and Reduction," *Philosophy of Science*, 46, pp. 533–558.

Robinson, W. S. (1988), *Brains and People: An Essay on Mentality and Its Causal Conditions*, Philadelphia: Temple University Press.

Russell, B. (1927), *The Analysis of Matter*, London: Kegan Paul.

Searle, J. (1992), *The Rediscovery of the Mind*, Cambridge, MA: MIT Press.

Schaffner, K. (1967), "Approaches to Reduction," *Philosophy of Science*, 34, pp. 137–147.

Schaffner, K. (1992), "Philosophy of Medicine," in M. Salmon, J. Earman, C. Glymour, J. Lennox, P. Machamer, J. McGuire, J. Norton, W. Salmon, and K. Schaffner (eds.), *Introduction to the Philosophy of Science*, Englewood Cliffs, NJ: Prentice Hall, pp. 310–344.

Shoemaker, S. (2007), *Physical Realization*, Oxford: Oxford University Press.

Smart, J. J. C. (1959), "Sensations and Brain Processes," *Philosophical Review*, 68, pp. 141–156.

Stoljar, D. (2006), *Ignorance and Imagination. The Epistemic Origin of the Problem of Consciousness*, New York: Oxford University Press.

Van Cleeve, J. (1990), "Emergence vs. Panpsychism: Magic or Mind Dust?" in Tomberlin, J. E. (ed.) *Philosophical Perspectives*, vol. 4, Atascadero, CA: Ridgeview Publishing Company, pp. 215–226.

Wimsatt, W. C. (1997), "Aggregativity: Reductive Heuristics for Finding Emergence," *Philosophy of Science*, 64 (4), Suppl. 2, S372–S384. Reprinted in M. Bedau and P. Humphreys (2008), pp. 99–110.

2

Philosophy of the Special Sciences

9

PHILOSOPHY OF LOGIC

Logic and Content—One Way to the Philosophy of Logic

Philippe de Rouilhan (Institut d'Histoire et de Philosophie des Sciences et des Techniques)

THE QUESTION OF the content of the different categories of expression of ordinary language is or should be one of the major questions of the philosophy of logic, and it would not even be an exaggeration to say that it is or should be the central question of the philosophy of what is called "philosophical logic." Readers are invited to enter into this question in order to be shown what the philosophy of philosophical logic is like, and thus, in a certain way, to be introduced to the philosophy of logic in general. To attain the same final goal, we could have begun with another question, that of demonstration and validity, for example, and taken the route of the philosophy of what is called "mathematical logic." If the first path was chosen rather than the second, it is not because we have a lower opinion of the latter than of the former. It is simply that, apart from being even more superficial than one necessarily is in engaging in this kind of exercise, a choice had to be made and we made it.[1]

---

[1] This chapter is an English translation of a revised version of the corresponding chapter of the original, French, edition of this book. Corrections are minor, except for the three following ones. First, I added a note (subsect. 2.3, n15) to present Kripke's new and devastating refutation of Quine's principle of exportation, pointed out to me by Serge Bozon once it was published in 2011. Second, Serge drove me to give a summary of Kripke's argument for the puzzling character of the disquotational principle (subsect. 3.2). Third, perplexities of François Rivenc led me to be more explicit about the place I give to formal ontology, syntax, and semantics, respectively, in the program of a content logic (subsect. 5.1. first para.). Last, there is of course this new, first note, which replaces the first two notes of French edition. There, I already thanked Serge for "avoir passé au peigne fin, comme il sait faire, l'avant-dernière version du texte." Thank you also to Claire O. Hill, whose fine English I could not keep myself from sullying with infelicities, as usual, in which I could see something of my own, abominable, English.

1. Introduction

First, a map of the field of logic in the broadest sense is made that assigns the disciplines mentioned in opening remarks their proper place and sheds some light on those remarks (subsect. 1.1). The question of content arises in the most acute manner for the joint analysis of singular terms and expressions of alethic or epistemic modality. Two versions of the paradox of the indiscernibility of identicals, up against which any such analysis must be measured, are presented (subsect. 1.2).

1.1 LOGIC IN THE BROADEST SENSE AND PHILOSOPHY OF LOGIC

Logic in the broadest sense may be characterized as the science aiming at *revealing the deep logical structure of statements* that is ordinarily hidden beneath their superficial grammatical structure and *correlatively evaluating the arguments involving such statements*, and the philosophy of logic as the part of logic devoted to the examination of the reasons liable to justify the choice of fundamental concepts, principles and methods of this science.

To be more precise, I shall say that the *division of labor* within the field of logic in the broadest sense can be achieved along two different lines. It can be achieved in accordance to whether the investigation *rather* deals with questions of foundations (an *approach* that is rather *philosophical, informal*), or *rather* aims at more or less complex results accessible in one given context or another and accepted as such (an *approach* that is rather *technical, formal*). The division of labor can also be accomplished according to whether the investigation concerns, be it formally or informally, objects of interest *rather* to mathematicians or *rather* to philosophers. Whence, by having the two lines intersect, one arrives at the double entry table below.

This table must be understood as reflecting neither clear-cut divisions, nor even just vague divisions, but gradual, therefore indeterminate, divisions, as indicated by the word "rather" used many a time. The labels sanctioned by use appear in bold type between quotation marks. This does not include the label "logic" in the broadest sense, because in current usage, "logic" is usually used in the narrower sense of "formal logic" or the even narrower sense of "mathematical logic." I cannot, however, imagine any science worthy of the name that does not imply, among its rights and its duties, that of philosophical reflection, and I do not abstain from speaking of "logic" plain and simple when it is a matter of logic in the broadest sense. It will be noted that the logic dubbed "philosophical," sometimes called "formal philosophy," is no less formal than the logic dubbed "mathematical" and that "philosophy of logic" does not merge any more with the one than with the other.

Insofar as the notion of *content* mentioned in the title of this chapter interests philosophers *rather* than mathematicians, one can say that this chapter has its place

TABLE 1

Logic in the broadest sense

Logic (logic in the broadest sense)	Approach *rather* technical (= formal): **"Formal logic"**	Approach *rather* philosophical (= informal): Philosophy of "formal logic," = "Philosophy of logic"
Objects concerning *rather* mathematicians	**"Mathematical logic"** (logic in the narrowest sense)	Philosophy of "mathematical logic"
Objects concerning *rather* philosophers	**"Philosophical logic"**	Philosophy of "philosophical logic"

in the last line and therefore belongs to "philosophical logic" or to the philosophy of that discipline. And insofar as the approach is philosophical *rather* than technical (or formal), one can also say that it fits into the last column and therefore belongs to the philosophy of logic. Finally, its place is at the intersection of the last line and the last column and therefore belongs to philosophy of "philosophical logic." But, as promised in its subtitle, this chapter on the philosophy of "philosophical logic" seeks to open a path leading to "philosophy of logic" in general.

1.2 THE PARADOXES OF THE INDISCERNIBILITY OF IDENTICALS

It is in the analysis of singular terms (proper names, demonstratives and other indexical expressions, descriptions and other complex singular terms) and expressions of alethic modality (such as possibility, impossibility, necessity, contingency) or epistemic modality (propositional attitudes, such as to believe, to know, or conceptual attitudes such as to seek) that the question of content arises most keenly. When it comes to singular terms, I shall essentially attach importance to proper names and to definite descriptions; when it comes to expressions of modality, to expressions of propositional attitude.

Naturally, numerous analyses, diversely inspired by the founding fathers of modern logic, Frege and Russell, have been proposed, and, as is often the case in the philosophical part of the sciences, whether it is a matter of the particular sciences or of logic (universal science), no consensus has been reached, but about a certain fundamental alternative that would constrain interested parties to choose between an analysis *à la* Frege and an analysis *à la* Russell. More concerned with proposing a panorama of a priori possible major options than with historical and exegetical faithfulness, I shall propose embarking on a purely rational course leading from a logic *à la* Frege (sect. 2) to a logic *à la* Russell (sect. 4), in the course of which I shall tackle a certain logic halfway between the two that I am surprised has not attracted the attention of logicians (sect. 3).

To tell the truth, the Frege and the Russell whose patronage I have just invoked are imaginary characters whom their historical namesakes neither knew, nor in which they would have willingly recognized themselves. I shall just say enough about these two thinkers and their successors for readers to be able to appreciate the extreme liberty I am taking in my rational reconstruction of their historical part.

I shall illustrate my words by drawing on examples of statement suggested by a situation imagined by Quine (1956):

> There is a certain man in a brown hat, whom Ralph has glimpsed several times under questionable circumstances on which we need not enter here; suffice it to say that Ralph suspects he is a spy. Also there is a gray-haired man, vaguely known to Ralph as rather a pillar of the community, whom Ralph is not aware of having seen except once at the beach. Now, Ralph does not know it, but these men are one and the same . . . . (Bernard J. Ortcutt, to give him a name) . . . .

The diverse logical analyses of the situation envisioned in sections 2-4 can be tested in diverse ways, in particular by comparison with paradoxical arguments (1) and (2) below:

(1a) Ralph believes that the man in a brown hat is a spy.
(1b) The man in a brown hat is the man seen at the beach.
∴ (1c) Ralph believes that the man seen at the beach is a spy.

The conclusion (1c) seems to follow from premises (1a) and (1b) in accordance with the principle of indiscernibility of identicals, or principle of substitutivity of identity:

> (SUBST) Co-referential singular terms are intersubstitutable in any (non-quotational) context *salva veritate*.

However, in the situation imagined by Quine, (1a) and (1b) are true, but (1c) is false. Here we have a first paradox of the indiscernibility of identicals, relative to a propositional attitude and definite descriptions (henceforth “the first Ralph paradox”).

Let us use our imaginations to complete the story told by Quine:

> Ralph learned that the man in a brown hat's name is Ortcutt and that the man seen at the beach's name is Bernard, but he still does not know that they are one and the same individual.

We obtain a second paradox of the indiscernibility of identicals, regarding a propositional attitude and, no longer definite descriptions, but proper names (henceforth, "the second Ralph paradox"):

(2a) Ralph believes that Ortcutt is a spy
(2b) Ortcutt is Bernard,
∴ (2c) Ralph believes that Bernard is a spy.

2. The Logic of Sense and Denotation (inspired by Frege, via Church and Quine)

Frege considered definite descriptions to be genuine and even paradigmatic singular terms. To solve the paradoxes of the indiscernibility of identicals regarding propositional attitudes, he was led to divide the content into sense and denotation. But he said nothing about statements in which quantifications and expressions of propositional attitude intertwined (subsect. 2.1). It would be up to Church, Frege's greatest Fregean successor, to do this and to construct what he would call "the logic of sense and denotation" (LSD) (subsect. 2.2). Quine, the greatest non-Fregean successor of Frege, would also have an opportunity to do this, in a completely different way and with completely different prospects than those of LSD, clearly giving, however, the idea of a remarkable variant of LSD (subsect. 2.3).

2.1 FREGE

One of the characteristic features of Frege's logic is that in it, with certain exceptions,[2] the definite descriptions of ordinary language are considered to be genuine singular terms, something which, through arguments based on the consideration of diverse problems—among them the one raised by the first Ralph paradox—induces a *division of the content* of definite descriptions into *sense* and *denotation*, if there is a denotation (see Frege 1891, 1892, 1892/1895). The denotation is the object described, which can be missing. The sense is the way in which this putative object is given by the description. Frege extends this

[2] That is to say, except for certain special uses such as, for example, when a definite description appears in the subject position of the verb "to exist," as, for example, in the statement "The largest natural number does not exist."

division of content, including the possibility that the denotation may be missing, to all singular terms (notably to proper names) and, finally, to all parts of discourse, all the categories of expression, in accordance with the following chart, characteristic, for the categories under consideration, of the logic of sense and denotation (LSD)[3]:

TABLE 2

Categories of LSD

Expression	Singular term	Predicative expression	Statement
Sense	Objectual concept	Property	Proposition
Denotation	Object	Extension	Truth-value

The distinction between sense and denotation enabled Frege to account for propositional attitude statements involved in the Ralph paradoxes. His idea was that, *logically*, the expression "Ralph believes" is not a fragment of an adverbial expression, "Ralph believes that," attached to the subordinate clause, let us say "the man in a brown hat is a spy." It is a predicative expression attached to a singular term, "that the man in a brown hat is a spy," denoting the proposition expressed by this statement. More precisely, let us designate the sense of an expression by using the expression resulting from bracketing it[4] (one could italicize it or systematically transform it typographically in one way or another).[5] This is how Frege analyzed the Ralph paradox:

---

[3] I remind readers that my Frege is not the historical Frege. (1) Independently of any terminological quarrel, he would have protested that it is not the extension—the set of objects falling under a predicative expression—that is the denotation of that expression, but what he called a "concept" (Frege 1891). The correction to Frege's theory that I am proposing in this regard is due to Church (1951a, p. 4; 1956, p. 13). (2) With an eye to escaping from the logical paradoxes, Church limited his ontology to a hierarchy of types based on a domain of *individuals*. As for me, I admit an infinitely vaster basic domain, that of *objects à la* Frege. I do not envision any other domain of quantification and I leave the question as to which is the best way to solve the paradoxes in question up in the air. This is why I speak of "object" and of "objectual concept" where Church spoke of "individual" and "individual concept." (3) The most surprising thesis of the chart—that the denotation of a statement is its truth-value—is definitely Frege's. It essentially results from the recognition of the definite descriptions of ordinary language as being genuine singular terms. Church showed that it would be just as compelling if one recognized other complex singular terms of ordinary language as genuine singular terms (cf. Church, 1943b, pp. 299–300).

[4] The brackets serve to paraphrase, not expressions of the form "the sense of . . . ," but expressions of the form "the objectual concept of . . . " ("the property of . . . ," "the proposition that . . . ," respectively) with a proper name or a definite description (a predicative expression, a statement, respectively) in the place of the ellipsis. Furthermore, I consider these expressions and their paraphrase to be singular terms and, correlatively, their denotation to be an object.

[5] Every device has its advantages and its drawbacks. Thus, bracketing has the advantage of being iterable, while italicization is not, *and the drawback of deceptively having the initial expression appear in the result of the transformation*, which italicization does not do (if one considers expressions of different styles to be different). One could be tempted to point out that this drawback is but the reverse side of an advantage that is a kind of "simulation" (in the sense of what will expressly be in question in subsect. 3.3) of ordinary language in the language of paraphrase, but I would reply that it is not worth the effort.

(3a) BEL_R([is-a-spy(the man in a brown hat)])
(3b) the man in a brown hat = the man seen at the beach
∴ (3c) BEL_R([is-a-spy(the man seen at the beach)]).

The principle behind the Fregean solution is the following. The definite descriptions flanking the equals sign in (3b) do *not really*, or *logically*, occur in (3a) and (3c). They no more (*logically*) occur there than in the singular term obtained from them by bracketing and that canonically denotes their sense. [the man in a brown hat], for example, is not the value at the man in a brown hat of a function that would be denoted by the brackets. What function could it be a question of? As Frege pretty much says, "there is no backward road from denotation to sense," so much so that the question of the (*logical*) intersubstitutivity of definite descriptions *salva veritate* does not arise. In other words, there is no matter for applying the principle (SUBST).[6]

This solution can be fine-tuned. Frege's idea is that, in the object clause of the verb "believes" of the original version (1) of the argument (3), the expressions have as their denotation what, customarily, would be their sense, and that holds in particular for definite descriptions. In order to be able to substitute one for the other *salva veritate*, they would customarily have to have not only the same denotation, but also the same sense. They would have to be synonymous—which they are not. To understand the more in-depth analysis, given below, which underlies this claim, it is necessary to realize that, for Frege, both here and in the historical Frege, but for terminological differences, a property—for example [is-a-spy]—is a function that, *applied* to an objectual concept—for example [the man in a brown hat]—has for value the proposition made from this property and this objectual concept, namely, in this case, [is-a-spy(the man in a brown hat)]. I note APP the logical functor of *application*, at which the historical Frege never stopped.[7] Thus, for example, APP([is-a-spy], [the man in a brown hat]) = [is-a-spy(the man in a brown hat)]. Whence the analysis announced:

(4a) BEL_R(APP([is-a-spy], [the man in a brown hat]))
(4b) [the man in a brown hat] = [the man seen at the beach]
∴ (4c) BEL_R(APP([is-a-spy], [the man seen at the beach])).

(The argument is valid, but the second premise is false, as is the conclusion.)

---

6 The point would be obvious if, instead of bracketing an expression, I had chosen a stylistic change, for example, italicization, to denote the sense of this expression: the descriptions in roman letters would *not even* occur *grammatically, graphically* in the italicized statements, and nothing could lead anyone to believe that they occurred there *logically* (cf. n. 5).

7 The need to introduce the logical functor "APP" here is due to the decision to treat expressions of the form "[. . .]" as singular terms (cf. n. 4). To form a singular term using two singular terms "[is-a-spy]" and "[Ortcutt]," a binary functor is necessary and "APP" is such a functor.

Without prejudging any objections that might be made to the Fregean analysis otherwise, it is in any case seriously incomplete. Frege did not envision the analysis of a quantified sentence like:

(5) There is someone of whom Ralph believes that he is a spy.

A famous argument, due to Quine, tends to show the problematic nature of quantifying in this manner, more precisely, its senselessness. This argument is closely related to the first Ralph paradox and runs as follows (this is not a quotation):

> What is (or what are) this (or these) individual(s) [this (or these) object(s)], of which statement (5) affirms the existence—what is it (or what are they) whose existence would make this statement true? There is at least one such thing, one is tempted to reply. It is the man in a brown hat, since Ralph believes that the man in a brown hat is a spy. But then, it is also the man seen at the beach, since they are both one and the same man. However, this cannot be the man seen at the beach, since Ralph *does not* believe that the man seen at the beach is a spy!

The argument is more forceful than is generally believed. It is an essential feature of statement (5) that the pronoun "he" occurs there in the object clause of the verb "believes" and has "someone" as antecedent there, and the paraphrase must, in one way or another, retain that feature. This is what the paraphrase

*(6) $(\exists x)\mathrm{BEL}_{\mathrm{R}}([\text{is-a-spy}(x)])$

would do in the simplest manner if the variable "x" could really occur in the argument of the belief predicate and thus be bound by the initial quantifier. It is against this paraphrase that Quine's objection is clear and devastating. It shows that, under the condition indicated, the truth conditions of *(6) would be indeterminate, *(6) would not mean anything determinate, and it would be the same for statement (5).[8] But then, what is the logical form of statement (5)?

Is it the uncertainty weighing on the logical form of statement (5) that kept Frege from engaging in the analysis of statements mixing quantification and propositional attitude? Be that as it may, it is necessary to take the variable "x" found in the argument position in paraphrase *(6) out of the scope of the brackets. There are, at first sight, two ways to do this. The first one, inspired by Church and faithful to the spirit

[8] The trap of pseudo-paraphrase *(6) goes hand in hand with the choice of bracketing a (well-formed) expression to designate its sense. Were one to use bold type, for example, in the place of bracketing, pseudo-paraphrase *(6) would be expressed by "$(\exists x)\mathrm{BEL}_{\mathrm{R}}(\textbf{is-a-spy(x)})$," with the variable "$x$" quantified vacuously and the bold letter "**x**" acting as intruder, and it would not enter anyone's mind to see a possible paraphrase of statement (5) there.

of Frege (see subsect. 2.2), preserves as naturally as possible the monadic character of the belief predicate:

(7) $(\exists x)\mathrm{BEL}_{\mathrm{R}}(\mathrm{APP}([\text{is-a-spy}], x)$.

The second one, proposed by Quine, and which Frege perhaps would not have cast aside (see subsect. 2.3), renounces this monadic character and appeals to a dyadic belief predicate:

(8) $(\exists x)\mathrm{BEL}_{\mathrm{R}}([\text{is-a-spy}], x)$.

2.2 CHURCH

Church[9] (1951b) expressly applies the Fregean division of content to variables of the language of paraphrase. To assign a complete value to a variable, it does not suffice to assign it a denotation (denotation-value). One must still assign it a sense (sense-value) to which corresponds this denotation. In short, one must assign it a content. And, since a content can be made up of a sense without a corresponding denotation, assigning a complete value to a variable can also be assigning it an objectual concept without corresponding object.[10] One can apply Church's idea (revised and corrected in this way, see n. 10) to expressions of ordinary language that correspond to the variables of the language of paraphrase, namely the anaphoric pronouns, then apply the Fregean idea of a systematic semantic shift of expressions to them when they are introduced in propositional attitude contexts, as Church (1943a) already suggested. As regards statement (5), if the antecedent "someone"—outside the object clause of the verb of belief—takes a first content as a value, the pronoun "he"—which represents that antecedent in the object clause—correlatively takes as value a second content the denotation component of which is the sense component of the first. To simplify the comparison with the alternative analyses of statement (5) presented later, one will henceforth act as if the antecedent were, not "someone," but "something" in the sense of "some object." As a first approximation, statement (5) must then be understood as expressing the existence of a content such that Ralph believes the proposition resulting from the application of the property of being a spy to the sense component of that content. In other words, without appealing to the idea of divided value of a variable any more than Church does in his formal construction of LSD (references given in n. 9), statement (5) must be understood as expressing the existence of an objectual concept such as Ralph believes the proposition resulting from the application of the property of being a spy to that objectual concept. But the original existential quantification

---

[9] Church is the inventor of the so-called logic of sense and denotation (1946, 1951a, 1973, 1974, 1993). His LSD was typed, the one I wish to give some idea of in this subsection is not.

[10] I distance myself here from Church, who, as Frege did for the language of science, excluded senses without denotation. For Frege, it was a question of principle, for Church, a matter of simplicity.

seems to imply, not only the existence of a certain objectual content, but also that of its object, it short, its non-vacuity. Finally, the faulty paraphrase *(6) of statement (5) is replaced by the new paraphrase (9), which is none other than statement (7) after explicit relativization of quantification to non-empty objectual concepts:

(9) $(\exists x \mid \text{is-a-non-empty-objectual-concept}(x))\text{BEL}_R(\text{APP}([\text{is-a-spy}], x))$.[11]

Thus, what seemed in statement (5) to be a *de re* propositional attitude, in which the belief was about the *res* (the man in a brown hat himself, i.e. the man seen at the beach himself), and one would believe of this *res*, independently of the objectual concept under which it is presented, that it possesses a certain property, is in reality a *de dicto* attitude (belief concerning the *dictum*, a certain proposition).

To tell the truth, this paraphrase is hardly satisfactory, as can be seen by an argument comparable to the one that R. Sleigh used to counter Quine's law of exportation (subsect. 2.3). Suppose a shortest spy exists and that Ralph believes this. If (let us say within the framework of a course on the philosophy of logic) one asked Ralph whether he believes that the shortest spy is a spy, there is no doubt as to how he would respond. He believes it too. But then the following statement is true:

(10) $\text{is-a-non-empty-objectual-concept}([\text{the shortest spy}]) \wedge \text{BEL}_R(\text{APP}([\text{is-a-spy}], [\text{the shortest spy}]))$,

and so are statement (9) and the original statement (5). Thus, simply assuming that the shortest spy exists and that Ralph believes in his existence, we have demonstrated that there is someone of whom Ralph believes that he is a spy. Paradox.

To see this, one must precisely arrive at the understanding that we have of this last statement, statement (5), as manifested in the use we would ordinarily make of it, because it is in reference to this use that we can and must test our analyses.[12] In ordinary usage, we would say that there is someone of whom Ralph believes that he is a spy only if the individual in question is given to him (Ralph) in a way that would allow him (Ralph) to give him away to the police, and would therefore enable the latter to look for this individual and, if need be, arrest him. It is clear that the objectual concept [the shortest spy] is not such a way of being given. From the almost trivial hypothesis according to which Ralph believes that the shortest spy is a spy, even admitting the existence of the shortest spy, one cannot draw the dramatic conclusion that there exists someone of whom Ralph believes that he is a spy. The Church inspired analysis is thwarted.

---

[11] Statement (9) is logically equivalent to the statement obtained by eliminating relative quantification in terms of absolute quantification, "$(\exists x)(\text{is-a-non-empty-objectual concept }(x) \wedge \text{BEL}_R(\text{APP}([\text{is-a-spy}], x)))$," but the intention here is not that the two statements should be synonymous for all that.

[12] Philosophy of logic thus sometimes intersects with philosophy of language, but it also sometimes takes liberties with respect to ordinary language that the philosophy of language condemns. The two disciplines do not merge with each other.

It would therefore be necessary to improve the paraphrase of statement (5) once again by placing a new constraint on the kind of objectual concept whose existence paraphrase (9) affirms. This objectual concept should be of a certain special kind for Ralph, a kind dependent on the context in which it is stated and of which considerations in the preceding paragraph give a vague idea. But, one does not see what the general rule could be that would determine the sense of a *de re* belief statement as a function of context in which it is stated. Be that as it may, the full comprehension of this kind of statement is not only a matter of syntax and semantics, it is also a matter of pragmatics. Regarding all that, the first indispensable reference is Kaplan (1968).

2.3 QUINE

Among the ideas cooked up by Quine (1956) in the course of his analysis of propositional attitude statements—"in the position of a Jewish chef preparing ham for a gentile clientele" (Quine 1977, in 1981, p. 116)[13]—is found the fundamental one of interpreting the quantification in statement (5), not as a quantification relativized to objectual concepts, but as an absolute quantification for the category of objects, ranging therefore over the class of all objects, whatever they may be, thus including non-conceptual objects. It is such, non-conceptual, objects that are in fact liable to make this statement true, and one of these objects, namely the man in a brown hat himself, i.e. the man seen at the beach himself, does indeed make it true. The idea, in other words, is to interpret Ralph's belief as a *de re* belief, and not as a *de dicto* belief. For that, one must, Quine explains, recognize an ambiguity in the verb *to believe*. The expression "Ralph believes" of statement (1a) gives rise to the monadic predicate "$\mathrm{BEL}_{\mathrm{R}}$" of statement (3a), predicate that I shall note $\mathrm{BEL}^1_{\mathrm{R}}$; while the same expression "Ralph believes" of statement (5) gives rise to the dyadic predicate "$\mathrm{BEL}^2_{\mathrm{R}}$" of the following new paraphrase, which is none other than statement (8), with the index "2" added:

(11) $(\exists x)\mathrm{BEL}^2_{\mathrm{R}}([\text{is-a-spy}], x)$.

Independently of any quantification, the contrast between *de dicto* modality and *de re* modality is highlighted well by the example of the two following statements of the new language of paraphrase:

(12) $\mathrm{BEL}^1_{\mathrm{R}}([\text{is-a-spy(the man in a brown hat)}])$,

(13) $\mathrm{BEL}^2_{\mathrm{R}}([\text{is-a-spy}], \text{the man in a brown hat})$.

The first, (12), says that Ralph believes that the man in a brown hat (*no matter what the object thus described may be*) is a spy; the second, (13), that Ralph believes of the man

[13] The ham, in this case, they are the intentional entities such as objectual concepts, properties, and propositions that Quine does not want to countenance in his ontology.

in a brown hat that he (*independently of the antecedent description and, generally, of any particular way of presenting the object so described*) is a spy. (The italicized indications in parentheses are only there to facilitate the proper understanding of what is said, but they are not part of it.) The monadic belief, in (12), is propositional, *de dicto*. The dyadic belief, in both (13) and (11), is relational, *de re*. Syntactically, everything seems clear, but there is something mysterious about the very sense of the *de re* belief. To ascribe to Ralph the *de dicto* belief in question, it suffices no doubt to ask him whether he believes that the man in a brown hat is a spy, or whether he assents to statement (12) and to apply a disquotational principle of the following kind (cf. Kripke, 1979):

> (DISQ) *If a normal, English speaker, on reflection, sincerely assents to "*p,*" then he believes that* p *(with, in place of the letter "*p*" any English statement devoid of systematic or accidental ambiguities).*[14]

However, to ascribe to Ralph the *de re* belief in question, what evidence could we provide?!

For a while, Quine, freely pursuing his exploration of the idea of *de re* belief, defended the thesis called the "law of exportation," which actually partially responded to this question:

> (EXPORT) De dicto *belief implies the corresponding* de re *belief.*

In particular, statement (12) implies statement (13). It was quite natural to wonder whether the two sorts of belief were linked by a relation of implication, but, in fact, the law of importation responds, at least partially, to the question about the sense of *de re* belief, in other words to that of its truth conditions. This law amounts to saying that the truth conditions of *de dicto* belief are truth conditions of *de re* belief. The sense of *de re* belief is thus at least partially elucidated by that of *de dicto* belief. But Quine soon had to abandon his thesis, an objection made by R. Sleigh (1968) having convinced him that it was false. This objection confirms the one I made above to the Church inspired paraphrase of quantification through propositional attitude expressions [subsect. 2.2, (9)], but it is directed at the law of exportation. Here it is, in a version equivalent to its original version.

On the one hand, the law of exportation manifestly lacks a premise of existence. For example, for statement (12) to imply statement (13), it manifestly lacks an existential premise: that the man in a brown hat exists, in other words:

(14) $(\exists!x)$(is-a-man-in-a-brown-hat(x)).

But, on the other hand, even weakened by adding a condition of existence, the law of exportation does not hold. Let us assume once again (subsect. 2.2) that the shortest

[14] Another, more reliable method, but a trickier one to implement, would obviously be that of betting odds.

spy exists and that Ralph believes it. Once again, Ralph believes that the shortest spy is a spy. The weakened law of exportation allows one to deduce that someone exists of whom Ralph believes that he is a spy. Once again, Ralph would be in a position to give the individual in question away to the police. Once again, paradox.[15]

Even weakened, the law of exportation is false. It needs to be weakened a second time, by adding a second supplementary hypothesis. Quine (1977) proposed the hypothesis:

(15) There is someone of whom Ralph believes that he is the man in a brown hat,

or, in the system of paraphrase of Quine (1956):

(16) $(\exists x)\mathrm{BEL}^2_{\mathrm{R}}$([identical to the man in a brown hat], x).

The proposal is convincing inasmuch as it seems fairly clear that, if there is someone of whom Ralph believes that he is the man in a brown hat, and that Ralph believes that the man in a brown hat is a spy, then there is someone of whom Ralph believes that he is a spy. [Hypothesis (15) makes (14) useless.] But, of course, over-weakened in this way, the thesis of exportation is of no help to us in grasping the sense of *de re* belief, since this hypothesis uses the expression of belief that it was precisely a matter of explaining.

Quine (1977) does not stop there. Following Hintikka (1962, p. 132) uncritically, he then reads statement (16) as a paraphrase of the statement:

(17) Ralph believes he knows who the man in a brown hat is,

and engages in a devastating critique. The notion of *believing one knows who someone is*, he notes to begin with, is *clearly dependent on context:* "Of itself the notion is empty" (Quine 1977, p. 121). He then contests the idea (p. 121) that one can still distinguish between admissible and inadmissible cases of exportation, and even between statements of *de dicto* belief and statements of *de re* belief, *except relative to context*. He finally denies (p. 122) that one can understand statements of *de re* belief, and even statements of *de dicto* belief, *except relative to context*. "I see the verb 'believe' even in its *de dicto* use

---

[15] In a paper published in the same year as the original, French, edition of the present book, Kripke (2011a) showed that the law of exportation has much more paradoxical consequences than had been believed (and had been made known) since Sleigh (1968), and in particular this one (adapted here from the story about Ralph): so long as one of Ralph's beliefs is wrong—and how could it be otherwise?— the law of exportation enables one to deduce from it that *Ralph believes of the Eiffel Tower that it is a spy*! Indeed, suppose that Ralph believes that p, where "p" abbreviates a false statement, and then use "c" to abbreviate the definite description "the object identical to Ortcutt, if p, and to the Eiffel Tower, if not p." It is clear that, deceived by his false belief that p, Ralph believes that c is Ortcutt, and therefore also that c is a spy (subsect. 1.2, last para.). Therefore (by exportation), Ralph believes of c that it is a spy and, therefore, also of the Eiffel Tower that it is a spy (since, actually, c is the Eiffel Tower).

as varying in meaningfulness from sentence to sentence" (p. 122). And what holds for belief obviously holds for propositional attitudes in general. On the way, as if to reassure the reader witnessing the destruction in progress, he had written: "At first, this seems intolerable, but it grows on one" (p. 121).

There is, however, a weak point in this destructive undertaking that prevents one from appreciating its outcome. It is his point of departure, with the intrusion of the notion of *knowing* or *believing to know who someone is.* As Kaplan would show (Kaplan 1986, p. 258–260), statement (17) is not an accurate reading of (16), something Quine would acknowledge (Quine 1986, p. 293), without for all that calling the rest into question.

3. The Logic of Meaning and Denotation (inspired by Quine's Variant of LSD, via Kripke and Kaplan)

In what is called here "the logic of meaning and denotation" (LMD), definite descriptions are still considered to be genuine singular terms, but proper names no longer have anything to do with descriptions. They are direct designators, and, in divided content, the sense merges with meaning. This analysis was already that of the early Russell (that of *Principles*, Russell [1903]), but it is found again here by taking an entirely different path, starting with the ideas of Frege and Quine discussed in subsection 2.3, and correcting them using Kripke's ideas about proper names and Kaplan's ideas about indexical expressions (subsect. 3.1). One comes up against the paradox of the indiscernibility of identicals with regard to propositional attitudes and to proper names (the second Ralph paradox), the solution to which is still uncertain (subsect. 3.2). For LSD, there was no adverbial expression of propositional attitudes. There is not any for LMD either, but at least it is possible, *to a certain extent*, to preserve the appearances of ordinary language in this regard (subsect. 3.3).

3.1 KRIPKE, KAPLAN

Church's logic was intended to complement Frege's through a certain analysis of quantification through a propositional attitude expression; see, for example, statement (5) and its paraphrase (9). In the considerations related above, suspending, for the sake of the argument, the extensionalism that made him otherwise reject intensional entities like objectual concepts, properties and propositions, Quine explored the possibility of attaining the same goal using an alternative analysis, see, for example, statement (5) and its paraphrase (11). Neither of these two analyses was completely satisfactory, but, independently of that, they both suffered, for want of having changed anything to it, from the same weakness as that of Frege as concerns the analysis of proper names and indexical expressions. Frege had analyzed these singular terms as possessing, like definite descriptions, a divided content. The second Russell (starting with "On Denoting") expressly analyzed proper names as "disguised definite descriptions," and, no more than Church did Quine have any objection to this.

However, the "descriptivist" analysis is completely unrealistic. This is what Kripke (1971, 1972) showed for proper names (notably by pointing out that speakers can very well use the proper name "Cicero," for example, only knowing that it is a matter of a Roman orator, and therefore not having at their disposal any definite description of this object, or any objectual concept of which it would be the object). And Kaplan (1977, published in 1989) did the same for indexical expressions (notably by drawing attention to the universally true character, in a sense, of the statement "I am here now," in spite of its being contingent, for which the analysis in question, no more than any other available up until then, could not account). The former worked out a new analysis of proper names as "rigid designators," and the latter, a new analysis of indexical expressions as "direct designators."

Kripke defined a *rigid designator* as designator that designates, or denotes, the same individual in all possible worlds in which this individual exists.[16] Kaplan defined a *direct designator* as a designator whose contribution to the proposition expressed by a statement in which it occurs is the very object that it designates, or denotes, so much so that this object itself is a constituent of this proposition. If a designator is direct, it is rigid. What it designates has its place once and for all in the proposition expressed by the statement considered independently of the consideration of any possible world. It designates, therefore, the same individual in all possible worlds *whatever they may be*. It is therefore rigid, and even rigid in a sense stronger than Kripke's. The converse, however, is not true. Certain definite descriptions are rigid without being direct designators (one will admit that "the smallest perfect number" denotes 6 rigidly, but not that 6 is a constituent of the proposition that the smallest perfect number is even). It is not even true, strictly speaking, that proper names themselves are direct designators owing to their rigidity. The concept of direct designator as defined above presupposes a theory of propositions as structured entities, having constituents, in short (borrowing the word from Cresswell, 1975, 1985), a theory of *hyperintensional* propositions, while the concept of rigid designator as defined above only presupposes possible worlds semantics, whose propositions are, one may say, *simple-intensions*, for which the question of constituents does not arise. But, even within the framework of a hyperintensional semantics, the temptation is great, in light of the second Ralph paradox (I come back to this below), to argue that proper names, however rigid they may be, are not direct designators. Kaplan (1977, pp. 497, 562), yet, affirm that they are. This thesis merits attention, be it only for the sake of testing it.

The propositions of Quine's variant of LSD are, as it happens, structured entities. We are therefore in a position simply to correct this logic to admit Kaplan's thesis that *proper names are direct designators*, or, what comes down to the same thing, that, *in the*

---

[16] This definition says nothing about what goes on the worlds in which the individual does not exist. In an exchange with Kaplan, going against certain statements that he had otherwise allowed to be published in his name, Kripke confirmed he did not want to commit himself on this point. On this ticklish matter, as well as on the relationship between rigid designation and direct designation that it is a question of in the rest of the text, see Kaplan (1977), pp. 492–497 and 569–571.

case of proper names, the sense is already the denotation. They are both the object denoted. I shall do this by speaking, no longer of "sense" (*Sinn*), but of "meaning," recalling the early Russell's logic (1903), which, within a different framework and without any argument comparable to those of a Kripke or of a Kaplan, considered proper names to be direct designators, and which, furthermore, as we have done up until now, considered definite descriptions to be genuine singular terms. Whence the following chart, characteristic (for the categories of expression that interest us) of what I call the "logic of meaning and denotation" (LMD):

TABLE 3

Categories of LMD

Expression	Singular term		Predicative expression	Statement
	Proper name	Definite description		
Meaning	Object	Objectual concept	Property	Proposition
Denotation	Object	Object	Extension	Truth-value

3.2 A LINK BETWEEN DE RE BELIEF AND DE DICTO BELIEF; THE PARADOX OF THE INDISCERNIBILITY OF IDENTICALS RELATIVE TO PROPOSITIONAL ATTITUDES AND PROPER NAMES

Let us take up the problem of the analysis of belief again where we left off (subsect. 2.3). Within this new framework, there is definitely a relation of implication between *de dicto* belief and corresponding *de re* belief, and even a relation of equivalence, at least *in the case of proper names*. For example, the statements:

(2a) Ralph believes that Ortcutt is a spy,

(18) Ralph believes of Ortcutt that he is a spy,

paraphrased respectively as follows:

(19) BEL^1_R([is-a-spy(Ortcutt)]),

(20) BEL^2_R([is-a-spy], Ortcutt),

are equivalent.

And, in the case of definite descriptions, things are not as simple, but there exists a link between *de re* belief and *de dicto* belief. For example, the statements:

(1a) Ralph believes that the man in a brown hat is a spy,

(21) Ralph believes of the man in a brown hat that he is a spy,
paraphrased respectively as follows:

(12) BEL^1_R([is-a-spy(the man in a brown hat)]),

(13) BEL^2_R([is-a-spy], the man in a brown hat),

are not, of course, equivalent, but, provided again (see subsect. 2.3, (14)) that the man in the brown hat exists, the latter statement, of *de re* belief, is equivalent to the following, *de dicto,* statement:

(22) BEL^1_R(APP([is-a-spy], the man in a brown hat)).

To simplify matters, let us remain with the case of proper names. The situation may be described in the following manner. Relatively to a proper name occurring within the scope of a propositional attitude expression, the distinction between *de dicto* attitude and corresponding *de re* attitude is, if not abolished, at least reduced to a distinction internal to a class of equivalent statements (relatively to the attitude in question).

But this victory over the obscurity of *de re* belief is a Pyrrhic victory, since, now, the statement of *de dicto* belief, with the *res* itself in the *dictum* (Ortcutt himself, instead of one of the objectual concepts of which he is the object, in the proposition object of the belief), has become perfectly obscure. This obscurity as such is well brought to light by the second Ralph paradox, no longer using definite descriptions, but proper names (cf. subsect. 1.2):

(2a) Ralph believes that Ortcutt is a spy
(2b) Ortcutt is Bernard,
∴ (2c) Ralph believes that Bernard is a spy.

The premises are true and the conclusion attributes to Ralph the belief that Bernard is a spy while, according to the story, he believes the negation of it. The major thing that is new with respect to the first Ralph paradox is that LMD is quite incapable of resolving the second one by proposing a paraphrase of it showing that the paradoxical inference is invalid. On the contrary, LMD formally validates the inference:

(23a) BEL^1_R([is-a-spy(Ortcutt)]),
(23b) Ortcutt = Bernard,
∴ (23c) BEL^1_R([is-a-spy(Bernard)]).

Need one see the second Ralph paradox as a sufficient reason to condemn LMD and, more generally, every logic extending the ideas of the theory of direct reference to proper names (notably, beyond LMD, that of section 4) and thus validating the principle of substitutivity (SUBST) for proper names and the inference that leads to the second Ralph paradox? Be that as it may, it should first be noted that the paradoxical character of the paradox does not depend only on (SUBST), but also on a further principle, say the disquotational principle (DISQ). At least it depends also on the latter *if*, as we may suppose, one knows that Ralph has such and such beliefs because he has had an opportunity to affirm or to assent to such and such sentences expressing them, notably "Orcutt is a spy" and "Bernard is not a spy," and all the conditions were met for applying (DISQ). As for (DISQ), now, Kripke (1979) shows that, *independently of* (SUBST) *for proper names and of the question as to whether the reference of proper names is descriptive or rigid or even direct, in and of itself* (DISQ) *is paradoxical!*

Kripke's argument is informal, it pertains to the philosophy of language. Here is a summary of the story. Pierre had always lived in France and only spoken his mother tongue, French, until the vicissitudes of life led him to move to England, to an unattractive area of London. There, remarkably, he learned how to speak English by the *direct method*, and nothing has ever driven him to identify the city he used to call "Londres" with the one in which he has been living and he naturally now calls "London." In France, he was told that "Londres est jolie" and since then he has believed it. Now that he is in London, he says to himself that "London is not pretty", he believes that London is not pretty. It is not that he has changed his mind, since, let us repeat, nothing has ever brought him to think that it was a matter of the same city. He still believes, as before, that London is pretty, but he now also believes that London is not pretty. To the fateful question and the only question genuinely puzzling in Kripke's eyes, "Does Pierre, or does he not, believe that London is pretty?", it seems one should answer by attributing obviously contradictory beliefs to Pierre, viz., on the one hand, that London *is* pretty and, on the other, that London is *not* pretty (even though Pierre certainly does not believe that London *is* pretty *and not* pretty). Such is Kripke's puzzle.

Kripke analyzes the puzzle in terms of an application of the *disquotational principle* (DISQ), already mentioned (subsect. 2.3), and of the *principle of translation*, (TRANS), according to which "*if a sentence of one language expresses a truth in that language, then any translation of it in any other language also expresses a truth (in that other language)*" (Kripke 1979, in 2011, p. 139). The latter principle is accompanied with conditions of application. The sentence must present no systematic or accidental ambiguity, which would deprive the principle of all intuitive plausibility; as for the translation, it must conform to our normal practice, even though the principle itself only retains the weak condition that the *truth-value* of sentences must be preserved. Kripke's analysis corresponds to the following presentation of the argument. On the one hand, by hypothesis, Pierre assents to "*Londres est jolie*"; thus, by (DISC) for French, *il croit que Londres est jolie*; thus, by (TRANS) from French into English, he believes that London is pretty. On the other hand, by hypothesis, Pierre assents to "London is not pretty"; thus, by (DISC) for English, he believes that London is not pretty. Eventually, Pierre comes at the same time to believe that London is pretty and to believe that London is not pretty.

Kripke says more than once that he does not see how to answer the fateful question cited above. His thesis is precisely that we are faced with a genuine puzzle, and he directs all his energy towards refuting numerous (fake) solutions that one might think about. One thing seems certain, it is that (DISC) or (TRANS) are open to challenge. In order to get rid of (TRANS), then, Kripke devises a second puzzle, a subtle variant of the first one. The whole story takes place in England. Since Pierre heard about "the famous pianist Paderewski," he has believed that Paderewski had musical talent. Later, he also heard about "Paderewski, the Polish nationalist leader and Prime Minister," but, being unaware that the two individuals involved were one and the same person, and fostering some prejudices against politicians, he has since then believed that Paderewski had no musical talent. To the new fatal question "Does Pierre, or does he not, believe that Paderewski had musical talent?," it seems that it should be answered by, one more time, attributing obviously contradictory beliefs to him.

The only difference of structure between the two puzzles is that, at the two stages of the new story, as Kripke convincingly argues, in spite of his epistemic gaps, Pierre has spoken not two different languages, but only one, the very same one in which contradictory beliefs are now being attributed to him. Thus, if it were to be a question of translation, then the translation in question should be homophonic. One might as well say that the question of translation does not arise and it is pointless to resort to (TRANS). So (DISQ) is the only principle that is still suspect, in and of itself it is a paradoxical principle, which is what needed to be proved.

Henceforth, what needs to be called into question in the second Ralph paradox, (SUBST) for proper names or (DISQ)? The situation is methodologically the same as that of König at the Heidelberg Congress in 1904, aspiring to refute Zermelo's well-ordering theorem by *reductio ad absurdum* by using the principles of the naïve semantics of definability. *Once the paradoxical character of these principles themselves was proved independently of the well-ordering theorem*, what needed to be called into question, the well-ordering theorem or the semantic principles?

The lesson that Kripke drew at the end of his analysis is that the question of the validity of the principle of substitutivity of proper names in propositional attitude contexts *salva veritate* is an open question. And the lesson that I draw from it here for LMD—which will also hold for the logic of section 4—is that the fact that this logic validates the principle of substitutivity for proper names is not, pending further investigation, an overwhelmingly compelling objection to it.

3.3 A SIMULATION OF MODAL OPERATORS

Whether in LSD or in LMD, acknowledging definite descriptions to be genuine singular terms and the division of content accompanying this naturally stand in the way of admitting the adverbial expression of modalities and their paraphrase in the form of what logicians customarily call *modal operators*. Examples of adverbial expression are: "Ralph believes that" and "Ralph knows that," formed out of the predicative expressions "Ralph believes" and "Ralph knows," respectively, and the subordinating conjunction "that." It is surprising to see certain logicians introduce both definite descriptions as singular terms and modal operators in their paraphrases of modal statement, as if they were completely unaware of the first Ralph paradox and of Quine's argument (subsect. 2.1).[17] It is true that, usually without warning or perhaps without even giving it a thought, these logicians seem to have abandoned the classical idea of logical paraphrase of statements of ordinary language that is supposed to reveal the logical structure of the original statement. Be that as it may, I have not abandoned it, and what they allow themselves to do, I forbid myself to do.[18]

However, the extension of the ideas of the theory of direct reference to proper names that led from Quine to LMD makes possible what would not be so in LSD: the *simulation* of modal operators, operation which does not go against the classical idea of logical paraphrase, since it is recognized as such and knowingly carried out. It is possible, for example, to simulate the operator "BEL^1_R-THAT," concerning any sentence whatsoever. If the sentence is closed, i.e. if it is a statement, it is easy:

(24) $\text{BEL}^1_R\text{-THAT } p \Leftrightarrow_{def} \text{BEL}^1_R([p])$,

where the letter "p" is a *schematic letter for a statement*. If the sentence is open, one can posit, for example (by limiting oneself, to simplify matters, to the case in which there is only one free variable):

(25) $\text{BEL}^1_R\text{-THAT is-a-spy(x)} \Leftrightarrow_{def} \text{BEL}^1_R\text{-THAT APP([is-a-spy], x)}$,

where the non-italicized letter "x" is a *schematic letter for a variable*.

But be careful! The schematic definition (25) must be understood literally, in the most restrictive manner: for each instance, the sentence on the left is defined by the one on the right, and no more than that. It is not in any way the definition of a new predicate, "BEL^1_R-THAT is-a-spy(. . .)," where the ellipsis could be replaced *ad libitum* by a singular term. It does not, *in and of itself*, justify any equivalence obtained by substituting anything other than a variable for the schematic letter "x" (and by eliminating the subscript "def" in the sign of definitional equivalence). That does not rule out finding, *from another source*, the justification for certain substitutions. For example, for "x" one can always substitute a proper name, for example "Ortcutt," but this is only because one already knows, *from another source*, that $\text{BEL}^1_R\text{-THAT is-a-spy(Ortcutt)} \Leftrightarrow \text{BEL}^1_R\text{(APP([is-a-spy], Ortcutt))}$. However, one cannot always substitute a definite description for "x," because, for example, it is not true, not even on condition that the man in the brown hat exists, that

[17] Examples: Carnap (1947); Hintikka (1957, 1962, § 6.6; 1969); Stalnaker and Thomason (1968); Thomason and Stalnaker (1968); Thomason (1969); Kaplan (1978); Salmon (1986, appendix C); Hughes and Cresswell (1996, chap. 18). For the significant reasons that they detail in the writings mentioned above, both Carnap and Hintikka are to be distinguished as concerns their method of paraphrase. Kaplan too, who in his Kaplan (1986) pleads for a certain amount of freedom with respect to the classical method, equivalent, in fact, to that which authorizes the operation of simulation.

[18] The choice of bracketing an expression to denote its sense or its meaning would prove an exception to this rule if I had not accompanied it with a warning, see n. 5.

BEL^1_R-THAT is-a-spy(the man in the brown hat) ⇔ BEL^1_R(APP([is-a-spy], the man in the brown hat)).

4. The Logic of Meaning (LM) (inspired by LMD, via Russell and A. Smullyan)

What stands in the way of authenticating the adverbial expression of modalities and from introducing full-fledged modal operators into the language of paraphrase? This is the analysis of definite descriptions inherited from Frege. In "On Denoting" (Russell 1905), Russell proposed an eliminative analysis of definite descriptions as singular terms that he would subsequently never call into question (subsect. 4.1). The adoption of his analysis leads to what is here called "the logic of meaning" (LM). This logic has much in its favor (subsect. 4.2), but it has its weak point (subsect. 4.3).

4.1 FROM THE ELIMINATIVE ANALYSIS OF DEFINITE DESCRIPTIONS TO THE LOGIC OF MEANING

One of the consequences of parsing definite descriptions as genuine singular terms and of the dividing of content that it implies may give cause for complaint. It is that the only expression of modalities recognized as genuine is their predicative expression. The logics envisioned up until now, whether LSD or LMD, do not recognize the adverbial expression of modalities. There is no modality operator in their language of paraphrase. They are not "*modal logics* properly so called" (to use a phrase *à la* Quine). Admittedly, within the framework of LMD, it is possible to *simulate*, to a certain extent, the modal operators, but it is precisely only a matter of a simulation.

To simplify matters, let us proceed as if the only singular terms that ordinary language contained were anaphoric pronouns, proper names, definite descriptions, and terms of the form "the objectual concept of . . . ," "the property of . . . ," or "the proposition that . . . ." An alternative, eliminative, analysis of definite descriptions as singular terms exists that enables one to avoid dividing the content and to do justice to the adverbial expression of modalities, and it is that of the second Russell (beginning with "On Denoting," 1905). According to the second Russell, definite descriptions are *not* genuine singular terms. They are disguised complex quantifiers. The paradigmatic example is:

(26) The King of France is bald,

which is analyzed:

(27) There exists an individual who is King of France and is the only one to be and who is bald.

The apparent subject of statement (26), the definite description "The King of France," ends up in the form of a complex quantifier, "There exists an individual who is King of France and is the only one to be and who." A nice way of abbreviating the new analysis (27) is to subscript the definite description of statement (26) and to indicate its scope as a quantifier there by assigning it the same subscript:

(28) (bald(the King of France)$_1$)$_1$.

If a definite description of ordinary language occurs in a sub-statement of a statement, the eliminative analysis of this description is not unique, and one must obviously take into account the context of this statement in order to dispel the ambiguity.

The eliminative analysis of definite descriptions as singular terms is subject to the same exception as their conservative analysis. It is inapplicable to cases in which the description is (grammatically) in the position of the subject of the verb to exist (cf. n. 2). But it is subject to another exception, which is proper to it (see subsect. 4.3). Furthermore, the eliminative analysis offers a new solution to the first Ralph paradox, the idea being that, once definite descriptions' appearance of being a singular term is dissipated, there is no reason to apply the principle of substitutivity, which led us from true premises to a false conclusion. As for the second Ralph paradox, the same considerations that applied to LMD (see subsect. 3.2) apply to LM.

There not being any reason to divide the content, the two-level semantics of LMD gives way to a single-level semantics, that of meaning, as in the following chart, characteristic (for categories of expression of interest to us) of what I call "the logic of meaning" (LM):

TABLE 4

Categories of LM

Expression	Proper name	Predicative expression	Statement
Meaning	Object	Property	Proposition

Let us give each one its due in making the transition from LMD to LM. The second Russell analyzed definite descriptions as derived, disguised quantifiers and, in *Principia* (with Whitehead), considered all expressions that are derived (through definition) to be "mere typographical conveniences" external to the logic's system (Whitehead and Russell, 1910, p. 11); A. Smullyan (1948) integrates them into the system. For LM, the difference is of little importance. What counts is the elimination of definite descriptions as singular terms and recognizing them as quantifiers. For the second Russell, ordinary proper names are disguised definite descriptions. For A. Smullyan (1947, p. 140), who seems to remember early Russell, one may conjecture that they are direct designators. On this point, LM is in early Russell's camp and, no doubt, in A. Smullyan's, as was LMD for the same reasons as it (subsect. 3.1), but no trace of these reasons is to be found, either in early Russell or, even less so, in A. Smullyan.

4.2 THE TWOFOLD ANALYSIS OF PROPOSITIONAL ATTITUDE STATEMENTS

Not only does LM account for the two modes, predicative and adverbial, of expression of modalities, and paraphrase the modal adverbs in the form of operators characteristic of modal logic properly so called, but distinguishing between monadic belief and dyadic belief becomes useless and modal expressions once again enjoy the univocity they had lost in LMD. The subordinating conjunction is paraphrased in the form of a genuine operator of nominalization, such that, contrary to what happened with the brackets in LMD (and in LSD), any sentence (open or closed) is really a part of the outcome of its nominalization. Thus, for example, statements:

(1a) Ralph believes that the man in a brown hat is a spy,
(5) There is someone of whom Ralph believes he is a spy,

can be analyzed in terms of the modal predicate "BEL_R" and of the nominalization operator "THAT":

(29) BEL_R(THAT(is-a-spy(the man in a brown hat)$_1$)$_1$),

(30) $(\exists x)$(BEL_R(THAT(is-a-spy(x)))),

or, just as well, in terms of the modal operator "BEL_R-THAT":

(31) BEL_R-THAT(is-a-spy(the man in a brown hat)$_1$)$_1$,

(32) $(\exists x)$(BEL_R-THAT(is-a-spy(x))).

4.3 THE PARTICULAR PROBLEM OF CONCEPTUAL ATTITUDE STATEMENTS

There is a problem, about which I have not spoken up to this point and with respect to which LM is less comfortably placed than its rivals. It is the analysis of statements of attitude whose object seems to have to be, not a proposition, but an objectual concept (cf. Church 1951b, n. 14).

The analysis of statements of attitude of this kind does not raise any particular problem, either for LSD or for LMS, since they recognize objectual concepts. Thus, for example (Church 1956, p. 8, n. 20, taken up by Kaplan 1975), the statement:

(33) Schliemann sought the site of Troy

can be analyzed as:

(34) $SOUGHT_S$([the site of Troy]).

But, in LM, this analysis is no longer available, and the eliminative analysis of the definite description as being disguised quantifier is inacceptable or impossible, depending on whether the scope of the quantifier in question is assumed to include or not include the expression of attitude.

For the example under consideration, (33), one could hope to work things out by appealing to a Quinean stratagem (Quine, 1956) that leads to the paraphrase:

(35) STRIVED-THAT(Schliemann finds the site of Troy).

But, even in passing over the fact that this paraphrase does not take into account that the verb *to find* is a verb of conceptual attitude (cf. Kaplan, 1986, p. 266) and the reflexive character, *de se*, of the propositional attitude that it should attribute to Schliemann [namely, of striving that *he* (*himself*) finds the site of Troy], Quine's stratagem is not always applicable. It is not, for example, to the statement "Schliemann is thinking of the site of Troy" (cf. Montague, 1960, mentioned by Kaplan, 1986, n. 102).

Another idea, not open to this criticism, is Church's (1951b, n. 14), which leads to the paraphrase:

(36) SOUGHT_S([is a site of Troy]),

but it no longer retains any trace of the unicity presupposed by the use of the definite article (singular) characteristic of definite descriptions (singular), and would therefore correspond, strictly speaking, to the statement "Schliemann sought *a* site of Troy" rather that the statement (33).

Finally, the best solution is Kaplan's (Kaplan, 1975), which leads to the paraphrase:

(37) SOUGHT_S([is a site of Troy and is the only one that is]),

eliminating the initial apparent attitude—whose object seemed to have to be an objectual concept formed from a property—in favor of an attitude whose object is the property obtained from the first by building into it the unicity of any object liable to possess it. Well done, but which one has won? LMD or LM?

5. Conclusion

One goes back over the ground covered, the goal pursued and the method utilized, the hyperintensional, universal, untyped character of the logics envisioned, their expressive force, the informal and hyperintensional character of the semantic considerations that led to them, the "principle of the name-relation" that oriented the paraphrasing, the naiveté assumed with respect to paradoxes (subsect. 5.1). One appraises the manner in which the three logics responded to four questions concerning the analysis of singular terms and expressions of modality, and so as (not) to finish one alludes to a logic *à la* Carnap constructed against the principles of the name-relation and contributing an affirmative answer to all the questions raised (subsect. 5.2).

5.1 RETROSPECTIVE CONSIDERATIONS

In this chapter, I have sought to show what the philosophy of logic is like by giving an idea of three content logics, LSD, LMD and LM, corresponding to diverse informal, syntactical or semantical analyses of singular terms (paradigmatically, proper names and definite descriptions) of ordinary language and, correlatively, of expressions of alethic or epistemic modality (paradigmatically, propositional attitudes). Beyond any heuristic considerations, a content logic, according to this idea is, in the first place, a (formal) ontology, viz., a (formal) theory of possible contents of expressions, considered in their own right, independently of any language. (Indeed, Church's logic of sense and denotation, for instance, did not extend beyond ontology.) Admittedly, a content logic should not, in principle, be restricted to an ontology, for, according to the idea in question, given such and such a part of ordinary language as paraphrased *within the framework of* this ontology—i.e., *within some extension of its language obtained by adding finitely many constants of individual, predicates and functors*—the logic in question should also include a (formal) syntax and semantics for that part *within this same framework.*[19] In particular, this logic should provide an explication of the relation of logical consequence, enabling one to validate ex post facto the informal considerations that proved crucial for the adoption of it. All the same, the elaboration of a content logic should begin with ontology, however naïve it may be. The method of exposition adopted in this chapter has consisted in showing how statements chosen as examples could be paraphrased within the framework of the ontology of LSD, LMD, or LM.

The ontology is in each case intended to be *universal*. When one writes a statement of the form "$(\forall x)$ . . . " or "$(\exists x)$ . . . ," with "x" as object variable, this is to speak of "*all* the objects," this "*all*" being taken in an absolute sense, without any implicit relativization, without the slightest mental reservation.[20] *Universalism* as such does not rule out the existence of multiple categories of variable, but here, the ontology has in each case a single category of variable, namely object variables. It is not that the entities other

[19] An analogy might help readers understand that program. If ordinary language were limited to its extensional part, then, independently of whether definite descriptions as singular terms were analyzed away or not, one would naturally be led to a "logic of denotation," whose corresponding ontology was a theory of individuals, and extensions (graphs, respectively) of possible, monadic or polyadic, predicates (functors, respectively). Such a theory would prove equivalent to a theory of (individuals and) sets. Given such and such a part of this limited ordinary language as paraphrased *within the framework of this set theory*—i.e. *within some extension of its language obtained by adding finitely many constants of individual, predicates and functors*—one could contemplate working out a (formal) syntax and a (formal) "denotational" semantics of that part *within the same framework*. On the way, the usual, extensional, first- or higher-order logic, in particular usual model theory, could be reconstructed and be at last given its natural and proper place, viz., *within the same framework* (for technicalities, see Rouilhan 2007, 2012).

[20] Well, that sentence by itself is not that clear and may be misleading. For more on the universalist point of view here assumed, see Rouilhan (2012), in particular, p. 561.

than individuals are excluded, it is that the former are values of the same variables as the latter, all are objects.

In this chapter, I have respected the principles that had, more often than not, guided, since Frege and Russell, logical analysis and the operation of paraphrase and that Carnap called the "principles of the name-relation" (1947, p. 98) (and that he only formulated to condemn their false obviousness and free himself of them, see below subsect. 5.2). These principles imply, in particular, that, if a singular term occurs *logically* in a statement of ordinary language, then that statement is about the denotation of that term, and that term is therefore replaceable in the case under consideration by any other term having the same denotation *salva veritate*. By contraposition, if a term is not replaceable in this way in one of its occurrences in a statement of ordinary language, this is because it occurs there, not *logically*, but only *grammatically* and it *hides* either another singular term, or something other than a singular term, something, in any case, that logical paraphrase is supposed bring to light. The same holds good, *mutatis mutandis*, for predicative and functorial expressions.

In the informal, syntactical or semantical, considerations presiding over the choice of such and such a logic, the intention has been to free oneself resolutely from the paradigm of possible worlds semantics and from its simple-intensions in order to contemplate, as Frege and Russell already had, finer intensions that, borrowing from Cresswell, one can call "hyperintensions." According to possible worlds semantics, the proposition expressed by a statement, for example, is the set of the possible worlds in which this statement is true, and two statements therefore express the same proposition if (and only if) they are equivalent in all possible worlds. According to hyperintensional semantics, the propositions are extra-linguistic entities structured like expressions, and, for two statements to express the same proposition, it is in no way sufficient for them to do this in the sense of possible worlds semantics. Hyperintensional semantics is much more demanding. It is sufficiently so that one can conceive of the logical form of a statement in close relation to the structure of the proposition that it expresses; speak of the constituents of a proposition and thus take the theory of direct reference of proper names literally; understand the affirmation that definite descriptions are not genuine singular terms as Russell understood it, namely that no constituent of the proposition expressed by a statement of ordinary language in which it occurs corresponds to a description; speak of the greater or lesser logical perfection (or imperfection) of a language and thus give credence to the idea of a language of paraphrase beyond its stenographic interest; and so forth.

In paraphrasing a statement of ordinary language in the framework of any one of the ontologies considered, I have used nominalization devices supposed to be available there to form, from an explicitly given singular term or predicative expression or functorial expression or statement, a singular term denoting the corresponding objectual concept or property or relation in intension or function in intension or proposition. For instance "[is a spy]" is a singular term denoting the property of being a spy, "[the man in the brown hat is a spy)]" a singular term denoting the proposition that the man in the brown hat is a spy. In all that, I have left the rules for the use of these singular terms in the dark, as if those rules could go without saying, and Russell's experience of paradox in set theory had not taught us anything. In fact, the careless use of these singular terms leads to paradoxes even harder to solve than Russell's paradox about sets. I am thinking of paradoxes *à la* Russell-Myhill here.[21] If I have proceeded in this way, it is simply because I do not take solving the great paradoxes to which the more or less naïve rules that naturally come to mind inevitably lead, for a reasonable preliminary to the study and use of such logics. The enterprise of knowledge, even in logic, never starts at the beginning.

5.2 PROSPECTIVE CONSIDERATION SO AS (NOT) TO FINISH

The chart below sums up the position of the three logics considered in this chapter on the questions that presided over their consideration.

TABLE 5

LSD, LMD, and LM on contentious questions

Contentious questions	LSD	LMD	LM
Are proper names direct designators?	No	Yes	Yes
Are definite descriptions singular terms?	Yes	Yes	No
Is the adverbial expression of propositional attitudes possible?	No	No (but simulation possible)	Yes
Is the predicative expression of propositional attitudes possible?	Yes	Yes	Yes

The preference for one or another of these logics is a function of semantical preferences concerning proper names, definite descriptions, or possible expressions of modality. It is remarkable that none of these logics corresponds to a positive response to the four questions raised. One might wonder whether it would not be possible to invent another logic enjoying this privilege, assuming that it is one, for a logic, of complying to the lessons of the grammar of ordinary language *in this way*. The answer is: certainly, and there is a logician to whom one can turn to for inspiration as to how to do it. It is Carnap (1947). Of course, a positive response to the first question would have assumed Carnap's having adopted semantics other than his own, in which the question of the directness of proper names could not have arisen. But Carnap's logic otherwise corresponded well to a positive response to the last three questions. In order to construct it, Carnap had to free himself resolutely from "principles of the

[21] For more about paradoxes of this kind, see Rouilhan (2004).

name-relation" (see above subsect. 5.1), with everything that that implied, especially the destruction of the classic idea of identity. Certain logicians also freed themselves of it after him, one might say, by simply dodging the requirements of the classic idea of logical paraphrase (see subsect. 3.3), but none of them had for all that rediscovered or taken up his idea. As for Carnap, not long after his discovery, yielding to Quine's critical injunctions, he himself abandoned his idea. It is not clear whether he was right to do so.[22]

References

Almog, J., Perry, J., and Wettstein, H. (eds.), 1989, *Themes from Kaplan*, Oxford: Oxford University Press.

Carnap, R., 1947, *Meaning and Necessity*, Chicago: Chicago University Press (2nd ed. 1956).

Church, A., 1943a, "Review of Quine 1943," *The Journal of Symbolic Logic*, 8, 45–47.

Church, A., 1943b, "Carnap's Introduction to Semantics," *The Philosophical Review*, 52, 298–305.

Church, A., 1946, "A Formulation of the Logic of Sense and Denotation (abstract)," *The Journal of Symbolic Logic*, 11, 31.

Church, A., 1951a, "A Formulation of the Logic of Sense and Denotation," in Henle, Kallen, and Langer 1951, 3–24.

Church, A., 1951b, "The Need for Abstract Entities in Semantic Analysis," *Contributions to the Analysis of Knowledge, Proceedings of the American Academy of Arts and Sciences*, 80, 100–112.

Church, A., 1956, *Introduction to Mathematical Logic*, vol. I, Princeton, NJ: Princeton University Press.

Church, A., 1973, "Outline of a Revised Logic of Sense and Denotation (Part I)," *Noûs*, 7, 24–33.

Church, A., 1974, "Outline of a Revised Logic of Sense and Denotation (Part II)," *Noûs*, 8, 135–156.

Church, A., 1993, "A Revised Formulation of the Logic of Sense and Denotation, Alternative (1)," *Noûs*, 27, 141–157.

Cresswell (M. J.), 1975, "Hyperintensional Logic," *Studia Logica*, 34, 25–38.

Cresswell (M. J.), 1985, *Structured Meanings: The Semantics of Propositional Attitudes*, Cambridge, MA, and London: MIT Press.

Davidson, D., and Harman, G. (eds.), 1972, *Semantics of Natural Language*, Dordrecht: D. Reidel.

Davis, J. W., Hockney, D. J., and Wilson, W. K. (eds.), 1969, *Philosophical Logic*, Dordrecht: D. Reidel.

Frege, G., 1891, Letter to Husserl of May 24, 1891, in Frege 1976, pp. 94–98.

Frege, G., 1892, "Über Sinn und Bedeutung," *Zeitschrift für Philosophie und philosophische Kritik*, 100, 25–50.

Frege, G., 1892/1895 "Ausführungen über Sinn und Bedeutung" (editor's title; written between 1892 and 1895), in Frege 1969, pp. 128–136.

Frege, G., 1969, *Nachgelassene Schriften* (ed. H. Hermes, F. Kambartel and F. Kaubach), Hamburg: F. Meiner.

[22] For more on Carnap's logic, with taking up again his ideas from a hyperintensional point of view, see Rouilhan (2002).

Frege, G., 1976, *Wissenschaftlicher Briefwechsel* (ed. G. Gabriel, H. Hermes, C. Thiel and A. Veraart), Hamburg: F. Meiner.

Hahn, L. E., and Schilpp, P. A. (eds.), 1986, *The Philosophy of W. V. Quine*, The Library of Living Philosophers, La Salle IL: Open Court.

Henle, P., Kallen, H. M., and Langer S. K. (eds.), 1951, *Structure, Method and Meaning. Essays in Honor of Henry M. Sheffer*, New York: The Liberal Arts Press.

Hintikka, J., 1957, "Modality as Referential Multiplicity," *Ajatus*, 20, 49–64.

Hintikka, J., 1962, *Knowledge and Belief*, Ithaca and London: Cornell University Press.

Hintikka, J., 1969, "Semantics for Propositional Attitudes," in Davis, J. W. Hockney, D. J., and Wilson, W. K. 1969, pp. 21–45.

Hughes, G. E., and Cresswell, M. J., 1996, *A New Introduction to Modal Logic*, London and New York: Routledge.

Kaplan, D., 1968, "Quantifying in," *Synthese*, 19, 178–214.

Kaplan, D., 1975, "How to Russell a Frege-Church," *The Journal of Philosophy*, 72, 716–729.

Kaplan, D., 1977, "Demonstratives: An Essay on the Semantics, Logic, Metaphysics, and Epistemology of Demonstratives and Other Indexicals," in Almog, Perry, and Wettstein 1989, pp. 481–563 (the text dated 1977; see also, in the same vol., "Afterthoughts," pp. 565–614).

Kaplan, D., 1978, "On the Logic of Demonstratives," *Journal of Philosophical Logic*, 8, 81–98.

Kaplan, D., 1986, "Opacity," in Hahn and Schilpp 1986, pp. 229–289.

Kripke, S., 1971, "Identity and Necessity," in Munitz 1971, pp. 135–164.

Kripke, S., 1972, "Naming and Necessity," in Davidson and Harman 1972, pp. 253–355 and 763–769 (reprinted *Naming and Necessity*, Oxford: Blackwell, 1972; rev. and enlarged ed. 1980).

Kripke, S., 1979, "A Puzzle about Belief," in Margalit 1979, pp. 239–283.

Kripke, S., 2011a, "Unrestricted Exportation and Some Morals for the Philosophy of Language," *in* Kripke (2011b), pp. 322–350.

Kripke, S., 2011b, *Philosophical Troubles. Collected Papers, vol. 1*, New York: Oxford University Press.

Lambert, K., 1969, *The Logical Way of Doing Things*, New Haven and London: The MIT Press/Bradford Books.

Lepage, F., Paquette, M., and Rivenc, F. (eds.), 2002, *Carnap aujourd'hui*, Montreal: Bellarmin and Paris: Vrin.

Margalit, A. (ed.), 1979, *Meaning and Use*, Dordrecht: D. Reidel.

Montague, R., 1960, "On the Nature of Certain Philosophical Entities," *The Monist*, 53, 159–194.

Munitz, M. K., 1971, *Identity and Individuation*, New York: New York University Press.

Quine, W. V. O., 1943, "Notes on Existence and Necessity," *The Journal of Philosophy*, 40, 113–127.

Quine, W. V. O., 1947, "The Problem of Interpreting Modal Logic," *The Journal of Symbolic Logic*, 12, 43–48.

Quine, W. V. O., 1956, "Quantifiers and Propositional Attitudes," *The Journal of Philosophy*, 53, 177–187.

Quine, W. V. O., 1977, "Intensions Revisited," *Midwest Studies in Philosophy*, 2, 5–11 (reprinted in *Theories and Things*, Cambridge, MA: Harvard University Press, 1981, 113–123).

Quine, W. V. O., 1986, "Reply to David Kaplan," in Hahn and Schilpp 1986, 290–294

Rouilhan (de), Ph., 2002 "Logiques du contenu," in Lepage, Paquette, and Rivenc 2002, pp. 317–343.

Rouilhan (de), Ph., 2004, "The Basic Problem of the Logic of Meaning (I)," *Revue internationale de Philosophie*, 58, 329–372.

Rouilhan (de), Ph., 2007, "La théorie des modèles et l'architecture des mathématiques" in P. Gochet et Ph. de Rouilhan, *Logique épistémique et philosophie des mathématiques*, Paris: Vuibert, 2007, pp. 39–114.

Rouilhan (de), Ph., 2012, "In Defense of Logical Universalism: Taking Issue with Jean van Heijenoort," *Logica Universalis*, 6, 553–586.

Russell, B., 1903, *The Principles of Mathematics*, London: George Allen & Unwin.

Russell, B., 1905, "On Denoting," *Mind*, 14, 479–493.

Salmon, N., 1986, *Frege's Puzzle*, Cambridge, MA: MIT Press.

Sleigh, R., 1968, "On a Proposed System of Epistemic Logic," *Noûs*, 2, 391–398.

Smullyan, A. F., 1947, Review of Quine 1947, *The Journal of Symbolic Logic*, 12, 139–141.

Smullyan, A. F., 1948, "Modality and Description," *The Journal of Symbolic Logic*, 13, 31–37.

Stalnaker, R. C., and Thomason, R. H., 1968, "Abstraction in First-Order Modal Logic," *Theoria*, 34, 203–207.

Thomason, R. H., 1969, "Modal Logic and Metaphysics," in Lambert 1969, pp. 119–146.

Thomason, R. H., and Stalnaker, R. C., 1968, "Modality and Reference," *Noûs*, 2, 359–372.

Whitehead, A. N., and Russell, B., 1910, *Principia Mathematica*, vol. I, Cambridge: Cambridge University Press (2nd ed. 1925).

10

PHILOSOPHY OF MATHEMATICS

Denis Bonnay (Université Paris Nanterre, IRePh and IHPST) and Jacques Dubucs (CNRS)

THE PHILOSOPHY OF mathematics occupies an exclusive position within philosophy of science. On the one hand, the importance of mathematics in contemporary science is such that, in principle, no philosophical inquiry into science can forgo reflection on the nature of mathematics and mathematical knowledge. So, on the horizon of philosophy of mathematics we find some of the fundamental questions of philosophy of science coming into play, questions such as the possibility of bringing the naturalization of epistemology program or the applicability of mathematics problem to fruition.[1] On the other hand, the methodology of mathematics seems far removed from the general methodology of science. To state it in a more caricatural fashion, the mathematician does not work in a laboratory. The classical issues of general philosophy of science which are applicable to the empirical disciplines such as, for example, the issue of confirmation, the question of the nature of causality or issues regarding theory change and sameness of reference, do not readily make sense in his world. When it comes to tackling the epistemology of mathematics, everything must be explained: what the activity of mathematicians consists of; in what sense it is a theoretical activity; what its objects are; what its methods are; how all of this fits into a global vision of science, including the natural sciences.[2]

[1] Are we obliged to think that there is something to explain, namely the "amazing" success of mathematized science? Or should it instead be said that there is no mystery, that mathematics is simply a set of tools?

[2] We wish to thank Christopher Robertson, who translated the French version of the present chapter and to acknowledge financial support from the Institut de Recherches Philosophiques (Université Paris Nanterre).

As was to be expected, philosophers of mathematics see eye to eye on almost none of this. Some consider that mathematics really is the study of objects which exist independently of us, that mathematical objects exist in just the same way as physical objects do, even if they are objects of a different kind. Others consider this to be nonsense, for them mathematical objects are simply useful fictions, inventions of ours, or they consider that mathematics only describes the most abstract properties of experience. Some consider that mathematical knowledge is *sui generis* knowledge of a purely intellectual nature. Others that it is indeed *sui generis* knowledge but that it relies on a form of intuition. Still others refuse to grant it its own place and don't wish to speak of mathematical knowledge unless it be integrated into the global edifice of science.

Connecting the answer to the ontological problem (what is it that mathematics studies?) to the answer to the epistemological problem (how is mathematical knowledge possible?) will be our underlying theme.[3] In the first section, we use the classical oppositions between empiricist, rationalist and critical approaches to set the stage and pose the question of mathematics' relationship with experience as well as the question of the respective roles of intuition and logical principles in mathematical knowledge. The second section presents a relatively detailed account of two antirealist programs which guarantee the success of forms of particular mathematical intuition against the retraction of the ontological independence of at least some parts of mathematics. Balancing this out, the third section presents arguments in favor of realism. Different forms of realism are discussed in the fourth section, particularly in terms of the status they accord to set theory. Having confronted the epistemological difficulties of various versions of mathematical realism (fifth section), the sixth section will be given over to the naturalist perspectives and to mathematical structuralism.

1. Mathematics—Between Logic and Intuition

1.1 TRUTHS OF REASON OR EMPIRICAL GENERALIZATIONS

Upon his valet's asking him what he believes, Don Juan replies, "I believe that two and two make four, Sganarelle, and that four and four make eight." Sganarelle may well go on to scorn the value of this fine belief; it nonetheless guards, and unanimously so, the character of cardinal belief Don Juan attributes it. Nothing could be more elementary than the proposition that two and two make four, nothing could be more certain than the truth of this proposition. It is remarkable that the difficulty in philosophy of

[3] The classical debates in philosophy of mathematics also deal with a group of specific questions concerning, for example, the nature of infinity, the nature of the continuum, the concept of computation, the notion of random process, or the question of which theory provides the best unified framework for contemporary mathematics. Some of these questions will be approached within the remit of the more general ontological and epistemological questioning we have adopted. Others, despite their intrinsic interest, did not find an appropriate place in this current presentation.

mathematics begins here, with the simplest truths of mathematics. Two and two do make four, but how do we know this?

Let us look at the classical response provided by a rationalist philosopher such as Leibniz. As elementary as the proposition may be, it is nevertheless not an entirely immediate truth; it must be demonstrated. To do this, Leibniz employs the definitions of the numbers (definition 1: 2 is 1 and 1, definition 2: 3 is 2 and 1, definition 3: 4 is 3 and 1, etc.) and a generally valid axiom, the principle of substitution of identicals (this is a valid axiom, according to Leibniz, insofar as it can be reduced to an identity principle). Here is the demonstration:

> "2 and 2 is 2 and 1 and 1 (def. 1)
> 2 and 1 and 1 is 3 and 1 (def. 2)
> 3 and 1 is 4 (def. 3.)
> Therefore (by the Axiom)
> 2 and 2 is 4—which is what was to be demonstrated."
> *New Essays On Human Understanding*, IV, VII, 10

The demonstration relies only on definitions and one axiom.[4] If it is correct, is a truth of reason, which does not depend in any way on experience and can be known a priori. Now, what Leibniz believes to have accomplished in the case of, he must also be capable of accomplishing for all mathematical truths. But problems arise before getting to any attempt at extending the strategy. As Frege (1884, §6) remarks, Leibniz' demonstration is incomplete: it implicitly uses the associativity of addition, allowing to go from the result of applying the definition of 1, namely $2+(1+1)$, to what is needed to apply the definition of 3, namely $(2+1)+1$. For the demonstration to be correct it would suffice to elucidate the function associativity fulfills in it. But to do this, the principle of associativity itself would have to be justified; there is no obvious way for doing this within the Leibnizian framework. What is needed would be reduction to a form of identity principle, and it is not a clear how such a reduction could be achieved.

Since the gap seems difficult to fill, let's try moving away from the rationalist approach to see the answer given by a radical empiricist like Mill. In *A System of Logic*, Mill contests the "simple definition" status of affirmations such as "3 is 2 and 1." The definition contains the assertion of a fact, namely that any totality composed of three elements can be divided into a totality of two elements and one other element: "The fact asserted in the definition of a number is a physical fact. Each of the numbers two, three, four, etc., denotes physical phenomena." (III, XXIV, 5) Mathematical notions are empirical notions, ("Two, for instance, denotes all pairs of things") and mathematical propositions are empirical propositions, though very general and just as abstract. From this point, a Millian response can be given to Leibniz' problem by proposing that the principle of associativity is an empirical principle, admittedly very general,

[4] The axiom that is used is the substitution of identicals.

but empirical all the same. The principle of associativity holds that when one aggregate can be divided into two aggregates—call the first one *a*—and the second of these aggregates is again divisible into two new aggregates, *b* and *c*, it is still possible to divide the initial aggregate into two aggregates, the first of which divides into two aggregates *a* and *b*, the second of which being the aggregate *c*. Radical empiricism, disposed to grounding mathematical truths on experience, doesn't meet with the rationalist's problem of having to explain, for every mathematical axiom, what makes it a truth of reasoning accessible independently of all experience. Nevertheless, radical empiricism meets with other problems. By reducing mathematical truths to empirical truths, it doesn't account for the apparent modal and epistemic properties of mathematical truths. Mathematical truths appear to be necessary and knowable independently of experience, in contrast to contingent empirical truths. This appearance may be illusory, but then the illusion too will need explanation. Furthermore, the distance between mathematical notions and experience renders the empiricist reduction difficult: as Frege would object, if an empirical denotation can be attributed to two, by speaking of aggregates made up of two things, what denotation will be attributed to zero?

1.2 A PURIFIED SENSIBLE INTUITION AT THE BASIS OF MATHEMATICAL JUDGMENTS?

Naively, it may be tempting to think that a good philosophy of mathematics should place itself somewhere between these extreme positions, embodied here for our purposes by Leibniz and Mill. On the one hand, there really does seem to be something like a mathematical experience at the core of the mathematician's activity that, furthermore, should be able to ground the validity of strictly mathematical principles like the law of associativity. On the other hand, this experience could not be of exactly the same nature as the experience which normally underpins our empirical generalizations; "2 +2 = 4" cannot be placed on the same level as "trees lose their leaves in the fall."

The temptation to seek a middle path does not mean this would be an easy task, nor does it mean such a path would lead anywhere. Kant's philosophy of mathematics is an attempt to explore this path, so let us see where it may lead us. Kant sought to identify a role for the intuition in mathematics, without its making mathematical truths depend on empirical content. In now famous passages from his *Critique of Pure Reason* and *Prolegomena*, Kant begins by advancing that mathematical propositions cannot be viewed as analytical propositions: there is something more in the concept of four than just the concept of the sum of two and two. For Kant, if we know that two and two makes four, it is because we leave the simple concept of the sum of two and two and turn to our intuition, by counting on our fingers, for example.

Again, the problem is understanding how we can base ourselves on an apparently empirical intuition to establish knowledge which is itself not empirical. In Kant's

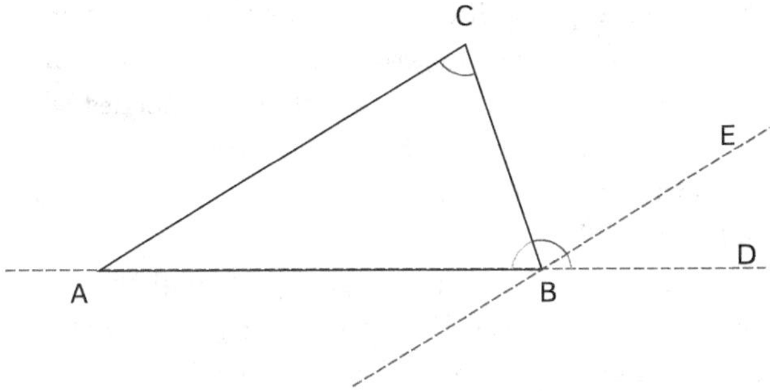

FIGURE 1 The sum of the angles of a triangle is equal to a straight angle.
Source : Wikipedia, License Creative Commons Attribution ShareAlike 3.0

terms, the problem is in understanding the possibility of synthetic a priori judgments and, in this instance, the possibility of synthetic a priori judgments grounded in the intuition. Kant's solution is to suppose the existence of a pure intuition, the pure intuition of the forms of sensibility. This form of sensibility idea relies on the distinction of two aspects of phenomena: their form, which corresponds to the manner in which phenomena are arranged relative to each other, and their matter, which corresponds to sensation. The forms of sensibility, namely time and space, are given a priori: they do not depend on the experience, instead they are the basis that renders experience possible.

Arithmetic relies on the pure intuition of time whereas geometry relies on the pure intuition of space. Though the link between arithmetic and geometry may seem to make sense thanks only to the specificities of Kant's elaboration of the relationships between consciousness and time, the link between geometry and space is clearly less problematic, and the Kantian philosophy of geometry benefits from a certain fidelity to geometers actual practices. As historians of mathematics have remarked, Euclid's postulates indicate construction possibilities: a circle can always be drawn out (empirically, with the help of a compass), a straight line can always be extended (empirically, with the help of a ruler). Correlatively, Euclidean geometric demonstrations rely on the realization of auxiliary constructions. For example, to show that the sum of the angles of a triangle is equal to a straight angle, we start with any triangle and draw a line parallel to one of the sides which passes through the point opposite this side. The demonstration then relies on reasoning on the initial figure and the auxiliary constructions carried out. In this instance, the reasoning will consist of using the properties of the angles formed by the newly drawn straight line and the lines extending the other two sides of the triangle (in the order of the demonstrations laid out in the *Elements*, these properties have already been demonstrated).

Mathematical intuition is just the intuition at play in these constructions, essential for successfully carrying out demonstrations. However, the contingent characteristics of the constructs are not and must not be brought into the demonstration, in which case a necessary geometric proposition would not have been demonstrated. Kant's idea is that the utilization of these constructions in proofs is legitimate because the only things retained in the demonstration are those properties which are based on what can be done in space, and not the empirical properties of the figures: only the pure part of the empirical intuition is relevant to the intuition which forms the basis for mathematical reasoning.[5]

The difficulties met by the Kantian philosophy of mathematics are matched by the force of its initial appeal. These difficulties partly result from the mysteries of the transcendental method: what are the forms of sensibility, why are they a priori, and what relationships do they maintain with the subject's empirical constitution? Supposing that psychology is capable of propping up the Kantian speculations,[6] how could the revelation of the links between awareness of time and numerical cognition possibly constitute an argument in favor of the a priori nature of mathematical knowledge?

The difficulties with a Kantian stance do not end there. By making his justification of mathematical truths depend on the structures of the experience, Kant preemptively resolves the problem of the application of mathematics: it is no wonder that the laws of arithmetic and geometry are applicable to phenomena located in time and space, since arithmetic and geometry just relate to the possible arrangements of phenomena in time and space. But Kant proves too much. He proves that Euclidean geometry applies to physical space. The later development of geometry would lead to the elaboration of different geometric systems, and the development of physics would lead to the choice of geometries other than the Euclidean for describing physical space. Explaining geometry's applicability can therefore not be equal to explaining that a given pure geometry is at the same time necessarily the geometry of physical space. It can only be equal to explaining that pure geometries provide physicists with the means needed to describe the geometry of physical space. In guaranteeing mathematical intuition against an intuition of the structures of the experience, in this instance the forms of sensibility, Kant connects pure mathematical theories to their applications more strongly than would seem wise to do so.

---

5 When "I construct a triangle, by the presentation of the object which corresponds to this conception [. . .] upon paper—in empirical intuition," I do so "without borrowing the type of that figure from any experience" and "the individual figure [. . .] is empirical; but it serves, notwithstanding, to indicate the conception, even in its universality, because in this empirical intuition we keep our eye merely on the act of the construction of the conception, and pay no attention to the various modes of determining it, for example, its size, the length of its sides, the size of its angles, these not in the least affecting the essential character of the conception" (*Critique of Pure Reason*, Transcendental Doctrine of Method, I,1).

6 We speak of "speculation" in the sense that the transcendental method, consisting of going back to the conditions of possibility of experience using reasoning, which is itself purely a priori, has not undergone the scrutiny of psychology. This objection to the method is certainly not meant as a criticism of Kant, psychology having not yet been formed in his day.

1.3 THE PURELY LOGICAL WAY

The aporia of Kant's strategy invite us to again ask what role intuition plays in mathematical demonstrations in general, notably in arithmetic. It is clear that intuition plays a pedagogical role. A child is taught that two and two makes four by using small batons. But is it really certain that this pedagogical role is doubled by an epistemic one? Do the justifications of arithmetic truths have anything to do with what is provided to us by some form of sensible intuition? One reason to respond negatively to this is the absolute generality of arithmetic. As Frege put it a century later when it was his turn to take on the problem of the intuition's place in arithmetical judgments, "The truths of arithmetic govern all that is numerable. This is the widest domain of all; for to it belongs not only the actual, not only the intuitable, but everything thinkable. Should not the laws of number, then, be connected very intimately with the laws of thought?" (1884, §14). In other words, if the validity of arithmetic is as general as that of logic, are we not obliged to envisage the possibility that arithmetic is just the result of the general laws of thought? That arithmetic be nothing other than pure logic?

How can we know whether intuition does or does not take part in arithmetical proofs? This can only be known if we manage to conduct a complete analysis of proofs in arithmetic. A framework must be developed in which a proof will appear as completed only if every axiom and every rule used has been made explicit, after which it must be determined whether the axioms and rules used for proofs in arithmetic are based on the intuition or if they are of a purely logical nature. In the first instance, this would require coming back to a Kantian or empiricist style solution. In the second instance, it would have been shown that arithmetic truths are a priori analytic, assuming that logical truths also are.

The program we have just outlined is Frege's. In this way, he thought he could refute the very starting point of the Kantian philosophy of mathematics, namely the synthetic nature of arithmetic truths. To see his program succeed, Frege attempted, with the highest of formal rigor, to explicitly lay out the concepts and laws of logic. In order to satisfy this necessity for rigor, the new logic was developed for an artificial language whose notation is designed to leave no room for ambiguity or uncertainty. We can now clarify the objectives pursued. It is first a question of defining the arithmetic notions with the sole aid of logical notions, and of then showing that the laws and rules used in arithmetic appear, via the translation allowed by the given definitions, as theorems of logic, that is, statements which can be demonstrated using only laws and rules of logic. Frege's philosophy of mathematics—known as logicism, since the central thesis is that arithmetic is nothing other than logic—thus leads to a task of a mathematical nature: effectively showing that arithmetic is reducible to logic. Though Frege's logicist position may be quite similar to that of Leibniz, this second dimension provides Frege with his originality: while Leibniz contented himself to suggesting, for example, that arithmetic truths were truths of reason, Frege's intention was to demonstrate that arithmetic truths are logical truths by showing that all the laws and rules used in arithmetic are reducible to purely logical laws and rules.[7]

An example of a typically arithmetic rule is the rule of induction: to show that all the whole numbers possess a certain property P (in logical notation, $\forall nPn$), it just needs to be shown that 0 possesses the property P, and that if any whole number n possesses the property P, then its successor $n+1$ also possesses this property (in logical notation, $P0 \wedge \forall n(Pn \rightarrow P(n+1))$. The content of this rule is more apparent when contrasted with the logical rule of universal generalization, stating that to show $\forall xPx$, one must first show Px without making assumptions on x. The universal generalization rule is, *prima facie* at least, a purely logical rule, it makes no assumption as to the kind of object involved and is valid without restriction. But it can be very difficult to show Px without some kind of assumption about x. The induction rule is easier to use: it suffices to show that $P0$ is definitely the case and that it is possible to pass from Pn to $P(n+1)$. Nevertheless, this rule is not, again *prima facie*, logical: there is a temptation to say that if this rule is valid then it is because of the structure of whole numbers, because every whole number is either 0 or else can be attained on the basis of 0 by executing the operation + 1 a sufficient number of times. Frege's stroke of genius consists of giving a definition of the whole numbers which makes the principle of induction appear as a logical theorem and not as a principle relying on other bases than logic (e.g. on our intuition of the structure of whole numbers).

What became of Frege's project? In 1902, while the second tome of *Grundgesetze*, intended to complete the logicist reduction, was on its way to print, Frege received a letter from Russell informing him of an inconsistency in the logical system he had developed. The now infamous law V was to blame: this principle appeared to be an unproblematic logical law dealing with the relationships between concepts and concept extensions (the extensions of two concepts F and G are identical if and only if every object that falls under F falls under G, and vice-versa). It allowed Frege to derive Hume's principle, stating that the number of F is equal to the number of G if and only if every object that falls under the concept F falls under the concept G and vice versa, with Hume's principle then being called on to derive arithmetic on the basis of the rest of the Fregean system.[8] But law V leads to worse still. It implies an unrestricted comprehension principle: for any formula $\varphi(x)$, there exists a y such that, for every x, $x \in y$ if and only if $\varphi(x)$. Let us now consider the formula $x \notin x$. Using the comprehension

[7] Frege lays the foundation stones for the new logic in his *Begriffschrift* (1879, Eng. trans. *Concept Script*, 1967). The logicist program is laid out in detail in *Grundlagen der Arithmetik* (1884, Eng. trans. *The Foundations of Arithmetic*, 1950), along with an idea of how it would be realized. The full realization of the reduction of arithmetic to logic is the object of *Grundgesetze der Arithmetik* (vol. 1, 1893 and vol. 2, 1903).

[8] Things work out better if we forgo law V and instead take Hume's principle as our fundamental principle. Boolos (1986) showed that from Frege's work one could extract what is called Frege's theorem, a derivation of Peano's axioms for arithmetic in second-order logic and using only Hume's principle. This result marks the beginning of neo-logicism, one possibility for its disciples being to advance the notion that Hume's principle is in fact a logical principle.

principle, we have a y such that $x \in y$ if and only if $x \notin x$. But then we have $y \in y$ if and only if $y \notin y$. Contradiction.

This is the failure suffered by Frege's logical system. Russell's paradox shows that at least some of the principles acknowledged to be logical by Frege must in fact be abandoned. It doesn't show that the logicist project is, as such, doomed to failure. Thus, following the discovery of the paradox, logicians and mathematicians developed new systems, designed to keep the paradox at bay, while allowing to account for large areas of mathematics. This trend explains the development of both Russell's type theory, which aims at excluding the paradox by authorizing consideration of formulas of type $x \in y$ only if x and y are of different levels (which is not the case when we write $x \notin x$), and Zermelo's set theory which imposes restrictions on the comprehension principle. These systems succeed in providing unified frames for the reconstruction of mathematics although, beyond that unification, the epistemological value of such a reconstruction is no longer clear. Some of the axioms in these theories, notably the infinity axiom included in both set theory and type theory, do not seem to be purely logical axioms (it is difficult to see in what way the existence of an infinity of objects could be a purely logical law).

The Kantian attempt to break the opposition between rationalism and empiricism, the logicist challenge to the role Kant bestows on the intuition in arithmetic proofs and, finally, the failure of classical logicism, constitute three steps which would determine the form that philosophy of mathematics took on in the 20th century. First, the (contested) question of mathematical intuition appeared as central, equally a matter of backing up mathematical practice which claims this intuition and of the difficulty in theorizing the formal intuition which would be involved. Second, and inversely, logic's role (also contested) in the formulation and establishment of theses in philosophy of mathematics was imposed. In a word, the foundational crisis which followed the discovery of Russell's paradox determined the foundationalist orientation of the programs that were to follow.[9]

2. Finitism and Intuitionism, Two Anti-realist Programs

2.1 THE CONSISTENCY QUESTION

Russell's paradox shows that intuitively reasonable principles are capable of leading to contradictions, that is, situations in which one can prove something and also prove its contrary, a fact which is obviously disturbing since, from an absurdity like *A and non-A* anything can be demonstrated (the *ex falso quod libet* principle). Thus it would be desirable to develop mathematical theories of whose non-contradictory nature we can be *certain*, that is, theories for which it can be proven that they cannot be used to simultaneously prove both a proposition and its negation. Here, the notion of proof

[9] By "foundationalist orientation" we mean the fact of trying to guarantee the solidity of the mathematical edifice, particularly its coherence.

extends its domain of application: what will be sought is to prove not only mathematical propositions of type $7+5=12$ (or other more complex ones . . .), but also "metamathematical" propositions of type "In the arithmetic theory T it cannot be proved that $0=1$." How can this be done? To prove that a proposition can be proven, proving it is all that is required, but to prove that it *cannot* be proven, it is not sufficient to simply not prove it!

What we have here is a variety of mathematical difficulty attested since ancient times, for instance with the "Delian" problem (how can the cubic temple to Apollo on the island of Delos be doubled?), with the trisection of an angle, or with the quadrature of a circle. None of these problems find their solution with just a ruler and compass, but the *proof* of their impossibility demanded a good two thousand years, the time it took for us to algebraically characterize the set of points constructible with a ruler and a compass (Wantzel, 1837). Another impossibility of considerable influence concerns the history of non-Euclidean geometries: Proving Euclid's fifth postulate ("there is at most one line that can be drawn parallel to another given one through an external point"). In the 19th century, this impossibility had been established by showing that a certain interpretation of the primitive terms of geometry transformed the theorems of hyperbolic geometry into theorems of *Euclidean* geometry: taking an open disc in the Euclidean plane, if "straight lines" are taken to mean the chords of this disc, then it is seen that the axioms of hyperbolic geometry can be demonstrated in Euclidean plane geometry (Beltrami, 1868). Supposing that hyperbolic geometry were inconsistent, that it proved both A and *non-A* for some A, it would be the case that the Beltrami style reinterpretation of the two propositions gave two contradictory theorems of Euclidean geometry. In short, if Euclidean geometry is consistent then hyperbolic geometry is too, putting an end to any attempts at proving the fifth postulate from the others. To obtain this kind of result, the usual or intuitive meaning of the primitive terms evidently needs to be set aside and *all* interpretations of these terms must be deemed legitimate as long as they satisfy the relevant axioms. In a word, a straight line is nothing other than an object which satisfies the axioms of geometry. Or rather, since these axioms simultaneously and "holistically" call on the notions of straight lines, points, etc., a legitimate interpretation of the set < straight lines, points, and so forth > is nothing other than a collective association of referents which satisfy the intended axioms. This is what Hilbert was referring to in his famous quip: "One must be able to say at all times—instead of points, straight lines, and planes—tables, chairs, and beer mugs."

Here we have the resolution, exclusively benefiting the *form* of mathematical language, of a tension between the *referential* and the *inferential* aspects of mathematics, a tension that M. Pasch (1882, p. 98) had formulated in the following way some years earlier: "If geometry is to be truly deductive, the process of inference must be independent in all its parts from the meaning of the geometric concepts, just as it must be independent from the diagrams. [. . .] In the course of deduction it is both permitted and useful to bear in mind the meaning of the geometric concepts that occur in it, but *it is not at all necessary*. Indeed, when it actually becomes necessary, this shows that there is a gap in the proof, and - if the gap cannot be eliminated by modifying the argument - that the premises are too weak to support it."

In other words, the "private" meaning the mathematician associates with the use of mathematical terms should have no decisive impact on the proofs: a proof is only truly a proof if it is capable of winning the support of someone who associates an entirely different meaning to these terms. The notion of proof must therefore be, in the modern parlance, "decidable": the question of whether a sequence of writing is or is not a proof is to be compared with the question of whether a travel-card is or is not valid, meaning that we must be able to arrive at an answer *mechanically*, in real time and by elementary procedures that require no hermeneutic ability relative to the meaning of the words contained therein. As A. Church (1956, p. 53 sq.) well remarked, if it weren't a simple matter of *routine* when it comes to deciding whether or not a sequence of formulas does or does not obey the rules of demonstration, then it would be the control of all assertions in the mathematical community which would become an impossible task, since the possibility would remain open at all moments and to anyone to declare themselves unconvinced by a proof.

And so we arrive at a "formalist" position, incorporating in itself several variants. In a minimal version, formalism consists of *suspending* the "expected" meaning, so to speak, of the symbols when verifying proofs; in a strong sense, it boils down to the position that mathematics *are*, in reality, a formal game involving formulas that are devoid of meaning, a game directly comparable to chess, for example (transitions from formula to formula that obey the rules of inference being equatable to "legal" moves on the chess board).

Naturally, it is possible to be formalist with respect to certain parts of mathematics but not others, if one imagines an opposition between "serious" mathematics, having intuitive content, and more speculative, risky mathematics where the intuition loses all capacity for control (in this latter domain is intended, for example, Cantor's theory of transfinite cardinals in which it is doubtful any intuitive control could operate). Hence, it is the partial (yet *strong* in the sense just given) kind of formalism that Hilbert defended in reserving the accolade of mathematics *having content* exclusively for elementary arithmetic. In a formalist perspective of this sort, it is not a matter of expecting of "transcendent" mathematics that they be *true* in a substantial sense, that is, that their statements adequately describe the properties of some domain of ideal *sui generis* objects. What is demanded is simply that they be consistent, that they lead to no contradiction.

To establish this consistency, the first thing to do is to *formalize* the theories, write them in a strictly defined formal language and use this language to specify the axioms and rules of inference. For a simple example of the consistency proofs we may rightfully wish for, let us consider the theory T whose language has as formulas all finite sequences of a and of b, which contains just one axiom, ab, and just one rule: from XY, make XYY (X and Y being any formula). In T, ab (the axiom!) and abb are proven (by applying the rule to ab), but ba cannot be proven. The proof of this fact is easily arrived at by recurrence or by "induction" (discussed previously) on the length of the proofs: the only axiom begins with *a* and the only rule of inference does not affect the start of the formulas to which it is applied. Therefore all provable formulas will begin with *a*, in such a way that *ba* can have no proof in *T*. In short, *T* is consistent because there is at least one formula, namely *ba*, which cannot be proven in T!

Let the difference be noted between this proof of consistency and the proof of consistency, mentioned earlier, for hyperbolic geometry (*HG*). The latter consisted of establishing that, if a formula *A* is provable in *HG*, then the proposition $A^{(B)}$, resulting from it in Beltrami's interpretation, is a theorem in Euclidean geometry (EG). So this was simply a *relative* proof of consistency, showing that if *HG* is contradictory, then EG also is. The proof of consistency for *T*, though, is not a semantic proof (no interpretation is sought for *T*'s language) and it seems to provide a result of "absolute" consistency, supposing the consistency of no other background theory. Yet this is clearly the kind of result expected of any attempt at founding mathematics: at risk of falling into an infinite regression it cannot, in this context, suffice to establish the consistency of one theory on the condition that another theory is also consistent, and so the consistency of some "ultimate" theory would have to be proven in some absolute, non-semantic way.

For the "ultimate" theory, it is of course arithmetic which claims the title. Geometry, *via* the system of "Cartesian" coordinates established since the 17th century, can be seen as a theory relative to the number systems which are used to describe the points (in general the system of real numbers is used). As for them, the real numbers, towards the end of the 19th century, had been defined in various ways as sets of rational numbers, the rational numbers themselves being obviously constructible as pairs of whole numbers. In this context, often defined as "Neo-Pythagorean" in reference to the ancient Pythagorean philosophy according to which whole numbers were the ultimate elements of the world's "furniture," it is evidently arithmetic which must have a proof of "absolute" consistency at its foundation. This is precisely why Hilbert presented the finalization of a proof of consistency for arithmetic as the second of the 23 problems he submitted for the consideration of mathematicians at the International Congress held at the Sorbonne in August 1900.

Can such a proof be obtained for this theory, akin to the procedure applied earlier to theory *T*? Hilbert attempted it in 1904 at the International Congress of Mathematics in Heidelberg by writing the axioms of arithmetic in such a way that all the formulas which it can demonstrate possess a certain morphological property, "homogeneity," that their negations do not possess. But does this really give an "absolute" proof of the consistency of arithmetic? As Poincaré remarked in his 1906 article "Mathematics and logic," the proof Hilbert proposed is run through with circularity, since the principle of recurrence, fundamental to arithmetic, is used in establishing the consistency of arithmetic: indeed, we reason by recurrence on the length of demonstrations to establish that the property of homogeneity, verified by the axioms (whose proof is of length 1) and hereditary by application of the rules of inference (used to go from a proof of length *n* to a proof of length $n+1$), is thus satisfied by all the theorems, whatever the length of their proof.

Hilbert acknowledged this by distinguishing, around 1920, between two sectors of mathematics, finitist and non-finitist, the consistency of the finitist part of mathematics being, so to speak, self-evident.

2.2. FINITISM

Stating what is obvious is not as easy as one might think. It is tempting to escape this difficulty by saying that nothing is obvious, that mathematics is a discipline completely devoid of presuppositions and that it must be reconstructed from a basis that is equally devoid of content. For example, by saying that mathematics is the offspring of logic, which, as we have seen, was Frege's opinion. But the problem with this supposedly content free logic is twofold.

On the one hand, a logic capable of spawning mathematics must already contain some amount of mathematics. As Hilbert said during the Congress at Heidelberg (1905, p 176), "Yet if we observe attentively, we realize that in the traditional treatment of the laws of logic certain fundamental notions from arithmetic are already used, such as the notion of set and, to a certain extent, that of number as well. Thus we find ourselves on the horns of a dilemma, and so, in order to avoid paradoxes, one must simultaneously develop both the laws of logic and of arithmetic to some extent."

On the other hand, and more fundamentally, logical reasoning itself presupposes an intuitive content:

> As a condition for the use of logical inferences and the performance of logical operations something must already be given in our faculty of representation, certain extra-logical concrete objects that are intuitively present as immediate experience prior to all thought. If logical inference is to be reliable, it must be possible to survey these objects completely in all their parts, and the fact that they occur, that they differ from one another, and that they follow each other, or are concatenated, is immediately given intuitively, together with the objects, as something that neither can be reduced to anything else nor requires reduction. (Hilbert, 1925, p. 376)

The objects in question are the logico-mathematical symbols, among which should first be presented the whole numbers conceived as simple sequences of bars: |, | |, | | |, etc. By comparing the objects | | | and | | it can be observed that in the first object the bar | appears one more time than in the second, an observation which can be written as "3>2". Similarly, the observation that | | | concatenated with | | gives rise to the same object as | | concatenated with | | | is expressed by $3+2=2+3$. Assertions of this type are thus not relative to abstract objects but to concrete sequences of | upon which diverse operations can be carried out such as the concatenation or addition or subtraction of an element. Hilbert's idea is that these assertions are so elementary as to be immediately justified. Mastery of them is presupposed by all acts of communication, to the point that anyone wishing to contest this principle would nevertheless have to have recourse to it in formulating their objection, be this only because of needing to identify as alike occurrences of certain words in their own as well as their adversary's discourse. In short it is a matter of recognizing that a certain type of elementary, combinatory reasoning, relative to material entities, is presupposed by mathematics themselves and, more fundamentally, by every action of thought or rational communication.

In this respect, mathematical formulas follow the same system as numbers and must equally be considered as concrete objects potentially made up of parts (symbols) capable of appearing in various places and of being re-identified as such. The competencies required by their syntactic or formal study are similarly presupposed by all mathematical activity. This is exactly what Bourbaki would assert years later:

> We do not propose to enter into a discussion of the psychological or metaphysical problems which underlie the use of ordinary language in such circumstances (for example, the possibility of recognizing that a letter of the alphabet is "the same" in two different places on the page, etc.). Moreover, it is scarcely possible to undertake such a description without making use of numeration. It is objected by some that the use of numbers in such a context is suspect, even tantamount to *petitio principii*. It is clear, however, that in fact we are using numbers merely as marks (and that we could for that matter replace them by other signs, such as colours or letters) and that we are not making use of any mathematical reasoning when we number the signs which occur in an explicitly written formula. We shall not enter into the question of teaching the principles of a formalized language to beings whose intellectual development has not reached the stage of being able to read, write and count. (1956, E.1.9–10)

In a word, with this being an essential characteristic of formalism: "in the beginning was the sign (*am Anfang ist das Zeichen*)" (Hilbert, 1922, p. 163). Naturally, discussion may arise over the genuinely "concrete" character of numbers or sequences of symbols conceived in this way. As A. Müller (1923) essentially says, if inequations are arbitrated based solely on sense perception, then we should indeed affirm that 3 is smaller than 2 in light of the comparison between ||| and | |. Also, numbers should instead be constructed as "types" of concrete bar sequences, that is, as entities which are more abstract than those advanced by Hilbert, or else as equivalence classes of equiform written symbols, thus independent of the vagaries and inessential differences with which their empirical realization is inevitably marked. Nevertheless, the essential is for the numerical symbols to not refer to ideal objects, and that the mathematical formulas not be the expression of thoughts: rather, the two constitute the primary material on which content-filled mental activity applies itself.

What then is the extension of this part of mathematics that neither needs to be nor indeed could be "founded"? Among its ranks can be counted all affirmations of type $7+5=12$, Boolean combinations (conjunctions, disjunctions, and negations) of these affirmations relative to determined numbers, but also their *generalizations by means of variables*, it being understood that a statement such as $\boldsymbol{a}+\boldsymbol{b}=\boldsymbol{b}+\boldsymbol{a}$ must simply be understood as a schema or prototype for affirmations of the same form in which ***a*** and ***b*** would have been replaced by determined numbers, with the justification of such general statements being reduced to the ability to justify, by intuitive, combinatory reasoning, any of their particular numerical cases.

As Hilbert and Bernays (1934) put it, finitism generalizes basic operations (concatenation and deletion of symbols) to operations which can be defined from them by "recursion." Thus, supposing that the two functions f (one variable) and g (three variables) are accepted from the finitist point of view, then finitism will accept the two variable function h defined as follows (this is the schema of primitive recursion):

$$h(0,m)=f(m)$$
$$h(n+1,m)=g(n,m,h(n,m)),$$

For example, if f is the constant map equal to 0 defined by $f(m)=0$, and if $g(n,m,k)=m+k$, then the new function h introduced by this is such that $h(0,m)=0$, $h(1,m)=g(0,m,h(0,m))=0+h(0,m)=m$ and, generally speaking, $h(n,m)=nm$ (this establishes that multiplication is a finitist operation). In the same way, a property φ will be considered finitist if its characteristic function is finitist in the sense just given. Thus, "being a prime number" is a finitist concept, since it is easy to show that the operation which associates | to any prime number and || to any number that is not prime, is a finitist operation.

In short, finitist mathematics contains all statements of type $\forall x_1 \ldots \forall x_n \varphi(x_1,\ldots,x_n)$, where the variables $x_1,\ldots x_n$ cover the domain of whole numbers constructed as shown earlier, or indeed any other domain of quasi-concrete entities like the formulas of a formalized language, and where φ is a property of such entities whose satisfaction or failing can be verified in each specific case by a simple mechanical and combinatory reasoning. "14 is an even number," "addition is an associative operation" $(\forall x \forall y \forall z[x+(y+z)=(x+y)+z])$, "all well-formed formulas of propositional logic contain an equal number of opening and closing parentheses" are just some such finitist statements, but also included are numerous mathematical propositions that are in no way trivial, such as "Fermat's last theorem" $\forall x \forall y \forall z \forall n[n>2 \rightarrow x^n+y^n \neq z^n]$. In the interpretation put forward by W.W. Tait (1981), these characteristics are a recommendation to consider the primitive recursive arithmetic (*PRA*), the system developed by Skolem (1923), as an appropriate formalization of the finitist fragment of arithmetic (this is also, essentially, "Language I" as defined by Carnap, 1937).

Finally, it should be noted (this feature, as the following sub-section will show, is of strategic importance to Hilbert's program) that the fundamental "meta-mathematical" notions are themselves of a finitist nature, which is consistent with the idea of founding mathematics with the help of supposedly incontestable concepts and methods. On the one hand, the basic morphological properties (the property of being a well-formed formula of some formal system, for example) are quite clearly finitist, in the sense given previously. On the other hand, and more essentially, "syntactic" properties also are. First among these is the notion of proof in a formal system: knowing whether a given sequence of formulas $\sigma_1,\ldots\sigma_n$ is or is not a proof can be judged exclusively by use of combinatory considerations. It is necessary only to verify, for each of these formulas, that it is an axiom of the system in question or that it results from formulas preceding it in the list $\sigma_1,\ldots\sigma_n$ in conformity with the system's rules of inference. In a word, "being a proof" is, for the sequence of formulas, a property of the same standing as "being prime" for numbers. On top of this, the very notion of consistency is finitist: to say that the theory *T* is consistent is to affirm that, in *T*, no proof contains "$0 = 1$" as a final formula, which, in light of what has just been seen, is obviously a finitist assertion.

Must a specific logic be constructed for these finitist mathematics? Hilbert always asserted that this would be pointless, differing in this from followers of the "intuitionistic" variety of anti-realism discussed in the following section. The reason it would be pointless has to do with the characteristic *instability* of finitist statements. The conjunction of two finitist statements makes another one, though this is not the case for their negation. Of course, the negation of *atomic* finitist statements (example: "14 is even") remains finitist, but not the negation of finitist *generalities*, since that negation is an existential statement whose justification may exceed the scope of combinatory reasoning capable of being brought to completion within a finite amount of time. Much more, a statement which is finitist may imply another which is not, as is the case with the following statements:

$$\text{(A)}\quad \forall p[p \text{ is a prime number} \rightarrow \exists p'(p < p' \leq p!+1 \text{ and } p' \text{ is a prime number})]$$

and

$$\text{(B)}\quad \forall p[p \text{ is a prime number} \rightarrow \exists p'(p < p' \text{ and } p' \text{ is a prime number})]$$

(A), the famous Euclidean theorem on infinite prime numbers, is a finitist statement: having proposed a prime number p, its justification consists in successively testing all the whole numbers n greater than p and less than $p!+1$ *until a prime number is found among them* (the important point here is that the restricted existential quantification is just a shortened way of writing out a long disjunction). In contrast, (B), while being implied by (A), is not a finitist statement, since the task of justifying it is not limited and its realization could carry on without limit should the statement turn out to be false.

This is why Hilbert proposes adding to the finitist statements (also called "real" statements) the statements which are not finitist (the "ideal" statements) in order to obtain a set of statements the laws of classical logic would leave stable. The consequences or negations of finitist statements could then be not finitist; however, they would still obviously be contained in the domain encompassing both the finitist and the ideal statements. The method which introduces ideal statements, and thus gives them a manner of legitimacy, is therefore absolutely comparable to the method in projective geometry which is behind the introduction of "ideal points" next to the normal points of the Euclidean plane. In projective geometry this is essentially a matter of expanding the generality and continuity of the laws. A statement like "any pair of distinct straight lines intersect in at most one point" is then no longer subject to any exceptions since the apparent anomaly presented by parallel lines is fixed by stipulating that two such lines will intersect "at infinity." Similarly, "imaginary numbers" are introduced in algebra to deal with exceptions to the principle stating that an equation of *n* degrees possesses exactly *n* roots (in this way, the body of real numbers is extended into a body of numbers that is "algebraically closed"). This comparison with the ideal elements of algebra or of geometry provides the key to the Hilbertian perspective on statements from the non-finitist part of mathematics. It would not be a matter of "giving them meaning" (as Hilbert wrote, "ideal propositions have no meaning in themselves," 1925, p. 216) or of considering them as describing a domain of *sui generis* abstract entities, but of introducing them purely by reason of usefulness or simplicity. In short, ideal statements are introduced for *instrumental* reasons: the admission of these ideal statements is the means Hilbert (1925) found for defining a mathematical zone that was constructive, incontestable and beyond all doubt (the finitist zone) without however having to give up the power of classical logic, which would be, as he put it, like depriving a boxer of his gloves or an astronomer of his telescope.

But it is not enough that the concepts of intersections at infinity, imaginary numbers, or ideal statements prove themselves useful. Their innocuousness must also be established. We must be assured that their admission will lead to no contradiction: "For there is a condition, a single but absolutely necessary one, to which the use of the method of ideal elements is subject, and that is *the proof of consistency*; or, extension by the addition of ideals is legitimate only if no contradiction is thereby brought about in the old, narrower domain." (Hilbert, 1925, p. 218) Taking into account the finitist character of the notion of consistency and of the thesis, long claimed by Hilbert to be self-evident, which states that we should be able to establish true finitist assertions by means of finitist methods, this allows a glimpse into one of the fundamental forms of Hilbert's foundational program: to give a finitist proof for the consistency of arithmetic.

2.3 CONSERVATIVITY AND CONSISTENCY

Providing a base for mathematics, we have seen, can signify the delineation of an incontestable, elementary, self-basing stratum on which the consistency of all mathematics could be established.

Nevertheless, there is another acceptation of the foundational enterprise, known about for a very long time, which rests on the ideal of *epistemic stability*. In short, if a certain stratum of the mathematical edifice is to be held as fundamental, the idea is that each and every problem concerning it be resolvable using only concepts and methods belonging to *that* stratum. It would indeed be unusual to qualify a domain as fundamental if its properties could only be established by means of extrinsic considerations or by introducing objects or properties from another level.

Historically, the first incarnation of the epistemic stability idea is probably the principle of "purity of methods" present in the mathematics of ancient Greece and the classical period. According to this methodological ideal, there is a natural "well-founded" stratification (without a decreasing infinite chain, that is, possessing a first layer preceded by no other) within which mathematical entities are arranged into ranks of increasing complexity. This ideal also states that justifications of propositions involving concepts of a given level should contain no concepts from any higher level. In other words, the principle of purity limits the array of notions that can be used in a proof. A mathematical proof is not merely an arrangement of arguments capable of upholding the rational conviction of its conclusion's truth: it must not just *reach for the nearest tool*, but rather it should only call on notions of a kind *befitting* its conclusion. As far as it seems, it must at least mention those notions mentioned in the statement of the proved proposition, but it should contain only these, or else notions of a related kind. The mathematical tradition is almost unanimously in agreement in denouncing proofs which infringe this precept by employing notions which are uselessly high up in the hierarchy.

Thus we have Pappus who condemned Archimedes, impure mathematician par excellence, in the following terms:

> It is no slight fault, it seems, for geometers who arrive at the solution to a problem by means of conical curves or linear sections or who, generally, solve it by means of a foreign kind (εξ ανοικειον γενουσ), as is the case with Archimedes who, in his book *On Spirals*, accepts a solid inclination though he be speaking of a circle—for it is possible to prove the theorem set down by Archimedes without having recourse to anything solid at all. (Pappus, p. 270, l. 28–33)

In the same way, Fermat, in the 17th century, cast a similar condemnation onto Descartes by writing that he

> offends pure geometry in his solution to a problem by reaching for curves which are too complex and of too high a degree, thus ignoring simpler and more fitting curves. For (. . .) this is no small fault in geometry, to solve a problem by improper means (*ex improprio genere*). (Fermat, 1643, p. 118)

Does there exist an elementary domain of mathematics whose epistemic stability is certain? The Greek delimitation (elementary to this is the geometry comprising Euclid's *Elements*, which deals with figures which can be drawn out with a ruler and compass, excluding conical sections and "mechanical" constructions) does not meet this condition: thus, as we explained, the problem of the quadrature of a circle, the statement of which is elementary (find a square of exactly the same area as a given circle), cannot be resolved with a ruler and compass. Hilbert clearly believed he had found such an epistemically stable elementary domain in the finitist part of mathematics, and his research program aimed precisely at *proving* that this was indeed the case. To establish this, it must thus be shown that if an ideal statement takes part in the proof of a real statement then there exists, for that same statement, a proof which does not contain the ideal statement in question.

Nowadays we are accustomed to formulating this property by using the notion of *conservativity*. Given a theory T and a theory T', an extension of the first (the language $L(T)$ of T is part of the language $L(T')$ of T' and every theorem in T is a theorem in T'), we say that T' is a *conservative extension of* T (or simply that it is *conservative on* T) if every formula of $L(T)$ which can be proven in T' is already provable in T. In other words, T' will certainly allow the demonstration of more theorems than T, since it is an extension of it, but it will only be qualified as conservative if it does not prove any new formulas of $L(T)$ that T does not already prove. What we meet again here is the notion of epistemic stability: a theory is epistemically stable if it possesses only conservative extensions, that is, if every proof for a statement of that theory's language can be purified, carried out within that theory itself, without the addition of any elements of "a foreign kind." With this vocabulary Hilbert's program, in the version formulated in terms of conservativity, asserts that the set of all mathematics {finitist statements + ideal statements} is a conservative extension of its finitist part.

It is worth noting that both versions of Hilbert's program (that referencing consistency as well as that relating to conservativity) are equivalent.

(a) Suppose first that a finitist proof for the consistency of arithmetic exists and let $\forall x\varphi(x)$ be a finitist statement possessing a transcendental proof (that is, a proof not limited to only finitist methods): then this statement is correct. For, if it wasn't, then there would exist a whole number a such that $\varphi(a)$ is false, or, to put it in a different way, such that $\neg\varphi(a)$ is true. However, a finitist statement of this kind, having no quantifier, is clearly provable when true. So we would be left with a contradiction in arithmetic because we could at the same time prove $\forall x\varphi(a)$ and $\neg\varphi(a)$. This is incompatible with the consistency of arithmetic for which, by hypothesis, a finitist proof can be given. This reasoning, whose core is the finitist proof of the consistency of arithmetic, is indeed a finitist proof of $\forall x\varphi(x)$. A finitist statement having some kind of proof therefore has a finitist proof. QED.

(b) From the opposite direction, let us suppose that arithmetic as a whole is a conservative extension of its finitist part and let us show that there exists a finitist proof for the consistency of arithmetic. The central argument hangs on the finitist nature of the very *notion* of consistency. Indeed, to say that an arithmetic theory T is consistent is to say that 0=1 cannot be proven in it, in other words, that no sequence of symbols σ is a proof for "*0=1*" in T. This last assertion is clearly finitist (it is a universal assertion attributing a decidable property to an assembly of symbols). Consequently, if the consistency of arithmetic had any proof at all then it would have a finitist proof (this is the conservativity hypothesis). However, a (trivial) semantic proof does exist for the consistency of Peano arithmetic: all the axioms are true (for example, there is no doubt that zero has no predecessor!), all the rules of inference go from true to true and thus (by recurrence) all the theorems y are true, in such a way that 0=1, which is false, cannot be a theorem of this arithmetic. The "finitization" of this semantic proof, which according to the conservativity hypothesis on the finitist part of mathematics is possible, is indeed the finitist proof sought for the consistency of arithmetic.

2.4 THE IMPACT OF GÖDEL'S INCOMPLETENESS RESULTS

A harsh blow was dealt to Hilbert's program by Gödel's (1931) incompleteness theorems, undoubtedly and justifiably the most famous theorems in logic. Let us take a closer look at the content of these theorems, starting with the first incompleteness theorem. Gödel shows that any decent theory of arithmetic is incomplete. What then is a *decent* theory? First and foremost, a decent theory is a consistent theory. An axiom system enabling the demonstration of just any statement, thus including 0 = 1, would be of little use. Second, a decent theory is such that its axioms are enumerable (what is called *recursively enumerable*). Again, a theory not having a systematic means for producing the axioms of that theory would be of little use. If this constraint is not satisfied, the property of being a proof relative to that theory will not be decidable (see subsection 2.1 on the decidability of proofs). What is a decent theory *for arithmetic*? All that is asked is that the theory enable the derivation of at least some number of elementary arithmetic truths or, to be more precise, it is asked that the theory be at least as powerful as elementary arithmetic.[10] What then is a *complete* theory? This is a theory that enables, for any statement in the language of that theory, either to prove that statement, or else to prove its negation.[11] How does Gödel manage to show that any theory T being consistent and recursively enumerable is also incomplete? The demonstration rests on the possibility of encoding the notion of proof in T within arithmetic (which is possible if T is at least as powerful as elementary arithmetic). We can construct a statement G_T, named Gödel's statement of T, which, relative to this encoding, says of itself that it is not provable in T. Thus it is shown that G_T, an arithmetic statement, is not provable in T unless the latter be inconsistent. Since G_T says of itself that it is not provable in T, and since it is in fact not provable in T, G_T is a true arithmetic theorem which is not provable in T.

The strength of Gödel's theorem is in its generality. It is not only that such and such a theory for arithmetic, the Peano axioms for example, is incomplete, in which

[10] The axioms of elementary arithmetic number seven. The first three axioms concern the successor function: 0 is the successor of no number, every number other than 0 is the successor of another number, and if two numbers have the same successor then they are equal. To this are added two axioms that give the recursive definition of addition and two axioms giving the recursive definition of multiplication. The schema of induction is not included.

[11] If we consider that arithmetic statements are either true or false, then an axiomatic theory whose aim is to allow the derivation of all true arithmetic statements into theorems should be complete. This point of view is not Hilbert's, who maintains that only finitist statements are meaningful.

case one could simply think that adding new axioms to the theory would be enough to complete it. Rather, it is every consistent, recursively enumerable theory containing elementary arithmetic which is incomplete. Adding new axioms thus provides no solution to the problem. So the first incompleteness theorem establishes the limits of formal methods.

Why is Hilbert's program jeopardized by this result? The "conservativity" version is the reason for this. There would have been no problem had only ideal statements been implicated by incompleteness. But this is not the case; G_T is a finitist statement (the encoding is such that the property of being a proof in T is primitively recursive in the sense seen previously). So G_T is an example of a finitist statement which is not provable by finitist methods (insofar as these finitist methods are covered by elementary arithmetic) but which is provable in a non-finitist theory (that which would be used to formally derive the result that G_T is true but not provable).

The second incompleteness theorem is presented as a refutation of the version of Hilbert's program concerning the consistency of arithmetic.[12] Based on encoding, an arithmetic statement Coh(T) can be constructed which expresses the consistency of the theory T. Gödel shows that Coh(T) is neither provable nor refutable in T. To prove the consistency of an arithmetic theory T, it is necessary to use a theory which is strictly stronger than the theory in question, and thus, in particular, stronger than finitist arithmetic when T contains finitist arithmetic. The aim of the project being to provide a finitist proof for the consistency of arithmetic in order to validate the use of non-finitist methods, it must therefore be abandoned.

Must we conclude that Hilbert's program is definitively refuted by the two incompleteness theorems? First of all, envisioning partial realizations does remain possible. Let us fix a certain interpretation of what non-finitist mathematics are, say primitive recursive arithmetic, and of what infinitist mathematics are, say second order arithmetic (a system rich enough to develop Real analysis). Simpson (1988) asks which portion of infinitist mathematics can be developed within sub-systems of second order arithmetic which are conservative on primitive recursive arithmetic relative to finitist statements. For example, Friedman (1976) shows such a result for WKL_0, a sub-system of second order arithmetic in which the induction schema is restricted.[13] WKL_0 enables to prove significantly more than primitive recursive arithmetic but is conservative on recursive arithmetic relative to finitist statements.[14]

In this context, it is also possible to contest the scope of Gödel's theorem, either by maintaining that the interpretation given for finitist methods is too restrictive, or by contesting the interpretation given for Hilbert's program. On the first point, Ackermann, in 1940, gave a demonstration for the consistency of arithmetic based

[12] The hypotheses appearing in the second incompleteness theorem are stronger than those used in the first. In particular, the predicate "provable in T" is indeed a provability predicate in Löb's sense of the term (1955).

[13] The induction schema is limited to Σ^0_1 formulas.

[14] If we identify the finitist statements of arithmetic with Π^0_1 formulas.

on transfinite induction.[15] Does such a demonstration count as a consistency demonstration by finitist methods? Ackermann is silent on this question, but he does point out that the functions used in his demonstration, even though they are not the kind of recursive functions usually used in finitist methods (specifically, they are not primitive recursive), well deserve the name "recursive function" to the extent that, for every particular number presented to them as an argument, they deliver a value at the end of a finite number of calculations. Gödel himself proposes extending the finitist methods by adding higher order functions to the standard primitive recursive functions (1958). On the second point, Detlefsen (1990), for example, has contested interpretations of Hilbert's program in terms of conservativity. For Detlefsen, there is no reason to demand, as is habitually done, that an ideal theory decide the *totality* of the finitist statements that can be formulated in its own language. It is more reasonable to hold to a simple *correction* demand relative to finitist statements, that is, to demand that among the finitist formulas provable in an ideal theory, none can be recognized as false through finitist means. This represents a weakened conservativity constraint whose satisfaction is not challenged by Gödel's first incompleteness theorem.[16]

2.5 INTUITIONISM

Philosophy of mathematics' objective is to give a faithful and cognitively plausible representation of the three elements at the core of mathematics: first, the objects the mathematician refers to; second, the formulas he uses; third, his own mental activity. Highlighting mathematical objects poses obvious ontological problems (what kind of objects are we dealing with?) as well as epistemological ones (how do we gain access to them?) From here, it may be tempting to avoid creating a hypostasis out of a domain of mathematical objects which enjoys independent existence and to explain what mathematics is on the unique basis of either mathematical language or else the mathematician's activity. Formalism, which we presented in detail with the help of Hilbert's program, constitutes such an attempt, focused on mathematical language. Intuitionism, which we shall present slightly less exhaustively, constitutes another attempt, this time focused on the mathematician's mental operations.

Intuitionism, like finitism, is a program at the crossroads of mathematics and the philosophy of mathematics. Resulting from the foundational crisis, it was born out of Brouwer's work at the beginning of the 20th century. Intuitionism, as opposed to all formalist approaches, asserts the prevalence of thought over language:

[15] The first proof of consistency was given by Gentzen (1936).

[16] Detlefsen also criticizes the interpretation of the second theorem of incompleteness in relation to Hilbert's program. In particular, he maintains that the hypotheses used, stronger than those in the first theorem, do not rule out the existence of "interesting" proof systems which may guarantee consistency.

> These mental mathematical proofs that in general contain infinitely many terms must not be confused with their linguistic accompaniments, which are finite and necessarily inadequate, hence do not belong to mathematics. (Brouwer, 1927)

Mathematics is not viewed as a theory whose objects could be sought outside of itself or whose linguistic formulation could be studied in itself, but rather it is viewed as an activity.[17] This activity consists in the constructions carried out by the mathematician. The truth of mathematical statements does not depend on some domain of independent objects but on these constructions. To say that an arithmetic statement is true is to say that it is possible to carry out certain constructions, or that certain constructions give such and such a result.

> An intuitionist accounts for the truth of $2+2=4$ by saying that if one constructs 2, constructs 2 again, and compares the overall result to a construction of 4, one sees they are the same. This construction not only establishes the truth of the proposition $2+2=4$, but is all there is to its truth. (van Dalen and van Atten, 2002)

This understanding of mathematical activity has consequences for the acknowledgment of the legitimacy, or lack thereof, of mathematicians' practices. Heyting, Brouwer's pupil, compares the definitions of two whole numbers k and l: k is defined as the biggest prime number such that $k-1$ is also prime, or $k=1$ if such a number does not exist, l is defined as the biggest primer number such that $l-2$ is also prime, or $l=1$ if such a number does not exist (Heyting, 1956). From a classical mathematics point of view, these two definitions are each as good as the other. From the point of view of intuitionistic mathematics, this is not the case. The first definition is acceptable, as we have a method for calculating k at our disposal (this calculation gives $k=3$). The second definition is not acceptable. It is not known whether or not twin primes exist in infinite number. An intuitionist rejects any definition of a whole number which does not provide the means of constructing that whole number.

The rejection of classical methods goes well beyond the matter of definitions, affecting everything up to logical principles. Intuitionists deny the universal validity of some of the principles of classical logic.[18] In the case of the finite, where, in principle, we can examine all relevant objects, a disjunction like $\forall x \phi x \vee \exists x \neg \phi x$ corresponds with an effectively *decidable* alternative: at the conclusion of a systematic search either it will have been verified for every object that it is a ϕ or else an object which is not one will have been found. However, in the case of the infinite, things are quite different.

[17] "Strictly speaking the construction of intuitive mathematics in itself is an action and not a science" (Brouwer, 1907).

[18] The rejection of a principle like the excluded middle $(\varphi v \neg \varphi)$ is not equivalent to affirming that this principle leads to contradictions since, on the contrary, the fact that the negation of the excluded middle is contradictory is accepted (see Brouwer, 1908).

We may well have a proof $\neg\forall x\phi x$, which shows that supposing all objects of the domain under consideration are ϕ leads to a contradiction, without this meaning we have a proof of $\exists x\neg\phi x$, a proof which would enable the construction of some object a of which it could be shown that it is not ϕ. Accepting the validity of $\forall x\phi x \vee \exists x\neg\phi x$, even for infinite domains, would be allowing ourselves to affirm the existence of objects which we have not constructed (by inferring $\exists x\neg\phi x$ from $\neg\forall x\phi x$). But if mathematical objects are nothing other than the result of the mathematician's activity then this is not legitimate: every demonstration of an existential statement like $\exists x\neg\phi x$ must be based on a construction of the object that is evidence of correction to the affirmation of existence.

Intuitionistic logic, formalized by Heyting (see Heyting, 1956, chap. 7 for a presentation), removes itself from classical logic, whose principles are typically justified by the truth conditions of the statements, by proposing rules which are justified by conditions of provability.[19] What an intuitionistic proof must be is defined by indicating how logical operations must be interpreted, in terms of proof. Giving a proof for $\varphi \vee \psi$ is defined as the fact of giving a proof for φ or giving a proof for Ψ. Giving a proof for $\varphi \rightarrow \psi$ is defined as the fact of giving a construction capable of transforming any proof for φ into a proof for Ψ. Intuitionistic logic is nothing other than the logic which results from this interpretation,[20] known by the moniker of the BHK interpretation.[21] On the one hand, it does not validate the principles which are valid from a classical point of view but problematic from the intuitionistic perspective, hence it does not enable derivation of the excluded middle. On the other hand, it possesses the "right" properties one would expect to find if every affirmation regarding mathematical objects must be guaranteed against the capacity of constructing these objects (like with the existence property,[22] if $\exists x\phi x$ can be demonstrated, then ϕa can also be demonstrated for some a).

The fundamental thesis of intuitionism is that the basis of mathematics consists of mental constructions. We have just seen, briefly, how this thesis led to a logical revisionism and also in what sense intuitionistic logic could be interpreted as a logic of constructions. But what are these mental constructions? First, these mental constructions must be viewed as the product of an ideal subject, and not as psychological realities which correspond to the mental states of some mathematician or another. Second, in dealing with the nature of constructions, Brouwer claims part of the Kantian heritage. Mathematical constructions are founded on the intuition of time and subjective time is seen as a dimension of consciousness necessary for the thought

[19] In classical logic, the validity of the excluded middle can be arrived at by showing that the definition given for the truth of statements relative to an interpretation guarantees that a statement ˘ is either true or false, so that, in either case, it will be seen that $(\varphi \vee \neg\varphi)$ is true. The epistemological value of such a demonstration is problematic.

[20] The meaning given here to the verb "result" is a bit loose, insofar as the BHK interpretation is an informal interpretation which is not precise regarding, for example, what is to be meant by "method enabling to . . ."

[21] The initials BHK refer, respectively to Brouwer, Heyting, and Kolmogorov.

[22] In the case where we are not only occupied with logic but also with a particular mathematical theory for which these logical rules are used, it must be shown that this existence property is preserved despite the extra axioms. This is indeed the case for intuitionist arithmetic, for example.

of any and all objects.[23] In Brouwer's formulation, the two "acts" at the foundation of intuitionistic mathematics consist, on the one hand, in acknowledging the role played by the perception of temporal change and, on the other hand, in acknowledging the possibility of giving rise to new mathematical entities, particularly with the help of infinite sequences whose members are chosen among a domain of already constructed mathematical entities. The whole numbers are constructed from the intuition of temporal change,[24] while acknowledging the possibility of giving rise to infinite sequences plays a crucial role in the construction of the real numbers. Indeed, the whole numbers having been constructed, Brouwer identifies the elements of the continuum with choice sequences of whole numbers. These infinite sequences represent, through encoding, intervals of rational numbers satisfying the Cauchy sequence condition. They may be given by a law enabling to calculate the nth element of the sequence, or they may be free sequences whose elements are not determined by a rule but are freely produced.[25] Analysis constructed on these bases diverges from classical analysis: it can be shown that every function on real numbers is continuous.[26] As this example borrowed from real analysis indicates, intuitionist mathematics is not some kind of diminished mathematics that would be obtained by subtracting certain contestable logical principles (like the excluded middle) from classical mathematics. It is an original mathematics in which certain classically false statements become central theorems.

On a mathematical level, constructive mathematics—that is, beyond the specificities of the Brouwerian approach, those mathematics which choose to interpret existence in terms of possibilities of construction—still constitute a vibrant tradition within mathematics today. Two striking examples of this are the work of Bishop (1967), which reveals how to develop a constructive analysis as rich as the classical one, and the development of type theory, following on from Martin-Löf (1975), which makes explicit

[23] As opposed to Kant, Brouwer holds no special role for spatial intuition because he considers the development of non-Euclidean geometries to have undermined the idea of a spatial intuition providing access to one unequivocal geometry. Our presentation of Brouwer's theories concerning the nature of mental constructions is primarily informed by van Atten (2004).

[24] Brouwer explains perception of a temporal change as "the falling apart of a life moment into two distinct things, one of which gives way to the other, but is retained by memory. If the twoity thus born is divested of all quality, it passes into the empty form of the common substratum of all twoities. And it is this common substratum, this empty form, which is the basic intuition of mathematics" (Brouwer, 1952, cited in van Atten, 2003, p. 4). The abstraction made on the basis of the temporal change of the simple "twoity" form counts for the construction of the two first whole numbers, the following whole numbers being constructed analogously.

[25] Troelstra (1977) is a detailed study of choice sequences and the conditions in which free-choice sequences either are or are not indispensable.

[26] The admission of free-choice sequences is problematic: for example, how does one calculate the value of a certain function for a real number when this number arises from an infinite choice sequence, which is thus never actually given in whole but which obeys no rule either. To resolve this difficulty, Brouwer adopts a continuity principle that guarantees that the value of a function on real numbers is determined, for every real number, by a finite number of elements from the choice sequence that gave rise to it. The theorum that all functions on real numbers are continuous, as well as other theorems from intuitionistic analysis that are not theorems of classical analysis, follows from this continuity principle.

the types attributed by our judgments to constructions. On a philosophical level, the Brouwerian theory of mental constructions is not the only possible philosophical foundation for intuitionist mathematics. Turning the classical anti-linguistic intuitionism perspective on its head, Dummett proposes founding intuitionism on a general theory of meaning.[27] The basis of this theory is a requirement of manifestability, in virtue of which, "The meaning of [. . .] a statement cannot be, or contain as an ingredient, anything which is not manifest in the use made of it, lying solely in the mind of the individual who apprehends that meaning" (Dummett, 1973, p. 216). A statement's meaning must therefore not be defined by reference to conditions which could, in principle, be satisfied *unbeknown* to the individual who apprehends that meaning: it must be identified with the statement's conditions of assertability (rather than with its truth conditions, for example).

It is not our place here to give the final word regarding formalist or intuitionist programs, so in concluding this section we will make do with a brief summary and a little bit of perspective. The type of intuition used by the formalists (perception of symbol types) is relatively non-problematic, but the mathematical results necessary for founding all mathematics on an intuition of this kind have not been obtained in the way Hilbert had hoped and, conversely, Gödel's negative results constitute an obstacle that every new attempt at updating the formalist program must get around. Following Brouwer and Heyting's work, intuitionist mathematics has undergone developments which suggest that it is no less rich or fruitful than classical mathematics. However, we can't help remarking that intuitionist mathematics remains a minority tradition within mathematics. Furthermore, the nature of Brouwerian intuition remains at least as mysterious as its Kantian counterpart.

Both programs we have just presented shared a quest to explain mathematics without resorting to the supposition of an objective reality to mathematics, independent of us. They subscribe to an anti-realist philosophy of mathematics. By contrast, the following sections are given over to realist conceptions whose objective is to take the existence of just such an objective reality seriously. Let us note, however, that finitism and intuitionism do not represent all possible forms of anti-realism. Indeed, further on we will have occasion to mention both Field's fictionalism and also nominalist structuralism.

3. Why Be a Realist?

3.1 SEMANTIC REALISM AND ONTOLOGICAL REALISM

To begin with, let us distinguish between two forms of mathematical realism. First we have semantic realism,[28] corresponding with the thesis that the truth or falsity of mathematical statements is an objective fact not dependent on us. Second, there is ontological realism, corresponding to the thesis that mathematical objects do exist

[27] In particular, see Dummett (1977).

[28] In speaking of semantic realism, a term sometimes employed is *truth-value realism*. See Shapiro (1997) and Linnebo (2009).

independently of us. Ontological realism seems to imply semantic realism:[29] if there are mathematical objects which are not dependent on us, then the truth or falsity of mathematical statements depends on these objects and does not, therefore, depend on us. If we wish to be "completely" realist, both theses must be accepted conjointly. Though let it be noted that semantic realism does not imply ontological realism. A Hilbertian formalist, for example, partially subscribes to semantic realism in being realist regarding the truth values of statements from finitist mathematics, since these statements are reinterpreted in such a way as to refer to mathematical symbols. It is even possible to reject ontological realism all the while accepting semantic realism for the full set of mathematical statements: this is the case for Hellman (1989) who gives a modal interpretation of mathematical statements without supposing a determined domain of mathematical objects to be fixed. In the following sections we will concentrate on that "full" realism which combines both the semantic and ontological forms. We will also come back to provide some required precision on the ontological realism thesis.

3.2 REALISM AND MATHEMATICAL PRACTICE

Why be a realist? The most obvious argument is based on a form of pre-theoretical realism. Spontaneously, we are tempted to say that $2+2=4$ is a true statement and that, if it is true, then this is thanks to the properties of the numbers referred to. The "we" spoken of here encompasses both mathematicians and non-mathematicians. For mathematicians, this naive realism is in part anchored in the practice of mathematical research. Indeed, trying to reveal a theorem which resists just is trying to reveal something about certain objects; the resistance or opacity itself suggests a certain objectivity which is independent of the researcher.[30] In Moschovakis's terms,

> The main point in favor of the realistic approach to mathematics is the instinctive certainty of almost everybody who has ever tried to solve a problem that he is thinking about 'real objects,' whether they are sets, numbers, or whatever, and that these objects have intrinsic properties above and beyond the specific

[29] However, Shapiro (1997) makes the remark that Tennant (1997) is an exception to this. Tennant's semantic anti-realism stems from the constraints he regards as weighing on the content of statements by reason of our using them, even though these statements refer to objects that exist independently of us.

[30] We write "suggests" because it is a case of feeling rather than argument. Gödel (1951) puts forward a genuine argument in favor of realism based on the difficulty of mathematics: if absolutely undecidable mathematical propositions exist, then the idea that mathematics are our own creation would be therein refuted, since a creator "necessarily knows all properties of his creatures" (1951, p. 16). If we are drawn more to the sentiment based on the difficulty of mathematics than to this argument, it is due to its being doubly limited. For one thing, the premise concerning the difficulty of mathematics needs strengthening. It is not simply a case of affirming that mathematics "resists," but that there are absolutely undecidable problems (something the theorems of incompleteness do not establish and Gödel does not take on). The other thing is that the second premise, supposing that the creator can in principle know all the properties of his creatures since "they can't have any others except those he has given to them," is entirely contestable. On the difficulty of mathematics and the meaning this takes on whether our position is realist or anti-realist, see Oumraou (2009).

> axioms about them on which he is basing his thinking for the moment. (1980, p. 605)

The argument is not only phenomenological. Mathematicians do not simply believe "without consequence," so to speak, that mathematical objects exist. They proceed as if mathematical objects did exist:[31] some of the principles used by mathematicians seem justified only if mathematical objects exist independently of us. If we consider that the practices of mathematicians are in proper order and that philosophers of mathematics have no right to interfere with them, what we get is an argument in favor of mathematical realism, in the sense that it is the only philosophical position that is consistent with these practices. Recourse to impredicative definitions is an example of this.[32] Impredicativity comes into play when a set M and a particular object m are defined such that m is an element of M and the definition of m depends on M. There are many mathematical definitions which are impredicative: this is the case, in analysis, for the definition of the upper limit of a set of real numbers.[33] As Gödel points out, impredicative definitions are not problematic so long as the ontological realism thesis is correct: "If, however, it is a question of objects that exist independently of our constructions, there is nothing in the least absurd in the existence of totalities containing members, which can be described (i.e. uniquely characterized) only by reference to this totality" (1944, p. 136). If, on the other hand, we consider in one way or another that mathematical objects are produced by definitions or constructed by mathematicians, then it is not possible to define an object on the basis of a totality which presupposes it. Mathematical realism can therefore boast not only of some vague faithfulness to the "philosophical" beliefs of mathematicians, but also, and above all, of faithfulness to their modes of reasoning.

Gödel (1953) also attempted to draw an argument out of the second incompleteness theorem. As we have seen, this theorem establishes the impossibility of using finitist methods to prove the consistency of sufficiently rich mathematical theories, such as arithmetic or set theory. Any conventionalist program aiming to reduce mathematics to a simple matter of symbol manipulation is therefore presented with the impossibility of establishing the consistency of the system of conventions it employs. This is

[31] Resnik (1980, p. 162) refers to those who accept the use of non-constructive methods in mathematics as "methodological Platonists." Shapiro (1997, p. 38) speaks of "working realism" with something similar in mind. Here we maintain that mathematicians' methodological platonism is an argument in favor of straight-up platonism.

[32] One philosophical-mathematical school in particular, predicativism, seeks to show how it is possible to do without impredicative definitions in developing mathematics. See Weyl's (1918) pioneering work and Feferman's recent (1988) theoretical-proof developments, which show that the majority of classical analysis can be elaborated without impredicativity.

[33] Real numbers being seen as Dedekind cuts, the least upper bound of a set X of real numbers, written lub(X), is the set of rational numbers which are elements of a real number belonging to X. lub(X) is defined as an element of the set R of real numbers, but its definition depends on R because, in the general case, X has itself been defined as a set of elements of R possessing some certain property (see Kleene, 1952, p. 43).

particularly problematic if it is a matter of maintaining that the syntactical systems used are without content, since an inconsistent system can be used to deduce any empirical proposition. In Gödel's terms,

> the scheme of the syntactical program to replace mathematical intuition by rules for the use of the symbols fails because this replacing destroys any reason for expecting consistency,[34] which is vital for both pure and applied mathematics. (1953, p. 346)

By contrast, the use of axioms can be justified by mathematical intuition. If we know that the axioms we use in arithmetic are true, because we know they correctly describe the properties of the mathematical objects constituting the whole numbers, then we know we can use arithmetic without fear of contradiction. In this perspective, the argument for realism would come from its ability to justify the trust we place in mathematics; this explains why we thought it pertinent to relate this argument to the previous one.

The scope of Gödel's argument is nevertheless limited.[35] First, if the argument works as a refutation of a certain anti-realist program whose aim is to reduce mathematics to a conventional manipulation of symbols, then that refutation is not enough to justify realism, especially not ontological realism. It remains entirely possible to replace the mathematical intuition of objects existing independently of us with other kinds of mathematical intuitions, for example, as already seen, with the intuition of mathematical objects of our own construction, or with empirical intuition (axioms being justified by the success of applications).[36] From this viewpoint, realism is but one among many solutions to the justification of axioms and consistency question, even once the domain of possible solutions has been reduced by the incompleteness theorem. Second, the argument is also problematic as a refutation of conventionalism in Gödel's sense. He supposes that conventionalism is only acceptable if the conventionalist can *show* that the rules he advances are consistent, and not simply if they are consistent. The conventionalist could refuse the burden of proof and retort that he does not have to mathematically prove that conventionalism cannot be refuted.

3.3 THE INDISPENSABILITY OF MATHEMATICS ARGUMENT

Both previous arguments can be overturned. We have maintained that if impredicative definitions are justified, and if we know that arithmetic is consistent,[37] then

[34] Gödel's target here is Carnap and other logical positivists such as Hahn or Schlick.

[35] Gödel's own position on the scope of the argument is not entirely clear. On the one hand, the article in which he exposes it is wholly concerned with a refutation of the syntactical program, and Gödel openly acknowledges that the role of mathematical intuition could be just as easily taken on by empirical intuition. On the other hand, Gödel declares that "the examination of the syntactical viewpoint; perhaps more than anything else, leads to the conclusion that there do exist mathematical objects and facts [. . .]" (1953, p. 337).

[36] In what way does the success of a mathematical theory's applications provide reasons for believing that the theory is consistent? In a certain way, belief in consistency may be based on the fact that no one has yet been able to derive a contradiction. Gödel also envisions, elliptically, the possibility of "a knowledge of empirical facts involving an equivalent mathematical content."

[37] Here we simplify the second argument, which is, as has been pointed out, more complex.

mathematical objects do exist independently of us. So, supposing that impredicative definitions are indeed justified, or that we do know that arithmetic is consistent, then it is possible to conclude that mathematical objects exist. But an anti-realist could maintain that the realist hypothesis is in fact the only possible reason we could have for considering impredicative definitions to be justified or arithmetic to be consistent. So the two previous arguments would beg the question. This anti-realist response is possible, even if it be questionable. It has been said of the argument we shall now present, the indispensability argument, that it was "the only non-question-begging argument against [nominalism]" as Field put it (1980, p. 4).

The indispensability argument deduces the existence of mathematical objects from the indispensability of mathematics to contemporary science. It is generally attributed to Quine (see especially 1953a and 1953b) and Putnam (1979);[38] here we give a rigorous presentation borrowed from Colyvan (2001, p. 11).

Premise 1	We have ontological commitment to all and only the entities that are indispensable to our best scientific theories.
Premise 2	Mathematical entities are indispensable to our best scientific theories.
Conclusion	We have ontological commitment to mathematical entities.

Premise 2 is an observation that feeds the methodological principle posed by premise 1. It affirms that, as a matter of fact, mathematics is an integral part of our best scientific theories. In this it has the weight of evidence behind it: theories of physics are entirely formulated using mathematical theories; consider the role of analysis in the formulation of mechanics, for example, or that of Hilbert spaces in quantum mechanics, or even that of Riemanian geometry in the theory of relativity.

Premise 1 itself follows from two methodological principles that Quine always defended, naturalism and holism of confirmation. The naturalist thesis is that the natural sciences are the ultimate judge in matters of truth and existence. It consists of abandoning the dream of a primary philosophy whose job it would be, using some method of its own, to have the final say on metaphysical and ontological questions. Instead, it asks for acknowledgment that science is the best guide we have, including when it comes to knowing what exists. When giving form to our best scientific theory of the world, if we notice that this theory posits the existence of quarks, then we are committed to admitting the existence of quarks. It would be pointless for metaphysicists to declare that quarks do not exist: their bold claim would have no value unless they could show that physics could be reconstructed without needing to speak in terms of quarks. How are things applicable to the theoretical entities postulated

[38] It is also possible to present a version of the indispensability argument which concludes in favor of semantic realism rather than ontological realism. Such a version would undoubtedly be closer to Putnam's conception. We can move from a version like this to an argument in favor of ontological realism by subsequently maintaining that it is proper to adopt a standard semantics for mathematical statements (see section 5.1).

by physics in turn also applicable to mathematical objects? This is where holism and confirmation come into play. The holistic thesis is that the data confirming a scientific theory do not confirm some part or other of the theory, but rather the whole theory. So if our best theory of the world is a theory whose laws simultaneously bring unobservable physical entities (such as quarks) and mathematical objects (like real numbers and functions on real numbers) into play, then the data that confirm this theory must be considered to confirm the existence of quarks just as much as they do the existence of real numbers and functions on real numbers.

The core of the indispensability argument is the refusal of a double ontological standard: mathematical entities, from the viewpoint of science, are on the same level as theoretical entities and we must therefore accord them the same ontological status. Putnam presents things in the following way, taking the law of universal gravitation as his example:

> One wants to say that the Law of Universal Gravitation makes an objective statement about bodies - not just about sense data or meter readings. What is the statement? It is just that bodies behave in such a way that the quotient of two numbers *associated* with the bodies is equal to a third number *associated* with the bodies. But how can such a statement have any objective content at all if numbers and 'associations' (i.e. functions) are alike mere fictions? It is like trying to maintain that God does not exist and angels do not exist while maintaining at the very same time that it is an objective fact that God has put an angel in charge of each star and the angels in charge of each of a pair of binary stars were always created at the same time! If talk of numbers and 'association'' between masses, etc. and numbers is 'theology' (in the pejorative sense), then the Law of Universal Gravitation is likewise theology. (1979, p. 74)

The role of Premise 2, in Putnam's example, is filled by the recourse to numbers and functions in the formulation of the law of universal gravitation: to speak of mass is to speak of a function which attributes a numerical value to bodies. But it must be noted here that in all generality, despite the "weight of evidence" and despite the examples, Premise 2 is open to question. Thus Field (1980) sets himself the task of defending a nominalist position by refuting the indispensability argument on the basis of a rejection of Premise 2. In this way Field intends to show that Newton's gravitational theory may be "nominalized": it is possible to reformulate its laws without using quantitative concepts and the mathematics which accompany their use. Field's demonstration has been contested on its relevance and on the possibility of generalizing it. Concerning the first point, Colyvan (2001) insists on the fact that, even if the nominalized theory is empirically equivalent to the initial theory, it does not follow that the nominalized theory is just as good as the initial one, considerations of simplicity and elegance playing an important role in our choices when it comes to theories. Even if a nominalization program could be finalized, it would still remain possible that the existence of mathematical entities be justified, to the extent that they are indispensable in the formulation of simple and elegant theories. Concerning the second point, the possibility of expanding the nominalization enterprise has been contested regarding quantum mechanics (Mallament, 1982) as well as the theory of relativity (Urquhart, 1990). The debate around Field's nominalist program is still active, quite in keeping with the philosophical importance a refutation of the indispensability argument would have.

4. Varieties of Platonism and Philosophy of Set Theory

4.1 WEAK PLATONISM AND STRONG PLATONISM

At the beginning of the previous section, we made the distinction between semantic realism and ontological realism and aimed to mobilize a "complete" realism, combining both semantic and ontological realism. It is now time to distinguish, within ontological realism, different forms it can take. Indeed, the arguments we have presented do not all justify the same "degree" of ontological realism. The arguments presented in section 4.1 take the form of an inference to the best explanation. Their conclusion is that it is desirable to suppose mathematical entities, these providing us with a bond liable to explain our mathematical knowledge. The indispensability argument presented in section 4.2 brings out no such bond with abstract entities. Supposing the existence of mathematical objects is the consequence of adopting mathematical theories, something which is underpinned by their integration into our best theory of the world. Mathematical entities are sort of like the projection of our theories, these not being built on the basis of some epistemic access to the objects they aim to describe. On this basis, it is fitting to distinguish what could be called a weak platonism and a strong platonism.[39] Weak platonism combines semantic realism and an "epistemologically neutral" ontological realism. Typically, this is Quine's position. Strong platonism combines semantic realism and an "epistemologically charged" ontological realism.

[39] Our distinction between weak platonism and strong platonism is epistemological, and one may be shocked to find an epistemological distinction called on in the characterization of ontological positions. In order to stay on ontological ground, we could have proposed a distinction based on the independence of mathematical objects (independent with respect to knowing subjects, their practices, their language, and their thought). This would be followed by saying that Quine's platonism, for example, is a weak platonism: mathematical objects are but the projections of our theories, to that extent their characterization depends on our theorization practices. Conversely, Gödel's platonism would be a strong platonism: mathematical objects are the elements making up a mathematical reality absolutely independent of our theoretical activity. Ontological characterization by independence, at least in the case of Quine and Gödel, would intersect with epistemological characterization precisely without its being necessary to part from ontological ground. Nevertheless, it seems to us that this speaking in terms of independence remains vague and that the epistemological distinction is clearer cut. Linnebo (2009) proposes a typology for ontological realism, dependent on whether it is simply the existence of mathematical objects, or else that these objects are also both abstract and independent, that is, being affirmed. In the same work, he admits that it is not really known what sort of thing a "non-independent" object would be.

Gödel undoubtedly proposed the most extreme version of strong platonism, but it is also the same position held by a logicist like Frege.[40]

Weak platonism and strong platonism agree in saying that mathematics speaks of mathematical entities which have an objective existence, just like physics is the study of physical entities, although the associated epistemologies vary greatly. For weak platonism, which does not place recognition of mathematical objects alongside the recognition of a specific mode of access to these objects, mathematical knowledge enjoys no particular privilege relative to all other knowledge. It is not certain, it is revisable, it is not a priori but rather depends on experience. Similarly, mathematical truths are not necessary, or in any case no more necessary than the principles of physics. Rather, strong platonism posits the existence of a mode of access specific to mathematical objects. In this way, Frege, in a posthumous text, evokes the existence, alongside sense perception, of a "logical source of knowledge"[41] (1924–1925). Remarkably, the characterization Gödel proposes for platonism calls on the perception of the mathematical reality;[42] thus platonism is presented as

> the view that mathematics describes a non-sensual reality, which exists independently both of the acts and [of] the dispositions of the human mind and is only perceived, and probably perceived very incompletely, by the human mind. (1951, p. 38)

Strong platonism thus proposes an epistemological regime for mathematics that is distinct from the natural sciences. Mathematical knowledge is a priori, in the sense that it is independent of sensual experience and, from a metaphysical viewpoint, it remains possible to attach some kind of necessity to mathematical truths which would not be attached to empirical truths.

[40] The logicist reduction does not equate to elimination of mathematical objects. Frege is a realist when it comes to logical objects. His intention is therefore not to show that there are no mathematical objects by reducing these objects to logical laws which have no content. Rather, he intends to show that mathematical objects are logical objects.

[41] Frege gives practically no positive characterization of our access to logical objects; the logical source of knowledge is simply distinguished negatively from sense-based perception and "geometrical" and "temporal" sources, no doubt making reference to the pure intuitions of space and time advanced by Kant. The laws of logic equating, for Frege, to the laws of thought, understood in a nonpsychological sense, this third source of knowledge could be assimilated with a reflexive capacity of thought to grasp the principles of its own inner workings, though all this remains highly speculative. Recently, Hale and Wright (2002) have advanced that logicist platonism could account for mathematical knowledge as purely conceptual knowledge within which the intuition never plays a key role.

[42] Frege and Gödel are both representatives of strong platonism. This does not, however, mean that they hold the same view on our mode of accessing mathematical objects. Frege would undoubtedly refuse to speak of mathematical *intuition*, while Gödel considers that an analogy exists between our rational grasp of mathematical concepts and our perceptual grasp of physical objects.

4.2 INTUITION AND SUCCESS

Let us come back now in more detail to our mode of access to the mathematical reality seen from the viewpoint of strong platonism, continuing thus to apply Gödel's conception.[43] The mathematical source of knowledge is thought of by analogy with sense perception. Mathematical intuition consists in a perception of mathematical reality:

> Despite their remoteness from sense experience, we do have something like a perception also of the objects of set theory, as is seen from the fact that the axioms force themselves upon us as being true. (1964, p. 529)

Gödel specifies that mathematical perception, like the perception of physical objects which is brought to awareness by sensations, is undoubtedly not an immediate form of knowledge. Nevertheless, Gödel gives no positive characterization of what would transmit awareness of mathematical perception. He contents himself to giving a negative argument stating that even our empirical ideas contain abstract elements which are "qualitatively distinct from sensations" (1964, p. 529) and whose origin cannot be in sensations. The objects of mathematical perception are instead determined as being concepts, Gödel thus affirms that the difference between sense perception and mathematical perception "consists solely in the fact that in the first case a relationship between a concept and a particular object is perceived, while in the second case it is a relationship between concepts" (1953/1959, V).

The intuition is not the only mode of access to mathematical truths recognized by Gödel. In the empirical domain, certain fundamental laws, whose content is not directly observable, are verified indirectly by their consequences; the same thing applies, in the mathematical domain, for axioms whose content escapes the intuition. Rather than establishing themselves by their obviousness, such axioms establish themselves by their "success" (1964, p. 522). Evoking the success of an axiom with a mind to justifying its adoption is standard practice for mathematicians. Take the example of the axiom of choice in set theory, affirming that for any collection of non-empty sets, there exists a function which chooses an element within each of these sets. A first mark of success is the possibility either of giving new demonstrations for previously known results or of generalizing these results. Hence, helped by the axiom of choice, it can be shown that for any set at least one of the two situations will apply: the elements can be grouped by pairs without omitting any single element, or the elements can be grouped by pairs omitting exactly one element. This result is both known and elementary in the scenario where the starting set is finite, the axiom of choice is needed when the starting set is infinite.[44] A second criteria for success is the possibility of resolving preexistent conjectures. Again, this criteria is satisfied by the axiom of choice

[43] For a detailed discussion of Gödel's platonism, see especially van Atten and Kennedy (2009), Parsons (1995), as well as Sabatier (2009).

[44] This argument is given by Sierpinski (1967).

which enables resolving the question of whether all sets can be well-ordered.[45] A third criteria is the capacity to systematize and simplify mathematical theory. Once again this is a scenario for the axiom of choice, in its capacity to lay the foundations of the cardinal numbers.[46] A fourth criteria, undoubtedly decisive in the case of the axiom of choice, is mathematicians' implicit use of the axiom: Zermelo (1908) shows that many developments concerning sets of real numbers, and particularly concerning Borel's set theory and projective sets, rely on the axiom of choice.

4.3 ADDING NEW AXIOMS

The difference between the realist and anti-realist positions, and the difference within the realist positions between weak platonism and strong platonism, distinguishes itself well when it comes to the attitude adopted when faced with results of independence and the question of whether these results do or do not call on an extension of the initial theory. Take the example of the continuum hypothesis (CH), the affirmation stating that the cardinal of the real numbers is the first uncountable cardinal or, in an equivalent wording, that every infinite subset of real numbers can form a bijection either with the set of integers or with the set of real numbers. This is a fundamental question for the theory of transfinite cardinals, which deals with mathematical objects "familiar" to non-set-theoretic mathematicians, namely integers and real numbers. Once Cantor had proven that the cardinality of the continuum was superior to the cardinality of the set of integers, it was natural to wonder "to what extent?": are there or are there not sets of greater cardinality than the set of integers and of lesser cardinality than the set of real numbers? The negative response to this question is the continuum hypothesis, formulated by Cantor as a conjecture in 1878. In 1940, Gödel showed that the negation of the continuum hypothesis is consistent with the standard Zermelo-Fraenkel set theory, including when the axiom of choice is present (ZFC). In 1964, Cohen showed that the continuum hypothesis is consistent with ZFC. On the basis of the axioms of ZFC, CH can neither be proved nor refuted, so that standard set theory leaves a fundamental question of cardinal arithmetic undecided. What must be concluded from this, and what must the set-theoreticians do about it? For those considering the notion of set to be defined conventionally by the axioms of ZF or ZFC, the undecidability result closes the debate over the acceptation of CH on the basis of our current notion of set. The development of rival set theories does, however, become possible. This was Church's spontaneous reaction:

> The feeling that there is an absolute realm of sets, somehow determined in spite of the non-existence of a complete axiomatic characterization, receives more of

---

[45] A relation R on a set A is well-ordered if R is an order and if every non-empty subset of A has a smallest element. The axiom of choice was explicitly put forward by Zermelo (1904) to show that all sets can be well-ordered, as Cantor supposed. Both hypotheses are in fact equivalent to each other.

[46] Tarski (1924) shows that the axiom of choice is equivalent to addition monotonicity for transfinite cardinals (if $m < n$ and $p < q$ then $m + p < n + q$).

> a blow from the solution (better, the unsolving) of the continuum problem than from the famous Gödel incompleteness theorems. [. . .] The Gödel-Cohen results and subsequent extensions of them have the consequence that there is not one set theory but many, with the difference arising in connection with a problem which intuition still seems to tell us must 'really' have only one true solution. (1966, p. 18)

The difference with the incompleteness theorems is that the result is not immediately interpretable as a consequence of intrinsic limitations of axiomatic methods. The proof of undecidability of Gödel's statement G(T) of a theory T stating about itself that it is not provable in T counts, upon reflection, as an (informal) proof of G(T). The same cannot be said for Gödel and Cohen's independence results which leave us with no indication as to what we should think of CH.

From a weak platonist point of view, the debate around CH is not entirely closed. ZFC is the classic set theory enabling the reconstruction of all mathematics used in science. The addition of new axioms to ZFC is justified if this addition leads to an improvement in our conceptual schema as a whole. Imagine that the addition of some axiom to ZFC enabled to prove previously unprovable theorems and that these theorems then found some application in some domain of the natural sciences. Imagine, moreover, that the axiom in question enabled, within ZFC, to derive the negation of CH. We should then consider that there is no set whose cardinal is strictly situated between the cardinal of the integers and the cardinal of the continuum. But this hypothetical scenario where applicable consequences appear is not the only scenario possible. The improvement of our conceptual scheme can also occur through simplification and ontological economy. For this reason, Quine himself leans toward CH:

> The main axioms of set theory are generalities operative already in the applicable part of the domain [of mathematics]. Further sentences, such as the continuum hypothesis and the axiom of choice, which are independent of those axioms, can still be submitted to the considerations of simplicity, economy, and naturalness that contribute to the molding of scientific theories generally. Such considerations support Gödel's axiom of constructibility '$V = L$'. It inactivates the more gratuitous flights of higher [transfinite] set theory, and incidentally it implies the axiom of choice and the continuum hypothesis. (1990, p. 95)

The axiom of constructibility says that the universe of all sets ("V") is nothing other than the totality of constructible sets ("L"). Constructible sets are built in steps, the steps in question being indexed by the ordinal numbers. L_0 is the empty set. $L_{\alpha+1}$ is the union of L_α and the definable subsets of L_α. If α is a limit ordinal, L_α is the union of L_β for $\beta<\alpha$. The axiom of constructibility is a minimality principle stating that only definable sets exist. If it is only a question of limiting our ontological commitments, it is natural to add 'V = L' to ZF since the constructible sets are sufficient in providing ZF with a model.[47]

From a strong realist's point of view, the debate around CH is definitely not closed. In the universe of all sets V, CH is either true or false. The independence results reveal the limits of our perception of V. To the extent that ZFC is the standard theory at some time *t*, ZFC represents all that is believed either explicitly (already demonstrated theorems) or implicitly (theorems awaiting demonstration) about the universe of sets by the mathematical community at time *t*. This knowledge is incomplete, and the proof of the independence of CH specifically shows that at time *t* we do not know (neither explicitly nor implicitly) what the status of CH is. Nevertheless, it is the job of set-theoreticians to push back the limits of this knowledge using the two criteria of truth constituted by mathematical intuition and success. Hence, in his 1947 article on the continuum problem,[48] Gödel maintains that new set theoretical axioms must be sought. Among the possible additions Gödel discusses, we find, notably, axioms of large cardinals. An example of an axiom of large cardinals is the axiom affirming the existence of inaccessible cardinals, and the addition of such an axiom obeys a maximality principle. An inaccessible cardinal is a set closed by the exponentiation operations and the limits of lesser cardinals. The idea behind the adoption of axioms of large cardinals is that nothing can exhaust the universe of all sets, or, to put it a little less metaphorically, that "the universe of all sets is beyond being captured by any closure condition on sets; instead, any such condition always closes off at a set" (Feferman, 1999).[49] Contrary to Gödel's wishes, research into the axioms of large cardinals did not lead to the formulation of axioms enabling to resolve CH. However, set-theoreticians have not given up: Woodin's program is the most famous contemporary attempt in this direction,[50] and Woodin (2002) certainly seems to subscribe to a platonist interpretation of his work.

Let us make a note of the meeting points: the "success" type criteria are just as admissible for weak platonism (in the way they make up part of good theorization practice and they are applied to ZF understood as a part of the system of science as a whole)

---

[47] On the contrary, if we consider V to be a realm of objects existing independently of us, there is no reason to consider that V is limited to L. For this reason, Gödel, who had introduced constructible sets to demonstrate the consistency of CH, in fact quickly turned his back on the axiom of constructibility (see Feferman, 1999).

[48] The article predates Cohen's result, although Gödel's philosophical and mathematical commitments are not dependent on it. Gödel moreover affirms that the undecidability of CH is the most probable hypothesis.

[49] Feferman lays out the motivations in favor of these axioms without, however, agreeing with them. He considers that "CH is an inherently vague problem" and that "there is no evidence on the basis of current logical work of the need for new axioms to settle such [arithmetical or finite combinatorial] problems" (1999).

[50] Woodin's idea is not exactly to show that a certain axiom that is remarkable by its evidence or its consequences can decide CH but, more indirectly, that *any* axiom having a certain effect (namely, to inactivate the forcing action up to the level of sets of size $\aleph_1$) can (negatively) decide CH. For an exoteric presentation of Woodin's program and its philosophical implications; see Dehornoy (2007).

as they are for strong platonism (in the way they are indications of the truth of axioms). But the application of the success criterion is not obvious, and it is remarkable to see that on this aspect Quine and Gödel diverge in their preference for extensions of ZF. For Quine, the axiom of constructibility commends itself by its success in making ZF ontologically economical. For Gödel, the axioms of large cardinals commend themselves both by their evidence and by their mathematical consequences.[51] Maybe this divergence bears witness to the fact that the underlying philosophical conception creates a bias on judgments concerning what a natural extension to ZF is. If the existence of sets results from our theories saying they exist, then minimality considerations will "naturally" win out (it is better to do more with less). If, on the other hand, the existence of sets is not posited by our theoretical activity but rather underpinned by a platonic universe of mathematical entities, then maximality considerations will "naturally" win out (all sets whose existence does not lead to contradiction exist).

Moreover, realist positions set themselves apart by maintaining that *one* correct response to the truth of CH question may potentially exist. Although this supposes that there is *one* universe of sets. In recent times, far more liberal versions of platonism have been put forward. Balaguer (1998) defends what he calls a "full-blooded Platonism" according to which any mathematical entity which could exist does in fact already exist.[52] If consistency is adequate in guaranteeing the possibility of existence, a notable result is the existence of both a universe of sets where CH is true and another where CH is false. Hamkins (2010) also defends the idea that a plurality of set-theoretical universes exists, based on the observation that "the most powerful set-theoretical tools developed in the past half-century are most naturally understood as methods of constructing alternative set-theoretical universes." Such versions of realism blur the boundaries with anti-realism, at least as far as the practical consequences which can be drawn from the opposition between realism and anti-realism are concerned.[53]

5. Reasons Not to Be a Platonist

5.1 BENACERRAF'S DILEMMA

The principal objection to platonism is that of epistemological access: how do we come to know the abstract entities whose properties make mathematical statements to be either true or false? This objection makes up one half of the dilemma laid out by

[51] We simplify here, insofar as Gödel himself acknowledges that this is not the case for all the axioms of large cardinals (see especially 1964, n. 20).

[52] More precisely clarifying the principle of maximality, which is constitutive of full-blooded platonism, does not go off without its own hitches (especially see Restall, 2003).

[53] Balaguer admits that full-blooded platonism and fictionalism are positioned back to back: "what I called the metaphysical project—the project of using considerations about mathematical theory and practice to solve the metaphysical problem of abstract objects—*does not work*" (1998, p. 158, emphasis ours). Hamkins seems more unilaterally platonist, insofar as he views methods such as forcing as means for *exploring* different set-theoretical universes.

Paul Benacerraf in his famous paper "Mathematical Truth" (1973): beyond the case of platonism, the problem Benacerraf poses takes the form of one of the major difficulties that philosophy of mathematics must solve. The dilemma is sparked by the meeting of two constraints which come into conflict. It involves

> (1) the concern for having a homogeneous semantical theory in which semantics for the propositions of mathematics parallel the semantics for the rest of language, and (2) the concern that the account of mathematical truth mesh with a reasonable epistemology. (1973, p. 661)

Let us see in more detail what these two constraints, the one semantic, the other epistemological, are. First of all, standard semantical theory tells us that a statement such as "there are twenty-five bridges over the Seine between the pont de Grenelle and the pont de Sully" is true if and only if there are twenty-five distinct objects which have a certain property, namely "being a bridge," and which are in a certain relationship with the pont de Grenelle and the pont de Sully, namely, "physically located between." If we adopt a similar semantical theory for mathematical statements, we must say that a statement like "there are twenty-five prime numbers between one and a hundred" is true if and only if there are twenty-five distinct objects having a certain property, namely "being a prime number," and which are in a certain relationship with one and a hundred, namely "appearing between."

Second, a reasonable epistemology says that for *x* to believe that *p*, a certain causal connection must exist between what *p* is about and the reasons *x* has for believing that *p*. If John believes that the platypus hibernates, but we know that John has never had either direct or indirect contact with platypuses (he has never seen one and has never received any information about them), then we can state, without a shadow of a doubt, that John does not know that the platypus hibernates, independently of and without knowing whether platypuses actually hibernate or not.

The dilemma occurs because it seems that satisfying desideratum (1) leads to not satisfying desideratum (2), and vice-versa. The conceptions that make truth a question of provability within a formal system satisfy (2); if being true is being provable, then it is enough to have a proof for a statement to know that this statement is true. But such conceptions do not satisfy the requisite (1): being true, generally speaking, is not simply a matter of being obtained as the last element in some sequence of symbols making up a formal proof. Platonism is in the opposite situation. If a universe of mathematical entities does exist, then it is certainly possible to provide a standard semantics for mathematical statements. These will be true if they describe this universe adequately, as has been pointed out. Desideratum (1) has been met. But it is (2) which is problematic: if mathematical objects are abstract objects located outside of space and time, what kind of connection could we have with these objects? According to (2), for a true belief to count as knowledge, it is necessary that what makes that belief true be causally responsible for that belief. But abstract objects, located neither in space nor in time, are causally inert.

To escape this dilemma, the Platonist must provide an explanation for the link between our cognitive faculties and known objects. But as Benacerraf points out, simply positing a mathematical intuition does not constitute an answer to the problem. Gödel imagined an analogy between mathematical perception and sense perception. Although, in the case of sense perception, we at least have the beginnings of an explanation for the link that exists between physical objects and the perceptual beliefs we hold about them; we can explain how physical objects create such and such a sensory impression, and the cognitive sciences take up the explanation of how sensory impressions create such and such a perception. No such schema exists in the case of mathematical perception. Worse still, if we accept the thesis of causal inertia of abstract objects, it seems that there could be no such connections even in principle.

The Benacerraf dilemma, or rather the half of it that deals with satisfying (1), at first seems to predominantly constitute an objection to strong platonism. Indeed, strong platonism maintains that a reasonable epistemology of mathematics must make our knowledge of mathematical truths depend on an epistemic access to objects whose existence has been confirmed. This is where the problems begin. A weak Platonist will refuse to take this step: she will wager a reasonable epistemology of mathematics against a reasonable epistemology of the entirety of our theory of the world, refusing to link our mathematical knowledge to mathematical objects. What we know about mathematical entities, we know it simply because the theories that systematize this knowledge are indispensable to science, and thus they are justified, in the same way as the rest of science, by science's various successes. Field (1989) proposed a version of Benacerraf's dilemma intended to be an objection not only against strong platonism but against weak platonism also.[54] According to Field, a reasonable epistemology of mathematics must explain the reliability of our mathematical knowledge. What Field means is that it is not enough to account for a belief in the existence of mathematical entities being justified, or for some specific beliefs about these entities being justified, there must also be "an account of the mechanisms that explain how our beliefs about these remote entities can so well reflect the facts about them" (1991, p. 26). Weak platonism can account for the justification of mathematical theories, but can it account for their reliability? It seems that the epistemological neutrality of weak platonism prevents it from doing so: precisely because mathematical objects are projected from our theories, there is no place for a mechanism explaining the compatibility between our mathematical beliefs and mathematical facts. We could state this in the following way. Weak platonism may think itself safe from objections based on the access problem since it is epistemologically neutral. Field would retort that this neutrality is also a problem. Strong platonism, epistemologically charged, proposes a debatable explanation for reliability. Weak platonism, epistemologically neutral, proposes no explanation for reliability whatsoever. Therefore, weak platonism performs no better than strong

[54] Another merit of Field's formulation is that it does not depend on the adoption of a causal theory of knowledge. See also Casullo (1992).

platonism. For the weak Platonist there remains the possibility of a deflationary response: the reliability of our mathematical beliefs would not need explanation for the precise reason that mathematical reliability is only the projection of theories which (in tandem with empirical theories) display their success.[55] In this sense, Field's criticism boils down to begging the question: he criticizes weak platonism for not giving an explanation which weak platonism chose not to give from the outset.

5.2 ARGUMENTS AGAINST WEAK PLATONISM

In a manner of speaking, weak platonism passes the Benacerraf dilemma test better than strong platonism. Though, for all that, choosing to prop up epistemology of mathematics entirely on a holistic epistemology brings about other difficulties which we will discuss now. The first difficulty is linked to the evidence of elementary mathematics. Weak platonism places mathematics on the same level as the most theoretical elements of the natural sciences. If we follow Quine's holistic conception, mathematical truths are statements situated at the "center" of our conceptual scheme, held back far from all direct confrontation with experience; the only justification they have is indirect, through the long chains of inferences that connect them to experience. As Parsons (1980) remarks, this allows Quinian realism to avoid certain excesses of Mill's empiricism (it is no longer necessary to interpret every mathematical statement as a certain empirical generalization, see section 1 of this chapter for more on this subject). However, this assimilation still does not do justice to the apparent specificities of mathematical truths. The most theoretical parts of science consist in bold hypotheses unifying sets of phenomena and aided by the simplest laws possible. But it is difficult to conceptually equate "2 + 2 = 4" and "$E=mc^2$". "2 + 2 = 4" is not a bold hypothesis, it is an elementary truth about whole numbers. For Parsons, the evidence of elementary mathematical truths can only be explained if we have a privileged access to these truths at our disposal:

> We are taking as a gross fact about arithmetic, that a considerable body of arithmetical truths is known to us in some more direct way than is the case for the knowledge we acquire by empirical reasoning. [. . .] What is more natural than the hypothesis that we have direct knowledge of these truths because the objects they are about are given to us in some direct way? (1980, p. 154)

Positing some privileged access to elementary mathematical truths implies turning away from the epistemological neutrality of weak platonism. The danger then is of falling back into the objections met by strong platonism. Another option is to reject the objection by refusing to accept that privileged epistemic access is the only explanation for this impression of obviousness. From an empiricist viewpoint, the latter

---

[55] For another response to the Benacerraf dilemma, from a Quinian viewpoint, see Steiner (1975).

would be but illusion. Mill had already highlighted that elementary arithmetic is "a truth known to us by early and constant experience" (1843, II, VI, §2). In parallel, a Quinian naturalist could maintain that it is the role of cognitive psychology to explain this impression of obviousness by enlightening us on the mechanisms of mathematical cognition.

At the other end of the chain, another difficulty involves the status of non-applied mathematics. If the ontological significance of mathematical theories is wholly derived from their use in the natural sciences, it follows that mathematical theories, or fractions thereof, which are not used in the natural sciences will have no such ontological significance. Here is what Quine says, following on from the issues in set theory we have discussed:

> So much of mathematics as is wanted for use in empirical science is for me on a par with the rest of science. Transfinite ramifications are on the same footing insofar as they come of a simplificatory rounding out, but anything further is on a par rather with uninterpreted systems. (1984, p. 788)

In other words, mathematics falls under a double regime. In the case of uninterpreted systems, mathematicians show that such and such theorems follow such and such axioms, but they do not show that these theorems are true, and there is no reason to suppose that entities having the properties described by the axioms do exist. In the case of interpreted systems, the axioms are about certain objects (whole numbers, real numbers, sets, etc.) and, by showing a theorem, the mathematician shows that something is true about these objects. But following the weak platonism logic, a mathematical system only acquires the status of interpreted system when it is applied, that is, when it is integrated into the totality of science: there are no interpreted systems which describe the properties of certain mathematical objects independently of some use of these systems in tandem with the theories of physics. The distinction between interpreted systems and uninterpreted systems is thus folded back onto the division between pure mathematics and applied mathematics. Quinian naturalism potentially leads to the introduction of ontological differences to areas where mathematicians place none.[56] As Leng points out, a mathematical theory which does not find its promised applications may see interest in it dropping, although this will not lead to its being considered as false or rejected. Applicability "will make no difference to how a mathematician goes about working in that theory" (2002, p. 408).[57] A Quinian naturalist could nevertheless reply that questions of ontology precisely exceed the competencies

[56] This assimilation is even more problematic given that the fraction of mathematics necessary for science is presumably limited. On the question of the extent of mathematics used in the natural sciences, see, as well as the debates already mentioned regarding Field's program, Feferman (1993).

[57] Leng takes the example of catastrophe theory. For a defense and development of the distinction between "recreational" mathematics and mathematics that constitute genuine knowledge, see Colyvan (1998, 2007).

of the mathematician, since such questions must always be posed relative to our best global theoretical system.

One last objection, partly developing on the previous one, is to remark that Quine and Putnam's position does not account for the fact that mathematicians turn to justification practices which are specific to them. As Maddy (2005) observes, Quine's naturalism is characterized by a bias towards empirical sciences. Quine considers that science must be seen as a totality, justified holistically by its empirical successes. Yet we can, on the contrary, be aware of and open to the diversity of disciplines which make up science. Mathematicians' method is not that of physicists. If the philosopher takes naturalism to consist of an abandonment of "first philosophy," by which we mean not attempting to be "cleverer" than the scientists, then the philosopher should definitely not attempt to be cleverer than the mathematicians in attributing standards of justification to mathematics which are foreign to it. This objection amounts to turning the double standard argument around. The indispensability argument was based on the idea that no double standard should be adopted in regards to the existential commitments of our theories. Maddy's objection to Quine's position is based on the idea that no double standard should be adopted in regards to respect for the methodology of scientists. One possible answer consists of contesting the existence of a gap between mathematical methods and natural science methods. Echoing Putnam, it can be said that "we have been using quasi-empirical and even empirical methods in mathematics all along" (1975a, p. 64). The example Putnam uses is the birth of analytical geometry. Descartes posits that one number—a real number—corresponds to each point on a straight line. This hypothesis was adopted because it turned out that it was rewarding, as much for pure mathematics as for applied mathematics (mechanics in this instance). These common points stretch out to cover the indirectly empirical elements of scientific methodology. Kitcher, for example, maintains that the unificationist theory of explanation enables simultaneously to account for mathematical explanations and also explanations from the natural sciences. Kitcher gives the example of the explanatory role played by the axiom system which characterizes a theory, in this case group theory:

> Similarly, the standard set of axioms for group theory covers both the finite and the infinite groups, so that we can provide derivations of the major theorems that have a common pattern, while the alternative set of axioms for the theory of finite groups would give rise to a less unified treatment in which different patterns would be employed in the finite and in the infinite cases. (1989, p. 437)

The unifying virtues which validate the choice of axioms of group theory are, in Kitcher's mind, exactly like the unifying virtues which validate, for example, the choice of the principles of mechanics. However, it's one thing to provide a reminder that the methodology of mathematics and that of the empirical sciences are not as far removed as we may believe, it's quite another thing to advance that no important difference between them is worthy of naturalists' attention. Kitcher, for example, may defend the idea that one and the same concept of explanation is just as valid in mathematics as it is in physics, this does not, however, imply that the facts to be explained, coming either from physics or from mathematics, are of the same nature. Similarly, the examples Putnam borrows from the history of mathematics show that applications outside of mathematics can also drive development within it. Still, it is quite clear that this is not always the case and that numerous theoretical developments in mathematics are based on purely mathematical considerations.

6. Naturalizing Platonism

6.1 DO WE SEE SETS?

Strong platonism goes too far: it posits a world of mathematical entities and a *sui generis* mathematical intuition to guarantee both the truth of classical mathematics and also our epistemic access to these truths. Weak platonism doesn't go far enough: the indispensability argument guarantees the truth of classical mathematics, but the differences between practices of justification in mathematics and in natural science are not acknowledged. So, it seems tempting to seek out a middle road which, while securing the truth of mathematics against their application in the sciences, would also account for the specifics of mathematics' own modes of justification. Notably, this would be a question of acknowledging the role played by a mathematical intuition that would be acceptable to the norms of naturalism. Maddy (1990) takes such a middle road in promoting a naturalized version of strong platonism. Maddy maintains that a naturalist philosophy of mathematics should not stop at the indispensability argument and that mathematical intuition is not incompatible with naturalism. Maddy's view is that there are no reasons, at least as far as sets are concerned, for dissociating mathematical intuition from perception. We do not only perceive colors, forms or objects, we also perceive sets of objects. Mathematical intuition, at least when it comes to set theory, would reside in our ability to perceive sets.

Let us examine the reasoning behind this proposed naturalization of mathematical intuition by taking up an example given by Maddy (1990). According to Maddy, when Steve opens the refrigerator door looking for the eggs needed for a recipe and sees three eggs sitting in the box, what he sees is just a set of three eggs. To say that Steve perceives a *set* of three eggs is to commit to several problematic points. First, it is acknowledging that sets exist (otherwise they could not be perceived). Second, it is admitting that impure sets (those not formed from the empty set but from sets of physical objects) have an "ordinary" spatio-temporal existence (the set of three eggs will cease to exist as soon as Steve cracks the first one). Third, the belief that there are three eggs must be a perceptual belief which is not based on inferences.[58] Fourth, the belief about the three eggs really is a belief about sets (and not about aggregates or

[58] If we consider belief to be inferential, then we must explain what justifies these inferences, as well as run the risk of regression, though this does not mean that this option is completely barred.

mereological sums, etc.). As reinforcement to the first and fourth points, the naturalist can call on the indispensability argument: in any case, we should suppose that sets exist since set theory is among our best scientific theories, and if we are to consider that it is sets that exist rather than aggregates or mereological sums then it is because it is set theory, rather than some aggregate theory or mereology, which is used in our best scientific theories. The second point is the price to be paid for the naturalization of platonism. To naturalize the set theory intuition, sets have to be allowed to occur in the physical world. The third point is the point where the naturalist commits herself as a naturalist: in order for Steve's belief that there are three eggs in the box to be open to consideration as a perceptual belief, Steve must be able to *perceive* three eggs or a set of three eggs. Perception transforms sensory data to provide us with a world of visible objects. For Steve to be able to perceive three eggs, perception must likewise transform sensory data to provide us with a world of visible sets.

> "[. . .] the hope is that something like what does the bridging in the case of physical object perception can be seen to do the same job in the case of set perception." (Maddy, 1990, p. 50)

Maddy speaks of hope; so, the naturalization of the set-theory intuition remains programmatic.

In the perspective of Quine's naturalism, unobservable entities, physical objects and mathematical objects all share the same status as myths, to use Quine's terms, developed in order to account for experience.[59] But mathematical objects remained, just like electrons or quarks, a *higher order* myth, destined to simplify the myth of physical objects. What distinguishes physical objects like apples or chairs, as first order myths, from unobservable entities like electrons or quarks, being higher myths, is that the first lot, as opposed to the latter, constitute a directly engaging part of our experience of the world. The myth is already developed on the perceptual level. Maddy's program consists of showing that, to a certain extent at least, sets are first order myths and not higher myths as Quinian orthodoxy would have it.

Sets can be considered as basic elements in contemporary mathematics, to the extent that all of these mathematics, in principle at least, are reconstructible within set theory. Nevertheless, it is far less obvious that the set intuition may be considered to be a basic element of mathematical intuition, or that every mathematical intuition be based on a set-like intuition. This is a problematic aspect of Maddy's naturalist approach, an aspect which resides in the promised articulation between the indispensability argument and the naturalization of the intuition. A priori there is no reason that the two should agree perfectly, that is, a priori there is no reason that what forms the basis of mathematics in our best theory of the world (namely, the set universe

[59] "A platonist ontology [. . .] is, from the point of view of a strictly physicalistic conceptual scheme, as much a myth as that physicalistic conceptual scheme itself is for phenomenalism. This higher myth is a good and useful one, in turn, insofar as it simplifies our account of physics" (Quine, 1953a).

studied within set theory) should simultaneously be the object of our mathematical intuition. After all, set theory is a late comer among mathematical theories and basing that theory on a perceptual ability seems more problematic than in the case of theories like arithmetic or geometry.

6.2 STRUCTURALISM AND INTUITION

A promising lead to follow would consist of generalizing Maddy's strategy by widening the naturalization of mathematical intuition program beyond set theory. The difficulty then resides in making our conception of what mathematical objects are (*qua* objects) compatible with a naturalist conception of the intuition of these objects underpinned by sensory perception. If numbers are to be objects like apples or chairs, it is only too clear that we do not perceive numbers. More positively speaking, what conception of the nature of mathematical objects must be adopted for it to be possible to partly base our epistemic access to these objects on sensory perception?

In philosophy of contemporary mathematics, structuralism—a now popular label regrouping positions which are in part heterogeneous—is the conception which seems the best able to form a response to the preceding question.[60] The idea common to the different forms of structuralism is that to do mathematics is to study structures, and that mathematical objects, such as numbers, are but positions within these structures. As Resnik puts it:

> in mathematics the primary subject matter is not the individual mathematical objects but rather the structures in which they are arranged. The objects of mathematics [. . .] are themselves atoms, structureless points or positions in structures. And as such they have no identity or distinguishing features outside of a structure. (1997, p. 201)

The structuralist conception enables us to underpin a mathematical intuition with perception, insofar as we are able to perceive not only objects but also patterns. Our perceptive link to mathematical structures resides in our capacity to perceive the manner in which physical objects are organized, insofar as systems of objects can instantiate mathematical structures. When a system of physical objects is organized in a manner which corresponds to certain structural properties of mathematical objects, our perception of this system of physical objects can inform us of the structural properties of mathematical objects. Resnik illustrates this point with the example of the elementary

[60] Structuralism is associated with authors such as Benacerraf (1965), Hellman (1989), Resnik (1997), Shapiro (1997), and, slightly more reservedly, Parsons (2008). Historically, Dedekind (1888) is often presented as the first structuralist. Evidence of this is the following characterization of natural numbers: "If in the consideration of a simply infinite system N set in order by a transformation φ we entirely neglect the special character of the elements; simply retaining their distinguishability and taking into account only the relations to one another in which they are placed by the order-setting transformation φ, then are these elements called *natural numbers* [. . .]" (1888, §73)

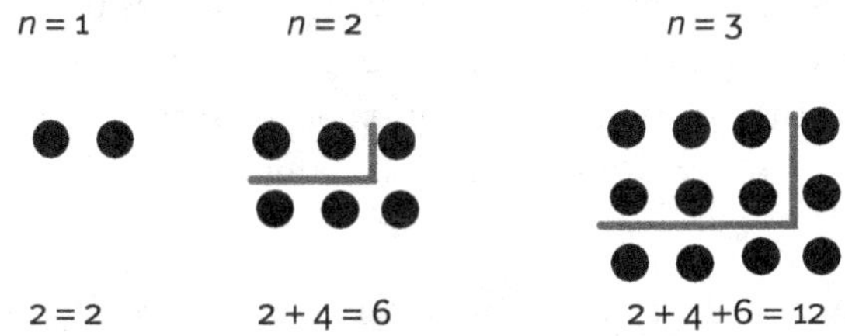

FIGURE 2 A visual proof that the sum of the first n even whole numbers is n(n+1)

arithmetic theorem stating that the sum of the first n even whole numbers is $n(n+1)$. This theorem can be formally demonstrated by deriving it from Peano's axioms, but it can also be given an intuitive proof based on the possibility of arranging the points representing the sum of the first n even whole numbers in a rectangle of length $n+1$ and of width n.

The crucial structural property here is that the nth rectangle is obtained by adding $2n$ points to the preceding rectangle, in such a way as the number of points of the nth rectangle give the sum of the first n even whole numbers.

The link between mathematical knowledge and perception must be qualified in at least two respects. First, the structuralists must not be accorded more than they have actually shown. The ability for "pattern recognition" is a supposition and a more detailed explanation of it is left to psychology.[61] Nevertheless, structuralists should not be too quickly satisfied with a naturalist division of labor under which the philosopher of mathematics' work would consist of showing in which sense mathematical objects are only positions in structures, whereas the psychologist's work would be to discover and explain the mechanisms of pattern recognition. It still remains to be established, on the one hand that the type of things mathematical structures are for the philosopher of mathematics is liable to be instantiated by the type of things patterns are for the psychologist and, on the other hand, that when our mathematical knowledge relies on empirical elements it is indeed pattern recognition, as we laid out, which comes into play.

Second, mathematical intuition underpinned by just perception could not constitute our only mode of access to mathematical structures. The discrete systems we perceive are finite systems which can only instantiate finite structures, but mathematics obviously does not study only finite structures. In the prior example, a crucial element of the demonstration is the supposition that it is always possible to iterate the arrangement of points into rectangles just like it is always possible to pass from one even number to the next. The nature of our ability to perceive sequences of drawn rectangles as patterns that can be stretched out has to this day never been explained.

[61] For example, Shapiro: "pattern recognition represents a sticky problem for psychology and cognitive science. There is no consensus among scientists as to how it works. Nevertheless, humans clearly can recognize at least some patterns" (1997, p. 12).

Structuralists like Shapiro (1997, chapter 4), alongside access through perception and pattern recognition, admit other modes of access to structures (typically, structures can be defined implicitly as satisfying certain axioms). Again, the articulation between these different modes of access demands explanation.

6.3 ARGUMENTS IN FAVOR OF STRUCTURALISM

Structuralism's merits do not consist exclusively in the epistemological reasons we have just developed. Indeed, the argument most frequently advanced in favor of structuralism is certainly not this epistemological one but rather a strictly ontological argument due to Benacerraf (1965). The starting point of the argument consists of remarking that within set theory there exist two normal ways of identifying natural numbers. The first way comes from Zermelo, the other from von Neumann. In both cases zero is identified with the empty set. Zermelo's procedure consists of identifying $n+1$ with the set whose only element is n. Zermelo's sequence of numbers is therefore given by $\varnothing$, $\{\varnothing\}$, $\{\{\varnothing\}\}$, $\{\{\{\varnothing\}\}\}$, etc. Von Neumann's procedure consists of identifying $n + 1$ with the set of its predecessors. Therefore, von Neumann's sequence of numbers is given by $\varnothing$, $\{\varnothing\}$, $\{\varnothing,\{\varnothing\}\}$, $\{\varnothing,\{\varnothing\}\{\varnothing,\{\varnothing\}\}\}$, etc. Benacerraf's argument is based on the fact that if numbers are objects "like others," then they must either be Zermelo's numbers or else von Neumann's numbers (or else other objects altogether). But if, for example, numbers are von Neumann's numbers, then it is true that 0 belongs to 3, though this will be false if numbers are in fact Zermelo's numbers. And if numbers are neither those objects identified by Zermelo nor those identified by von Neumann, but indeed some other objects, then they will certainly have still more distinctive properties possessed neither by Zermelo's nor von Neumann's numbers.[62] The problem lies in the fact that choosing between Zermelo's or von Neumann's numbers, or between them and some totally different system, does not make sense. Likewise, deciding whether 0 does or does not belong to 4 does not make sense. If we do not wish to choose between von Neumann's whole numbers and Zermelo's, then we are tempted to say that they are both equally good candidates, to the extent that they both instantiate the structure of natural numbers. Or, to word it as a formula, whole numbers are neither Zermelo's whole numbers nor von Neumann's but rather they are what these have in common, namely certain structural properties.

Beyond Benacerraf's argument, structuralism can also claim compatibility with mathematicians' actual practices (Reck and Price, 2000). Thus, mathematicians study the structural properties of the entities which interest them while disregarding

[62] Benacerraf's argument is first an argument against the thesis that numbers are sets of some kind. The extension of this argument into a more general one establishing that numbers can be neither sets, nor any other sort of object in the normal sense, is more problematic. What is meant by "objects in the normal sense" is vague and would have to be made clear and precise for us to be able to evaluate the hypothesis crucial to the argument's correction, namely, that whatever the objects in the normal sense that may be chosen, problems linked with the supplementary properties (independent of the structure of whole numbers) of these objects will be encountered.

the specific nature of these same entities. Many moments can be found which bear witness to this. Down the history of mathematics we see it notably through works on reduction. Several definitions of real numbers as sets of rational numbers have been proposed (such as Dedekind cuts or equivalence classes in Cauchy sequences). Just like in Benacerraf's example, no definition is better than any other since, in any case, the real numbers defined have the expected structural properties, so it makes no difference whether the specific nature of such and such a real number be identified with this set of rational numbers and not some other one. Abstract algebra gives us another striking example: we take groups, rings and fields in order to study their general properties and then class them. Many different systems can instantiate a group structure. Each time, the only important thing is the properties these systems have *qua* groups (rings or fields). Similarly, in mathematical logic, the languages used in formalizing mathematical theories are such that two structures of interpretation which are isomorphic will satisfy the same statements. It is remarkable that this applies just as well for classical first order languages, for the extensions of first order logic by the addition of new quantifiers, as it does for higher order logics. If being isomorphic implies satisfying the same statements, then only the properties of structures preserved through isomorphisms count. That is, only the "structural" properties of structures.[63]

6.4 VARIETIES OF STRUCTURALISM

The structuralist position, as we have thus far presented it, remains under-determined. We said that mathematics studies structures before it studies objects, in the sense that only the structural properties of objects are pertinent to the truth or falsity of mathematical statements. We did not say what the structures studied were, nor what was the relationship between objects and structures. One way of broaching the subject would be to ask what makes a mathematical statement true. Let us consider, for example, the statement φ from arithmetic language, "there are infinitely many prime numbers." According to a first variant of structuralism, φ is true if and only if the structure of the whole numbers makes φ true. By "structure of the whole numbers," what is to be understood is

> a single *abstract structure*, the pattern common to any infinite collection of objects that has a successor relation, a unique initial object, and satisfies the induction principle. (Shapiro, 1997, p. 72)

This variant of structuralism is known by the name "*ante rem* structuralism" (Shapiro, 1997, by analogy to the quarrel of universals), "pattern structuralism" (Reck and Price,

[63] Though the formula may sound a little tautological, this does not mean the fact of it is trivial: if we can speak of *structures* of interpretation this is precisely because the only thing that counts is the structural properties of interpretations.

2000) or "non-eliminativist structuralism" (Parsons, 2008).[64] *Ante rem* structuralism distinguishes itself from the strong platonism presented earlier in that, for example, it does not acknowledge the number 2 as having an independent existence. 2 is only a position in the structure of the natural numbers. It does, however, agree with strong platonism in admitting that what mathematics is occupied with (structures and not objects) exists independently of any instantiation (the structure of the integers exists even if it is instantiated by no system of physical objects).

According to a second variant of structuralism, φ is true if and only if every infinite system which makes the axioms of arithmetic hold true[65] also makes φ hold true. The structure of the integers makes φ hold true. This variant is known by the name of "*in rebus* structuralism" (Shapiro, 1997), or "eliminativist structuralism" (Parsons, 2008) and it is a version of Reck and Price's (2000) "universalist structuralism." The idea is to not hypostasize mathematical structures existing independently of the systems which exemplify them and to interpret mathematical statements as universal affirmations concerning all systems of a certain kind. *In rebus* structuralism is not (at all) a form of platonism, since neither the mathematical objects nor structures exist independently of the systems which exemplify them. A dangerous consequence is that if no physical system exists to exemplify the structures which are the subject of such a mathematical theory, then all the statements of the mathematical theory in question will be true. For example, if there is no physical system which makes the axioms of arithmetic hold true then, trivially, no physical system making the axioms of arithmetic hold true is liable to falsify a statement of arithmetic.

A third variant of structuralism aims at conserving the spirit of eliminativist structuralism while also providing a solution to the problem just raised. This time we say that φ is true if and only if, for every possible system S, if S makes the axioms of arithmetic hold true, then S also makes φ hold true. This is a modal variant of structuralism known, precisely, by the name "modal structuralism" and developed in detail by Hellman (1989). The idea is that, even if infinitely many objects do not actually exist, so that no real system can make the axioms of arithmetic hold true, infinitely many objects and systems making the axioms of arithmetic hold true could still exist. Consequently, arithmetic truth is not suddenly trivialized in the absence of real infinite systems. Modal structuralism is not *prima facie* a strong platonism, since it does not admit mathematical structures existing independently of the systems which

---

[64] On the nuances to be drawn on the identification between *ante rem* structuralism and non-eliminativist structuralism, see Parsons (2008, p. 52).

[65] We purposely leave undetermined what is to be understood by "the axioms of arithmetic." If it is to be understood as Peano's second-order arithmetic, which characterizes the structure of integers up to isomorphism, then truth in the eliminativist structuralism sense will be equivalent to truth in the sense of non-eliminativist structuralism, on condition that there be at least one infinite system that makes the axioms in question hold true. If it is to be understood as Peano's first-order arithmetic, which admits models that are not elementarily equivalent, then truth in the eliminativist structuralism sense will not be equivalent to truth in the non-eliminativist structuralism sense.

exemplify them. Nevertheless, an exact evaluation of the ontological engagements of modal structuralism depends on the analysis that would be made of modalities.

Besides the question of the exact ontological interpretation given for structuralism, the problem of its application to a theory such as set theory also arises.[66] Set theory takes on the role of a background theory in which it is possible to define the systems that instantiate the various mathematical structures studied, as has already been said regarding the natural numbers and real numbers. But then what about sets themselves taken as mathematical objects? Must they also be seen as positions in a structure, that of the set-theoretical universe? While mathematics has given us the habit of seeing integers or real numbers as a structure liable to exemplification through different systems, this does not apply in the case of set theory: we don't have (or we have less) the habit of interpreting the membership relationship by another relationship between objects which are not sets. Above all, providing a structuralist interpretation for the set-theoretical universe is problematic, insofar as set theory is used in defining what a structure is, as is done in model theory. Faced with this difficulty, several solutions are possible. We can consider the notion of structure to be a primitive notion and, following Shapiro, envisage a structure theory which would fit alongside set theory. Such a solution is certainly not very economical. Another option would be to make an exception for set theory and only adopt a structuralist interpretation for the other theories.

7. Conclusion

Philosophy of mathematics quite easily (if not indeed, too easily) allows itself to be described as the battleground for conflict between several schools of thought. These divisions are partly inherited from the philosophical tradition (realism *vs* nominalism, as well as platonism vs. aristotelianism). Though they also partly find their origin in the developments of logic (logicism) or in the reaction to the foundational crisis (finitism, intuitionism). They are also determined by the more general theoretical choices engaged by contemporary philosophy in its entirety (naturalism). In this introduction, while presenting these different frameworks, we have sought to show how the two tasks incumbent to philosophy of mathematics have been articulated at each stage: first is a strictly epistemological task to account for mathematical knowledge in terms of what it may or may not have in common with other scientific knowledge, then comes the ontological task of accounting for what mathematical objects are, or, more broadly, what it is that mathematics studies.

We will conclude by saying a couple of words about what seem to us to be the major challenges in philosophy of mathematics. Concerning the epistemological task, at least three elements have already been identified which seem to direct the formation of mathematical knowledge. First, a certain mathematical intuition whose links with perception and acceptability from a naturalist point of view are problematic. Second,

[66] On this question, see Parsons (2008, chap. 4).

general theoretical criteria, such as consistency, simplicity, or unifying power, whose impact is real but whose sufficiency in explaining what mathematics is can be doubted. Third, the application of mathematical theories outside of mathematics, something which plays an important role in ontological discussions but whose epistemological significance is less clear. One of the first challenges for philosophy of contemporary mathematics is to clarify the functioning of these different modes of mathematical development, to say if they also constitute modes of justification and to explain how, should this turn out to be the case, these different modes of justification coexist.

Regarding the ontological task, the challenges certainly vary according to whether an anti-realist or a realist perspective is adopted. In the first case, the matter often comes down to showing that it is possible to be anti-realist, and this matter in turn partly depends on mathematical realizations: a paradigmatic example of this is Field's program and the nominalist reconstruction of science. In the second case, what is at stake, it seems, is the elaboration of some notion of object that would be adequate for mathematical objects, in the sense that it would account for their ontological specificity and that it could be integrated into an explanation of the modes of mathematical justification. Notably, this is the reason we chose to present structuralism on the basis of considerations about the naturalization of mathematical intuition.

Two striking characteristics of recent time which we have already encountered in passing during this exposition but to which we deem it fitting to come back in concluding are, exterior to philosophy of mathematics, advances concerning our understanding of mathematical cognition and, within philosophy of mathematics, closer and more acute attention to the actual practice of mathematics. On the first point, a remarkable example is the case of arithmetic cognition, through the development of subtle hypotheses on the different cognitive systems at work, their having or not having a symbolic character or an innate origin (see Dehaene 1997). On the second point, the study of diagrammatic reasoning and the role played by visualization constitutes yet another remarkable example (see Mancosu et al., 2005). The integration of these new elements into the general epistemological and ontological perspectives we have developed here is the ultimate challenge we set out to highlight.

References

Ackermann, W. (1940) "Zur Widenspruchsfreiheit der Zahlentheorie," *Mathematische Annalen*, 117 (1), 162–194.

Balaguer, M. (1998) *Platonism and Anti-platonism in Mathematics*, Oxford: Oxford University Press.

Beltrami, E. (1868) "Saggio di interpretazione della geometria non-euclidea," *Giornale di Mathematische*, 6, 285–315.

Benacerraf, P. (1965) "What Numbers Could Not Be," *Philosophical Review*, 74, 47–73.

Bishop, E. (1967) *Foundations of Constructive Analysis*, New York: McGraw-Hill.

Boolos, G. (1986) "Saving Frege from Contradiction," *Proceedings of the Aristotelian Society*, 87, 137–151.

Bourbaki, N. (1956) *Éléments de mathématique. Livre 1: Théorie des ensembles*, Paris: Hermann, Actualités scientifiques et industrielles, English trans. *Elements of Mathematics, Theory of Sets*, Paris: Hermann/Reading: Addison-Wesley, 1968.

Brouwer, L. E. J. (1907) "Over de grondslagen der wiskunde," PhD thesis, University of Amsterdam, English trans. *in* A. Heyting (ed.), *L. E. J. Brouwer. Collected Works I. Philosophy and Foundations of Mathematics*, Amsterdam: North-HollandAmsterdam, pp. 11–104.

Brouwer, L. E. J. (1908) "De Onbetrouwbaarheid der logische Principes, Door L. E. J. Brouwer," *Tidjschrift voor Wijsbegeerte*, 2, 152–158.

Brouwer, L.E.J. (1927) "Über Definitionsbereiche von Funktionen," *Mathematische Annalen* 97, 60–75.

Carnap, R. (1937) *The Logical Syntax of Language*, London: Routledge and Kegan Paul.

Casullo, A. (1992) "Causality, Reliabilism, and Mathematical Knowledge," *Philosophy and Phenomenological Research*, 52 (3), 557–584.

Church, A. (1956) *Introduction to Mathematical Logic*, Princeton, NJ: Princeton University Press

Church, A. (1966) "Paul J. Cohen and the Continuum Problem" *in* G. Petrovsky (ed.), *Proceedings of the International Congress of Mathematicians*, Moscow: Izdatel'stvo "Mir," 1968, pp. 15–20.

Colyvan, M. (2001) *The Indispensability of Mathematics*, Oxford: Oxford University Press.

Colyvan, M. (2007) "Mathematical Recreation Versus Mathematical Knowledge," *in* M. Leng, et al. (eds.), *Mathematical Knowledge*, Oxford: Oxford University Press, pp. 109–122.

Van Dalen D., and van Atten, M. (2002) "Intuitionism," *in* D. Jacquette (ed.), *A Companion to Philosophical Logic*, Oxford: Blackwell, 2002, pp. 513–553.

Dedekind, R. (1888) "Was Sind und was sollen die Zahlen?," Braunschweig: Vieweg, English trans. by W. W. Beman (2012), *Essay on the Theory of Numbers*, New York: Dover.

Dehaene, S. (1997) *The Number Sense*, New York: Oxford University Press.

Dehornoy, P. (2007) "Au-delà du forcing: la notion de vérité essentielle en théorie des ensembles," *in* J. B. Joinet (ed.), *Logique, dynamique et cognition*, Paris: Publications de la Sorbonne (2007), pp. 147–170.

Detlefsen, M. (1990) "On an alleged refutation of Hilbert's Program Using Gödel First Incompleteness Theorem," *Journal of Philosophical Logic*, 19 (4), 343–377.

Dummett, M. (1973) "The Philosophical Basis of Intuitionistic Logic," in *Truth and Other. Enigmas*, Cambridge, MA: Harvard University Press, pp. 215–247.

Dummett, M. (1977) *Elements of Intuitionism*, New York: Oxford University Press.

Feferman, S. (1988) "Weyl Vindicated: Das Kontinuum Seventy Years Later," *repr. in* S. Feferman (1998), *In the Light of Logic*, New York: Oxford University Press, pp. 249–283.

Feferman, S. (1993) "Why a Little Bit Goes a Long Way: Logical Foundations of Scientifically Applicable Mathematics," *Proceedings of the Philosophy of Science Association*, 2, 442–455.

Feferman, S. (1999) "Does Mathematics Need New Axioms?," *American Mathematical Monthly*, 106, 99–111.

Fermat, P. de (1643) *De solutione problema tum geometricorum per curvas simplicissimas et unicuique problematum generi proprie convenientes dissertatio tripartita*, in *Oeuvres*, ed. Ch. Henry and P. Tannery, Paris: Gauthier-Villars, 1891, t. 1.

Field, H. (1980) *Science without Numbers: A Defence of Nominalism*, Oxford: Blackwell.

Field, H. (1989) *Realism, Mathematics and Modality*, New York: Basic Blackwell.

Frege, G. (1879) *Begriffsschrift, eine der arithmetischen nachgebildete Formelsprache des reinen Denkens*, Halle: Nebert, English trans. by S. Bauer Mengelberg as "Concept Notation: A Formula Language of Pure Thought, Modelled upon That of Arithmetic," in J. van Heijenoort (ed.) (1967), *From Frege to Gödel: A Sourcebook in Mathematical Logic*, 1879–1931, Cambridge, MA: Harvard University Press.

Frege, G. (1884) *Die Grundlagen der Arithmetik, eine logisch-mathematische Untersuchung über den Begriff der Zahl*, Breslau: W. Koebner, English trans. by J. L. Austin (1974), *The Foundations of Arithmetic: A Logic-Mathematical Enquiry into the Concept of Number*, 2nd rev. ed., Oxford: Blackwell.

Frege, G. (1893, Band 1 and 1903, Band 2) *Grundgesezte der Arithmetik, begriffsschriftlich abgeleitet*, Band 1, Iéna: Pohle, English trans. by P. A. Ebert and M. Rossberg (2013), *Basic Laws of Arithmetic*, Oxford, Oxford University Press.

Frege, G. (1924–1925) "Sources of Knowledge of Mathematics and Natural Sciences," *in Posthumous Writings*, English trans. by P. Long and R. White, Chicago: University of Chicago Press, 1979, pp. 267–274.

Friedman, H. (1976) "Systems of Second Order Arithmetic with Restricted Induction," I, II (résumés), *Journal of Symbolic Logic*, 41, 551–560.

Gentzen, G. (1936) "Die Widerspruchfreiheit der reinen Zahlentheorie," *Mathematische Annalen*, 112, pp. 493–556, English trans. *in* G. Gentzen (1969), *The Collected Papers of Gerhard Gentzen*, Amsterdam: North-Holland, pp. 132–213.

Gödel, K. (1931) "On Formally Undecidable Propositions of Principia Mathematica and Related Systems I," *in* S. Feferman et al. (eds.) (1986), *Kurt Gödel Collected Works*, vol. I, Oxford: Oxford University Press, pp. 145–195.

Gödel, K. (1944) "Russell's Mathematical Logic," reprinted *in* S. Feferman et al. (eds.) (1990), *Kurt Gödel Collected Works*, vol. II, Oxford: Oxford University Press, pp. 119–141.

Gödel, K. (1951) "Some Basic Theorems on the Foundations of Mathematics and Their Implications," Josiah Willard Gibbs Lecture, American Mathematical Society, *in* S. Feferman et al. (eds.) (1995), *Kurt Gödel Collected Works*, vol. III, Oxford: Oxford University Press, pp. 304–323.

Gödel, K. (1953/1959,V) "Is Mathematics Syntax of Language?," *in* S. Feferman et al. (eds.) (1995), *Kurt Gödel Collected Works*, vol. III, Oxford: Oxford University Press, pp. 356–362.

Gödel, K. (1958) "Uber eine bisher noch nicht benützte Erweiterung des finiten Standpunktes" *in* Feferman et al. (eds.) (1995), *Kurt Gödel Collected Works*, vol. II, Oxford: Oxford University Press, pp. 240–251.

Hale, B., and Wright, C. (2002) "Benacerraf's Dilemma Revisited," *European Journal of Philosophy*, 10 (1), pp. 101–129.

Hamkins, J. (2010) "The Set-Theoretic Multiverse: A Model-Theoretic Philosophy of Set Theory," delivered at the conference *Philosophie et Théorie des Modèles*, Paris, June 2010.

Hellman, G. (1989) *Mathematics without Numbers*, Oxford: Oxford University Press.

Heijenoort, J. van (ed.) (1967) *From Frege to Gödel. A Source Book in Mathematical Logic, 1897-1931*, Cambridge, MA: Harvard University Press.

Heyting, A. (1956) *Intuitionism, An Introduction*, Amsterdam: North–Holland.

Hilbert, D. (1904), "Über die Grundlagen der Logik und der Arithmetik," *Verhandlungen des dritten Internationalen Mathematiker-Kongresses in Heidelberg vom 8. bis 13. August 1904*, Leipzig: Teubner, 1905, pp. 174–185. Quoted from the English translation in J. van Heijenoort (ed.) (1967), pp. 129–138.

Hilbert, D. (1905) "Über die Grundlagen der Logik und der Arithmetik," *in* A. Kratzer (eds.), *Verhandlungen des dritten Internationalen Mathematiker-Kongresses in Heidelberg vom 8. bis 13. August 1904*, Leipzig, Tuebner, pp. 174–185.

Hilbert, D. (1922) "Neubegründung der Mathematik (Erste Mitteilung)," *Abhandlungen aus dem Mathematischen Seminar der Hamburgischen Universität*, 1, p. 157–177

Hilbert, D. (1925) "Über das Unendliche," Conference given at Münster, June 4, 1925, *Mathematische Annalen*, 95-1926, pp. 161–190.

Hilbert, D., and Bernays, P. (1934) *Grundlagen der Mathematik*, vol. 1, Berlin: Springer Verlag, French trans. M. Guillaume et al. (2001), Paris: L'Harmattan.

Kitcher, Ph. (1989) "Explanatory Unification and the Causal Structure of the World," *in* Ph. Kitcher and W. Salmon (1989), *Scientific Explanation*, Minnesota Studies in the Philosophy of Science, vol. XIII, Minneapolis: University of Minnesota Press, pp. 410–505.

Kleene, S. C. (1952) *Introduction to Metamathematics*, Amsterdam: North-Holland.

Leng, M. (2002) "What's Wrong with Indispensability?," *Synthese*, 131, 395–417.

Linnebo, Ø. (2009) "Platonism in the Philosophy of Mathematics," *The Stanford Encyclopedia of Philosophy*, Zalta, E.N. (ed.), http://plato.stanford.edu/archives/fall2009/entries/platonism-mathematics/.

Löb, M. H. (1955) "Solution of a Problem of Leon Henkin," *Journal of Symbolic Logic*, 20, 15–118.

Maddy, P. (1990) *Realism in Mathematics*, Oxford: Oxford University Press.

Maddy, P. (2005) "Mathematical Existence," *Bulletin of Symbolic Logic*, 11 (3), 351–376.

Malament, D. (1982) "Review of Field's *Science without Numbers*," *Journal of Philosophy*, 79, 523–534.

Mancosu, P., Jorgensen, K. F, and Pedersen, S. A. (eds.), (2005) *Visualization, Explanation and Reasoning Styles in Mathematics*, Synthese Library, Berlin: Springer.

Mill, J. S. (1843) *A System of Logic*, London: John William Parker.

Moore, G. H. (1982) *Zermelo' s Axiom of Choice*, New York: Springer.

Moschovakis, Y. (2009) *Descriptive Set Theory*, Providence, RI: American Mathematical Society.

Müller, A. (1923) "Über Zahlen als Zeichen," *Mathematische Annalen*, 90, pp. 153–158

Pappus, (1876–1878) *Collectionis quae supersunt*, F. Hultsch (ed.), Berlin: Weidmann.

Parsons, Ch. (1980) "Mathematical Intuition," *Proceedings of the Aristotelian Society*, 80, 145–168.

Parsons, Ch. (1995) "Platonism and Mathematical Intuition in Kurt Gödel's Thought," *Bulletin of Symbolic Logic*, 1 (1), 44–74.

Parsons, Ch. (2008) *Mathematical Thought and Its Objects*, Cambridge, MA: Cambridge University Press.

Pasch, M. (1882) *Vorlesungen über neuere Geometrie Vorlesungen über der neuere Geometrie*, Leipzig: Teubner (quoted from 2nd ed., 1912).

Poincaré, H. (1906) "Les mathématiques et la logique," *Revue de métaphysique et de morale*, 14, 294–317.

Putnam, H. (1979) "Mathematical Truth," *in Mathematics Matter and Method: Philosophical Papers, vol. 1*, 2nd ed., Cambridge: Cambridge University Press, pp. 60–78.

Quine, W. V. O. (1953a) "On What There Is," *in From a Logical Point of View*, Cambridge, MA: Harvard University Press, pp. 1–19.

Quine, W. V. O. (1953b) *From a Logical Point of View*, Cambridge, MA: Harvard University Press.

Quine, W. V. O. (1984) "Review of Charles Parsons' *Mathematics in Philosophy*," *Journal of Philosophy*, 81, 783–794.

Quine, W. V. O (1986) "Reply to Charles Parsons" *in* K. Hahn and P. Schilpp (eds.), *The Philosophy of W. V. Quine*, La Salle: Open Court, 396–404.

Quine, W. V. O. (1992) *The Pursuit of Truth*, Cambridge, MA: Harvard University Press.

Oumraou, L. (2009) *Pourquoi les mathématiques sont-elles difficiles?* Paris: Vuibert.

Reck, E., and Price, M. P (2000) "Structures and Structuralism in Contemporary Philosophy of Mathematics," *Synthese*, 125, 341–383.

Resnik, M. D. (1980) *Frege and the Philosophy of Mathematics*, Ithaca, NY: Cornell University Press.

Resnik, M. D. (1997) *Mathematics as a Science of Pattern*, Oxford: Oxford University Press.

Restall, G. (2003) "Just What Is Full-blooded Platonism?," *Philosophia Mathematica*, 11 (1), 82–91.

Sabatier, X. (2009) *Les formes du réalisme mathématique*, Paris: Vrin.

Shapiro, S. (1997) *Philosophy of Mathematics: Structure and Ontology*, Oxford: Oxford University Press.

Sierpinksi, W. (1967) "L'axiome du choix," *Notre Dame Journal of Formal Logic*, 8 (4), 257–266.

Simpson, S. G. (1988) "Partial Realizations of Hilbert's Program," *Journal of Symbolic Logic*, 53 (2), 349–363.

Skolem, T. (1923), "Begründung der elementare Arithmetik durch die rekurrierende Denkeweise ohne Anwendung scheinbarer Veränderlichen mit unendlichem Ausdehnungsbereich," *in Videnskapsselskapets skripfter*, I. *Matematisknaturvidenskabelig klasse*, vol. 6., English trans. in van Heijenoort (1967), pp. 302–333.

Steiner, M. (1975) *Mathematical Knowledge*, Ithaca, NY: Cornell University Press.

Tarski, A. (1924) "Sur quelques théorèmes qui équivalent à l'axiome du choix," *Fundamenta Mathematicae*, 5, 147–154.

Tait, W. W. (1981) "Finitism," *Journal of Philosophy*, 78, 524–546, reprinted *in* Tait (2005), pp. 21–42.

Tait, W. W. (2005) *The Provenance of Pure Reason. Essays in the Philosophy of Mathematics and Its History*, New York: Oxford University Press.

Troelstra, A. (1977) *Choice Sequences: A Chapter of Intuitionistic Mathematics*, Oxford: Clarendon Press.

Urquhart, A. (1990) "The Logic of Physical Theory," *in* A. D. Irvine (ed.), *Physicalism in Mathematics*. Dordrecht: Kluwer, pp. 145–154.

Van Atten, M. (2004) *On Brouwer*, Belmont: Wadsworth/Thomson Learning.

Van Atten, M., and Kennedy, J. (2009) "Gödel's Modernism: On Set-Theoretic Incompleteness," *in* S. Lindström et al. (eds.), *Logicism, Intuitionism and Formalism: What Has Become of Them?*, Berlin: Springer, pp. 303–356.

Wantzel, P.-L. (1837) "Recherches sur les moyens de reconnaître si un problème de géométrie peut se résoudre avec la règle et le compas," *Journal de mathématiques pures et appliquées*, 2, 366–372.

Weyl, H. (1918) *Das Kontinuum, Kritische Untersuchungen über die Grundlagen der Analysis*, Leipzig: Veit, English trans. S. Pollard and T. Bole (1994), *The Continuum: A Critical Examination of the Foundation of Analysis*, Mineola: Dover Publications.

Woodin, W. H. (2002) "The Continuum Hypothesis and the Omega-Conjecture," *Coxeter Lectures*, Fields Institute, Toronto, Novembre 2002, recordings are available online at http://www.fields.utoronto.ca

Zermelo, E. (1904) "Proof That Every Set Can Be Well-Ordered," English trans. in J. van Heijenoort (ed.) (1967), pp. 139–141.

Zermelo, E. (1908) "A New Proof of the Possibility of Well-Ordering," English trans. in J. van Heijenoort (1967), pp. 183–198.

11

PHILOSOPHY OF PHYSICS

Anouk Barberousse (Sorbonne Université)

SOME DECADES AGO, philosophy of physics used to be the main part of philosophy of science. Nowadays, it is still a highly developed and very active part of this field, even though philosophy of science has diversified a lot. The reason of philosophy of physics being central to philosophy of science is that philosophy of science as we know it today was born as a reflection on physics at the end of the 19th century. This reflection was originally worked out by so-called scientists-philosophers like Ludwig Boltzmann, James Clerk Maxwell, Henri Poincaré, Pierre Duhem and a few others. These physicists were deeply concerned by the rapid development of theoretical physics and looked for secure foundations for the practice of their discipline. As a result, they engaged in a reflection on their own theoretical activity, asking questions like: What should be called a "principle"? A "model"? What should the role of mathematics be in the investigation of empirical phenomena?

What is left from this origin? As this book illustrates, philosophy of science is by no means restricted to a reflection on theoretical physics. This dramatic change raises the question of the current place of philosophy of physics within philosophy of science. Has physics become a "special science" on the same footing as other empirical disciplines, or does it still have a more fundamental status within philosophy of science? Some physical theories being fundamental does not mean that philosophy of physics should also be more fundamental than other parts of philosophy of science. This is the main background assumption of this chapter whose aim is to present some specific features of philosophy of physics among other "special sciences."

A complete review of the field is not an option. We shall take the question of the role of mathematics in the investigation of physical phenomena as our leading theme. This choice will be vindicated in the first section. In the second section, we shall briefly come back to the 19th-century origins of philosophy of physics in order to present the most general questions related to physics and its use of mathematics: What is measurement? Can physical theories tell us anything about the hypothesis that the world is deterministic? We shall then focus on probabilities, a mathematical tool loaded with lots of epistemological questions. The last section deals with the role of computers in physics.

1. The Nature of Physics, from the Point of View of the Philosophy of Physics

Philosophers of physics mainly focus on theories. They view physics as a domain of inquiry whose architecture is determined by theories and the relations they entertain with each other. Such a strong theoretical structure, if not without shortcomings, distinguishes physics from other disciplines. As a result, philosophers of physics spend a lot of time and energy to investigate into the nature of physical theories and inter-theoretical relationships (see, e.g., Cushing, 1998; Sklar, 1992; Torretti, 1999).

This inquiry has two aspects. The first one pertains to the general philosophy of science. The main questions are: What is a scientific theory in general? What are inter-theoretical relationships in general? As philosophy of science first emerged as a reflection about physics, these questions were first answered in the context of physics and only detached from it after a while. The second aspect is specific to physics. Here, the inquiry focuses on the details of physical theories in order to capture their logical structure and/or their metaphysical implications (like in Maudlin, 2007). This has given rise to very active subdisciplines within philosophy of physics such as the philosophy of space-time (Earman, 1989; Friedman, 1983; Sklar, 1974) and the philosophy of quantum mechanics and its more recent developments, as in particular quantum field theory (Albert, 1994; Fine, 1996; van Fraassen, 1991; Jammer, 1974; Maudlin, 1994; Ruetsche, 2011). They both inquire into the meaning of the fundamental concepts of these theories and into the mathematical structures they involve. With respect to inter-theoretical relationships, the debates have focused on the reduction relationship and its various meanings, with a strong emphasis on the relationship between statistical mechanics and thermodynamics (Albert, 2000; Sklar, 1993; but see Batterman, 2001, 2002, for a different perspective).

What has just been described is undoubtedly the heart of current philosophy of physics. Its main focus is on mathematical structures, seen as the heart of physical theories, and their metaphysical implications. More recently, philosophers of science have been willing to include models within their objects of investigation, realizing that models, and not just theories, play a major role in physicists' activity. The philosophy of modelling emerged as a result from the turn toward more practical aspects of scientific activity, sometime called the "practical turn." Whereas the philosophy of modelling is by no means restricted to physics, many philosophers of physics have taken part in this development. The last section of this chapter bears witness of this new stage in philosophy of physics.

Before entering into a few topics exemplifying the methods that are currently used in philosophy of physics, it is worth wondering whether it is fair to include the philosophy of modelling and computer simulation in a chapter devoted to philosophy of physics. There is a decision to be made with this respect. Either philosophy of science only bears on the products of scientists' activity, or it also includes this activity itself. As philosophy of science has widen its scope as to include scientists' activity, it seems fair to consider physicists' specific activity, which is largely focused on modelling, as belonging to the objects of philosophy of physics.

2. Fundamental Questions about Physics

At the end of the 19th century, physics began to look similar to the field as we know it today. A handful of partly competing and partly complementary theories were available and the imperfections of the mathematics of the day were felt as a hindrance to the development of some of these theories. The most ancient and revered theory was (classical) mechanics, whose predictive and explanatory power was considered a model for other theories. Thermodynamics was competing with mechanics on two grounds: its domain of application, which was supposed to be universal, and its confirmation basis, as the principles of thermodynamics appeared as suffering from no counter-example. (Classical) electromagnetism was still developing and tended to reach the micro-domain, at the scales of the molecules and atoms. The price of this development was mathematical complexity. Other theoretical endeavors directed toward the micro-domain, like spectroscopy, opened new fields of experimental work, but were difficult to related to mainstream theories.

In the midst of these difficulties, some physicists and physically-inclined mathematicians, like Poincaré, felt the need to better understand what was fundamental to this rapidly growing domain of investigation: physics. They wanted to go beyond the usual practices and assumptions of their time and raise basic questions about why, and to what extent, physical theorizing could be meaningful and fruitful. The nature of measurement was one of Poincaré's main object of philosophical investigation.

Accordingly, the present section begins with a brief review of the questions raised by the nature of measurement operations, which are at the very bottom of every physical theory. Investigation of these questions have developed throughout the 20th century and we will have a look at this development. With the rise of quantum mechanics in the 1902s, another general question emerged, which also became a major theme for philosophy of physics: are physical processes all deterministic, as earlier physicists usually supposed them to be, or are some of them intrinsically random? The end of this section is thus devoted to the notion of determinism. In between, there is a small section on precision, a major virtue achieved by the mathematical investigation of empirical phenomena.

2.1 MEASUREMENT

One of the major, common assumptions about physics is that this is an essentially quantitative domain of investigation. The physical phenomena have to be described by quantitative propositions. Words of the common, everyday languages are supposed to be not enough. For sure, the use of mathematics allows for well acknowledged and highly praised benefits, such as precision and predictive ability. The latter is particularly invaluable in applied science. But how is the assumption that physics should be quantitative justified? The very first activity by which observed phenomena are given a quantitative description is measurement. Through this activity, a property that has been identified as important beforehand is associated with both a number and a measurement unit. Under what conditions is this association justified? This question has been extensively treated within a long tradition beginning with Poincaré (1902) and Duhem, going on with Carnap (1966), and culminating with Patrick Suppes (1989) on the formal side, and Hasok Chang (2004) and Bas van Fraassen (2008) on the historical and epistemological side.

Measurement is a nice example of a fundamental topic in the philosophy of physics. The formal clarifications teach us a lot about the conditions under which a physical property can be linked with a number, but do not answer all the relevant questions. As Chang and van Fraassen point out, measurement also raises a historical and epistemological puzzle. Let us briefly review each side of the question.

2.1.1 Defining Measurement

Measurement is often viewed as the attribution of a numerical value, associated with a measurement unit, to a physical quantity. However, simply viewing measurement as an attribution procedure will not do. The complex assumptions and conventions that are the actual foundations of the association between a physical property and a numerical value would then be overlooked. Before proceeding to such an association, it is first necessary to identify, in the phenomenon of interest, the property to be measured. This means that the phenomenon has already been analyzed in order to be able to recognize the property in its various realizations. A property is usually not identified alone, but by the role it plays with respect to other properties. We often forget these assumptions because we focus on simple examples like length or mass measurement. In these cases, it seems pretty obvious that they are important properties playing many roles in physical phenomena. However, in the case of spin for instance, it is necessary to understand both that it is an important property and how it relates to other physical properties in order to measure it.

Another way to define measurement is to focus on the physical interaction between the measurement instrument and the investigated phenomenon that is involved in most measurement operations. This implies to precisely define the involved physical contact and the conditions under which it produces results that can justifiably be called valid "measurement results." For instance, one may rely on the notion of congruence in order to define length measurement with a ruler, and then on the notion of stability of the results when the measurement act is iterated in similar circumstances. This approach highlights the role of measurement instruments as a means toward the discovery of relations between theories and phenomena, because they are designed by theories on the one hand, and in contact with the phenomena on the other. However, the required physical contact needs not be as simple as congruence is. As an example, let us take temperature measurement. Although it looks almost as simple as length measurement, note that it takes some time for the height of the mercury column to stabilize in the thermometer after it has been put into contact with the investigated system. How long has one to wait for the hoped for stabilization? The answer to this question is not immediately given when one uses a thermometer; it requires further investigation and is by no means obvious. It should also be emphasized that the measurement procedure with a thermometer presupposes that the system's temperature is itself stable, which is equally difficult to assess in the context of an actual measurement operation.

2.1.2 Two Perspectives on Measurement

There are two ways to better appreciate the complexity of measurement procedures and to describe them as generally as possible. The first way is to consider the role of measurement in the (static) context of well-established theories; the second is to focus on the role of measurement in the (dynamical) establishment of theories. In the first case, the relationships among quantities of interest are well-known, whereas in the second, it is impossible to rely on well-accepted scientific statements in order to define relationships among properties.

When measurements are made in the context of well-established theories, they are the primary means to take phenomena into account in physical theories. They relate the theories' general statements with the specific situations in which phenomena occur, that is, with *hic et nunc* tokens. It is entirely different when there is no theory, or not enough theories. For instance, before the constitution of thermometry, people tried to pair sensations of heat with numbers by relying on the mediation of some liquid's increase of volume, without however being fully aware of the regularities governing the relationships between heat and the behavior of fluids. A major obstacle was that without a thermometer, it is impossible to be sure that the temperature of such or such material system is stable. There is a formidable epistemic circle here that has been analyzed in the past by Ernst Mach (1896), and more recently by Hasok Chang (2004): how is it possible to design thermometers if in order to do so, it is necessary to already dispose of thermometers?

This general circle divides into smaller circles. How is it possible to be sure that melting water and boiling water are actual fixed points enabling one to define a well-behaved temperature scale? How can one assess that the specific heat of mercury does not vary too much with temperature in order to ensure that the regular graduations of common thermometers actually reflect a regular increase in temperature? It was impossible to get out of uncertainty about these matters during the long establishment of thermometry from the seventeenth to the 19th century. Warrants could only be reached when thermodynamics and, later, statistical mechanics have been settled in the middle of the 19th century. Without the theoretical foundation they provide, the construction of thermometers relied on uncertain principles and empirical contingencies.

2.1.3 Measurement as Representation

When analyzing measurement within the framework of well-established theories, it appears as a partial, numerical representation of a natural phenomenon or system. It is a piece of scientifically useful information about the investigated system or phenomenon. A measurement result is thus a relational object for, being a representation, it both entertains relations with the represented object and with the user of the representation, that is, the scientist wanting to gain information about the investigated system or phenomenon. Besides, the representation's user also possesses concepts and hypotheses allowing her to situate the measurement result within a meaningful theoretical framework. Therefore, the measurement result represents the investigated system both in the user's theoretical framework and as possessing such or such numerical property. Moreover, measurement is also intentional for it is performed in order to gain information about a phenomenon or system.

What does it mean to represent something as possessing such or such numerical property? First, talking of "a numerical value" is an idealization since the measurement instrument provides the user with a numerical value (associated with a measurement unit) plus an error interval. Second, as seen above about the construction of thermometers, many conditions have to be satisfied in order for the measurement act to give the hoped for results. Let us now come back to the example of length measurement. Several assumptions are necessary unless the ruler put in congruence with the object to be measured will give wrong results. For instance, one has to assume that the ruler's length does not vary when moved in space. This example shows that the assumptions underlying measurement operations do not only include what we know, or believe to know, about length itself, but also other pieces of knowledge, here, about the properties of the ruler's material. More generally, measurement operations rely on empirical hypotheses, usually about the regular behavior of the components of the measurement instrument. This is the reason why standards are so important in the construction of measurement instruments.

The problem of coordination

This problem has been intensively investigated by Mach (1900), Poincaré (1902), and Reichenbach (1928). It can be expressed as follows: How is it possible to anchor physical properties, as defined by physical theories, in particular phenomena? To put it in other words: What gives empirical content to measurement results, which represent properties that have been defined within a theoretical framework? The task is to specify the conditions under which measurement operations actually allow one to gain empirical information. There is a seemingly intuitive answer to this question: the measurement operation has to identify the "physical correlate" of measurement. Now this identification is only possible within a well-establish theoretical framework because the representational function of measurement results is theory-laden. It is only in the context of admitted theories and the classifications they provide that certain physical interactions are identified as being able to correctly represent (part of) the investigated phenomena.

A first necessary condition for an operation of measurement to yield adequate empirical information is that a scale of measurement has been established. This cannot be done unless one relies on some contingent, empirical regularity. For instance, as mentioned above, in order to establish a length scale, one relies on the assumption that the ruler will not shrink when moved around (one can adapt this condition to other instruments than rulers). Another condition has been discussed in details by Poincaré (1902): one has to rely on an equality criterion, itself based on a physical property of the measurement instrument. In the case of length measurement, the equality criterion is the following: one has to assume that the space between two graduations on the ruler does not vary with temperature, pressure, displacement, and so forth. In other words, the length between two graduations at some instant of time has to remain identical in time. How can such an equality criterion be established? As Poincaré emphasizes, it is impossible to rely on strictly empirical regularities, because one would have to use some already available instrument in order to establish these very regularities. This would lead to infinite regress. How can one escape this regress? According to Poincaré, the only way out, in the case of length measurement, is to decide to accept the assumption that the physical properties of the ruler do not change when the surrounding physical conditions change. Without this decision, length measurements are simply impossible to perform. For Poincaré, this is an argument in favor of the claim that conventions play a fundamental role in science.

Van Fraassen has given a different analysis of the problem of coordination. According to him, the rules and principles on which one relies in order to solve the problem of coordination, and which are introduced in order to define particular measurement operations, cannot be formulated outside of a scientific and historical context in which other measurement operations are already accepted as valid and used on a regular basis. He agrees with Poincaré in saying that there can be no starting point with no presupposition at all that could solve the problem of coordination. To put it in other words, there is no independent access to the parameters to be measured: the parameters to be measured are progressively identified, through some historical process during which one has to use, in order to identify the parameters, the very measurement practice about which one hopes that it will stabilize and become a well-established measurement practice.

Van Fraassen's claims are in direct opposition to other epistemological claims about the nature and role of measurement results, as Bridgman's operationalism (1916) or Carnap's views about protocol statements (1932). According to both Bridgman and Carnap, measurement results are absolutely free of any theoretical presupposition. For them, measurement being detached from theoretical hypotheses is the necessary condition of the epistemic value of empirical science. However, as it is already clear from Quine's (1951) criticism, and is even clearer from van Fraassen's analysis, it is impossible to implement an "observational language" that would contain no theoretical element at all. This ideal absolutely contradicts what we know of actual scientific development. Carnap's and Bridgman's views about measurement are therefore too idealized to be of any use in analyzing actual, scientific measurement operations.

2.2 PRECISION

Once the physical properties that are considered important in the investigation of some physical phenomenon can be associated with numbers, it is possible to try and find out regular relations among these numbers. These relations may take any mathematical form; this is the place where the expressive power of mathematics enters the scene. The game to play in physics is indeed to find out regular relations among quantities, or "regularities." Physics has made a leap forward when Newton, Euler, d'Alembert, Laplace, Lagrange, and others succeeded in making calculus meet the needs of physics. With calculus, the set of possible regularities was dramatically enlarged and the representative power of physics accordingly increased.

Differential equations are the main tools by which change is represented with the help of calculus. They represent how various quantities change as time goes by. This representation is powerful because it allows for the computation of the value of any variable figuring in the equation at any time, given its value at some initial time and the values of boundary conditions. Differential equations thus look like a magic wand because they allow for the knowledge of any future state if the knowledge of some past state of the system under investigation is given, and if it is possible to compute the solution of the equation. Let us examine these two conditions in turn. This will allow us to better appreciate the powers of this magic wand.

Within the context of the application of calculus, knowledge of a past state of the system under investigation and of its boundary conditions takes the form of a list of values for the variables of interest. As said above, the "variables of interest" are the physical quantities that have been acknowledged as important in order to describe the behavior of the system. In order to know their values, it is necessary to measure them, or to compute them from other elements that are already known about the system. Either way, the main difficulty is to obtain sufficiently precise values to ensure that the computed solutions of the differential equations may themselves be precise enough. This difficulty arises when any mathematical statement is applied to empirical phenomena, but has especially dramatic consequences when calculus is applied. This is because mathematical equations are only valid, *stricto sensu*, for exact values of the variables they contain. However, exact values are outside the realm of measurement or previous computations. Therefore, some discrepancy is bound to appear between the target solution of the equation and the computed one, as the target solution only holds in the realm of mathematical, exact values, whereas the computed equation can only contain approximate, measured values.

Even though this difficulty is both pervasive and well-known, its implications are not always fully taken in account when analyzing the so-called efficiency of mathematics in the study of empirical phenomena (Wigner 1960). For sure, mathematics, and especially calculus, are efficient as they allow for the representation and computation of past, present, and future states of empirical systems, if certain conditions are satisfied. However, it is important not to forget that these conditions are rather stringent. For instance, if the past state of the system under investigation is not known precisely enough, it is useless to try to apply the differential equation describing its behavior as time goes, because the computation would only provide meaningless results. Duhem provides us with a nice illustration of how the application of calculus sometimes leads to nonsense: "It may be that the problem of the stability of the solar system is meaningless for the astronomer . . . One cannot go through the various and difficult deductions of Celestial Mechanics and Mathematical Physics without fearing that some of them be sentenced to sterility" (Duhem 1906, second part, chapter 3).

2.3 DETERMINISM

As we have just seen, when it is possible to know the initial and boundary conditions of the investigated system precisely enough, then differential equations allow one to predict the system's state at any time in the future (provided that Cauchy theorem holds). The prediction holds within the limits of the validity of the description of the initial state and the computed solutions. As most differential equations used in physics are difficult, or even impossible to compute exactly, the predictions are almost always approximate (see Humphreys 2004 for a detailed discussion of the difference between computability in principle and computability in practice). However, within these limits, differential equations provide one with knowledge of the future. Even more so, they allow one to know the state of the investigated system moment by moment, that is, to follow its transformations step by step.

Let us take a simple example: free fall on Earth as represented in classical mechanics. Generally speaking, the motion of a body is described in classical mechanics by the differential equation $\mathbf{f} = m\mathbf{a}$, where $\mathbf{f}$ is the sum of the forces acting on the body, m its mass and $\mathbf{a}$ its acceleration. Acceleration is represented as the second derivative of its position $\mathbf{x}$ relative to time t: $\mathbf{a}(\mathbf{x}) = d^2\mathbf{x}/dt^2$. In the free fall case, the only force acting on the body is the gravitational force, because the resistance of air is neglected. In order to know the velocity of the body at any instant, one has to integrate the equation $\mathbf{f} = m\mathbf{a}$, which means that for each t_i of interest, one has to compute $v(t_i) = \int \mathbf{a}(t)dt$ between t_o et t_i

The possibility to compute the state of a system at any instant seems to presuppose that the succession of states of the investigated system is knowable in advance, in other terms, that they are pre-determined. This is the reason why differential equations are closely linked to the concept of determinism. Let us now examine some aspects of this concept and the related thesis.

It is first important to distinguish between two deterministic claims: the metaphysical and the epistemic claims. The metaphysical claim has two parts, each of which being equally important. According to the first part of the claim, the state of the investigated system can be completely described at each time by the values of the relevant physical quantities. These physical quantities are defined independently of the knowledge human beings can acquire about them. The second part of the claim is that the state of the system at any time fully determines the total succession of its future states. As the investigated system can be the universe itself, the metaphysical claim admits of a fully general version.

The epistemic claim is defined relative to the knowledge someone can acquire about a system (possibly the universe). Here, "someone" may be a human being, but also a "demon," as suggested by Laplace (1814). The omnipotent intelligence of Laplace's demon is able to grasp all the connections among the system's elements. A system's being deterministic, in the epistemic sense, amounts to its being thoroughly accessible to a (hypothetical) perfect intelligence in both its static and dynamic aspects. This claim involves the conviction that the system's evolution (or even the whole universe) is correctly described by mathematical laws. More precisely, it involves the conviction that the system's evolution is adequately describable by differential equations. Therefore, the epistemic claim may itself involve a metaphysical component, as the thesis that the evolution of any system, *a fortiori* the whole universe, can be correctly described by a set of differential equations probably goes beyond the scope of physics.

Let us investigate the implications of the epistemic claim. The claim that a system's evolution can be represented by a set of differential equations is often (but not always) associated with the belief that no objective chance event can possibly affect it. Thus, relative to this system, what we call "chance" is only a symptom of our incomplete knowledge. This belief can be extended to the whole universe: if its evolution is deterministic (in the epistemic sense), objective chance is either an illusion or does not have any effect on its evolution. Such a position might have been plausible until the discovery of radioactivity and quantum phenomena. It had a strong influence on epistemology until the beginning of the 20th century because it was defended by Kant in his *Metaphysical Foundations of Natural Science* (1786) and then by the neo-Kantians, like Ernst Cassirer, who profoundly renewed epistemology in the first decades of the 20th century.

It took quite a long time (more than a century, from the mid-19th century on) before physicists managed to tame the mathematical representation of chance and to dispose of mathematical tools for the representation of chancy events that were as powerful as the tools they had for the representation of non-chancy events. Before these tools have been elaborated, it was a reasonable strategy to represent the investigated systems as if they were deterministic, because of the predictive power associated with differential equations.

As clear from the previous paragraphs, the notion of determinism (both in its metaphysical and epistemic sense) is different from both notions of causality and necessity (see chap. 3). Claiming that the universe is deterministic does not amount to saying that each event has a cause or that it obeys necessary laws, but to saying that its occurrence depends on all preceding events. The main difference between determinism, causality and necessity is that determinism involves temporally ordered but reversible events. On the contrary, the notion of causality settles an asymmetrical relations between two events. Therefore, when Laplace claims that the state of the universe at any point in time is the cause of all its future states, he gives up the common notion of causality. Russell (1913) goes a step further and claims that the notion of causality has to be given up altogether and replaced with the notion of determinism (in the epistemic sense) as defined above. On the other hand, claiming that every event results from the instantiation of a necessary law does not imply that this law is deterministic, as necessary laws need not be expressed by differential equations.

This brief discussion of the distinction between determinism, causality, and necessity can be supplemented with a presentation of the relation between the notions of determinism and law of nature. For sure, determinism is no essential component of laws of nature, for several laws take on a statistical form, and others seem of intrinsically probabilistic nature, like the laws of radioactivity. However, deterministic laws enjoy a special status because they possess at the highest degree two valuable features: simplicity and informational content. These features are even claimed by some philosophers, like David Lewis (1983), to distinguish laws from other universal sentences. Deterministic laws are indeed especially simple for they encompass an infinity of past, present, and future states in a single formula. Whereas simplicity and informational content are usually conflicting properties of sentences, they nicely dovetail with one another in deterministic laws.

As mentioned above, the natural systems that are described by deterministic laws may not be deterministic in any metaphysical sense of the term. In the 1950s, Reichenbach has further shown that it is always controversial to claim that any natural system is actually, and objectively, deterministic. In order to analyze Reichenbach's proposal, let us first introduce the definition of determinism as proposed by Russell (1913):

> A system is said to be "deterministic" when, given certain data, $e_1, e_2, \ldots, e_n$, at times $t_1, t_2, \ldots, t_n$ respectively, concerning this system, if E_t is the state of the system at any time t, there is a functional relation of the form $E_t = f(e_1, t_1, e_2, t_2, \ldots, e_n, t_n)$. (1913)

This definition has a surprising consequence. According to it, claiming that the universe is deterministic provides the reader with no information at all as it amounts to saying that the total state of the universe at time t can be expressed by a function of t, without specifying the function.

It is however possible to transform Russell's definition in a new one requiring that time does not appear as a factor in the evolution of the system. For a system to be deterministic, there must exist a function f such that $\forall t, \forall b>0, s(t+b) = f(s(t), b)$, where s is the trajectory of the system within its phase space (i.e., the set of all its possible states). f has an important symmetry: it is invariant under time translation, or periodic: $\forall t$ and $\forall t'$, if $s(t) = s(t')$ then $s(t+b) = s(t'+b)$

Nevertheless, periodicity is not enough to define determinism, because it does not remove the possible trajectories in phase space that are not instantiated, whereas the notion of determinism involves that these possible, but never instantiated, trajectories should be disposed of. In other terms, the system's evolution should not depend on where the system is at a given time. In formal terms: $\forall t, \forall b>0, \forall s', s'(t+b) = f(s'(t), b)$. The symmetry of the last formula makes every possible trajectory in phase-space periodic.

From this analysis, two conclusions may be drawn about the attribution of determinism to a natural system:

(i) First, it is important to keep in mind that when certain properties are selected to provide an adequate description of the system (like, e.g., the temporal positions and velocities of the system's elements), there is no guarantee that they are actually relevant to a better understanding of the system's behavior. Let us examine a case in which the selected properties are misleading. First, recall that the trajectories in phase space are defined by the selected properties, supplemented by an evolution equation. Now it may happen that these trajectories have more symmetries than the initial situation. This is the case when the trajectories are time-reversible when the investigated phenomenon is not. However, the solutions cannot have more symmetries than the initial situation unless there is some flaw in the representation of the system and its behavior.

(ii) The above mentioned definition of determinism is based on symmetries of possible states and trajectories. This implies that it applies to types of systems. As a result, when the description of a natural system satisfies the definition, it can be said to belong to a deterministic type, but it would be fallacious to say that it is deterministic in itself for a deterministic type may include indeterministic subtypes, as shown by Reichenbach (1956). Therefore, the question whether a natural system is deterministic has no univocal answer. This significantly reduces the metaphysical import of the notion of a deterministic system. Reichenbach's result implies that what is deterministic is the model (the representation), not necessarily the physical process it represents.

Even though the use of deterministic representations does not indicate in itself whether the investigated phenomena are deterministic or not, they have been, and still are, popular. The reason for this popularity may be that they allow for the construction of an image of the phenomena that is relatively simple: an image in which many symmetries are revealed and exploited in the models.

In particular, the use of deterministic representations allow physicists to satisfy a requirement that is commonly accepted and has been formulated by Pierre Curie as follows: "When certain causes produce certain effects, the symmetries of the causes should also be found in the produced effects" (1894). It is possible to rephrase Curie's principle without the notion of cause by using Russell's function f. In this way, the principle can be interpreted as a commonly used methodological tactics, especially when constructing deterministic models. It consists in looking for solutions to the investigated problem that do not add any symmetry to the initial description of the problem. This amounts to choosing models with as many symmetries as possible. When we try to know whether a system is deterministic, we thus apply Curie's strategy.

The above elucidation of the notion of determinism, based on Russell's and Reichenbach's analyses, allows one to purge it from its unclear metaphysical component in order to remain with a perfectly clear notion. By doing so, one ends up with a model-relative notion instead of a notion pertaining to the description and understanding of the universe (van Fraassen 1985; for a more advanced discussion on determinism, see Earman 1986).

3. The Meaning of Probability in Physics

In the preceding section, quantum phenomena, including radioactive decay, have been mentioned as forbidding deterministic modelling. Quantum phenomena are currently viewed as the domain in which chance has to be integrated into the models. This is why probabilistic representations are unavoidable in quantum mechanics. This is not to say that probabilistic functions are always interpreted as representing objectively chancy events. Their minimal interpretation is that they represent our inability to predict the result of certain measurements. The aim of this section is to review some uses of probability in physics and their related problems. Probability is not only used in quantum mechanics. Before the development of this theory, it has been used in statistical mechanics. The questions it raises in this theory are still open.

3.1 PROBABILITY IN STATISTICAL MECHANICS

The founders of statistical mechanics, Maxwell and Boltzmann, wanted to base the study of thermal phenomena on the theory they conceived as the most scientifically secure one, namely (classical) mechanics. Before they engaged in this project, the regularities exhibited by thermal phenomena had been gathered within thermodynamics. Thermodynamics has a strong explanatory power and its domain of application is very large, but it only deals with measurable quantities (it is called a "phenomenological" theory for that reason). In order to provide a mechanical explanation to thermal phenomena, it is necessary to leave the domain of phenomenological theories and to rely on a theoretical hypothesis whose plausibility was still questionable in the middle of the 19th century: the atomic hypothesis. According to the atomic hypothesis, matter is made of moving microscopic particles. Thus, in order to provide a mechanical explanation to thermal phenomena, one had to refer to quantities to which there was no empirical access, like the mass and velocity of molecules. In statistical mechanics, the laws of mechanics are applied to these microscopic and hypothetical quantities in order to infer the macroscopic properties of the macroscopic systems they are part of, like gas samples.

On the way from the motion of molecules to the macroscopic properties of fluids, one encounters two problems. First, there are much too many molecules for a scientist to represent all their individual motions. (For instance, in 22.4 liters of air, at normal temperature and pressure, there are $6.02.10^{23}$ molecules). This is the reason why it is only possible to study the motion of very large sets of molecules with the help of statistical notions, among which the notion of probability.

At the beginning, the introduction of probabilistic hypotheses at the scale of molecular motions did not seem to raise any special difficulty. In 1860, Maxwell was content with postulating that when two molecules collide, all scattering directions are equiprobable. However, this hypothesis is only justified when molecules are "hard spheres" (i.e., perfectly elastic spheres, the collision between which causing no loss of energy). But Maxwell was not concerned by this question for he was struggling with the justification of statistical laws (see Maxwell 1873, 1875). At the time, when deterministic modelling was triumphant, scientific theorizing was thought to primarily look for certainty. Statistical laws apparently required giving up this search for certainty. Several decades were needed to convince physicists that statistical laws are as legitimate as dynamical laws expressed by differential equations.

Today, the situation is entirely different. The first major difference is that the concept of probability is now a honorable mathematical concept to use in physics. In the 19th century, there was no proper theory of probability, but only a set of ill-founded recipes. In 1933, the mathematical theory of probability has been axiomatized by Kolmogorov (see von Plato, 1994, for details about the creation of "modern probability"). As a result, the necessary precautions to use this theory in order to investigate empirical phenomena are now well-known. For instance, it is required that the set of events on which the probability function is defined be carefully defined in order to avoid erroneous application of the probability concept. Generally speaking, it is never easy or straightforward to apply a mathematical theory to empirical phenomena. In the case of probability theory, the conditions at which the mathematical concepts are applicable to empirical situations are more often than not counter-intuitive. For instance, when probability theory is applied to continuous quantities, it is first necessary to define a measure allowing to state the conditions at which two probabilities are equal. In some cases, the choice of such or such measure is difficult to justify. Now the measure plays an important role because it determines the values of the probability functions that are used to describe the system. Several questions relative to the application of probability theory to empirical systems in the framework of statistical mechanics are still open.

Among the open questions still facing statistical mechanics, the origin of irreversibility is probably the most debated one (Albert 2000, Price 1996, Sklar 1993). Let us now present its main characteristics. As said above, statistical mechanics relates two scales of phenomena: the macroscopic scale of thermal phenomena and the microscopic scale of molecular motions. Thermal phenomena are described by irreversible laws, for according to the second principle of thermodynamics, when a system is out of equilibrium at some initial time, it is bound to evolve to an equilibrium state if is isolated from outside influence. And when an isolated system is at equilibrium, there is no way for it to lose its equilibrium for a long (macroscopic) time, unless someone adds energy to it, in which case it is not isolated anymore. Here, the equilibrium state is defined in the following way: at equilibrium, the macroscopic quantities that are considered as relevant for the macroscopic description of the system (such as temperature, pressure, volume) do not vary with time. On the contrary, molecular motions are described by the reversible laws of mechanics. In classical statistical mechanics, these are the laws of Newtonian mechanics, and in quantum statistical mechanics, it is Schrödinger equation, which is also reversible. This means that when you replace the time parameter t by -t in the dynamical equations, the system goes back to its initial state. This operation, though imaginary, is physically meaningful for there is no physical impossibility for the system to go back to its initial state: even if, generally speaking, it is impossible to go back to the past, it is not impossible that the system naturally evolve so as to be in the very same state in which it was at the initial time. Here the state of the system is defined by the positions and velocities of its component molecules.

The above shows that the two scales that are considered relevant for the description of thermal phenomena obey highly different kinds of laws for the macroscopic laws are irreversible whereas the microscopic ones are reversible. At the beginning of statistical mechanics, this difference was not viewed as a problem. It emerged as a formidable one when it was realized that in order to go from the microscopic, reversible laws to the macroscopic, irreversible laws, it is necessary to introduce some supplementary hypothesis on top of mechanical laws applied to molecular motion. This means that the sole laws of mechanics are not enough to account for thermal phenomena. Let us now briefly examine how this problem has emerged and has been recognized as an important one in the history of statistical mechanics. This will allow us to capture the heart of the problem, which is still a major problem at the frontier between physics and the philosophy of physics (see Uffink, 2007, for a thorough review of the historical side of this question).

In the first paper in which Boltzmann tried to design a mechanical explanation of the second principle of thermodynamics (Boltzmann 1872), the shift from the reversible laws of mechanics to the irreversible laws of thermodynamics was achieved by the introduction of the hypothesis called *Stosszahlansatz*, or "hypothesis about the number of collisions." According to this hypothesis, the number of collisions of a given type is proportional to the number of molecules with kinetic energy x and the number of molecules with kinetic energy x'. This amounts to supposing that the velocities of any two colliding molecules are independent of one another. This hypothesis was considered so benign by Boltzmann that he did not even point out that it contradicts Newton laws of motion according to which, in an isolated system, all molecular motions are interdependent because of gravitational forces. However, the *Stosszahlansatz* does contradict Newton's laws, their reversibility, and associated determinism because it amounts to neglecting intermolecular forces.

Because the *Stosszahlansatz* was not perceived, at first, as a novel, anti-Newtonian, and anti-deterministic hypothesis, the mechanical explanation of the irreversible laws of thermodynamics could be considered complete. However, it soon appeared that it is strictly impossible to explain irreversible phenomena by relying on reversible laws. So the shift from reversible, microscopic laws to irreversible, macroscopic laws is paradoxical, as first pointed out by Loschmidt (hence the name "Loschmidt's paradox" to designate this problem).

What is the solution to Loschmidt's paradox, also called "irreversibility" paradox? Isn't there any way to go from reversible to irreversible laws? Yes, there is, if one is willing to accept statistical explanations. The explanation relying on the *Stosszahlansatz* or any other probabilistic hypothesis cannot be said to be strictly mechanical, but it counts as a statistical explanation because it allows one to describe the collective motion of large sets of molecules. This kind of explanation is not the only possible one and is still open to discussion. For instance, it may be asked on what grounds the introduction of a statistical hypothesis like the *Stosszahlansatz* may be based: why should this hypothesis be chosen?

3.2 PROBABILITY IN QUANTUM MECHANICS

The domain of thermal phenomena was the first one to which probability theory was applied; as briefly presented above, it raised, and still raises, difficult questions. However, the questions raised by the application of probability theory are perhaps even more difficult in the case of quantum phenomena. Most developments in the philosophy of quantum mechanics are devoted to these problems. Their origin lies in the following fact, which has been brought to light since the very beginnings of quantum mechanics: whereas the evolution of isolated quantum systems is described by a differential equation (Schrödinger equation), this equation is no longer valid as soon as one tries to make a measurement on the system. More precisely, one has to use another, different formalism each time the quantum system interacts with a macroscopic system (among which measurement devices). The result of this interaction is described by probabilities. This is called the "measurement problem" for quantum mechanics, lying at the origin of several competing proposals to interpret the twofold formalism of quantum mechanics.

The measurement problem is both physical and philosophical and runs through the history of quantum mechanics (Jammer, 1974). It can be presented in the following way: how can we fill the gap between the evolution of isolated quantum systems and their evolution when they interact with a macroscopic system, as described by quantum mechanics? This question emerged as a problem threatening the foundations of quantum mechanics themselves in the "EPR paper" (Einstein, Podolski, Rosen, 1935). The aim of this paper is to show that quantum mechanics cannot be considered a complete physical theory because it contains no element explaining (not to mention filling) the gap between the two ways of describing the evolution of quantum systems. Therefore, according to Einstein, Podolski, and Rosen, quantum mechanics is unable to justify the presence of probabilistic predictions on top of its fundamental equation, which is deterministic.

Einstein's main target in the EPR paper is Bohr's hypothesis that any measurement operation on a quantum system disturbs its evolution, the perturbation being unpredictable and impossible to analyze with the help of the theory's components. (Bohr defended his hypothesis against the EPR paper the same year). According to Bohr (1935), the only possible attitude for the physicist is to submit herself to this fact: it is the only way to continue the enterprise of physics. On the contrary, according to Einstein, there is no physically valid reason to admit that this perturbation is both necessary and impossible to analyze, and further that it must play so important a role in the understanding of the theory. To put it briefly, Einstein is blaming Bohr for introducing metaphysical elements in quantum mechanics through the back door.

In order to show that Bohr's conception is not satisfactory, Einstein imagined a thought experiment. He described a quantum process that had not been observed at the time, but whose possibility seemed to be a consequence of quantum mechanics. Here is the starting point. Imagine a setup in which two quantum systems, let us say electrons, so interact as to keep up their relative positions along the axis of motion and as to keep the total momentum of the whole system along this axis equal to zero. Let us now make a first hypothesis, the separability hypothesis, according to which the two systems are separated in a specific sense: when a measurement is performed on one of the systems, the other is not influenced by the measurement interaction and maintains its identity (in the sense that it does not undergo any perturbation). This hypothesis seems straightforward because it is valid for classical systems. Even more so, it is one of the main pillar of classical theories. We now make a second hypothesis, the locality hypothesis, about the measurement act, according to which it is purely local: it can be performed on one of the systems without making any difference in the other. This amounts to saying that the systems do not interact with one another when a measurement is performed on one of them. Again, this second hypothesis is straightforward when classical systems are concerned. Let us now imagine that the quantum systems interact at initial time and then move apart so as to respect the above conditions. When they are spatially separated, a measurement is performed on one of them, for instance one measures its position. If one accepts the two above hypotheses and the laws of quantum mechanics, then one has to admit that the measurement on one system immediately provides us with information on the position of the other system as well, because both are described by the same wave function. It is as if the first system "knew" the other's state. Bohr claims that one has to accept this counter-intuitive result whereas Einstein argues that it is the symptom of something missing in the theory.

Einstein and his co-authors draw several conclusions from this thought experiment. These conclusions are still debated today. As mentioned above, the main conclusion is that the description of a quantum system as it is given by quantum mechanics cannot be considered complete. As a result, another theory has to be elaborated in order to provide quantum mechanics with firm ground.

Let us examine how Einstein arrives at this conclusion. The role of his argument has been so important in the development of the philosophy of quantum mechanics that it is worth looking at it in details. Its goal is to determine whether the two following statements are logically compatible (this is a good example of a philosophy of science question, for it is usually assumed that scientific statements are clear and that their logical status is univocal, whereas this is a case in which a hard work has to be done in order to settle logical questions about scientific statements):

(1) Quantum mechanics is an incomplete physical theory for it cannot simultaneously describe all relevant aspects of the systems falling in its domain.
(2) Two quantities whose operators do not commute (this means that they do not act in the same way when they are applied in a different order, like position and momentum of a same particle) cannot simultaneously possess objective reality.

The thought experiment shows that (1) and (2) are incompatible. Therefore, one of them is false. According to Einstein, (2) is false; therefore, (1) is true. He claims that the thought experiment shows that both position and momentum of a particle have objective reality because it is possible to measure to any degree of precision either position or momentum of one electron in the thought experiment.

Most physicists remained unconvinced by Einstein's argument. They thought, and still think, that the EPR thought experiment does not show that quantum mechanics is incomplete, but rather that either the hypothesis of separability or the hypothesis of locality is false, or that both are false. Today, the EPR thought experiment is viewed as an especially clear way to demonstrate the major feature of the quantum world, entanglement. Entanglement can be informally characterized in the following way: within most quantum processes, the quantum states of systems that were actually separated at initial time get so intertwined as to allow for the emergence of a new system within which it is impossible to discriminate the initial systems. In the EPR thought experiment, the two electrons get entangled and make up a new, single system. Entanglement is the source of the statistical predictions of quantum mechanics. When a system is in an entangled state, the result of a measurement act on one of its properties can only be predicted statistically. The EPR thought experiment clearly shows that entanglement is incompatible with the joined truth of the hypotheses of separability, locality, and completeness of quantum mechanics.

In 1951, David Bohm imagined another experimental setup allowing one to achieve a better grasp of entanglement in order to clarify the relationships between separability, locality, and completeness of quantum mechanics. In 1964, John Bell showed that, contingent on the validity of certain assumptions, among them locality and "realism" (in this context, a theory is said "realist" when each element in the formalism corresponds to an "element of reality" as in Einstein's phrasing), the statistical correlations that can be measured in the course of an EPR experiment must satisfy a set of constraints called "Bell's inequalities." More precisely, a quantum theory is said to be "realist" when it postulates, besides the quantum state, a "complete state" containing hidden variables which determine the total set of measurement results on the system (see Fine 1982 for a useful clarification of the implications of Bell's theorem).

What is the impact of Bell's theorem? To answer this question, it is important to realize that predictions based on quantum mechanics violate Bell's inequalities. In the 1960s and 1970s, other theorems have been proven by Bell and other physicists demonstrating that no physical theory satisfying the "realism" and locality conditions can agree with the statistical predictions of quantum mechanics. These theorems are usually interpreted as implying that it is impossible to give quantum mechanics an interpretation that would be both local and "realist."

The EPR saga did not stop with the discussions on the implications of Bell's theorem. In the 1980s, actual laboratory experiments have been performed that aimed at realizing Einstein's thought experiment. It is generally admitted that actual EPR-type experiments do violate Bell's inequalities (see Aspect, Grangier, Roger, 1982 and Aspect, Dalibard, Roger, 1982). Most experiments that have been performed since then also confirm the violation of Bell's inequalities by quantum systems; however, their interpretation is an area of endless controversy. They are likely to suggest that quantum mechanics is a complete theory, because there does not seem to be any "complete state" above and over the quantum state, but nevertheless a theory in which the locality hypothesis does not hold, because of entanglement. Nevertheless, some adopt another interpretation, according to which quantum mechanics is neither local nor complete. This interpretation is defended by David Bohm, following Louis de Broglie's work. According to this interpretation, it is important to explain observed correlations instead of being content with Bell's negative results (see for instance Goldstein, 2001).

4. Physics with Computers

If philosophy of physics is to investigate what physics actually is and how physicists develop their theories and models, it is bound to acknowledge that the use of computers is now pervasive in all parts of physics. Everyday work has become unconceivable without the help of machines. This is a tremendous change from a practical point of view. The question arises whether this major shift is also important from a conceptual point of view. When talking with elderly physicists who have experienced this shift within their career, it becomes clear that the "phenomenology of physics," so to speak, the way it is perceived by working physicists, has dramatically changed in the past decades. But has physics changed as well?

At first sight, it seems that the heart of physics, at least as seen from the philosophical point of view, is only slightly affected by the computational revolution. This domain is still highly structured by a few theories whose relationships are well-known, if not without problems. The mathematics allowing to write down the fundamental equations of these theories has been established for a long time, even though the resolution methods undergo important developments. For these reasons, it seems that computers are not likely to transform physics when it is understood as a set of theories, some fundamental and the others derived from the fundamental ones.

However, as mentioned in the introduction of this chapter, philosophy of physics has recently widen its scope so as to include the study of physicists' activity besides the study of their results. Now, it is clear that the availability of cheap and powerful computers has no less invaluable practical consequences than the existence of super computers. Further in this section, we shall investigate some of these consequences. Beforehand, we shall have a brief, introductory look at physical models, as intermediaries between physical theories and computers (this section is a brief summary, applied to physics, of elements from chap. 5).

4.1 PHYSICAL MODELS

As we shall see, computer simulations are closely linked to their underlying models, that is, to the sets of equations describing the systems of interest. It is therefore important to understand the specificity of modelling by contrast with theory construction (a seminal paper on this topic is Redhead, 1980). The first, obvious difference is that when you endeavor to build up a physical theory, you want to identify regularities that hold in a very large domain of phenomena—maybe in the whole universe. When you build up a model, on the contrary, your activity is oriented (i) toward a specific phenomenon and (ii) toward a specific aim. For instance, you build up a model in order to study the aerodynamic properties of this type of car in order to improve its fuel consumption.

A second major difference between theory and model construction is that usually, you rely on already existing theories in order to design your model. You may thus use fundamental equations as ingredients in your model, but then, you have to transform them in order to make sense of the phenomenon you want to represent within their framework. Modelling is mostly about equation transformation, either to adapt the initial equations to the target phenomenon or to achieve computable equations, for the solutions of fundamental equations are seldom easily computable. The main methodological difficulty of modelling is to assess to what extent the purported transformations are justified (Morrison, 1998). We shall see further in this section how this question is magnified in the context of computer simulations.

4.2 COMPUTER SIMULATIONS

One of the first questions that has been raised about the pervasive use of computers and computer simulations in physics is whether it indicates a shift in scientific methodology (Frigg and Reiss, 2009; Humphreys, 2004, 2009; Morrison, 2009; Parker, 2008, 2009; Winsberg, 2001, 2003). When computers were not available, scientific methodology, although a topic of lively debates, was generally considered to include two items, experiments and theory construction. The debates mainly focused on the relations between theories and the results of experiments. Now that a third item, computer simulation, has been introduced, the question arises where to locate it on the methodological map. Are computer simulations *in silico* experiments, or investigations into the mathematical consequences of theories? There does not seem to be any simple answer to this question for some computer simulations are commonly used as experiments when experiments in the lab or field observations are not possible for physical, economical, or ethical reasons, whereas other simulations are clearly investigations into the mathematical properties of the solutions of model equations.

In order to shed some light on the current debate, it may be useful to specify what a computer simulation is. Unfortunately, there is no consensus even on this basic point. A computer simulation can be seen from several, complementary points of view. On the one hand, as mentioned above, a computer simulation is usually based on an underlying model that is made of a series of equations; the simulation helps scientists solve these equations and explore the mathematical features of the solutions. On the other hand, it can also be used to explore the behavior of a natural system by visualizing it on a computer screen even though no experiment is performed on it. Some insist on the first point of view and consider computer simulation as a means to compute solutions to model equations whereas others view simulations as processes (Hartmann, 1996, Parker, 2009) taking place in machines that are somehow able to mimic natural processes.

For sure, some computer simulations are used as processes mimicking the natural phenomenon one wants to investigate. In this case, the possibilities of visualizing are especially helpful because they enable scientists to follow various aspects of the investigated phenomenon on screen in a way that is usually impossible *in concreto*. However, this is only one aspect of the use of simulations. Even though visualization and other experiment-like features of simulation are highly relevant to what scientists do with them, at the bottom, simulations are (very) long series of lines of code. These lines of code are written to compute the solutions to some equations. But it would be naive to think that the computer program that gives rise to a simulation only consists in machine-based recipes to compute solutions to differential equations. To put it briefly: on the one hand, the machine does not understand differential equations, and on the other hand, the simulation's user does not understand the simulation's direct outputs.

Let us investigate these two points further. First, the machine only "understands" series of 0s and 1s. This has several consequences. The first is that it is not possible to simply provide it with a differential equation and initial and boundary conditions in order to obtain a solution. This information has to undergo a tedious series of transformations in order to be readable by the machine. Second, the machine is unable to deal with the mathematical properties of continuity that are embedded in differential equations. It can only deal with discrete mathematical expressions and thus provide approximate solutions to differential equations. Therefore, among the many transformations the initial model has to undergo in order to be used as the basis of a simulation, the shift from differential equations to finite difference equations is a major one. It is not the only one, though. Many further approximations and idealizations are needed for the program to compute what it is meant to compute (Winsberg 2001, 2003). As mentioned above, a computer simulation usually yields large amounts of outputs which are unusable as such and thus have to be transformed in turn. The most common transformations at this point are done by further programs for statistical analysis and/or visualization.

At the end of all these transformations, what is left from the initial model? When everything goes fine, the simulation's user ends up with the solutions she was looking for. But how can she assess that everything went fine? Several obstacles stand on the way. The first one is that, usually, the simulation's user is not the programmer. Now division of labor does not facilitate epistemic control. The programmer is likely to use many tricks enabling her to obtain at least some outputs; however, the user may not be able to understand their effects or to assess whether they are justified. The division of labor has lately taken another, more distributed form, as many elements in computer programs can be obtained online. In this case, it is even more difficult to check the validity and effects of their inclusion into the main program. Second, the large computer programs from which simulations originate are epistemically opaque. This means that it is impossible to check the validity of all their components because of their length, unreadability, and complexity, in the sense that the various components are often interdependent. To put it briefly, computer simulations are meant to provide scientists with the solutions of their models in various situations, but the latter cannot assess whether they actually fulfil their function: they have no usable means to check whether the introduction of simplifications, idealizations, approximations, and so on, are harmless. Of course, when it is possible to compare the simulation's results with experiment results, the above-mentioned problems are lessened. However, simulations are mainly used in the investigation of phenomena on which experiments are difficult or impossible to perform.

What is the upshot of the above-described features of computer simulations? It is that the main methodological problem related to computer simulations is not whether they are closer to theory than to experiments, but their validation. It requires a renewed conception of the epistemic procedures leading one to consider some result as scientifically acceptable. This is the field where the epistemology of computer simulation is likely to be fruitful.

References

Albert, D. (1994) *Quantum Mechanics and Experience*, Cambridge, MA: Harvard University Press.

Albert, D. (2000) *Time and Chance*, Cambridge, MA: Harvard University Press.

Aspect, A., Dalibard, J., and Roger, G. (1982) "Experimental Test of Bell's Inequalities Using Time-Varying Analyzers," *Physical Review Letters*, 49(25), pp. 1804–1807.

Aspect, A., Grangier, P., and Roger, G. (1982) "Experimental Realization of Einstein-Podolsky-Rosen-Bohm Gedankenexperiment: A New Violation of Bell's Inequalities," *Physical Review Letters*, 49(2), pp. 91–94.

Batterman, R. (2001) *The Devil in the Details: Asymptotic Reasoning in Explanation, Reduction, and Emergence*, New York: Oxford University Press.

Batterman, R. (2002) "Intertheory Relations in Physics," *The Stanford Encyclopedia of Philosophy*, Edward N. Zalta (ed.), http://plato.stanford.edu/entries/physics-interrelate/.

Bell, J. S. (1964) "On the Einstein-Podolsky-Rosen Paradox," *Physics*, 1, pp. 195–200, reprinted in Bell (1987).

Bell, J. S. (1987) *Speakable and Unspeakable in Quantum Mechanics*, New York: Cambridge University Press.

Bohm, D. (1951) *Quantum Theory*, New York: Prentice Hall.

Bohr, N. (1935) "Can Quantum-Mechanical Description of Physical Reality Be Considered Complete?," *Physical Review*, 48, pp. 696–702.

Boltzmann, L. (1872) "Weitere Studien ûber das Wärmegleichgewicht unter Gasmolekülen," Wiener Berichte, 66, pp. 275–370, *repr. in* L. Boltzmann (1909), *Wissenschaftliche Abhandlungen*, Vol. I, II, and III. F. Hasenöhrl (ed.) Leipzig, reissued New York: Chelsea, 1969, Vol. I, paper 23.

Bridgman, P. W. (1916) "Tolman's Principle of Similitude," *Physical Review*, 8, pp. 423–431.

Carnap, R. (1932) "Über Protokollsätze," *Erkenntnis*, 3, pp. 215–228.

Carnap, R. (1966) *Philosophical Foundations of Physics*, London: Blackwell.

Chang, H. (2004) *Inventing Temperature: Measurement and Scientific Progress. Oxford Studies in the Philosophy of Science*, New York: Oxford University Press.

Curie, P. (1894) "Sur la symétrie dans les phénomènes physiques. Symétrie d'un champ électrique et d'un champ magnétique," *Journal de Physique*, 3e séries, T. III, pp. 393–417.

Cushing, J.T. (1998) *Philosophical Concepts in Physics: The Historical Relation between Philosophy and Scientific Theories*, Cambridge: Cambridge University Press.

Duhem, P. (1906) *La théorie physique, son objet, sa stucture*, Paris: Chevalie et Rivière éditeurs, English translation Tr. Philip P. Wiener (1962), *The Aim and Structure of Physical Theory*, New York: Atheneum.

Earman, J. (1986) *A Primer on Determinism*, Dordrecht: Reidel.

Earman, J. (1989) *World Enough and Space-Time: Absolute vs. Relational Theories of Space and Time*, New York: MIT Press, Bradford Books.

Einstein, E., Podolsky, B., and Rosen, N. (1935) "Can Quantum-Mechanical Description of Physical Reality Be Considered Complete?," *Physical Review*, 47, pp. 777–780.

Fine, A. (1982) "Some Local Models for Correlation Experiment," *Synthese*, 50(2), pp. 279–294.

Fine, A. (1996) *The Shaky Game: Einstein, Realism and the Quantum Theory*, 2nd edition, Chicago: University of Chicago Press.

Friedman, M. (1983) *Foundations of Space-Time Theories: Relativistic Physics and Philosophy of Science*, Princeton, NJ: Princeton University Press.

Frigg, R., and Reiss, J. (2009) "The Philosophy of Simulation: Hot New Issues or Same Old Stew?," *Synthese*, 169(3), pp. 593–613.

Goldstein, S. (2001) "Boltzmann's approach to statistical mechanics," *in* J. Bricmont et al. (eds.), *Chance in Physics: Foundations and Perspectives*, Lecture Notes in Physics, 574, New York: Springer-Verlag, pp. 39–54.

Hartmann, S. (1996) "The World as a Process. Simulations in the Natural and Social Sciences," *in* R. Hegselmann, U. Mueller, & K. Troitzsch (eds.) (1996) *Modelling and simulation in the social sciences from the philosophy of science point of view*, pp. 77–100.

Humphreys, P. (2004) *Extending Ourselves. Computational Science, Empiricism and Scientific Method*, Oxford: Oxford University Press.

Humphreys, P. (2009) "The Philosophical Novelty of Computer Simulation Methods," *Synthese*, 169, pp. 615–626.

Jammer, M. (1974) *The Philosophy of Quantum Mechanics*, New York: Wiley

Laplace, P. S. (1814) *Essai philosophique sur les probabilités*, Paris.

Lewis, D. (1983) "New Work for a Theory of Universals," *Australasian Journal of Philosophy*, 61, pp. 343–377.

Mach, E. (1883) *Die Mechanik in ihrer Entwicklung Historish-kritisch dargestellt*, Leipzig: F.A. Brochkhaus

Mach, E. (1896) *Die Prinzipien der Wärmelehre*, Leipzig: Barth. English translation B. McGuinness (ed.), *Principles of the Theory of Heat (Historically and Critically Elucidated)*, trans. T.J. McCormack, P.E.B. Jourdain, and A.E. Heath, with an introduction by M.J. Klein. Dordrecht: Reidel. This translation is from the 2nd German edition, 1900.

Maudlin, T. (1994) *Quantum Non-Locality and Relativity: Metaphysical Intimations of Modern Physics*, Oxford: Basil Blackwell; 2nd ed., 2002; 3rd ed., 2011.

Maudlin, T. (2007) *The Metaphysics within Physics*, Oxford University Press.

Maxwell, J. C. (1873) "Molecules," *Nature*, 8, pp. 437–441, repr. in W. D. Niven (ed.), *The Scientific Papers of James Clerk Maxwell* (1890/1961), vol. II, pp. 361–378, New York: Dover.

Maxwell, J. C. (1875) " Atom," *Encyclopaedia Britannica*, 9th edition, vol. 3, pp. 36–49, repr. in W. D. Niven (ed.) *The Scientific Papers of James Clerk Maxwell (1890/1961)*, vol. II, pp. 445–484, New York: Dover.

Morrison, M. (1998) "Modelling Nature: Between Physics and the Physical World," *Philosophia Naturalis*, 35, pp. 65–85.

Morrison, M. (2009) "Models, Measurement and Computer Simulation: The Changing Face of Experimentation," *Philosophical Studies*, 143(1), pp. 33–57.

Parker, W. S. (2008) "Franklin, Holmes and the Epistemology of Computer Simulation," *International Studies in the Philosophy of Science*, 22(2), pp. 165–183.

Parker, W. S. (2009) "Does Matter Really Matter? Computer Simulations, Experiments and Materiality," *Synthese*, 169, pp. 483–496.

Plato, J. von (1994) *Creating Modern Probability*, Cambridge: Cambridge University Press.

Poincaré, H. (1902) *La Science et l'Hypothèse*, Paris: Flammarion, English translation W. J. Greenstreet (1952), *Science and Hypothesis*, New York: Dover Publications.

Price, H. (1996) *Time's Arrow and the Archimedean Point*, Oxford: Oxford University Press.

Quine, W. V. O. (1951) "Two Dogmas of Empiricism," *The Philosophical Review*, 60, pp. 20–43.

Redhead, M. (1980), "Models in Physics," *British Journal for the Philosophy of Science*, 31, 145–163.

Reichenbach, H. (1928) *Philosophie der Raum-Zeit-Lehre*, Berlin/Leipzig: Walter de Gruyter, English translation Maria Reichenbach and John Freund (1958), *The Philosophy of Space and Time*, London: Routledge & Kegan Paul.

Reichenbach, H. (1956) *The Direction of Time*, Berkeley: University of California Press

Ruetsche, L. (2011) *Interpreting Quantum Theories: The Art of the Possible*, Oxford: Oxford University Press.

Russell, B. (1913) "On the Notion of Cause," *Proceedings of the Aristotelian Society*, 13, pp. 1–26.

Sklar, L. (1974) *Space, Time and Spacetime*, Berkeley: University of California Press.

Sklar, L. (1992) *Philosophy of Physics*, Boulder: Westview Press, Dimensions of Philosophy Series.

Sklar, L. (1993) *Physics and Chance: Philosophical Issues in the Foundations of Statistical Mechanics*, Cambridge: Cambridge University Press.

Suppes, P., Krantz, D.M., Duncan Luce, R., and Tversky, A. (1989) *Foundations of Measurement*, vol. 1. New York: Academic Press.

Torretti, R. (1999) *The Philosophy of Physics*, Cambridge: Cambridge University Press.

Uffink, J. (2007) "Compendium of the Foundations of Classical Statistical Physics," in J. Butterfield and J. Earman, eds., *Handbook for Philosophy of Physics*, Elsevier. Available online at http://philsci-archive.pitt.edu/archive/00002691/.

van Fraassen, B. (1985) *Laws and Symmetry*, Oxford: Clarendon Press.

van Fraassen, B. (1991) *Quantum Mechanics. An Empiricist View*, Oxford: Clarendon Press.

van Fraassen, B. (2008) *Scientific Representation. Paradoxes and Perspectives*, Oxford: Clarendon Press.

Wigner, E. (1960) "The Unreasonable Efficiency of Mathematics in the Nature Sciences," *Communications on Pure and Applied Mathematics*, 13, pp. 1–14.

Winsberg, E. (2001) "Simulations, Models and Theories: Complex Physical Systems and Their Representations," *Philosophy of Science*, 68 (Proceedings), pp. 442–454.

Winsberg, E. (2003) "Simulated Experiments: Methodology for a Virtual World," *Philosophy of Science*, 70, pp. 105–125.

12

PHILOSOPHY OF BIOLOGY

Thomas Pradeu (ImmunoConcept, UMR5164, CNRS & University of Bordeaux)

Introduction

"Philosophy of biology" refers to the critical examination of the conceptual, theoretical and methodological foundations of today's life sciences. Although biology, contrary to an erroneous notion, was not entirely absent from the logical positivists' preoccupations (Byron, 2007), philosophy of biology, as such, remains a recent field of study; for the most part its founders, practically all of them Anglo-Saxons, are still alive today. Important founders include David Hull (1935–2010), Michael Ruse (born in 1940), and Elliott Sober (born in 1948). A testimony to the influence of these founders, philosophy of biology has been dominated by questions about evolution, something which is still true today, as testified, in particular, by the contents of the main textbooks in the field (Sterelny and Griffiths, 1999; Sober, 1984, 1994, 2006; Rosenberg and McShea, 2008; Godfrey-Smith 2014). This situation is probably undergoing a slow change as interest in questions of molecular biology, developmental biology, microbiology and immunology become more and more established, but the domination of evolutionary issues in philosophy of biology remains extremely strong. Philosophy of biology has been particularly structured and institutionalized around one journal, *Biology and Philosophy*. Founded in 1986 by Michael Ruse, it was directed by him until 2000, when Kim Sterelny took over the reins. Since 2017, the editor-in-chief has been Michael Weisberg.

Philosophy of biology has a paradoxical status. Its problems are unquestionably of a philosophical bent (e.g., "What is an individual?," "Does such a thing as human nature exist?"), and yet it does not seem, not essentially at least, to be structured by the fundamental problems of philosophy of science (e.g., What is a theory, a law, a model? What is a scientific explanation?) So, even though it has been one of the most active fields within philosophy of science since the end of the 1980s, philosophy of biology could seem to be little representative of general philosophy of science. For example, the textbooks of Sterelny and Griffiths (1999) and Sober (1984, 1994, 2006) mentioned above tackle almost none of the questions typically associated with general philosophy of science. At its outset, philosophy of biology was in equal measure built on and also in opposition to the foundations of general philosophy of science. The building on tendency is illustrated, in particular, by Ruse (1973), who applies traditional problems of the philosophy of science to biology, even placing himself in the continuation of logical empiricism, although sometimes with a critical attitude (see Hull's most enlightening analysis, 1977). The second, opposing, tendency is clearly apparent in Hull (1969, 1974), who is of the opinion that philosophy of biology must be built in large part in opposition to general philosophy of science of his time, seen as being excessively dominated by just one science, physics, and also, above all, by certain problems rooted in logical empiricism and an excessively analytical conception of philosophy of science, which Hull attacks for its idealistic views (Hull, 1969, 1988, 1989b).

After the 1960s, philosophy of biology's autonomy with respect to general philosophy of science has tended to increase even further. Indeed, philosophy of biology has progressively broken free of the big questions of philosophy of science as its growing specialization has advanced. Of course, some influential philosophers of biology were really involved in general philosophy of science, or else had fields of interest that stretched beyond biology before becoming interested in the life sciences (see Rosenberg 1985; Sober, 1984, pp. ix–x). The latter have applied to the living world some classical questions of philosophy of science, notably questions about the theoretical status of the theory of evolution, or about reductionism. However, during the 1990s, a steady trickle of specialists in philosophy of biology began to appear who distanced themselves from general philosophy of science and its questions while drawing closer to biologists. Kim Sterelny and Paul Griffiths are two typical examples of this trend in philosophy of biology.[1]

In this presentation, I hope to show the diversity of the problems posed in philosophy of biology by drawing attention to seven of them. The first, regarding the status of the theory of evolution, is undoubtedly the closest to typical questions from within general philosophy of science. Second, I will ask what is meant by the idea of adaptation in biology, when it is said, for example, that an organism is "well adapted" to its

---

[1] This is how Sterelny and Griffiths lay out the approach they took in their manual *Sex and Death*: "One option would be to use biological examples to stalk general issues in philosophy of science—the nature of theory change, causation, explanation, and prediction . . . . that is definitely not the book we have written. This book is very much focused and the conceptual and theoretical problems generated by the agenda of biology, rather than pursuing a philosophy of science agenda through biological examples" (Sterelny and Griffiths, 1999, p. xi). See also Sterelny (1995).

environment, or that an organ is "well adapted" to its function. This will lead us to a third problem regarding the basis biologists lean on when they turn to a functional and, seemingly at least, finalist vocabulary in speaking of an organ's or a trait's "function." Based on these first three steps, which will enable us to lay out several crucial aspects of the theory of evolution, we will consider, fourth, the problem which has undoubtedly occupied philosophers of biology the most since the birth of their discipline, that is, the units of selection problem: On which biological entities (genes, genomes, cells, organisms, groups, species, etc.) does natural selection operate? Fifth of our problems, though evolution may be the dominant theme in philosophy of biology since its inception, the issues relative to organism development (by which is meant the changes an individual organism undergoes from its conception to its maturity—or, according to some, to its death—) are in the process of establishing themselves as another major theme that we will analyze in turn. One of the objectives of research into development is to answer certain questions that the theory of evolution passes over, for example, questions concerning the legitimacy of speaking about a genetic programming of organism development and how the regulation of this development occurs. As we can see, the subject of development enables us to connect specifically evolutionary questions to those more tied in with cellular and molecular biology, these, in turn, being ever more studied by philosophers of biology. In the sixth section, we will look at the question of reductionism, which, within the context of contemporary philosophy of biology, primarily consists of asking whether or not it is possible to reduce macromolecular biology to molecular biology. In the seventh section, I will draw attention to some recent work that, contrary to the dominant trend in philosophy of biology, does not focus on evolution only. Following analysis of these problems, and by way of conclusion, I will come back to the question of the relationships held between philosophy of biology, general philosophy of science, and biology itself.

1. The Status of the Theory of Evolution

The theory of evolution is generally considered to be the foundation to every proposition in biology, as well as the primary, if not unique, biological theory. What then, precisely, does "the theory of evolution" mean?

The aim of the theory of evolution is to explain modifications in species over time; their adaptations and their diversification. Darwin was not the first to put forward an explanation for this phenomenon, nor to speak of species evolution (this idea can be found in Lamarck, in Erasmus Darwin, etc.). Nonetheless, Darwin (1859) advanced two decisive theories: common descent (captured in a species tree), that is, the assertion that today's organisms are descended from common ancestors; and natural selection, according to which there is a process of variation and then of differential survival and reproduction among organisms (the "struggle for existence" leading to the "survival of the fittest," to use the expression Darwin would eventually borrow from Spencer). So what we call the "theory of evolution" is a set of propositions initially put forward by

Darwin and then, between the 1920s and 1950s, solidified around the central ideas of common ancestry and natural selection by those active in the "Modern evolutionary synthesis" (Mayr and Provine, 1980). However, as much in Darwin's case as in the case of the Modern Synthesis, speaking of the theory of evolution causes problems.

First, can we truly speak of *the* theory of evolution? According to Mayr (1982), Darwin does not propose one but five theories: evolution as such, common descent, gradualism (the idea that species evolution occurs by means of cumulated minor modifications and not by "leaps"), population speciation (the idea of a continuity between population and species, a population of living creatures which undergoes variation being considered as a "nascent species"), and natural selection. Each of these theories met with a different fate. In particular, common ancestry was very quickly accepted by biologists following the publication of *The Origin of the Species*, while natural selection was neither well understood nor widely accepted in Darwin's own lifetime. Even though Darwin held to each of them, taken together they did not constitute a unified theoretical structure (Mayr, 1982). Furthermore, precisely as a result of this plurality of ideas in Darwin, they were on the verge of being abandoned at the turning of the 20th century: following work which had rediscovered Mendel's "laws" of heredity, a certain tension arose between gradualism and speciation (Bowler, 1983; Gayon, 1998). Darwin was in the dark regarding the mechanism behind variation in individuals, contenting himself to simply observe the phenomenon. But to his eyes, it was clear that the variations were gradual and not saltatory. The first "geneticists" found the mechanism of variation in what they called "mutations" but, according to them, mutations were quite precisely leaps and not gradual modifications: for de Vries, in particular, species appeared suddenly following one of these mutations (Allen, 1969). The Darwinian theory of gradualism and natural selection thus found themselves strongly rejected (Bowler 1983; Gayon, 1998). The first step of the Modern Synthesis (corresponding roughly to a period between the 1920s and 1930s) was the unification of genetics and Darwinism, primarily under Fisher's (1930) influence. Fisher showed that mutations, whose effects are generally limited, are perfectly compatible with Darwinian gradualism and in fact account for the variating mechanism so desperately sought since Darwin's day. It would, however, be erroneous to think that the Modern Synthesis led to a unified theory of evolution. The second step of the Modern Synthesis (roughly from the 1930s to the 1950s) involved the aggregation of various disciplines of biology (zoology, botany, systematics, etc.) around a "solid core" of hypotheses (Mayr and Provine, 1980). So the Modern Synthesis came about more as a result of a sociological convergence (the unification of practically all the branches of biology on the basis of principles relative to evolution) than by the formulation of one theory of evolution (Gayon, 1998, p. xiv).

Nevertheless, can we take the common principles all biologists have accepted since the Modern Synthesis and use them to deduce propositions for the "theory of evolution"? This leads us to our second question: can we really speak of a *theory* of evolution? Concerning Darwin's own ideas, we should perhaps speak not so much of a veritable theory as of a descriptive generalization which created a paradigm (in the sense of an exemplary model, widely imitated afterward) for understanding species evolution, at least in as far as common descent is concerned (Gayon, 1998). Nevertheless, it has often been emphasized (e.g., Ghiselin, 1969; Lewens, 2007a; Sober, 2011) that Darwin had complied to the canons of theory construction of his time, and in particular to the views of Whewell and Herschel. Concerning the theory of evolution as it has been presented since the Modern Synthesis, philosophers of science have attempted to determine whether or not it constitutes a veritable theory. Many are the philosophers who have doubted its validity as a theory, their primary argument being that biology, since it is a "historical" science, cannot formulate laws, and hence cannot offer theories in a nomological sense (Smart, 1963; see also Beatty, 1995). Most of Smart's arguments are invalid and rely on a false understanding of biology (Ruse, 1973; Hull, 1969, 1977): contrary to his claims, biology deals not with such and such albino mouse but with processes of a much wider scope, such as the conditions for the expression of recessive genes, crossing-overs, and the notion of geographically isolated populations—which all are processes about which generalizations are possible. On the other hand, it is undeniable that biological entities are spatiotemporally situated within an evolutionary history: for example, a biological species is a historical entity, the product of an evolutionary history, and not a class of objects open to abstract generalization which disregards spatiotemporal conditions, as is standard in physics. Consequently, formulating laws of biology, that is, general abstract propositions, at the level of historical entities, seems impossible. But for Hull, for instance, biology can formulate laws about entities that are not defined genealogically, in particular at levels of organization higher than particular taxa (Hull, 1978, pp. 353–354). From this point of view, the claim that biology could not offer laws at all is misleading (Hull, 1976, 1978). It is, however, difficult to assess the scope of the claim that laws cannot be formulated about historical entities: in the context of that debate, doesn't physics run the risk of isolating itself from the majority of other empirical sciences, all "historical" in the sense we have defined, such as biology, geology, and the social sciences? If physics be the only science capable of formulating laws, should it remain a model for philosophy of science in general? Furthermore, certain branches of physics, like astronomy, also deal with historical entities. If the future reveals all empirical sciences to be "historical," wouldn't we have to soften our stipulation that a science must necessarily produce (spatiotemporally unrestricted) laws? Alternatively, we could suggest other, more "relaxed," conceptions of what a "law" is. Finally, the implicit assertion stating that a science cannot advance any theories once it advances no laws, must be handled with caution, as it depends on one particular vision of theories which, we shall now show, is not only not well suited to biology, but is also not the only vision of scientific theories possible.

In the 1970s, philosophers of biology brought precision to the debate around the problem of the theory of evolution's being a genuine theory or not by posing the following question: If the theory of evolution is a theory, is this in the "syntactic" or "semantic" sense of the term? According to the syntactic conception, best expressed by Hempel (1965), a theory is a hypothetico-deductive system in which, based on just a few axioms, one must be capable of deducing a large number of propositions.

According to the semantic conception, defended in particular by van Fraassen (1972) and Suppe (1977), a theory is a collection of models that must serve as the representation of empirical phenomena. In the semantic conception, to describe a theory is to present a class of models and to specify the manner in which those models reflect the phenomena. It quickly became apparent that the theory of evolution was not a theory in the syntactic sense of the term. Several biologists (M. B. Williams, 1970; Lewontin, 1970) and philosophers (Ruse, 1973) have attempted an axiomatization of the theory of evolution, but this has led more to uncovering the theory of evolution's "structural core" than it has to a veritable axiomatization: by means of a method exemplified by Lewontin (1970), they pushed themselves to defining the minimal conditions a population of individuals must meet to be said to evolve by natural selection (for a fresh look at these questions, see Godfrey-Smith, 2007 and 2009). The most enthusiastic advocates for an axiomatization of the theory of evolution finally ended up showing that this effort could only be partial (M. B. Williams, 1981). Several philosophers of biology have defended the view that if the theory of evolution is a "theory," then it is so in the semantic rather than the syntactic sense of the term (e.g., Thompson, 1983; Lloyd, 1993 [1988]): it can be interpreted as a collection of models that must serve as the representation of empirical phenomena (important discussions of models in recent philosophy of biology, with a different perspective than that of Thompson or Lloyd, include Godfrey-Smith, 2006 and Weisberg, 2006). Were a consensus to emerge regarding the semantic conception of scientific theories (but see Ereshefsky, 1991), the oft repeated claim that the theory of evolution is not really a theory would have to be just as quickly flatly rejected. The work carried out by philosophers and biologists on the structure of the theory of evolution by natural selection since the end of the 1980s (e.g., Lloyd, 1988; Gould, 2002) sets its goal precisely as the clear definition of these models and the conditions for their testing.

2. Adaptation

According to many evolutionists, the aim of the theory of evolution is not so much to explain species modifications in general as the fascinating complexity of their traits and their amazing adaptation to their environment. The principal goal of the theory of evolution, in other words, would be in this view to account for adaptive complexity. Darwin (1859) himself stands behind this position,[2] which he illustrates through several examples, among which we find the recurring example of the woodpecker. In observing a woodpecker's beak, how could one not conclude that it was perfectly "adapted" to the bird's goal, which is to grab the insects from within the cracks in bark? For Darwin, and for many contemporary biologists as well, the question of adaptation can be viewed

[2] Darwin describes his aim as to show "how the innumerable species inhabiting this world have been modified, so as to acquire that perfection of structure and coadaptation which most justly excites our admiration" (Darwin 1859, Introduction).

as being the "atheistic" inheritor of William Paley's natural theology: "The main task of any evolutionary theory is to explain adaptive complexity, i.e. to explain the same set of facts which Paley used as evidence of a Creator" (Maynard-Smith, 1969, p. 82). According to Paley (1802), a man going ashore on an island and finding a watch could not but deduce that the island was or had been inhabited, since such a complex artifact could not be the product of random chance; similarly, when we see a living being, its complexity (far superior to that of any artifact) is such that its origin in divine creation, and not random chance, cannot be doubted. Darwin, having eagerly taken the lectures on natural theology at Cambridge, takes on Paley's problem but asserts that it is a natural force without design, natural selection, which explains the adaptive complexity of living beings. Likewise, for Dawkins (1986), Paley poses the right problem, it's just that the watchmaker is "blind": natural selection is a force without design but which nevertheless explains the design-like appearance of things. In fact, natural selection appears as an optimizing force in respect to a given environment: when only the fittest survive and reproduce, natural selection explains the correct adaptation of organisms to their environment, right down to the most fascinating of consequences, like in the case of the woodpecker. It is also what the phenomenon of fixation illustrates, where an allele correlated to an advantageous trait goes from being rare to being ubiquitous (or "fixed") within a population. Of course, the adaptation process always depends on a given environment: when the environment changes, those organisms who were the best fitted to the previous environment will probably no longer be so with respect to the newly changed one. But as long as the environment remains relatively stable, adaptations resulting from natural selection can be transmitted along the generations.

The above passage could seem to be nothing more than a simple, uncontroversial description of the adaptive effects of natural selection. However, the adaptation debate has been one of the most charged areas in philosophy of biology since the 1970s. Given that the terms of this debate have not always been sufficiently clear, I will try to define adaptation and its related concepts as best as possible before presenting the bases of attacks on "adaptationism" (adaptationism, analyzed further on, is the position stating that living beings, thanks to natural selection, are perfectly adapted to their environment).

How should the concept of adaptation, seeming at once central to the theory of evolution as well as reliant on a firmly intuitive interpretation, be defined? Taking our lead from Lewens (2007b), we can begin by envisaging an informal definition of adaptation: an adaptation is just a trait which seems guided towards a design but whose existence is in fact the result of natural selection (G. C. Williams, 1966; Dawkins, 1986). However, a definition like this excessively hangs the notion of adaptation on Paley's theological outlook: it seems quite unacceptable to suggest that the identification of adaptations depend on a discipline which is unanimously viewed as unscientific and whose clearest echoes are found in the modern thesis of "intelligent design." To more precisely and more robustly determine what adaptations are, Sober (1984, p. 208), picking up from Brandon (1978) and Burian (1983), proposed an explicitly historical definition: "A is an adaptation for task T in population P if and only if A became prevalent in P because there was selection for A, where the selective advantage of A was due to the fact that A helped perform task T." An adaptation is thus a trait whose presence and persistence within a given population is the result of its contribution to the overall fitness of the organisms that possessed this trait in the past (it is, therefore, important not to confuse adaptation and fitness, as the remainder of this section shall show in detail). This definition may seem circular: it would seem difficult to accept that the Darwinian revolution consisted of explaining adaptation by means of natural selection if adaptation were to be defined as the product of natural selection. However, the circularity vanishes when we assert that Darwin meant for traits such as the vertebrate eye or the woodpecker's beak, and for behaviors such as certain instincts to be seen as the consequences of natural selection's specific action. In an equivalent but more precise manner, the Darwinian position can be reformulated by saying that Darwin showed how the "correct adaptation" of living beings towards their present environment is explained away as being the product of the past actions of natural selection. Even though a non-historical definition has been put forward, stating that an adaptation, in the present, is, "a phenotypic variant that results in the highest fitness among a specified set of variants in a given environments" (Reeve and Shermann, 1993, p. 9),[3] the historical definition is nevertheless the dominant one (Brandon, 1990).

On the basis of this historical definition, philosophers of biology have proposed several useful conceptual distinctions for understanding adaptation. The most important is that between an adaptation, a trait selected in the past because it increased the fitness of its bearer, and an adaptive trait, a trait that increases its bearer's fitness in the present. An adaptive trait is not necessarily an adaptation, and vice-versa. Imagine, for example, that the woodpecker's beak presently enables it to nest in drain pipes and that this increases its fitness: this would be an "adaptive trait" but not, however, an "adaptation" in the historical sense, that is, the product of repeated natural selection in the woodpecker species' past. Conversely, very thick plumage in certain woodpeckers could be the product of natural selection's past action and yet no longer be "adaptive" in a climate where the temperature had increased significantly. Similarly, the distinction must be made between an adaptation, the result of a process (the woodpecker's beak, for example), and adaptation in the general sense, that is, the process that lead to this result itself. In most cases, philosophers of biology have the result, and not the process, in mind. What precedes illustrates the fundamental difference between the concepts of adaptation and of expected fitness: the mechanism of natural selection probabilistically predicts that individuals with higher fitness will survive and reproduce, while adaptation is the name given to the result of the immediate selection process. The definition of adaptedness (Brandon, 1990) enables us to highlight the non-tautological character of the natural selection hypothesis: this is not worded to say that the fittest survive by defining the fittest as those who survive, rather it defines

[3] In this definition, a "phenotypic variant" refers to a particular trait possessed only by a subset of living beings of a given species (for example, those woodpeckers who possess a much stronger beak than their kin) in a given environment.

those traits which increase the probabilities of survival and reproduction in those who possess them (Mills and Beatty, 1979). The consequence of this is that those individuals with the highest expected fitness (Burian, 1983; Brandon, 1990 refers to adaptedness) may not necessarily be those who survive and reproduce the best, that is, who display the highest realized fitness.

Let us look now at some criticism "adaptationism" has received. In what has remained one of the most famous articles in biology and philosophy of biology over the last decades, Gould and Lewontin (1979) denounced this adaptationism, that is, biologists' "panglossian" attitude, which, like Voltaire's Pangloss, sees every biological trait as proof that all is for the best in the best of all possible biological worlds. Gould and Lewontin, using numerous examples, show that this attitude is extremely widespread in the biology of their time. In their view, adaptationism occurs in two steps: (1) the atomization of the organism into traits, each of which is described as a structure optimally conceived for its "function" by natural selection; (2) each trait seeming, in fact not to be perfectly adapted to its function, the adaptationist explains that each organism is the best trade-off possible between the different environmental demands to which it is subjected. Dennett (1995), without doubt the most audacious or, depending on one's own opinion, most naive of the adaptationists, openly admits this "panglossianism," which he believes to be inherent to the argument for natural selection (on this matter, see also Dupré, 1987).

In reality, the criticisms drawn together by Gould and Lewontin seem today to be as passionate as they are insufficiently unraveled. Here, aligning myself in part with the remarkable clarifications put forward by Godfrey-Smith (2001) and Lewens (2009), I distinguish three overlapping problems in their article.

(1) Is natural selection the only, or even the primary, evolutionary mechanism? Clearly, natural selection is not the only mechanism of evolution: random genetic derivation (that is, modification due to random sampling of allelic frequencies from one generation to the next for a given population),[4] in particular, plays an important role in species evolution, especially in cases of small population sizes. Other evolutionary mechanisms are also acknowledged by many biologists (allometry, for example, the correlation between the size and the shape of an organism). Gould and Lewontin insist on development constraints and on the "Baupläne" which limit natural selection's action possibilities and thus innovation. Darwin himself vehemently asserted that natural selection was the primary but not the

[4] Imagine a population of woodpeckers. The allele A (possessed by woodpeckers with a very strong beak) has a frequency p. The frequency of this allele A in all possible descendant populations of woodpeckers will average p. However, the realized woodpecker descendants constitute only a sample selection from within these possible descendants, so that the actual realized frequency of allele A could, in fact, differ from p. Thus, independently of the effects of natural selection, the frequency of a given allele changes from one generation to another simply by the effect of random sampling.

only evolutionary mechanism (see, in particular, a famous letter of Darwin, published in Nature: Darwin, 1880). In asserting that natural selection is not the only mechanism of evolution, Gould and Lewontin can rest assured that all biologists will agree with them. However, the authors add that, at time of writing, these other mechanisms of evolution had not yet drawn sufficient attention, something which was undoubtedly true in 1979 but we can affirm is less so today (especially since Kimura, 1983). Most importantly, Gould and Lewontin pose two real problems which still remain pertinent now: (1) What exactly are the other mechanisms of evolution? (2) Which part of evolution must be attributed to each mechanism (selection, derivation, etc.)?

(2) Do mechanisms other than natural selection enable explanation of adaptive complexity? In their criticism of adaptationism, Gould and Lewontin do not sufficiently differentiate this question from the previous one. If their position in respect to (1), where they affirm that natural selection is not the only adaptive mechanism, seems, today at least, to meet with consensus, with this second question they place themselves in opposition to the great majority of biologists (first among these being G. C. Williams, Dawkins, and Maynard-Smith). Gould and Lewontin show that the "correct adaptation" of organisms to their environment can sometimes be explained by other mechanisms than natural selection, like phenotype plasticity, for example (what Mary-Jane West-Eberhard more recently describes as "the ability of an organism to react to an environmental input with a change in form, state, movement, or rate of activity"; West-Eberhard, 2003, p. 34). The difficulty lies in precisely demonstrating in what measure these mechanisms are indeed evolutionary, that is, liable to ancestral transmission, making possible the accumulation of adaptive effects down through the history of the species, rather than their just being processes involving only the individual (ontogeny). Gould and Lewontin do not sufficiently rise to this but, once again, recent work is carrying their intuitions forward by showing, notably, the evolutionary effects of phenotypic plasticity (West-Eberhard, 2003). Another, partly related possibility is epigenetic adaptation (Jablonka and Lamb, 2005). As of yet, no consensus has emerged on these matters. For now, the vast majority of biologists deem that, if adaptive complexity is what we are trying to explain, then the best explanation available to us is certainly natural selection (though see West-Eberhard, 2003 and Müller, 2007).

(3) Must every biological trait be explained in terms of adaptation? In my view, this is the heart of Gould and Lewontin's article. This third question is obviously linked to the first two, but it is important to point out how it is distinct from them also. The first question is rightfully removed from the question of adaptation itself, being rather concerned with the problem of natural selection's importance in evolution. The second question,

> by contrast, contains the admission that adaptation is a fundamental biological phenomenon, therefore necessitating explanation, and asks if the best explanation for adaptation is the mechanism of natural selection. In the third question, the idea that one of the most fundamental characteristics of living creatures is their "adaptation" is challenged and, along with it, the claim that one of biology's primary tasks is to explain this adaptation. To use one of Gould and Lewontin's most famous examples, many biologists state that the existence of the Tyrannosaurus's small sized forelimbs is a puzzle: what purpose could possibly be served by limbs so short they don't even reach the mouth? What could their "function" have been? Gould and Lewontin respond that relentlessly seeking a solution to these limbs' "adaptedness" to nature may be of little benefit: (i) they must first and foremost be seen as being inherited from organs existing in the Tyrannosaurus's ancestors; (ii) present usefulness and adaptation, in the sense of "product of natural selection," must not be confused (in this they agree with the point established above). In other words, Gould and Lewontin maintain that a great many biological traits are not "well adapted." The dominant interpretation is that Gould and Lewontin's article is merely a useful, though little consequential (in terms of actual biology) warning against the excesses of seeing adaptation everywhere. And yet, in reality, the article contains far more than that (Godfrey-Smith, 2001), particularly when it is seen in relation to Lewontin's argument that the construction metaphor must replace the adaptation metaphor in contemporary biology (Lewontin, 1978; this argument has recently seen a highly remarkable, though sometimes contested, resurgence in interest in the wake of Odling-Smee et al., 2003): he proposes research avenues to biologists for the exploration of processes other than just adaptation in the practice of biology, as well as for new perspectives on the living world.

To conclude this section, I come back to the most fundamental question. Is adaptation the most important element of the living world, that which biology must therefore explain as a priority? Joining Gould and Lewontin, we seem in our right to be doubtful. What is certain is that those supporting this proposition must put forward far more solid arguments than those which have thus far been formulated, of which Dawkins's (1986, p. 303) is the most typical: "Large quantities of evolutionary change may be non-adaptive, in which case these alternative theories may well be important in parts of evolution, but only in the boring parts of evolution . . ." Not only could the exact contrary be asserted but, in any case, the argument of the scientific interest bears no weight when dealing with a question about the reality of the living world (Godfrey-Smith, 2001). As Lewens (2007b) shows, the fact of many biologists focusing on the phenomenon of adaptation seems, in reality, to be largely evidence for the genealogy of the Darwinian theory, itself stemming from Paley's natural theology (this includes the likes of Dawkins, Dennett, Grafen, Maynard-Smith, etc.). It may well be doubted that there exist one fundamental question in biology. As for evolutionary biology, if we had to ascribe just one fundamental question to it, then, following Ghiselin (1983), this would be "What happened?" that is, "What is the history of life?" and not the riskier and indeed very problematic, "How can life's amazing adaptation be explained?"

3. Functions and Teleology in Biology

Contrary to physics or chemistry, biology seems to employ a teleological vocabulary: do we not say, for example, that the "function" of the heart is to circulate blood, or that the heart is "for" circulating blood? The question arises of how such statements should be understood, and whether or not the presence of teleological vocabulary is problematic in an experimental science such as biology.

Nagel (1961) considered the use of teleological vocabulary to be a serious obstacle should biology wish to gain real scientific legitimacy, compared to the example of physics which had slowly but surely freed itself of teleology. He also suggested replacing functional statements with ordinary causal statements, interpreting cause as necessary condition. For example, the expression "the function of the heart is to circulate blood" would have to be replaced by the expression "the heart is a necessary condition for the circulation of blood." However, as shown by Larry Wright in a founding article (1973), this suggestion fails because it does not allow for the distinction of two cases that biologists imperatively wish to distinguish. For example, hemoglobin is at once a necessary condition for the redness of blood as well as for the transportation of oxygen; nevertheless, biologists will state that its "function" is to transport oxygen, never that it is to give blood its red color. In other words, while Nagel had perfectly exposed the problem, unfortunately his solution didn't fit.

Philosophy of biology, largely relying on philosophy of mind, has enabled decisive progress on the question of biological functions. This progress has led to what could be called, following Godfrey-Smith (1993), a "consensus without unity," in light of the fact that two clearly distinct uses of the term "function" exist in biology today, the etiological and the systemic uses. This distinction has clarified the functions debate considerably.

According to the etiological conception, of which Wright (1973) and Neander (1991) are among the principal representatives, the statement "the function of the heart is to circulate blood" means "the heart was selected in the past for its capacity to circulate blood." This conception is, on the one hand, fundamentally historic and, on the other hand, immediately correlated to the idea of adaptation by natural selection met in the previous section: a function is any trait that is the product of its positive contribution, in the past, to the fitness of the members of a species. The etiological conception seems to dominate in the function debate, the majority of philosophers of biology of the last twenty-five years or so having aligned themselves with it. One of the most notable is Karen Neander who has proposed defining a function, quite simply, as a "selected effect" (Neander, 1991). One of the main reasons for the success of this conception is that it seems to meet the demand for a "naturalization" of the teleological wording found at the center of philosophical reflection on functions. For example, Gayon (2006, p. 482) writes: "When the biologist uses the notion of function, his interest is not only in the actual effect of some apparatus or process. He is not only occupied with what it does, but also with what it is supposed to do." The etiological conception defines a functional norm relative to a certain type of organisms (for example, all vertebrates or all zebras, etc.); in so doing, it allows for the assertion that a heart which does not fulfill the function for which hearts were selected therefore does not function "normally": it doesn't do what it is "supposed to do." One of the possible objections to the etiological conception is that it may run the risk of a certain "adaptationism" if it begins to see all traits as "functions." However, this is only a possible, rather than a necessary, consequence of the etiological conception. Another objection to this conception is the difficulty it has, with its own specific vocabulary, in accounting for the "adaptive" character of an innovation: a newly appeared trait increasing the fitness of its bearer (and, in the future, of its descendants) cannot be said to have a "function," in the etiological sense, since it is not the product of an evolutionary history. Nevertheless, it is likely that biologists would still choose to speak of this as a "function." And yet, a simple conceptual clarification, such as that between adaptation and adaptive trait, could probably dissipate this difficulty.

The second conception, called the "systemic" conception, differs greatly, because its eye is not on the past, and it is not reliant on the theory of evolution by natural selection. The systemic conception is founded on the present analysis of a biological mechanism. According to Cummins (1975), functions are not effects that explain why something is there but effects that contribute to the explanation of more complex abilities and dispositions within a system they partake in. In other words, Cummins's starting point is the delimitation of a biological "system," the organism being available to analysis within several systems (circulatory, nervous, respiratory, etc.), themselves divided into characteristic capacities which, in turn, can be analyzed as organs and structures participating in the realization of that ability. For example, in speaking of the respiratory system, we can talk of its capacity to transport food, oxygen, waste, and so forth, and in the context of that systemic capacity, we can say that the heart is capable of pumping, which means that this is its "function," in the systemic sense of the word (Cummins, 1975, p. 762). The systemic conception offers the advantage of being equally applicable to living beings as to artifacts or technical systems, with Cummins even proposing that biological functions be understood following the analysis model for the functioning of an assembly line. Furthermore, in assuring that one and the same function can be accomplished by various structures having diverse evolutionary histories, it accounts for the fundamental difference struck in biology of evolution between homologies (organisms having similar traits due to their common genealogical origin) and evolutionary convergences (similar traits not due to a common genealogical origin).

The systemic conception prolongs and enriches what Mayr (1961) advanced under the title of "functional biology"—that part of biology which poses "how?" type questions, in opposition to the biology of evolution which poses "why?" type questions. This explains that the systemic conception dominates physiology and experimental biology. As for the etiological conception, it is situated exactly on the evolutionary biology side ("why?" questions). It must, however, be made clear that within the context of the debate on biological "functions" the term "functional biology" must be avoided, since it sweeps away the distinction between the systemic and etiological conceptions.

The systemic conception is not teleological, it accounts for the causal contribution of a mechanism to a system it forms a part of, it does not attempt to say what ends traits have, nor what they are "supposed to do." It is therefore "mechanistic," one of the possible extensions of that conception being precisely the recent interest in philosophy of biology for the notion of mechanism (Machamer, Darden, and Craver, 2000; Craver, 2007; Bechtel, 2005). Consequently, it does not answer the demand often formulated with regard to the notion of function, the question of what something is "supposed to do." For this reason, the systemic conception often meets with the same criticism that Nagel had already faced: not accounting for the normativity of the notion of function, nor, correlatively, for the possibility of dysfunction. This is undoubtedly the most serious objection that can be leveled at this conception, though it is unlikely to be a fatal one. First of all, quite simply, we cannot criticize the systemic conception for not doing precisely what it wishes to not do, namely, giving an answer to the question of what a trait is "supposed to do" (Cummins, 1975, p. 757, n. 13). Second, some dysfunctions at least can be understood "systemically," for example when we analyze a disease as a series of causal contributions to the overall effect on the system. In this way we can, for example, explain the development of an autoimmune disease by saying that it results from a dysfunction of the immune system (which in these cases stops performing what is generally considered to be its function, defending the integrity of the organism), but we can also explain it by detailing the cellular and molecular mechanisms which lead to this pathological state—by showing, for example, in what way the number of regulatory cells in the organism have decreased, why there is cross-matching with a pathogen, and so forth (Pradeu, 2010a). In addition, the systemic conception may resort to a statistical definition of norms, where what is normal would be defined simply as what is the most commonly occurring.

Finally, it appears that both conceptions are operational and that each of them is dominant within one or the other of the two main branches of biology (evolutionary biology and "mechanistic" biology, in the broad sense). It may be deemed regrettable that the term "function" find itself split into two such different meanings. Some philosophers have attempted to bring these two meanings for "function" together under one single definition. The most remarkable of these attempts is Kitcher's (1993), who proposes uniting them under the umbrella concept of "design." However, his attempt did not convince Godfrey-Smith (1993), and it can in fact be claimed that one of the most clear advancements due to philosophy of biology is the firm assertion that there does exist two distinct concepts of function. In the interests of complete clarification, it would perhaps be useful to reserve the term "function" for just one of the two ideas analyzed here, though such a reform of functional vocabulary seems unlikely given that, on the one hand, biologists are quite firmly attached to it and that, on the other hand, few actual incompatibilities between them have arisen.

4. The Units of Selection Debate

Unquestionably, the most intense and impassioned of all debates in philosophy of biology these last forty years has been the units of selection debate. To a lesser, yet still considerable, degree it has also drawn in the biologists themselves. With hindsight, it could be said that the fierce tensions identified with it were at least partly created by a lack of clarity in the initial wording of the problem. The most significant contributions to the debate have come from biologists but the most important clarifications were the work of philosophers.

The starting point is the aforementioned problem with the structure of the theory of evolution by natural selection (TENS). Following on from Mary Williams's (1970) work, Lewontin (1970) showed that the structure of the TENS made it applicable to a wide variety of entities, and not to organisms only: any population made up of entities characterized by variation, differential fitness, and the heredity of that fitness can be said to evolve by natural selection. Lewontin's question, in what was the first text entitled "Units of selection," is this: "Which entities are capable of evolution by natural selection?" His response covers a wide spectrum, not only individual organisms, as is generally asserted, but also a whole hierarchy of biological entities: genes, organelles, cells, organisms, populations, species, even as far as ecosystems and prebiotic molecules (see also Lewontin, 1985).

In publishing *The Selfish Gene*, Dawkins (1976) set the debate on units of selection into full swing, in the sense that, from that point onwards, huge numbers of biologists and practically every philosopher of biology felt the need to have their say on the matter. Dawkins's thesis, inspired by the views of George C. Williams (1966), Hamilton, and, in part, by certain actors of the Modern Synthesis (Mayr, 2004), is called "genic selectionism" or the "gene-centered view of evolution." Dawkins is largely responsible for the confusion which reigned over this debate for several decades because he uses the same term as Lewontin, that is, "unit of selection," despite their dealing with completely different problems. Dawkins's question is: "To whose benefit does natural selection occur?" For Dawkins, genes are the real units of selection as they are the real beneficiaries of natural selection and its effects. His argumentation can be summed up in four steps: (i) the most important biological phenomenon, adaptive complexity, can only be understood over long evolutionary periods of time; (ii) however, placed beside such lengthy time periods, organisms are very ephemeral beings, they are, as famously suggested by Dawkins, "like clouds in the sky or dust-storms in the desert" (Dawkins, 1976, p. 34); (iii) in contrast, the genes contained within these organisms are transmitted between generations with great fidelity, they are what truly persist on the evolutionary scale, making the accumulation of discrete adaptations possible; (iv) consequently, the theory of evolution by natural selection applies not so much to organisms as to those entities which truly span the ages, the genes (for a corroborative philosophical analysis of Dawkins's position, see Sterelny and Kitcher, 1988; see also Lloyd's 2005 rebuttal).

The gene-centered thesis brings confusion to the units of selection debate as it tends to be presented as a response to Lewontin's question, while in actual fact it responds to a different one. Were it a response to Lewontin's question it would be this: natural selection operates exclusively, or at least primarily, on the genic level of life. As a response to Dawkins's question, it is this: genes are the true beneficiaries of the action of natural selection. As well as the re-apparition of Dawkins's aforementioned "adaptationism' (meaning that the essential question for Dawkins is that of adaptive complexity), the main difficulty is that Dawkins doesn't define the problem he claims to answer with sufficient clarity. This ensuing confusion is then carried on by numerous biologists, each one chipping in on the "units of selection" debate, even though it is still not known precisely what question they are actually responding to.

Certain philosophers of biology have played a decisive role in this debate: since the beginning of the 1980s, a handful of them have brought considerable clarification to it (several results of this clarification are detailed in Brandon and Burian, 1984).

One of the most useful clarifications came from David Hull (1980, 1981, 1988, particularly p. 407 sq.) Hull proposes the differentiation of two biological entities involved in the evolutionary process: the replicator, being "an entity that passes on its structure directly in replication" (i.e., an entity that is faithfully copied), and the interactor, being "an entity that interacts as a cohesive whole with its environment in such a way that this interaction causes replication to be differential" (i.e., an entity that natural selection acts on directly; Hull, 1988, p. 408). Although some philosophers have recently criticized the idea that all evolutionary processes can be understood through application of the replicator/interactor distinction (Godfrey-Smith, 2009), this distinction has nevertheless quite certainly contributed to clarifying the units of selection debate.

Hull makes it clear that the best replicators, given our current knowledge on the subject, are genes (though this does not mean that they are the only ones; see, for example, Sterelny, 2001) and that therefore the real "units of selection" debate is actually a debate concerning only interactors (Hull, 1992; Lloyd, 1988; Gould, 2002). When the debate is taken at this level, Dawkins's response is not quite so convincing. Admittedly, many biologists find the idea Dawkins popularized to be heuristically useful (e.g., Grafen and Ridley, 2006), but this does not change the fact that the dominant response to Hull's clarified problem (following directly on from Lewontin's 1970 suggestions) is that a hierarchy of interactors exists, within which the most clearly defined level is that of the organism, with genes only sometimes having the possibility to be interactors. Indeed, the organism is probably the best example of an interactor since it is on the organism's phenotypic traits that natural selection primarily operates (e.g., Mayr 1963, 2004; Gould, 1980, 2002; Hull, 1988, although the latter is equally insistent that organisms are not the only interactors). Dawkins partially recognized this point in developing his "extended phenotype" conception (Dawkins, 1982). However, for Dawkins the real entity on which natural selection operates is not the organism as such but rather the assembled collection of phenotypic traits on which genes exercise their influence, that is, the "extended phenotype," something which can go far beyond the boundaries of the organism itself. For example, in the case of a parasite, the nervous system of the host organism can constitute part of the parasite's extended phenotype, as parasitic genes efficient in influencing the nervous systems of their hosts will have been selected by natural selection (Dawkins, 1982, p. 216; for a critical evaluation of the evolution of Dawkins's ideas, see Hull, 1988; Gould, 2002).

Other philosophers have put forward useful distinctions that at least partially corroborate Hull's. Brandon (1982) expands on Hull and states that the interactors debate should be called the "levels of selection" debate, while the replicators debate be called the "units of selection debate"; Burian proposes a similar distinction (see Brandon and Burian, 1984). Sober (1984, pp. 97–102) distinguishes between selection of (what is conserved after natural selection has happened: referring to effects) and selection for (the reason for which natural selection happened: referring to causes). As for the biologist Niles Eldredge (1984), he proposes distinguishing two classes of living entities, one containing genealogical entities (which pass along information through replication of a structure, typically genes, local populations, and species) and the other containing ecological entities (characterized by stable structure and homeostasis, typically proteins or ecosystems).[5] The biological entity that best occupies both classes, being at once a genealogical and an ecological entity, is the organism.

Using these conceptual clarifications we can in turn provide a historical one. One of the primary sources for confusion in the units of selection debate is the heated discussions around the question of "group selection" and whether it can or cannot occur. But there are several ways to understand this question. Wynne-Edwards (1962) approaches it this way: can groups be the beneficiaries of adaptation? It is this question that Williams (1966), Maynard-Smith (1976), and Dawkins (1976) all answer in the negative, hence their flat-out rejection of group selection. However, if the question of group selection is understood rather as, "Can groups be interactors?" (that is to say, can natural selection occur on the group level?), then the arguments of Maynard-Smith, of Williams and of Dawkins become to a large extent ineffective (see, in particular, Wilson and Sober 1989), as all but the latter have come to acknowledge (see Maynard-Smith, 1987, p. 123 and Williams, 1992). So, the confusion between the interactor and the beneficiary questions stems largely from this grand debate in the 1960s and 1970s (on all these points, see Lloyd, 2007). Most surprising in this story is that there is a whole tradition of renowned biologists having openly posed the level of selection question, totally independently of the "beneficiary" of evolution question which they had considered to be irrelevant (Lewontin, 1970; Wright, 1980). Dawkins falls into the prolongation of this interactor/beneficiary confusion and spreads the debate further, into the units of selection domain. He does, however, add a third confusion, between long-term survival and adaptation: for him, cumulative adaptation is so definitely the major phenomenon

[5] That is, a system of auto-regulatory processes.

of evolution that the beneficiary of evolution could only be something which survives over extremely long periods along the evolutionary process. However, no real demonstration for this claim is given. According to Gould (2002), the gene-centered view relies on a false comprehension of the theory of evolution by natural selection, and more precisely on a confusion between book-keeping (which refers to the counting of certain hereditary attributes' differential augmentation) and evolution's causality (the mechanism that produces relative reproductive success). Evolutionary causality occurs at the interactor, and not the replicator, level. Furthermore, Dawkins says that to be a unit of selection an entity must possess sufficient stability; this is quite true, but, precisely, organisms do last long enough to act as units of selection in Darwinian processes, thus they do possess the "sufficient stability" required to be counted as evolutionary individuals. Lasting for vast periods of time, up to several millennia, is not a necessary condition for evolution by natural selection. The process of evolution by natural selection does not require perfectly faithful transmission, only the influencing of future generations' biological make up (often genetic). In opposition to genic selectionism, Gould argues for a "hierarchical perspective" on evolution (Gould, 2002; see also Gould and Lloyd, 1999, and Brandon, 1988). According to this perspective, evolution occurs on several levels of natural life (genes, genomes, organelles, cells, organisms, species, etc.), all understood as interactors.

One of the extensions of the hierarchical perspective of evolution is the so-called "multi-level" debate on selection. The following question, in particular, seems pressing: if natural selection operates simultaneously on several levels of life, for example, on an organism and on the cells that make up that organism, wouldn't tensions exist between these levels? (Buss, 1987 offers the founding approach to this question.) Could it not happen that cell lineages would favor their own fitness at the expense of the organism containing them and its fitness? The example of cancer cells shows that this phenomenon is certainly possible. Work on multi-level selection, mainly inspired by the pioneering work of Buss (1987), has flourished since the 1990s (e.g., Maynard-Smith and Szathmary, 1995; Michod, 1999; Okasha, 2006; Godfrey-Smith, 2008, 2009). Samir Okasha's book *Evolution and the Levels of Selection* (Okasha, 2004) has played a very useful role in assessing and clarifying the debate, thanks, in particular, to the distinction (suggested by Damuth and Heisler, 1988) between "multilevel selection 1" (MLS1, in which a collective's fitness is defined as the average fitness of the particles within the collective) and "multilevel selection 2" (MLS2, in which a collective's fitness is defined as the expected number of offspring collectives contributed to the next generation). One of the results of this work has been to highlight the existence of particular levels in the hierarchy of living things, levels where the competition at lower levels is suppressed, thanks to numerous mechanisms. The best examples of these particular levels could be the multicellular organism (Buss, 1987; Michod, 1999; Pradeu, 2013) and the "superorganism" (Wilson and Sober, 1989; Bouchard, 2013; Haber, 2013)—two notions that, according to some, should be conflated (Queller and Strassmann, 2009).

Behind the scenes, so to speak, the units of selection debate also poses a metaphysical question regarding the distinction of biological individuals (Hull, 1978, 1980, 1981,

1989a, 1992; Gould, 2002). The criteria generally employed to circumscribe individuals are stability, cohesion, discretion, and continuity. From the theory of evolution by natural selection's point of view, a whole hierarchy of levels of individuality exist (gene, cell, organism, species, etc.) Species, for example, are "individuals" in that they are spatio-temporally defined entities, and not classes of individuals. This means that a species is defined genealogically and not by some intrinsic properties which would be common to all its members (Ghiselin, 1974; Hull, 1976, 1978). Nevertheless, the biological entity that best satisfies all the criteria for being a biological individual in the sense of an interactor is probably the organism (Eldredge, 1984; Hull, 1978; Gould, 2002), an observation that, coming in the wake of numerous criticisms leveled at the privileging of the organism biological level (particularly following Dawkins's remarks, 1976), could now lead the organism back to its central position. In addition, it is likely that the articulation of evolutionary and physiological (in particular immunological) criteria of individuality will strengthen the view of organisms as highly individuated entities (Pradeu, 2010b, 2012).

Recently, Peter Godfrey-Smith (2009) has offered a different picture of biological individuality. Godfrey-Smith defines "Darwinian individuals" as members of a "Darwinian population," which itself is defined as a population of entities characterized by variation, heredity and differential fitness. On this basis, Godfrey-Smith suggests to distinguish several components of Darwinian individuality, and several degrees of this individuality, which has led him to propose a renewed conception of biological individuality (Godfrey-Smith 2013).

Frédéric Bouchard has suggested a very different view, according to which one must not give too much weight to the process of reproduction in the Darwinian theory of evolution, and one should instead focus on the process of persistence (of which reproduction would just be an instantiation). Fitness, Bouchard suggests, is often more a question of determining which entities live longer than others than which entities reproduce more than others. Therefore, Bouchard proposes a re-definition of fitness on the basis of this idea of differential persistence of lineages (Bouchard, 2010).

As all these discussions make clear, the debate over levels of selection and biological individuality is far from closed in biology and philosophy of biology. It will certainly continue to foster fruitful discussions in the near future.

5. From Egg to Adult, from Egg to Death: Development in Organisms

Development is the construction of a novel organismal form. Development is commonly, but not indisputably, thought as the set of processes that accompany life from the egg stage to sexual maturity. Although development did not get a lot of attention from the first philosophers of biology, it has today become the subject of intense research (e.g., Oyama, 2000[1985]; Amundson, 1994; Gilbert and Raunio, 1997; Griesemer, 2000; Oyama, Griffiths, and Gray, 2001; Brigandt, 2002; Burian, 2005;

Laubichler, 2007; Laubichler and Maienschein, 2007; Love, 2008; Pradeu et al., 2011; Minelli and Pradeu, 2014).

An important problem is the connection of development and the notion of information, which plays a crucial role in molecular biology. Generally it is said that genes bear information, in that they "encode" for the synthesis of precise proteins, according to some maybe even for the expression of phenotypic traits (Monod, 1971; Jacob, 1973; see also Sarkar, 2004, and Maynard-Smith, 2000), a point of view that has been analyzed critically by several philosophers (especially see Sarkar, 1996; Oyama, 2000; Godfrey-Smith, 2004; Godfrey-Smith and Sterelny, 2007). In developmental biology, the debate has become focused around the question of whether or not genes contain all the information necessary for the formation of an embryo and the adult organism, even whether this formation is not "programmed" by the genes, as many biologists had claimed between the 1970s and 1990s (among the most influential see Mayr, 1969a; Monod, 1970, 1971; Jacob, 1973; Gilbert, 1992; Wolpert, 1994) and as certain philosophers of biology believe today (Rosenberg, 1997, 2007). According to the genetic program hypothesis, genes contain all the information which, once "read," enables the realization of a complete individual organism. The difficulty lies in the fact that nothing allows us to isolate any particular meaning for the term "information" that would make it specifically applicable to genes but not to other developmental factors (epigenetic, environmental, etc.), as the partisans of developmental systems theory (DST) have shown (Oyama, 2000; Griffiths and Gray, 1994; Griffiths, 2001; Oyama, Griffiths, and Gray, 2001; Griffiths and Stotz, 2013). Certain philosophers (Oyama, 2000, 2009; Francis, 2003) even advance excellent arguments for considering that the very notion of information carries too much risk (notably the risk of anthropomorphism) to be allowed a place in biology.

The work of clarification on the notion of information in developmental systems theory has been accompanied by questioning into the temporal and spatial boundaries of development (Pradeu et al., 2011). From a temporal point of view it seems preferable to say that development doesn't simply stop at adulthood but that in reality it continues throughout life as a continuous constructive interaction with the environment (Gilbert 2013). From a spatial point of view, DST, following on from Lewontin (1983), rejects the theory that the organism is simply a product of the self-actualization of internal potentialities (an idea which is really just a contemporary form of preformationism: see Lewontin, 2000), and asserts that it comes into being through incessant interaction with its environment. This is where the idea that what actually develops is the system made up of the organism and its environment comes from (Oyama, 2000; Oyama, Griffiths, and Gray, 2001; Griffiths and Gray, 2004). This insistence on the interactions between the developing organism and its environment aligns with Scott Gilbert's so-called eco-evo-devo perspective (connecting ecology, evolution, and development; Gilbert, 2001, 2002, 2006; Gilbert and Epel, 2009) and its meeting point with niche construction (Laland, Odling-Smee, and Gilbert, 2008).

Important questions about development, yet to be examined in detail, include the causality of development, the formulation of general principles of embryogenesis, and the exact role played by theories in developmental biology (see Minelli and Pradeu, 2014).

Regarding developmental biology's place within the life sciences, something previously unnoticed became evident in the 1980s: developmental biology had, to a great extent, been neglected during the Modern Synthesis of the 1920s to 1950s (Hamburger, 1980). "Evo-devo" is the name given to the domain dedicated to connecting developmental biology with biology of evolution. The evo-devo field, thus called, is a recent one. Its sources are generally considered to be a few articles and publications from the 1980s and 1990s (particularly, Raff and Raff, 1987; Hall, 1992; Raff, 1996; Gilbert, Opitz, and Raff, 1996); its institutionalization in research programs and journals (*Evolution and Development Journal of Experimental Biology Part B*) primarily took place on the cusp of the 1990s to 2000s. However, attempts to bring together results from biology of evolution and embryology, now known as developmental biology, have a long history, particularly in the 20th century (especially Waddington, 1940; Gould, 1977), but going further back also (see Gilbert, Opitz, and Raff, 1996; Minelli, 2003; Amundson, 2005; Laubichler and Maienschein, 2007).

The principal problems posed in evo-devo are as follows (Laubichler, 2007; Müller, 2007):

1. The origin and the evolution of developing systems. Even though development seems to be both stable and robust over time, in fact developmental mechanisms change with evolution. It is these changes that are studied in the scope of this first problem. The notions of module and correlatively of modularity[6] have acquired decisive importance in this research (for an overview, see Müller, 2007).
2. The homology problem. How does one determine what counts as a homology and explain the emergence of homologies in the course of evolution (Brigandt, 2002; Griffiths, 2006; Griffiths, 2007)?
3. The relationship between genotype and phenotype. The claim, long held in population genetics, that development does not influence the correspondence between genotype and phenotype (the idea that development could be seen as a sort of "black box") can no longer be

[6] A module is a subsystem within the developing system (the latter may be an organism, a cell, etc.), characterized by intense interactions between its constituent parts, relative independence with respect to the system as a whole, an auto-regulatory capacity, redundancy (the same effect can be obtained in various ways), and persistence throughout evolution (some modules are found, sometimes in differing forms, in various species, some of which are in no way closely related). The module is to be found at an intermediary level, between easily individuated entities (such as cells in the case of an organism) and the level of the system in its totality (for example, the organism itself). An oft described example of a module is gene networks, with their regulatory systems. See for example von Dassow and Munro (1999).

accepted today. Research into phenotypic plasticity (West-Eberhard, 2003) is one way of posing the genotype-phenotype relationship problem anew.

4. Developmental constraints on phenotypic variations (Maynard-Smith et al. 1985; Amundson, 1994). The problem here is determining in what way development limits and constrains the range of possibilities for phenotypic variations.
5. The role of the environment in development and evolution. This role, long looked over, is considered to be absolutely crucial today (Gilbert and Epel, 2009; the role of symbiosis in development seems especially important: McFall-Ngai, 2002; Pradeu, 2011; McFall-Ngai et al., 2013).
6. The origin of evolutionary innovations. With genes (particularly regulatory genes like Hox) being highly persistent through evolution, it is necessary to look to other explanatory factors than just genes for an explanation of the manifest phenotypic differences between the species. Many consider that the explanation resides in the developmental modifications of gene regulation networks, but it still remains difficult to define with any precision what is to count as an "evolutionary innovation" (Müller and Wagner, 1991; Müller, 2007). Several developmental biologists claim that the theory of evolution resulting from the Modern Synthesis does not offer an explanation for evolutionary innovation and that this explanation must rather be provided by developmental biology, in opposition to the "classical" view (Gilbert, Opitz, and Raff 1996; Gilbert, 2006).

There is almost full consensus in affirming that the years to come will see evo-devo becoming one of the most dynamic fields within biology and one of the most exciting for philosophy of biology (Hull, 2002; Amundson, 2005; Laubichler, 2007). It is nevertheless not easy to tell whether or not this field will profoundly modify the acquired knowledge of the Modern Synthesis as its followers regularly and insistently claim it will. Most likely, evo-devo will neither replace nor erase the Modern Synthesis, but rather complete it, and decisively so (Arthur, 2002; Hull, 2002; Amundson, 2005; Minelli, 2010).

6. Reductionism and the Gene Concept

Though it may have enthralled the first philosophers of biology, as a result of logical positivism's influence, the problem of biology's reduction to physical chemistry seems now to belong to the past. There is full consensus regarding physicalism (ontological reductionism), which states that all biological processes are nothing other than physicochemical. There is also almost full consensus regarding explanatory anti-reductionism, that is, the assertion that we cannot adequately explain biological processes by means of physicochemical theories and terms. These questions recently resurfaced during debates on the notion of emergence applied to biology (see for example, Wimsatt, 2007, and, for a general overview, Bedau and Humphreys, 2008; on the related notions of self-organization and complexity, see Kauffman, 1993) but not in a way that challenged this double consensus.[7]

The real issue now concerns the possibility of an internal explanatory (theoretical) reductionism in biology, and more precisely the possibility of reducing macromolecular biology to molecular biology (e.g., Rosenberg, 2007). According to the reductionists, all biological explanations must be completed, amended, clarified by more fundamental explanations coming from molecular biology. Discussion on this reductionism has been focused on the possibility of reducing Mendelian genetics to molecular genetics. The word gene, originating from the term pangene, has a very loose meaning within Mendelian genetics: it refers simply to a factor of heredity. Mendelian genetics is a theory of hybridization and transmission; it is interested in genetic differences, which are correlated to the possession of this or that trait. Following the discoveries made by molecular biology in the 20th century, in particular the discovery of the double helix structure of DNA in 1953, the question arose as to whether it would be possible to reduce Mendelian genetics to molecular genetics. In molecular genetics, which is a theory of development and not a theory of heredity, the gene is an encoding sequence of nucleotides for the synthesis of a protein (Hull, 1974). The question of reducing Mendelian genetics to molecular genetics attracted practically all of the first philosophers of biology. There is a relative consensus to responding to this question in the negative because genetic processes are just far too complex to envisage identifying a Mendelian gene with some particular continuous sequence of nucleotides (see for example Hull, 1974; Kitcher, 1984; Mayr, 2004. See also, however, Schaffner, 1967; Ruse, 1971; Rosenberg, 1985 2007; Waters, 1990).

One of the most beneficial consequences of this debate has been the testing of the term "gene" itself. Indeed, it turned out that, contrary to popular belief, it was extremely difficult to precisely answer the question "What is a gene?" (Falk, 2000; Keller, 2000). Griffiths and Stotz (2007, 2013) distinguish three definitions for gene: the instrumental gene (a "Mendelian factor," i.e. a variable which takes part in the Mendelian transmission of a phenotypic trait), the nominal gene (referencing the nucleotide sequences similar to those which were studied at the time of molecular biology's discoveries in the 1950s to 1970s, such as sonic hedgehog, for example), and the classical molecular gene (a sequence of nucleotides which determines the structure of biological products, typically proteins) which has today become the post-genomic gene (the complex set of elements carrying out the function believed to have been carried out by the molecular gene). All three of these definitions are useful, but their coexistence suggests that it has become indispensable, for biologists speaking about genes, to specify which signification they are intending.

---

[7] Let us simply say that a property is said to be "emergent" relative to a system (an organism, for example) if it is not reducible to the properties possessed by the constituent parts of that system (for example, the organism's cells). On the distinction between ontological and epistemological emergence, see for example Wimsatt (2007).

To conclude on this point, is it possible to reduce macromolecular biology to molecular biology? If the disciples of this reduction continue to put the emphasis on the idea that it is necessary to complete macromolecular explanations with molecular explanations (constituting a weak form of "reduction"), as Rosenberg (2007) seems to be doing more and more, then, given the ever more repeated affirmation of a need to connect various modes of explanation within contemporary biology (Lewontin, 2009; Morange, 2009), we may consider that a consensus on this matter is beginning to emerge.

7. Philosophy of Biology beyond Evolution

As was noted at the beginning of this chapter, the field of philosophy of biology is dominated by evolutionary issues. A significant illustration of this fact is that, from 2008 to 2012, 64% of the articles published in the journal *Biology and Philosophy* concerned evolution (Pradeu, 2017). This focus on evolution is easy to understand, as evolution raises fundamental philosophical questions about the nature of species, the status of human beings in the living world, essentialism, individuality, and so on.

But this almost exclusive attention paid to evolutionary biology in philosophy of biology could also become problematic. Philosophers of biology have always aimed at working in close connection with biologists and at reflecting on the "real" and current biological sciences. The difficulty is that the great majority of the articles currently published in biology are not about evolution (although, of course, they accept the theory of evolution as an essential background). In the same period (2008–2012) during which *Biology and Philosophy* published 64% about evolution, the *Proceedings of the National Academy of the Sciences of the USA (PNAS)*, one of the major scientific journals in the world, had 6% of its biological papers put in the "evolution" section. The most exciting and discussed breakthroughs in today's biology concern fields such as neurobiology, cancerology, immunology, microbiology (and especially virology), or the renewed "omics" studies in molecular biology (genomics, proteomics, metabolomics)—all fields that have been almost entirely neglected by philosophers of biology.

Yet the situation has started to change. A growing number of philosophers of biology have become interested in neurobiology (e.g., Craver, 2007), microbiology (e.g., O'Malley and Dupré, 2007; Dupré and O'Malley, 2009; O'Malley, 2013), immunology (Tauber, 1994; Pradeu, 2012), systems biology (e.g., Green 2015), or in the "omics" detailed in recent molecular biology (e.g., Griffiths and Stotz 2013). If philosophers of biology want to maintain their wish to remain closely connected to biology as it is actually done today, it is likely that the young generations in the field will be increasingly attracted by these domains and the often highly philosophical questions that they raise.

8. Conclusion

I have presented in this chapter some of the major problems raised by today's philosophy of biology. Due to space limitation, many important issues and domains have not been analyzed, including the evolution of humans, the human mind,[8] and the possibility to speak of a "human nature."[9] To conclude, I would like to come back to the problem with which we started this chapter, namely the ties that exist between philosophy of biology and general philosophy of science.

In 1969 two seminal articles appeared, one by a philosopher (Hull, 1969), the other by a biologist (Mayr, 1969b). The first mourned the fact that a philosophy both specific to biology and well instructed on biological findings had not yet emerged; the second affirmed that "philosophy of science" more fittingly suited the moniker "philosophy of physics," and called for a rejuvenation of philosophy of science through the embracing of the magnificent advances accomplished in the life sciences. Almost 50 years on, the status report, as I see it, is this: Hull's wishes have been answered, better than he could have hoped, while Mayr's are still far from fruition.

Philosophy of biology, in accordance with Hull's wish, has today become a well-structured and flourishing philosophical domain with its own journals, academic circle, and so forth. It can even be viewed as a genuine example for all philosophies of science (in saying this, we certainly don't mean the only example) in at least two respects. First, it has enabled real progress to be made, as much from the philosophical as the scientific point of view. Second, it is characterized by genuine collaboration and dialog with scientists, the best example of which is that the journal *Biology and Philosophy* not only hosts frequent contributions from biologists but also is regularly cited in scientific journals. Several biologists have made major contributions to philosophy of biology (Dawkins, Gould, Lewontin, Maynard-Smith, and Mayr, in particular). Philosophers of biology have played, and continue to play, an important role in biology, something that is quite exceptional in philosophy of science.[10] Several biologists have openly acknowledged this, like Gould, for example, when he affirmed that philosophers have brought remarkable clarification to the biological debate on units of selection (Gould 2002, p. 598). From this point of view, we can highlight the contrast between Hull's original discourse (1969, p. 259), where he says that philosophers had not yet contributed to biology but that they could and should do so, and what he shows in Hull (2002), which is the fact that this contribution has become a reality.

However, in parallel, philosophy of biology has established a quite strong autonomy with respect to general philosophy of science, with less and less importance being given over to the latter's fundamental problems, often considered to be too dependent on its particular conditions of development (logical positivism, the physics model, etc.), and more and more attention being given to the grand problems of general philosophy (What is an individual? What are the entities that make up the world? Where

---

[8] These issues, often situated at the frontier between philosophy of cognitive science, psychology, and philosophy of biology, are well-represented in a journal like Biology and Philosophy. For a very stimulating example of a recent work situated at this frontier, see Sterelny (2012).

[9] These issues are at the crossroads between philosophy of biology and ethics. See for example Wilson (1975, 1978), Hull (1986), Francis (2003), and Ayala (2009).

[10] Physicist Richard Feynman supposedly said that, in his view, philosophy of science was no more useful to science than ornithology is to birds. Philosophy of biology clearly demonstrates that he was mistaken.

is the frontier between man and animal? Can we explain the origins of morality? Are humans free or determined? Can we speak of such a thing as "human nature"?). And so, philosophy of biology, unquestionably a well-structured domain posing classical philosophical problems, has not yet sufficiently lead to a rejuvenation of general philosophy of science, and thus seems to have failed in assuaging Mayr's (1969b) regrets for the discipline.

However, there are many signs indicating that a new phase is now taking shape, a phase where, precisely, general philosophy of science undergoes a partial re-creation thanks to the contribution of philosophy of biology (see for example Hull, 1988; Craver, 2005; Godfrey-Smith, 2006; Wimsatt, 2007; Sober, 2008; Stotz and Griffiths, 2008; Woodward, 2010). We can but impatiently await the fruit of this rejuvenation.

References

Allen, G. E., 1969, Hugo de Vries and the reception of the "mutation theory." Journal of the *History of Biology*, 2(1), 55–87.

Amundson, R., 1994, Two concepts of constraint: Adaptationism and the challenge from developmental biology. *Philosophy of Science* 61(4), 556–578.

Amundson, R., 2005, *The Changing Role of the Embryo in Evolutionary Thought*, Cambridge: Cambridge University Press.

Arthur, W., 2002, The emergent conceptual framework of evolutionary developmental biology. *Nature*, 415, 757–764.

Ayala, F., 2009, What the biological sciences can and cannot contribute to ethics. In Ayala, F. & Arp, R. (eds.).

Ayala, F., & Arp, R. (eds.), 2009, *Contemporary Debates in Philosophy of Biology*, Oxford: Wiley-Blackwell.

Barberousse, A., Morange, M., & Pradeu, T., 2009, *Mapping the Future of Biology. Evolving Concepts and Theories*, Boston Studies in the Philosophy and History of Science, 266, Dordrecht: Springer.

Beatty, J., 1995, The evolutionary contingency thesis. In G. Wolters & J. G. Lennox, (eds.), *Concepts, Theories, and Rationality in the Biological Sciences*. Konstanz: Universitätsverlag Konstanz, and Pittsburgh: University of Pittsburgh Press, 45–81 .

Bechtel, W., 2005, *Discovering Cell Mechanisms*, Cambridge: Cambridge University Press.

Bedau, M., & Humphreys, P., 2008, *Emergence: Contemporary Readings in Philosophy and Science*, Cambridge, MA, MIT Press.

Bouchard F., 2010, Symbiosis, lateral function transfer and the (many) saplings of life. *Biol Philos* 25, 623–641.

Bouchard, F., 2013, What is a symbiotic superindividual and how do you measure its fitness? In F. Bouchard and P. Huneman (eds.) *From Groups to Individuals: Perspectives on Biological Associations and Emerging Individuality*. Cambridge, MA: MIT Press, 243–264.

Bowler, P. J., 1983, *The Eclipse of Darwinism: Anti-Darwinian Evolution Theories in the Decades around* 1900, Baltimore: Johns Hopkins University Press.

Brandon, R., 1978, Adaptation and evolutionary theory. *Studies in the History and Philosophy of Science*, 9, 181–206.

Brandon R., 1982, The Levels of Selection. *PSA: Proceedings of the Biennial Meeting of the Philosophy of Science Association* 1982, 315–323.

Brandon, R., 1988, The levels of selection: A hierarchy of interactors. In H. Plotkin (ed.), *The Role of Behavior in Evolution*, Cambridge, MA: MIT Press, pp. 51–71.

Brandon, R., 1990, *Adaptation and Environment*, Cambridge: Cambridge University Press.

Brandon, R., & Burian, R. (eds.), 1984, *Genes, Organisms and Populations: Controversies Over the Units of Selection*, Cambridge, MA: MIT Press.

Brigandt, I., 2002, Homology and the origin of correspondence. *Biology and Philosophy*, 17(3), 389–407.

Burian, R., 1983, Adaptation. In M. Greene (ed.), *Dimensions of Darwinism*, New York and Cambridge: Cambridge University Press, pp. 287–314.

Burian, R., 2005, *The Epistemology of Development, Evolution, and Genetics*. Cambridge: Cambridge University Press.

Buss, L., 1987, *The Evolution of Individuality*, Princeton, NJ: Princeton University Press.

Byron, J. M., 2007, Whence philosophy of biology? *British Journal for the Philosophy of Science*, 58(3), 409–422.

Craver, C., 2005, *Beyond reduction: Mechanisms, multifield integration, and the unity of science*. Studies in History and Philosophy of Biological and Biomedical Sciences 36, 373–396.

Craver, C., 2007, *Explaining the Brain: Mechanisms and the Mosaic Unity of Neuroscience*, Oxford, Oxford University Press.

Cummins, R., 1975, Functional analysis. *The Journal of Philosophy*, 72, 741–764. Reprinted in Sober, E., (ed.), 1994.

Damuth, J., & Heisler, I. L., 1988, "Alternative formulations of multi-level selection." *Biology and Philosophy*, 3, 407–430.

Darwin, C., 1859, *The Origin of the Species*. London: John Murray.

Darwin, C., 1880, Sir Wyville Thomson and natural selection. *Nature*, 23, 32.

Dassow (von), G., & Munro, E., 1999, Modularity in animal development and evolution: Elements of a conceptual framework for EvoDevo. *Journal of Experimental Zoology B (Mol Dev Evol)*, 285, 307–325.

Dawkins, R., 1976, *The Selfish Gene*, Oxford: Oxford University Press.

Dawkins, R., 1982, *The Extended Phenotype*, Oxford: Oxford University Press.

Dawkins, R., 1986, *The Blind Watchmaker*, New York: Norton.

Dennett, D., 1995, *Darwin's Dangerous Idea*, New York: Simon and Schuster.

Dupré, J., 1987, *The Latest on the Best: Essays on Evolution and Optimality*, Cambridge, MA: MIT Press.

Dupré, J., & O'Malley, M., 2009, Varieties of living things: Life at the intersection of lineages and metabolism. *Philosophy and Theory in Biology*, 1, 1–25.

Eldredge, N., 1984, Large-scale biological entities and the evolutionary process. *Proceedings of the Biennial Meeting of the Philosophy of Science Association* 1984, 2, 551–566.

Ereshefsky, M., 1991, The semantic approach to evolutionary theory. *Biology and Philosophy*, 6, 59–80.

Falk, R., 2000, The gene: A concept in tension. In P. Beurton, R. Falk, & H-J. Rheinberger (eds.), *The Concept of the Gene in Development and Evolution*. Historical and Epistemological Perspectives, Cambridge: Cambridge University Press, 317–348.

Fisher, R. A., 1930, *The Genetical Theory of Natural Selection*, Oxford: Clarendon Press.

Francis, R., 2003, *Why Men Won't Ask for Directions: The Seductions of Sociobiology*, Princeton, NJ: Princeton University Press.

Gayon, J., 1998, *Darwinism's Struggle for Survival: Heredity and the Hypothesis of Natural Selection*, Cambridge: Cambridge University Press.

Gayon, J., 2006, Les biologistes ont-ils besoin du concept de fonction? Perspective philosophique. *Comptes Rendus Palevol.*, 5, 479–487.

Ghiselin, M., 1969, *The Triumph of the Darwinian Method*, North Andover, MA: Dover.

Ghiselin, M., 1974, A radical solution to the species problem. *Systematic Zoology*, 23, 536–544.

Ghiselin, M., 1983, Lloyd Morgan's canon in evolutionary context. *Behavioral and Brain Sciences*, 6, 362–363.

Gilbert, S. F., 2001, Ecological developmental biology: Developmental biology meets the real world. *Developmental Biology*, 233, 1–12.

Gilbert, S. F., 2002, The genome in its ecological context. *Annals of the New York Academy of Science*, 981, 202–218.

Gilbert, S. F., 2006, The generation of novelty: The province of developmental biology. *Biological Theory*, 1(2), 209–212.

Gilbert, S. F., 2013, *Developmental Biology*, 10th ed. Sunderland, MA: Sinauer Associates.

Gilbert, S. F., & Epel, D., 2009, *Ecological Developmental Biology*, Sunderland, MA: Sinauer Associates.

Gilbert, S. F., Opitz, J. M., & Raff, R. A., 1996, Resynthesizing evolutionary and developmental biology. *Developmental Biology*, 173, 357–372.

Gilbert, S. F., & Raunio, A. M., (eds.), 1997, *Embryology: Constructing the Organism*. Sunderland, MA: Sinauer Associates.

Gilbert, W., 1992, Vision of the grail. In D. J. Kevles and L. Hood (eds.), *The Code of Codes*, Cambridge, MA: Harvard University Press, pp. 83–97.

Godfrey-Smith, P., 1993, Functions: Consensus without unity, *Pacific Philosophical Quarterly*, 74, 196–208. Reprinted in D. Hull & M. Ruse (eds.), 1998, pp. 280–292.

Godfrey-Smith, P., 2001, Three kinds of adaptationism. In S. Orzack & E. Sober (eds.), 2001, *Adaptationism and Optimality*, Cambridge: Cambridge University Press.

Godfrey-Smith, P., 2004, Genes do not encode information for phenotypic traits. In Hitchcock, C., (ed.), *Contemporary Debates in Philosophy of Science*, Malden, MA: Blackwell, pp. 275–289.

Godfrey-Smith, P., 2006, The strategy of model-based science. *Biology and Philosophy*, 21, 725–740.

Godfrey-Smith, P., 2007, Conditions for evolution by natural selection. *The Journal of Philosophy*, 104, 489–516.

Godfrey-Smith, P., 2008, Varieties of population structure and the levels of selection. *British Journal for the Philosophy of Science*, 59, 25–50.

Godfrey-Smith, P., 2009, *Darwinian Populations and Natural Selection*, Oxford: Oxford University Press.

Godfrey-Smith P., 2013, Darwinian Individuals. In Bouchard F, Huneman P (eds.), *From Groups to Individuals: evolution and emerging individuality*, Cambridge, MA: MIT Press, pp. 17–36

Godfrey-Smith, P., 2014, *Philosophy of Biology*, Princeton, NJ: Princeton University Press.

Godfrey-Smith, P., & Sterelny, K., 2007, Biological information. *Stanford Encyclopedia of Philosophy* (online).

Gould, S. J., 1977, *Ontogeny and Phylogeny*, Cambridge, MA: Belknap Press.

Gould, S. J., 1980, *The Panda's Thumb*, New York: Norton.

Gould, S. J., 2002, *The Structure of Evolutionary Theory*, Cambridge, MA: Harvard University Press.

Gould, S. J., & Lewontin, R., 1979, The Spandrels of San Marco and the Panglossian Paradigm: A critique of the adaptationist programme, *Proceedings of the Royal Society of London B* 205, p. 581–598. Reprinted in Sober, E., (ed.), 2006.

Gould, S. J., & Lloyd, E., 1999, Individuality and adaptation across levels of selection: How shall we name and generalize the unit of Darwinism? *Proceedings of the National Academy of Sciences USA* 96(21), 11904–11909.

Grafen, A., & Ridley, M. (eds.), 2006, *Richard Dawkins: How a Scientist Changed the Way We Think*, Oxford: Oxford University Press.

Green S., 2015, Revisiting generality in biology: systems biology and the quest for design principles. *Biol Philos*, 30, 629–652.

Griesemer, J., 2000, Development, culture and the units of inheritance. *Philosophy of Science*, 67 (Proceedings), S348–S368.

Griffiths, P., 2001, Genetic information: A metaphor in search of a theory, *Philosophy of Science*, 68(3), 394–412.

Griffiths, P., 2006, Function, homology and character individuation, *Philosophy of Science*, 73(1), 1–25.

Griffiths, P., 2007, The phenomena of homology, *Biology and Philosophy*, 22(5), 643–658.

Griffiths, P., & Gray, R., 1994, Developmental systems and evolutionary explanation, *Journal of Philosophy*, 91, 277–304. Reprinted in D. Hull & M. Ruse (eds.), 1998.

Griffiths, P., & Gray, R., 2004, The developmental systems perspective: Organism-environment systems as units of development and evolution, in M. Pigliucci & K. Preston (eds.), *Phenotypic Integration: Studying the Ecology and Evolution of Complex Phenotypes*, Oxford & New York: Oxford University Press, pp. 409–430.

Griffiths, P., & Stotz, K., 2007, Gene, in D. Hull & M. Ruse, 2007 (eds.), pp. 85–102.

Griffiths, P., & Stotz, K., 2013, *Genetics and Philosophy: An Introduction*, Cambridge: Cambridge University Press.

Haber, M., 2013, Colonies are individuals: Revisiting the superorganism revival. In F. Bouchard & P. Huneman (eds.), *From Groups to Individuals: Perspectives on Biological Associations and Emerging Individuality*, Cambridge, MA: MIT Press, pp. 195–217.

Hall, B. K., 1992, *Evolutionary Developmental Biology*, New York: Chapman and Hall.

Hamburger, V., 1980, Embryology and the modern synthesis in evolutionary theory, in E. Mayr & W. B. Provine (eds.), pp. 97–112.

Hempel, C. G., 1965, *Aspects of Scientific Explanation*, New York: The Free Press.

Hull, D., 1969, What philosophy of biology is not. *Journal of the History of Biology*, 2(1), 241–268.

Hull, D., 1974, *Philosophy of Biological Science*, Englewood Cliffs, NJ: Prentice-Hall.

Hull, D., 1976, Are species really individuals? *Systematic Zoology*, 25, 174–191.

Hull, D., 1977, A logical empiricist looks at biology. *British Journal for the Philosophy of Science*, 28(2), 181–189.

Hull, D., 1978, A matter of individuality. *Philosophy of Science*, 45,335–360. Reprinted in E. Sober (ed.), 2006, pp. 363–386.

Hull, D., 1980, Individuality and selection. *Annual Review of Ecology and Systematics*, 11, 11–332.

Hull, D., 1981, Units of evolution: A metaphysical essay. In U. J. Jensen & R. Harré (eds.), *The Philosophy of Evolution, Brighton*, England: Harvester Press, pp. 23–44. Reprinted in R. N. Brandon & R. M. Burian (eds.), 1984, pp. 142–160.

Hull, D., 1986, On Human Nature, *Proceedings of the Philosophy of Science Association*, 2, 3–13. Reprinted in D. Hull & M. Ruse (eds.), 1998, pp. 383–397.

Hull, D., 1988, *Science as a Process: An Evolutionary Account of the Social and Conceptual Development of Science*, Chicago: University of Chicago Press.

Hull, D., 1989a, *The Metaphysics of Evolution*, Albany, Ithaca, NY: State University of New York Press.

Hull, D., 1989b, A function for actual examples in philosophy of science. In M. Ruse, (ed.), *What the Philosophy of Biology Is: Essays Dedicated to David Hull*, Dordrecht, Holland: Kluwer Academic Publishing, pp. 313–324. Reprinted in Hull, D., *Science and Selection: Essays on Biological Evolution and the Philosophy of Science*, Cambridge: Cambridge University Press, 2001.

Hull D., 1992, Individual. In E. F. Keller, E. A. Lloyd (eds.) *Keywords in Evolutionary Biology*. Cambridge, MA: Harvard University Press, pp. 181–187.

Hull, D., 2002, Recent philosophy of biology: A review. *Acta Biotheoretica*, 50, 117–128.

Hull, D., & Ruse, M. (eds.), 1998, *The Philosophy of Biology*, Oxford: Oxford University Press.

Hull, D., & Ruse, M. (eds.), 2007, *The Cambridge Companion to the Philosophy of Biology*, Cambridge: Cambridge University Press.

Jablonka, E., & Lamb, M. J., 2005, *Evolution in Four Dimensions: Genetic, Epigenetic, Behavioral, and Symbolic Variation in the History of Life*, Cambridge, MA: MIT Press.

Jacob, F., 1973 [1970], *The Logic of Life: A History of Heredity*, trans. Betty E. Spillmann. New York: Pantheon Books.

Kauffman, S., 1993, *The Origins of Order: Self-Organization and Selection in Evolution*, Oxford: Oxford University Press.

Keller, E. F., 2000, *The Century of the gene*. Cambridge, MA: Harvard University Press.

Kimura, M., 1983, *The Neutral Theory of Molecular Evolution*, Cambridge: Cambridge University Press.

Kitcher, P. S., 1984, 1953 and all that: A tale of two sciences. *Philosophy of Science*, 93(3), 335–373.

Kitcher, P. S., 1993, Function and design. *Midwest Studies in Philosophy*, 18(1), 379–397. Reprinted in D. Hull & M. Ruse (eds.), 1998, pp. 258–279.

Laland, K., Odling-Smee, J., & Gilbert, S. F., 2008, EvoDevo and niche construction: Building bridges. *Journal of Experimental Zoology (Mol Dev Evol)*, 310(B), 1–18.

Laubichler, M., 2007, Evolutionary developmental biology. In D. Hull & M. Ruse (eds.), pp. 342–360.

Laubichler, M., & Maienschein, J., 2007, *From Embryology to Evo-Devo*, Cambridge, MA: MIT Press.

Levins, R., & Lewontin, R., 1985, *The Dialectical Biologist*, Cambridge, MA: Harvard University Press.

Lewens, T., 2007a, *Darwin*, London and New York: Routledge.

Lewens, T., 2007b, Adaptation. In D. Hull and M. Ruse (eds.), 2007, pp. 1–21.

Lewens, T., 2009, Seven kinds of adaptationism. *Biology and Philosophy*, 24(2), 161–182.

Lewontin, R., 1970, Units of selection. *Annual Review of Ecology and Systematics*, 1, 1–18.

Lewontin, R., 1978, Adaptation. *Scientific American*, 239(9), pp. 156–169. Reprinted in a slightly different version in Levins, R., & Lewontin, R., 1985, pp. 65–84.

Lewontin, R., 1983, The organism as the subject and object of evolution, *Scientia*, 118, 63–82. Reprinted in Levins, R. & Lewontin, R., 1985. pp. 86–106.

Lewontin, R., 2000, *The Triple Helix*, Cambridge, MA: Harvard University Press.

Lewontin, R., 2009, Carving nature at its joints. In A. Barberousse, M. Morange, & T. Pradeu (eds.)

Lloyd, E., 1993, *The Structure and Confirmation of Evolutionary Theory*, Princeton, NJ: Princeton University Press, 1st ed. 1988.

Lloyd, E., 2005, Why the gene will not return. *Philosophy of Science*, 72, 287–310.

Lloyd, E., 2007, Units and levels of selection. In D. Hull & M. Ruse, (eds.), 2007, pp. 44–65.

Love, A. C., 2008, Explaining the Ontogeny of Form: Philosophical Issues. In S. Sahotra A. Plutynski (eds.) *A Companion to the Philosophy of Biology*. Blackwell Publishing Ltd., pp. 223–247

Machamer, P., Darden, L., & Craver, C., 2000, Thinking about mechanisms. *Philosophy of Science*, 67(1), 1–25.

Maynard-Smith, J., 1969, The status of neo-Darwinism. In C. H. Waddington (ed.), *Towards a Theoretical Biology*, Edinburgh: Edinburgh University Press, pp. 82–89.

Maynard-Smith, J., 1976, Group selection. *Quarterly Review of Biology*, 51, 277–283.

Maynard-Smith, J., 1987, How to model evolution. In Dupré, J. (ed.), *The Latest on the Best: Essays on Evolution and Optimality*, Cambridge, MA: MIT Press, pp. 119–131.

Maynard-Smith, J., 2000, The concept of information in biology. *Philosophy of Science*, 67, 177–194.

Maynard-Smith, J., Burian, R., Kauffman, S., et al. (1985), Developmental constraints and evolution. *Quarterly Review of Biology*, 60(3), 265–287.

Maynard-Smith, J., & Szathmary, E., 1995, *The Major Transitions in Evolution*, Oxford New York: W. H. Freeman Spektrum.

Mayr, E., 1961, "Cause and effect in biology. *Science*, 134, 1501–1506.

Mayr, E., 1963, *Animal Species and Evolution*, Cambridge, MA: Harvard University Press.

Mayr, E., 1969a, Comments on "Theories and hypotheses in biology." In R. S. Cohen and M. W. Wartofsky (eds.), *Boston Studies in the Philosophy of Science*, Vol. 5, pp. 452–456. Dordrecht: Springer. Reprinted under the title Theory formation in developmental biology, in E. Mayr (1976), *Evolution and the Diversity of Life. Selected Essays*, Cambridge, MA: Harvard University Press, pp. 377–382.

Mayr, E., 1969b, Footnotes on the philosophy of biology. *Philosophy of Science*, 36, 197–202.

Mayr, E., 1982, *The Growth of Biological Thought*, Cambridge, MA: Harvard University Press.

Mayr, E., 2004, *What Makes Biology Unique*, Cambridge: Cambridge University Press.

Mayr, E., & Provine, W. B. (eds.), 1980, *The Evolutionary Synthesis*, Cambridge, MA: Harvard University Press.

McFall-Ngai, M. J. 2002, Unseen forces: The influence of bacteria on animal development. *Developmental Biology*, 242(1), 1–14.

McFall-Ngai, M., et al. 2013, Animals in a bacterial world, a new imperative for the life sciences. *Proceedings of the National Academy of Sciences*, 110(9), 3229–3236.

Michod, R., 1999, *Darwinian Dynamics: Evolutionary Transitions in Fitness and Individuality*, Princeton, NJ: Princeton University Press.

Mills, S., & Beatty, J., 1979, The propensity interpretation of fitness. *Philosophy of Science*, 46, 263–286.

Minelli, A. 2003, *The Development of Animal Form*, Cambridge: Cambridge University Press.

Minelli, A. 2010, Evolutionary developmental biology does not offer a significant challenge to the neo-Darwinian paradigm. In F. J. Ayala & R. Arp (eds.), *Contemporary Debates in Philosophy of Biology*, Chichester: Blackwell, pp. 213–226.

Minelli, A. & Pradeu, T. 2014, *Towards a Theory of Development*, Oxford: Oxford University Press.

Monod, J., 1971 [1970], *Chance and Necessity: An Essay on the Natural Philosophy of Modern Biology*, Translated from the French by Austryn Wainhouse. New York: Knopf.

Morange, M., 2009, Articulating different modes of explanation: The present boundary in biological research. In A. Barberousse, M. Morange & T. Pradeu (eds.), pp. 15–26.

Müller, G.B., 2007, Evo-devo: extending the evolutionary synthesis. *Nature Reviews in Genetics*, 8, 943–949.

Müller, G. B., Wagner, G. P. (1991) Novelty in Evolution: Restructuring the Concept. *Annual Review of Ecology and Systematics* 22:229–256. doi: 10.1146/annurev.es.22.110191.001305

Nagel, E., 1961, *The Structure of Science*, New York: Harcourt Brace.

Neander, K., 1991, The teleological notion of function. *Australian Journal of Philosophy*, 69, 454–468.

Odling-Smee, J., Laland, K., & Feldman, M., 2003, *Niche Construction. The Neglected Process in Evolution*, Princeton, NJ: Princeton University Press.

Okasha, S., 2004, *Evolution and the Levels of Selection*, Oxford: Oxford University Press.

O'Malley, M. A. 2013, Philosophy and the microbe: A balancing act. *Biology and Philosophy*, 28, 153–159.

O'Malley, M. A., & Dupré, J. 2007. Size doesn't matter: Towards a more inclusive philosophy of biology. *Biology and Philosophy*, 22(2), 155–191.

Oyama, S., 2000 [1985], *The Ontogeny of Information*, Durham, NC: Duke University Press.

Oyama, S., 2009, Compromising positions: The minding of matter. In A. Barberousse, M. Morange, & T. Pradeu (eds.), pp. 27–45.

Oyama, S., Griffiths, P., & Gray, R. (eds.), 2001, *Cycles of Contingency*, Cambridge, MA: MIT Press.

Paley, W., 1802, *Natural Theology—or Evidence of the Existence and Attributes of the Deity Collected from the Appearances of Nature*, 2nd ed. (1827), Oxford: J. Vincent.

Pradeu, T., 2010a, Peut-on attribuer une fonction au système immunitaire? In J. Gayon & A. de Ricqlès (eds.), *Les Fonctions: des organismes aux artefacts*. Paris: Presses Universitaires de France, pp. 261–275.

Pradeu, T., 2010b, What is an organism? An immunological answer. *History and Philosophy of the Life Sciences*, 32, 247–268.

Pradeu T. 2011, A mixed self: The role of symbiosis in development. *Biological Theory*, 6(1), 80–88.

Pradeu, T., 2012, *The Limits of the Self: Immunology and Biological Identity*, New York: Oxford University Press.

Pradeu, T., 2013, Immunity and the emergence of individuality. In F. Bouchard & P. Huneman (eds.), *From Groups to Individuals: Perspectives on Biological Associations and Emerging Individuality*, Cambridge, MA: MIT Press.

Pradeu T., 2017, *Thirty years of Biology & Philosophy: Philosophy of which biology?* Biology & Philosophy 32(2), 149–167.

Pradeu T., Laplane L., Morange M., Nicoglou A., & Vervoort M. 2011. The boundaries of development. *Biological Theory*, 6(1), 1–3.

Queller, D. C., & Strassmann, J. E. 2009, Beyond society: The evolution of organismality. *Philosophical Transactions of the Royal Society B: Biological Sciences*, 364(1533), 3143–3155.

Raff, R., 1996, *The Shape of Life: Genes, Development and the Evolution of Animal Form*, Chicago: University of Chicago Press.

Raff, R. A., & Raff, E. C. (eds.), 1987, *Development as an Evolutionary Process*, New York: Alan R. Liss.

Reeve, H. K., & Sherman, P. W., 1993, Adaptation and the goals of evolutionary research. *Quarterly Review of Biology*, 68, 1–32.

Rosenberg, A., 1985, *The Structure of Biological Science*, Cambridge: Cambridge University Press.

Rosenberg, A., 1997, Reductionism redux: Computing the embryo. *Biology and Philosophy*, 12, 445–470.

Rosenberg, A., 2007, Reductionism (and antireductionism) in biology. In D. Hull & M. Ruse (eds.), 2007, pp. 120–138.

Rosenberg, A., & McShea, D. W., 2008, *Philosophy of Biology. A Contemporary Introduction*, New York: Routledge.

Ruse, M., 1971, Reduction, replacement, and molecular biology. *Dialectica*, 25, 38–72.

Ruse, M., 1973, *The Philosophy of Biology*, London: Hutchinson University Press.

Sarkar, S., 1996, Decoding 'coding': Information and DNA. *BioScience*, 46, 857–864.

Sarkar, S., 2004, Genes encode information for phenotypic traits. In C. Hitchcock (ed.), *Contemporary Debates in Philosophy of Science*, Malden, MA: Blackwell, 259–274.

Schaffner, K., 1967, Approaches to reduction. *Philosophy of Science*, 34, 137–147.

Smart, J. J. C., 1963, *Philosophy and Scientific Realism*, London/New York: Routledge & Kegan Paul/ Humanities Press.

Sober, E., 1984, The nature of selection. *Evolutionary Theory in Philosophical Focus*, Cambridge, MA: MIT Press. 2nd ed., Chicago: University of Chicago Press, 1993.

Sober, E. (ed.) 1984, 1994, 2006, *Conceptual Issues in Evolutionary Biology*, Cambridge, MA: MIT Press.

Sober, E., 1994, *From a Biological Point of View: Essays in Evolutionary Philosophy*, Cambridge: Cambridge University Press.

Sober, E., 2008, *Evidence and Evolution: The Logic Behind the Science*, Cambridge: Cambridge University Press.

Sober, E., 2011, *Did Darwin Write the Origin Backwards?*, Amherst, NY: Prometheus Books.

Sterelny, K., 1995, Understanding life: Recent work in philosophy of biology. *The British Journal for the Philosophy of Science*, 46(2), 155–183.

Sterelny, K., 2001, Niche construction, developmental systems, and the extended replicator. In S. Oyama, P. E. Griffiths & R. D. Gray (eds.), *Cycles of Contingency. Developmental Systems and Evolution*, Cambridge, MA: MIT Press.

Sterelny, K., 2012, *The Evolved Apprentice: How Evolution Made Humans Unique*, Cambridge, MA: MIT Press.

Sterelny, K., & Griffiths, P., 1999, *Sex and Death: An Introduction to the Philosophy of Biology*, Chicago: University of Chicago Press.

Sterelny, K., & Kitcher, P., 1988, The return of the gene. *The Journal of Philosophy*, 85, 339–360. Reprinted in D. Hull & M. Ruse (eds.), 1998, pp. 153–175.

Stotz, K., & Griffiths, P., 2008, Biohumanities: Rethinking the relationship between biosciences, philosophy and history of science, and society. *Quarterly Review of Biology*, 83(1), 37–45.

Suppe, F. (ed.), 1977 [1974], *The Structure of Scientific Theories*, 2nd ed., Urbana: University of Illinois Press.

Tauber, A. I., 1994, *The Immune Self: Theory or Metaphor?*, Cambridge: Cambridge University Press.

Thompson, P., 1983, The structure of evolutionary theory: A semantic approach. *Studies in the History and Philosophy of Science*, 14, 215–229.

van Fraassen, B. C., 1972, A Formal Approach to the Philosophy of Science. In R. Colodny (ed.), *Paradigms and Paradoxes*, Pittsburgh: University of Pittsburgh Press.

Waddington, C. H., 1940, *Organisers and Genes*, Cambridge: Cambridge University Press.

Waters, C. K., 1990, Why the antireductionist consensus won't survive the case of classical mendelian genetics. In A. Fine, M. Forbes, & L. Wessells, (eds.), *Proceedings of the Biennial Meeting of the Philosophy of Science Association*, Vol. 1: Contributed Papers, pp. 125–139. Reprinted in E. Sober (ed.), 2006, pp. 283–300.

Weisberg, M., 2006, Forty years of "the strategy": Levins on model building and idealization. *Biology & Philosophy*, 21, 623–645.

West-Eberhard, M. J., 2003, *Phenotypic Plasticity and Evolution*, Oxford: Oxford University Press.

Williams, G. C., 1966, *Adaptation and Natural Selection*, Princeton, NJ: Princeton University Press.

Williams, G. C., 1992, *Natural Selection: Domains, Levels, and Challenges*, Oxford: Oxford University Press.

Williams, M. B., 1970, Deducing the consequences of evolution: A mathematical model. *Journal of Theoretical Biology*, 29, 343–385.

Williams, M. B., 1981, Similarities and differences between evolutionary theory and the theories of physics, *Proceedings of the Biennial Meeting of the Philosophy of Science Association (1980)*, Volume 2: Symposia and Invited Papers, pp. 385–396.

Wilson, D. S., & Sober, E., 1989. Reviving the superorganism. *Journal of Theoretical Biology*, 136(3), 337–356.

Wilson, E. O., 1975, *Sociobiology, the New Synthesis*, Cambridge: Belknap Press.

Wilson, E. O., 1978, *On Human Nature*, Cambridge, MA: Harvard University Press.

Wimsatt, W., 2007, *Re-Engineering Philosophy for Limited Beings*, Cambridge, MA: Harvard University Press.

Wolpert, L., 1994, Do we understand development? *Science*, 266, 571–572.

Woodward, J., 2010, Causation in biology: Stability, specificity, and the choice of levels of explanation. *Biology & Philosophy*, 25(3), 287–318.

Wright, L., 1973, Functions. *Philosophical Review*, 82(2), 139–168. Reprinted in E. Sober (ed.), 1994, pp. 27–47.

Wright, S., 1980, Genic and organismic evolution. *Evolution*, 34, 825–843.

Wynne-Edwards, V. C., 1962, *Animal Dispersion in Relation to Social Behavior*, Edinburgh: Oliver & Boyd.

13

PHILOSOPHY OF MEDICINE

Élodie Giroux (Jean Moulin Lyon 3 University and Lyon Institute for Philosophical Researches) and Maël Lemoine (University of Bordeaux, Immunoconcept UMR5164)

1. What Is Philosophy of Medicine?

Since the 1970s it appears to be the case that, in the aftermath of "the philosophy of biology," a distinct disciplinary field having its place within the philosophy of science has progressively developed through its interest in questions specific to medicine.[1] A theme arising in several collections of texts (Caplan, Engelhardt, and McCartney, 1981; Humber and Almeder, 1997; Caplan, McCartney, and Sisti, 2004) has developed around a philosophical analysis of health, disease and illness concepts and of the scientific nature of medicine. Particularly in native English-speaking countries, research, institutions and specialized reviews like *Man and Medicine* (1975), *The Journal of Medicine and Philosophy* (1976) and *Theoretical Medicine and Bioethics* (in 1977, *Metamed*) were created.[2] Yet far from being as consensual as the philosophy of biology,

[1] This section was finished in 2013. This rapidly growing field of philosophy of science has since provided a wealth of contributions, papers, and handbooks. Due to the editorial process, it was not possible to update it.

[2] From 1957, a specialized journal concerning biology and medicine was created: *Perspectives in Biology and Medicine*. Sections were consecrated to the philosophy of medicine in the *Philosophy of Science Association* (1976) and the *Philosophy of Science* (1977). Since 1975, a collection directed by S. Spicker and T. H. Engelhardt and published by Reidel then Kluwer has been dedicated to this field, and other journals have been created: e.g., *The Journal of Medicine and Philosophy* (1976), *Metamedicine* (1977; which became *Theoretical Medicine* in 1979 and then *Theoretical Medicine and Bioethics*), *Medicine, Health Care, and Philosophy* (1998), and *Studies in History and Philosophy of Biological and Biomedical Sciences* (1998).

the definition, limits, and very existence of this domain, and in particular, its belonging to—or at least its proximity to—either the philosophy of science or ethics, were more recently the subject of an abundant debate (Caplan, 1992; Wulff, 1992; Pellegrino, 1998; Engelhardt, 2002; Ten Have, 1997; Stempsey, 2004; Stempsey, 2008) which continues (Pellegrino, 1976; Pellegrino, 1986; Engelhardt, 1976a; Engelhardt, 1986). Several reasons can explain this questioning. First of all it is appropriate to underline that the relations between philosophy and medicine largely preceded the 1970s. Medical anthropology in Germany from the beginning of the 20th century (Weizsäcker, 1987), the Polish school (Löwy, 1990) with in particular the work of Ludwig Fleck published in German in 1935, then the French school of historical epistemology, principally represented by Georges Canguilhem (Canguilhem, 1978[1943]), Mirko Grmek, and Michel Foucault (Foucault, 1963), had already contributed to giving themes and orientations to this domain. The philosophy of medicine oscillated between three major orientations: anthropological, epistemological and ethical (Ten Have, 1997). Most of all, it is difficult to define and delimit the actual contours of medicine given the number and diversity of its sub-disciplines. Is it appropriate to include public health, psychiatry and nursing? Would it then not be better to use the more global term "health care" rather than "medicine"? What is the goal of medicine? Treating, curing and preventing illness and disease? Improving health? Prolonging life? Added to these difficulties relating to unity and finality are those concerning the nature of medical knowledge between practice, technology and theory, but also between human and social sciences, and natural science. Whatever the case may be one can justly argue that philosophical interest in medicine is the occasion for fundamental epistemological reflection on the relation between theory and practice (Canguilhem, 1978; Grene, 1976). Besides, and to finish, since the 1970s, ethical and bioethical questions have been foremost to such an extent that some have spoken "of the moralization of the philosophy of medicine" (Ten Have, 1997, p. 105). It is notably this preponderant place of ethics which led Caplan (1992) to not only question the definition of the field but also its very existence.

As a matter of course, we can understand the diversity of definitions, from the most expansive and large, to the most specific. Edmund Pellegrino (1976) and Arthur Caplan (1992) advocate a narrow definition but in a very different sense. Pellegrino (1976, 1986, 1998) began by distinguishing the "philosophy of medicine" from three other modes of relation between philosophy and medicine: (1) "Medical philosophy" includes the informal and literary reflections of doctors regarding their clinical experiences; (2) "philosophy *and* medicine" incorporates mutual considerations of problems common to both disciplines, "each retains its identity and enters as a distinct discipline into independent and autonomous dialogue with the other" (1998, p. 321);[3] and (3) "philosophy in medicine" which consists in the application of reflexive tools from philosophy as a whole to medical problems. Pellegrino refers for example to the principle-based ethics

[3] For a recent illustration, see Johansson and Lynøe's introduction (Johansson and Lynøe 2008). The content of Marcum's *Introductory Philosophy of Medicine* also registers itself in this type of relationship between philosophy and medicine and/or in that designated by "philosophy in medicine" (Marcum, 2010).

of Beauchamp and Childress. But for him, the "philosophy of medicine" in the strict sense is the discipline which examines the conceptual foundations of the clinical encounter between patient and doctor.[4] Caplan's alternative view is that the *philosophy* of medicine must be a sub-discipline of the philosophy of science and separate from bioethics: "as such its primary focus is epistemological not ethical, legal, aesthetic, or historical." Its objective is "the epistemological, metaphysical and methodological dimensions of medicine; therapeutic and experimental diagnostic, therapeutic and palliative" (1992, p. 69). But he concludes that defined as such it is inexistent, and he calls for its development.

Others prefer to adopt a large definition of the philosophy of medicine which fits with what Pellegrino calls "philosophy in medicine," at the same time relativizing the place of bioethics, which has become a very multi- and interdisciplinary field and has acquired autonomy in relation to philosophy (Jonsen, 1998; Hottois, 2004; Carson and Burns, 1997). For Engelhardt and Schaffner (1998), the philosophy of medicine includes "philosophical inquiries within medicine" and encompasses "those issues in epistemology, axiology, logic, methodology and metaphysics generated by or related to medicine" (Engelhardt and Schaffner, 1998, p. 268).[5] Others again contest the very pertinence of a philosophy of medicine and consider it preferable to position henceforward in a plurality of approaches to medicine; which amounts to what is called the "medical humanities."

But for several years, a new impulsion has been given to the analysis of epistemological, methodological, and metaphysical questions, apparently accomplishing Caplan's wish and legitimizing the domain of fertile and promising analysis in the bosom of the philosophy of science (Kincaid and McKitrick, 2007; Gifford, 2011). As well as the long predominant and central thematic of the concepts of disease, illness and health, we are, in effect, witnessing the deployment of traditional questions from the philosophy of science, renewed in the special field of medicine: the causality and the *explanation* of disease (Nordenfelt and Lindahl, 1984; Thagard, 1999), *theories* in biomedical science (Schaffner, 1993; Thompson, 2011a; Kazem Sadegh-Zadeh, 2011) and the status and nature of proof in medicine, and the relationship between theory and practice at the heart of Evidence-Based Medicine (John Worrall, 2007a; Howick, 2011). For this chapter we retain a restrained sense of the philosophy of medicine comparable to that in the handbook directed by Gifford: a field that "encompasses the topics connected to the philosophy of science that arise in reflection upon medical science" (2011, p. 1). We begin by presenting the long time dominant analysis of health, illness and disease

---

4 "Philosophy of medicine seeks to understand the nature and phenomena of the clinical encounter, i.e., the interaction between persons needing help of a specific kind relative to health and other persons who offer to help and are designated by society to help" (Pellegrino, 1998, p. 327).

5 If Engelhardt and Erde (1980) devote a whole section to bioethics in their 1980 description of the philosophy of medicine, in 1998 it is only present in a subordinate manner to questions of an epistemological, logical, and methodological nature (Engelhardt & Schaffner, 1998, p. 264).

concepts, before presenting the question of causal analysis and experimentation in medicine, and, to finish, that of clinical reasoning.

2. The Concepts of Health and Disease: Naturalism versus Normativism

2.1 ONTOLOGY AND NORMATIVITY

Two types of question have been asked on the nature of disease. The first, dominant until the mid-20th century, bears on the nature of the individual entities we call "diseases" (tuberculosis, AIDS, cancer, etc.): do there exist individual diseases as real entities, or just the ill person? The second, more recent, bears on the belonging of our concepts of health and disease to the field of biological facts or to the field of human values. The actors of these two debates share elements in response to these questions, which are thus narrowly linked.

In the first case, the question is to know whether diseases refer to natural and real units of classification. The question derives from an ancient opposition between two conceptions of the nature of disease. According to the ontological conception, disease is a thing, distinct from the organism in which it is or on which it acts. Thus diseases exist, and it is quite probable that a natural classification of them is possible. During a time when living beings began being identified to natural classes according to the taxonomic model of species, some, like Sydenham (1624–1689) then Linné (1707–1778) and Boissier de Sauvages (1706–1767), applied this same method to diseases and accordingly developed medical nosology, a discipline therefore which studies the characteristics of diseases, aiming to classify them into individual and discrete entities. According to the *physiological conception*, disease is rather a process which affects the organism, for example, its equilibrium or its functioning. The disease entities of the nosology are consequently less sure to coincide with natural classes, because a complete comprehension of the pathological process is necessary to be certain that a distinction of classes does not correspond in reality to superficially different manifestations of one underlying process. Claude Bernard is one of those who have strongly criticized the ontological presuppositions of nosology. In line with François Broussais (1772–1838), he maintains that only the distinction between the normal and the pathological is significant, and that in itself it is not a difference of nature but rather of degrees. This opposition between ontological and physiological accounts of disease partially recoups the more general opposition between realists and nominalists (Faber, 1930; Cohen, 1955; Temkin, 1963; Engelhardt, 1975) and that, more recently, between realists and constructivists (Simon, 2011).

The psychiatrist and philosopher Lawrie Reznek has proposed one of the most complete philosophical analyses to date concerning the two levels of the question of the nature of disease: first, the level of "disease-status," where the question concerns the nature of the distinction between disease and health, and, second the level of "disease-identity," i.e. the nature of the distinction between individual diseases (Reznek, 1987; Reznek, 1995). According to Reznek, it is this question on the nature of diseases, more precisely, the fact that they are or are not natural kinds, which allows us to determine if the concept of disease is value-free or value-laden. In effect, for the latter, our concept of disease is value-free if and only if disease is a natural kind, i.e. a condition is or is not a disease by virtue of its nature. Reznek defends a value-laden concept of the disease-status compatible with a value-free account of disease-identity: contrary to disease, specific disease entities such as tuberculosis or Down's syndrome are natural kinds in the sense that each of those entities shares a common explanatory nature.

Since the 1970s the second type of questions have largely dominated and been the source of an abundant controversy between naturalists and normativists.[6] They concern the general concepts of disease and health: are they value-free concepts? Is there any objective and natural way of drawing the distinction between disease and health and of defining them? Canguilhem (1978 [1943]) has introduced this topic, on the one hand, attacking two objectivist conceptions *taken separately*—that of Claude Bernard functional and physiological, and that founded in the approach of the statistician Adolphe Quételet, statistical and empiricist of the norm as means—and, on the other hand, introducing and defending the existence of a biological and individual "normativity." The contemporary debate, mainly Anglo-American, develops on, and distances itself from, these analyses (Giroux, 2010). The context and methodology are different: on the one hand, the "French style" in philosophy of science which integrates historical and philosophical approaches, and on the other "conceptual analysis" inherited from analytical philosophy. Above all, normativity is not here biological but much rather social or cultural. The American philosopher Christopher Boorse defends a Bio-statistical Theory of disease (BST) that articulates these two functional and statistical conceptions subject to isolated criticism from Canguilhem. He has thus renewed and reset the possibility of a value-free and objective concept of disease. His BST has been fundamental in the emergence and the actual development of the controversy. Before presenting Boorse's theory, it is beholding to say a few words about the context in which this controversy emerged.

2.2 OPPOSITION TO THE BIOMEDICAL CONCEPT OF DISEASE

In the first half of the 20th century, a set of characteristics today associated to the idea of scientific medicine or biomedicine came into being: professional organization, pathological and organ specialization, the association between biological sciences and analysis of pathological mechanisms, hospital teaching, recourse to experimental modelling, and, analysis and statistical management of the health of populations (Gaudillière, 2002; Gaudillière, 2006). A biological and statistical conception of disease tends to predominate. It is in reaction to this said "biomedical" conception that a

[6] There are multiple ways to characterize this controversy (objectivism vs. constructivism; reductionism vs. subjectivism; neutralism vs. normativism; etc.). We retain the most usual designation. For a description of the diversity in this debate, see (Hofmann, 2001).

certain number of criticisms were formulated from the 1950s on which were to be the linchpin of the said "normativist" theories concerning health concepts.

First of all, in proposals comparable to those by Canguilhem, some denounce the illusion that there are grounds to consider that the pathological could be defined as a simple deviation from a statistical normality objectively (King, 1954; Murphy, 1966; Offer and Sabshin, 1966): other norms, social and subjective, inevitably come into account. Statistical normality is neither necessary (there are diseases which are statistically common: atherosclerosis) nor sufficient (there are rare states which are healthy: blood group B, ginger hair, exceptional intelligence, etc.). Then, a body of studies produced by historians, sociologists,[7] anthropologists, philosophers, doctors, and psychiatrists concerned with controversial states such as alcoholism (Szasz 1972), homosexuality (Green, 1972), menopause (Barnes, 1962), masturbation (Engelhardt, 1974), or aging (Engelhardt, 1977; Caplan, 1981) demonstrate the historical and social relativity of judgements governing the decision to classify a state as normal or pathological. Some even denounce the ideological nature of this categorization: the supposedly scientific and objective medical argumentation used to promote social or moral norms (Sedgwick, 1973; Foucault, 1976; Engelhardt, 1976b; Margolis, 1976).

Besides, correlatively to this critique of the biomedical concept of disease, a concept which in reality was not well-defined, other concepts like "bio-psychosocial" (Engel, 1960) or "ecological" (Dubos, 1959) have been proposed. Equally expressed have been a critical analysis of "bio-power" (Foucault, 1994), which supposedly constitutes medicine and public health, and the excessive medicalization of life which leads to an "expropriation of health" (Illich, 1976). The 1960s anti-psychiatry debate questioned the validity of the extension of the biomedical concept into the mental domain and hence its univocity for the somatic and the mental realms (Szasz, 1960). Still more devastating was the study led by Rosenhan (Rosenhan, 1973), which raised doubts regarding all objective justifications for psychiatric internment. Moreover, the World Health Organization, in the preamble to its 1946 Constitution, considerably enlarges the scope of health, defining it in positive terms as "a state of complete physical, mental and social well-being and not merely the absence of disease or infirmity."[8]

2.3 Christopher Boorse's Bio-Statistical Theory (BST)

It is in this context that Christopher Boorse takes on the challenge of elaborating a value-free definition of health and disease, precisely with the goal of avoiding the

---

[7] The beginnings of the sociology of medicine date back to the 1950s. The sociologist Talcott Parsons played a central role in the emergence of the theorization of the social dimension of illness and disease through, notably, analyses of the social role of the ill person and also of the doctor (Parsons, 1951, 1958, 1975).

[8] Preamble to the Constitution of the World Health Organization as adopted by the International Health Conference, New York, June, 19–22, 1946; signed on July 22, 1946 by the representatives of 61 nation states (Official Records of the World Health Organization, no. 2, p. 100) and effective on April 7, 1948.

relativism which for him necessarily emanates from normativism. His theory has notably underpinned naturalist approaches in bioethics (Daniels, 1985). It is exposed in a series of three seminal articles (Boorse, 1975; Boorse, 1976a; Boorse, 1977), then taken up again and defended in two articles published in 1987 and 1997 incorporating slight modifications in response to objections (Boorse, 1987; Boorse, 1997).

Two theoretical commitments enable Boorse to put aside the criticism of the biomedical concept formulated by the normativists. The first (1975) consists in proposing the distinction between a theoretical concept and a practical concept.[9] This distinction enables Boorse to validate everything the normativists say in considering that their theses only concern practical concepts, but not the theoretical one. It is solely the latter he undertakes to define. He restrains his analysis to the concept of Western medical science, that is to say, physiology. The notion of disease that he defined is very broad and does not correspond to everyday use: it includes "such conditions as injuries (broken arm, dog bite), deformities (club foot, cleft palate), static abnormalities (foreign bodies in the stomach), functional impairments (blindness, deafness), poisonings (arsenic or alcohol intoxication), environmental effects (sunburn, heatstroke, frostbite), and various other phenomena (starvation, drowning)" (1997, p. 41); in reality it is "the pathological," in its somatic and psychological dimensions (1997, p. 7). This concept is "analytical" in the sense that it is valid first and foremost for the parts, traits or processes of the organism. Its conceptual analysis also attempts to account for the use of this concept in relation to animals and plants (1977, p. 565) and, concerning humans, aims at being as valid for somatic diseases as for mental disorders. Boorse argues that at the theoretical level, that is to say in pathology, the demarcation between the normal and the pathological stems effectively from a factual judgment which does not necessitate recourse to values or norms which are social or individual. Western medicine rests first of all on the idea that "the normal is the natural—that health is conformity to a 'species design'" (1997, p. 7). Next, health and disease are opposite and exclusive concepts. To define one of these concepts is enough to define the other. Let us note that Boorse defines theoretical health in a negative way as the absence of disease, he does not exclude the possibility of a positive concept of health, as that which is "beyond the absence of disease" (1977, pp. 553–554, 568 ss), but it is no longer a theoretical concept.

Thus, pathology would use a theoretical concept of disease which is independent of clinical practice, and so the values and norms which this introduces. Boorse maintains in effect that the practical concepts (clinical—diagnostic and therapeutic—and social) articulate themselves on the theoretical concept by adding evaluative criteria, and not the opposite.[10] Founding the theoretical concept on the clinical concept, which

---

9 In his first article (Boorse, 1975), the distinction between theoretical concept and practical concept covers that between "disease" and "illness." Subsequently, Boorse redefined "illness" as a systemic disease, i.e., a theoretical concept, and he substituted the simple dichotomy for a "multilevel approach" (1997, pp. 11–13).

10 "Diagnostic normality" is the absence of a clinically detectable pathological condition, and "therapeutic normality," the absence of a diagnostic abnormality worthy of treatment.

integrates normative elements such as suffering and negativity associated to the experience of disease, is to expose oneself to relativism and the difficulty of taking stock of the fact that, on the one hand, we consider as pathological numerous asymptomatic or pre- and infra-clinical conditions (hemophilia, hypertension, numerous cancers, etc.), and on the other hand, that some pathological conditions could be desired (infertility, cowpox). This distinction enlightens differences between medical and profane conceptions of disease. Every disease in the theoretical sense does not necessarily entail the presence of a disease in the practical sense—it is possible to have a disease in the theoretical sense without feeling ill; however, for there to be a disease in the practical sense, the presence of a disease in the theoretical sense is necessary. The theoretical definition could be seen as delimitating the field of medicine. But in reality medicine is concerned with many conditions that are not pathological (pregnancy, contraception). Thus the practical importance of a theoretical definition of disease is limited: these concepts are "far from setting all clinical or social questions" (1997, p. 99). Nonetheless, the aim of Boorse in making explicit the theoretical definition of disease is to deliver "a bedrock requirement to block the subversion of medicine by political rhetoric or normative eccentricity" and "to avoid false presumptions caused by calling something a disease (e.g. masturbation)" (1997, pp. 99–100). But it remains that we must be able to describe in a value-free manner the significance of this theoretical concept used by physiologists.

His second theoretical commitment consists in conceding to the normativists that statistical normality is in effect neither necessary nor sufficient to define disease. He enumerates seven views, which reappear frequently in the literature on concepts of health: (1) value, (2) treatment by physicians, (3) statistical normality, (4) pain, suffering, discomfort, (5) disability, (6) adaptation, (7) homeostasis. He shows that none of them provides a necessary or sufficient condition for a definition (1977, pp. 543–550). The statistical criterion remains nonetheless a fundamental component of his definition. In his view the articulation of *statistical normality* with a non-normative concept of *biological function* overcome the difficulties of the statistical criterion. Before developing how this articulation operates and before presenting the four theses with which he summarizes his theory, two notions have to be detailed: *biological function* and *reference class*. The definition of *function* refers back to questions debated in the philosophy of biology (see the preceding chapter), which Boorse distinguishes from the philosophy of medicine. His analysis of function is pre-supposed by his theory of health but this latter does not depend on it, having validity with other theories of the biological function, like that proposed by Wakefield (Boorse, 1997, pp. 8–11). He defends a value-free analysis of function that he sees as a causal contribution to a goal in a teleological system (Boorse, 1976b; Boorse, 2002). Organisms are goal-directed systems in the sense that Sommerhoff and Nagel (Sommerhoff, 1950; Nagel, 1961) had tried to characterize, that this teleological orientation of living organisms is an objective property. And "the structure of organisms shows a means-end hierarchy with goal directedness at every level" (Boorse, 1977, p. 556). In physiology, "the highest-level goals are the organism's survival and reproduction." The biological function of an organ, a trait or a process is then defined as a contribution by it to the individual survival and reproduction: the function of the heart is to pump blood, that of the lungs, to breathe, and so on. Besides, physiological function statements are relative to what Boorse calls "species design." They "describe species or population characteristics, not any individual plant or animal." The clinical judgment which bears on individual health consists in an evaluation of that patient's health regarding theoretical health or typical functional normality. More precisely, the functional statements of physiology are relative to a fraction of a species: the reference class is relative both to sex and to age. Indeed there are many variations in normal physiology between males and females, young and old. Physiology compares individuals of the same sex and in the same age range. For Boorse, the functional organization of individuals of a same age and of the same sex is uniform enough to enable us to distinguish diverse reference classes. These reference classes are defined statistically as well as the normal level of the efficiency of a function.

Health is then defined as the normal functioning, i.e. the ability to perform physiological functions with at least a statistical typical efficiency. As for a disease, it is a reduction of the functional ability below the typical efficiency that characterizes the species norm of the organism. Here is the definition of the theoretical concepts of health as proposed in 1997 (pp. 7–8):

1. The reference class is a natural class of organisms of uniform functional design; specifically, an age group of a sex of a species.
2. A normal function of a part or process within members of the reference class is a statistically typical contribution by it to their individual survival and reproduction.
3. A disease is a type of internal state which is either an impairment of normal functional ability, i.e. a reduction of one or more functional abilities below typical efficiency, or a limitation on functional ability caused by environmental agents.
4. Health is the absence of disease.

If the concepts of *function* and *reference class* are fundamental in this definition, it is on *statistical normality* that rests the demarcation between the normal and the pathological. More precisely, it is from the population distribution of the function's efficiency that a limit is established.

> Normal functioning in a member of the reference class is the performance by each internal part of all its statistically typical functions with at least statistically typical efficiency, i.e. at efficiency levels within or above some chosen central region of their population distribution (. . .). Abnormal functioning occurs when some function's efficiency falls more than a certain distance below the population mean. (Boorse, 1977, p. 559)

The pathological is identified with statistical *sub*normality rather than with abnormality. The hyper- or hypo- diseases (e.g. hyperthyroidism) are inadequate denominations. The disease is not the clinical value measured here (the rate of Thyroid Stimulating Hormone). In effect, to determine health what matters is not the concrete process that makes the physiological contribution but the level of the function's efficiency. Thus for thyroid disease, the hyper- and hypo-thyroidism are pathological because in both cases, the level of efficiency is well below the norm. The problem of the so called extremal disease would be here resolved.

Ever since the 1977 article, Boorse points out the principal limitations of his theory for at least two classes of recognized diseases: "structural diseases" and "universal diseases" (pp. 565–568). In effect, some structural anomalies may be compatible with a normal functioning like "congenital absence of the appendix, calcification of the pineal gland, minor deformities such as those of the nose or the ear, or even, perhaps, some internal tumors." But for Boorse, even if it is not in conformity to present medical usage, it would be more pertinent to exclude these structural anomalies from the concept of disease. What's more, if health is defined as statistical normality of function, how can we account for statistically typical dysfunctions or what he calls "universal diseases," such as arterial thickening after a certain age, lung irritation, and benign hypertrophy of the prostate in old men? Concerning the first two, Boorse considers it surprising that medicine does not apply here age-relativity to normality: "the puzzle is why old age is not always seen as a stage with its own statistical norms of healthy functioning" (1977, p. 567); in this case these two conditions would simply be counted as normal for the old persons. This would not prevent them from being subject to medical care. In effect, as we have said previously, medicine is not exclusively concerned with illness and disease. To resolve the problem of lung irritation, Boorse adds a clause according to which if the disease is not a reduction of one or more of normal functional ability, it can be "a limitation on functional ability caused by environmental agents." But he conceded later that this clause raised too many difficulties, presupposing particularly the possibility of clearly distinguishing the internal and external causes. He preferred to abandon it considering that it only concerned "an infinitesimal part of the field (medical recognized disease)" (1997, p. 86). Most important, in Boorse's view, as long as we haven't found another view of comparable explanatory power to BST, these anomalies in his theory are not sufficient to invalidate it (1977, p. 568; 1997, pp. 99–100).

2.4 CRITICISMS OF THE BIO-STATISTICAL THEORY

According to Boorse, one can distinguish two principal lines of criticism, one stemming from the philosophy of biology, the other from the philosophy of medicine (1997). The philosophers of biology opposed his claim to define normality on a purely biological base. This is tantamount to identifying, in a questionable manner and without argument, theoretical normality and statistical normality (Wachbroit, 1994). The very idea of biological normality (Amundson, 2000), and of the possibility of identifying natural traits proper to each member of a same species, has little biological founding, as much in the field of taxonomy and in the definition of species, as in that of genetics. The modern synthesis of the theory of evolution teaches us that neither homogeneity nor qualitative resemblance, but heterogeneity and ascent are relevant in this domain (Hull, 1978; Sober, 1980). Besides, an important criticism concerns the pertinence of Boorse's non-normative and anhistoric concept of the physiological function (Neander, 1991) and, in particular, the limitation of the organism's goals to those of survival and reproduction. Biologists describe numerous other states of organisms that are not limited to these two goals, as for example a non-reproductive sexuality or the fact of eating for pleasure, and so forth (Ereshefsky, 2009). As such, it appears difficult to determine, in a purely theoretical manner, that these two goals are those of human organisms (Brown, 1985; Schaffner, 1993).

In the philosophy of medicine, one encounters some of these latter criticisms (Engelhardt, 1996), but generally speaking, as he himself justly remarks (1997, p. 6), it is his concept of disease which has been criticized rather than his analysis of it. Counterexamples have mostly been used to demonstrate the limitations of his biostatistical definition. The principal objections have consisted in claiming that it is illusory to define a dysfunction without having recourse to a value-judgement, or that dysfunction is neither necessary nor sufficient to distinguish the normal from the pathological. The example of homosexuality has been used against BST to underline the theoretical insufficiency of dysfunction to define disease, but also its inadequacy with what is considered pathological in Western medicine (Ruse, 1981; Ruse, 1997; Nordenfelt, 1995, pp. 131–139; Wakefield, 1992). In effect, according to BST, homosexuality is a dysfunction as it prevents the normal contribution of the individual organism to reproductive function. That said it appears inappropriate to consider it as pathological. Following a decision by the American Psychiatric Association in 1973 homosexuality was removed from the *Diagnostic Statistical Manuel of Mental Disorders II*, after its introduction in 1968, and this change was not motivated by a modification in medical knowledge but by an evolution of our value judgments relative to this state (Bayer, 1981, pp. 101–154). Jerome Wakefield's definition of mental disorder defends the idea that dysfunction is necessary but not sufficient. He defines mental disorder as a "harmful dysfunction," adding thus a normative component i.e. harm, allowing, in particular, homosexuality to be excluded from the pathological domain (Wakefield, 1992). But is it satisfactory to admit that biological dysfunction, despite its insufficiency, remains necessary? If, for example, female orgasm has no biological function, it seems appropriate to consider the absence of this mechanism as pathological (Reznek, 1987, p. 131).

In the responses he addresses to these counterexamples, Boorse maintains that homosexuality is a disease in the theoretical sense. In the case of orgasm, the objection rests on a misunderstanding of his theory whose precise aim is to distinguish the pathological from the treatable, and which therefore does not claim to define all that medicine can and should treat. That the incapacity to have orgasms is not pathological (and nor therefore an illness or disease in the proper sense) does not imply that it cannot be the object of medical attention in the same way as contraception (1997, pp. 92–94). But as previously highlighted, by insisting on this disjunction between a concept of health and the domain of medical practice, one tends to considerably reduce the practical interest of defining theoretical health, which then seems relevant only to the biologist (Engelhardt, 1996, p. 202).[11] At most we can say that a definition marks out a domain in which therapeutic action is less controversial (Boorse, 1997, pp. 98–99).

Another criticism has concerned the difficulty for BST to take into account the dynamic and evolving relationship of the organism to its environment. In effect, the question of physiological adaptation, as well as that of disease caused by environmental factors, have led its definition to include the notion of statistically normal environment (1997, pp. 83–84) which raises important theoretical difficulties (Lorne, 2004, p.93). More recently there has been more criticisms on the actual validity of Boorse's *analysis* of the concept of disease. Regarding the question of environmental relativity, Kingma has pointed out the necessity to index functions against situations to account for dynamic physiological functions. But to her it leads to excluding several diseases that result from specific environmental factors (Kingma, 2010). Concerning the physiological goals of the organism (survival and reproduction), apart from the fact that they can oppose each other, as in the case of giving birth, which may be life threatening, the organism can well no longer have any reproductive function but continue to suffer from illness or disease (Schwartz, 2007b). As for the notion of reference class, it is doubtful that it is a pure and simple statistical and empirical abstraction effected from the human population as its delimitation appears to require a value-judgment (Cooper, 2002; Lorne, 2004; Kingma, 2007; Giroux, 2009). Moreover, one can question more fundamentally the very pertinence of the frequency criteria: can we really maintain that certain states, whose consequences are negative, are *normal* because they are *frequent* for a certain class of reference? It is what Peter Schwartz calls the "common diseases problem" and which concerns many modern pathological conditions like obesity, cardiovascular diseases and certain cancers, but also certain forms of premature senility like Alzheimer's disease, and all the diseases associated with aging and life-style.[12]

2.5 ALTERNATIVE CONCEPTIONS

2.5.1 The Numerous Varieties of Normativism

Normativism has many varieties. Boorse (1975, p. 51) has proposed a minimal division into "strong normativism," i.e. "the view that health judgments are pure evaluations

[11] Benditt (2007), while defending a naturalist account of disease inspired by Boorse, points that we risk being led to conclude that the implications of a naturalist concept for the medical practice are very small or non-existent.

[12] Schwartz (2007b) shows that in reality the frequent pathologies that pose a problem for BST are not so much those that are universal (besides, do such pathologies exist in reality?) or very frequent (more than 50%), but those which are sufficient to surpass the 5% limit, that is to say, between 10% and 20% of a reference class.

without descriptive meaning," quite a rare position in philosophical literature, and "weak normativism that allows such judgments a descriptive as well as a normative component." But a large diversity is due mostly to the fact that various senses are given to the notion of norm (epistemic, descriptive, prescriptive, evaluative, naturalistic, etc.) and diverse types of values (objective, subjective, social, moral, cultural) are pre-supposed. Usefully, clarifications have recently been made (Simon, 2007).

Tristram Engelhardt, an American physician and philosopher, was one of the representatives of normativism with whom the controversy developed more particularly in the 1970s and 1980s, before going on to contribute specifically to the development of bioethics (Engelhardt, 1996; Engelhardt, 2002). Its conception is difficult to classify. Far from adopting the method of conceptual analysis, and rather sensitive to the historical and critical approaches to medicine, and to the idea, attributed to Fleck and Kuhn, that there is no strict separation between facts and values, he underlines the probably insurmountable ambiguity, and the heterogeneity of health and disease concepts.[13] The concept of disease has a function and a significance that is indissolubly normative and descriptive, evaluative and explanatory. For him, there cannot be a pathological judgement of dysfunction without reference to human goals and interests (Engelhardt, 1974; Engelhardt, 1975; Engelhardt, 1976b; Engelhardt, 1984) and the implicit goal in calling a set of phenomena a disease is to enjoin to medical intervention. In his conception of disease, which Boorse (1987, p. 368) baptized the "3-D" theory, at least three types of judgement are required: those relative to a *Disability*, to aesthetics *(Deformity)* and to subjective experience *(Discomfort)*: "what medicine addresses as diseases is a cluster of physiologically or psychologically based problems with function, freedom from pain, and bodily form" (1984, p. 31). The concept of disease is both aesthetic and ethical but more fundamentally, it is a pragmatic concept "whose truth is found in action directed to the elimination of illness and toward the establishment of health" (1975). Indeed, calling a condition a disease rather than a demonic possession or a simple exhaustion is to make an explanatory move that allows a medical intervention: "the concept of disease is a general scheme for explaining, predicting, and controlling dimensions of the human condition" (1975). If this analysis has the merit of highlighting the multiplicity of implicated norms and values, it faces the risk of circularity between the definition of disease and the definition of medicine, and above all, it leaves unresolved the vagueness of the frontiers between, one the one hand, political, social and subjective devaluation, and, on the other hand, medical devaluation: "the concept of disease has fuzzy borders with moral concepts."

Other approaches which come explicitly from a weak normativism have defended the idea of an "objective value" in a bid to avoid the pitfalls of relativism, ideological usage and moralism denounced by Boorse in normativism; and to take account of the

[13] "Perhaps the concept of disease indicates a family of conceptually consanguineous notions. That is, the concept of disease may be basically heterogeneous, standing for a set of phenomena collected together out of diverse social interests, not on the basis of the recognition of a natural type or a common conceptual structure" (Engelhardt, 1975).

fact that most illnesses and diseases are communally identified enough, as such, beyond cultural diversities, societies, and even medical specialities. A first version of this normativism rests on the idea that in the medical domain there are shared and pertinent universally objective values. Clouser, Culvert, and Gert's (1981) starting point is that having a disease is first having something wrong with one's self. The notion of evil or harm is the common feature of those of death, pain and disability frequently associated with disease. What is harmful can be *objectively* determined in the sense that there is agreement on it by all rational persons. A second version, the view that function statements are value-laden, recalls Aristotle. Health and disease, according to James Lennox, should not be viewed as predicates in judgments of approval or disapproval, but rather in judgments of function contribution to life (Lennox, 1995). There is some empirical basis for judging of the success or failure of a system in achieving its goal. Health refers "to that state of affairs in which the biological activities of a specific kind of living thing are operating within the ranges which contribute to continued, uncompromised living."[14] But both of these forms of normativism are widely criticized for being too inclusive.

Another declination of weak normativism could be characterized as the combination of several descriptive and normative criteria that are jointly necessary and sufficient for a condition to be a disease. A "hybrid" form is the "harmful-dysfunction analysis" defended by Wakefield: it articulates a biological and normative component. His analysis has had a towering influence on American mental health professions and the philosophy of psychiatry. It can almost appear as the articulation of Boorse's theoretical and practical concepts, aside the fact that his concept of biological function rests on an etiological and not systemic account. Thus it would appear to be more pertinent to classify it with BST as being part of the "dysfunction-requiring accounts" (Schwartz, 2007a). Another form that joins what Cooper (2007) calls "messy accounts" and in which her own definition falls (2002) employs a number of conditions to define disease. Reznek (1987), who considers that the presence of a dysfunction is not a necessary criterion, suggests a group of criteria: abnormality, harm (a disease is a bad thing to have), and the necessary and appropriate character of the medical treatment.[15]

2.5.2 Action-Based Accounts

The most developed alternative theories of BST are those that ground their accounts of health in action theory and that propose a definition on the basis of categories of ability and disability: health is a kind of ability to act and illness a disability or lack

[14] This form of normativism can be interpreted as an extension of the notion of "biological normativity" introduced by Georges Canguilhem. See also (Grene, 1978).

[15] Here is his definition: "A has a pathological condition C if and only if C is an abnormal bodily/mental condition which requires medical intervention and for which medical intervention is appropriate, and which harms standard members of A's species in standard circumstances" (1987, p. 167).

of action (Fulford, 1989; Nordenfelt, 1995).[16] Nordenfelt shares with Boorse the same method of conceptual analysis and the project of defining a concept of health that can take into account our diverse usages, for the mental and the somatic, but also extending to humans, plants, and animals.[17] He underlines that his objective is to "sharpen the borders" of concepts of health (1995, p. 11), but also to bring a certain number of clarifications on their relations to other associated concepts like happiness, morality, legality, decency, excellence (talent, intelligence, strength, creativity), and on the relation between health and environment (1995, pp. 4–6). To Nordenfelt such a clarification has importance for clinical medicine and health care as much as for philosophy of science.

The interest in the notions of ability and disability is to include those of function and dysfunction while closely linking the latter to their consequences on human life. In running counter to Boorse, Nordenfelt gives the priority to commonsense holistic concepts of health and illness, over the scientific and analytical notion of disease. Founding his analysis on analytic action theory, Nordenfelt gives the less objectionable priority to the notion of disability over that of suffering. He also gives priority to the concept of health over that of illness or disease, providing in his view a better adequation with the importance attributed to health promotion in contemporary medicine (Nordenfelt, 2004), and he highlights its positive dimension in our contemporary societies and in our social and political institutions. It is from a positive and holistic health concept, considered as the logically prior concept, that all other health-concepts derive. The complement of health is illness. Disease is a technical concept defined in relation to illness: it is a bodily or mental process which tends to cause illness and to compromise health (1995, pp. 109–110). As with BST, this distinction between disease and illness allows us to become aware of divergence between medical and ordinary conceptions: diseases do not always cause illness in their bearers, and an individual can feel in good health despite having a disease.

Thus, we can accordingly expose with more precision the contents of the concept of health. Here is the definition: "A is completely healthy, if and only if A is in a mental and bodily state which is such that A has a second order ability, given accepted circumstances, to realize the states of affairs which are necessary and together sufficient for A's minimal happiness in the long run" (Nordenfelt, 2000, p. 93). The concept of health is a relational concept consisting of three fundamental elements: the agent's ability, the vital goals, and the accepted environment. Ability consists in "that kind of possibility for action which is determined by factors internal to the agent's body or mind" (1995, p. xiv). At first glance close to BST's functional ability, it demarcates itself however through a greater placing in environmental perspective. This environment is not defined statistically, as with Boorse, but rather in reference to the cultural

[16] There are many other propositions sharing this perspective (Whitbeck, 1978; Whitbeck, 1981; Pörn, 1984; Pörn, 1993; Agich, 1983; Agich, 1997).

[17] Two books treat reciprocally on the application of his concept of health to mental health (Nordenfelt, 2007) and to animals (Nordenfelt, 2006). The application to plants is considered analogical.

and social norms of a given society. Therefore a relativity and normativity inherent to the concept of health is linked to the natural and cultural environment. Nonetheless normativity and relativity do not necessarily entail relativism. To what degree is it possible to affirm that an individual can be healthy in a given environment and ill in another? Nordenfelt illustrates this through the example of a political refugee who, to provide for his family, was able to farm in his homeland, but in the country to which he emigrates is unable to have this activity and lifestyle. For all that can one say that he was healthy in his homeland and became ill in his host country? To avoid this kind of affirmation, Nordenfelt introduced the distinction between "first-order ability" and "second-order ability." A person can *actually* be unable (first-order) to perform a particular action but nonetheless be *potentially* able (second-order) to perform it. This effectively signifies that a person can acquire through a training program the ability (first-order) to do this action. Second-order ability is thus compatible with a first-order inability, but inversely this is not the case. The notion of second-order ability brings us closer to "the biologically founded capabilities of man"; this is what is implied in the definition of health (1995, pp. 49–53).

But what is one to understand by "vital goals"? Nordenfelt begins by demonstrating the limitations of two main theories. On the one hand, the one which rests on the notion of "basic human needs" tends to reduce them to survival and reproduction as in BST (1995, pp. 57–65), and on the other hand, that which defines them in terms of the goals set by the agent himself, which leads to an excess of subjectivism and relativism (pp. 65–76).[18] A better definition is based on minimal happiness, or more broadly, on minimal welfare. The general idea is that these vital goals are those for which the accomplishment is both necessary and sufficient for a minimal welfare. This notion (such as the notion of accepted environment) cannot be completely defined in descriptive terms (1995, p. 79). But the evaluation is one of welfare, that is an "evaluation sui generis"—to be kept distinct from moral evaluation—and which is conducted with as much rigor as a scientific investigation. The values and culture which the individuals of a society implicitly share enable a certain consensus on the minimum degrees of well-being and these degrees can also be–and already are—made explicit and decided in the social policy frameworks of each country. The theoretical content of the definition of health proposed by Nordenfelt eventually remains very sketchy, which probably is an insurmountable consequence of the project to elucidate what is the unique basic health concept. Nevertheless, this theory prolongs and deepens the relevance of some of Canguilhem's theses and renews the debate on health concepts through its opening of interesting new perspectives (Giroux, 2010).

---

[18] This subjective definition of vital goals has, among other things, the fault of not accounting for the use of animal and plant health concepts in the incapacity to decide on their goals. According to Nordenfelt, these accounts are those used by Whitbeck (Whitbeck, 1978, 1981) and Pörn (Pörn, 1984, 1993) in their theories of health.

2.6 ASSESSMENT AND PERSPECTIVES

In the debate on health and disease concepts, it appears that strict naturalism, which consists in arguing that the question of values is external to medicine and/or that practical medicine is no more than applied biology, is commonly recognized as no longer relevant.[19] Effectively, for Boorse, if a value-free concept of disease can be defined, it is nevertheless insufficient to apprehend disease in all its meanings. Practical and normative concepts are necessary for clinical medicine. Moreover, Thomas Schramme considers that the BST can be interpreted as part of a weak form of normativism (Schramme, 2007).[20] From the mental health viewpoint, Dominic Murphy defends "a revisionist objectivism" which integrates a normative dimension compatible with a value-free concept of disease (Murphy, 2006). One can also consider that the controversy in reality opposes weak and strong forms of normativism, i.e. those who maintain that it is possible to distinguish normative and non-normative elements and those who, on the contrary, affirm that their intricacy is essential (Khushf, 2007). The question arising is how to determine their nature and status (Kincaid and McKitrick, 2007).

Also, an evolution of the debate is a displacement of the analysis in two principal directions: we focus less on whether or not the concept of disease is value-laden, than on the utility and the nature of the analysis. These past few years, have come to light definitions presenting themselves as modified versions of BST. Mahesh Ananth (2008) defends an "evolutionary-homeostatic" concept of *physical* health. The importance he accorded to the notion of homeostasis would allow a response to the difficulties of BST with regard to the taking into account of the environment. Schwartz (2007b) modifies the formulation of the organism's physiological goals: "survival *or* reproduction" rather than "survival *and* reproduction." And, to resolve what he calls the "common diseases problem," he introduces a "negative consequences" criterion, that is to say, the effect which a given level of functioning has on the organism. Moreover, in the context of the framework for the construction of disease ontologies, others are taking an interest in defining disease. The objective is to provide an adapted definition of this ontological project, itself linked to the need for systematizing and standardizing diagnosis and disease vocabulary to make disease data computable in bioinformatics (Williams, 2007). Following debates, in the general philosophy of science, on the status of natural kinds and on what natural kind realism could amount too, one also witnesses a renewal of the analysis on the nature of disease and of individual diseases (Sulmasy, 2005). In order to accommodate biological species which plausibly lack essential properties,

[19] In his reflection on the reducibility of medicine to biology, Kenneth Schaffner defends the possibility of partial reductions but underlines that in medicine this question leads to value problems and in particular to the possibility of ethical naturalization (Schaffner, 1992, pp. 341–343).

[20] Schramme (2007) maintains the necessity of elaborating a medical concept that is independent of values while recognizing their importance. As with Ananth, BST should better be interpreted as a "descriptive normativism" (2008, p. 41).

some non-essentialist accounts of natural kind have been developed which could be relevant to embrace the nature of diseases (Murphy, 2006; Cooper, 2007).

Second, faced with an absence of consensus on a viable and satisfactory definition, the utility of an analysis of health concepts has itself become an object of reflection. It has been argued that we have no need for a disease concept particularly with regard to decision-making concerning the treatment and judgment of responsibility (Hesslow, 1993). More directly, what is criticized is also the recourse to so-called conceptual analysis for these concepts, a method principally adopted by Boorse, Nordenfelt, and Wakefield, but also Reznek and Cooper (Worrall and Worrall, 2001; Murphy, 2006; Nordby, 2006; Ereshefsky, 2009; Lemoine, 2013). Do we not have to deal with a multiplicity of irreducible disease concepts (Simon, 2007)? For Schroeder (Schroeder, 2013), there has been too much focus on defining a state of health rather than defining a relation (healthier than); he thus argues for the possibility and relevance of a comparativist theory of health. If the philosophical analysis of the concept of disease is seen as pertinent, other conceptions of the definition are also suggested in which one renounce to find a descriptive and essentialist definition with necessary and sufficient criteria (Sadegh-Zadeh, 2000, 2008; Schwartz, 2007a). It then becomes important to specify the objective of the analysis: is it to serve as a foundation for clinical practice through the determination of what is to be treated and what is not? Or is it more modestly to define and clarify the notions and their logical relations? If the explicit goal of the philosophical analysis is to modify and specify our concepts, is it appropriate to seek a definition common to both science and medical practice with regards to a greater coherence, or, on the contrary, to maintain and reinforce the distinction between different concepts? These questions are associated with the one which concerns the role of philosophy in medicine.

3. Problems in the Medical Sciences

Medicine deals with disease scientifically by way of description and explanation. Description raises one major issue, classifications (3.1). Explanation in turn raises the question of causality.[21] In medicine, causal research is twofold: bench research (3.2) and clinical research (3.3). Some causes of diseases can be studied in bench research; others are best studied in clinical research. Both bench and clinical research encounter the problem of multifactorial diseases in their own field (3.4). This last problem leads to the difficulty of interpreting causality as being either mechanistic or probabilistic (3.5).

[21] To simplify the exposition, we separate here the questions of disease classification and disease explanation/causation. But they are strongly intricate, since disease causation or disease explanation can be a strong and relevant criterion for classification and definition of individual diseases. See for instance (Nordenfelt 2000).

3.1 PROBLEMS IN THE CLASSIFICATION OF DISEASES

As said before, normativists have opposed to naturalists that both disease in general and individual diseases are not natural kinds. The focus now is not anymore on the demarcation between health and disease but on the identity of specific diseases and on the criteria of their classification. The focus is therefore on disease entities, not on their status as diseases: does the similarity we perceive between different cases of disease correspond to robust, natural facts? On this question too, there are naturalist and normative stances. Yet being a normativist on health and disease does not entail being a normativist on disease entities, as both Reznek and Cooper advocated. As Cooper emphasized,

> Whether a condition is a disorder is partly a value judgement, but the distinctions between types of disorder might still depend solely on psychological and biological facts. If this were the case then the domain of mental disorders would be analogous to the domain of weeds. Weeds are unwanted plants, thus whether a daisy is a weed is at least in part a value judgement. Still, the distinctions between kinds of plants generally considered weeds are fixed by the nature of the world. Botanical facts make it the case that daisies and thistles are genuinely distinct types of plant. (Cooper, 2005, 45)

Conversely, naturalism about health and disease does not necessarily entail naturalism about disease entities. Boorse, for instance, is committed to a naturalistic view on health and disease, but does not particularly endorse the current classification of diseases:

> the BST is aimed at the demarcation problem, not at nosology. It seeks to say what is disease, not to individuate diseases. Individual disease entities should be defined and classified on whatever is the most scientifically convenient basis (Boorse, 2007, 67).

At the very least, this implies indifference toward the question, which, if possible indeed, is an argument toward the independence of the two questions, albeit a weak one. That said, current classification is one thing, possible classifications are another. Claude Bernard also advocates such a position. According to him, only disease is a natural fact to the experimentalist's eye, whereas disease entities lead only to classifications that have no ground in nature, but could be useful.

The history of medicine in the 18th and 19th centuries provided a very striking example of changing classifications. "Descriptive" or "phenomenological" criteria classify diseases on the basis of the co-occurrence of symptoms: these were prominent during the 18th century, and are not inexistent today, particularly (but not exclusively) in psychiatry. Pathological criteria classify diseases according to the localization of lesions. Foucault studied this approach, prominent in the first half of the 19th century, in *The*

Birth of the Clinic: An Archaeology of Medical Perception. He also showed how physicians discovered the limits of the *révolution anatomopathologique*, that is, that numerous conditions leave no anatomical or morphological trace.

Etiological criteria classify diseases on the basis of causes of pathological conditions. They revealed powerful in the late 19th century with the discovery of infectious diseases. If a specific bacterium or virus could be associated to each disease as its one and unique cause, any condition, with or without lesion, would be in its proper class, as naturally carved out as species of bacteria or viruses are distinct. Nevertheless, the limits of this approach soon appeared. First, the experimental isolation of the germ is a thorny issue: the so-called Henle-Koch postulates, which define the conditions of a proven infection, are sometimes impossible to apply (see section 3.4.). Second, it has been sensed very early, and proven many times since, that the mere presence of a germ in an organism is not sufficient for a disease to occur. The organism and its environment also provide necessary conditions of a disease, whether it is labeled "infectious" or not, which leads to the question of multifactorial diseases, obviously a major issue for etiological criteria. Another interesting case of etiological definition of disease is that of so-called genetic diseases. Is that a natural kind with crisp criteria, demarcating genetic from non-genetic diseases? The received view in medicine has it that the distinctive character of genetic diseases is "being caused by one or several genes." Of course, it soon came to be known that environment plays a major role in these diseases. Philosophers thus questioned the meaning of "cause" in the case of genetic diseases, and discarded all satisfactory interpretations: it cannot be a sufficient condition of genetic diseases because environmental conditions are necessary to any disease, and it cannot be a specific difference between those affected with a disease and the rest of the population, because this depends on the choice of a contrastive population. The general conclusion was that "genetic disease" is not a natural class (Hesslow, 1984; Gifford, 1989; Smith, 2001; Magnus, 2004).

Contemporary nosology mixes phenomenological, anatomical, physiological and microbiological criteria together with immunological and genetic criteria (Wulff, Pedersen, and Rosenberg, 1986), sometimes even therapeutic criteria. They are generally consistent with each other, but in some cases, several non-overlapping classifications may coexist, as is the case for leukemia: for instance, whereas the Rai-Binet classification of forms of chronic lymphoid leukemia was based on anatomic and pathological criteria together with life expectancy, the MIC classification is based on morphologic, immunologic and cytogenetic criteria. Oncologists and hematologists use them depending on various reasons, and there seems to be no way to capture the best of both in a unitary classification. As disease classification therefore seems pragmatic, this sometimes prompts bouts of anti-realism at least at the level of individual diseases, as noted earlier, but it might be considered a nice case of "promiscuous realism" (Dupré, 1995) all the same: that is, many different clusters of properties provide justified, but non-equivalent, classifications.

The terms "realism" and "anti-realism" have been endowed with many senses in the philosophy of medicine. An exhaustive review of these meanings and of arguments pro and contra is provided by Simon (2011). Realism about types seems to consist in the defense of essentialist natural kinds, that is, the view that types of disease could "carve nature at its joints" thanks to determinate necessary and sufficient conditions, that diseases are "eternal" natural facts and that they can take part in explanations and predictions. As opposed to realism, anti-realism consists in the view that human interests, not real-world features, identify disease tokens and group them into types, a view advocated by Claude Bernard, and also more recently by (Whitbeck, 1977; Engelhardt, 1975; Severinsen, 2001). This debate has of late been very prolific about mental disorders. Two plausible reasons to that are that most mental disorders are about deviant behaviors, which is not an argument, but a serious clue that they are normative constructs, and that no mental disorder is consensually defined by anything else than symptoms and their conjunction to date.

Antipsychiatry has long provided an influent turnkey template for papers on specific mental disorders: first, describe moral, social or economic influences on a construct, and then conclude in favor of an anti-essentialism. Masturbation (Engelhardt, 1974), hyperkinesia (Conrad, 1975), hysteria (Szasz, 1984), and, more recently, premenstrual syndrome (Richardson, 1995) and social anxiety (Lane, 2007), have been successes on this agenda. More recently, some have shifted direction towards a more moderate position, considering either that investigating the construction of a disorder such as post-traumatic stress disorder is theoretically separable from examining its naturalness (Young, 1997); or that natural facts such as autism are also social constructs (Hacking, 2000); or that part of a psychiatric construct as major depressive disorder is indeed a natural fact, and part of it, a mere scientific error (Horwitz and Wakefield, 2007).

Nearly all the more radical contributions are historical drawings of the birth of disorder such and such: they relate it to the social, cultural and historical contingencies of its appearance. Nevertheless, appeal to history is not specifically anti-realistic: as a matter of fact, a traditional argument in favor of the naturalness of a psychiatric entity is historical continuity. Depression, after all, seems to have been considered a disease since Hippocratic times. Yet this also is questionable. Radden, for instance, does not consider it plausible that "depression" denotes what "melancholy" used to denote (Radden, 2003). Yet she acknowledges that with a strong theory of what depression consists in, not just a simple description of what it looks like, the question might be solved differently. In this, she accepts an in-principle prominence of natural approaches to the status of a disease entity. Putnam has famously proposed the principle of "benefit of doubt": if medicine was to discover that what we used to call "multiple sclerosis" is in fact not at all what we used to consider it was, we would still consider that "multiple sclerosis" is nevertheless what former physicians referred to (Putnam, 1975, 310–311). Yet it is a difficult question to settle: whereas "consumption" and "phthisis" are not considered natural kinds anymore, "chlorosis" is sometimes considered to have been hypochromic anemia, sometimes not. Is diabetes really the same entity as it used to be when polyuria and polydipsia were the cardinal signs?

What when different conditions, previously thought of as different diseases, come to be considered the same disease? Strikingly, history is sometimes considered an argument pro or contra diseases as natural kinds, and sometimes, it seems to presuppose the notion of a given disease as a natural kind.

On the social side, Hacking asserted the possibility for a mental disorder to be both a natural kind and a social construction (Hacking, 2000). The behavior of autistic children can be influenced by the way they are classified, and at the same time, there may be one or several neurobiological processes justifying the naturalness of the grouping of pathology P in which child autism possibly consists. Nevertheless, some mental disorders are probably mere social constructs.

Zachar has defended the view that psychiatric kinds are not natural kinds because there cannot be essentialistic kinds, that is, defined by necessary and sufficient conditions and thereby discrete (Zachar, 2000). Instead, they are dimensional, prototypal and fuzzy. However, this does not mean that they are arbitrary—which they are not according to Zachar –, nor even that they are impractical. On the contrary, as "practical kinds," they fit with treatment, are useful as prognostic tools, design groups for research, coordinate clinics with biology, and so on (Zachar, 2002). According to Cooper, who defends a non-essentialist definition of natural kinds, some psychiatric kinds are likely to be natural kinds in the sense that that they share similar (not necessarily identical) determining properties. Determining properties should be, according to her, theoretically important properties. Other psychiatric kinds probably are partial kinds, that is, share common processes at some crucial stages of the disorder, not all stages. Their naturalness does not exclude border fuzziness, but consists in clustering in a multidimensional quality space (Cooper, 2005). As natural kinds, some mental disorders allow for what she calls "natural-history based explanations" (Cooper, 2007), that is, explanations (or predictions) of an individual's behavior by its belonging to a class.

Many, probably not all, difficulties of classification in psychiatry may also be found in medicine in general: historicity, fuzziness, dimensionality hold for somatic diseases too (e.g., diabetes, epilepsy, hypertension). Some diseases with known etiology or pathophysiology are difficult to establish as natural kinds. Yet a fundamental reason of the difficulty to establish that some diseases, such as mental disorders, are natural kinds, lies in the absence of a clear and sharp pathophysiology and etiology. For that reason, causal research in medicine has been considered an important part of the solution to the problem of natural kinds.

3.2 CAUSAL RESEARCH IN MEDICINE: BENCH RESEARCH

Medical knowledge mainly deals with the causal knowledge of diseases, that is, not only which factors cause which diseases, but also, which cause resistance, immunity, recovery, and the success or failure of cures. Importantly, medicine also contains the causal knowledge of the inner workings of disease processes. Two sources of causal knowledge are biological or "bench" research, and clinical research.

Bench research consists in experimentation on either in vitro or in vivo models. Experimentation on in vivo, animal models increasingly attracts philosophers' attention. In his own time, Canguilhem had already investigated the question of experimentation on animals (Canguilhem, 1965). A distinction must be made between experimenting on animal models in biological science and in medical science. Philosophers have first focused on the former, dealing with problems of phylogeny, species definition and genetic determination: the aim is generally to understand evolution better or to acquire a rough picture of a general process in living beings, such as the role of genes (Schaffner, 1998; Ankeny, 2001; Weber, 2005). Focusing on animal models in biomedical science, the question is very different. Animal systems are not explored for themselves, but rather as artificial surrogates for human systems: they are therefore manipulated, genetically or environmentally, to resemble human diseases. This has many consequences of importance.

First, it matters whether animal models are indeed considered instances of the same disease as humans suffer from, or just heuristic tools. LaFollette and Shanks thus distinguish between causal analogue models (CAM) and hypothetical analogical models (HAM): whereas the former are perfectly analogous to humans as to causes of a disease, the latter only prompt the formation of hypotheses about the human disease. LaFollette and Shanks argue that animal models are HAM, not CAM. The main reason is that whatever shared properties between humans and a given animal are, and numerous as they may be, it is still impossible to draw the conclusion that another property associated with them in the animal model will necessarily be present in humans also (LaFollette and Shanks, 1995).

Second, animal research in medicine seems more demanding on conditions of external validity for the experiment than in biology; external validity concerns whether conclusion drawn from an experimental context can be generalized beyond this context and beyond the population who participated in the study. The reason is that treatments generally are in perspective, and side effects, however small, matter much. Steel proposed an account of "animal extrapolation" in biomedical science (Steel, 2008). Obviously, the additional difficulty is that the experimental population does not belong to the same species as the target population. As opposed to LaFollette and Shanks though, Steel does not consider disanalogies to be a major obstacle to extrapolation. But he establishes a number of conditions for it to be conclusive, namely

- Ideal intervention: the elimination of other causal effects than the intervention on the target variable, of the intervention on other variables, and of the system's variables on the intervention (randomization is a way to approach this condition, which hardly obtains in biomedical science).
- Disruption principle: in the tested population, the effect is absent if and only if every possible link between cause and effect is disrupted. This is a powerful principle from which many experimental principles follow, such as the notion

that if an intervention is known to disrupt a causal pathway and the effect is still observed in the population, then there must be another causal pathway to the same effect.

- Additional knowledge: the knowledge of what plausibly and importantly differs in base model and target. Several philosophers have noted the importance of the knowledge of standard animal models per se (Weber, 2005; Ankeny and Leonelli, 2011; Meunier, 2012). For instance, biologists have accumulated much knowledge about the particular physiology of *Mus muscularis*: among other things, metabolism differs greatly in mice and humans, an important point to consider when testing drugs.
- Comparative process tracing conditions: parts of the mechanisms involved in the model and the target are more or less similar. Where there are more dissimilarities, lies the best chance that the model fails. When these admittedly most important differences are known to be small, the chances are better that the animal disease is a good equivalent of the modeled disease.
- Level of precision of the claim: extrapolation in a particular case is possible or not depending on the level of precision of the claim at hand. The difficulty is increasing, depending on the nature of the causal claim. From easiest to most difficult, claims are: "x is relevant to disease y," "x has a causal effect on y," "x causes/impedes y," "the effect on y increases with the dose of x on an interval [a;b]."

In addition to these necessary conditions, Steel mentions further conditions, some of which have been considered necessary in biology, but are merely facilitating conditions in biomedical research:

- Contextual unanimity, i.e., the fact that there is no subpopulation in the target human population for which the causal claim does not hold or is reversed. An example Nancy Cartwright emphasized is the causal relevance to thrombosis of both birth control pill and pregnancy: birth control medication is a neat risk factor for a subpopulation of women who will not be pregnant, but decreases risk for the rest of the population because it decreases the probability of being pregnant, a condition where the risk of thrombosis is much higher (Cartwright, 1989).
- Consonance: the absence of conflicting mechanisms of the intervention on the system, such that either one or the other may prevail depending on the subject (for instance, radiotherapy is known to both treat cancer and increase the risk of cancer, and serotonin reuptake inhibitors to worsen depression in some subjects because it equally stimulates antagonistic regions of the brain).
- Modularity: the possibility of altering the functioning of components independently from one another (Weber, 2005) allows for experimenting

on subsystems in a model organism similar to human subsystems, notwithstanding important animal/human differences in other subsystems.
- Phylogenetic closeness (Wimsatt, 1998): the closer two species, the more similar they are on the whole and the easier extrapolation is.

Some of these conditions are specific to either medicine or biology, and some are more important in one than in the other, which strongly suggest that animal experimental models in biomedicine have specific features. Another entirely different argument has recently been proposed in favor of the specificity of biomedical research relative to biological research. Pathological mechanisms, as bench research tries to establish them, might be different in nature as compared to physiological mechanisms. According to Mauro Nervi, mechanisms of some diseases at least are not just non-functioning mechanisms (Nervi, 2010): they are not considered to be pathological just because they fail to fulfill one of the normal functions. They are pathological mechanisms per se, that is, they display specific properties as such: outcome variability, ambivalence and dependence on a range, according to Nervi. Advocating this view on the theoretical independence of pathology does not imply supporting precisely these three specific properties: both claims have been discussed (Moghaddam-Taaheri, 2011). The theoretical independence and originality of pathophysiology is a crucial question in demarcating philosophy of biology and philosophy of medicine.

3.3 CAUSAL RESEARCH IN MEDICINE: CLINICAL RESEARCH

Experiments on model organisms have proven necessary, but not sufficient in causal research in medicine, and particularly therapeutic research. Inspired by the experimental design in agronomy, a method developed by the statistician Ronald Fisher (1935), other tools have been developed since the middle of the 20th century, based on new clinical study designs, the controlled comparison of well-defined groups of individuals, and statistical techniques. These techniques have been increasingly recognized as relevant not only for improving the design of an epidemiological study but also for drawing valid inference from these studies. The randomized clinical trial (RCT) more particularly renewed ideas on what good experiments and sound evidence consist in (Marks, 2000; Fagot-Largeault, 2003). Experimenting on humans outside of the laboratory demands to be very rigorous on controlling the potential bias in the design of the study.

Since the attempt to assess the effect of streptomycin on tuberculosis by sir Bradford Hill, a British statistician and physician, randomized clinical trials (RCT) have been progressively considered the gold standard of clinical experiments and their results, the best evidence for efficient treatments. They consist in assessing the probability and the efficacy of a new treatment relative to a standard one. The crucial point is that there is no significant difference between the test (treated) and control (non-treated) group save for the treatment itself. Ronald Fisher had introduced the procedure of randomization as a method "by which the test of significance may be guaranteed against corruption by the causes of disturbance which have not been eliminated" (1947, 19). The test of significance permits to measure the risk that the association investigated is due to chance. Bradford Hill, a (former) student of Fisher, applied the randomization procedure to clinical trials. In what has become the RCT, he added the double-blind procedure (neither the physician nor the patient knows which group the patient belongs to) to the randomized allocation of the subjects to groups (treatment or placebo). This method of allocation has then been considered and used as the best one to guarantee and secure the equivalence of the treatment and control groups with regard to confounding factors that may influence the outcome of the study.

Since the deliberate exposure to a presumed disease factor would be unethical, this kind of controlled randomized experiment cannot be used in etiological research. Instead, scientists resort to epidemiological observational studies. The most important in epidemiology are case-control studies, a comparison of cases to control subjects, and prospective cohort studies, a follow-up comparison of individuals exposed and non-exposed to the factors under scrutiny. These studies allowed the identification of risk factors such as hypertension, tobacco or hypercholesterolemia for cardiovascular diseases, and contributed to modify the medical conception of disease (Aronowitz, 1998; Giroux, 2008).

The problem of inference from these observational studies is that it is inevitably much exposed to errors and biases: biases due to potential defects in the design of execution of a study and confounding variables. Epidemiologists distinguish selection biases and information biases (Hennekens, Buring, and Mayrent, 1987; Elwood, 1992). Selection biases intervene in the constitution or allocation of samples. It introduces a difference between the characteristics of the people selected for the study and the characteristics of those who were not. Information biases pertain to how data are collected and measured. It occurs when classifications of disease or exposure are not valid. This kind of errors can be introduced by the observer, by the study individual, or by the instruments used to make the measurement. Confounding occurs when an estimate of the association between and exposure and an outcome is mixed up with the real effect of another exposure on the same outcome, the two exposures being correlated: for instance, tobacco consumption is such a factor in studies about the relationship between alcohol consumption and lung cancer, because it is frequently associated with both separately. The resort to statistical tools such as multivariate analysis constitutes an important, if limited, means to quantitatively estimate the independent effects of several factors on the outcome and to control against confounding. Besides, the test of statistical significance and calculating confidence intervals allow for assessing the role of chance. But the scientific and philosophical question remains: to what extent is it possible to conclude from statistical association to causation in the context of observational studies?

It is often assumed in methodological debates that only experimental clinical studies such as RCTs, not observational studies, can be conclusive on causal relations for clinical decision and practice. Only the manipulation of the variable under study in clinical experimental conditions would allow for control strictly speaking, which observational studies can only approach. In etiological research, observational studies would therefore be merely heuristic and suggest hypotheses that either biological experiments or RCTs (if possible) would confirm as causal relations. Evidence-based medicine, a very influential movement born in the 1990s, proposes a gradation of levels of evidence where RCTs come first, followed by observational studies. "Expert opinion without explicit critical appraisal, or based on physiology, bench research or 'first principles'" comes last (http://www.cebm.net). This hierarchy is controversial. In particular, some methodologists and philosophers of science contest three claims.

First, the superiority of experiment over observation and the very distinction between these two kinds of scientific analysis are questionable. Bradford Hill and Jerome Cornfield (Hill, 1953; Cornfield, 1954) both considered that there is no difference in nature, only in degree, between observation and experiment. Moreover, they emphasized the inappropriateness of "bench research" kind of experiment to the study of human phenomena medicine consists in. Because it secludes its object from its natural environment, it has more defaults than observational inquiries. According to them, etiological observational studies provide data and analyses otherwise inaccessible. They belong to what has been called "quasi-experiments" (Campbell and Stanley, 1963), a kind of study that strongly resembles experiments except for the randomized allocation of the variable under study.

Second, the supposed virtues of randomization are discussed (Urbach, 1985; Worrall, 2002; Worrall, 2007b).[22] Worrall (2002) concludes that among the many virtues attributed to randomization in RCTs, only the avoidance of selection bias, which occurs when clinicians assign patients to the treatment or to the control group, is real. In particular, randomized studies do neutralize some confounding factors, known and unknown, but it is not true that they could rule out all possible confounding factors: the samples are never large enough to do that. Moreover he points out that alternative methods for preventing selection bias are possible and equally effective. In the same perspective, some Bayesian analyses of experiments have been proposed (Urbach, 1985), from which the notion of Bayesian trials has emerged (Teira, 2011).

Third, it has often been opposed that bench research sometimes discovers causal effects, the size of which is sufficient to be self-evident. Jeremy Howick, otherwise a strong advocate of EBM, has recognized the fact and pleaded for a minor revision of the hierarchical principle (Howick, 2011). Although RCTs are not always necessary, due to the size of the observed effect, nor rule out all possible confounders, RCTs are more likely to give the strongest evidence in most cases. From a different, but related perspective, some philosophers have defended the view that the results of fundamental research were already implicitly taken into account in randomized clinical trials, but also, that only strongly theoretical models, not just empirical bunches of facts, could support causal claims (Thompson, 2011b).

[22] As opposed to Fisher, Jerzy Neyman and Egon Pearson thought that randomized allocation was not the only means to reproduce laboratory control conditions, avoid selection biases, and allow for the test of significance (Gigerenzer et al. 1990, 90–106).

3.4 CAUSAL INFERENCE IN MULTIFACTORIAL DISEASES

Both bench and clinical causal research in medicine have soon encountered the same problem of multifactorial processes, which many diseases seem to involve, mainly chronic diseases such as cancer, diabetes, Alzheimer's disease, and many mental disorders.

Multifactorial diseases contrast with simpler etiological models, of which the most elegant may be Henle-Koch's postulates for infectious diseases: a bacteria or virus is the cause of a disease if and only if (1) it is always present in cases of a disease, (2) it is never present in disease-free individuals, and (3) it can be isolated from a diseased subject and inoculated to a disease-free individual (who therefore turns ill). Here, the definition of a cause of a disease is a necessary and sufficient condition of the disease, which makes it predictable and perfectly manipulable (Evans, 1976; Carter, 2003). Another example of such simple models is so-called mendelian diseases, that is, diseases that seem to follow simple mendelian inheritance mechanisms.

Those models were soon revealed as either too simple or impracticable. Koch's postulates are too stringent and they were later relaxed (Evans, 1993), and even the expression "mendelian diseases" seems not to be used so often now. The fact is for instance that many subjects are healthy despite being carriers of a bacteria (e.g., *Vibrio choleriae*), a virus, or a gene (notion of penetrance): in those cases, the bacteria, virus, or gene is a necessary but insufficient condition of the disease. Besides, some diseases seem to group many different causal pathways (as previously noted). In the case of cancer, it seems that several mixes of genetic mutations and environmental exposures can often (but not in all cases) lead to the same result. Such conditions are considered multifactorial diseases, because it seems impossible to identify one specific cause.

In the 1960s, the risk factor approach was devised to deal with multiple causation of chronic diseases. Several multifactorial models were then proposed by epidemiologists such as the causal web (MacMahon and Pugh, 1970) in which the occurrence of different exposures is required for the occurrence of a disease, none of which is necessary. Risk factors are neither necessary nor sufficient; they just increase the probability of a disease. Smoking increases the probability of having cancer, obesity and alcohol consumption too, all three, much more so.

A controversy arose about the causal interpretation of risk factors, and particularly, about the causal role of tobacco in lung cancer which strongly stimulated an enduring discussion on the logic of causal inference (e.g. Susser, 1973; Susser, 1991; Rothman and Greenland, 2005). The statistician Joseph Berkson defended the view that the positive association between the two could be the result of a selection bias (Berkson, 1958), and Ronald Fisher was wary of a possible confounding factor such as a genetic factor (Fisher, 1959); both advocated the search of an underlying biological mechanism (Parascandola, 2004; Berlivet, 2005). In 1964, the Report of the advisory committee to the surgeon general of the American Public Health Service on smoking and health asserted that couples with other data, epidemiological studies provide decisive elements to establish the causal significance of the association. Several criteria were used for judgment of causality, no one of which is an all-sufficient basis. Five criteria were used, to which Bradford Hill added four others (Hill, 1965). The resulting list is often referred to: (1) the strength of the association, (2) the consistency of the association (across various studies), (3) specificity, (4) temporal relationship, (5) biological gradient, (6) plausibility, (7) coherence with the natural history of the disease, (8) experimental evidence, and (9) analogy with other diseases. These constitute only convergent clues, not necessary conditions for a causal judgment.

Such set of clues for causal inference tend to be used in bench research too, where the complexity of causal processes is often so overwhelming that simple causal hypotheses and analysis seem not to fit anymore. Genomics in particular, all –omics more generally, seem to plead for the use of complex mathematical models that seem to make causal processes in a disease a black box, and traditional functional explanations inappropriate (Dupré, 2011; Gross, 2011). When dozens of effects of a protein usually occur with no conceivable link between them whatsoever, it seems awkward indeed to explain it by postulating function(s).

3.5 CAUSALITY IN MEDICINE: MECHANISTIC OR STATISTICAL?

To think more clearly about causal inference, it seems useful to have a notion of what is claimed when a "cause" is invoked (Parascandola and Weed, 2001). It seems at first sight that the mechanistic notion of causality should be prominent in biological research, the statistical, in epidemiological research. To be sure, on the one hand, traditional interventionist or manipulationist conceptions of causal relevance, as applied to experimentation on biological mechanisms (Craver, 2007), and on the other hand the probabilistic notion of causality some epidemiologists (Elwood, 1992; Lagiou, Adami, and Trichopoulos, 2005) prefer, support this view. Yet surprisingly, deterministic approaches seem to be more influential. Rothman's Sufficient-Component Cause Model (Rothman, 1976), which is based on Mackie's model (Mackie, 1965) of INUS conditions (Insufficient but Non-redundant part of an Unnecessary but Sufficient condition) and John Stuart Mill's multifactorial conception of causes, allows for the integration of the determinist and necessary/sufficient terms for thinking of causation into a multifactorial account. In this model, a cause of an individual case of disease occurrence is defined as a set of conditions that are both necessary and sufficient, in the five circumstances, for the disease to occur. There is no cause which is necessary and sufficient for its effect but there are "component causes" which are necessary parts of a "sufficient cause," i.e. a complete causal mechanism that is defined as a set of minimal conditions and events that inevitably produce disease. At least thus, do we observe a tension over causation in contemporary epidemiology between probabilistic risk factors and deterministic approach to causal mechanisms (Parascandola, 2011).

Moreover, Hill's criteria appear to mix mechanistic and probabilistic aspects of causality (Russo and Williamson, 2007). To many, statistical associations as established by epidemiological studies are only preliminary research, whereas mechanistic and biological considerations on temporality, biological plausibility, coherence, experimental data and analogies with known diseases, are the core of the causal interpretation. To Wesley Salmon (Salmon, 1984), as to defenders of the physical and mechanistic view of causation (Glennan, 1996; Machamer, Darden, and Craver, 2000), the continuity of stages as well as a complete map of interactions are necessary for causal assertion. Yet due to the pragmatic dimension of medicine, it is often impossible to wait for such knowledge to decide and intervene. Moreover, mechanistic knowledge is not sufficient: there need to be some evidence that the cause makes a difference to its effects and concerning the strength of the statistical association. For instance, it has been shown that p53 is a gene involved in the causal pathway between tobacco consumption and lung cancer. But such information is useless if we do not have information concerning the actual existence of the association in populations and its strength (Thagard, 1998; Parascandola, 1998; Russo and Williamson, 2007).

Mechanistic and probabilistic approaches to causality therefore either compete or collaborate. Some consider that their relevance depends on the levels of observation, mechanistic approaches to individuals, probabilistic and statistical approaches to population (Parascandola and Weed, 2001). Others think that this plurality just reflects the necessity to use various causal models in epidemiology (Greenland and Brumback, 2002). Some philosophers defend a unified concept of causality in medical sciences, transcending the mechanistic and probabilistic accounts and the monist and pluralist theories of causality. Russo and Williamson (2007) want to show that scientists use a single notion of cause in health science besides the fact that both probabilistic and mechanistic approaches are necessary in causal inference: probabilistic approach identifies causal relationships, whereas mechanistic evidence explains it. They make a distinction between *types of evidence*, of which mechanistic and probabilistic aspects are irreducibly heterogeneous to one another, and *the causal relation itself*, which is unique. There could then be two types of evidence for causal assertion and yet only one causal relation. They defend a dual-faceted epistemic theory of causality. In their view, the causal relation is not an ontological entity, it is epistemological: it should be identified with the causal beliefs of an omniscient rational agent. Causality is thus determined by causal epistemology. The duality or plurality in the types of evidence is thus compatible with a unified epistemic account of causality.

Another way of dealing with this problem of the interpretation of causality is to make a move from the concept of causation toward the concept of causal explanation. Thagard shows that for each disease type such as cancer, infectious diseases, autoimmune diseases, etc., there is a unified system of disease explanation schemas (Thagard, 1999, pp. 20–36), but that there is no unified system of explanations for all diseases. In each system of schemas, nodes are connected by the causal relation inferred on the basis of very different kinds of considerations: statistical associations, alternative causes, and mechanisms (Thagard, 1999, pp. 113–117).

In his analysis of causation in epidemiology, Broadbent (2009) considers that too much effort has been directed toward how to infer causation. At least in epidemiology, the main question should rather be explanation and prediction. Relying on Peter Lipton's contrastive theory of causal explanation (Lipton, 1990), he argues for a contrastive model of causation in epidemiological explanation. Lemoine considers that the notion of causality itself is relative to particular beliefs defining what should count as a sufficient explanation (Lemoine, 2011). As there are many sets of such beliefs in medicine, each defining what he calls an "explanatory value," there also are irreducibly different concepts of causality, not only mechanistic and epidemiological, but also, pharmacological, psychiatric, and maybe others.

4. How Rational Is Clinical Reasoning?

All of medical knowledge does not consist in classificatory and causal knowledge. Since the early 1960s, clinical reasoning per se, including diagnostic, prognostic and therapeutic reasoning, has been an object of scientific investigation for clinicians. Either descriptive or normative approaches aim at improving it. To philosophers, clinical reasoning is an interesting epistemological case. It involves all traditional kinds of knowledge (Russell, 1912). First, it involves knowledge of things, by acquaintance (knowing that *this* is cyanosis for instance), and by description (knowing what *carphology* looks like). Second, it involves knowledge of truths (cyanosis is a sign of hypoxia). It also involves various kinds of what is called "tacit knowledge" (Polanyi, 1962), among which, practical as well as theoretical knowing-how. Clinical reasoning also seems to mix them at each step and often to support the replacement of one by the other. Besides, it cannot be considered the strict application of a general knowledge of diseases to particular patients: the nature of clinical reasoning is intrinsically different. The recognition of this fact lies at the heart of the emergence of "clinical epidemiology," and of the application of logical and mathematical models in decision theory to clinical judgment.

Both the emergence of these scientific domains and contemporary epistemology thus contribute to renew the ancient distinction between art and science of medicine. As an art, clinical reasoning is a skill, often remains implicit, and relies on intuition, epistemic virtue, heuristics, biases, and narrative reasoning. As a science, clinical reasoning can always be explicit, resorts to measurement, relies on statistical and probabilistic inferences, and has the structure of a decision tree.

4.1 THE ART OF CLINICAL REASONING

In the past, intuition (flair) has been much celebrated as an achievement of expert clinicians. Its content is often thought of as fuzzy or allusive. It refers either to a hidden cognitive process, which can be illuminating as well as misleading, or to the incomprehensible appearance of the right solution to a problem (in which case it is illuminating by definition). Among contemporary physicians, some conclude that intuition, as the mysterious appearance of the truth, does not exist since intuition in the first sense is sometimes misguided; others, that intuition as a hidden cognitive process is always right, because in the second sense, intuition is the appearance of the truth. It is not tantamount to the same thing to study hidden cognitive processes in normal diagnosticians and in exceptional diagnosticians. As Dreyfus's model of skill acquisition suggests, it might be the case that the cognitive processes at work are not the same at all (Dreyfus and Dreyfus, 1980): whereas the novice relies on the strict application of a few rules, the expert might rely on a very sophisticated "repertoire of situational discriminations," which is very difficult, if not impossible, to reduce to algorithms. One should not confound the impossibility to simulate clinical reasoning with algorithms with the impossibility to describe it. Based on the notion of heuristics and biases in decision (Kahneman, Slovic, and Tversky, 1982), it is sometimes possible to infer the various rules of the thumb clinicians follow from the observation of their behavior. They seem to be arbitrarily and selectively oriented toward a limited set of hypotheses, which, interestingly, account both for clinicians' mistakes and achievements. This approach is called the information processing theory (Elstein, Shulman, and Sprafka, 1978).

Part of what is labeled "intuition" in clinical reasoning amounts to what Michael Polanyi called "tacit knowledge" (Polanyi, 1962), that is, a kind of know-how that cannot be phrased. Tacit knowledge is not limited to practical skills, such as applying the right pressure with the hands in a diagnostic or therapeutic procedure, but also extends to perceptual and intellectual skills, such as, using contextual (Pantazi, Arocha, and Moehr, 2004) or social and non-verbal communicational data (Førde, 1998), perceiving signs of anomalous movements, heartbeats or respiration, and examining the most relevant small set of alternative diagnoses. Some have argued that tacit knowledge is an irreplaceable part of clinical judgment, and that it is irreducible to explicit knowledge (Goldman, 1990; Goldman, 1991; Braude, 2009; Braude, 2012), although this does not amount to denying the usefulness of the latter.

As clinical reasoning is an ability that improves through exertion, it can be thought of as an intellectual virtue (Marcum, 2009). In particular, Aristotle's phronesis (wisdom, practical reason, prudence) consists in the ability to exert one's judgment in particular instances by applying general knowledge as one sees fit (Montgomery, 2005).

Some consider narrative reasoning to be an essential part of the art of clinical judgment. It involves rational interpretation of the patient's story and experience rather than inferential reasoning. A story is intelligible as an agent's actions and aspirations in the context of the agent's natural and interpersonal world (Mattingly, 1998). Illness is not only a natural fact, it is also a human event. Although it may sometimes provide clues as to the patient's condition and what the doctor could do for him, it is nonetheless counterintuitive both to describe clinical judgment and to analyze its commonsense meaning by resorting to such notions. There might be a slip from a descriptive to a normative point of view, as many think that good clinicians should endorse such approach. In *Rational Diagnosis and Treatment* (Wulff, 1981), Henrik Wulff considers the introduction of hermeneutics and ethics into clinical reasoning as a critical assessment rather than a description.

4.2 THE SCIENCE OF CLINICAL REASONING

In 1954, an influent book by Paul Meehl had suggested that traditional clinical reasoning is less reliable than statistical inferences (Meehl, 1954). The sense that explicating the inner logic of clinical reasoning could make it more precise, reliable and valid (Murphy, 1976; Wulff, 1981; Engelhardt, 1979) emerged together with the notion that artificial algorithms could outperform natural clinical reasoning. In both cases, explicit reasoning was opposed to implicit reasoning. The question is empirical which of these is best for clinical reasoning. From a conceptual point of view, some important issues, as Sober (1979) listed them, were these:

1. Can the implicit art of clinical judgment be explicated and reduced to explicit logical protocols, and is the result a faithful reflection of what clinicians do?
2. Does clinical judgment take idiosyncrasies into account in a way that scientific explicit reasoning cannot in principle?
3. Is clinical judgment irreducibly qualitative, and are quantitative scientific methods more accurate and more reliable because they are quantitative?
4. Is there background information, such as gestalts and the perception of emotions, the extraction of which cannot be laid into explicit processes? (Sober, 1979)

As opposed to Meehl, Alvan Feinstein defends the irreplaceability of clinical judgment by statistical inference as instantiated in experimental science, but claims that clinical judgment can be made explicit. The result would be an improvement of clinical science through clinical taxonomy, formalization and standardization thanks to statistical and mathematical tools. First, he emphasizes the scientific function of clinical data as observed at bedside, a function shunned because of "hard" data from physiopathology and biomedical science. He distinguished data relevant to diseases (morphological, chemical, etc.), data relevant to the patient characteristics and environment, and data resulting from the interaction of the first two, that is, symptoms, as lived by the patient, and signs, as observed during the examination (Feinstein, 1967, p. 24–25). This third kind of data can be made reliable through the statistical control of variability and the accuracy of measurement. It is indispensable, for instance, as an indicator of severity, or as the only source of complementary information that can improve our knowledge of the natural history and various forms of a disease. From this point of view, clinical science is as basic a science as biomedical science (Feinstein, 1967, pp. 381–390; Feinstein, 1983). The hardening of so-called soft data through standardization and quantification of observations results in new disciplines such as "clinical epidemiology," a phrase accepted nowadays (Fletcher, 2001), despite its fuzziness and Feinstein's own reservation about it (Feinstein, 1985, vi) and "clinimetrics," a less successful term he also coined (Feinstein, 1987). Feinstein's works often stand at the boundaries of science and philosophy of science; his thought is complex, his prose, full of neologisms.

Feinstein also reacted against the blind application of the result of RCTs to particular cases, as a substitute for clinical judgment. For all its defenders' claims to the contrary (Sackett et al., 1996, p. 71), it has often been objected to EBM that it is ill-equipped for individual decision making (Daly, 2005).[23] EBM has often been associated with frequentist approaches. The latter approaches seem particularly adequate for populations, and Bayesian approaches, for individual, because of variability and the resulting uncertainty (Ledley and Lusted, 1959; Suppes, 1979; Fagot-Largeault, 1982). In a Bayesian approach to an individual decision, the result of RCTs is used as one determinant among others. Systematic reviews of RCTs and meta-analyses, as provided by EBM or the Cochrane Collaboration, serve that purpose. Initial probabilities can also be revised in the light of the results of successive diagnostic tests. This provides a quantitative approach to the reasoning itself, not only to the initial data of the reasoning.

An important question about clinical science is a relevant and useful decomposition of clinical reasoning into consistent stages. Many have proposed various global models of clinical reasoning (Ransohoff and Feinstein, 1976; Dowie and Elstein, 1988; Jenicek, 2003). Indeed, clinical problems have structures, which can be expressed through decision trees. Diagnostic trees constitute a major part, but not all, of these flow charts, inasmuch as diagnosis too, not only therapeutics, consists in a series of decisions or a course of actions. Didactic examples are the decision tree of the course of action in the face of acute abdominal pain, and the decision tree of the diagnosis of an icterus. Probabilities and utilities are often assigned to all branches of the decision tree, in order to decide which has the highest utility. Nonetheless, several problems have been raised (see Marcum, 2010):

1. Estimation of probabilities appropriate to the particular case of the patient is difficult (Ransohoff and Feinstein, 1976).
2. Estimation of utilities is often arbitrary (Ransohoff and Feinstein, 1976).
3. These estimations have to include non-medical information too, and there is often too much of it.
4. Decision trees have to be exhaustive but can rarely be (Ransohoff and Feinstein, 1976).
5. Clinical reasoning is often too large for decision trees to be useful, so that "pruning" may become necessary (Kassirer, 1976): yet although there are principles to it, pruning is often arbitrary.

A last theoretical question is whether fuzzy theory is a better means for either capturing what clinicians actually do or improving what could be done in clinical decision making. Kazem Sadegh-Zadeh, a philosopher and a founder of *Theoretical Medicine and Bioethics* as well as *Artificial Intelligence in Medicine*, has explored the

[23] See *Perspectives in Biology and Medicine*, 2005, vol. 48, no. 4 and *Journal of Evaluation in Clinical Practice*, 2010, vol. 16, no. 2. See also (Rothwell, 2007).

possibilities of its application to diagnosis, nosology, prognosis (Sadegh-Zadeh, 1999, 2000, 2001).

5. Conclusion

Philosophy of medicine is a rapidly growing field. The discussion over concepts of health and disease has become less prominent, as conceptual analysis is not the only philosophical approach to it anymore. Cognitivist and social approaches, but also, philosophy of causality, of statistics and epistemology all contributed to its renewal. Nonetheless, just as the question of the independence of medicine as a science is still raised, the question of the independence of the philosophy of medicine is too. Many philosophers of science had contributed to specific questions in the philosophy of medicine in the past, without considering themselves as philosophers of medicine. Some younger philosophers of medicine now take the existence of the field, including, but not limited to the conceptual analysis of health and disease, for granted. It may be all that it takes for an emergent field in philosophy to exist as such.

References

Agich, George J. 1983. "Disease and Value: A Rejection of the Value-Neutrality Thesis." *Theoretical Medicine* 4 (1): 27–41.

Agich, George J. 1997. "Toward a Pragmatic Theory of Disease." In *What Is Disease?* edited by James M. Humber and Robert F. Almeder, 219–246. Biomedical Ethics Reviews 14. Totowa, NJ: Humana Press.

Amundson, Ron. 2000. "Against Normal Function." *Studies in History and Philosophy of Science Part C: Studies in History and Philosophy of Biological and Biomedical Sciences* 31 (1): 33–53.

Ananth, Mahesh. 2008. *In Defense of an Evolutionary Concept of Health Nature, Norms, and Human Biology*. Aldershot, England and Burlington, VT : Ashgate.

Ankeny, Rachel A. 2001. "Model Organisms as Models: Understanding the 'Lingua Franca' of the Human Genome Project." *Proceedings of the Philosophy of Science Association* 2001 (3): S251–S261.

Ankeny, Rachel A., and Sabina Leonelli. 2011. "What's So Special about Model Organisms?" *Studies in History and Philosophy of Science* Part A 42 (2): 313–323.

Aronowitz, Robert A. 1998. *Making Sense of Illness: Science, Society and Disease*. 1 edition. Cambridge, U.K.; New York, NY: Cambridge University Press.

Barnes, Allan C. 1962. "Is Menopause a Disease." *Consultant* 2: 22–24.

Bayer, Ronald. 1981. *Homosexuality and American Psychiatry: The Politics of Diagnosis*. Princeton, NJ: Princeton University Press.

Benditt, Theodore. 2007. "Normality, Disease, and Enhancement." In *Establishing Medical Reality: Essays in the Metaphysics and Epistemology of Biomedical Science*, edited by Harold Kincaid & Jennifer McKitrick, 13–21. Dordrecht, Netherlands: Kluwer Academic Publishers.

Berkson, Joseph. 1958. "Smoking and Lung Cancer: Some Observations on Two Recent Reports." *Journal of the American Statistical Association* 53 (281): 28–38.

Berlivet, Luc. 2005. "'Association or Causation?' The Debate on the Scientific Status of Risk Factor Epidemiology, 1947–c. 1965." Clio Medica/The Wellcome Series in the *History of Medicine* 75 (1): 39–74.

Boorse, Christopher. 1975. "On the Distinction Between Disease and Illness." *Philosophy and Public Affairs* 5 (1): 49–68.

Boorse, Christopher. 1976a. "What a Theory of Mental Health Should Be." *Journal for the Theory of Social Behaviour* 6 (1): 61–84. doi:10.1111/j.1468-

Boorse, Christopher. 1976b. "Wright on Functions." *The Philosophical Review* 85 (1): 70–86.

Boorse, Christopher. 1977. "Health as a Theoretical Concept." *Philosophy of Science* 44 (4): 542–573.

Boorse, Christopher. 1987. "Concepts of Health." In *Health Care Ethics: An Introduction*, 377. Philadelphia, PA: Temple University Press.

Boorse, Christopher. 1997. "A Rebuttal on Health." In *What Is Disease?* edited by James M. Humber and Robert F. Almeder, 1–134. Biomedical Ethics Reviews 14. Totowa, NJ: Humana Press.

Boorse, Christopher. 2002. "A Rebuttal on Functions." In *Functions: New Essays in the Philosophy of Psychology and Biology*, edited by Ariew André, Cummins Robert, and Perlman Mark, 63–112. Oxford: Oxford University Press.

Braude, Hillel D. 2009. "Clinical Intuition versus Statistics: Different Modes of Tacit Knowledge in Clinical Epidemiology and Evidence-Based Medicine." *Theoretical Medicine and Bioethics* 30 (3): 181–198.

Braude, Hillel D. 2012. *Intuition in Medicine: A Philosophical Defense of Clinical Reasoning.* Chicago: University of Chicago Press.

Broadbent, Alex. 2009. "Causation and Models of Disease in Epidemiology." *Studies in History and Philosophy of Biological and Biomedical Sciences* 40 (4): 302–311.

Brown, W. Miller. 1985. "On Defining 'Disease.'" *Journal of Medicine and Philosophy* 10 (4): 311–328.

Campbell, Donald T., and Julian Stanley. 1963. *Experimental and Quasi-Experimental Designs for Research*. 1st ed. Belmont, CA: Wadsworth Publishing.

Canguilhem, Georges. 1965. *La Connaissance De La Vie*. Paris: J. Vrin.

Canguilhem, Georges. 1978. *On the Normal and the Pathological.* Dordrecht, Netherlands: Kluwer Academic Publishers.

Caplan, Arthur L. 1981. "The Unnaturalness of Aging: A Sickness unto Death?" In *Concepts of Health and Disease: Interdisciplinary Perspectives*, edited by Arthur L. Caplan, H. Tristram Engelhardt, and James J. McCartney, 725–737. Menlo Park, CA: Addison-Wesley.

Caplan, Arthur L. 1992. "Does the Philosophy of Medicine Exist?" *Theoretical Medicine* 13 (1): 67–77.

Caplan, Arthur L., H. Tristram Engelhardt, and James J. McCartney. 1981. *Concepts of Health and Disease: Interdisciplinary Perspectives*. Menlo Park, CA: Addison-Wesley.

Caplan, Arthur L., James J. McCartney, and Dominic A. Sisti. 2004. *Health, Disease, and Illness: Concepts in Medicine*. Washington, DC: Georgetown University Press.

Carson, Ronald A., and Chester R. Burns. 1997. *Philosophy of Medicine and Bioethics: A Twenty-Year Retrospective and Critical Appraisal*. Vol. 50. New York: Springer.

Carter, Kay Codell. 2003. *The Rise of Causal Concepts of Disease: Case Histories*. Farnham, UK: Ashgate Publishing.

Cartwright, Nancy. 1989. *Nature's Capacities and Their Measurement*. Oxford/New York: Oxford University Press.

Clouser, K. Danner, Charles M. Culver, and Bernard Gert. 1981. "Malady: A New Treatment of Disease." *Hastings Center Report* 11 (3): 29–37.

Cohen, Henry. 1955. "The Evolution of the Concept of Disease." *Proceedings of the Royal Society of Medicine* 48 (3): 155.

Conrad, Peter. 1975. "The Discovery of Hyperkinesis: Notes on the Medicalization of Deviant Behavior." *Social Problems* 23 (1): 12–21.

Cooper, Rachel. 2002. "Disease." *Studies in History and Philosophy of Science Part C: Studies in History and Philosophy of Biological and Biomedical Sciences* 33 (2): 263–282.

Cooper, Rachel. 2005. *Classifying Madness: A Philosophical Examination of the Diagnostic And Statistical Manual of Mental Disorders*. New York: Springer.

Cooper, Rachel. 2007. *Psychiatry and Philosophy of Science*. Montreal: McGill-Queen's University Press.

Cornfield, Jerome. 1954. "Statistical Relationships and Proof in Medicine." *The American Statistician* 8 (5): 19–23.

Craver, Carl F. 2007. *Explaining the Brain: Mechanisms and the Mosaic Unity of Neuroscience*. Oxford: Oxford University Press, Clarendon Press.

Daly, Jeanne. 2005. *Evidence-Based Medicine and the Search for a Science of Clinical Care*. Vol. 12. Berkeley: University of California Press.

Daniels, Norman. 1985. *Just Health Care*. Cambridge: Cambridge University Press.

Dowie, Jack, and Arthur S. Elstein, eds. 1988. *Professional Judgment: A Reader in Clinical Decision Making*. Cambridge: Cambridge University Press.

Dreyfus, Stuart E., and Hubert L Dreyfus. 1980. *A Five-Stage Model of the Mental Activities Involved in Directed Skill Acquisition*. Berkeley: Operations Research Center, University of California, Berkeley.

Dubos, René Jules. 1959. *Mirage of Health: Utopias, Progress and Biological Change*. New York: Harper and Brothers.

Dupré, John. 1995. *The Disorder of Things: Metaphysical Foundations of the Disunity of Science*. Cambridge, MA: Harvard University Press.

Dupré, John. 2011. "Emerging Sciences and New Conceptions of Disease; Or, Beyond the Monogenomic Differentiated Cell Lineage." *European Journal for Philosophy of Science* 1 (1): 119–131.

Elstein, Arthur S., Lee S. Shulman, and Sarah A. Sprafka. 1978. *Medical Problem Solving: An Analysis of Clinical Reasoning*. Cambridge, MA: Harvard University Press.

Elwood, J. Mark. 1992. *Causal Relationships in Medicine: A Practical System for Critical Appraisal*. Oxford: Oxford University Press.

Engel, George L. 1960. "A Unified Concept of Health and Disease." *IRE Transactions on Medical Electronics* 1: 48–57.

Engelhardt, H. T., Jr. 1974. "The Disease of Masturbation: Values and the Concept of Disease." *Bulletin of the History of Medicine* 48 (2): 234–248.

Engelhardt, H. T., Jr. 1975. "The Concepts of Health and Disease." In *Evaluation and Explanation in the Biomedical Sciences*, edited by H. T. EngelhardtJr. and S. F. Spicker, 125–141. Dordrecht, Holland: D. Reidel Publishing Company.

Engelhardt, H. T., Jr. 1976a. "Is There a Philosophy of Medicine?" *PSA: Proceedings of the Biennal Meeting of the Philosophy of Science Association*, 94–108.

Engelhardt, H. T., Jr. 1976b. "Ideology and Etiology." *Journal of Medicine and Philosophy* 1 (3): 256–268.

Engelhardt, H. T., Jr. 1977. "Treating Aging: Restructuring the Human Condition." *Extending the Human Life Span: Social Policy and Social Ethics*. Washington, DC: National Science Foundation, 35–40.

Engelhardt, H. T., Jr. ed. 1979. *Clinical Judgment: A Critical Appraisal. Proceedings of the Fifth Trans-Disciplinary Symposium on Philosophy and Medicine Held at Los Angeles, California, April 14–16, 1977*. New York: Springer.

Engelhardt H. T., Jr. 1984. "Clinical Problems and the Concept of Disease." In *Health, Disease, and Causal Explanations in Medicine*, 27–41. New York: Springer.

Engelhardt H. T., Jr., 1986. "From Philosophy and Medicine to Philosophy of Medicine." *The Journal of Medicine and Philosophy* 11: 3–8.

Engelhardt, H. T., Jr. 1996. *The Foundations of Bioethics*. New York: Oxford University Press.

Engelhardt H. T., Jr. 2002. *The Philosophy of Medicine and Bioethics: Framing the Field*. Dordrecht, Holland: Reidel.

Engelhardt, H. T., Jr., and Edmund L. Erde. 1980. "Philosophy of Medicine." In *A Guide to the Culture of Science, Technology and Medicine*, 364–461. New York: Free Press.

Engelhardt, H. T., Jr., and K. F. Schaffner. 1998. "Philosophy of Medicine." In *Routledge Encyclopedia of Philosophy*, edited by E. Craig, 264–269. New York: Routledge.

Ereshefsky, Marc. 2009. "Defining 'health' and 'disease.'" *Studies in History and Philosophy of Science Part C: Studies in History and Philosophy of Biological and Biomedical Sciences* 40 (3): 221–227.

Evans, Alfred S. 1976. "Causation and Disease: The Henle-Koch Postulates Revisited." *Yale Journal of Biology and Medicine* 49 (2): 175.

Evans, Alfred S. 1993. *Causation and Disease. A Chronological Journey*. 1st ed. New York: Springer.

Faber, Knud. 1930. *Nosography: The Evolution of Clinical Medicine in Modern Times*. New York: Paul B. Hoeber.

Fagot-Largeault, Anne. 1982. *Médecine et Probabilités*. Paris: Didier Erudition.

Fagot-Largeault, Anne. 2003. "Preuve et niveau de preuve dans les sciences biomédicales." In *La Vérité Dans Les Sciences*, edited by Jean-Pierre Changeux, 215–236. Paris: Odile Jacob.

Feinstein, Avan R. 1967. *Clinical Judgment*. Philadelphia, PA: Lippincott Williams & Wilkins.

Feinstein, Alvan R. 1983. "An Additional Basic Science for Clinical Medicine: I. The Constraining Fundamental Paradigms." *Annals of Internal Medicine* 99 (3): 393–397.

Feinstein, Alvan R. 1985. *Clinical Epidemiology: The Architecture of Clinical Research*. Philadelphia, PA: W.B. Saunders Company.

Feinstein, Alvan R. 1987. *Clinimetrics*. New Haven, CT: Yale University Press.

Fisher, R. A. 1947. *The Design of Experiment Oliver and Boyd*. 4th ed. Edinburgh: Oliver and Boyd.

Fisher, Ronald. 1959. *Smoking: The Cancer Controversy: Some Attempts to Assess the Evidence*. Oxford: Oliver and Boyd.

Fisher, Ronald. 1935. *The Design of Experiments*. Vol. 11. Oxford: Oliver & Boyd.

Fletcher, R H. 2001. "Alvan Feinstein, the Father of Clinical Epidemiology, 1925–2001." *Journal of Clinical Epidemiology* 54 (12): 1188–1190.

Førde, Reidun. 1998. "Competing Conceptions of Diagnostic Reasoning: Is There a Way Out?" *Theoretical Medicine and Bioethics* 19 (1): 59–72.

Foucault, Michel. 1963. *Naissance de La Clinique*. Paris: P.U.F.

Foucault, Michel. 1976. *Histoire de La Folie À L'âge Classique*. Vol. 9. Gallimard.

Foucault, Michel. 1994. "La Naissance de La Médecine Sociale." *Dits et Écrits*, 2: 215–224.

Fulford, K. 1989. *Moral Theory and Medical Practice*. Cambridge: New York: Cambridge University Press.

Gaudillière, Jean-Paul. 2002. *Inventer La Biomédecine: La France, l'Amérique et La Production Des Savoirs Du Vivant, 1945–1965*. Paris: La Découverte.

Gaudillière, Jean-Paul. 2006. *La Médecine et Les Sciences: XIXe-XXe Siècles*. Paris: La Découverte.

Gifford, Fred. 1989. "Complex Genetic Causation of Human Disease: Critiques of and Rationales for Heritability and Path Analysis." *Theoretical Medicine and Bioethics* 10 (2): 107–122.

Gifford, Fred. 2011. *Philosophy of Medicine*. Vol. 16. Amsterdam: Elsevier.

Gigerenzer, Gerd, Zeno Swijtink, Theodore Porter, Lorraine Daston, John Beatty, and Lorenz Krüger. 1990. *The Empire of Chance: How Probability Changed Science and Everyday Life*. Reprint ed. Cambridge: Cambridge University Press.

Giroux, Élodie. 2008. "Enquête de cohorte et analyse multivariée: une analyse épistemologique et historique du rôle fondateur de l'étude de Framingham." *Revue d'Epidémiologie et de Santé Publique* 56 (3): 177–188.

Giroux, Élodie. 2009. "Définir objectivement la santé : une évaluation du concept bio statistique de Boorse à partir de l'épidémiologie moderne." *Revue philosophique de la France et de l'étranger Tome* 134 (1): 35–58. doi:10.3917/rphi.091.0035.

Giroux, Élodie. 2010. *Après Canguilhem : Définir La Santé et La Maladie*. Paris: Presses Universitaires de France—PUF.

Glennan, Stuart S. 1996. "Mechanisms and the Nature of Causation." *Erkenntnis* 44 (1): 49–71.

Goldman, G. M. 1990. "The Tacit Dimension of Clinical Judgment." *Yale Journal of Biology and Medicine* 63 (1): 47–61.

Goldman, G. M. 1991. "Clinical Judgment: Will HAL Take over by 2001?" *Hospital Practice* 26 (5A): 7.

Green, Richard. 1972. "Homosexuality as a Mental Illness." *International Journal of Psychiatry*. 10 (1): 77–98.

Greenland, Sander, and Babette Brumback. 2002. "An Overview of Relations among Causal Modelling Methods." *International Journal of Epidemiology* 31 (5): 1030–1037.

Grene, Marjorie. 1976. "Philosophy of Medicine: Prolegomena to a Philosophy of Science." In *PSA: Proceedings of the Biennial Meeting of the Philosophy of Science Association*, (2): 77–93.

Grene, Marjorie. 1978. "Individuals and Their Kinds: Aristotelian Foundations of Biology." In *Organism, Medicine, and Metaphysics*, edited by S. F. Spicker, 121–136. New York: Springer.

Gross, Fridolin. 2011. "What Systems Biology Can Tell Us about Disease." *History and Philosophy of the Life Sciences* 33 (4): 477–496.

Hacking, Ian. 2000. *The Social Construction of What?* Cambridge, MA: Harvard University Press.

Hennekens, Charles H., Julie E. Buring, and Sherry L. Mayrent. 1987. *Epidemiology in Medicine*. 1st ed. Boston: Little, Brown and Co.

Hesslow, Germund. 1984. "What Is a Genetic Disease?" In *Health, Disease and Causal Explanation in Medicine*, edited by Lennart Nordenfelt and B. I. B. Lindahl, 183–193. Dordrecht, Holland: Reidel.

Hesslow, Germund. 1993. "Do We Need a Concept of Disease?" *Theoretical Medicine and Bioethics* 14 (1).

Hill, A. Bradford. 1953. "Observation and Experiment." *New England Journal of Medicine* 248 (24): 995–1001.

Hill, Austin Bradford. 1965. "The Environment and Disease: Association or Causation?" *Proceedings of the Royal Society of Medicine* 58 (5): 295–300.

Hofmann, Björn. 2001. "Complexity of the Concept of Disease as Shown through Rival Theoretical Frameworks." *Theoretical Medicine and Bioethics* 22 (3): 211–236.

Horwitz, Allan V., and Jerome C. Wakefield. 2007. *The Loss of Sadness: How Psychiatry Transformed Normal Sorrow into Depressive Disorder*. 1st ed. New York: Oxford University Press.

Hottois, Gilbert. 2004. *Qu'est-Ce Que la Bioéthique?* Paris: Librairie Philosophique J. Vrin.

Howick, Jeremy. 2011. *The Philosophy of Evidence-Based Medicine*. Oxford: Wiley-Blackwell, BMJ Books.

Hull, David L. 1978. "A Matter of Individuality." *Philosophy of Science* 45 (3): 335–360.

Humber, James M., and Robert F. Almeder. 1997. *What Is Disease?* Vol. 14. Totowa, NJ: Humana Press.

Illich, Ivan. 1976. *Medical Nemesis: The Expropriation of Health*. New York, Pantheon Books/ Random House.

Jenicek, Milos. 2003. *Foundations of Evidence-Based Medicine*. Boca Raton, London, New York, Washington, DC: The Parthenon Publishing Group/CRC Press.

Johansson, Ingvar, and Niels Lynøe. 2008. *Medicine & Philosophy: A Twenty-First Century Introduction*. Berlin: Walter de Gruyter.

Jonsen, Albert R. 1998. *The Birth of Bioethics*. Oxford: Oxford University Press.

Kahneman, Daniel, Paul Slovic, and Amos Tversky. 1982. *Judgment Under Uncertainty: Heuristics and Biases*. Cambridge: Cambridge University Press.

Kassirer, J. P. 1976. "The Principles of Clinical Decision Making: An Introduction to Decision Analysis." *Yale Journal of Biology and Medicine* 49 (2): 149–164.

Khushf, G. 2007. "An Agenda for Future Debate on Concepts of Health and Disease." *Medicine, Health Care and Philo*sophy 10: 19–27.

Kincaid, Harold, and Jennifer McKitrick. 2007. *Establishing Medical Reality: Essays in the Metaphysics and Epistemology of Biomedical Science*. Dordrecht, Netherlands: Kluwer Academic Publishers.

King, Lester S. 1954. "What Is Disease?" *Philosophy of Science* 21 (3): 193–203.

Kingma, Elselijn. 2007. "What Is It to Be Healthy?" *Analysis* 67 (294): 128–133.

Kingma, Elselijn. 2010. "Paracetamol, Poison, and Polio: Why Boorse's Account of Function Fails to Distinguish Health and Disease." *British Journal for the Philosophy of Science* 61 (2): 241–264. doi:10.1093/bjps/axp034.

LaFollette, Hugh, and Niall, Shanks. 1995. "Two Models of Models in Biomedical Research." *Philosophical Quarterly* 45 (179): 141–60.

Lagiou, Pagona, Hans-Olov Adami, and Dimitrios Trichopoulos. 2005. "Causality in Cancer Epidemiology." *European Journal of Epidemiology* 20 (7): 565–574.

Lane, Christopher. 2007. *Shyness: How Normal Behavior Became a Sickness*. 1st ed. New Haven, CT: Yale University Press.

Ledley, R. S., and L. B. Lusted. 1959. "Reasoning Foundations of Medical Diagnosis: Symbolic Logic, Probability, and Value Theory Aid Our Understanding of How Physicians Reason." *Science* 130 (3366): 9–21.

Lemoine, Maël. 2011. *La Désunité de La Médecine. Essai Sur Les Valeurs Explicatives de La Science Médicale*. Paris: Hermann.

Lemoine, Maël. 2013. "Defining Disease beyond Conceptual Analysis: An Analysis of Conceptual Analysis in Philosophy of Medicine." *Theoretical Medicine and Bioethics* 34 (4): 309–325. doi:10.1007/s11017-013-9261-5.

Lennox, James G. 1995. "Health as an Objective Value." *Journal of Medicine and Philosophy* 20 (5): 499–511.

Lipton, Peter. 1990. "Contrastive Explanation." *Royal Institute of Philosophy Supplement* 27: 247–266.

Lorne, Marie-Claude. 2004. "Explications Fonctionnelles et Normativité: Analyse de La Théorie Du Rôle Causal et Des Théories Étiologiques de La Fonction." Paris, EHESS. http://www.theses.fr/2004EHES0069.

Löwy, Ilana. 1990. *The Polish School of Philosophy of Medicine: From Tyfus Chalubinski (1820–1889) to Ludwik Fleck (1896–1961)*. New York: Springer.

Machamer, Peter K., Lindley Darden, and Carl F. Craver. 2000. "Thinking About Mechanisms." *Philosophy of Science* 67 (1): 1–25.

Mackie, J. L. 1965. "Causes and Conditions." *American Philosophical Quarterly* 2 (4): 245–264.

MacMahon, Brian, and Thomas F. Pugh. 1970. *Epidemiology: Principles and Methods*. Boston: Little Brown & Co. Published in Great Britain by J. & A. Churchill, London http://www.cabdirect.org/abstracts/19712703347.html.

Magnus, David. 2004. "The Concept of Genetic Disease." In *Health, Disease and Illness: Concepts in Medicine*, edited by Arthur L. Caplan, James J. McCartney, and Dominic A. Sisti, 233–242. Washington, DC: Georgetown University Press.

Marcum, James. 2010. *An Introductory Philosophy of Medicine: Humanizing Modern Medicine*. New York: Springer.

Marcum, James A. 2009. "The Epistemically Virtuous Clinician." *Theoretical Medicine and Bioethics* 30 (3): 249–65. doi:10.1007/s11017-009-9109-1.

Margolis, Joseph. 1976. "The Concept of Disease." *Journal of Medicine and Philosophy* 1 (3): 238–255.

Marks, Harry M. 2000. *The Progress of Experiment: Science and Therapeutic Reform in the United States, 1900–1990*. 1st edition. Cambridge and New York: Cambridge University Press.

Mattingly, C. 1998. "In Search of the Good: Narrative Reasoning in Clinical Practice." *Medical Anthropology Quarterly* 12 (3): 273–297.

Meehl, Paul Everett. 1954. *Clinical versus Statistical Prediction: A Theoretical Analysis and a Review of the Evidence*. Minneapolis: University of Minnesota Press.

Meunier, Robert. 2012. "Stages in the Development of a Model Organism as a Platform for Mechanistic Models in Developmental Biology: Zebrafish, 1970–2000." *Studies in History and Philosophy of Science Part C* 43 (2): 522–531.

Moghaddam-Taaheri, Sara. 2011. "Understanding Pathology in the Context of Physiological Mechanisms: The Practicality of a Broken-Normal View." *Biology and Philosophy* 26 (4): 603–611.

Montgomery, Kathryn. 2005. *How Doctors Think: Clinical Judgment and the Practice of Medicine*. 1st ed. New York: Oxford University Press.

Murphy, Dominic. 2006. *Psychiatry in the Scientific Image*. Cambridge, MA: MIT Press.

Murphy, E A. 1966. "A Scientific Viewpoint on Normalcy." *Perspectives in Biology and Medicine* 9 (3): 333–348.

Murphy, Edmond A. 1976. *Logic of Medicine*. Baltimore: Johns Hopkins University Press.

Nagel, Ernest. 1961. *The Structure of Science: Problems in the Logic of Scientific Explanation*. New York: Harcourt, Brace & World.

Neander, Karen. 1991. "Functions as Selected Effects: The Conceptual Analyst's Defense." *Philosophy of Science* 58 (2): 168–184.

Nervi, Mauro. 2010. "Mechanisms, Malfunctions and Explanation in Medicine." *Biology and Philosophy* 25 (2): 215–228.

Nordby, Halvor. 2006. "The Analytic–synthetic Distinction and Conceptual Analyses of Basic Health Concepts." *Medicine, Health Care and Philosophy* 9 (2): 169–180.

Nordenfelt, Lennart. 1995. *On the Nature of Health: An Action-Theoretic Approach*. 2nd Revised ed. Dordrecht, Netherlands: Kluwer Academic Publishers.

Nordenfelt, Lennart. 2000. *Action, Ability and Health*. Dordrecht, Netherlands: Kluwer Academic Publishers.

Nordenfelt, Lennart. 2004. "The Logic of Health Concepts." In *Handbook of Bioethics: Taking Stock of the Field from a Philosophical Perspective*, edited by G Khushf, 205–222. Dordrecht, Netherlands: Springer.

Nordenfelt, Lennart. 2006. *Animal and Human Health and Welfare: A Comparative Philosophical Analysis*. Wallingford, England: CABI.

Nordenfelt, Lennart. 2007. *Rationality and Compulsion: Applying Action Theory to Psychiatry*. Oxford: Oxford University Press.

Nordenfelt, Lennart, and B. I. B. Lindahl. 1984. *Health, Disease, and Causal Explanations in Medicine*. Dordrecht, Netherlands: Reidel.

Offer, Daniel, and Melvin Sabshin. 1966. *Normality: Theoretical and Clinical Concepts of Mental Health*. New York: Basic Books.

Pantazi, Stefan V., José F. Arocha, and Jochen R. Moehr. 2004. "Case-Based Medical Informatics." *BMC Medical Informatics and Decision Making* 4 (1): 19. doi:10.1186/1472-6947-4-19.

Parascandola, M. 1998. "Epidemiology: Second-Rate Science?" *Public Health Reports* 113 (4): 312–320.

Parascandola, M., and D. L. Weed. 2001. "Causation in Epidemiology." *Journal of Epidemiology and Community Health* 55 (12): 905–912. doi:10.1136/jech.55.12.905.

Parascandola, Mark. 2004. "Skepticism, Statistical Methods, and the Cigarette: A Historical Analysis of a Methodological Debate." *Perspectives in Biology and Medicine* 47 (2): 244–261. doi:10.1353/pbm.2004.0032.

Parascandola, Mark. 2011. "Causes, Risks, and Probabilities: Probabilistic Concepts of Causation in Chronic Disease Epidemiology." *Preventive Medicine* 53 (4): 232–234.

Parsons, T. 1975. "The Sick Role and the Role of the Physician Reconsidered." *Milbank Memorial Fund Quarterly. Health and Society* 53 (3): 257–278.

Parsons, Talcott. 1951. "Social Structure and Dynamic Process : The Case of Modern Medical Practice." In *The Social System*, 428–479. Glencoe, IL: The Free Press.

Parsons, Talcott. 1958. "Definitions of Health and Illness in the Light of American Values and Social Structures." In *Patients, Physicians, and Illness: A Sourcebook in Behavioral Science and Health*, edited by E. Gartly Jaco, 165–187. New York: The Free Press.

Pellegrino, Edmund D. 1976. "Philosophy of Medicine: Problematic and Potential." *Journal of Medicine and Philosophy* 1 (1): 5–31.

Pellegrino, Edmund D. 1986. "Philosophy of Medicine: Towards a Definition." *Journal of Medicine and Philosophy* 11 (1): 9–16.

Pellegrino, Edmund D. 1998. "What the Philosophy Of Medicine Is." *Theoretical Medicine and Bioethics* 19 (4): 315–336.

Polanyi, Michael. 1962. *Personal Knowledge: Towards a Post-Critical Philosophy*. London and New York: Routledge & Kegan Paul.

Pörn, Ingmar. 1984. "An Equilibrium Model of Health." In *Health, Disease and Causal Explanations in Medicine*, edited by Lennart Nordenfelt and B. I. B. Lindahl, 3–9. Dordrecht, Holland: Reidel.

Pörn, Ingmar. 1993. "Health and Adaptedness." *Theoretical Medicine* 14: 295–303.

Putnam, Hilary. 1975. *Mind, Language, and Reality*. Cambridge: Cambridge University Press.

Radden, Jennifer. 2003. "Is This Dame Melancholy?: Equating Today's Depression and Past Melancholia." *Philosophy, Psychiatry, & Psychology* 10 (1): 37–52.

Ransohoff, D. F., and A. R. Feinstein. 1976. "Editorial: Is Decision Analysis Useful in Clinical Medicine?" *Yale Journal of Biology and Medicine* 49 (2): 165–168.

Reznek, Lawrie. 1987. *The Nature of Disease*. London and New York: Routledge & Kegan Paul.

Reznek, Lawrie. 1995. "Dis-Ease about Kinds: Reply to D'Amico." *Journal of Medicine and Philosophy* 20 (5): 571–584.

Richardson, J. T. 1995. "The Premenstrual Syndrome: A Brief History." *Social Science & Medicine* 41 (6): 761–767.

Rosenhan, David. 1973. "On Being Sane in Insane Places." *Science*, 179: 250–258.

Rothman, Kenneth J. 1976. "Causes." *American Journal of Epidemiology* 104 (6): 587–592.

Rothman, Kenneth J., and Sander Greenland. 2005. "Causation and Causal Inference in Epidemiology." *American Journal of Public Health* 95: S144–S150.

Rothwell, Peter M. 2007. *Treating Individuals: From Randomised Trials to Personalised Medicine*. Amsterdam: Elsevier Health Sciences.

Ruse, Michael. 1981. "Are Homosexuals Sick?" In *Concepts of Health and Disease: Interdisciplinary Perspectives*, edited by Arthur L. Caplan, H. Tristram Engelhardt Jr., and James J. McCartney, 245–272. Reading, MA: Addison-Wesley.

Ruse, Michae. 1997. "Defining Disease. The Question of Sexual Orientation." In *What Is Disease?* edited by James M. Humber and Robert F. Almeder, 137–171. Totowa, NJ: Humana Press.

Russell, Bertrand. 1912. *The Problems of Philosophy*. London: Williams and Norgate.

Russo, Federica, and Jon Williamson. 2007. "Interpreting Causality in the Health Sciences." *International Studies in the Philosophy of Science* 21 (2): 157–170.

Sackett, D. L., W. M. Rosenberg, J. A. Gray, R. B. Haynes, and W. S. Richardson. 1996. "Evidence Based Medicine: What It Is and What It Isn't." *British Medical Journal (Clinical Research Ed.)* 312 (7023): 71–72.

Sadegh-Zadeh, Kazem. 1999. "Fundamentals of Clinical Methodology: 3. Nosology." *Artificial Intelligence in Medicine* 17 (1): 87–108.

Sadegh-Zadeh, K. 2000. "Fuzzy Health, Illness, and Disease." *Journal of Medicine and Philosophy* 25 (5): 605–638.

Sadegh-Zadeh, Kazem. 2008. "A Prototype Resemblance Theory of Disease." *Journal of Medicine and Philosophy* 33: 106–139.

Sadegh-Zadeh, Kazem. 2011. *Handbook of Analytic Philosophy of Medicine*. Dordrecht, Holland: Springer.

Salmon, Wesley. 1984. *Scientific Explanation and the Causal Structure of the World*. Princeton, NJ: Princeton University Press.

Schaffner, Kenneth F. 1992. "Philosophy of Medicine." In *Introduction to the Philosophy of Science*, edited by M. Salmon, 310–345. Englewood Cliffs, NJ: Prentice Hall.

Schaffner, Kenneth F. 1993. *Discovery and Explanation in Biology and Medicine*. Chicago: University of Chicago Press.

Schaffner, Kenneth F. 1998. "Model Organisms and Behavioral Genetics: A Rejoinder." *Philosophy of Science* 65 (2): 276–288.

Schramme, Thomas. 2007. "A Qualified Defence of a Naturalist Theory of Health." *Medicine, Health Care and Philosophy* 10 (1): 11–17.

Schroeder, S. Andrew. 2013. "Rethinking Health: Healthy or Healthier Than?" *British Journal for the Philosophy of Science* 64 (1): 131–159.

Schwartz, Peter H. 2007a. "Decision and Discovery in Defining 'Disease.'" In *Establishing Medical Reality*, edited by Harold Kincaid and Jennifer McKitrick, 47–63. Philosophy and Medicine 90. Dordrecht, Netherlands: Springer.

Schwartz, Peter H. 2007b. "Defining Dysfunction: Natural Selection, Design, and Drawing a Line." *Philosophy of Science* 74 (3): 364–385.

Sedgwick, P. 1973. "Illness—Mental and Otherwise." *Hastings Center Studies* 1 (3): 19–40.

Severinsen, Morten. 2001. "Principles Behind Definitions of Diseases: A Criticism of the Principle of Disease Mechanism and the Development of a Pragmatic Alternative." *Theoretical Medicine and Bioethics* 22 (4): 319–336.

Simon, Jeremy. 2007. "Beyond Naturalism and Normativism: Reconceiving the 'Disease' Debate." *Philosophical Papers* 36 (3): 343–370.

Simon, Jeremy R. 2011. "Medical Ontology." In *Philosophy of Medicine*. Amsterdam: Elsevier.

Smith, Kelly C. 2001. "A Disease by Any Other Name: Musings on the Concept of a Genetic Disease." *Medicine, Health Care and Philosophy* 4 (1): 19–30.

Sober, Elliott. 1979. "The Art and Science of Clinical Judgment." In *Clinical Judgment: A Critical Appraisal*, edited by Hugo Tristram EngelhardtJr., Stuart F. Spicker, and Bernard Towers, 29–44. Philosophy and Medicine 6. Dordrecht, Netherlands: Springer.

Sober, Elliott. 1980. "Evolution, Population Thinking, and Essentialism." *Philosophy of Science* 47 (3): 350–383.

Sommerhoff, G. 1950. *Analytical Biology*. London: Oxford University Press.

Steel, Daniel. 2008. *Across the Boundaries: Extrapolation in Biology and Social Science*. London and New York: Oxford University Press.

Stempsey, William E. 2004. "The Philosophy of Medicine: Development of a Discipline." *Medicine, Health Care and Philosophy* 7 (3): 243–251.

Stempsey, William E. 2008. "Philosophy of Medicine Is What Philosophers of Medicine Do." *Perspectives in Biology and Medicine* 51 (3): 379–391.

Sulmasy, Daniel P. 2005. "Diseases and Natural Kinds." *Theoretical Medicine and Bioethics* 26 (6): 487–513.

Suppes, Patrick. 1979. "The Logic of Clinical Judgment: Bayesian and Other Approaches." In *Clinical Judgment: A Critical Appraisal*, edited by Hugo Tristram EngelhardtJr, Stuart F. Spicker, and Bernard Towers, 145–159. Philosophy and Medicine 6. Dordrecht, Netherlands: Springer. http://link.springer.com/chapter/10.1007/978-94-009-9399-0_10.

Susser, Mervyn. 1973. *Causal Thinking in the Health Sciences: Concepts and Strategies of Epidemiology*. New York: Oxford University Press.

Susser, Mervyn. 1991. "What Is a Cause and How Do We Know One? A Grammar for Pragmatic Epidemiology." *American Journal of Epidemiology* 133 (7): 635–648.

Szasz, T. S. 1972. "*Bad Habits Are Not Diseases. A Refutation of the Claim That Alcoholism Is a Disease*." Lancet 2 (7767): 83–84.

Szasz, Thomas S. 1960. "The Myth of Mental Illness." *American Psychologist* 15 (2): 113–118.

Szasz, Thomas S. 1984. *The Myth of Mental Illness: Foundations of a Theory of Personal Conduct*. Revised ed. New York: Harper Perennial.

Teira, David. 2011. "Frequentist versus Bayesian Clinical Trials." In *Philosophy of Medicine*, edited by Fred Gifford, 255–297. Handbook of the Philosophy of Science. Amsterdam: Elsevier.

Temkin, Owsei. 1963. "The Scientific Approach to Disease: Specific Entity and Individual Sickness." *Scientific Change: Historical Studies in the Intellectual, Social and Technical Conditions for Scientific Discovery and Technical Invention, from Antiquity to the Present*, edited by A. C. Combie, 629–647. London: Heinemann.

Ten Have, Henk. 1997. "Form Synthesis and System to Morals and Procedure: The Development of Philosophy of Medicine." In *Philosophy of Medicine and Bioethics: A Twenty-Year Retrospective and Critical Appraisal*, edited by Ronald A. Carson and Chester R. Burns, 105–123. Dordrecht, Netherlands: Kluwer. http://link.springer.com/content/pdf/10.1007/0-306-48133-2_7.pdf.

Thagard, Paul. 1998. "Explaining Disease: Correlations, Causes, and Mechanisms." *Minds and Machines* 8 (1): 61–78.

Thagard, Paul. 1999. *How Scientists Explain Disease*. New ed. Princeton, NJ: Princeton University Press.

Thompson, R. Paul. 2011a. "Models and Theories in Medicine." In *Philosophy of Medicine*, edited by Fred Gifford, 115–136. Handbook of the Philosophy of Science 16. Amsterdam: Elsevier.

Thompson, R. Paul. 2011b. "Models and Theories in Medicine." In *Philosophy of Medicine*, edited by Fred Gifford, 115–136. Handbook of the Philosophy of Science 16. Amsterdam: Elsevier.

Urbach, Peter. 1985. "Randomization and the Design of Experiments." *Philosophy of Science* 52 (2): 256–273.

Wachbroit, Robert. 1994. "Normality as a Biological Concept." *Philosophy of Science* 61 (4): 579–591.

Wakefield, J C. 1992. "The Concept of Mental Disorder. On the Boundary between Biological Facts and Social Values." *American Psychologist* 47 (3): 373–388.

Weber, Marcel. 2005. *Philosophy of Experimental Biology*. Cambridge: Cambridge University Press.

Weizsäcker, Viktor von. 1987. *Gesammelte Schriften in Zehn Bänden: 7: Allgemeine Medizin—Grundfragen Medizinischer Anthropologie: BD 7*. 1st ed. Frankfurt: Suhrkamp Verlag.

Whitbeck, Caroline. 1977. "Causation in Medicine: The Disease Entity Model." *Philosophy of Science* 44 (4): 619–637.

Whitbeck, Caroline. 1978. "Four Basic Concepts of Medical Science." *PSA: Proceedings of the Biennial Meeting of the Philosophy of Science Association* 1978 (January): 210–222.

Whitbeck, Caroline. 1981. "A Theory of Health." In *Concepts of Health and Disease: Interdisciplinary Perspectives*, edited by Arthur L. Caplan, H. Tristram Engelhardt, and James J. McCartney, 611–626. Reading, MA: Addison-Wesley.

Williams, Neil. 2007. "The Factory Model of Disease." *Monist* 90 (4): 555–584.

Wimsatt, William C. 1998. "Simple Systems and Phylogenetic Diversity." *Philosophy of Science* 65 (2): 267–275.

Worrall, Jennifer, and John Worrall. 2001. "Defining Disease: Much Ado about Nothing?" In *Life Interpretation and the Sense of Illness within the Human Condition*, 33–55. New York: Springer. http://link.springer.com/chapter/10.1007/978-94-010-0780-1_3.

Worrall, John. 2002. "What Evidence in Evidence-Based Medicine?" *Proceedings of the Philosophy of Science Association* 2002 (3): 316–330.

Worrall, John. 2007a. "Evidence in Medicine and Evidence-Based Medicine." *Philosophy Compass* 2 (6): 981–1022.

Worrall, John. 2007b. "Why There's No Cause to Randomize." *British Journal for the Philosophy of Science* 58 (3): 451–488.

Wulff, Henrik R. 1981. *Rational Diagnosis and Treatment: Introduction to Clinical Decision-Making*. Édition: 2nd rev. edition. Oxford and Boston: WileyBlackwell.

Wulff, Henrik R. 1992. "Philosophy of Medicine—from a Medical Perspective." *Theoretical Medicine* 13 (1): 79–85.

Wulff, Henrik R., Stig Andur Pedersen, and Raben Rosenberg. 1986. *Philosophy of Medicine: An Introduction*. St. Louis, MO: Blackwell Scientific Publications.

Young, Allan. 1997. *The Harmony of Illusions: Inventing Post-Traumatic Stress Disorder*. 1st ed. Princeton, NJ: Princeton University Press.

Zachar, Peter. 2002. "The Practical Kinds Model as a Pragmatist Theory of Classification." *Philosophy, Psychiatry, & Psychology* 9 (3): 219–227.

Zachar, Peter. 2000. "Psychiatric Disorders Are Not Natural Kinds." *Philosophy, Psychiatry, & Psychology* 7 (3): 167–182.

14

PHILOSOPHY OF SOCIAL SCIENCES

Jon Elster (Columbia University) and Hélène Landemore (Yale University)

1. Introduction

This chapter on the philosophy of social sciences addresses problems and themes in the social sciences, where the latter are understood in the specific sense of sciences that have (or that should have) the following minimal characteristics: their object of study is human behavior, and they follow a certain number of methodological principles, including: (1) a marked effort towards analytical clarity; (2) the investigation of causal explanations through the formulation of causal laws or at least causal *mechanisms*; and (3) a subscription to a form of methodological individualism, if an amended one, which puts at the heart of social science the notion of *choice*. We assume that the object criterion and the first methodological criterion of analytic clarity are self-explanatory. In this chapter, we develop the significance and the implications of the two latter methodological principles. Beyond these common traits, styles and methodologies remain varied, and by no means do we suppose that the social sciences can be considered as anything other than a multidisciplinary field. It is a field, moreover, traversed by the tension between two major types of social sciences: qualitative social sciences, which include case studies and narrative approaches, and quantitative social sciences, which include approaches that make use of modeling techniques and statistical analysis.[1]

[1] This dichotomy between qualitative and quantitative sciences is, of course, imperfect. The technique of modeling is not properly quantitative and, inversely, in sociology one finds statistical analyses without an explicative mathematical theory behind them. To the extent that this opposition is relatively transparent and widespread, we nonetheless invoke it here for pragmatic reasons.

After a brief first section discussing the boundaries of what can legitimately be called "social sciences," we discuss three principal themes.[2] The first theme raises the question of the status of laws in the social sciences and, in particular, that of "consequence laws," otherwise known as functionalist explanations. The second theme takes up methodological individualism, as compared to holistic approaches. The last theme concerns hypotheses of rationality and self-interested motivations which, more and more often, figure in social scientific explanations.

In addressing these themes, we take the opportunity to establish three points that seem important to us. First of all, against the Weberian tradition, we defend the idea that there is no difference between so-called hermeneutic approaches and explanatory approaches. Put differently, in our view, to interpret is to explain. Another point concerns social science's reductionist ambition. We believe that far from being destructive, reductionism is enriching and that it is necessary for the social sciences to strive to tie their explanations to concepts and approaches from the natural sciences, in particular biology and neurophysiology. We underline, finally, the importance of the cognitive turn taken by the social sciences since the fundamental works of Kahneman and Tversky on "heuristics" and "cognitive biases," which have notably complicated the opposition between rational and irrational beliefs in showing that certain irrational beliefs are not attributable to the influence of the passions (see notably Kahneman and Tversky, 1974).

2. The Boundaries of the Social Sciences

It is necessary at this point to say a word on the boundaries that can be delineated around the multidisciplinary field of the social sciences. The disciplines traditionally classified within the "humanities"—philosophy (at least as a history of ideas), ancient languages, literature, literary criticism, and certain branches of law—are excluded a priori by the aforementioned criteria of object and method. Although these disciplines have as their object of study the human being in all of its complexity, they are only incidentally interested in the question of choice and do not directly pursue the investigation of causal laws. On the other hand, economics, sociology, political science, certain branches of law (as in the recent discipline of "law and economics," more developed in the Anglo-Saxon world than in continental Europe), history, information and communication sciences, psychology and anthropology, are included within the field of social

[2] For a more general introduction to the themes of the philosophy of the social sciences, see for example Alan Ryan, *Philosophy of Social Sciences* (1979), the anthology by the same author entitled *The Philosophy of Social Explanation* (1973), the anthology by Michael Martin and Lee McIntyre, *Readings in the Philosophy of Social Science* (1994), or Alexander Rosenberg's *Philosophy of Social Science* (1995). On the debates opposing "naturalists" and "interpretivists," see James Bohman, *New Philosophy of Social Science* (1991), David Braybrooke, *Philosophy of Social Science* (1987), Daniel Little, *Varieties of Social Explanation* (1991), Martin Hollis, *The Philosophy of Social Science* (1994), and Jon Elster, *Explaining Social Behavior* (2007).

sciences, insofar as they satisfy, at least minimally, the criteria formulated above. We insist in particular on the artificial and dated character of the distinction between history and social science. Even if causal laws and causal mechanisms are not always explicitly mentioned in qualitative approaches, they do underlie the selection and the description of the reported events and facts. Conversely, quantitative approaches strive to demonstrate statistical regularities that serve as causal explanations (unfortunately often incomplete in our view).

We also exclude, at the two extremes of qualitative and quantitative social sciences, that which we will take the liberty of characterizing as obscurantisms, respectively soft and hard. These exclusions are not entirely symmetric, insofar as the byproducts of hard obscurantism are a lesser evil compared to the by-products of soft obscurantism.[3]

2.1 SOFT OBSCURANTISM

"Soft" obscurantism is closer to literary criticism or even to literature itself (about which the authors offer no value judgment, except to say that it does not constitute social science) than to the qualitative empirical research that literary criticism often claims to be. Despite their creativity and possible power of suggestion, postmodernism, postcolonial studies,[4] subaltern studies,[5] and Kleinian or Lacanian deconstructionism, among others, fit into this category because, and insofar as, they represent sectarianisms based more on common linguistic reflexes than on an investigative principle of universal, inclusive rationality and a search for truth. To the extent that it is impossible to falsify that which does not expose itself to falsification, we refer back to Sokal and Bricmont (1997) and Sokal, Bricmont and Hochstedt (2005) for a definitive proof by absurdity.

2.2 HARD OBSCURANTISM

Hard obscurantism characterizes one part of the research conducted in the "quantitative" social sciences to the extent that measuring exercises, data analysis and modeling

---

[3] Thus hard obscurantism, more characteristic of the American university milieu, does less damage there than does the soft obscurantism that has long paralyzed the French social sciences.

[4] Postcolonial studies represent a current of thought that appeared during the 1970s after Edward Said's (1978) critique of Western constructions of the Orient in his classical work *Orientalism*. Globally, the term refers to studies of the interactions between European nations and their former colonies in the modern era.

[5] Subaltern studies are a historiographic current from India associated with postcolonial studies and postmodernism. This current, developed during the 1980s, attempts to write, or rather to rewrite, the postcolonial or postimperial history of the societies of the region of South Asia and, more generally, of developing countries, from a point of view centered on popular masses rather than on elites. The term "subaltern" (a reference to the work of Antonio Gramsci) applies to persons and, by extension, groups of persons of inferior status and rank, whether because of their sex, ethnicity, religion or social class. The founder of subaltern studies is Ranajit Guha, the author of a monograph entitled *Elementary Aspects of Peasant Insurgency in Colonial India* (1983).

have no more than a thematic relation to the reality of the human phenomena that they are meant to explain or predict. This critique targets, in part, rational choice theory, including theories of individual decision-making, game theory and social choice, as well as certain practices of the statistical branches of economics and political science.

As far as statistical analysis is concerned, it is worth emphasizing the fact that it is closer to a profession or a technique, in which precision and value increase with practice and experience, than a science properly understood. In this sense, the quantification of social phenomena, and the simple act of identifying a regularity in a mass of data, are not proof of good social science. There are many things at play, indeed, in the selection of variables and the interpretation of data. Moreover, it is necessary to denounce the common practices of adjusting curves to the data ("curve fitting"), or of selectively choosing data to fit them to curves ("data snooping").[6] The measuring rod against which practitioners of statistical analysis in social science should assess their results is prediction, or "postdiction," which consists of predicting one part of the observations obtained on the basis of the analysis performed on another part of the same observations. One way (perhaps impractical) to prevent *post hoc* curve or data manipulation would be to force researchers to present their datasets and hypotheses to the journals in which they intend to publish their results, say, two years before the submission of those results.

Modeling is, the majority of the time, a deductive exercise that begins with hypotheses and concludes with predictions.[7] It exists in at least two forms in the social sciences: rational choice theories and, more and more frequently, evolutionary models that do not presuppose rational or even intentional responses. We shall limit ourselves here to touching on certain limits of rational choice theory models.[8]

2.3 IS RATIONAL CHOICE THEORY THE SCIENCE OF CHOICE?

Rational choice-type modeling is dominant in economics, where it has and will continue to be applied far afield of the traditional domain of consumer and producer behavior. At present there exists an economics of suicide, marriage, religious practice and, more generally, all activity that implies, in one form or another, a choice. The tendency is towards the disappearance of the conceptual difference between intentional action and rational action. If an opportunity for choice exists, it is assumed that it will be exercised in a rational way. Does rational choice theory amount to a science of choice, as it aspires to be?

First, it is worth mentioning that for some of its more eminent practitioners (Reinhart Selten and Ariel Rubinstein, for example), rational choice theory's primary

[6] Data snooping is different from data mining, which can occasionally be justifiable. See Freedman, 2009, p. 64.

[7] Still, simulation has come to play an increasing role in modeling in the social sciences, notably in economics, as with the Santa Fe school, which studies, for example, the functioning of artificial markets.

[8] For a more thorough study, see Elster (2007, chapter 12) and Landemore (2004); for a defense of the "virtual causality" at work in rational choice theories, see Pettit (2004).

goal is not to predict or to explain the behavior of real individuals, but rather like a type of "science fiction" to study of ideally rational agents (even if theirs is a limited rationality) who have never existed, nor ever will.[9] As for social choice theory (the axiomatic study of voting mechanisms), twenty years ago, one of the most important journals in economics, *Econometrica,* established a moratorium on the publication of articles within this domain, judged to be excessively mathematical and lacking any evident link to reality.

Next, there are a number of problems with the propositions advanced by rational choice theorists. For example they attribute to agents motivations and cognitive capacities that they do not have, such as a disposition towards exponential time discounting, or the faculty of resolving complex equations or probability calculations at the instant of decision-making. Other limitations of these models have to do with their treatment of the uncertainty surrounding real actions by real agents, or the hypothesis underlying these models that deviations from the norm of rationality are temporary, or cancel out in the aggregate. If the goal of social science should continue to be the explanation of real social phenomena, useless complications introduced by certain currents of rational choice theory seem to us a failure in so far as they do not lead to a plausible description or to any prediction.

2.4 THE COGNITIVE TURN

In this regard, one must welcome the salutary turn brought by the foundational work of D. Kahneman and A. Tversky on judgment under conditions of uncertainty (1974), which shows how human decisions are guided by all sorts of "heuristics" and often distorted by cognitive "biases" that are more or less anchored in human beings. This work stands at the forefront of what today is called "behavioral economics," a subfield of economics that studies and documents the limits of human rationality and integrates the contributions of psychology and cognitive sciences to neo-classical economic models.[10] Despite criticism to the contrary by the separate tradition of bounded rationality, developed after Herbert Simon's works, the theory founded by Kahneman and Tversky is not just a simple complication of rational choice theory, in the sense in which utility maximization might be performed under increasing

---

[9] See for example Ariel Rubinstein's afterword (entitled "Final Thoughts") in *Modeling Bounded Rationality,* MIT Press, 1998. Here, Ariel Rubinstein defends his abstract and mathematical approach of "bounded rationality," which was initially theorized by Herbert Simon (1978) in hopes of turning the field of economics in a more empirical and psychological direction (such as that of Kahneman and Tversky). Rubinstein responds to Herbert Simon's objections by asserting that, for him, the goal of economic science is neither to predict nor to prescribe human behavior. In so doing, he makes an epistemological thesis out of the disciplinary habits omnipresent in economic theory.

[10] This is not the same as behaviorist economics, far from it. Behaviorism is, in brief, the doctrine by which human psychological events must be explained by means of observable behaviors and objective phenomena, rather than by intentions, beliefs, desires, and other non-observable mental states. Behavioral economics, by contrast, moves away from a certain behaviorism present in traditional economics (in revealed preference theory, for example) in order to recover the concepts of folk psychology.

uncertainty constraints and with more limited cognitive capacities.[11] Of course, a certain number of non-Bayesian theories of risky and uncertain choice, such as prospect theory, remain theories of optimization. But, in general, behavioral economics calls on entirely different resources for human choice than an optimizing rational calculus, such as "heuristics" and emotions.[12] An important result established by the scholarship in this domain merits mention: the idea that the sources of irrationality are not just "hot" (to be found in the emotions, also known as the "passions") but equally "cool"—to be found in the existence of systematic cognitive biases, such as the phenomena of *anchoring* or *framing*.[13,14] For these reasons and more, it seems to us that rational choice theory fails in its pretention to be the science of choice. A charitable interpretation renders it at best one of the tools of a pluralist, or, in Phillip Pettit's terms, "ecumenical" methodology.

3. The Status of Laws in the Social Sciences

Having defined the "explicative" mission of the social sciences, we now present the two major types of laws in social science: causal laws and consequence laws. Judging by the great indeterminacy and almost non-existent predictive value of social scientific causal laws, we suggest that the goal of social sciences could usefully be brought back to the more humble objective of identifying causal "mechanisms" underlying facts, events, and human choices. As for consequence laws, we defend their importance in social science insofar as they meet two conditions. First is the necessity of establishing the existence of a *feedback loop* between the beneficial consequences of a behavioral pattern and the decision by the individual to undertake the actions that this pattern entails (whether the motivation is conscious or unconscious). Another requirement is that of explaining the first occurrence of the behavior yielding beneficial consequences by something other than those beneficial consequences, namely by a causal law or a causal mechanism.

---

[11] See Gigerenzer and the Berlin ABC Group for a critique of behavioral economics as an optimization under constraints and a defense of the program of "bounded rationality" as a different and more promising theoretical alternative in terms of description and explanation of human choices (see for example Gigerenzer et al., 2001, and Gigerenzer, 2008).

[12] As we see it, the crucial difference between behavioral economics and bounded rationality theory is that the first continues to measure cognitive "biases" in relation to the ideal rationality of homo economicus while the second explicitly (and in some ways problematically) frees itself from this normative anchoring.

[13] The anchoring phenomenon corresponds to the tendency of human beings to rely too heavily on a particular piece of information, often because it is the only one available, in making a decision or forming a belief. For example, my belief about the number of people living in the city of London may be anchored in my knowledge of the number of inhabitants of the city I live in. If my point of comparison is a city larger than London, I should have the tendency to overestimate my response, and if it is smaller, to underestimate it.

[14] The framing phenomenon corresponds, for example, to the fact that people respond differently to the same question depending on the way it is formulated.

3.1 EXPLANATION IN THE SOCIAL SCIENCES

The principal mission of the social sciences is that of explaining social phenomena. In this regard, we maintain that every explanation is causal, in the sense that to explain a phenomenon (an *explanandum*) is to specify an antecedent phenomenon that caused it. According to Hume's billiard-ball model, one can thus say than a certain event A is caused by a certain event B, in the same way that a billiard ball is pushed by another ball.

As a first, rough approximation, one can say that social scientists aim to produce causal explanations on the model of the causal explanations found in the natural sciences. They aim to identify causal laws from which the explanandum can be logically deduced. According to this deductive-nomological (or hypothetico-deductive) model, social scientists choose a theory, that is, a set of mutually interrelated causal propositions, then specify a hypothesis that applies the theory to a given question, and finally show that the *explanandum* follows logically from the hypothesis.[15] A causal explanation thus understood will be more or less accepted depending on how well, relative to rival explanations, it can accommodate already observed facts as logical consequences, as well as help predict "new facts," that is fact observed after the causal explanation has been formulated.

This conception of an explanation, which is both causal and nomological, would seem to apply equally to interpretations, that is, to "intentional" explanations of behavior, insofar as these intentions may be causes of human action. Hence, on that view, there is no reason to distinguish between explanation and interpretation. Even if, following Weber, one has often contrasted *Verstehen* and *Ekläre*n as the respective task of the human or "spiritual" sciences (*Geistewissenschaften*) and that of the natural sciences (*Naturwissenschaften*), we think that, for the social sciences as we have defined them above, interpretation is never other than a type of causal explanation. If the task of natural science is not, to paraphrase Weber, to "interpret" the behavior of cells, it is the task of social science to explain the behavior of individuals by interpreting it. Interpretation is none other than a particular application of a deductive-nomological model. In brief, to interpret is to explain, through causes.

Incidentally, this thesis also applies a priori to the interpretative approach of works of art, particularly literary (Elster, 2007, chap. 14). It is common in literary criticism not to consider the intention of the author relevant, which gives commentators great freedom of interpretation, sometimes to the point of rendering the exercise somewhat gratuitous (or creative, depending on the point of view). We propose that a hypothesis positing the intentionality of the creator of a work, as it so happens a text, can nonetheless serve as a theoretical anchor for the plausibility of commentary on a work, defining a "fact of the matter" with respect to which proposed interpretations are more or less plausible, and not merely more or less coherent. In literature, as in social science, the interpretative principle should thus be anchored in a hypothesis

[15] See chapter 1 on explanation.

about the intentionality of the principal agent. For example, the simple fact that a text is consistent with a numerological regularity should not allow the reader to think that the author was conscious of and wanted her readers to perceive it, any more than "data snooping" and "curve fitting" in the social sciences authorize us to believe that observed regularities have a causal significance. One can parallel the opposition between the kingdom of efficient causes and the kingdom of final causes in Leibniz, each of which offers, independently of the other, a satisfactory explanation for human action (Leibniz, 1969, p. 588) and two explanatory logics to which every text can be subjected.[16]

3.2 CAUSAL LAWS

The natural sciences, particularly physics and chemistry, offer law-based explanations. The majority of laws in the natural sciences describe the evolution of a system over the course of time: the truth of the first proposition at a given moment makes it possible to infer the truth of the second proposition at a later moment. For example, when we know the positions and the speed of the planets at a given instant t, we can deduce and predict, thanks to the laws governing the movement of planets, their position at time $t + n$. This type of law is deterministic: given the antecedents, there is only one possible consequent.

The social sciences have always had the ambition of producing causal laws in the same way as the natural sciences. Unfortunately, social science offers few examples of such laws. In general, the social sciences predict that one antecedent can generate several consequences, and even that several antecedents can generate all these consequences. Put differently, the ideal model of a sole or multiple antecedents producing a sole consequence is rarely found in the social sciences, hence the indeterminate character of their predictions.

Some efforts have been made to model this type of relation, with multiple potential antecedents and consequences, by means of statistical methods. From this perspective, the social scientific ambition might simply be to establish general laws, to wit, purely statistical "macro laws" with little or even no predictive value.[17] Kincaid, for example, cites in support of this position the case of laws in biology that limit themselves

---

[16] To choose an example, one can find two different logics behind the actions of young Pip who, at the start of Charles Dickens's *Great Expectations*, comes to the aid of a criminal in flight. The first explanation lies in the logic of verisimilitude: it is out of fear that the young boy obeys the orders of the ex-convict. The other explanation refers to the structural demands of the plot: Pip's help is necessary so that Magwitch will remember it and later try to reimburse the debt he owes him and thus provide the spark of the plot (Rimmon-Kenan, 1983, pp. 17–18).

[17] Weak predictive value is characteristic of laws that specify at least the "sign" of the expected phenomenon, even if they do not predict the exact amplitude of the change. Thus, the law of supply and demand in economics predicts that as the price of a good increases, with all else equal, the demand for the good will diminish, although it does not specify how much the total decrease will be. See chapter 15 on philosophy of economics.

to establishing relationships of correlation between two phenomena "controlling" for potential rival causal factors at the level of methods of statistical regression (Kincaid, 1994). Even so, statistical explanations establishing correlations are incomplete in and of themselves, because ultimately one must rely on causal intuitions, not only regarding the mechanism at work, but also regarding the variables that must be controlled for. Without this, we are dealing with no more than correlations without explicative value. In effect, correlation is not causation. Thus, a law cannot be called explicative or causal except insofar as it identifies a precise mechanism accounting for the relation between a particular event and its presumed consequence.

In the face of social science's failure to produce causal laws that are not purely statistical and that could have real predictive value, Jon Elster has proposed assigning social sciences the humbler tasks of identifying "mechanisms" and building itself up as a toolbox of such mechanisms. A mechanism is defined by Elster as a frequent and easily identifiable causal pattern that is triggered in generally unknown conditions and with indeterminate consequences. The property of a mechanism is to explain, but not to predict.[18] Some well-established examples of mechanisms are the reduction of cognitive dissonance by the "sour grapes" effect, through which one ceases to find desirable that which one could not obtain, or motivated beliefs, which make us believe that which is convenient for us. Many mechanisms are expressed by popular proverbs, such as "Out of sight, out of mind," or "Opposites attract," or "Like father, like son." These mechanisms often have mirror opposites. One can thus match up the preceding mechanisms with their inverses: the "forbidden fruit" effect, which arouses the desire for that which one cannot have, and the mechanisms expressed by the proverbs, "Absence makes the heart grow fonder," "Birds of a feather flock together," or "The miser's son is a spendthrift." Mechanisms, then, are of two types, which we can call A and B. Type A-mechanisms produce a particular effect to the exclusion of another: thus, the "sour grapes" mechanism and its opposite, the "forbidden fruit" mechanism, make an object more or less, but not more *and* less desirable at the same time. Type B-mechanisms, on the other hand, yield two simultaneous effects whose net effect is not possible to determine: for example, a mechanism like the "tyranny" effect, at work as a government intensifies repression of its opponents, produces at the same time hatred and fear among individuals, which can in turn bring about submission or revolt depending on which effect overcomes the other.

One can easily combine "atomic" mechanisms, to call them thus, into more complex causal mechanisms. Let us imagine that one seeks to explain the impact of democracy on the importance of religion in a given country. Over the centuries, elites have claimed that the disappearance of political authority would result in the weakening of religious authorities, by a spillover effect. Inversely, Tocqueville always maintained that democratic peoples would seek in religion a compensation for the loss of political authority, by a compensation effect. According to him, critics of democracy were

[18] On the question of the asymmetry between explanation and prediction, see chapter 1 on explanation.

mistaken in considering only the opportunities opened by the loss of political authority, and not the desires towards those opportunities, whereas it is possible that an increase in the field of opportunities might not be accompanied by a comparable increase in desires. The two mechanisms—the spillover effect and the compensation effect—can thus be combined to form a general mechanism which can be formulated in the following manner: if the influence of democracy on religion is mediated more by the compensation effect than by the spillover effect,[19] then democratic societies will be religious; if the negative effect of democracy on desires (mediated by religion) is sufficiently strong to trump the positive effect of democracy on opportunities, then democratic citizens will behave in a moderated way.[20]

3.3 CONSEQUENCE LAWS

A second type of laws in the social sciences consists of explaining a phenomenon, not by an antecedent phenomenon but by a subsequent phenomenon, hence the name "consequence laws,"[21] given to a certain type of functional explanations or explanations by function. Not all functional explanations, however, deserve the title of consequence laws.

Functional explanations that limit themselves to indicating the production of beneficial consequences and merely assume that these consequences suffice to explain the behavior that has them for consequences are not scientific. When the *explanandum* is a unique event or fact, this type of explanation fails for one evident metaphysical reason: a cause should precede its effect and an event cannot be caused by a later event. To take an example from biology, one cannot explain the appearance of a neutral or a dangerous mutation by the fact that that mutation was a precondition for another, advantageous mutation.

When the *explanandum* is an institution or a recurring behavioral pattern (and not a one-time action or behavior), the functional explanation may or may not be valid. Insofar as an explanation does not specify a particular feedback mechanism detailing the corresponding impact of the consequence of a behavior on that behavior, we must hold this explanation to be invalid. Certain anthropologists, for example, have maintained that revenge-seeking behavior against one's enemies can have various types of beneficial consequences, from population control to the provision of an alternative punitive mechanism in countries in which the State is weak.[22] But even supposing that these beneficial consequences are, in fact, produced, it is possible that they might have occurred in a fortuitous or accidental manner. To prove that these advantages are not accidentally produced, and that they effectively reinforce the revenge-seeking behavior that caused them, it is necessary to demonstrate the

[19] Tocqueville (1993) speaks of a "carrying" of the effect of one sphere onto another.
[20] See Elster (2009a) for a more in-depth study.
[21] According to G. A. Cohen's terminology (1982), with which we nonetheless differ in substance.
[22] See Elster (2007, chapter 22), for additional examples.

existence of a feedback mechanism. Nevertheless, we should note that even if this feedback mechanism is proven to exist, the explanation is still not complete as long as the initial occurrence of the behavior is not explained by anything other than that mechanism, that is to say, either by a causal law or by a mechanism.

In the end, a viable functional explanation can be defined as follows:

> An institution or a behavioral pattern X is explained by its function Y for the group Z if, and only if
>
> (1) Y is an effect of X.
> (2) Y is beneficial for Z.
> (3) Y is not intentionally pursued by the agents that produce X.
> (4) Y (or, at least, the causal relation between X and Y) is not recognized by the actors of Z.
> (5) Y supports or reinforces X by a retroactive causal loop that goes through Z.
> (6) A distinct mechanism W explains the initial production of X.

If it seems doubtful that a perfect functional law exists in contemporary social science, let us consider two that come close the ideal. The first, in economics, is the explanation, by economists of the Chicago School, of firm profit-maximizing behavior as a result of "natural selection" of firms by the market. Here,

X = behavior rules guiding firms' actions
Y = profit maximization
Z = firms
W = technological innovation

According to the functionalist explanation, only the firms in group Z that unconsciously follow behavioral rules X, whose unintended result is to assure maximization of profit Y, survive competition in the market. The behavioral rules in question spread among the firms in group Z, whether because surviving firms absorb the others, or through imitation. The initial occurrence of behavior X is produced following event W, for example a technological innovation in managing production. The only problem with this apparently complete functional explanation is that it is difficult to conceive an analogy to natural selection in the world of firms that is sufficiently precise as to yield refined predictions.

Another example of a successful functionalist explanation can be borrowed from political science. It consists of Morris Fiorina's explanation of the excessive growth of American bureaucracy as a result of the fact that the career of Congress members benefits from the unplanned growth of that bureaucracy. Here,

X = growth of governmental agencies (which obtain their budges from Congress and which respond, consequently, to Congressional demands for help with their voters)

Y = reelection of members of Congress who please their voters by intervening with governmental agencies on their behalf
Z = members of Congress
W = ?

Because they spend more time at the service of their voters, members of Congress often delegate, even if unintentionally, decision-making power and resources to agencies in such a way that voters interact more and more with governmental agencies. Thus, a feedback effect from representatives' careers on the growth of bureaucracy is produced in two ways:

1. The growth of the bureaucracy results in more demands on the part of voters and, in consequence, more occasions for members of Congress to seek a role as mediators.
2. Playing the role of mediator deters members of Congress from their legislative and supervisory roles, such that they end up delegating more decision-making power to administrative agencies.

The result is the selective survival of the most "adapted" members of Congress, that is, those whose electors and interest groups can assure them enough votes to raise them from the marginal reelection threshold. New members of Congress learn by example that service to voters pays off at the moment of elections.

We note that, in Fiorina's analysis, there is not one, but two mechanisms that produce the feedback loop, or reinforcement effect, between Y and X. The initial mechanism W, which leads to the first occurrence of X, is not specified, but one could imagine a plausible explanation like the external shock of war, for example, bringing about the initial increase in the number of government agencies and their personnel.

When all of the criteria are satisfied but criterion (4)—a lack of awareness among agents that behavior X is beneficial to them—is missing or disappears as a result of increased awareness, it often makes more sense to talk about "filter explanations." [23] The process is that of "artificial selection" where intelligent agents are capable of filtering mutations in the most advantageous manner: accepting an unfavorable mutation allowing access to a global maximum in the long term, and refusing a beneficial mutation that does not lead to a local optimum. In this case, members of Congress can continue to improve their chances of reelection by intervening with administrative agencies on behalf of their voters.

Finally, as long as criteria (1) to (4) and (6) are met, but not criterion (5), one should speak of an explanation as an "invisible hand" phenomenon.[24] The self-interested exchanges between my butcher and me produce an optimal situation for the two of us

[23] See Hardin (1980, p. 756).
[24] See Hardin (1980, p. 756).

without, nonetheless, there being reinforcement of his behavior or mine thanks to the beneficial effects of our respective egotisms. In this case,

X = behavioral rules aiming to maximize my individual profit.
Y = production of a Pareto optimal situation.[25]
Z = economic agents.

Let us note that here the invisible hand *explanandum* is Y (the Pareto optimal situation), not X. We are dealing here with a classic causal law and not a consequence law, of which functional or filter explanations are a type.

3.4 THE FUTURE OF THE SOCIAL SCIENCES

Are the social sciences' aspirations to predictivity, determinism, and precision in its predictions capable of ever being fulfilled?

The incorporation of neuroscientific discoveries into the social sciences will, in the future, no doubt place psychology on more solid foundations and permit the resolution of certain current controversies. It has been suggested, for example, that brain scanners confirm or at least support the recent hypothesis advanced by economists (for reasons of mathematical simplicity) that individuals have a quasi-hyperbolic discount function of time and not a hyperbolic one.[26]

There are two reasons why the social sciences are currently incapable of predicting or explaining in the strict sense. One is that, because of certain beliefs and preferences, action may to a certain degree remain indeterminate, that is to say, unpredictable. In the decision-making under conditions of strong uncertainty or complexity, people resort to all sorts of decision-making rules, too numerous to ensure the determination of one particular outcome.

The second reason is our limited comprehension of preference-formation mechanisms. Individuals are subject to different competing inclinations whose relative strengths, in a given situation, are indeterminate. If someone threatens you, will it elicit a "fight" or a "flight" response? If a country transitions from dictatorship to democracy, will its citizens, now liberated from one form of political authority, also reject religious authority or, on the other contrary, seek it out with greater ardor? We are for the most part unable to answer these questions in advance, even if we can identify the mechanisms at work after the fact.[27] It is certainly difficult to identify the conditions that set off ("triggering conditions") these reactions.[28] One example is the

---

[25] A Pareto optimal situation is a situation in which it is impossible to improve the well-being of one person without diminishing that of another person.

[26] See McLure et al. (2004 and 2007).

[27] The fact that the idea of a mechanism leaves a great margin of indeterminacy does not imply that we should take this as evidence of an objective indeterminism, call it "liberty" or "free will" or any other name. On this point, we think that the approach of the social sciences must remain agnostic.

[28] See Elster (2007, chapter 2) for more details.

case of an individual who overpays for a seat to a mediocre performance. It is a situation of cognitive dissonance,[29] in which we might expect that the effect by which the individual takes her desires for realities ("wishful thinking") will trump the effect by which she changes her beliefs. Given that the individual cannot easily convince herself that she has not spent a significant amount, one can predict that she will choose instead to decide that the show is exceptional. According to the writer Arthur Miller, the increase in ticket prices is thus what explains the multiplication of standing ovations for Broadway shows.

4. Methodological Individualism and the Question of Reductionism

Methodological individualism (MI) consists in affirming that social phenomena should be explained with reference to the choices, desires, and beliefs of individuals and not with reference to supra- or infra-individual entities, for example, institutions or genes. In the first section, we will elaborate on this definition by offering what is in our opinion its most plausible and most defensible interpretation, notably by dissociating MI from absurd interpretations like atomism or related, potentially correlated, but conceptually independent positions, such as political or ethical individualism. We will respond to the objection that methodological individualism is not a valid methodological principle because it ignores the existence of collective phenomena that defy explanations in terms of individual rationality. We take seriously, though, certain ideas developed by theorists of what one might call the collective mind ("we-thinking") and, more generally, the field of social epistemology. Finally, we examine the relation between MI, the reductionist ambition in the social sciences and, in particular, psychological reductionism. As we see it, MI implies reductionism in the social sciences, but the question of whether the ideal is to formulate explanations of individual choice in the language of revealed preference theory or in that of folk psychology, that is, the beliefs and internal desires of the individual, remains an open question. We lean towards the latter position.

4.1 DEFINITION

MI is a principle both central to and hotly contested within social science. The two great controversies surrounding its definition (which took place during the 1950s and 1980s, respectively) have at least had the merit of clarifying some points and positions that we take for granted in what follows.[30]

---

[29] A cognitive dissonance situation is a situation in which there is a tension between what one knows to be true and what one wishes to be true.

[30] For the precise genealogy and details of these (Anglo-Saxon) controversies, see the article "Methodological Individualism" in the *Stanford Encyclopedia of Philosophy* (Heath, 2009).

Methodological individualism signifies that in principle, explanations in social sciences should make reference exclusively to individuals and to their actions.[31] In this, MI is opposed to methodological holism, which attempts to explain social phenomena by reference to aggregates like the State, the nation, the family, or the firm. Contrary to what has been suggested by Durkheim and reaffirmed, albeit with various amendments, by what one might call the French school of sociology (represented by Marcel Mauss, Pierre Bourdieu,[32] and Louis Dumont, among others), there are no "social facts" that act in the world and move individuals, no more than exist social objects with intentions different from those of the individuals of which they are composed.[33]

One can advance at least two reasons for which the social sciences cannot assume that aggregates are unified actors.

A. The Problem of Aggregating Individual Preferences into Coherent Social Preferences

This problem has been formalized by Arrow's impossibility theorem (1950), according to which there is no social choice function conforming to reasonable criteria that permits aggregating individual preferences into social preferences, so long as there are three or more options. These reasonable criteria are: universality or the non-restriction of the domain of preferences (the requirement that the social function be defined by the total profile of logically possible preferences), non-dictatorship (according to this criterion, no individual can impose his preferences, independently of the preferences of others), unanimity (which demands that, when all individuals have the same preferences, the social choice function should match these preferences to society); independence of irrelevant alternatives (according to which the relative ranking of two alternatives depends solely on their relative position for the individual, and not on the ranking of third alternatives; if one considers only a subset of options, the function should not lead to another ordering of the subset). Arrow's theorem is itself a generalization of Condorcet's paradox, which references the possibility of preference cycling in elections, that is, the fact that any alternative can be chosen, depending on the pairs between which the choice is structured at the start. For Condorcet, no *simple* system could guarantee this coherence. Arrow proved that, conditional on acceptance of the four hypotheses just mentioned, there is *no* system guaranteeing the requisite coherence.

The implication of Arrow's impossibility theorem is less dramatic than it has been made out to be, notably regarding the possibility and meaningfulness of democracy.[34]

[31] The principle of methodological individualism formulated by Popper is the following: "[T]he task of social theory is to construct and to analyze our sociological models carefully in descriptive or nominalist terms, that is to say, in terms of individuals, of their attitudes, expectations, relations, etc.—a postulate which may be called 'methodological individualism'" (Popper, 1945, p. 136).

[32] See Bourdieu (1979).

[33] See Quinton 1975, p. 17.

[34] See, for example, Mackie (2003) for a refutation of the objections to the possibility of democracy that the political scientist Riker (1988) derives on the basis of Arrow's theorem.

Nonetheless, the theorem conclusively establishes that there is no unique, unambiguous translation of individual wills and preferences into the "will" or preference of the group that they constitute. In terms of predictive power, the implications of Arrow's theorem for MI are twofold. First, it means that it is necessary to specify the mode of aggregating individual preferences before it is possible to talk about a collective will or decision. Second, it is necessary to ensure that given a particular mode of aggregating individual preferences, the results will not be cyclical (for example, by verifying that preferences are actually single-peaked). If this condition is not met, one can predict nothing, and it is difficult to give content to the notion of group will.

B. The Collective Action Problem

How can one be assured that collective action will take place when agents have, or may have, potentially divergent private interests? The Prisoner's Dilemma embodies this problem at the small-group level, whereas the Tragedy of the Commons illustrates it in the case of collective action involving a large number of people.[35,36] The collective action problem arises regardless of the presence of strategic interactions—that is, regardless of whether a particular individual's action has an impact on the well-being of others or not. In both types of situations, even though each individual has an interest in all others behaving in the common interest, each one also has an individual interest in "free-riding" themselves, on the condition that the others do not do the same. This problem, very important in social science since the foundational work of Paul Samuelson (1954), Anthony Downs (1957), Mancur Olson (1965), and Garret Hardin (1968), has made it impossible straightforwardly to attribute wills or intentions to collective entities or institutions like the "proletariat" or "big capital." The very possibility of collective action must overcome the problem of individual incentives and, in particular, the free-rider problem. The advantage of applying MI to the social sciences is that of being able to avoid analysis that commits the error of postulating an intention where there is no intentional actor.[37]

In cases where practitioners in social science invoke supra-individual entities in the way ordinary people commonly do, this may be an example of inconsequential linguistic approximation, or of an inevitable alternative in the absence of data or more

---

[35] The Prisoner's Dilemma is a famous example in game theory, in which the players are two prisoners whose dilemma consists of having to decide if it is better to confess to the police for a crime committed or to keep quiet. Because the prisoners are interrogated separately by the police, they cannot coordinate deliberately on the optimal strategy, which would consist of each keeping quiet (in this case, each receiving the minimum sentence). Since they are in doubt, and since each has an incentive to confess if the other keeps quiet (in which case the prisoner who confesses is released by the police while the other prisoner is heavily punished), they are both rationally led to confess, condemning each to a heavy sentence. In this game, it is hypothesized that each player tries to maximize his own utility.

[36] The tragedy of the commons (public goods) formalizes a situation in which several people are competing for access to a limited resource (for example, natural resources). Each has an individual interest in overconsuming the resource in question, which leads to the disappearance of the public good.

[37] Elster (1982), p. 452, and Elster (1989b).

individualistic theories. Thus, insofar as social aggregates are the object of individual beliefs or desires, one cannot always substitute them for co-extensive individual referents, any more than the truth of the sentence, "He believes that Venus is the morning star" implies that of "He believes that Venus is the evening star" (although Venus is, indeed, both). In the sentence, "The United States fears Iran," the reference to a collective entity can be broken down into assertions about the fears of individual Americans. The first part of the sentence makes no sense unless it is broken down in this way. The second component, on the other hand, Iran, resists attempts at such decomposition. That which Americans fear is in fact a collective entity with its own goals, not a particular collection of Iranian citizens with heterogeneous goals aggregated at the national level and set in action by particular individuals. Thus, one can make this minimal concession to methodological holism: insofar as individuals have beliefs and desires about social aggregates, these should be part of explanations of their behavior. On the other hand, it is not scientific to ascribe desires and beliefs to supra-individual entities. Thus, MI does not mean that social science can in principle eliminate all references to social entities, collectives, or systems. To the extent that these concepts are part of current vocabulary, they are indispensable for analysis, which is not the same thing as saying that they can be used as explanatory factors.

In this regard, it is important to emphasize that MI should not be confused with atomism, that absurd position that no one, aside perhaps from Leibniz, has ever defended, according to which the world is made up of individuals with no relation to each other. Nor should MI be confused with methodological atomism, which would have us disregard the existence of interdependence among individuals. MI has often been suspected of not being aware of the interrelational and intersubjective dimension of social phenomena, and of being incapable of registering social interactions, for which a holistic approach would seem more appropriate. This is a misunderstanding. Indeed, MI is a natural fit for considering relationships among individuals in a way that is not conceivable for holism, in that the holist actually erases the differences between individuals, and thus their relational potential. Game theory, for example, is in fact neither feasible nor useful as a tool for studying strategic relations between individuals unless it is grounded on methodological individualism, and not holism.

4.2 THE ANTI-SINGULARIST OBJECTION

The position defended here has been called "singularist" by Philip Pettit (who borrows the term from Margaret Gilbert). According to Pettit, methodologically speaking, the correct position for the social sciences to take is anti-singular. According to the definition given by Pettit, anti-singularists deny that a group's actions coincide with—that is, are reducible to—individual action. The idea is that groups may satisfy conditions by which any behavioral or response center can be taken as of that of a person endowed with an intention, and even a spirit. Groups can be organized so as to present a behavioral model inviting explanation in terms of beliefs or desires, insofar as these states of intentionality do not simply reflect the presence of corresponding states among their members. Groups can also be organized in a way such as to make it possible to hold them responsible for certain actions, in the same way that we do hold people responsible who have certain intentional attitudes or act according to them.[38]

The reasons advanced by Pettit in favor of anti-singularism (and, more generally, what he calls "individualist holism") are based on an analysis of the problem known in law as the "doctrinal paradox," and renamed by Pettit as the "discursive dilemma." This dilemma reveals the different results that can occur when a group decides to take a decision using a procedure based on the premises of the question at stake or a procedure based on its conclusions.

From this possible tension, Pettit deduces the necessity of recognizing the existence of a different level of intention from that of individuals, namely that of "integrated collectivities." Leaning on recent works of philosophy defending the existence of collective subjects (Gilbert, 1989; French, 1984; Bratman, 1993a and 1993b; Searle, 1995; and Tuomela, 1995 and 2007), Pettit maintains that certain groups are "going to display all the functional marks of an intentional subject and (. . .) there is no reason to discount those marks as mere appearances."[39] According to him, the burden of proof rests on those who deny the existence of these collective intentional subjects. Pettit also refutes the principal objection to the idea of the collective intentional subject, namely that such an entity requires postulating a domain emerging ontologically from these groups—which would lead to count the group in addition to each of its members. Pettit agrees, in fact, with the idea that "if we replicate how things are with and between individuals in a collectivity—in particular, replicate their individual judgments and their individual dispositions to accept a certain procedure—then we will replicate all the collective judgments and intentions that the group makes."[40]

It seems to us that, in this last comment, Pettit considerably trivializes his initial holist proposition. At the end, Pettit proposes both that group intentions are "real" in a different but analogous sense to the reality of individual intentions, but that insofar as scientific explanation is concerned, one can still fully reduce collective intentions and judgments to the "way in which things happen to individuals and between them at the heart of a collectivity." In the end, it is difficult to see how the methodology this seems to imply for the social sciences is different from the singularist position to which Pettit claims to be opposed.

4.3 Miller's Objections

Let us turn now to two strong objections to MI, which we borrow from Richard Miller. The first objection is the following:

1. MI allows us to explain social phenomena but only in terms of psychological dispositions, when one should also, or rather instead, take into account "objective interests."

[38] Pettit (2004).
[39] Pettit (2004), p. 182.
[40] Pettit (2004), p. 184.

Miller asks us to consider, for example, the capitalist who regards bourgeois interest as coinciding with the national interest because such a belief suits his own desires and personal goals. If we were to consider the psychological inclinations of such an individual, we might guess that they would not correspond with objective interests that actually motivate the identification of bourgeois interest with national interest. In effect, the capitalist lies to himself in telling himself that the accumulation of profit by some serves the national community in its entirety.

Such a critique rests, we believe, on a confusion between, on the one hand, the distinction between psychological dispositions and objective interests and, on the other hand, the distinction between conscious and unconscious interests. The capitalist described in the example reduces cognitive dissonance in a way that permits him to reconcile his desire to see his interests fulfilled and his desire that these be driven by more noble motivations. Contrary to what Miller's objection suggests, MI is perfectly capable of recognizing the difference between these two contradictory desires.

Next, Miller raises the following objection:

2. MI tends to confuse explanations of a phenomenon and descriptions of its causes.

According to Miller,[41] in the explanation for the First World War, MI mistakenly focuses on the assassination of Archduke Franz Ferdinand, because in his view, any incident in the economic and political context of the era would have set off the powder keg. Here, one can give two responses. First, one can point out that a structuralist explanation is an individualist interpretation of a certain sort. True, one can give a structuralist explanation for the First World War, insisting on the fact that, given the economic and social context of the era, anyone might have set things off, not just Archduke Franz Ferdinand's assassin. This structuralist explanation is still individualist, however, because it comes back to a single individual, even if an unspecified one, as the cause of the war. The difference between this structuralist explanation and a non-structuralist one is that in the first case, the individual is not a specific individual, but a variable capable of taking on individual values (if the assassin had failed, someone else might have taken his place). The problem with this type of "structuralist laws" is that they are difficult to prove with certainty.

A second response would be to point out that proving causality and proving causation are two different things. One can explain an event as it really happened or argue that the event was inevitable, but these are two different objectives. We think that, in order to explain the First World War, one cannot dispense with the investigation of effective causes, like the archduke's assassin. It might also be interesting to try to model the necessity of an event like the First World War using a probabilistic model predicting that, given the circumstances, a certain type of incident would necessarily

[41] Miller (1978).

(given a probability distribution of this type of incident) have set things off. One analogy might be the case of a rickety bridge over which pedestrians are crossing one by one. Given, on the one hand, the characteristics of the bridge and, on the other, the average distributions of the pedestrians' weights and the force with which each one strikes the bridge while walking, it is necessarily true that there will come a day in which some pedestrian causes the collapse of the bridge, even if it is impossible to predict which day in particular.[42] Such a model would prove the necessity of the bridge's collapse. But this would still not be enough to explain it.

4.4 IS MI INCAPABLE OF EXPLAINING IRRATIONAL MASS PHENOMENA?

MI justifies itself in part because of the difficulties raised by the idea of "collective action," as in the case of the prisoner's dilemma or the tragedy of the commons. As we have already noted, before we can talk about collective action, we have to take account of individual incentives in a plausible way. From another angle, in reality, we observe forms of collective action that seem to defy explanations by individual rationality: the fact of voting, following rules, etc. Don't these observations defy the individualist approach? To this, one might respond that we take care to distinguish between methodological individualism and the hypothesis of individual rationality. Contrary to one overly popular account,[43] the two are, in effect, conceptually different. Mass phenomena can be reduced to a combination of irrational *and* individual behaviors. Nothing in this, then, puts MI in doubt.

In the last section of this chapter, we will examine the limits of the individual rationality hypothesis, but we should mention here two interesting approaches, one in philosophy and one in game theory, which can be considered as reintroducing holistic principles of explanation. Both attempt to think at the level of collective thought/give some consideration to the nature of collective thought.

In France, Vincent Descombes has recently defended, against methodological individualism, a form of holism inspired by Wittgenstein and Hegel.[44] In this holistic frame, Descombes proposes to entirely rethink the conceptual difference between a simple collection and a whole, developing a conception of "collective individuals," [45] which owes much to the tools of modern logic. Descombes considers that methodological individualism's main error, at least in its Popperian version,[46] comes in too quickly reducing certain types of "wholes" to abstract logico-mathematical sets, and in its incapacity to consider the intermediate and real (not abstract) category of

[42] This example does not involve a cumulative cause (each pedestrian worsens the bridge's condition), in which case we would have a sorites paradox. As it happens, a sole pedestrian is the cause of the collapse.

[43] Illustrated, for example, by the article by Heath (2009).

[44] There are numerous and explicit references to Wittgenstein in Descombes. At the end of *The Mind's Provisions* (2001), the idea of the "real totality" suggests the influence of the Hegelian concept of the objective spirit.

[45] See Descombes (2000), (2001/2002), and (2004).

[46] Popper (1945).

collective individuals. Thus, for Descombes, "if a society could be assimilated to a set of individuals, individualist reduction would be possible. But that would require that this society be an abstract object, not a real totality." [47]

For Descombes, collective individuals are not simple collections of individuals with no relationship to each other (for example, an arbitrary list of white objects or a list of department employees who went on vacation in Japan), nor the logico-mathematical set to which these individuals can be reattached (whiteness, the set of employees who spent their vacations in Japan). Collective individuals are characterized by the type of concrete relation that exists among their members. Thus, for Descombes, a group of employees who are friends and who visited Japan together has a real existence, irreducible to either the list of employees who have visited Japan (since nothing about the latter category tells us that they made the trip together), or to the abstract category of the "set of employees who have visited Japan." For Descombes, it is the existence of such attributes like "traveling in a group" that allows us to go from a collective of individuals to a collective individual.

The difference between a collection of individuals and a collective individual is that only the collective individual can be the subject of a different predicate than the individuals, that is, the subject of irreducible predicates, since every predicate that applies to a simple collection of individuals also applies to each individual, independent of the others. Thus, "in order for the group (constituted by a ministerial position) to go from Paris to Tokyo, normally its members have to move from Paris to Tokyo. In order for the group to be greeted by the mayor, each of its members must be greeted by the mayor." [48] For Descombes, "collective individuals [. . .] are beings with an unimpeachable/inalienable status, provided that we are careful not to confuse them with collections of individuals or with sets of individuals."[49]

Even though Descombes' position seems to be posed as an explicit alterative to methodological individualism, seeming sometimes to align itself with the social holism of thinkers like Peter Winch or Louis Dumont,[50] it is not certain that this philosophical position is actually in tension with the pragmatic and metaphysically agnostic position that we defend below. For us, the task of methodological individualism in the social sciences is not to respond to the question Descombes raises, namely "to give a satisfactory metaphysical status to collective individuals." [51] To the extent that Descombes himself attempts to preserve the autonomy of the subject and places himself at a distance from the structuralist tradition, his adversary seems to us to be a variety of

[47] Descombes (2001/2002, pp. 46–47).

[48] Descombes (2001/2002, pp. 127–ss130). One cannot predict that a group member will be greeted by the mayor except when it is true that the group is (even if Descombes insists on the fact that it is neither necessary nor sufficient that all group members travel and be received by the mayor in order for the group to travel and be greeted by the mayor). It is in this sense that the predicate is irreducible.

[49] Descombes (2001/2002), p. 125.

[50] Two thinkers who insist on maintaining the division between natural sciences and social sciences. See Winch (1958) and Dumont (1991).

[51] Descombes (2001/2002), p. 63.

individualism that comes close to ontological nominalism, more so than the properly *methodological* individualism that we favor. We are still uncertain of the practical implications for social science of Descombes' philosophical position.[52]

The late game theoretician Bacharach also laid down the foundations for an apparently non-individualistic approach to game theory.[53] Faced with the difficulty of explaning cooperation among individuals where rational choice theory would have predicted non-cooperation, Bacharach strove to develop the idea of "we-thinking" (the fact of thinking as a "we"), that is, the reasoning of an individual who thinks as if he was part of a larger unit. Instead of asking himself: "Is it to my advantage not to throw this greasy paper/trash on the floor or to go vote?" each individual asks himself: "What actions should we choose in order to improve our collective well-being?" For us, what Bacharach was attempting to formalize was nothing other than magical thinking, which consists of an individual confusing the diagnostic or symptomatic value of his individual action for a causal action. In the examples given, the causal action was a priori nonexistent, because the fact that I decide not to throw my trash on the ground or to go vote cannot of itself cause the corresponding decisions of other citizens. This attempt to formalize group intention is not necessarily in contradiction with MI, but rather, with the hypothesis of self-interested individual rationality (see earlier discussion). That said, if we interpret "we-thinking" not as an example of magical thinking, but as something else,[54] Bacharach's attempt at an explanation becomes perhaps an attempt to return to explanatory holism. We leave the question open.

4.5 MI, POLITICAL INDIVIDUALISM, ETHICAL INDIVIDUALISM, AND THE QUESTION OF FREE WILL

As a general matter, we have to now emphasize the point that MI is a position regarding social scientific *method*, not a metaphysical, ontological, or even political or ethical position. Methodological individualism is thus distinct from political individualism, or ethical individualism. Political individualism is defined by Schumpeter as the position according to which "freedom contributes more to the development of the individual and that of the society than anything else."[55] According to Schumpeter, PI and MI are independent of each other, in the sense that any combination of accepting or rejecting one and the other is both possible and coherent.

[52] A priori these should be nonexistent for Descombes, as is seemingly suggested by the philosopher's remark on the separation between logical and social sciences, which he believes Popper mistakenly ignored: "Before going further, it is important to note that logic, in and of itself, cannot tell us what there is in the world. It does not form part of the debate on individualism and holism in the social sciences, contrary to what Popper's account suggests" (Descombes 2001/2, p. 65).

[53] Bacharach (2006).

[54] See Susan Hurley (1990).

[55] Schumpeter (1908, p. 90), English translation p. 58.

Ethical individualism is the meta-ethical position according to which theories must be formulated exclusively in terms of concepts defined at the level of the individual, whether they be concepts of individual wellbeing, individual rights or individual autonomy. EI excludes ethical theories invoking supra-individual or non-individual concepts as fundamental moral notions. One example of an ethical theory based on a supra-individual concept is the idea of a public policy with the goal of achieving equality between the sexes or equality between nations, even if the cost is greater inequality between individuals themselves. An example of a non-individual theory is the idea that politics should aim to protect nature or to encourage the growth of scientific knowledge, independently of the damage done to the rights or the well-being of human beings. This position, once again, is logically independent of MI.

4.6 REDUCTIONISM

In reality, methodological individualism is simply the consequence of a more general thesis on the validity and importance of the reductionist program in social science. Reduction consists of explaining phenomena situated at one level of the scientific hierarchy in terms of phenomena situated at a lower level. Reductionist programs have been criticized at two levels: whether because they are not considered feasible, or because they are not considered desirable. As for the non-feasibility hypothesis, what one could call "Durkheim's error,"[56] every day it continues to be falsified by the fruitful relations cultivated between, for example, economics and psychology or between psychology and different branches of biology (genetics, physiology, developmental biology, and evolutionary biology). This development bears some resemblance to the earlier falsification of the belief that the living world could never be explained by chemistry.

As for desirability, it seems to us that as reductionism has been the motor of progress in science, one cannot plausibly make a stand against it, except in premature, approximate, or speculative forms of it. Premature reductionism is that which does not (yet) possess the means to its ambitions, as illustrated by the failure of early efforts to design satisfactory automatic machine translators. Approximate reductionism is exhibited by scientists who propose to explain a particular behavior in biological terms, even though what probably needs explaining in such terms is the capacity or the tendency that such a behavior may instantiate. The same goes for explanations of the behavior of political actors in terms of "territorial imperatives," comparable to that of animals. Finally, speculative reductionism is that of "just-so stories," which offer a possible explanation of a given behavior without showing that that behavior has actually emerged because of the proposed reasons. Sociobiology and the closely related discipline of evolutionary psychology offer numerous examples of this sort of speculative reductionism, for example when one explains the fact of lying to oneself as

[56] Durkheim (1895) believed, in fact, that what he called "social facts" had a unique reality, independent of individual actions, and could not be reduced to these last, and even less to a lower level of explanation such as the biological.

a capacity that appeared thanks to the adaptive advantages it secured,[57] or postpartum depression in women as a bargaining tool within the family.[58]

4.7 PSYCHOLOGICAL REDUCTIONISM

The importance of reductionism in science and particularly in the social sciences having been posited, one can still raise the following question: if the ultimate goal of social sciences is to reduce human behaviors to their most basic foundations (their "rock-bottom explanations," following Watkins, 1957), why then stop at the level of the individual as opposed to that of a gene or of an atom? Is methodological individualism condemned to dissolve into the reductionist ambition?

The response, we think, is unambiguous. Yes, MI aspires to dissolve itself into the reductionist aim, once the "bridge" between social sciences and natural sciences is firmly established. But to the extent that such a junction is still far from being realized, the level of the individual remains privileged because it is, at the present time, the only level of explanation to which we have access and at which the proposed explanations have been found convincing.

This conclusion allows us, in passing, to take a position on the controversy raised in the 1950s over MI's "ontological" or "metaphysical" presuppositions. Critics such as Leon Goldstein (1958) and, later on, Steven Lukes (1968) argued that MI was an indirect way to defend an individualist metaphysical or ontological position, perhaps indeed suggested by Watkin's interpretation of MI as the proposition that the ultimate constituents of the social world are individuals (Watkins, 1957, p. 105). For us, MI does not defend a privileged position of the individual for any reason other than pragmatic considerations related to the present advancement of the social sciences. As concerns the ultimate foundations of the social world, MI is agnostic.

Another question that might be raised is the following: what is meant exactly by "the level of the individual"? Does it refer to sticking to the observable "surface" of individuals, as with the economic theory of revealed preferences? Or must we go further, "below the surface," so to speak, of agents, in order to discover observable beliefs and preferences? Here, one could give two responses. The first is that a purely behavioral approach like revealed preference theory entirely rolls back the notion of preferences for that of choices, eliminating the role of beliefs about the available options. Still, it is evident that it is possible to choose an option which one does not prefer, for example if one does not know that one's preferred option is a possible option. Revealed preference theory ignores this possibility, in part because it fails to consider the importance of the beliefs that condition human choices. Moreover, in failing to consider the role of beliefs, it makes game theory theoretically impossible.[59] Revealed preference theory can therefore lead to the absurd case of being unable to tell the difference between a prisoner's

[57] Trivers (2002).

[58] Hagen (1999, 2000).

[59] For a more complete critique, see Hausman (2000).

dilemma situation, in which the option to free ride is the dominant strategy, and an "insurance game" in which cooperation is the best strategy provided that others cooperate, but in which the players believe that other players have the same preferences as in a prisoner's dilemma scenario. In this latter case, mutual suspicion incites the players to free ride. Even if the equilibrium in the prisoner's dilemma and this particular type of "insurance game" marked by pluralistic ignorance is the same in the end (free ride, free ride), it is still clear that these are two very different situations. Revealed preference theory is nonetheless unable to make the distinction.

The second response is pragmatic. It might perhaps be legitimate to stick to the surface of the individual and to revealed preference theory if the latter was undeniably successful in predicting and explaining behaviors. But this is far from being the case. Given that the instrumentalist position is not justified, we are thus entitled to appeal to concepts of folk psychology, which as much as possible, permit us to open the black box of human actions, considering mental states and processes, as well as desires and beliefs. Turning to folk psychology presents clear problems of method, but not insurmountable ones. Just as historians draw on cross references between actions, public declarations of intent, and confessions made under the veil of secrecy or years later in memoires or letters, it is possible, to a certain extent, to gain access to real and individual intentions.

One question we leave open is whether it will still be relevant to use folk psychology once the link is established with neurobiology and other more fundamental sciences. It is possible that, even if all human choices might one day be explained in the language of neurobiology, these choices will remain unintelligible (as they are unintentional), unless they are accompanied by a description, in the terms of basic psychology, that puts the spotlight on intentions.

We now pose the final question. If the level of basic psychology offers greater immediate intelligibility for us human beings, shouldn't we admit that there is a privileged "interpretative" level of social sciences at which introspection and empathy are possible? We cannot understand cells, just as we cannot really understand institutions and groups. On the other hand, we can truly understand our neighbor (and even his dog). To this, one might respond that if introspection and empathy are privileged sources of hypotheses, verifying these hypotheses can take place at any level of the scientific analysis. One can privilege the interpretative level as a point of departure, that is, in order to search for hypotheses, but not to find answers. In other words, from a scientific point of view, it is more important that the answers be true, than that they necessarily be immediately intelligible or intuitive.

4.8 HYPOTHESES ON RATIONAL AND SELF-INTERESTED BEHAVIOR

Much more than methodological individualism, it is undoubtedly the rationality and self-interest hypotheses, two fundamental hypotheses of economics, that we should relax in order to explain a certain number of phenomena.[60]

[60] For an in-depth treatment of these questions, see Elster (2009b and 2010).

According to the rationality hypothesis, the rational individual maximizes some indeterminate objective function, subject to a constraint of coherence. The self-interest hypothesis specifies the egoistic or egocentric nature of the individual action. In spite of the frequent empirical falsification of these two hypotheses, a large number of social scientific researchers, principally in economics or political science, persist in using them, in the name of simplicity or parsimony. To the extent that, to paraphrase Tolstoy, we can say that every rational or self-interest actor is rational or self-interested in the same way, while every irrational or selfless actor is irrational or selfless in his or her own way, it seems preferable to try to explain behavior in well-defined terms rather than risk falling into the potential arbitrariness of explanations that renounce them.

Still, it seems to us that when the rationality and self-interest hypotheses—the principal traits of *homo economicus*—are not verified, we should abandon them. The rationality hypothesis is empirically contradicted by the demonstration of cognitive biases, which have been itemized by behavioral economics. Similarly, the self-interested behavior hypothesis is frequently contradicted by a large number of altruistic acts performed by agents, often at very high cost to themselves (this is also true of voting, according to certain interpretations, but also of triply anonymous gifts and kamikaze acts).

Let us make three points. First, it is generally more costly to give up on the rationality hypothesis than on that of self-interest. Next, it is not easy to dissociate the effects of the rationality hypothesis from those of the self-interest hypothesis. Finally, it is not evident that the inverse of the self-interest hypothesis, namely, selflessness, has a meaning.

A. Why It Is More Costly to Give Up on the Rationality Hypothesis

The rationality and self-interest hypotheses are logically independent from one another. Thus, the rationality hypothesis does not imply the self-interest hypothesis, and vice versa. The hypothesis of self-interested or egoistic motivation may be combined with the rationality hypothesis to lead to a particular case of rationality, even an important case, but there is no methodological reason to privilege it. Inversely, self-interested behavior may be irrational in that the agent does not apply the most adequate means for pursuing his egoistic desires. Nonetheless, there is a certain asymmetry between the two hypotheses, to the extent that rationality is also a norm that human beings seek out over its opposite, irrationality, whereas self-interest is a purely contingent motivation from an empirical point of view, since we do not always have reason to privilege personal interest. The rationality norm constitutes a permanent counterweight to irrational tendencies, which is not the case with self-interest. For explicative purposes, it is therefore more useful to preserve the idea of maximizing a utility function, even if the maximized object includes the wellbeing of others, than it is to preserve the idea that the object of the attempted action is individual interest.

B. Why It Is Not Always Possible to Dissociate the Effects of the Rationality Hypothesis from Those of the Self-Interest Hypothesis

According to the so-called Duhem-Quine thesis, scientific hypotheses are not presented to the world one by one, but en bloc and simultaneously.[61] Thus, when rational choice theory comes face to face with counter-examples, it is not evident if these counter-examples refutes the rationality hypothesis, the self-interest hypothesis, or both. Let us take, for example, the voting paradox. It seems that one could explain it in at least three different ways: voting as a rational but selfless act; as a self-interested but irrational act; or as a self-interested and also irrational act. How to know which of these interpretations is the right one?

Even when an experiment is devised in order to test out a precise hypothesis, a negative result does not conclusively refute that hypothesis, since it is possible that the error lies in one of the auxiliary hypotheses adopted implicitly or explicitly by the researcher. Let us take the following experiment: a negotiation in which two agents make each other successive offers and counteroffers over how to divide up a sum of money which diminishes with each successive period of the negotiation. In the first period, Agent I (Paul) proposes to Agent II (Marie) a division of a sum total of 5 dollars. In the second period, Marie can accept that offer, or reject it and make a counteroffer, in which case the total to be shared goes from 5 to 2.5 dollars. Finally, in a third period, Paul can either accept this counteroffer, or decide on a division of 1.25 dollars, to which the sum of money has been reduced.

We note that, even when the decision tree involves three nodes, the process may stop at the second if Marie accepts the initial division proposed by Paul.

Let us suppose that the two agents are rational, self-interested, and have perfect information about the other agent. In this case, reasoning by backward induction leads Paul to propose (3.75, 1.25) and Marie to accept his proposition.[62] This constitutes "the equilibrium" of the game, that is, a point of stability for two rational, self-interested and informed agents.

In reality, when two subjects have to carry out this negotiation, the mean offer made by Agent I is (2.89, 2.11). This is clearly a more generous proposition towards Agent II than the equilibrium predicted by backward induction. Should we interpret generous offers by real agents as proof of irrationality, selflessness, or a combination of the two? We could indeed imagine that individuals are rational and selfless by choice, hoping to portray themselves as fair or manifesting a form of put-on/assumed altruism. We could imagine that they are self-interested but irrational and that, for example, they lose their composure when faced with a charming smile from Agent II. Finally, we could imagine that agents are perfectly rational and self-interested, but fear that Agent II is

---

[61] See chapter 2 on confirmation and induction.

[62] We start by asking what Paul would do if the last node was ever reached, the response obviously being that he would propose the division (1.25, 0). This fact constitutes a constraint on Marie's decision at the second node, because Paul will reject any division that gives him less than 1.25. At the same time, the fact that Marie can obtain for herself 1.25 minus epsilon by offering Paul 1.25 plus epsilon constitutes a constraint on Paul's decision at the first node. If Paul offers Marie a sum $5 - x < 1.25$, she will make a counteroffer $(2.5 - y, y)$ such that $2.5 - y > 1.25$ (and thus more advantageous for Paul than what he could obtain by refusing) and $y > 5 - x$ (thus more advantageous for Marie than Paul's offer).

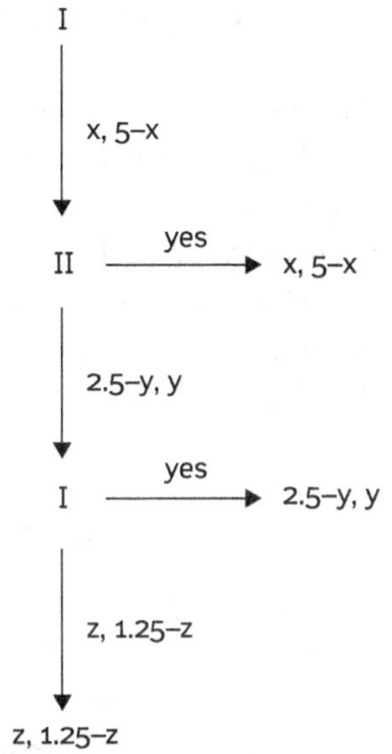

FIGURE 1 A decision tree for Marie and Paul

herself irrational, or that, subject to what we call "selflessness through negligence" and ready to retaliate against an offer that they consider too stingy, they quit, suffering the financial consequences themselves at the end. If Agent I fears that Agent II will reject the equilibrium offer, he will make a more generous offer, hoping to avoid rejection, but this would violate neither the rationality hypothesis nor that of self-interest.

C. Do Selfless Actions Exist?

Even if the self-interest hypothesis is less necessary, to economic analysis in particular, than the rationality hypothesis, a problem arises from the fact that, contrary to the irrationality hypothesis, which is well-established for certain actions, the self-interest hypothesis is not necessarily plausible, nor, in particular, easy to verify. Selflessness refers to motivation detached from all personal interest. One example of this might be *triply anonymous* donations to charitable works, in which neither the identity of the recipients, nor of the organizer of the charity, nor of the public is known. An example of a triply anonymous donation is given by the case of a person who deposits a 100-dollar bill in the donation box of an empty church. The problem is that it is always possible to say, even in this example, that the charitable action is motivated not by real selflessness but by the desire to please God and save one's soul, or the desire to win internal plaudits from one's own conscience. The question posed, then, is based on Kant's model of the existence of a good intention: does there exist, in this depraved world, one single authentically selfless action? One can imagine an even more convincing paradigm than the case of the triply anonymous donor, that of the anonymous, atheistic kamikaze (unless he experiences—a possibility, according to Kant, we are unable to exclude—a feeling of self-satisfaction at the instant preceding his death that would ruin the hypothesis of complete disinterest).

We propose that there exist three forms of authentic selflessness: selflessness in fact, selflessness by choice, and selflessness through negligence. The first corresponds, broadly speaking, to the disinterestedness of the judge, who has no personal stake in the question of which he is an a priori impartial arbitrator. The second corresponds to the selflessness of the altruist, who consciously chooses to pursue the interests of others above his own. Finally, the third corresponds to that of the revenge-seeking individual whose passion carries him back to self-interest.

5. Conclusion

Clearly, it is difficult, even dangerous, to make predictions about the future of a disciplinary field. In view of the impasses reached by social science over the course of the twentieth century, and in light of more recent debates, it seems nonetheless possible to offer the following forecast. The future of social science probably lies in a resolutely reductionist program, in the sense defended in this article. This reductionist program rejects the distinction between interpretation and explanation, is anchored is a non-dogmatic methodological individualism (that is to say, a methodological individualism that is essentially pragmatic and agnostic on the ultimate foundations of the social world) and aims to fuse the concepts of folk psychology with those of natural science. The reductionist aim does not mean that the level of analysis at which explanation makes sense for us—the "interpretative" level, according to the classical distinction—does not have a heuristic advantage over other levels concerning the formulation of questions and hypotheses. Investigating the responses to these questions, however, cannot be limited to this particular level of analysis.

To the extent that the reductionist aim today is at the order of a regulative ideal more than an empirical reality, it seems to us that the social sciences would be better off, for now, cultivating a certain epistemological modesty. This modesty would in part imply abandoning, at least temporarily, the search for general laws and instead concentrating on the clarification of a certain number of fundamental hypotheses, like those of rationality or self-interest, and on the collection of mechanisms.

References

Arrow, K. (1950) "A Difficulty in the Concept of Social Welfare," *The Journal of Political Economy*, 58 (4), pp. 328–346.

Bacharach, M. (2006) *Beyond Individual Choice: Teams and Frames in Game Theory*, Princeton, NJ: Princeton University Press.

Bohman, J. (1991) *New Philosophy of Social Science: Problems of Indeterminacy*, Cambridge, MA: MIT Press.

Bourdieu, P. (1979) *La Distinction: Critique sociale du jugement*, Paris: Les Editions de Minuit.

Bratman, M. (1993a) "Shared Intention," *Ethics* 104, pp. 97–113.

Bratman, M. (1993b) *Faces of Intention*, Cambridge, MA: Cambridge University Press.

Braybrooke, D. (1987) *Philosophy of Social Science*, Englewood Cliffs, NJ: Prentice-Hall.

Cohen, G.A. (1982) "Functional Explanation, Consequence Explanation, and Marxism," *Inquiry* 25(1), pp. 27–56.

Descombes, V. (2000) "Philosophie des représentations collectives," Available at: http://classiques.uqac.ca/contemporains/descombes_vincent/philo_representations_collectives/philo_repres_coll.html

Descombes, V. (2001/2002) "Les Individus Collectifs," *Revue du MAUSS*, 18, pp. 305–337. Available at: http://www.cairn.info/article.php?ID_ARTICLE=RDM_018_0305

Downs, A. (1957) *An Economic Theory of Democracy*, New York: Harper.

Dumont, L. (1991) *Essais sur l'individualisme*, Paris: Seuil.

Durkheim, E. (1895) *Les règles de la méthode sociologique*, Paris: Flammarion, 1988.

Elster, J. (1979) *Ulysses and the Sirens*. Cambridge: Cambridge University Press.

Elster, J. (1982) "The Case for Methodological Individualism," *Theory and Society*, 11, pp. 453–482.

Elster, J. (1989a) *Nuts and Bolts for the Social Sciences*, Cambridge: Cambridge University Press.

Elster, J. (1989b) "Marxism and Individualism," in Dascal, M. and Gruengard, O. (eds.), *Knowledge and Politics. Case Studies in the Relationship Between Epistemology and Political Philosophy*, Boulder: Westview Press, pp. 189–206.

Elster, J. (2007) *Explaining Social Behavior. More Nuts and Bolts for the Social Sciences*, Cambridge: Cambridge University Press.

Elster, J. (2009a) *Alexis de Tocqueville, the First Social Scientist*, Cambridge: Cambridge University Press.

Elster, J. (2009b) *Le Désintéressement: Traité critique de l'homme économique*, Vol. 1, Paris: Seuil.

Elster, J. (2010) *L'Irrationalité: Traité critique de l'homme économique*, Vol. 2, Paris: Seuil.

Freedman, D. (2009) *Statistical Models: Theory and Practice*, Cambridge and New York: Cambridge University Press.

French, P.A. (1984) *Collective and Corporate Responsibility*, New York: Columbia University Press.

Gigerenzer, G. (2008) *Rationality for Mortals: How People Cope with Uncertainty*, Oxford: Oxford University Press.

Gigerenzer, G., and Selten, R. (eds.) (2001) *Bounded Rationality: The Adaptive Toolbox*. Cambridge, MA: MIT Press.

Gigerenzer, G., Todd, P.M. et al. (eds.) (1999) *Simple Heuristics That Make Us Smart*. Oxford: Oxford University Press.

Gilbert, M. (1989) *On Social Facts*, Princeton, NJ: Princeton University Press.

Goldstein, L. (1958) "The Two Theses of Methodological Individualism," *The British Journal for the Philosophy of Science*, 9, pp. 1–11.

Guha, R. (1983) *Elementary Aspects of Peasant Insurgency in Colonial India*, Delhi: Oxford University Press India.

Hagen, E.H. (1999) "The functions of postpartum depression," *Evolution and Human Behavior*, 20, pp. 325–359.

Hagen, E.H. (2000) "Depression as Bargaining: The Case Postpartum," *Evolution and Human Behavior* 23 (5), pp. 323–336.

Hardin, G. (1968) "The Tragedy of the Commons," *Science*, 162, pp. 1243–1248.

Hardin, R. (1980) "Rationality, Irrationality and Functionalist Explanation," *Social Science Information*, 19, pp. 755–772.

Hausman, D. (2000) "Revealed Preference, Belief, and Game Theory," *Economics and Philosophy*, 16, pp. 99–115.

Heath, J. (2009) "Methodological Individualism" *in* Zalta, E. (ed.), *The Stanford Encyclopedia of Philosophy (Summer 2009 Edition)* URL = <http://plato.stanford.edu/archives/sum2009/entries/methodological-individualism/>.

Hollis, M. (1994) *The Philosophy of Social Science: An Introduction*, Cambridge: Cambridge University Press.

Hurley, S. (1990) *Natural Reasons*, Oxford: Oxford University Press.

Kahneman, D. and Tversky, A. (1974) "Judgment under Uncertainty: Heuristics and Biases," *Science*, 185, pp. 1124–1131.

Kincaid, H. (1994) "Defending Laws in the Social Sciences," in Martin, M. and McIntyre, L.C. (eds.), *Readings in the Philosophy of Social Science*, Cambridge, MA and London: MIT Press, pp. 111–130.

Landemore, H. (2004) "Politics and the Economist-King: Is Rational Choice Theory the Science of Choice?," *Journal of Moral Philosophy* 1, pp. 177–197.

Leibniz, G.W. (1969) *Philosophical Papers and Letters: A Selection*, Loemker, L.E. (ed.), Dordrecht: Kluwer Academic Publishers.

Little, D. (1991) *Varieties of Social Explanation: An Introduction to the Philosophy of Social Science*, Boulder: Westview Press.

Lukes, S. (1968) "Methodological Individualism Reconsidered," *The British Journal of Sociology*, 19(2), pp. 119–129.

Mackie, G. (2003) *Democracy Defended*, Cambridge, MA: Cambridge University Press.

Mauss, M. (1923–1924) *Essai sur le don. Forme et raison de l'échange dans les sociétés archaïques*, Paris: PUF, reprint 2007.

Martin, M. and McIntyre, L.C. (eds.) (1994) *Readings in the Philosophy of Social Science*, Cambridge, MA: MIT Press.

McClure, S.M., Laibson, D., Loewenstein, G. and Cohen, J.D. (2004) "Separate Neural Systems Value Immediate and Delayed Monetary Rewards," *Science*, 306, pp. 503–507.

McClure, S.M., Ericson, K.M., Laibson, D., Loewenstein, G. and Cohen, J.D. (2007) "Time Discounting for Primary Rewards," *Journal of Neuroscience*, 27(21), pp. 5796–5804.

Miller, R. (1978) "Methodological Individualism and Social Explanation," *Philosophy of Science*, 45(3), pp. 387–414.

Olson, M. (1965) *The Logic of Collective Action: Public Goods and the Theory of Groups*, Cambridge, MA: Harvard University Press.

Pettit, Ph. (2004) "Groups with Minds of Their Own," in Schmitt, F. (ed.), *Socializing Metaphysics*, New York: Rowman and Littlefield, pp. 167–193.

Popper, Karl (1945) "The Poverty of Historicism III," *Economica*, 11: 69–89.

Quinton, A. (1975) "Social Facts," *Proceedings of the Aristotelian Society*, 75, pp. 1–27.

Riker, W. (1988) *Liberalism Against Populism: A Confrontation Between the Theory of Democracy and the Theory of Social Choice*, Prospect Heights, IL: Waveland Press.

Rimmon-Kenan. S. (1983) *Narrative Fiction: Contemporary Poetics*, London and New York: Methuen.

Rosenberg, A. (1995) *Philosophy of Social Science*, Boulder: Westview Press.

Rubinstein, A. (1998) *Modeling Bounded Rationality*, Cambridge: MIT Press.

Ryan, A., ed. (1973) *The Philosophy of Social Explanation*, Oxford: Oxford University Press.

Ryan, A. (1979) *Philosophy of Social Sciences*, London: McMillan.

Samuelson, P. (1954) "The Pure Theory of Public Expenditure," *Review of Economics and Statistics*, 36, pp. 387–389.

Schumpeter, J. (1908) *Das Wesen und der Hauptinhalt der theoretischen Nationalokonomie*, Leipzig: Duncker & Humbolt; Eng. trans. by Mc Daniel, B. *The Nature and Essence of Economic Theory*, New Brunswick and London: Transaction Publishers, 2010.

Searle, J. (1995) *The Construction of Social Reality*, New York: Free Press.

Simon, H. (1978) "Rationality as Process and as Product of Thought," *American Economic Review*, 68, pp. 1–16.

Sokal, A. and Bricmont, J. (1997) *Impostures intellectuelles*, Paris: Odile Jacob.

Sokal A., Bricmont J. and Hochstedt, B. (2005) *Pseudosciences et postmodernisme*, Paris: Odile Jacob.

Tocqueville, A. (1993) *De la Démocratie en Amérique*, Paris: Garnier-Flammarion.

Trivers, R. (2002) *Natural Selection and Social Theory: Selected Papers of Robert Trivers*, Oxford: Oxford University Press.

Tuomela, R (2007) *The Philosophy of Sociality: the Shared Point of View*, Oxford: Oxford University Press.

Tuomela, R. (1995) *The Importance of Us: A Philosophical Study of Basic Social Notions*, Stanford Series in Philosophy, Stanford: Stanford University Press.

Watkins, J.W.N. (1957) "Historical explanation in the social sciences," *British Journal for the Philosophy of Science* 8, pp. 104–117.

Winch, P. (1958/1990) *The Idea of a Social Science and Its Relation to Philosophy*, 2nd ed., London: Routledge and Kagan.

15

PHILOSOPHY OF ECONOMICS

Mikaël Cozic (Paris-Est Créteil, Institut Universitaire de France and IHPST)

1. Introduction

1.1 THE PHILOSOPHY OF ECONOMICS

Economic science occupies a large area in our daily life: its concepts, statistics, predictions, if not its theories themselves, are all transmitted to the public at large and contribute significantly to economic and political behavior.[1] Still, the epistemological status of economics never fails to spark debate. For example, economics will be criticized for hiding its inability to predict or advise behind sophisticated mathematical constructions,[2] for basing itself on an inadequate understanding of man and society,[3] or indeed for surreptitiously propagating a questionable ideology. It is certain

[1] I heartily thank Philippe Mongin for his commentary and advice on two successive versions of this chapter. I also thank Jean Baccelli, Denis Bonnay, and Bernard Walliser for their comments. The content is in large part taken from the notes of my class "Philosophie de l'économie," given in collaboration with Philippe Mongin at the Ecole Normale Supérieure of Paris (2007–2010), and I would like to express my gratitude to the students who followed this somewhat unusual class and facilitated its improvement. Thanks also to Christopher Roberston, who helped me to translate the original version of this chapter. Part of this work was done at the Institut d'Etudes Cognitives (Paris, ENS Ulm) with the support of the ANR-10-LABX-0087 IEC and ANR-10-IDEX-0001-02 PSL* grants.

[2] On the history of the mathematization of economics, see Ingrao and Israel (1990) regarding the general equilibrium theory.

[3] See the article "Rational Fools" in the collection Sen (1987): "The *purely* economic man is indeed close to being a social moron. Economic theory has been much preoccupied with this rational fool decked in the glory of his *one* all-purpose preference ordering. To make room for the different concepts related to his behavior we need a more elaborate structure."

that economic science is *singular*, particularly among the other social sciences from which it seems, in method, so different. This in part explains why the philosophy of economics (and notably the methodology of economics) is practically as old as the discipline itself and has punctuated its whole development. Economics raises a diverse array of philosophical questions. Three main fields can be distinguished within philosophy of economics (Hausman, 2008c). (1) Like all scientific disciplines, economics is the object of epistemological and methodological discussions; this first field is generally referred to as the *methodology of economics*. (2) To the extent that among the fundamental theories of contemporary economics is to be found, in some guise or another, the assumption that economic agents behave in a rational manner, economics raises questions belonging to the *theory of action and rationality*. (3) Finally, to the extent that economics provides concepts and principles for the appraisal of institutions, states and economic processes, its questions are also part of normative philosophy and, more specifically, *moral and political philosophy*. These three fields make up the subject matter of *Economics and Philosophy* (Cambridge University Press), the leading international journal, founded by D. Hausman and M. MacPherson in 1985. Methodology is the specific subject of the *Journal of Economic Methodology* (Routledge) created in 1994.[4]

1.1 "POSITIVE" ECONOMICS

The present chapter deals with the methodology of economics, conceived of as the branch of philosophy of science dedicated to economics. Numerous economists participate in the evaluation of policies and socio-economical institutions. If we approach economics with a philosopher of science's eye it is because a part of economists' goals, attitudes and contributions seems at first glance to obey an epistemic regime similar to that of the sciences. The assumption, generally implicit, upon which most economic methodology rests is that these goals, attitudes and contributions of economics are sufficiently *separable* from its normative dimensions for us to evaluate and analyze them using the tools and criteria of philosophy of science. This assumption is closely linked to that famous and still widespread distinction between *positive economics* and *normative economics*: it is positive economics that is the philosopher of science's preferred subject. The distinction dates back to the trichotomy between "positive science," "normative science" and "art" introduced by Keynes senior (1890/1917): the first is a "body of systematized knowledge concerning what is," the second a "body of systematized knowledge relating to criteria of what ought to be," while the third is a "system of rules for the attainment of a given end."

In making the assumption of separability explicit, we do not wish to suggest that questions attached to the distinction between positive and normative in economics are already solved or even easy to solve, nor that the assumption itself is self-evident. The distinction between positive and normative is, in the literature, inextricably linked to

[4] See also Davis et al. (1998) *The Handbook of Economic Methodology*.

the role of economists' value judgments, and in particular to the question of *axiological neutrality*: is it possible, or is it desirable, that economists "*qua* economists" refrain from asserting value judgments (the formulation is Mongin's [2006a])? Robbins (1932/1935), who is largely responsible for introducing the distinction between facts and values into economic literature, responds in the positive to both parts of the question.

Opposing this, others have maintained that

(T1) Economics cannot (in any of its domains) be axiologically neutral.

From this point of view, even economic contributions ordinarily qualified as "positive" would be run through with value judgments. In supporting (T1), one asserts that economics (and, generally speaking, other social sciences and humanities, see Part VII of Martin and McIntyre [1994] is run through with value judgments in a manner, or to extents, which differentiate it from the natural sciences. Thus (T1) directly threatens the working hypothesis upon which the core of the methodological literature is based.

What prompts such an argument is that economics is concerned with elements to which we spontaneously attach value judgments—think, for example, of revenue distribution or poverty. On the basis of this analysis, which is difficult to contest, a partisan of (T1) such as Myrdal (1958) can develop his position by concluding (a) that the economist's value judgments are *inevitably* expressed through (i) the selection of questions posed, (ii) the kinds of answers given and (iii) the evaluation of these answers. He can also conclude (b) that economic concepts *necessarily* carry an evaluative dimension. Conclusion (a) leads doubly to confusion. For one thing, it combines heterogeneous phenomena. The fact, for example, that the economist's values guide him in (i), the selection of the questions which he will attempt to resolve, does not imply that these questions (and the answers they call for) are not "factual." For another thing, (a) doesn't do justice to the distinction between the *assertion* of value judgments and the *influence* of value judgments on the formation and evaluation of factual judgments.[5] As for conclusion (b), it is, according to Mongin (2006a), a false generalization of a partially correct truth. The economist's conceptual toolbox contains many evaluative concepts, starting with the concept of rationality, but it also contains genuine non-evaluative ones.

For the reasons which have just been indicated, and for others beside, the (T1) thesis is difficult to uphold. Rather, the focus of discussion is on the examination of the different components of the axiological neutrality thesis. The claim of neutrality presupposes that judgments of fact and value judgments can be easily and unambiguously distinguished. The philosophical examination of this presupposition is closely linked with contemporary debates on "fact-value entanglement" (see e.g. Putnam, 2002) and requires a thorough conceptual analysis of judgment categories and of their linguistic expressions. Certainly, this examination is one of the important tasks on the

[5] Hausman & McPherson (2006, chap.3) contains two examples of interference between value judgment and positive economics.

current philosophy of economics agenda,[6] and one of the most grueling, as it demands the establishment of communication between abstract philosophical considerations and an economic tradition which has independently developed its own reflexive tradition.

Economics, sometimes referred to as the "dismal science" (Carlyle), is often poorly understood and little loved among philosophers. Before beginning our methodological reflections, we will very briefly present some notions of economics. Often, one looks to the 18th century when locating the birthplace of modern economic science, particularly to the works of Cantillon (*Essai sur la nature du commerce en général*, 1730), Hume, and, above all, Adam Smith (*An Inquiry into the Nature and Causes of the Wealth of Nations*, 1776). It is relatively easy to name the kind of things that have been of priority interest to economics since that time: production, consumption and the exchange of goods, revenues, currency, employment, and so forth. In contrast, it is more difficult to give a less extensional, more general characterization.

Certain attempts nevertheless remain influential.[7] Mill (1848) discusses the idea, dominant in the 19th century, according to which

(T2) Economics is the science of wealth.

where, by wealth, is to be understood anything that has a use or is pleasant, and which has an exchange value. (In the same order of ideas, economics is sometimes defined as the science whose object is material welfare.) For Mill, that definition is not restrictive enough since, in principle, it includes all disciplines that deal with diverse forms of wealth and the factors which influence them (agronomics, meteorology, geology . . .). So Mill (1836) proposes defining economics as "the science which traces the laws of such of the phenomena of society as arise from the combined operations of mankind for the production of wealth, in so far as those phenomena are not modified by the pursuit of any other object." Economic science, from among all individual motivations, would take only the desire for wealth into account, disregarding all others. It doesn't rely on any thesis saying that this motivation is the *only* one, but its purpose is to study the social effects of this motivation without considering the other ones. We could summarize this idea as follows:

(T3) Economics is the science of the effects of the desire for wealth considered in and of itself.

[6] Mongin (2006a) attempts something of this sort. The author pleads in favor of a position of "weak non-neutrality" according to which (i) the economist can (and must) assert value judgments and (ii) these value judgments are abundant and difficult to distinguish, in principle and in practice, from factual judgments. See also Reiss (2008), who equally exploits the idea that certain concepts simultaneously carry an evaluative and a non-evaluative dimension, as well as Sen's distinctions (1970, chap.5) regarding value judgments, notably "basic" and "non-basic" judgments.

[7] For a historical contextualization of the definitions of economics, consult Backhouse and Medema (2009).

The favored objects of economics, which we have already mentioned (production, consumption and exchange of goods, etc.), are, in this perspective, phenomena of which the desire for wealth is, one supposes, the overriding factor. Often this "substantial" definition of economics is contrasted with the "formal" (and no less influential) definition given by L. Robbins (1932/1935): according to him, the science of economics owes its unity and specificity to the fact that it studies certain types of behavior, *choices* under constraint. The agent choosing has limited means at his disposal and he must allocate these across several end goals, consequently he must *sacrifice* the fulfillment of some of these goals to the benefit of others. Thus,

(T4) "Economics is the science which studies human behavior as a relationship between ends and scarce means which have alternative uses."

This definition has been reused frequently up to the modern day, for example in Stiglitz and Walsh's manual (2000). It intrinsically links economics to the theory of choices made by economic agents. This implies that economics has a scope which, in principle, greatly surpasses the subjects it traditionally favors. (T4), latterly, has sometimes been specified with the addition of the assumption stating that, in these choice situations, agents behave rationally ("instrumental" rationality), and potentially coupled to the assumption stating that they form beliefs about their environment rationally ("cognitive" rationality).

Economics is marked by the existence, alongside a dominant or orthodox orientation, of heterodox schools. The *divisions* of dominant economics are relatively well defined. In general one distinguishes (i) *macroeconomics* from (ii) *microeconomics*. (i) Macroeconomics, which in its separate form is often traced back to Keynes' *The General Theory of Employment, Interest and Money* (1936), deals with the national output, unemployment rate, inflation, balance of trade, etc. (see for example Blanchard, 2017). Thus, it deals with *economic aggregates* and is interested notably with the way in which economic policy (fiscal and monetary policy) can influence the values of these aggregates. The macroeconomic *theory* typically proceeds by making assumptions about the relationships between these aggregates; for example, by supposing that the aggregate consumption C of a national economy is an (increasing) function of aggregated disposable income Y_D, which is equal to the total income Y from which we subtract taxes T. The hypothesis thus obtained is $C=C(Y-T)$, which, in the Keynesian theory of the "multiplier," is specified linearly: $C=c_0 + c_1 (Y-T)$ where c_1, taken between 0 and 1, is called the marginal propensity to consume. (ii) As for microeconomics, its starting point is the behavior of economic agents (typically, firms and consumers) and, on the basis of assumptions regarding this behavior, it proposes to explain and predict the resulting collective phenomena (see for example Mas-Colell et al., 1995). (iii) Sometimes an extra branch is added to these main two areas, *econometrics*. Appearing in the 1930s, it is dedicated to the statistical estimation of micro- and macroeconomic relationships—for example, the estimation, for a given type of good and population, of the manner in which that population's demand for that good varies according to its price—as well as to the testing of models coming from these two branches. Macroeconometrics, in particular, contributes to the economic forecasting of national aggregates and to the simulation of the effects of public policies.

The methodological discussions which follow will have a dominant, though not exclusive, application in microeconomics. Microeconomics proceeds from a method characteristic of the contemporary economic approach which grants a central position to mathematical theories and models and relies primarily on two fundamental assumptions: (a1) the *rationality* of economic agents, and (a2) the *equilibrium* of the system formed by their interactions. We will clarify both of these assumptions in turn.

(a1) Economics begins with *agents* who evolve in a certain material and institutional environment and who, generally speaking, are not designated individuals but functionally specified categories: the consumer (in fact, the household) who buys goods on the markets, and the firm who produces the goods that it sells to the consumers. Economic models start with assumptions about the agents' behavior; these are supposed to *specify*, for a given class of agents and for the environment in which they evolve, the general assumption of rationality. Hence, consumer theory makes the following assumptions:

> (c1) The agent has transitive and complete preferences over various "bundles of goods," represented by vectors $x=(x_1, \ldots, x_N)$ where x_1 is the quantity of good 1, . . . , and x_N the quantity of good N. Transitivity and completeness are stated thus: for any x, y, z, if the agent prefers x to y and y to z, then she prefers x to z; for any x, y, she either prefers x to y or y to x.
>
> (c2) The set of all bundles of goods between which the agent can *choose* is determined by her wealth w and by the standard prices of each good $p=(p_1, \ldots, p_N)$: the total price of a bundle of goods must be less than or equal to w, that is to say $x_1 . p_1 + \ldots + x_N . p_N \leq w$.
>
> (c3) The consumer chooses the basket of goods which she prefers among those which fall within the budgetary constraints defined in (c2).

Assumption (c3) determines the consumer's *demand* $x = x(p,w)$ on the basis of her preferences and of the constraints (price and resources) that she encounters. For every good *n*, the consumer demands a quantity $x_n(p,w)$ of that good. Assumption (c3) justifies our speaking of "optimization" or "maximization" models of behavior. Optimizing models are not found exclusively in microeconomics: contemporary macroeconomics has massive recourse to it and, through borrowing, they have spread also to other social sciences.

(a2) Once these assumptions about the economic agents have been made, the question of their interaction arises. At this stage, the assumption of equilibrium is introduced to assure *compatibility* between the behaviors of the different agents. For example, when we consider the market for a certain good *n* produced by certain firms and bought by certain consumers, assuming perfect competition, the concept of equilibrium means a state of equality between supply and demand for this good, this

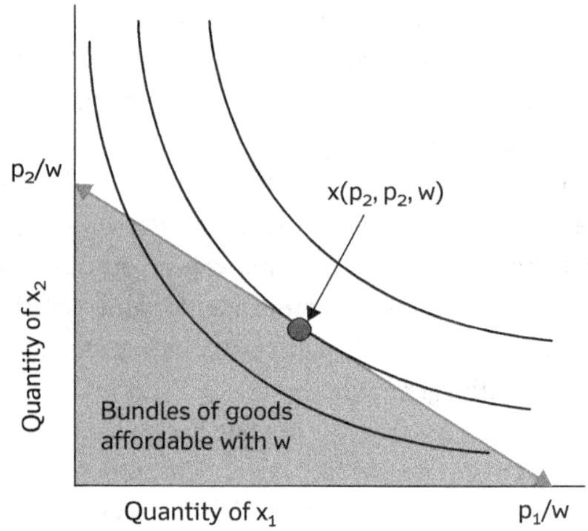

FIGURE 1 Graphic representation of the consumer's choice. Given a budget w and the prices p_1 and p_2, the affordable bundles of goods (the "consumption set") form the shaded triangle closed by the budgetary line. So-called indifference curves, generally supposed to be convex, join up the bundles of goods, between which the consumer is indifferent. The optimal choice $x(p_1, p_2, w)$ is the point of tangency of the budget line and the tangential indifference curve.

coordination taking place by means of the good's price: p_n is such that the sum of the individual demands for n is equal to the sum of all supplies of n. The *existence* of an equilibrium is not obvious, particularly when there are numerous goods and numerous agents on the market. One of the traditional fields of microeconomics, the theory of general equilibrium, specifically studies the conditions for the existence of equilibria in such a situation. Models relying on the assumption of equilibrium generally remain silent in regards to the *mechanism* leading to equilibrium, and they typically deploy their predictions and explanations by determining the way in which states of equilibrium are affected from the outside. For example, they will look at the way in which the introduction of a sales tax, which alters the demand of a good, will modify the price and the equilibrium quantity of that good, and for that they compare the states of equilibrium from before and after the introduction of the tax. *Comparative statics* is the name given to the exercise which consists in studying the effect of an exogenous change on the resulting equilibrium (Samuelson, 1947, p. 8; see Figure 2). Economic theory also relies extensively on the notions of equilibrium elaborated by game theory, which is a general theory of strategic interactions, that is, individual actions which are rationally determined relative to the actions of other agents. The fundamental notion is the *Nash equilibrium*: the actions of each individual are such that it is in no one's interest to change their action unilaterally; in other words, other individuals' actions being determined, one's own action is optimal.

The relationship between micro- and macroeconomics is itself the subject of important methodological discussions that we will not elaborate on in this chapter. Many of these revolve around the question of the *microfoundation* of the macroeconomy, i.e. around the question of knowing whether it is possible or desirable to *reduce*

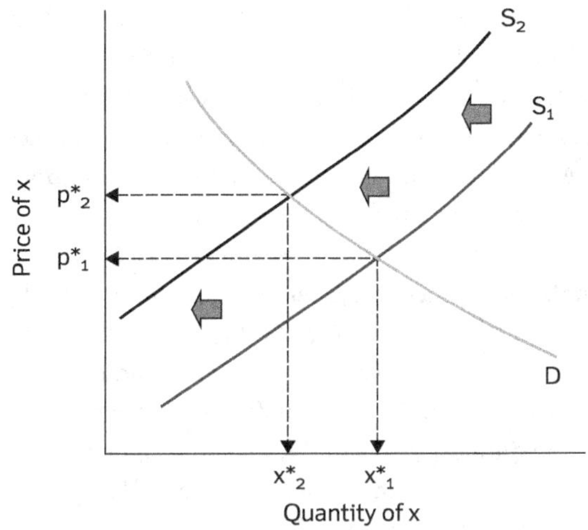

FIGURE 2 Graphic representation of the market equilibrium for good *x*. Curve D represents the demand aggregate. Curve S_1 represents the initial supply aggregate. The intersection (p^*_1, x^*_1) of the two curves is the point of equilibrium. If, following the increase in price of some factor of production, for example, the supply curve moves to S_2, a new equilibrium obtains (p^*_2, x^*_2). Thus the quantity exchanged decreases while the price increases.

macroeconomics to microeconomics (see in particular, Malinvaud, 1991, and Hoover, 2001b, chap. 3). That question is partly related to the question of methodological individualism in social sciences (see chapters 8 and 14 of the present volume).

1.2 THE METHODOLOGY OF ECONOMICS

Modern development of economics has been constantly accompanied by reflections regarding the discipline's method, object and scope.[8] Methodology today is largely the concern of specialists and the impact of epistemological assertions on economic research is less than it may have been in past decades. Economists are not always kind to professional "methodologists" (Samuelson, 1992, p. 240: "Those who can, do science; those who can't, prattle about its methodology"). Those who did take an interest, sometimes actively, in methodology left themselves open to similar "kindness" in return, (Hausman, 1992b: "If one read only their methodologies, one would have a hard time understanding how Milton Friedman and Paul Samuelson could possibly have won Nobel Prizes"). Methodological discussions still have important current relevance, as witnessed by the lively debates concerning so-called behavioral economics and neuroeconomics (see subsection 7.3).

It is difficult to present economic methodology in an analytic fashion, markedly differentiating the principal questions of contention: indeed, these questions are very

[8] Elements of the history of economic methodology can be found in Blaug (1980/1992, section II) and Hausman (1992a).

closely linked to each other. For that reason we will follow the dominant trend, which consists of approaching the field by way of the principal *doctrines* which animate it. Nevertheless, we will attempt to work out a question or preoccupation common to the area of methodology. This question dates back to Mill who, according to Hausman (1989), posed himself this problem: how can an empiricist methodology be reconciled with the manner in which economic science is built and practiced? In particular, how can empiricism be reconciled with the apparent falsity of the assumptions of economic theories and the small importance which *seems* to be accorded to the confrontation between the theories and the empirical data? *Mill's problem* ends up being more general even than the author's own empiricism, spreading beyond empiricism as a philosophical position: when we ponder on the realism of economic assumptions, on economists' sensitivity to empirical data or, further still, on the progress of economics, it is often because we wonder whether economics obeys the methodological standards of an empirical science—supposing that such standards exist. This *generalized problem* of Mill's is at the very heart of a large number of the reflexive discussions on economics. It explains the particular interest, in philosophy of economics, for some of the "great" questions of general philosophy of science, like the demarcation between the sciences and the non-sciences, the relationship between theory and experience, the nature of scientific progress, etc.

The formulation of Mill's generalized problem could lead one to believe, incorrectly, that economic methodology consists of comparing a would-be unified and homogeneous discipline, economics, to methodological standards which would be the object of a consensus and which would characterize what it is to take on a field of study scientifically. This is clearly not the case. For one thing, though economics does perhaps have a stronger discipline identity than other social sciences, it is still marked by significant disagreements and by the existence, alongside one dominant or orthodox orientation, of heterodox schools, Marxist or institutionalist, for example. Secondly, as could have been expected, the degree of consensus is still far lower on the side of philosophy of science. The hope, legitimately harbored during the middle of the twentieth century, of providing simple, consensual and universal criteria for scientific methodology or for scientific progress has been largely abandoned today. Resulting from this evolution, analyses of Mill's problem can vary noticeably from one methodologist to another, and broad intuitions regarding the nature of scientific method are taken on with more flexibility and less certitude than they may have been in the past.

We won't deal with these two complications symmetrically: we will give room for expression to a diversity of epistemological points of view, but, as is often the case in methodological literature, we will concentrate first and foremost on what we have called dominant or orthodox economics (even if we do tackle distinct programs of research at the end of the chapter).

Two lines of approach will be pursued: the first ("Millian themes") will address in sections 2, 3, and 4 those positions we can relate to the ideas of J. S. Mill, pioneer of economic methodology and representative of the English empiricism of the 19th century. We will begin with Mill's famous deductive method and with his Anglo-Saxon successors (section 2) to discuss their current ramifications and, in particular, current neo-Millian views (sections 3 and 4). In the second approach ("Neo-positive themes"), we will broach the methodological views close to certain epistemological trends in sections 5 and 6, in particular neo-positivism and logical empiricism, which held center stage in the philosophical scene between the 1920s and 1950s, and which, in economic methodology, ousted the Millian tradition during the 1930s. Section 5 is dedicated to the contributions of P. Samuelson and to refutationist ideas, section 6 to M. Friedman's famous theses. We complete our review by reviewing contemporary discussions on experimental economics, behavioral economics, and neuroeconomics in section 7.

2. Mill's Deductivism

2.1 THE DEDUCTIVE METHOD

The *deductive* conception of economics has its origin in the methodological writings of J. S. Mill (1836, 1843), and it is later found (with some differences of greater or lesser importance) in Cairnes (1857, 1875),[9] J. N. Keynes[10] (1890/1917; even if Keynes often presents himself as seeking to reconcile deductivists with their adversaries), and maybe even in L. Robbins (1932/1935). We present it in detail not only because it dominated economic methodology for close to a century, but also because certain modern economic philosophers, like D. Hausman (1992a), claim considerable allegiance to it.

Mill (1836) distinguishes two principal methods in empirical sciences: the a posteriori (or inductive) method, and the *a* priori (or deductive) method. The first essentially consists in detecting regularities in the empirical data and then proceeding by generalizing inference.[11] The data in question *directly* concerns the proposition to be established; in the simplest case, if the proposition has a universal conditional form ("All *P* are *Q*"), these findings could be positive instances of this (an entity or an example which is simultaneously *P* and *Q*). The second method consists in reasoning *deductively* on the basis of prior assumptions. The procedure breaks down into three steps (Mill, 1843, book III, chapter XI):

(s1) The assumptions are first formulated and established *inductively*.
(s2) The consequences of these assumptions are extracted by deduction.
(s3) These consequences are compared to the available empirical data (see earlier discussion).

One fact must be insisted upon: the assumptions forming the starting point of the reasoning are themselves established by generalizing inference (or else deducted from other assumptions, these having been established by generalizing inference).

[9] On the differences between Mill and Cairnes, see Hands (2001), p. 27.
[10] John Neville Keynes (1852–1949) is the father of John Maynard Keynes (1883–1946).
[11] Which, in philosophical jargon, is also called "enumerative induction." See also Cairnes (1857/1875), p. 41.

The term *a priori*, which, since Kant, more frequently refers to the property of certain propositions to be justifiable independently of experience, can thus be misleading. The a priori method is, in reality, an *indirect* method of induction. In contrast with the a posteriori method, the target propositions are not established by generalizing inference. But induction plays an indirect role since it is by this means that are established the assumptions on the basis of which the propositions of interest are deduced. In the case we are occupied with, the assumptions are the fundamental propositions of economic science. Mill is quite evasive about their exact content. He mainly evokes the psychological law according to which a greater gain is preferable to a lesser gain (Mill, 1843, book VI, chapter IX, §3, p. 901), all the while affirming that economics considers mankind as "occupied solely in acquiring and consuming wealth" (Mill, 1836, p. 382). Alongside other commentators from classical economics, Cairnes mentions the efficient search for individual advantage as well as the law of diminishing marginal return to the soil (1857/1875, p. 41). With Robbins, a reference not from classical but neoclassical economics, the first fundamental assumption is that agents are able to arrange their options according to their preferences; the second is the law of diminishing returns, which could, he says, be reduced to the assumption stating that there exists more than one factor of production (on the justifications for the law of diminishing returns, consult Mongin (2007) who criticizes them all as being inaccurate).

2.2 WHY RESORT TO THE DEDUCTIVE METHOD?

The deductive method is not unique to economics. According to Mill, it is this method which we employ, for example, in mechanics. It imposes itself on the economist because

(T5) The a posteriori method is not viable in the economic domain.[12]

(T5) means that induction cannot directly establish the propositions targeted by economics. The inapplicability of the a posteriori method comes from two fundamental characteristics of economics: it is a *non-experimental* science of *complex* phenomena. The empirical data of economics emerges essentially from *observation* and not from experimentation.[13] For the deductivists, such data does not generally allow one to proceed inductively (or a posteriori) because of the intrinsic *complexity* of the phenomena in question:[14] too many factors interact simultaneously for us to ever hope to directly extract any robust regularities or causal relations.[15] If, for example, we wanted to establish that restrictive and prohibitive commercial legislation influenced national wealth, what would be needed in order to apply what Mill calls the "Method of Difference" would be to find two nations which were identical in everything except

[12] Mill (1836), p. 50; Keynes (1890/1917), p. 13.
[13] Mill (1836), p. 51; Keynes (1890/1917), pp. 85–88; Robbins (1932/1935), p. 74.
[14] Cairnes (1857/75), p. 43; Keynes (1890/1917), pp. 97–98.
[15] Mill (1836), p. 55; Keynes (1890/1917), p. 98.

their commercial legislation.[16] If we wish to proceed via direct induction, only experimentation is capable of untangling the complexity of the economic phenomena, but this is excluded.[17] So we cannot hope to justify economic propositions a posteriori.

For disciples of the deductivist approach, the fundamental assumptions are established, inductively, by introspection (Mill, 1836, p. 56) or by observation raised to the level of induction. These are the "indisputable facts of experience"[18] which require no supplementary empirical investigation.[19] Thus, for Cairns, contrary to the physicist, "the economist starts with a knowledge of ultimate causes" (1857/1875, p. 50). So confidence in economic theory results from the confidence its assumptions inspire, a particular kind of confidence, as it is put in the following characteristically deductivist assertion:

(T6) The propositions of economic theory are true only hypothetically—it is also said that they hold abstractly, in the absence of disturbing causes,[20] or *ceteris paribus*.[21]

The propositions of economic theory are not true *simpliciter*. Here we have something which contrasts, in appearance at least, with the preceding statements about the obviousness of economic assumptions. There are two ways to resolve that tension. (i) The first consists in restricting (T6) to the *conclusions* of economic theory, which Cairns does.[22] The objection we can formulate in this case is that if the reasoning were deductively correct and if the premises were true *simpliciter,* then, equally, the conclusions would be too. Cairnes, however, maintains that this may not be the case,[23] since the premises, even if they are true, are nevertheless *incomplete*: they don't describe *all* the factors that could affect the phenomena in question. To develop an analogy with mechanics: the parabolic movement of a body may be "deduced" from the laws of movement and gravitation, which are true; however, the movements of bodies do not necessarily trace a parabola—frictions with the air, for example, disturb the trajectory. In this way we could pass deductively from propositions that are true *simpliciter* to others that are not. The analogy is not very convincing: to deduce the parabolic form

[16] Mill (1843), VI, VII, §3.

[17] Mill (1843), VI, VII, § 2 and Cairnes (1857/75), pp. 43–44. One is struck by the similarity between these Millian stances and those of contemporary economists. See, for example, Malinvaud (1991), pp. 346–347.

[18] Robbins (1932/1935), p. 78. It is doubtful, on this point, that Robbins's position be equatable to those of Mill's or Cairnes's: von Mises's *apriorism* wields considerable influence over Robbins. Robbins (1938) brings up an interesting perspective: he seems to want to preserve a kind of neutrality between apriorism and empiricism. For him, the most important is that the two grant a very high level of certitude to the fundamental propositions of economics.

[19] Robbins (1932/1935), p. 79; Keynes (1890/1917), p. 13.

[20] Keynes (1890/1917), p. 14.

[21] Keynes (1890/1917), p. 101.

[22] Cairnes (1857/1875), p. 39: ". . . the conclusions of economic policy do not necessarily represent real events."

[23] Cairnes (1857/1875), pp. 38 sq.

of the trajectory, the assumption that gravity is the *only* force acting must be made, but this last assumption is false. When approaching the issue in this way, the deductivists tend to mix the logico-semantic and the causal domains, the latter being essentially thought of in analogy with *forces* and their combination through vector addition in classical mechanics. Mill himself (cited by Cartwright, 1989, p. 173) seems to see reasoning in mechanics as causal and *non-monotonic* reasoning: what lends itself to being inferred from an assumption will not necessarily lend itself to being inferred from this same assumption joined to another. Rather than saying that assumptions that are true *simpliciter* can result in false conclusions, deductivists should say that premises that correctly describe some factors but assume that no other is present can be false (when some of these other factors are in fact present). (ii) The second way to resolve the tension, the only tenable one in our view, consists of applying (T6) for *all* economic propositions, including the assumptions. According to this interpretation, economics is a thoroughly *inexact* science. Whichever theoretical option they choose, deductivists nevertheless agree on the fact that, hypothetical or not, the premises retained in economic theory are not arbitrary.[24] First of all, they describe authentic factors that influence economic phenomena.[25] Second, these selected factors must be among the most important.[26]

2.3 THEORY AND EXPERIENCE ACCORDING TO THE DEDUCTIVE METHOD

From here on let us concentrate on the most contested step in the deductive method, step (s3), which deals with the comparison between the conclusions of the theory and the empirical data. Divergences between the two must be expected: even if the premises of economic reasoning deal with the *principal* causes of economic phenomena, they do not mention *all* the causes which can perceptibly influence them. For example, deductivists mention customs and moral or religious convictions as factors which could interfere with the desire for wealth. This raises the following question: which attitude must be adopted when the conclusions resulting from theory diverge from empirical data? The deductivist replies that comparison with experience lets us know whether we have omitted any important disturbing causes.[27]

The specific content of this reply varies from author to author. For Mill, taking account of these "disturbing causes" belongs to the domain of applied economics and not to economics *stricto sensu*. For Keynes, on the other hand, disregarding all factors apart from the desire for wealth allows us to deliver a "first approximation" which can at times be first rate. But "neither the conception of the economic man nor any other abstraction can suffice as an adequate basis upon which to construct the whole science

[24] Even so, see Mill (1836), p. 46.

[25] Keynes (1890/1917, p. 104) speaks of *verae causae*.

[26] Cairnes, (1857/1875), p. 31 speaks of "leading causes." See Mill (1836), p. 38; Mill (1843), p. 901; and Keynes (1890/1917), p. 60.

[27] Mill (1836), p. 64.

of economics."[28] To resolve many economic questions, the simplistic *homo economicus* theory must be enriched and opened to the other social sciences. This difference has perhaps limited epistemological importance in comparison to the strong and debatable assertions which unite deductivists:

(T7) The divergences between empirical data and economic theory should not incite rejection of the fundamental assumptions.

(T8) Every proposition that is false *in concreto* can be transformed into a true proposition which does take the disturbing factors omitted by the first analysis into account.[29]

(T7) seems justified by the fact that the fundamental assumptions are supposed to be *already* justified by induction. (T8) is difficult to clarify as it mixes logico-semantic and causal concepts. It is once again mechanics, more precisely the *vector sum of forces*, which serves as a model: if a force was omitted from the initial description, that description must be rectifiable by adding the omitted force to those that were mentioned. In contrast, for Mill, chemical phenomena do not obey this composition of causal factors. Economic phenomena are thus closer to mechanical phenomena than to chemical phenomena: they are phenomena where the "Composition of Causes" applies (Mill, 1843, book III, chapter VI, §1), this being a generalization of the composition of forces in mechanics. Neither of these two positions is self-evident. (T7) seems excessively conservative from the perspective of contemporary philosophy of science as it definitively immunizes the basic economic assumptions against empirical questioning. As for (T8), it is clearly weakened by the absence of analogy between mechanics and the social sciences, where there is not any principle of composition of causes similar to the vector sum of forces. The justification that Mill gives for this in asserting that "human beings in society have no properties but those which are derived from, and may be resolved into, the laws of the nature of individual man" (1843, book VI, chapter VII, §1) is too rushed and imprecise to be efficient (see Hausman, 2001). Even if a law like the "Composition of Causes" did exist for social phenomena, economics is not destined to use it systematically: it is occupied with phenomena where the causal factors it traditionally retains among its assumptions (like the desire for wealth) are dominant. Contrary to other scientific domains, the deductive method in economics is *partial*.[30]

---

[28] Keynes (1890/1917), p. 61.

[29] Mill (1836), p. 47.

[30] Hausman (1992a), pp. 145–146. Note that among the assumptions on behavior that are made in contemporary economics, one may distinguish between, on the one hand, those pertaining to agent's rationality (e.g., the transitivity of preferences), which are studied by decision theory, and, on the other hand, those making specific hypothesis on the content of preferences (e.g., that agents prefer larger commodity bundles or larger sums of money). In conventional models, this second set of assumptions typically captures the idea that agents are self-interested. Depending on one's view of the respective entrenchment of these two sets of assumptions in economics, one obtains distinct senses in which economics can be seen as partial. In the strongest sense, it studies the influence of rationality *and* self-interest.

3. Economics as an Inexact and Separate Science

Among the works which have dominated economic methodology for the last 25 years or so, *The Inexact and Separate Science of Economics* (1992a) by Daniel Hausman features incontestably. The author formulates and defends a neo-Millian view of contemporary microeconomics which he calls the "theory of equilibrium."[31] The "theory of equilibrium" hangs on a handful of fundamental laws: the laws of consumer theory, theory of the firm, and also on the assertion that markets arrive rapidly at a situation of equilibrium (where the prices of goods are such that the aggregated supplies and demands balance out). For Hausman, the fundamental assumptions of that theory (for example, the transitivity of consumer preferences, or the maximization of profit for firms) are *inexact* laws. The economic analysis is developed essentially through exploration of their consequences and confidence in the implications of the theory comes more from the confidence placed in the assumptions than from empirical testing.

Hausman's exact position is relatively complex, notably because it combines elements of Millian exegesis, descriptive methodology of contemporary economics, and also normative methodology. We can describe it by indicating the main ideas he identifies in Mill: the position stating that economic laws are inexact; his defense of the deductive method; and the idea that economics is and must be "separate" from the other social and human sciences. Hausman's idea is made up of three components which we will examine one by one: an *enrichment* of the inexactness of economic laws thesis, a *revision* of the deductive method, and a *rejection* of the separation thesis.

3.1 ENRICHMENT OF THE INEXACTNESS POSITION

The assumptions of (micro-)economic theory do not, according to Hausman, have the same status as the fundamental laws of nature: rather they are inexact laws.[32] Thus, he proposes a semantic *and* epistemological analysis of inexactness which breaks down into an analysis of (1) the truth conditions and (2) the justification conditions of the *ceteris paribus* propositions. In his view, an economic assumption like the transitivity of consumer preferences must be understood as "*ceteris paribus*, consumer preferences are transitive" (1992a, chap. 8).

(1). Let us consider propositions of the form

Ceteris paribus, all P are Q.

[31] The place in economics of what Hausman calls the "theory of equilibrium" is the subject of an informed examination in Backhouse (1998), chap. 17. Incidentally, the article also lets us position the two fundamental assumptions (rationality and equilibrium) exposed in subsection 1.2.

[32] This does not preclude that in certain branches of the natural sciences, including physics, laws just as inexact as those in economics can be found.

A semantics for propositions of this form must authorize exceptions to the proposition which is within the scope of the clause: it must be possible for an entity to be *P* without being *Q*, and for "*ceteris paribus*, all *P* are *Q*" to be nevertheless true. The natural idea, taken up by Hausman, is that the *ceteris paribus* clause expresses a domain restrictor (implicit and context-dependent). Let us suppose that we can explicitly formulate that restriction with the predicate *S*: then "*ceteris paribus*, all *P* are *Q*" is true iff

All *P* and *S* are *Q*

is true. The compatibility of this analysis with the deductive method is not self-evident, as Hausman in essence remarks: if the restrictors can vary depending on which proposition they are applied to, then the application of deductive reasoning to a set of propositions does not lend itself to easy interpretation; in other words, the *logic* becomes singularly complicated.[33] Why, in these conditions, should one interpret economic proposition by adding implicit *ceteris paribus* clauses? Hausman's reply hangs, in large part, on what we may call his *nomocentrism*:

> Theorists use basic economic 'laws' to try to explain economic phenomena. They cannot regard them as mere assumptions, but must take them as expressing some truth, however rough. Otherwise, their attempts to use them to explain economic phenomena would be incomprehensible.(p. 139; see also Hausman, 2009)

In other words, laws are required to account for the *explicative* ambitions of economics.

(2). Let us now move on to the *epistemology* of the *ceteris paribus* propositions: in what conditions are we justified in believing of a *ceteris paribus* proposition that it is a law? This is not a trivial matter: for some, these clauses are suspect as they allow one to indefinitely keep falsifiers of the propositions they concern at bay. If we consider a proposition like "*ceteris paribus p*," Hausman puts forward the following four necessary conditions of justification:

(j-i) proposition *p* (unmodified by the clause) must be *lawlike*. In philosophical literature, the term *lawlike* (or *nomological*) is used to speak of a proposition which has all the characteristics of a law, except perhaps that of being true[34]. This condition is natural when we take into account the preceding semantic analysis and the commonly envisaged criteria for characterizing something as lawlike.

---

[33] For example, if the restrictions vary from one proposition to another, then we cannot conclude, in total generality, "*ceteris paribus*, all *P* are *R*" from "*ceteris paribus*, all *P* are *Q*" and "*ceteris paribus*, all *Q* are *R*." Indeed, it is incorrect to conclude "All *P* which are *S′* are *R*" from "All *P* which are *S′* are *Q*" and "All *Q* which are *S* are *R*."

[34] Hence, all laws are nomological, and every true nomological proposition is a law.

(j-ii) *p* must be *reliable,* i.e. largely true in its field of application once certain precise interferences have been taken into account.

(j-iii) proposition *p* must be *refinable,* i.e. we must be able to add qualifications to it which make it more reliable, or reliable in a wider domain.

(j-iv) the proposition must be *excusable,* i.e. the major interferences which allow us to explain the instances when *p* is false must be known.

According to Hausman, the propositions which make up the "theory of equilibrium" are candidates for the title of inexact law. And certain among them, like the assumption of diminishing marginal rates of substitution in consumer theory, or that of diminishing returns in the theory of the firm, would be *good* candidates[35]. It is useful to point out that this is not the case for all propositions to be found in theories of the domain. For example, the proposition stating that goods are infinitely divisible is not lawlike. Hausman calls these unlawlike falsities *simplifications* and proposes a series of acceptance conditions for them analogous to, but distinct from, (j–i)–(j–iv).[36]

3.2 REVISION OF THE DEDUCTIVE METHOD

According to Hausman, if economists do subscribe to a method, it is not exactly Mill's one: they don't accept the (T7) thesis, according to which divergences between empirical data and economic theory should never incite the rejection of economic theory (nor any part of it). In other words, economists, perhaps despite appearances, distance themselves from the dogmatism of the original deductive method. It is nevertheless true that they are reticent, when faced with empirical anomalies, to question their theories. Yet they often have good reason not to. On the one hand, the essential part of their empirical data comes from uncontrolled observation and is not easily compared to the *ceteris paribus* propositions. On the other hand, economic theory, to make empirical predictions, resorts to numerous auxiliary assumptions, assumptions in which economists often have far less confidence than they do in the fundamental assumptions, and that they are thus more inclined to reject. In these conditions, in the case of a conflict with empirical data, it is not unreasonable to blame one or the other of the auxiliary assumptions rather than one of the fundamental ones. This situation makes fundamental assumptions poorly falsifiable from a methodological point of view. Hausman proposes a revision of the deductive method which is supposed to be at once methodologically acceptable and compatible with economists' practices (Hausman, 1992a, p. 222):

---

[35] Let us remember that in contemporary microeconomic theory, the first of these has given way to the assumption of convexity of consumer preferences, and the second to the assumption of convexity of production sets. See, e.g., Mas-Colell et al. (1995), pp. 44 and 133.

[36] Hausman (1981), p. 142.

(s1′) Formulate credible and convenient *ceteris paribus* generalizations concerning the operation of the relevant causal factors.

(s2′) Deduce from these generalizations, initial conditions, and simplifications, etc., predictions concerning the relevant economic phenomena.

(s3′) Test the predictions.

(s4′) If the predictions are correct, consider the whole as confirmed.[37] Otherwise, attempt to explain the failure by comparing the assumptions based on explanatory success, empirical progress and pragmatic virtues.

3.3 REJECTION OF THE SEPARATION THESIS

Are we to conclude from what precedes that, in economics, all's well that methodologically ends well? Hausman replies in the negative. Indeed, in his view it is another important component of economists' practice that is at fault; the idea that economics should be conceived of as a *separate science*. According to that conception, (1) economics is defined by the causal factors it accounts for, (2) its domain is the one where its causal factors predominate, (3) the laws of these factors are already reasonably well known, and (4) it accounts for its domain in an inexact yet unified and complete fashion (1992a, pp. 90–91). From this point of view, economics would be a unified and general science of economic phenomena which borrows nothing from other disciplines.

Some important methodological consequences follow on from the conception of economics as a separate science: among them, the idea that particular intervening assumptions are only legitimate if these assumptions (in the best of cases) derive from fundamental assumptions, or are at least *compatible* with them. If this is not the case, then these assumptions are readily considered as ad hoc. This, according to Hausman, is what drives economists to a form of dogmatism. The statement is justified notably by the study of economists' reaction to the famous phenomenon of *preference reversal*. At the beginning of the 1970s, the psychologists Slovic and Lichtenstein conducted the following experiment: when two subjects are asked to directly give their preferences between two monetary lotteries *H* and *L* (for example: *H* gives a 99 in 100 chance of winning €4 and a 1 in 100 chance of losing €1; *L* gives a 1 in 3 chance of winning €16 but a 2 in 3 chance of losing €2), the majority states a preference for *H* over *L*. Yet when the subjects are asked to assign *minimum selling prices*, the majority assigns a higher minimum selling price to *L* than to *H*![38] Hausman's interest is in economists' reaction

[37] This part of (s4′) is a reflection of the Millian inspiration: it is the initial confidence in the fundamental assumptions that justifies considering the whole as confirmed. A liberal Popperian, who would accept the *ceteris paribus* clauses, would further demand independent tests. We are indebted to Philippe Mongin for this remark.

[38] We recommend the collection Lichtenstein and Slovic (2006) for more on this fascinating phenomenon.

to preference reversal. This reaction was to quite quickly admit that this was a case of authentic empirical anomaly for preference theory, though without going so far as to question the theory's central role. The alleged reason for this hangs on an attachment to economics as a separate science. Grether and Plott (1979), for example, assert, "No alternative theory currently available appears to be capable of covering the same extremely broad range of phenomena." Hausman judges this assertion to be characteristic of partisans of the separation thesis.

In summing up what precedes, we can compare the perspectives of Mill, economists (in Hausman's view), and of Hausman himself concerning Mill's three principal ideas about economic methodology: (a) all agree on the inexact nature of economic laws; (b) Hausman and the economists accept a revised version of the deductive method that authorizes the modification of the fundamental assumptions relative to the empirical data; (c) Mill and the economists are attached to economics as a separate science, something Hausman criticizes. There seems to be a certain tension in Hausman's attempt to defend the economists' methodological practice while also criticizing their conception of economics as a separate science. Hausman (1997) recognizes this tension and delimits the precise part of the economists' methodological practice with which he agrees: the usual empirical data has connections too distant from economic theory for them to maintain decisive relations of confirmation or disconfirmation.

3.4 DISCUSSION

The importance Hausman accords to *ceteris paribus* propositions found echoes in philosophy of the special sciences during the 1990s and 2000s.[39] His position, and other analogous positions, were discussed and contested. Before getting to these criticisms, it is indispensable to point out that the philosophers of science participating in these discussions *interpret* the propositions of some special science as *ceteris paribus* propositions without the representatives of the discipline having openly affirmed the corresponding *ceteris paribus* clauses beforehand. Economics is special: *ceteris paribus* clauses have been explicitly used there for a long time, going back at least to Petty's *Treatise of Taxes and Contributions* (1662, quoted by Reutlinger et al., 2014). It was greatly popularized by A. Marshall. In his *Principles of Political Economy* (1890/1920, see notably V, 5, §2), he makes use of them to signify that, in studying a phenomenon, certain factors can be deliberately put aside. Marshall is interested, for example, in the demand function $x_n(p_n)$ for a particular good *n*, this function being constructed to depend only on the price p_n of this good, as it occurs on the market.[40]. But an individual's demand obviously depends on more factors than just the price of the good in question, be it on his resources, on the price of other goods, etc. These

[39] For an introduction to this literature, see Reutlinger et al. (2014).

[40] See Figure 2, *infra*.

supplementary factors are thus considered to be fixed while we authorize the variation of the price n. Economists' use of *ceteris paribus* clauses has itself been the subject of methodological discussions (see Hausman, 1992b, chap.11), notably because, along with those just mentioned, supposedly exogenous variables (like resources) are mixed with supposedly endogenous variables (the price of other goods to n). A more general theory of demand than Marshall's would take into account the interdependence of prices by contradicting the assumptions stating that the prices of the other goods do not vary.

However, we leave these questions aside to come back to the interpretation of economic propositions as being implicit *ceteris paribus* propositions. Woodward (2002) criticizes this kind of view for its latent nomocentrism. He rejects the idea that laws are necessary to the scientific legitimacy of a discipline or to its explanatory capacities. Following in the steps of Earman and Roberts (1999), he also criticizes the analyses of the truth conditions of *ceteris paribus* propositions such as the one proposed by Hausman. These analyses would be at risk of trivialization: if the system studied is determinist, then it must always be possible to find conditions expressed by S which are by themselves nomologically sufficient for Q and therefore such that "All P and S are Q" is true. Refining the analysis by demanding that neither P nor S be individually nomologically sufficient for Q may lead to consequences which are no less counter-intuitive.

The possibility of confirming or disconfirming *ceteris paribus* propositions, which Hausman defends and analyzes with the conditions (j-i)–(j-iv), is often challenged, for example by Earman and Roberts (1999) and Earman, Roberts, and Smith (2002). They assert, in essence, that when conditions like (j-ii) and (j-iv) are satisfied, we learn the nature and the limits of a statistical relationship without there necessarily being convincing reasons to infer the existence of a law. Besides this, if Hausman is conscious of the "danger of trivialization" present in the conditions (j-ii) and (j-iv), an abusive use of which could lead to the justification of "laws" which clearly should not be laws, this danger could be judged to be too great. This is particularly true of condition (j-iv) which demands an explanation for the counter-examples only a posteriori.

Revisiting the major arguments of his 1992 publication, Hausman (2009) considers that his work may have been marked by the potentially exaggerated role he accorded to laws. It seemed to him that the primary task of philosophy of economics was to understand if, and in what sense, the fundamental propositions of economic theory could be analyzed as laws. Influenced by the recent work of J. Woodward and others on causality,[41] Hausman now intends to organize his methodological contributions on the basis of this latter concept: it is preferable to conceive of economic generalizations as causal claims rather than as inexact laws.

---

[41] See the chapters 1 and 3 of the present volume.

3.5 CETERIS PARIBUS CLAUSES, FOLK PSYCHOLOGY, AND PROGRESS IN ECONOMICS

Before moving on to other works inspired by Mill but which start by placing causality and causal powers at the center of their analysis, it is worth pausing on the ideas of A. Rosenberg. Last in a long series of publications dedicated to economics, *Economics: Mathematical Politics or Science of Diminishing Returns* (1992) accepts both the Millian position of inexactness and its contemporary re-reading, by Hausman, in terms of implicit *ceteris paribus* clauses. We will nevertheless see that, in other regards, he paints quite a different picture of economic science.

Rosenberg's first contribution to philosophy of economics was his book *Microeconomic Laws. A Philosophical Analysis* (1976). It speaks about the nature of the general propositions of microeconomics,[42] and, more precisely, discusses whether or not those propositions that deal with agent behavior can be assimilated to the laws (or nomological propositions) of the natural sciences. Rosenberg's central argument, novel at the time, is that the concepts brought into effect by microeconomic generalities, and the explanatory role that these can play, bring them closely alongside folk psychology, that is, the way in which we habitually explain actions in terms of beliefs and desires.[43] As philosophers of action have highlighted, one of the essential characteristics of our common explanation of action is that the *explanans* appear as a *reason* to undertake the *explanandum.* Going against a tradition often associated with the writings of Wittgenstein and once influential in philosophy of action and the social sciences, Rosenberg maintains that this characteristic does not prevent microeconomic propositions from being causal. Thus, he subscribes to the position, known as causalist and notably maintained by D. Davidson (1980), according to which the reasons for an action can be its causes (Rosenberg, 1975, sec. II; 1976, chap.4 and 5). Another important argument of Rosenberg's (1976) hangs on the assertion that microeconomic propositions are not only causal but also *nomological.* Indeed, they satisfy the generality, the regularity and the necessity which are supposed to be the particularity of laws. According to Rosenberg's view, "there [is] no conceptual obstacle to microeconomic theory's status as a body of contingent laws about choice behavior, its causes and consequences" (1992, p. xiii).

Between the end of the 1970s and the beginning of the 1990s, Rosenberg developed supplementary theses which present that conclusion in a less favorable light:

(T9) Economics does not manifest significant predictive progress in the long term.

Rosenberg considers it an epistemological empiricist commitment that a scientific discipline *must* manifest predictive progress in the long term (1992, p. 18), without which

---

42 The analysis of these propositions is the subject of discussions approaching from other angles than their nomological properties; Mongin (2006b, 2007) discusses their status with regard to the distinctions between analytic and synthetic, and *a priori* and *a posteriori.*

43 Economics certainly borrows from other fields of expertise, whether scientific or otherwise. We can reconcile this to Rosenberg's view by formulating the hypothesis that it is the borrowing from folk psychology that calls for philosophical clarification.

its "cognitive status" as an empirical science becomes problematic. He defends this demand and thinks it is accepted by many economists. But (T9) asserts that it is not satisfied in economics, which is different.[44] The discipline would essentially produce "generic predictions," in other words, "predictions of the existence of a phenomenon, process, or entity, as opposed to specific predictions about its detailed character" (1992, p. 69). The problem, in his view, is not that economics produces generic predictions, but that it seems *incapable* of producing anything else. Why, despite real efforts, does it find itself in this situation? Rosenberg's response is once again based on the bringing together of the conceptual arsenal of microeconomics and the "folk" explanation of action. The two domains share a recourse to *intentional states* (or "propositional attitudes" in philosophy of mind terms), such as beliefs and desires. According to Rosenberg, "the intentional nature of the fundamental explanatory variables of economic theory prohibits [an] improvement [of its predictive power]" (1992, p. 149); in other words,

(T10) The reason for the failure of economics as an empirical science lies in the recourse it has to intentional states.

The same supposedly crushing reason leads Rosenberg to uphold an even stronger proposal: economics *cannot* truly improve its predictive power. Economics as an empirical science, therefore, does not suffer due to a conceptual problem but because it rests on a false hypothesis that it shares with folk psychology, according to which, "the categories of preference and expectation are the classes in which economic causes are to be systematized" (1983). These categories "do not describe 'natural kinds,' they do not divide nature at the joints." This is manifest in the "problem of improvability": if one views the theory of choice on which economics rests as a set of nomological propositions relating intentional states and behaviors and if these intentional states can only be measured through the observation of behavior with the help of this theory, it is hard to see how to improve our predictions in this framework—be it by improving our measurement of intentional states or by considering a better theory. This is how (T10) explains and justifies (T9).

Rosenberg paints an unsparing and contested (see, for example, Hoover, 1995) portrait of economics: its predictive failure is such that the discipline lends itself better to being conceived as a kind of "formal political philosophy" (1992, chap.7) or applied mathematics (1992, chap. 8). Though not accepting this reduction, Hausman shares a part of Rosenberg's pessimism.[45] The reasons for economics' mitigated success are not to be found in its psychological roots but in the fact, already highlighted by Mill, that economic phenomena are *complex and unpredictable.*[46]

[44] See the counter-examples proposed by Hoover (1995), pp. 726–727.

[45] In his view, "[. . .] scientific methods have not worked very well for economists and . . . they are unlikely to work well . . . The best methods of knowledge acquisition . . . have their limits and . . . one should not expect much of economics." (1992b, pp. 99–100).

[46] Rosenberg (2009) later goes back on his own arguments.

4. Tendencies, Capacities, and Idealizations in Economics

4.1 TENDENCIES AND CAPACITIES

Hausman is not the only contemporary philosopher of science to claim allegiance to Mill. Cartwright (1989) defends an idea of causality, influential today in philosophy of the natural sciences, which she reads in his work. For Mill, the fundamental assumptions of economics are tendency laws: not in the sense that they would be generally speaking true, but in the sense that what they express is at work even when other causes disturb their effect:

(T11) A causal law doesn't only describe what is happening *in the absence* of disturbing factors; it says what *tends* to happen regardless of the disturbing factors which may be present.

The introduction of tendencies notably preserves the laws' universal scope. Nancy Cartwright brings them down to what she calls *capacities*. The capacity of a system or device is the property they have to produce certain characteristic results. Thus gravity is a capacity of attraction that bodies have in virtue of their mass and which results in characteristic movements. According to Cartwright, many causal statements, scientific or otherwise, are attributions of capacities: "[. . .] the laws of electromagnetic repulsion and attraction, like the law of gravity, and a host of other laws as well, are laws about enduring tendencies or capacities." This holds not only in the natural sciences: social sciences typically presuppose the existence of capacities too. For example, what would justify resorting to idealizations, the importance of which is widely recognized in modern science, is the assumption that the capacities at work in the ideal cases are *also* at work in the real situations. As for the economic sciences, much of the work in econometrics would rely on the assumption, implicit or not, that some factor (let's say, price) influences, in a *stable* and *measurable* manner, some other factor (let's say, demand). Generally speaking, econometrics occupies an important place in Cartwright's work (1989) because of its philosophically "refined" procedures of causal inference. If Millian economic methodology inspires Cartwright's general philosophy of science, it is however difficult to draw a systematic conception of economic science from her writings, and this despite the enduring interest she displays for the subject (2007, 2009).

4.2 ECONOMIC MODELS AND IDEALIZATIONS

These recent contributions concern the function of theoretical economics models and, more precisely, the persistent problem of their unrealism (see also section 6 on M. Friedman). Economists recognize and claim a fundamental role for these models,[47]

[47] See the letter of July 4th 1938 from J. M. Keynes to Harrod: "Economics is a science of thinking in terms of models joined to the art of choosing models which are relevant to the contemporary world." More recently, Krugman (2009, p. 18) affirms: "The only way to make sense of any complex system, be it global

whose lack of realism is manifest. Economists are sometimes accused of studying the imaginary worlds that the models describe rather than the real world itself. Economic methodology converges toward contemporary discussions, very much alive in general philosophy of science, around that concept (see Frigg and Hartmann, 2009 and chapter 5 in this volume).

Cartwright thinks that physics models lack no less realism than economic ones and that the lack of realism objection is not the right one. Economic models are situated, at first glance, among the methodologically respectable family of *Galilean idealizations*[48] (McMullin, 1985): procedures by which, theoretically or experimentally, a cause is isolated from other causes which could disturb the effect that it produces. For Cartwright, Galilean idealization allows a capacity to be fully exercised and consequently allows the scientist to understand the causal contribution it brings *in general*. From this point of view, the lack of realism is not a *problem*, but rather a *means*: "frequently, what we are doing in this kind of economic theory is not trying to establish facts about what happens in the real economy but rather, following John Stuart Mill, facts about stable tendencies" (2007, p. 221). Which we can reword thus:

(T12) An essential part of economic modeling is destined to isolating causal factors so that their effects can be studied separately.

It is a position defended in a different philosophical setting by U. Mäki (see Mäki, 2009c).

For a partisan of (T12), the question to be asked is whether economic modeling *succeeds* in this enterprise of isolation. Cartwright (2007, 2009) gives a reserved response. Indeed, many idealizations present in economic models are not Galilean but instead consist in supplementary assumptions pertaining to the "structure" of economics. This claim is illustrated by contemporary macroeconomic models like Lucas's (1972).[49] In such a model, individuals live for two periods, are of equal number in each generation, all produce goods that cannot be stocked, cannot pass the goods they possess to the following generation, etc. According to Cartwright, the economist needs these supplementary assumptions because the fundamental principles on which these models rest, typically specifications of assumptions (a1) and (a2) (rationality and equilibrium), are too few to result in interesting conclusions. But as a result, the guarantee that the conclusions could be exported to other circumstances—as Galilean idealization would have it—is lost. Economic models would thus be "overly constrained." The situation would be far more favorable in physics where one can rely on fundamental

---

warming or the global economy, is to work with models—simplified representations of that system which you hope help you understand how it works."

[48] Discussion of the properties of economic models permanently enlists the notion of idealization. For a classification of the different types of idealization, see Walliser (2011), chap. 3, sec. 2.

[49] R. E. Lucas Jr. (1972), "Expectations and the Neutrality of Money," *Journal of Economic Theory*, vol. 4, pp. 103–124.

principles in far higher number.[50] To sum up, with economic models, "the worry is not just that the assumptions are unrealistic; rather, they are unrealistic just the wrong way" (2009, p. 57).[51]

4.3 DISCUSSION: MODELS AS "CREDIBLE" WORLDS

The question of knowing whether, and how, models such as those found in economics allow us to acquire knowledge about relevant aspects of the world is particularly debated in philosophy of economics today. For example, according to R. Sugden (2000, 2009), special theoretical models[52] like Akerlof's "market for lemons" (1970)[53] do not aim at disregarding causal factors which supposedly exist. More generally, they don't have the ambition of providing firmly grounded knowledge about the capacities at work in these phenomena. Instead, they should be seen as counter-factual worlds which, by virtue of their similarities with the real world, can convince us of the plausibility of certain conjectures concerning it. For example, the market for lemons model makes plausible the proposition stating that, all other things being equal, an asymmetry of information about the quality of the goods being exchanged tends to reduce the volume exchanged (see Table 1).

Sugden particularly puts the accent on the *abductive* use of economic models: logical exploration of the model shows that in the counter-factual world it describes, some factor F (the asymmetry of information) induces some economic phenomena (e.g., low volumes exchanged). If the model presents relevant similarities with the real world, and if analogous economic phenomena are observed in the real world, then the model makes plausible the explanation of these phenomena by a factor analogous to F. The inductive force of these kinds of reasoning, according to Sugden, lies in the similarity between the real world and the worlds described by the models: these must be *credible* given what we believe about our real world. In this view, "[. . .] the model is not so much an abstraction from reality as a parallel reality. The model world is not constructed by starting with the real world and stripping out complicating factors: although the model world is simpler than the real world, the one is not a *simplification* of the other" (Sugden, 2000).[54]

* * *

[50] The contrast between economics and physics would, in reality, demand much deeper examination. It is not obvious that in physics the fundamental principles are sufficient for avoiding the "overly-constrained" problem once we move away from the discipline's "central core." We thank B. Walliser for his remarks on this point.

[51] For an elaboration of this idea, see also Reiss (2013, chap. 7), who proposes several dimensions along which differ truly Galilean idealizations and those found in economic models.

[52] The models Sugden is interested in belong to those which, in an article which anticipates current discussion on models, Gibbard and Varian (1978) call "caricatures." These are simple models which are applied to economic situations in a "casual" way: they must "explain aspects of the world that can be noticed or conjectured without explicit techniques of measurement," in contrast to the models which are applied in an econometric way. Gibbard and Varian's central argument is that these models are not conceived to be approximations of the economic reality, but as *deliberate exaggerations* of certain of its characteristics.

[53] Akerlof (1970), pp. 488–500.

[54] Hoover (2001a) also discusses Cartwright's ideas about economics and its models. The angle of attack varies to the one we have presented here and favors macroeconomics and econometrics.

TABLE 1

Akerlof's market for lemons (1970)

The automobile market brings the members of two groups into play. The members of group 1 possess N cars whose quality x is uniformly distributed between 0 and 2. Their utility function is given by $U_1 = M + \Sigma^n_{i=1} x_i$ where M stands for the consumption of other goods and x_i is the quality of car i. The members of group 2 do not have cars. Their utility function is given by $U_2 = M + \Sigma^n_{i=1} 3/2\, x_i$. (Thus, members of group 2 attach more value to these cars and it is expected that some trade will take place.) The respective revenues of the two groups are noted Y_1 (which includes any possible revenue made from the sale of cars) and Y_2. All agents maximize their expected utility. The price (unique) of the automobiles is p while the price of the "other goods" is 1. The information is asymmetrical: the members of group 1 have knowledge of the cars' quality, those of group 2 know only their average quality μ. According to these assumptions, the members of group 1 will be inclined to sell a quantity $S(p) = p.N/2$ of cars if $p \leq 2$ and the average quality of the cars exchanged will thus be $\mu = p/2$. In these conditions, the global demand $D(p, \mu)$ will be null and no automobile will be exchanged: the members of group 2 knowing μ are only inclined to buy at the price ¾ p. (Intuitively: owners of high-quality cars won't have any interest to sell and given that only low-quality cars are available, buyers are not willing to pay p.)

If, on the other hand, the members of group 1 also only have knowledge of the average quality of the automobiles, i.e. if the information is imperfect yet *symmetrical*, then equilibria will exist where the volumes exchanged will be non-null.

Millian deductivism is largely defensive: its intention is to explain *and* justify the epistemological particularities of economics. In its original form, it immunizes the fundamental assumptions of economic theory, since the comparison between empirical data and theoretical predictions would not be engaged in evaluating them. This view has always aroused unwillingness, even stretching to economists' way of doing things, insofar as they seemed to conform to the deductive method. We will not trace the detailed history of its decline through the 1930s and 1940s. Two factors undoubtedly played an important role, factors which can be considered in either a disciplinary or more conceptual manner. From the side of economics, it appeared doubtful, notably in theory of the firm, that fundamental assumptions like the maximization of profit should enjoy the obviousness that certain Millians were crediting them with.[55] In this way, confidence in the propositions of economic theory becomes difficult to rationalize if we suppose that it results primarily from confidence in these

[55] See what we say further on about the historical context of M. Friedman's *Essay*.

assumptions. Moreover, on the side of philosophy of science, this period saw the diffusion of ambitious and demanding visions of scientific knowledge, notably those from within neo-positivism, which ousted the older views like Mill's. The Millian solution to Mill's "generalized" problem revealed itself to be inadequate: it assessed economics through the lens of defective methodological standards, standards that *in any case* economics couldn't manage to satisfy. Sections 5 and 6 of our chapter are dedicated to a methodological tradition which we can liken, but only to a certain extent, to the neo-positivist views. It is not just a matter of *variants* of neo-positivism, since refutationism will be included which, in its Popperian version, was vigorously opposed to the Vienna Circle. Rather it is a matter of notions, directly influenced or otherwise, which take up certain fundamental positions, starting with the determining importance for theory evaluation of the comparison between its predictions and empirical data.[56]

5. Paul Samuelson, Revealed Preference Theory, and Refutationism

5.1 REVEALED PREFERENCE THEORY

We will start with the methodological views vindicated or implemented by Paul Samuelson during the 1930s and '40s (from the revealed preference theory to the *Foundations of Economic Analysis*, 1947). On Samuelson's methodology, see Mongin (2000a, section III), to whom this section is deeply indebted, and also Wong (1978/2006). Samuelson certainly didn't "apply" neo-positivist ideas to economics. But many of his methodological options or convictions relate to them. We will give just one example, which we shan't come back to: Samuelson was attached to the ideal of the *unity of science*, as witnessed by his Nobel Prize acceptance speech which he dedicated specifically to the unifying role of maximization, in economics as among the sciences. We will concentrate our study on two of Samuelson's major projects, which also happen to be closely linked: revealed preference theory and the search for "operationally meaningful theorems" of economics.

Revealed preference theory is the result of a research program on the microeconomic consumer theory, launched by Samuelson at the close of the 1930s, and that many (including Samuelson) consider to have been completed by Houthakker (1950). With Samuelson (1938a), the objective attached to this program is to allow economics to do without the "residual traces of the concept of utility" found in contemporary consumer theory, developed on the basis of the concept of preferences (or ordinal utility, see Hicks and Allen, 1934). Hicks and Allen (1934), in the wake of Pareto's arguments, had proposed replacing Marshall's consumer theory, which relied on a notion of cardinal

[56] Popper (1963/1989, p. 54) formulates and defends "the principle of empiricism which asserts that in science, only observation and experiment may decide upon the acceptance or rejection of scientific statements, including laws and theories." It is this kind of principle that unites the ideas developed in this second part of the paper.

utility,[57] with a theory which would make do with ordinal utility (or with preferences, to use more recent terminology):

> It is necessary, in any theory of value, to be able to define just what we mean by a consumer of 'given wants' or 'given tastes'. In Marshall's theory (like that of Jevons, and Walras, and the Austrians) 'given wants' is interpreted as meaning a utility function, a given intensity of desire for any particular collection of goods. This assumption has made many people uncomfortable, and it appears from Pareto's work that it is not a necessary assumption at all. 'Given wants' can be quite adequately defined as a given *scale of preferences*; we need only to suppose that the consumer has a preference for one collection of goods rather than another, not that there is ever any sense in saying that he desires the one collection 5 per cent more than the other, or anything like that.[58]

However, the concepts of utility and preference are considered as psychological *and* non-observational, in contrast to choice behavior, which is supposed to be observable. For Samuelson, a consumer theory based only on behavior, thus "more directly based upon those elements which must be taken as *data* by economic science," is "more meaningful in its formulation" (1938a, p. 71).

These initial motivations of revealed preference theory seem to belong to a kind of timid eliminationism with regard to non-observational concepts: to rely on a theory formulated exclusively in terms of observational concepts is (all things being equal) *a* progress, not a *sine qua non* condition to the field's scientificity. The approach is not always understood in this way. For example, for Malinvaud (1972/1985), who is not one of its defenders, it belongs to a stronger eliminationism which he puts like this: "the scientist must not introduce non-operational concepts into [her] theories which do not lend themselves to objective observation." The discipline's history itself decided on this by creating a coexistence between the consumer theory of Hicks and Allen and the study of behavioral properties put forward by Samuelson.

5.2 Discussion of Revealed Preference Semantics

Revealed preference theory call for other, less historical, remarks.

(1) The theory's eliminationist motivations underwent a similar fate in economic methodology to that of eliminationism in general philosophy of science: the elimination of theoretical concepts is considered as neither desirable nor, more often, practicable. Economics has the particularity that, for some of its central theories (such

---

[57] Simply put, a numerical function on options is an ordinal utility function if it represents only the way in which the individual ranks her options in terms of her preferences; it is cardinal if it also represents the *intensity* of these comparisons.

[58] Hicks (1939, pp. 17–18). Certain economists nevertheless think that the two notions of preference and ordinal utility do not coincide: it would also be possible to "cardinalize" preference (see d'Aspremont and Mongin, 1998).

as consumer theory), elimination does seem possible: it can be shown that Hicks and Allen's version, which contains theoretical concepts, and Samuelson's version, which contains only observational (or supposedly observational) concepts, are in fact equivalent. As Mongin (2000b) underlines, this epistemic situation is not without its advantages since the theory formulated in observational language allows not only for the characterization of all the testable consequences of the initial theory, but also for the containment of any potential refuters of that theory.

(2) Moreover, revealed preference theory can be associated with a *semantics* for the concept of preference which largely exceeds the theory itself: in that perspective, preferring option x to option y *signifies* choosing x rather than y when the two options are available. Despite regular warnings from economic philosophy dating back to Sen, economists persist in incorrectly distinguishing that vague and dubious semantics from the theory which, as we have just seen, is precise and tenable. In contrast to the latter, which is barely discussed at all any more, the former continues to play an important methodological role; in particular it inspires Gul and Pesendorfer's (2005/2008) hostile anti-neuroeconomics manifesto. The open defenders of revealed preference essentially maintain that

(T13). The only legitimate or necessary notion of preference in economics is the notion of revealed preference.

Sen (1973) was the first to distinguish himself by rejecting (T13). First of all, it wouldn't be tenable to see in revealed preference theory an attempt to *eliminate* the concept of preference: if we completely deprive ourselves of it then we also lose any possible source of justification for the assumptions of the new theory. If we dismiss that first interpretation, we are still left with the revelation hypothesis which states that preferences are directly expressed in choices. Yet, again according to Sen, an individual's choices are not rigidly linked to her preferences; in forming an assumption of this sort we run the risk of muddling the preferences revealed by choices, the genuine individual preferences, and other motives which also influence choices all into one and the same concept. Sen was followed by Mongin and d'Aspremont (1998) and Hausman (1992, 2000, and 2008), who maintained that "economics cannot function without a subjective notion of preference, which does not and cannot stand in any one-to-one relationship with choices" (2008, p. 132). Hausman imagines several objections: (a) The first is that preferences, in the usual sense, are not expressed in choices without assumptions about the agents' beliefs. (b) Economics doesn't only relate preferences to objects of choice, nor even to hypothetical choices. It borrows from game theory, where preferences relate to possible *consequences* of the interaction between agents, as well as from social choice theory where, according to the model established by Arrow (1951), preferences relate to abstract states of society. Concerning game theory, we can think of its elementary predictive task as being the prediction of *choices* between feasible strategies on the basis of beliefs and preferences about the possible *consequences*. (c) Finally, the theoretical apparatus of economics and decision theory would lose its explanatory power if we adopted the semantics of revealed preference: it would be a matter only of recording the behavioral generalizations without once looking at the causal factors responsible for this behavior.

5.3 SAMUELSON'S "OPERATIONALLY MEANINGFUL THEOREMS"

As Houthakker (1950) was already pointing out, Samuelson doesn't always attach his theory to an exclusive methodological motivation. In Samuelson (1950), it is no longer a question of *eliminating* the residual traces of the utility concept of consumer theory but of obtaining the "full empirical implications for demand behavior of the most general ordinal utility analysis." One of the objectives of the *Foundations of Economic Analysis* (1947) is precisely to derive "operationally meaningful theorems." These are hypotheses a "about empirical data which could conceivably be refuted, if only under ideal conditions,"[59] Samuelson wants to show that economics, and consumer theory in particular, do indeed entail operationally meaningful theorems.[60] For example, if a consumer obeys conventional theory (in terms of preferences), then she must conform to the Weak Axiom of Revealed Preference (WARP) according to which, for all price vectors p, p′ and budgets w, w′:

(a) if the consumer doesn't choose the same basket of goods in conditions (p,w) and (p′,w′) (i.e. $x(p,w) \neq x(p',w')$), and
(b) if she can buy the basket of goods x(p′,w′) in conditions (p,w),

then she cannot buy x(p,w) in conditions (p′,w′)—in other words, x(p,w) exceeds the budget w′ when prices are p′.

The axiom is better understood if we introduce the concept of preferences on top of that of choice: if the consumer doesn't choose the basket of goods chosen for (p′,w′) in conditions (p,w), even though she has the means to, this means that she *prefers* the basket she chooses, and the choice observed in conditions (p′,w′) must be compatible with that same preference; so x(p,w) must not be affordable given her budget. Often the relation "x is *revealed preferred* over y" is *defined* by the property that the consumer demands the basket of goods x even though both the prices and her budget allow her equally to demand y. Thus the Weak Axiom comes down to demanding that the relation ". . . revealed preferred over . . ." be asymmetrical. These refutable consequences give birth to what economists call the non-parametric tests of consumer theory (see

[59] See Samuelson (1970, p. 10): "From the beginning I was concerned to find out what *refutable* hypotheses on the observable facts on price and quantity demanded were implied by the assumption that the consumer spends his limited income at given prices in order to maximize his ordinal utility."

[60] Two answers are given to the question of knowing which pressures on consumer behavior are implied by the theory. (i) Slutsky's substitution matrix must be symmetrical, negatively semi-defined, and the demand function homogeneous to degree 0 relative to prices and to revenue. (ii) The demand function must obey the Strong Axiom of revealed preference. The second answer is the result of revealed preference theory.

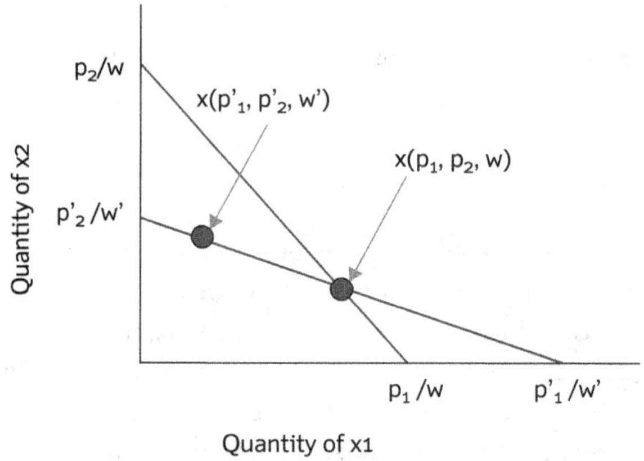

FIGURE 3 Violation of the weak axiom of revealed preference. The consumer does not choose the same basket of goods in conditions (p_1, p_2, w) and (p'_1, p'_2, w); she can buy $x(p'_1, p'_2, w)$ in conditions (p_1, p_2, w); but she can also buy $x(p_1, p_2, w)$ in conditions (p'_1, p'_2, w).

Varian, 1982, 1992, chap. 8, 12). It is important to point out that we are dealing with an idealized notion of refutability. What we can directly observe at a given moment *t*, is at the very most a consumer's demand (given the prices and her budget). For the demands of the consumer x(p,w) at *t* and x(p',w') at *t*′ to conflict with the Weak Axiom, we must suppose that the consumer's preferences, or her demand function, remain *stable* between t and *t*′. If we really want to conduct tests with natural data, then hypotheses about the identification of the consumers, the identification of the goods, the separability of present and future demands, etc., must also be made, and account must also be taken for the fact that these data are finite, whereas the demand function x(p,w), by definition, covers a continuum of situations (see Chiappori, 1990).

5.4 REFUTABILITY AND REFUTATIONISM

The determination of the refutable consequences of theories plays a crucial role in a refutationist approach to science. Refutationism wielded great influence over economic methodology with Samuelson's *Foundations*, but already it had inspired the strictly methodological work of Hutchison, *On the Significance and Basic Postulate of Economics* (1938), and it finds a rebirth through the seminar, "*Methodology, Measurement and Testing in Economics*" (M²T) at the London School of Economics (Archibald, Lancaster, Lispey).[61] The work of M. Blaug (1980/1992) is the current methodological manifestation of this. Unlike Samuelson, whose philosophical sources are poorly identified, all these authors are influenced by the Popperian version of refutationism which makes refutability the criterion of demarcation between science and non-science, and makes refutation the means by which our scientific theories are evaluated.

[61] See Lipsey (2008). Klappholz and Agassi (1959) can be associated with the same group.

At the meeting point of Samuelson's research program and Popperian ideas, several members of the M²T, during the 1960s, explored the refutable consequences of various contemporary economic models (see Mongin, 2005). It was already appearing from *Foundations* that, following the ordinary distinction of what is observable from what is not, the refutable consequences of economic theory were to be found in qualitative comparative statics: the interest, then, is on the sign of variation of an endogenous variable when an exogenous parameter varies. It turns out that the variables and parameters have to maintain very particular relationships for the signs of variation of the former to be unequivocally determined by the variations of the latter and so that, consequently, refutable consequences can be reached. Archibald (1965) arrives at the conclusion that, "It seems unfortunately to be the case that the general qualitative content of maximizing models is small, if not trivial." For a refutationist wanting to turn refutability into a criterion for scientificity while holding on to parts of the economic theory in question, this conclusion is discouraging. The question of the refutable consequences of economic theories has an interest which exceeds refutationism, on top of which we would like to add some elements concerning more recent microeconomic models.[62]

(1) After the Second World War, theoretical economics progressively adopted the model of expected utility as a reference for individual decision made under uncertainty, that is, when the decider is not in a position, for all possible actions, to know what the consequence of that action will be. According to this model, an action's value is the sum of the products of all the values of the action's possible consequences multiplied by the probability of their occurring. Thus, when uncertainty is already probabilized, the options can be identified with probability distributions (economists speak of "lotteries") and the model posits that the decision maker prefers lottery *P* over lottery *Q* if and only if

$$\sum_{c\in C} P(c).u(c) \geq \sum_{c\in C} Q(c).u(c)$$

We have noted *P(c)* the probability of obtaining consequence *c* if lottery *P* is chosen, and *u(c)* the utility the agent attaches to *c*. This model imposes a property of "independence," according to which option *P* is preferred over option *Q* if and only if the probability mixture of *P* with some other option *R* is preferred over the probability mixture of *Q* with the same *R*, and in the same proportions.[63] This proposition is considered to be refutable, and, in certain situations, individuals *seem* to violate the axiom of independence.[64] The reservation is important, as the situation resembles a Duhemian problem, see Mongin (2009). The expected utility model is thus refutable and, according to the general view, *refuted* too. A vast program of collective research

62 For the sake of space, we leave aside Lakatos's influence on economic methodology.

63 By definition, the α-mixture of lotteries *P* and *R*, written $\alpha P \oplus (1-\alpha)R$ assigns the probability $\alpha P(c) + (1-\alpha)R(c)$ to a consequence *c*. It is easily verified that $\alpha P \oplus (1-\alpha)R$ is also a lottery.

64 These cases of alleged refutation match well known paradoxes, such as Allais' paradox (1953).

among economists and psychologists, still ongoing, allowed for the elaboration of decision models for uncertainty that are compatible with the observed anomalies. For the moment, the most convincing models are typically *generalizations* of the expected utility model, which make one lose in refutable content what is gained in empirical validity. As such, refutationism is safe only on first analysis (again, see Mongin, 2009).

(2) A second innovation of contemporary economics, even more recent, is the massive reliance on game theory. The question arises, once again, of knowing whether the theory is refutable or not. Several economists and philosophers of economics have looked into this question in recent times (Weibull, 2004; Hausman, 2005; Guala, 2006). Game theory works by constructing "solution concepts" which select, for a set I of participants and for a given strategic configuration G, certain action profiles noted $S(G) \subseteq \times_{i\in I}(A_i)$ where A_i is the set of actions available to the individual i. At first glance it seems easy to think up a situation which would be disadvantageous for such a solution concept: (a) we observe individuals interacting according to G; (b) the actions $\underline{a} \in \times_{i\in I}(A_i)$ selected by these individuals do not belong to $S(G)$. Hence, it is often considered that Nash's equilibrium (recalled in subsection 1.2) is jeopardized in situations which reproduce the Prisoner's Dilemma (Table 2): experimentally, individuals tend to "cooperate" rather than "defect."[65]

In this perspective, the refutability of game theory seems to pose no particular problem. Moreover, it would be variable according to the games since, with some of them, the solution concept involved is incompatible with many action profiles, something which is not the case with others. Several remarks are necessary, however.

First let us note that we have supposed game theory lends itself to the customary game of scientific assumptions, even though it is not obvious that this be the case when it is proposing solution concepts. For many specialists, it thus more defines *norms* for comparison with observed actions and not strictly speaking assumptions. It is only within certain applications that the theory appears to want to expose itself. Here, straight away, we see a difference with the theory of individual decision. But let's pursue anyway, supposing an empirical interpretation of the theory.

We must then bring attention to the fact that our provisional conclusion, according to which the refutability of the theory seems unproblematic, is based on the assumption that the individuals do indeed take part in game G. What is open to testing then, is simultaneously (ai) the assumption that, in situation G, individuals obey the solution proposed by game theory, and (aii) the assumption that they play game G. This second assumption cannot be directly assessed, if only because the individuals' preferences, supposed to be unobservable, take part in the definition of what a game is. Consequently, when it is observed that the profile of selected actions $\underline{a}$ is incompatible with $S(G)$, we can, in principle, point the figure at (aii) rather than (ai), that is to

[65] It is easy to verify that the action profile (defect, defect) is a Nash equilibrium: the best option for a player, presuming that the other player is defecting, is to do the same. Moreover, this equilibrium is unique.

TABLE 2

The Prisoner's Dilemma

		Player 2	
		Cooperate	Defect
Player 1	Cooperate	(3, 3)	(0, 4)
	Defect	(4, 0)	(1, 1)

Each player has the choice between cooperating and defecting. To each action profile corresponds, in the grid, the vector of utilities for the two players. Thus, the profile where each player cooperates induces a utility of 3 for each player.

TABLE 3

Prisoner's Dilemma's Game Form

		Player 2	
		Cooperate	Defect
Player 1	Cooperate	($3, $3)	($0, $4)
	Defect	($4, $0)	($1, $1)

say, we can question whether the individuals are really playing game *G*. Let's suppose, for example, that the subjects are put in the following situation: each of them have the choice between two possible actions and, according to the actions chosen, they obtain the vectors of *monetary gain* given in Table 3.

Table 3 doesn't describe a game, as the individuals' preferences are not specified. If the subjects do not defect, it will be possible to save the theory by maintaining that they didn't play the game described by Table 2. It could, for example, be maintained that the preferences of a subject *i* are not increasing functions of her monetary gain. This (natural) idea, inspires much work in experimental game theory which associates situations like those described in Table 3 with games where individuals' preferences take the monetary gain of the other players into account.

Coming back to the general discussion, the essential difficulty resides in the fact that it is awkward to test (aii) independently. From that, one may conclude, like Hausman (2005), that "economists can often learn more by using game theoretic anomalies to study the factors influencing preferences rather than by treating them as disconfirming game theory." Guala (2006) recognizes these methodological difficulties, but maintains that constraints from decision theory on the revelation of the players' preferences impose certain limits on the flexibility of game theory which, as a result, is refutable—*and refuted* by certain recent experiments.

6. Milton Friedman and the "Realism" of Assumptions

6.1 THE CONTEXT

The most famous contribution to contemporary methodology certainly remains "The Methodology of Positive Economics" (1953) by Milton Friedman. This article was read and widely discussed, not only by philosophers of economics, but also by economists themselves. Commentaries are legion and continue to multiply: Nagel (1963), Simon (1963), Mongin (1988, 2000a), Musgrave (1981), Blaug (1980/1992), Hausman (1992b), Mäki (2009a). Friedman's essay was interpreted in many ways: refutationist, conventionalist, instrumentalist, realist, etc. In fact it is doubtful whether the *Essay* presents any single coherent epistemology.[66] The article can be seen as an attempt at reconciling economic methodology and philosophy of science, as they were practiced at the time. It was largely taken to be a *defense* of economic practice in the face of the most tenacious objections it encounters and, in particular, in the face of the objection, which we have already discussed in reference to Mill, according to which economic theory is based on excessively *unrealistic* assumptions. So we shouldn't be surprised that Friedman's theses were favorably received by certain economists.[67]

Before looking at these theses, it is worthwhile placing them in their historical context. Indeed, the article follows after one of the major internal controversies of the discipline, the marginalist controversy in the theory of the firm which had developed just after the Second World War. The theory of the firm that we know today took root progressively during the 1930s (see Mongin, 2000a). At the end of this time period, several researchers attempted to test its fundamental assumption—of profit maximization—independently of its consequences, by directly posing questions to company heads. The results of these questionnaires, in regards to methods of price fixation and hiring, seemed to utterly contradict that assumption. If, as the Millian tradition would have it, we consider that confidence in economic theory stems from confidence in its assumptions, the situation becomes problematic. Friedman proposes another way of thinking about the assessments of the theory of the firm and of economic theories in general, a way of thinking which, ultimately, would vindicate them against objections based on the implausibility or falseness of its assumptions.

6.2 FRIEDMAN'S THESES

Many reconstructions of Friedman's theses have been proposed. We will opt for this one:

(T14) A (scientific) theory must be primarily judged by the correctness of its predictions (pp. 4, 9–10, 15, 31)

[66] Mongin (2000a), Mäki (2009b).

[67] To take just one example, the introductory manual of Stiglitz & Walsh (2000, p. 123) rejects criticisms of consumer theory's lack of psychological realism in much the same way as Friedman.

(T15) A theory must not be judged by the "realism" of its assumptions (pp. 14, 41)

(T16) A theory affirms that everything happens *as if* its assumptions were true (pp. 17–19, 40)

(T17) If a theory is important and significant, then its assumptions are not "realistic" (p. 14)

Theses (T14) and (T15), baptized "F-Twist" by Samuelson (in Archibald et al., 1963), are the most important, we won't really discuss the other two. Nagel (1963) and Mäki (2009b) highlight the ambiguity of the "*as if*" in (T16), the latter showing that, in certain passages (pp. 19–20), use of the phrase is clearly instrumentalist, while, in others (p. 40), the author draws on realism by suggesting that everything happens *as if* certain ideal conditions were satisfied. As for (T17), the thesis is particularly elaborated by Mongin (1988) who discerns in it both a banal interpretation and an unreasonable interpretation with the help of the neo-positivist definitions of scientific theories.

The first thesis (T14) rests upon a notion of prediction that Friedman sees in an epistemic and not temporal fashion: *P* is the prediction of a certain theory at a moment *t* if *P* follows on from the theory, potentially enriched by auxiliary assumptions, and if *P* is an empirical proposition *whose truth value is not yet known at t*. Consequently, *P* can just as well concern a future phenomenon (prediction in the strictest sense) as a phenomenon which has already occurred (retrodiction). Friedman seems to see only a pragmatic difference between prediction and explanation, i.e. to explain is to predict something we know has occurred.[68] In reality, he limits the field of prediction by adding that a "theory is to be judged by its predictive power for the class of phenomena which it is intended to 'explain'" (1953, p. 8). In other words, the theory's surface domain, that to which it seems to apply, must be distinguished from its *target domain*, that which matters in its empirical evaluation; and (T14) becomes:

(T18) A theory must be (primarily) judged by the correctness of its predictions relative to its target domain.

To the question of knowing what economic theory's target domain is, two main responses are conceivable:

(a) The first, which corresponds with Friedman's examples, consists in maintaining that it includes the behavior of economic agents, but not their mental states or processes. Certainly, the best illustration is to be found in the article that Friedman wrote with Savage in defense of expected utility theory and which it is worth quoting at length:

> The hypothesis does not assert that individuals explicitly or consciously calculate and compare expected utilities. Indeed, it is not at all clear what such an

[68] See chapter 1 of the present volume.

> assertion would mean or how it could be tested. The hypothesis asserts rather that, in making a particular class of decisions, individuals behave *as if* they calculated and compared expected utility and *as if* they knew the odds. The validity of this assertion does not depend on whether individuals know the precise odds, much less on whether they say that they calculate and compare expected utilities or think that they do, or whether it appears to others that they do, or whether psychologists can uncover any evidence that they do, but solely on whether it yields sufficiently accurate predictions about the class of decisions with which the hypothesis deals.[69]

(b) The second response consists of maintaining that the target domain contains only aggregated variables, such as prices or the quantities of goods exchanged. This interpretation dates back at least to Machlup (1967), for whom the target domain would be made up of "*mass behaviors*." He matches this interpretation to a restriction to predictions of comparative statics (see *infra*), a restriction not read in Friedman.

The first thesis, (T14), modified in (T18), serves as a foundation for the second, (T15), directly pointed against objections to economists' practices. The scope of the response depends on the notion of "realism" employed, something which is far from unequivocal with Friedman. Several commentators have sought to clarify this.[70] The two most common interpretations are: (i1) realism as exhaustivity (in this sense a set of assumptions is unrealistic when it doesn't say *everything* about relevant objects); (i2) realism as truth (in this sense a set of assumptions is unrealistic when some of the assumptions are false), or with very high probability of being true.

The premise of Friedman's argument in favor of (T15) is that a set of scientific assumptions is *necessarily* unrealistic. The question to be asked then is whether this set is realistic *enough*, despite everything else, to satisfy the economist's epistemic objectives. It is here that (T18) intervenes: the only standard we possess for judging the previous question is the empirical correctness, relative to the target domain, that the hypotheses authorize. There is no intrinsic criterion for deciding whether a set of assumptions is a "good approximation" or not. Just as it is pointless to abstractly debate the realism of the law of falling bodies—this depends on the kind of context in which predictions of the law are expected—so it is pointless to criticize the central assumptions of economic theory for the reason that they do not accurately describe economic agents' reasoning, or even their individual behavior. The strength of the argument will obviously depend on the meaning given to the notion of realism. If we go far (i1), then the premise is trivial, as Nagel (1963) remarks, and the part of the conclusion dealing with unrealism will be too. On the other hand, if "realism" is to be understood in the (i2) sense, the premise is far more contestable.

69 Friedman and Savage, 1948, p. 298

70 Notably Nagel (1963), Musgrave (1981), Mäki (2000).

Perhaps, to obtain a non-trivial methodological argument, the sequence must be understood in still a different way. In essence, Hausman (1992b) proposes passing by the intermediary conclusion (C), which differs subtly from (T18):

(T18) A theory must be (primarily) judged by the correctness of its predictions relative to its target domain.

(C) The only test for judging a theory consists of directly determining whether it provides correct predictions relative to its target domain.

(T15) A theory must not be judged by the "realism" of its assumptions.

6.3 DISCUSSION

It is difficult to give an overview of the objections that have been brought against Friedman's theses. We will concentrate on Hausman's objection (1992b) which focuses on the argument we have just exposed. The passage from thesis (T18) to the intermediary conclusion (C) is not, in his view, legitimate. Indeed, let's consider the parallel thesis about the purchasing of a second-hand car:

(T18′) A good second-hand car is reliable, economical and comfortable.

(C′) The only test for knowing whether a second-hand car is a good second-hand car consists in directly determining whether it is reliable, economical and comfortable or not.

(T15′) Everything that can be found out by opening the hood of the second-hand car and inspecting its various component parts is irrelevant to its evaluation.

The conditions mentioned in (T18′) must be understood to be necessary and sufficient conditions in assuring the parallel with (T18). This last thesis would be convincing were it possible to know all the road performances, past and future, of a second-hand car. Then we wouldn't need to "look under the hood." In the same way, for somebody who, like Friedman, accepts (T18), if all the empirical performances, past and future, of a theory could be known, we would have everything necessary for its evaluation. But the point that Hausman puts forward is that *we are not* in such an epistemic situation. The inspection of a theory's "components" can be an important resource when, for example, we want to extend the theory to new situations, or when we have to react to empirical difficulties.

It is not certain, however, that Hausman's objection quite does justice to a strong intuition discernible behind Friedman's proposals and theses and which consists in giving prominence to the *division of labor* between the special sciences (specifically, economics and psychology). For example, it would have the consequence, in the case of microeconomics, of defending the stylization of psychological description by justifying it with the fact that a keener description is a job for psychologists, while the economist must concentrate on the consequences for collective phenomena. It is not surprising then, that theses of a Friedmanian bent frequently reappear in current methodological discussions about behavioral economics and neuroeconomics (see above) which raise, implicitly at least, the question of division of labor between economists and psychologists.

7. Experimental Economics, "Behavioral" Economics, and Neuroeconomics

7.1 EXPERIMENTAL ECONOMICS AND ITS OBJECTIVES

For a long time, the dominant view was that economics was exclusively a science of *observation*, and not an *experimental* science. But since the last 40 years or so, experimental economics has been progressively developing.[71] The Sveriges Riksbank Prize in Economic Sciences (known as the "Nobel Memorial Prize") 2002, bestowed on the experimenters D. Kahneman and V. Smith, bears witness to this development and to its recognition by the economics community. The number and variety of experimental work is now considerable, as shown by the *Handbook of Experimental Results* by Smith and Plott (2008) or the *Handbook of Experimental Economics* by Kagel and Rott (1995). Indeed, the experiments are just as much about individual decision and the markets as they are about strategic interactions. Moreover, they can either be laboratory *or* field experiments. In laboratory conditions, subjects evolve within a context (set by the task they must accomplish, the information they may receive, the goods they consider, etc.) that is largely artificial, while in the field, one is closer to a natural environment.[72] We can also make distinctions among field experiments. Harrison and List (2004) distinguish those which are "framed," where the context is natural in one or several of its aspects and where subjects know they are participating in an experiment, from those which are truly "natural," where they do not have such knowledge. They also distinguish field experiments from *social experiments*, where a public institution, in its action, undertakes a rigorous statistical procedure so as to understand the effects of certain factors it can control, and from *natural experiments*, where one *observes* variations which occur without the experimenter's intervention, but whose structure approaches that of the controlled variants.

The experiments may pursue differing objectives. We can distinguish at least three:[73]

(oi) Testing a preexisting theory—for example, we have already mentioned the experimental tests of expected utility theory.

(oii) Discovering unknown phenomena.

(oiii) Exploring questions of economic policy.[74]

---

[71] Readers can initiate themselves on experimental economics with Friedman & Sunder (1994).

[72] Furthermore, subjects of laboratory experiments are very often students of the universities where they take place.

[73] See Roth (1995), pp. 21–22 and Sugden (2005).

[74] Thus, in 1993, when the Federal Communications Commission, an American government agency, was wondering by which mechanism it would be wisest to allocate new telecommunications permits, experimenters were called upon to test various propositions (see Guala, 2005, chap.6).

In the past, experimenters themselves have often put the accent on objective (oi), that is, the test of economic theories. Today, more emphasis is put on the partial autonomy of experimentation with regard to economic theory: experimenters often introduce variations relative to factors which economic theory does not take into account, and allow themselves to be guided by local and informal hypotheses regarding the importance of such and such a parameter (see e.g. Guala, 2005, p. 48).

7.2 METHODOLOGICAL QUESTIONS

The methodological questions raised by experimental economics are many, and have recently been the object of some monographs (Guala, 2005; Bardsley et al., 2010).[75] Some of these questions concern the specifics of experimental economics, like the systematic use of financial motivations, which differentiates it from other experimental human sciences like psychology. In market experiments, which concern the coordinating role of that institution, financial motivations serve to experimentally *control* certain individual characteristics like the value assigned to options. Smith's "induced value theory" (1976) is the canonical formulation of this use.[76]

As we have already recalled, one of the objectives commonly assigned to experimentation is the *testing* of those economic theories which lend themselves to it. What is highlighted then is that the experimental approach makes possible empirical testing whose results are far more unequivocal than those that could be obtained from natural data. The *confirmational impact* of experimental data is, however, not easily gauged, and this divides economists. Economic theories are, indeed, largely thought of as seeking to predict and explain "real" phenomena. From this point of view, the relevance of their empirical adequacy in artificial contexts is in no way obvious: why, for example, should a theory which is confounded by experimental data suffer the same fate if applied outside of the laboratory? The way in which we conceive the confirmational impact of experimenting depends on two factors: (1) on the *domain* assigned to the economic theories, and (2) on the response given to the question of *external validity* or parallelism (see notably Starmer, 1999b; Guala, 2005; Bardsley et al., 2010), that is to say, the question of knowing what is allowed to be inferred concerning real economic phenomena on the basis of experimental phenomena. If we go as far as including laboratory behaviors in the domain of economic theories, then regardless of the actual response given to the question of external validity, the confirmational impact of the experimentation will already be notable: a theory confounded by experimental data will be a theory confounded in its domain. Experimental economist Ch. Plott's point of view can be read in this way:

> General models, such as those applied to the very complicated economies in the wild, must apply to simple special cases. Models that do not apply to the simple cases are not general and thus cannot be viewed as such . . .

[75] See also the special edition "On the Methodology of Experimental Economics" of the *Journal of Economic Behavior and Organization*, 73(1), January 2010.

[76] See Friedman and Sunder (1994), pp. 12–15 for a synthesis presentation.

> Theories that predict relatively poorly in the laboratory are either rejected or refined. Models and principles that survive the laboratory can then be used to address questions about the field. (Plott, 1991, p. 905)

Inversely, if experimental phenomena are excluded from the domain of economics and if it is thought that there are large differences between real and laboratory behaviors, then the confirmational impact of the data resulting from the laboratory will be extremely limited. We will now add a few separate remarks about the domain of economics and about external validity.

(1) The positions concerning the question of what belongs to the domain of economic theories cannot be reduced to the opposition between those who exclude laboratory behavior and those who don't. Thus Binmore (1999) limits relevant experiments to those where (a) subjects are faced with "simple" problems, (b) their motivations are "adequate" and (c) the time given to them to adjust their behavior to the problem is "sufficient." Symmetrically, he also limits the application of economic theories *in the field* to those situations which obey analogous conditions. This is not self-evident: among the phenomena generally considered to be relevant to the domain of economics feature situations which are complex or whose stakes are low or which offer little opportunity for learning (Starmer, 1999a). Moreover, it is not obvious that all economic theories must maintain the same relationship with experimental data. It can, for example, be considered that if consumer theory's job is to account for behavior in the field, and not in the laboratory, then the abstract theory of decision has a more universal scope and experimental data *must* be involved in its evaluation. The very notion of domain quite certainly calls for clarification. A first effort in that direction was carried out by Cubitt (2005) who distinguishes

(i) The *base domain*: the set of phenomena to which the theory applies without ambiguity[77]
(ii) The *intended domain*: the set of phenomena that the scientist intends to explain or predict with her theory,[78]
(iii) The *test domain*: the set of phenomena which can be legitimately considered for the testing of the theory

Cubitt maintains that these three domains should not coincide and, in particular, that the test domain not be limited to the intended domain. Specifically in the situation with which we are interested, one may recognize that the experimental situations do not belong to (ii) while maintaining that some of them at least belong to (iii). This assertion will not get a detailed argument but can be justified by calling

[77] For example, we can consider expected ("objective") utility theory to apply unambiguously to choices between bets on the color of balls randomly removed from various boxes, the proportion of balls of each color in each box being known.

[78] For example, we can consider the purchase of insurance policies as belonging to the domain targeted by expected utility theory.

on the external validity of the experimental phenomena, to which we now turn our attention.

(2) In which conditions can one "export" results obtained in the laboratory to the field? Guala (2005) asserts, in essence, that the inference from laboratory to field must take place on a case by case basis, and by a rigorous accounting of information about the experiments and about the natural domain of application. The objective is to ensure that the two contexts have enough relevant causal factors in common to allow for reasoning out, by analogy, from the laboratory to the field. According to Guala, it is essentially with a view to exploiting the analogy that these experiments have an interest for economists: the experimental situations are not so much *components* of the specific domain of economics (natural economic phenomena, what Cubitt calls the intended domain) as they are *representations* of the domain which enable it to be understood, along with models or simulations. Borrowing from contemporary literature on models, Guala sums up his idea by affirming that experiments are "mediators" between the domain of economics and the hypotheses we can form about it (pp. 209–211).

7.3 On the Border between Economics and Cognitive Science: Behavioral Economics and Neuroeconomics

Experimental economics is often associated with two other movements, both of which also make extensive use of experimentation: (1) so-called behavioral economics and (2) neuroeconomics.

(1) The qualifiers "experimental" and "behavioral" are often used in an interchangeable fashion, but perhaps wrongly so. While experimental economics consists of studying economic phenomena with the help of controlled experiments, behavioral economics defines itself as a project that "increases the explanatory power of economics by providing it with more realistic psychological foundations" (Camerer and Loewenstein, 2004). This project involves much experimentation, but it also relies on taking into account natural data and revisiting the psychological and behavioral assumptions which orthodox economics rests upon. Decision theory, game theory and the auxiliary assumptions which economists often rely upon (like the assumption stating that individual preferences increase with monetary gain) are subject to particular attention. This project is largely motivated by dissatisfaction with regards to orthodox economics and by the anti-Friedmanian working hypothesis:

(T19) An improvement in the assumptions made concerning economic agents will result in a significant improvement in economic science.

Behavioral economics typically proceeds by generalization or modification of received assumptions and in this sense it constitutes a "soft" heterodoxy. Hypothesis (T19) is empirical and behavioral economics is undoubtedly too fragmented for us to be able to evaluate it at this stage. If it does seem to go against the Friedmanian thesis (T15), which states that a theory must not be judged by the realism of its assumptions, the conflict may be only apparent. Some of its followers have the paradoxical ability of remaining loyal to the thesis grounding (T15), thesis (T18), according to which a theory must be judged by the correctness of its target domain predictions, all the while considering that an improvement in the psychological realism of economic theory is the *means* of obtaining the best predictions. Others, on the contrary, reject (T18) and consider that economic theory *must* be founded on plausible psychological principles, whether or not that results in a significant predictive improvement. (T19) can thus mask different epistemological motivations. Moreover, the reference to psychology and to psychological realism is not free of ambiguity. Certainly, the disciples of behavioral economics are opposed to the separation of economics and psychology, of the sort that Robbins (1932/1935), for example, defended.[79] But if we judge on the basis of the most striking works of behavioral economics, it is not a question of applying or being inspired by a preexisting cognitive psychology of decisions, nor even of approaching economic behavior through the use of the concepts and methods of cognitive psychology. Nor is it a question, generally, of opening the "black box" of mental states and processes that traditional economics, in its timidness, would leave closed: numerous hypotheses or theories within the domain are neither more nor less "psychological," in this sense, than the traditional theories in economics. What more certainly unifies various work within the domain is the conviction that, in many situations, the models used by traditional economics in describing agents' behavior is systematically erroneous. The call for "psychological realism" largely consists of *taking account of* these empirical anomalies through theoretical revision. This attitude has consequences for the discipline which are still difficult to evaluate. In defending the recourse to assumptions which, sometimes significantly, move away from the standards of rationality, behavioral economics also disturbs the traditional organization of economics, and particularly the communication between positive economics and normative economics, which to a considerable extent rests upon agents' individual rationality, understood in the traditional fashion.

(2) Neuroeconomics, born at the beginning of the 2000s, has the objective of exploring the neural bases of economic behavior. To do this, it employs the methods and tools of contemporary neuroscience, notably functional magnetic resonance imaging (see Glimcher and Fehr, 2008/2014 for an encyclopedic state of the art). For example, in McClure et al. (2004) subjects have to choose between two options with varied delayed monetary rewards. The first option (*sooner-smaller*) yields the sum R after a delay d, and the second (*later-larger*) the sum R' after a delay d', with $d < d'$ (where d is today, in two weeks, or in one month) and $R < R'$. The authors show that (a) the limbic system is preferentially activated when the first option involves an immediate reward (d = today), (b) the parietal and prefrontal cortex is engaged uniformly by the task (irrespective of the value of d), and (c) greater activity of the

[79] Robbins (1932/1935), pp. 83 sq. For this reason Robbins criticizes the attitude of Gossens, Jevons, or Edgeworth. Bruni and Sugden (2007) trace the divorce between scientific psychology and neo-classical economics back to Pareto.

parietal and prefrontal cortex is associated with choosing the second option rather than the first.

In seeking to enlighten the study of certain social phenomena through neurobiology, neuroeconomics cannot help but raise questions linked to the reductionism dealt with in the chapter "Philosophy of the social sciences." Methodology's first interest is in what the neurosciences could bring to economics, and particularly in the more specific question of the relationship between neural data and models of choice, taking the following affirmation from F. Gul and W. Pesendorfer as its target:

(T20) Neural data can neither confirm nor disconfirm the models of decision that economics makes use of.

Gul and Pesendorfer develop several arguments to support their statement (see Hausman, 2008). If some rely more particularly on the semantics of revealed preference, all highlight the fact that the traditional models of decision are *silent* on the cognitive side of things (see Cozic, 2012) and that, consequently, these models imply no testable restriction on the direct observations we could collect of individuals' deliberative processes. As the defenses and objections collected by Caplin and Schotter (2008) attest, there is today a striking absence of consensus regarding (T20) and the arguments which are supposed to justify it. These debates explain why, though economists may not doubt the interest of neuroeconomics for the cognitive neurosciences, they are often more skeptical regarding its fecundity in the treatment of the traditional questions of economics (see Camerer, 2007, Bernheim, 2009).

8. Conclusion

We have placed our chapter on economic methodology under the banner of *Mill's generalized problem*: does economic science obey the methodological standards of an empirical science? This question, implicitly or explicitly, has oriented a large portion of epistemological reflection about the discipline since the beginning of the nineteenth century. It is time now to see what can be learned from the principal responses this problem gave rise to.

- The judgments concerning economics as an empirical science cover an extremely wide spectrum: some defend the core of neo-classical economics' achievements, others are of the opinion that economists are guilty of insufficiently testing their theories, still others that the conceptual framework in which they work is doomed to failure.
- Unequivocally unfavorable diagnostics of economics are today, it seems, rather in the minority among specialists in economic methodology. The dominant impression is that these are based on either overly rigid epistemological foundations or overly partial considerations of the discipline's accomplishments. Unequivocally favorable diagnostics are not abundant

either. A majority would no doubt be in agreement that several episodes or tendencies of contemporary economics were manifestations of excessive dogmatism.

- There is virtual unanimity *against* the methodological view which is certainly the most frequently cited by the economists of the last few decades, this being Friedman's view.
- Recent evolutions in economics, which assign increased roles to theoretical diversification, to interdisciplinary openness and to attention to empirical data (experimental or otherwise), are positively welcomed by most methodological analyses of the discipline.

Let us come back to the status of economic methodology. We said it in the introduction: methodological standards do not exist today which could be the object of a consensus among philosophers of science and whose mere application to economics would suffice to appraise its scientific legitimacy. The methodological contributions we have chosen to present draw no radical conclusions, in the sense that they reckon a normative reflection about economics as an empirical science to still be possible,[80] even if, with the most recent of them, their way of reaching that reflection encounters significant reorientation, in accordance with an increased attention to the economist's actual procedures and a larger distance with respect to the doctrines which animated general philosophy of science in the twentieth century. This is not the only reaction possible. The absence of consensus sometimes joins forces with an absence of expertise argument to the effect that it falls on the experts (the economists), and not on the methodologists, to judge their own work,[81] to nourish a skepticism with regard to any normative ambition in methodological inquiry. In response, some, particularly severe with the application of philosophy of science to economic methodology, have abandoned the normative project and sometimes even the conceptual tools of philosophy of science. This is the case with works belonging to the "rhetoric of economics" school of thought initiated by McCloskey (1985/1998),[82] which propose studying, armed with the tools of rhetoric and literary criticism, the way in which economists *persuade* themselves.[83] In our view, and that of many philosophers of economics, these absence of consensus and absence of expertise arguments have limited reach,[84] even if they call

[80] This attitude is, for example, openly declared by Rosenberg (1992, pp. xii–xiii) or Hausman (1992a, sec. 14.3).

[81] See McCloskey (1985/1998), p. 139.

[82] McCloskey (1985/1998, chap. 9) paints a particularly severe picture of the "modernism" in epistemology, which largely covers the themes we have labeled "positivist."

[83] See Hands (2001, pp. 257–258) for a detailed bibliography of these studies. These meta-methodological questions exceed the boundaries of philosophy of economics and are at the heart of the rivalry that exists between philosophy of science and what is known as "science studies."

[84] See notably the criticisms of Blaug (1992), Hausman (1992a, pp. 262–269), Rosenberg (1992), Hoover (1995). Other meta-methodological elements of discussion can be found in Hands (2001), Kincaid et Ross (2009b), Guala (2009). The latter defends an "instrumental" normative methodology that is supposed to constitute a middle road between a "categorical" normative methodology, which evaluates

(and rightly so) for the methodologist to consider her abstract principles as fallible and to base her judgments on a deep and charitable understanding of the economist's work. Moreover, in such remarkably active domains as philosophy of physics, of biology or of cognitive science, the specialists carry on the project of critical reflection of the object of their discipline.

Let us finish by pointing out some tendencies and some gaps within current economic methodology.[85] These will be easier brought out if we pursue a comparison with the other domains of philosophy of the special sciences. These other domains grant less importance to Mill's generalized problem and perhaps greater attention to their own objects. They often seek to clarify the fundamental concepts and principles of their discipline by evaluating their coherence, as much within the discipline as with the rest of our knowledge. In this way these reflections take on an allure which is (a) more specialized and (b) more *ontological* than the majority of the contributions we have presented. (a) The tendency toward specialization is already at work in methodology of economics. Section 7, dedicated to experimental economics, behavioral economics, and neuroeconomics, will certainly have made the reader aware that, though the most recent debates often remain tied to Mill's problem, they do move toward more specific questions which are dealt with in a more autonomous fashion. In this regard, we must admit that, because of length constraints, we were not able to do justice to questions like those of causality in economics,[86] of econometric reasoning,[87] or regarding the links between micro- and macroeconomics.[88] (b) Philosophy of economics as a positive science is still widely dominated by methodological preoccupations.[89] One could consider as fruitful for it to develop its ontological inclination, all the more so that economics spans from infra-individual properties (the mental states of actors) to supra-individual entities such as organizations or institutions. It could benefit, on the one hand, from the progress of philosophy of mind and the cognitive sciences, and, on the other hand, from recent studies in philosophy of social sciences which are interested in the status of collective

---

scientific activity from the perspective of abstract and supposedly universal prescriptions, and the abandoning of normative methodology. The idea is to have the methodologist evaluate the scientific practices he is studying from the perspective of the objectives the scientist pursues. In this way, Guala defends the idea that the normal practices of experimental economics are justifiable *if* the objective pursued is the discovery of robust causal relations (rather than universal laws).

[85] Kincaid and Ross (2009b) give another point of view, more controversial, about recent tendencies in economic methodology, which they name the "new philosophy of economics." In their view, this differentiates itself favorably from the works that dominated economic methodology from the 1970s to the 1990s—those of Blaug, Hausman and Rosenberg in particular—which attached themselves too exclusively to the theoretical core of neo-classical economics and which approached it with philosophical concepts that are now obsolete.

[86] See Hoover (2001a), Reiss (2008, chap. 7–9).

[87] For an introduction, see Reiss (2013, chap. 10). See also Hendry (1980), Malinvaud (1991, chap. 12 and 13), Spanos (2006, 2012).

[88] See Kirman (1992), Hoover (2001b, chap. 3, 2009).

[89] This domination is also visible linguistically: "methodology of economics" in fact refers to the collected research falling under the category of philosophy of economics as a positive science.

entities and properties. As an example, the clarification of a concept as fundamental to economic analysis as "the market" is far less straight-forward than it may seem. Finally, there are two characteristics of contemporary economics which call for a supplementary effort of analysis: the considerable development of its theoretical apparatus, and the internal coexistence of positive *and* normative preoccupations. On the one hand, we certainly haven't reached a satisfactory degree of clarification of the norms and objectives which have required this development of the theoretical apparatus of economics. This is the case, for example, with the status of the theory of general equilibrium. Advances on this question probably necessitate a better understanding of the general nature of *theoretical progress*. On the other hand, the juncture between positive economics and normative economics, and notably the role of individual rationality in the communication between the two types of research, still largely remains to be clarified.

References

Akerlof, G. (1970) "The Market for 'Lemons': Quality Uncertainty and the Market Mechanism," *The Quarterly Journal of Economics*, 84(3), pp. 488–500.

Allais, M. (1953) "Le comportement de l'homme rationnel devant le risqué: critique des postulats et axiomes de l'école américaine," *Econometrica*, 21(4), pp. 503–546.

Archibald, G.C. (1965) "The Qualitative Content of Maximizing Models," *The Journal of Political Economy*, 73(1), pp. 27–36.

Archibald, G.C., Simon, H. and Samuelson, P. (1963) "Discussion," *American Economic Review*, 53(2), pp. 227–236.

Arrow, K. (1951/1963) *Social Choice and Individual Values*, 2nd ed., Yale University Press.

Backhouse, R. (ed.), (1994) *New Directions in Economics Methodology*, London: Routledge.

Backhouse, R. (1998) *Explorations in Economic Methodology: From Lakatos to Empirical Philosophy of Science*, London: Routledge.

Backhouse, R., and Medema, S. (2009) "On the Definition of Economics," *Journal of Economic Perspectives*, 23(1), pp. 221–233.

Bardsley, N., Cubitt, R., Loomes, G., Moffatt, P., Starmer, C., and Sugden, R. (2010) *Experimental Economics: Rethinking the Rules*, Princeton, NJ: Princeton University Press.

Bernheim, D. (2009) "On the Potential of Neuroeconomics: A Critical (but Hopeful) Appraisal," *American Economic Journal: Microeconomics*, 1(2), pp. 1–41; abridged version in Glimcher, P., Camerer, C., Fehr, E. and Poldrack, R. (eds.) *Neuroeconomics*, Amsterdam: Elsevier, chap. 9, pp. 115–126.

Binmore, K. (1999) "Why Experiments in Economics?," *The Economic Journal*, 109(453), pp. F16–F24.

Blanchard, O. (2017) *Macroeconomics*, Harlow: Pearson Education, 7th ed.

Blaug, M. (1980/1992) *The Methodology of Economics*, Cambridge: Cambridge University Press.

Boland, L. (1979), "A Critique of Friedman's Critics," *Journal of Economic Theory*, 17, pp. 503–522.

Boumans, M., and Morgan, M.S. (2001) "Ceteris paribus Conditions: Materiality and the Application of Economic Theories," *Journal of Economic Methodology*, 6(3), pp. 11–26.

Bruni, L., and Sugden, R. (2007) "The Road Not Taken: How Psychology Was Removed from Economics, and How It Might Be Brought Back," *The Economic Journal*, 117, pp. 146–173.

Cairnes, J. E. (1857/1875) *The Character and Logical Method of Political Economy*, 2nd edition, London: Macmillan.

Camerer, C. (2007) "Neuroeconomics: Using Neuroscience to Make Economic Predictions," *The Economic Journal*, 117, C26–C42.

Camerer, C. F., and Loewenstein, G. (2004) "Behavioral Economics: Past, Present, Future" *in* Camerer, C.F., Loewenstein, G., and Rabin, M. (eds.) *Advances in Behavioral Economics*, Princeton, NJ: Princeton University Press, pp. 3–51.

Caplin, A., and Schotter, A. (ed.) (2008) *The Foundations of Positive and Normative Economics*, Oxford: Oxford University Press.

Cartwright, N. (1989) *Nature's Capacities and Their Measurement*, Oxford: Oxford University Press.

Cartwright, N. (2007) *Hunting Causes and Using Them. Approaches in Philosophy and Economics*, Cambridge: Cambridge University Press.

Cartwright, N. (2009) "If No Capacities, Then No Credible Worlds, but Can Models Reveal Capacities," *Erkenntnis*, 70, pp. 45–58.

Chiappori, P-A. (1990) "La théorie du consommateur est-elle réfutable?" *Revue économique*, 41(6), pp. 1001–1025.

Cozic, M. (2012) "Economie sans esprit et données cognitives" *Revue de philosophie économique*, 13(1), pp. 127–153.

Cubitt, R. (2005) "Experiments and the Domain of Economic Theory," *Journal of Economic Methodology*, 12(2), pp. 197–210.

Davidson, D. (1980), *Essays on Actions and Events*, Oxford: Clarendon Press.

Davis, J., Hands, D. W., and Mäki, U. (eds.) (1998) *The Handbook of Economic Methodology*, Cheltenham: Edward Elgar.

Earman, J., and Roberts, J. (1999) "*Ceteris Paribus*, There Is No Problem of Provisos," *Synthese*, 118, pp. 439–478.

Earman, J., Roberts, J., and Smith, S. (2002) "*Ceteris Paribus* Lost," *Erkenntnis*, 57, pp. 281–301.

Friedman, D., and Sunder, S. (1994) *Experimental Methods. A Primer for Economists*, Cambridge: Cambridge University Press.

Friedman, M. (1953) *Essays in Positive Economics*, Chicago: University of Chicago Press.

Friedman, M., and Savage, L. (1948) "The Utility Analysis of Choices Involving Risks," *The Journal of Political Economy*, vol. 56(4), pp. 279–304.

Frigg, R. and Hartmann, S. (2009) "Models in Science," *The Stanford Encyclopedia of Philosophy (Summer 2009 Edition)*, Edward N. Zalta (ed.), http://plato.stanford.edu/archives/sum2009/entries/models-science/.

Gibbard, A., and Varian, H. R. (1978) "Economic Models," *The Journal of Philosophy*, 75(11), pp. 664–677.

Glimcher, P., and Fehr, E. (2008/2014) *Neuroeconomics. Decision Making and the Brain*, Amsterdam: Elsevier.

Grether, D., and Plott, C. (1979) "Economic Theory of Choice and the Preference Reversal Phenomenon," *American Economic Review*, 69, pp. 623–638

Guala, F. (2005) *The Methodology of Experimental Economics*, New York: Cambridge University Press.

Guala, F. (2006) "Has Game Theory Been Refuted?" *The Journal of Philosophy*, C 3(5), pp. 239–263.

Guala, F. (2009), "Methodological Issues in Experimental Design and Interpretation," *in* Kincaid and Ross (2009a).

Hands, D. Wade (2001) *Reflection without Rules. Economic Methodology and Contemporary Science Theory*, Cambridge: Cambridge University Press.

Harrison, G., and List, J. (2004) "Field Experiments," *Journal of Economic Literature*, 42(4), pp. 1009–1055.

Hausman, D. (1981) *Capitals, Profits, and Prices*, New York: Columbia University Press.

Hausman, D. (1989), "Economic Methodology in a Nutshell," *The Journal of Economic Perspectives*, 3(2), pp. 115–127.

Hausman, D. (1992a) *The Inexact and Separate Science of Economics*, Cambridge University Press.

Hausman, D. (1992b) *Essays on Philosophy and Economic Methodology*, Cambridge: Cambridge University Press.

Hausman, D. (1997), "Theory Appraisal in Neoclassical Economics," *Journal of Economic Methodology*, 4(2), pp. 289–296.

Hausman, D. (2000), "Revealed Preference, Belief, and Game Theory," *Economics and Philosophy*, 16, pp. 99–115.

Hausman, D. (2001) "Tendencies, Laws and the Composition of Economic Causes," *in* Mäki, U. (ed.), *The Economic World View*, Cambridge: Cambridge University Press, pp. 293–307.

Hausman, D. (2005) "'Testing' Game Theory," *Journal of Economic Methodology*, 12, pp. 211–223

Hausman, D. (ed.), (2008a) *The Philosophy of Economics. An Anthology*, 3rd ed., Cambridge; Cambridge University Press.

Hausman, D. (2008b), "Mindless or Mindful Economics: A Methodological Evaluation," *in* Caplin and Schotter (2008), chap. 6.

Hausman, D. (2008c) "Philosophy of Economics," *in* Zalta (ed.), *The Stanford Encyclopedia of Philosophy* (Fall 2008 Edition), http://plato.stanford.edu/archives/fall2008/entries/economics.

Hausman, D. (2009) "Laws, Causation and Economic Methodology" *in* Kincaid and Ross (2009a), chap. 2.

Hausman, D., and McPherson, M. (2006) *Economic Analysis, Moral Philosophy and Public Policy*, Cambridge: Cambridge University Press.

Hendry, D. (1980) "Econometrics: Alchemy or Science?" *Economica*, 47, pp. 387–406.

Hicks, J. R. (1939) *Value and Capital*, Oxford: Oxford University Press.

Hicks, J. R., and Allen, R. G. (1934), "A Reconsideration of the Theory of Value," *Economica*, 1(1), pp. 52–76.

Hoover, K. D. (1995) "Why Does Methodology Matter for Economics?" *The Economic Journal*, 105(430), pp. 715–734.

Hoover, K. D. (2001a) *Causality in Macroeconomics*, Cambridge: Cambridge University Press.

Hoover, K. D. (2001b) *The Methodology of Empirical Macroeconomics*, Cambridge: Cambridge University Press.

Hoover, K. D. (2006) "The Methodology of Econometrics," *in* Mills, T. C., and Patterson, K. (eds.) *Palgrave Handbook of Econometrics*, vol. 1, chap. 2, pp. 61–87, Basingstoke: Palgrave Macmillan.

Hoover, K. D. (2009) "Milton Friedman's Stance: the Methodology of Causal Realism," *in* Mäki (2009a), chap. 12.

Houthakker, H. (1950) "Revealed Preference and the Utility Function," *Economica*, 17, pp. 159–174.

Hutchison, T. W. (1938) *The Significance and Basic Postulates of Economic Theory*, London: Macmillan.

Hutchison, T. W. (1994), "Ends and Means in the Methodology of Economics," *in* Backhouse (1994), pp. 27–34.

Ingrao, B., and Israel, G. (1990) *The Invisible Hand. Economic Equilibrium in the History of Science*, Cambridge, MA: MIT Press.

Kagel, J. H., and Roth, A. E. (eds.) (1995) *The Handbook of Experimental Economics*, Princeton, NJ: Princeton University Press.

Keynes, J. N. (1891), *Scope and Method of Political Economy*, London: Macmillan.

Kincaid, H., and Ross, D. (eds.) (2009a) *The Oxford Handbook of the Philosophy of Economics*, Oxford: Oxford University Press.

Kincaid, H., and Ross, D. (2009b) "The New Philosophy of Economics," *in* Kincaid and Ross (2009a), chap. 1.

Kirman, A. (1992) "Whom or What Does the Representative Individual Represent?" *Journal of Economic Perspectives*, 6(2), pp. 117–136.

Klappholz, K., and Agassi, J. (1959) "Methodological Prescriptions in Economics," *Economica*, 26, pp. 60–74.

Kolm, S-C. (1986), *Philosophie de l'économie*, Paris: Seuil.

Krugman, P. (2009), *The Return of Depression Economics and the Crisis of 2008*, New York: W. W. Norton.

Lichtenstein, S., and Slovic, P. (2006) *The Construction of Preference*, Cambridge: Cambridge University Press.

Lipsey, (2008) "Positive Economics" *in* Durlauf, S. and Blume, L. (eds.) *The New Palgrave Dictionary of Economics*, 2nd ed., Basingstoke: Palgrave Macmillan, http://www.dictionaryofeconomics.com/article?id=pde2008_P000130.

Machlup, F. (1967) "Theories of the Firm: Marginalist, Behavioral, Managerial," *The American Economic Review*, LVII(1), pp. 1–33.

Mäki, U. (2000) "Kinds of Assumptions and Their Truth: Shaking an Untwisted F-Twist," *Kyklos*, 53(3), pp. 317–336.

Mäki, U. (ed.) (2009a) *The Methodology of Positive Economics. Reflecting on the Milton Friedman Legacy*, Cambridge: Cambridge University Press.

Mäki, U. (2009b) "Unrealistic Assumptions and Unnecessary Confusions: Rereading and Rewriting F53 as a Realist Statement," *in* Mäki (2009a), pp. 90–116.

Mäki, U. (2009c) "Realist Realism about Unrealistic Models," *in* Kincaid and Ross (2009a), pp. 3–54.

Malinvaud, E. (1972/1985) *Lectures on Microeconomic Theory*, 2nd ed., Amsterdam: North-Holland.

Malinvaud, E. (1991) *Voies de la recherche macroéconomique*, Paris: O. Jacob.

Marshall, A. (1890/1920) *Principles of Political Economy*, London: Macmillan.

Martin, M., and McIntyre, L. C. (eds.) (1994) *Readings in the Philosophy of Social Science*, Cambridge, MA: MIT Press.

Mas-Colell, A., Whinston, M., and Green, J. (1995) *Microeconomic Theory*, Oxford: Oxford University Press.

McCloskey, D. (1985/1998) *The Rhetoric of Economics*, Madison: University of Wisconsin Press.

McClure, S. M., Laibson, D., Loewenstein, G., and Cohen, J. D. (2004) "Separate Neural Systems Value Immediate and Delayed Monetary Rewards," *Science*, 306, pp. 503–507.

McMullin, E. (1985) "Galilean Idealization," *Studies in History and Philosophy of Science*, 16(3), pp. 247–273.

Mill, J-S. (1836) "On the Definition of Political Economy and on the Method of Investigation Proper to It," *in* Robson, J. M. (ed.) (1967), *The Collected Works of John Stuart Mill, Volume IV—Essays on Economics and Society Part I*, Toronto: University of Toronto Press.

Mill, J-S. (1843) *A System of Logic Ratiocinative and Inductive, Being a Connected View of the Principles of Evidence and the Methods of Scientific Investigation, in The Collected Works of John Stuart Mill, Volume VII and VIII*, Robson, J. M. (ed.) (1974), Toronto: University of Toronto Press.

Mill, J-S. (1848) *The Principles of Political Economy with Some of Their Applications to Social Philosophy, in The Collected Works of John Stuart Mill, Volume II*, Robson, J. M. (ed.) (1965), Toronto: University of Toronto Press.

Mingat, A., Salmon, P., and Wolfelsperger, A. (1985) *Méthodologie économique*, Paris: PUF.

Mongin, Ph. (1988) "Le réalisme des hypothèses et la *Partial Interpretation View*," *Philosophy of the Social Sciences*, 18, pp. 281–325.

Mongin, Ph. (1999) "Normes et jugements de valeur en économie normative," *Social Science Information*, 38 (4), pp. 521–553.

Mongin, Ph. (2000a) "La méthodologie économique au XXème siècle. Les controverses en théorie de l'entreprise et la théorie des préférences révélées," *in* Béraud, A., and Faccarello, G., *Nouvelle histoire de la pensée économique*, tome 3, Paris: La Découverte, chap. 36.

Mongin, Ph. (2000b) "Les préférences révélées et la formation de la théorie du consommateur," *Revue économique*, 51(5), pp. 1125–1152.

Mongin, Ph. (2005) "La réfutation et la réfutabilité en économie," *miméo*

Mongin, Ph. (2006a) "Value Judgments and Value Neutrality in Economics," *Economica*, 73, pp. 257–286.

Mongin, Ph. (2006b) "L'analytique et le synthétique en économie," *Recherches économiques de Louvain*, 72, pp. 349–383.

Mongin, Ph. (2007) "L'apriori et l'a posteriori en économie," *Recherches économiques de Louvain*, 73, pp. 5–53.

Mongin, Ph. (2009) "Duhemian Themes in Expected Utility Theory" *in* Brenner, A., and Gayon, J. (eds.), *French Studies in the Philosophy of Science*, New York: Springer, pp. 303–357.

Mongin, Ph., and d'Aspremont, Cl. (1998) "Utility Theory and Ethics" *in* Barbera, S., Hammond, P., and Seidl, Ch. (eds.), *Handbook of Utility Theory*, vol. 1, Dordrecht: Kluwer, pp. 371–482.

Musgrave, A. (1981) "'Unreal Assumptions' in Economic Theory: The F-Twist Untwisted," *Kyklos*, 34, pp. 377–387.

Myrdal, G. (1958) *Value and Social Theory*, London: Routledge.

Nagel, E. (1963) "Assumptions in Economic Theory," *American Economic Review*, 53(2), pp. 211–219.

Popper, K. (1963/1989) *Conjectures and Refutations: The Growth of Scientific Knowledge*, London: Routledge.

Plott, Ch. (1991) "Will Economics Become an Experimental Science?," *Southern Economic Journal*, 57, pp. 901–919.

Putnam, H. (2002) *The Collapse of the Fact/Value Dichotomy and Other Essays*, Cambridge, MA: Harvard University Press.

Reiss, J. (2008) *Error in Economics*, London: Routledge.

Reiss, J. (2013) *Philosophy of Economics. A Contemporary Introduction*, London: Routledge.

Reutlinger, A., Schurz, G., and Hüttermann, A. (2014) "*Ceteris Paribus* Laws," *The Stanford Encyclopedia of Philosohy* (Winter 2014 Edition), E. N. Zalta (ed.), http://plato.stanford.edu/entries/ceteris-paribus/.

Robbins, L. (1932/1935) *An Essay on the Nature and Significance of Economic Science*, London: Macmillan.

Robbins, L. (1938) "Live and Dead Issues in the Methodology of Economics," *Economica*, 5(19), pp. 342–352.

Rosenberg, A. (1975) "The Nomological Character of Micro-economics," *Theory and Decision*, 6(1), pp. 1–26.

Rosenberg, A. (1976) *Microeconomic Laws. A Philosophical Analysis*, Pittsburgh: University of Pittsburgh Press.

Rosenberg, A. (1992) *Economics—Mathematical Politics or Science of Diminishing Returns*, Chicago: University of Chicago Press.

Rosenberg, A. (2009) "If Economics Is a Science, What Kind of Science Is It?" *in* Kincaid and Ross (2009a), chap. 3.

Roth, A. (1995) "Introduction to Experimental Economics," *in* Kagel and Roth (1995), pp. 3–109.

Samuelson, P. (1938a) "A Note on the Pure Theory of Consumer's Behaviour," *Economica*, 5(17), pp. 61–71.

Samuelson, P. (1938b) "The Empirical Implications of Utility Analysis," *Econometrica*, 6(4), pp. 344–356.

Samuelson, P. (1947) *Foundations of Economic Analysis*, Cambridge, MA: Harvard University Press.

Samuelson, P. (1950) "The Problem of Integrability in Utility Theory," *Economica*, 17(8), pp. 355–385.

Samuelson, P. (1970) "Maximum Principles in Analytical Economics," Nobel Prize speech, reprinted in *Collected Scientific Papers of Paul A. Samuelson*, Vol. 3, Cambridge, MA: MIT Press.

Samuelson, P. (1992) "My Life Philosophy: Policy Credos and Working Ways," *in* Szenberg, M. (ed.), *Eminent Economists. Their Life Philosophies*, Cambridge: Cambridge University Press, pp. 236–247.

Sen, A. (1970) *Collective Choice and Social Welfare*, San Francisco: Holden-Day.

Sen, A. (1973) "Behaviour and the Concept of Preference," *Economica*, 40, pp. 241–259.

Sen, A. (1987) *On Ethics and Economics*, Oxford: Blackwell.

Smith, V. L. (1989) "Theory, Experiment and Economics," *The Journal of Economic Perspective*, 3(1), pp. 151–169.

Smith, V.L., and Plott, C. (eds.) (2008) *Handbook of Experimental Economics Results*, Amsterdam: North-Holland/Elsevier.

Spanos, A. (2006) "Econometrics in Retrospect and Prospect," *in* Mills, T.C., and Patterson, K. (eds.), *Palgrave Handbook of Econometrics*, vol. 1, Palgrave, chap. 1, pp. 3–60, Basingstoke: Palgrave Macmillan.

Spanos, A. (2012) "Philosophy of Econometrics," *in* Mäki, U. (ed.) *Philosophy of Economics*, *in* Gabbay, D., Thagard, P., and Woods, J. (eds.) *Handbook of the Philosophy of Science*, Oxford and Amsterdam: Elsevier, pp. 329–393.

Starmer, C. (1999a) "Experimental Economics: Hard Science or Wasteful Tinkering," *Economic Journal*, 453, F5–F15.

Starmer, C. (1999b) "Experiments in Economics: Should We Trust the Dismal Scientists?," *Journal of Economic Methodology*, 6, pp. 1–30.

Stiglitz, J., and Walsh, C. (2000) *Principles of Microeconomics*, 3rd ed., New York: W. W. Norton.

Sugden, R. (2000) "Credible Worlds: The Status of Theoretical Models in Economics," *Journal of Economic Methodology*, 7(1), pp.1–31.

Sugden, R. (2005) "Experiments as Exhibits and Experiments as Tests," *Journal of Economic Methodology*, 12(2), pp. 291–302.

Sugden, R. (2009) "Credible Worlds, Capacities and Mechanisms," *Erkenntnis*, 70, pp. 3–27.

Varian, H. (1982) "The Nonparametric Approach to Demand Analysis," *Econometrica*, 50(4), pp. 945–974.

Varian, H. (1992) *Microeconomic Analysis*, 3rd ed., New York: W. W. Norton.

Walliser, B. (2011) *Les fonctions des modèles économiques*, Paris: O. Jacob.

Weibull, J. (2004) "Testing Game Theory" *in* Huck, S. (ed.) *Advances in Understanding Strategic Behavior*, New York: Palgrave, 2004, pp. 85–104.

Wong, S. (1978/2006) *Foundations of Paul Samuelson's Revealed Preference Theory*, 2nd ed., London: Routledge.

Woodward, J. (2002) "There Is No Such Thing as a *Ceteris Paribus* Law," *Erkenntnis*, 57, pp. 303–328.

16

PHILOSOPHY OF COGNITIVE SCIENCE

Daniel Andler (Université Paris-Sorbonne)

COGNITIVE SCIENCE APPEARS as an articulated group of research programs whose aim is to constitute a science of the mind. In some respects it is a discipline much like any other, and the philosophy of cognitive science resembles the philosophy of other particular sciences. But in other ways cognitive science is very different from most other disciplines or groups of disciplines, and as a result the philosophy of cognitive science differs somewhat from, for example, the philosophy of physics, the philosophy of biology, or the philosophy of economics.[1]

One might suppose that the main difference has to do with the plural label—the cognitive sciences—that is sometimes used to refer to the field. This difference does play a certain role, and it explains why the philosophy of cognitive science is similar in some respects, for example, to the philosophy of the social sciences. But the unity of a set of disciplines is a matter of degree, and no assessment can be offered that does not already commit us to theoretical hypotheses regarding the field under scrutiny. As a first approximation, one may say that physics is more unified than biology (also often referred to as the life sciences), that the social sciences are much less unified than the life sciences, and that cognitive science is situated between the latter two in this regard. Thus a relative absence of unity is not what gives cognitive science its

[1] Many thanks to Jeff Lewis, who provided a translation of the French original, which the author then revised and updated.

philosophical singularity, although the plurality of disciplines that make up cognitive science is something for a philosopher to reflect on.

Cognitive science is also quite young, another distinguishing feature that accounts for the fact that many people have a somewhat blurred perception, and see it as precarious and immature. In this regard, philosophy can take on the task of explaining this science, and of defending it as philosophy defended the emerging physical sciences at the time of the Scientific Revolution—although the defense strategy will be very different, and not only because of the temporal distance involved.

As I see it though (and this is up for discussion), the essential difference has to do with a *persistent uncertainty regarding the object* of cognitive science (which we term cognition as a matter of convention), which goes hand in hand with a certain *interpenetration of cognitive science and philosophy*, with several areas of philosophy each playing a part.

However that may be, the fact is that the philosophy of cognitive science is a burgeoning area of scholarship, with boundaries that are ill-defined. Whether one is still within them, or has entered the territory of another branch of philosophy, or even that of one of the positive sciences, is often moot. Such questions of the demarcation may seem of limited importance: research programs are individuated by problems and their interrelations, far more than by the labels we attach to them in the process of organizing institutions and students' curricula. However, philosophers do disagree on the role they should or must play in relation to cognitive science. It may thus be useful to have at least an approximate idea of the relative position of the main areas of philosophical activity concerning the subject of cognition, given that these main areas of inquiry (which are referred to in a variety of ways, as we shall see) are pursued by whole cohorts of philosophers, in greater number than any other branch of the philosophy of science, and whose output, in diversity and in quantity, beggars the imagination.

A geography of the field will only be sketched out at the end of the present chapter however: better to start by trying to get a sense of some of its central topics. Suffice it to say at this point that I will choose to focus on matters that belong to the philosophy of cognitive science in the sense where such topics as the notion of an organism, the concept of a function, or molecular reduction belong to the philosophy of biology. In other words, as befits a chapter in a volume devoted to philosophy of science, this one gives priority to issues typical of philosophy of science, while only briefly touching on more general philosophical problems. It is hardly necessary to observe that I make no attempt to provide an overview of the philosophy of cognitive science: I can only sample the field, and try to make sure that the sample is a representative one.

Nor will the reader find a mini-encyclopedia of cognitive science. One doesn't expect a chapter on the philosophy of mathematics to provide a mini-encyclopedia of mathematics, and *mutatis mutandis* the same goes for philosophy of biology or philosophy of physics. Cognitive science is young, to be sure, but it now rests on a vast library of works, both introductory and advanced, general and specialized. Philosophers of cognitive science can now be excused from tasks which they may have felt compelled to assume at first, such as recounting the history of their subject, or trying to provide simplified expositions of it.

1. The Structure of the Mind: A Program for Research

1.1 FROM GALL TO FODOR

1.1.1 The Notion of an Architecture of the Mind and the Project of a Faculty Psychology

Our point of departure is a question that has occupied cognitive science since the time a philosopher, Jerry Fodor, one of the principal theoreticians of cognitive science, formulated some 35 years ago the hypothesis of a modular architecture of the mind (Fodor, 1983). The initial insight, the first scientific formulation of which is attributed to Franz Gall in the early 19th century, was simple. The mind is considered as a collection of specialized mental powers, or faculties. Gall's idea led him and his disciple Spurzheim into what is today considered a disaster of pseudo-science: phrenology, roughly the theory of bumps on the cranium. For example, aptitude for mathematics might be explained by overdevelopment of a specific part of the cerebral cortex, causing an anatomical deformity at the corresponding point on the skull, which would be called the "mathematics bump," and so on for an entire list of "faculties" (27 to be exact, 19 shared with animals and 8 specific to humans) all derived from a largely speculative psychology, further distorted by certain anthropological or anthropometric prejudices prevalent during that time period (Gall and Spurzheim, 1810–1819).

What interests us today are not the errors of phrenology but the grain of truth that might have been contained in the speculations of Gall and Spurzheim. In fact, one may retrospectively credit them with formulating a research program that would end up taking in a large portion of contemporary cognitive science. This program is based on the responses to three central questions:

(1) Knowing that the human mind is capable of accomplishing tasks of great variety and complexity, does it follow that it is composed of parts, and what are they?
(2) If we maintain that the mind is produced by a dedicated system within our organism, to wit, the brain, or to speak more precisely the central nervous system (CNS), what are the relationships, first, between the mind (seen as the repertory of mental or psychological functions) and the CNS, and second between the faculties (the components of the mind) and the parts of the CNS?
(3) If it can be confirmed that the mind is composed of parts that correspond to the parts that make up the CNS, how can we explain the (real or apparent) ability the mind has to confront an indefinitely varied set of situations, which cannot be handled by the skills of a single fundamental

faculty? And correspondingly, How can we account for the introspective feeling that the mind is essentially one?

It might be thought that such general questions are after all too vague, or only rhetorical—that they simply will not prove fruitful terrain for scientific inquiry. We shall see that this is not the case. However, to begin with, it will be helpful to get clearer about the framework in which these questions can acquire meaning.

1.1.2 The Mind as Laborer

The first of our questions leads us to consider the mind, in the first instance, as an entity that carries out tasks. A sewing machine sews, a plow turns the soil, the heart pumps blood throughout the body, bees make honey, a student multiplies 13 by 17; and likewise the mind busies itself getting countless things done. On the other hand, whatever we may understand by the term "mind," we need not endow it with any of the mysterious qualities that would induce us to translate "esprit" in French as "spirit" rather than "mind," to realize that it presents itself to us as altogether different than a laborer. It presents itself rather as a "mental flux" (the flow of thoughts), and also as the "seat" of consciousness; or it might be said to present itself as an inner eye, or perhaps the "Cartesian theatre" mockingly referred to by the philosopher Daniel Dennett (Dennett, 1978, 1991). For William James, the central topic of psychology was our "conscious mental life," and its goal was to provide an account consistent with human physiology (James, 1890).

Some remarks are in order. First, these two conceptions—laborer *vs.* mental flux—are incompatible only insofar as each is construed as providing the essence or the core of the mind. One viewpoint can in fact be subordinated to the other: the mind as a conscious flux can be set to carry out a task, as happens when the mind of the student (as opposed to his liver or his legs) is given the task of multiplying 13 by 17; conversely we can easily imagine that the mind, seen as a capacity for accomplishing tasks (we will presently offer a less clumsy expression), is also a locus for secondary phenomena that are manifest in our personal experience in the form of a mental flux of conscious thoughts, or the form of a mental theatre in which appearances follow one another.

Still, the task-carrying view of the mind appears at first glance the more restrictive one, corresponding to purely deliberative episodes in our mental life; making this the core of the mind is a bold theoretical move, reminiscent of similar moments in the emergence of a science. Thus Galileo and Descartes's conception of motion launched an "impoverished," thin science of dynamics, cut free from the "rich," thick conception of dynamics inherited from Aristotle. Such a position starts out with some degree of legitimacy insofar as it is taken as a conjecture or as a bet, and it may gain credibility to the extent that it gives rise to a fruitful or progressive research program.

Third, one should expect (as was the case with the concept of motion in physics) that the sense in which the mind "carries out tasks" will undergo significant changes as the program unfolds. One starts out with examples such as solving simple formal problems, discovering the cause or the agent responsible for some typical event of daily life, translating a simple text, planning a simple action, and so forth; the means typically used by the mind in accomplishing tasks such as these belong to logic (taken in a broad sense). But cognitive science is not held to this paradigm; as it turns out it fairly quickly went beyond it. This two-step procedure—restriction to a subset of commonsense concepts followed by the lifting of restrictions from conceptions of common sense, or of our metaphysical heritage—is at work in the creation of every science and should come as no surprise (again physics is a prime example). In the case of cognitive science, due to its relative youth and porous boundaries with philosophy and common sense, this process by which an emerging science constitutes its object must be emphasized and recalled as often as necessary, for it is not well understood outside the field, and gives rise to disputes that are usually based on misunderstandings.

Finally, philosophers are not obliged to accept the restrictive position—the reduction of mind to a task-accomplishing system, or, briefly, the position of mind as laborer—without an independent scrutiny, even if they accept the legitimacy of these hypotheses as a conjectural or even speculative starting point. This is brought home to us all the more strongly today because it is being called into question, not from viewpoints outside cognitive science, but from within, by participating scientists and philosophers who hold that cognitive science should in one way or another break out of the conceptual framework in which it was originally constituted (I briefly return to this issue in section 2.2.4).

1.1.3 The Brain and the Mind

Let us move on to the second question that comes from Gall's framework. For Gall, as for his predecessors and his materialist contemporaries, this had to do with the idea that the productions of the mind are, in a sense, also productions of the brain. In what sense? Physicians, and philosophers in their wake, have been content with the metaphor of a "seat." To think of the brain as the seat of thought (in opposition to Aristotle who located thought in the heart) was to believe that without a brain, thinking is impossible, and that a brain lesion generally leads to an alteration of thought. Nonetheless, it was clear from the start that the brain does not "produce," in a causal sense, anything but cerebral events or episodes, of a biological, electrical and chemical nature, capable of triggering in turn certain motor events. But thinking (productions or manifestations that are characteristic of the mind) is not something of a biological, chemical, or electrical, or again motor nature. We have hit on a version of the mind-body problem, which philosophers and the first representatives of scientific psychology were attempting to solve, or dissolve: Gall's idea appeared to hold out not a solution, but a way of getting around the problem, given that none of the proposed solutions appear to be capable of generating consensus. This may seem surprising, because the notion of a correspondence between a particular faculty of the mind and a specific area of the brain appears logically to depend on the notion of correspondence between the mind as a whole and the brain as a whole. How can we understand the significance of the fact that one part of the cortex produces mathematical thought, when we don't understand what it means to say that the brain produces thinking? But this is how one may hope to get over the difficulty. Let us suppose that we have succeeded in showing (i) that all mental processes are elementary "moves" that belong to a particular faculty, or a rule-governed combination of such moves belonging to different faculties; (ii) let us further suppose that to each faculty there corresponds a specific, dedicated section of the brain; (iii) and further, that for every combination of elementary mental processes there is a specific transformation of the cerebral substrate. In that case, we might be led to conclude that (iv) there exists between thoughts, that is, the entire group of the productions of the mind, on one hand, and the states and transformations of the brain on the other, a kind of isomorphism, and that (v) in strictly scientific terms, this empirical correspondence is sufficient to meet the needs of explanation and prediction, rendering superfluous the unavoidably diverse metaphysical conceptions that had been and would be proposed in order to explain and to make use of this notion of correspondence. We may observe that there is a similarity between this way of handling a metaphysical problem by means of science and the solution offered by structural realism to the general question of scientific realism: following a path opened up by Poincaré (1905; which was to some extent anticipated by Comte; see for example Comte, 1848/1998), the contemporary supporters of structural realism, such as John Worrall (Worrall, 1989) believe that science can only identify a system of relations between entities in the world, and that science should give up trying to determine the ultimate nature or essence of those entities. One may speak of a kind of structuralism inherent in the neuropsychology of faculties such as Gall outlines it, which finds its most general and its most precise expression, as we shall see, in the functionalist conception that is still the basic frame of reference of cognitive science.

Yet at the same time, this outline of the modularist solution, or rather dissolution of the mind-body problem, may be a Pyrrhic victory. For once we have neatly bundled all the mental happenings we can think of into a manageable number of functionalities that are well enough defined to be put in 1-1 correspondence with brain areas, we must ask where has the mind gone? Surely, it might be argued, the mind is more than any particular one of these functionalities, but also more than the collection of them all. Isn't the mind precisely that which escapes all specialization? If not the mind, what is it that activates in an appropriate way the various specialized faculties? So now we have reached the third of our initial questions. There are various possible responses. We can propose that the mind is nothing over and above a set of *combinations* of specialized processes, emphasizing that such combinations can produce an infinite variety of hybrid thoughts (involving several specialized components). It then falls on us to examine what is left of modularity if we permit all combinations between the productions of different modules. Or we can accept outright that part of thought escapes modularity, even if the latter is enriched by the play of appropriate combinations. In either case, the initial plan—establishing modularity as a scientific fact and using that fact to circumvent the mind-body problem—seems to run into trouble. We can then hope to overcome the problems, or else decide that they are serious enough to make us return to the basic hypotheses we have depended on up to this point in order to formulate the questions that Gall's program raises.

So we already have a series of questions to consider as we look back on Gall's project for a faculty psychology, or as we might express this today, a modularist conception of the functional architecture of the mind—and all before we have even begun to unpack the fundamental concepts of cognitive science. In this chapter we will see how these concepts make this problematic more precise, and its content more determinable.

1.1.4 The Two Stages of the Mind according to Fodor

Let's go back to Fodor. According to him,[2] the mind is made up of two kinds of processes; on one side there are specialized and autonomous faculties, which he calls *input systems*; on the other side there are so-called *central systems* that ensure the "fixation of beliefs," that is, the outcome of cognitive processes in the form of the conscious acceptance of a proposition such as "A red book is on the table." There are in fact a great variety of conscious mental states that can be characterized by a "propositional attitude": acceptance (which can be graded), doubt or rejection, fear or hope . . . concerning a real or supposed state of affairs, which is expressed in a language, such as our own mother tongue or perhaps, as we shall see shortly, a "resident" language. The central processes postulated by Fodor lead the mind toward states of this type, on the basis of data furnished by input systems, whose function according to Fodor is to "present the world to thought": they are the processes of perception, plus language, or at least the automatic components of the processing and production of spoken language. These processes are local, specialized, and only handle certain kinds of information; they are "insulated" in the sense that they have no access to information that is external to their base, and thus cannot make use of it; they are automatic and swift; they exhibit characteristic profiles with regard to acquisition and loss in the case of brain lesions or other diseases; they are at least approximately localizable in the brain and have an innate component. These properties make the modules accessible to scientific inquiry; and in fact cognitive science has

[2] Fodor, alone, neither invented nor reinvented the contemporary context of the notion of modularism or the modularist hypothesis. He only worked out a systematic theory of it, using the resources of cognitive science and conceptual analysis; he took the risk of offering an explanation of the uneven achievement of cognitive science going as far as to set a principled limitation to what it can hope to accomplish. I mention this for two reasons. First, I want to make clear that this chapter does not aim at historical exactness, and the references cited are given solely as general points of reference. Second, Fodor's contribution to the question of modularity is a typical example of "cognitive philosophy," as I'll be using the term toward the end of this chapter.

made notable progress in the theorization of modular cognitive processes. On the other hand, science encounters obstacles when it attempts to account for central processes. According to Fodor, cognitive science has in fact made no progress at all in this area, and at the time he predicted that it never would (he stuck to his pessimistic view thereafter; see Fodor, 2000). His argument is based on a comparison with the theory of scientific confirmation (see chapter 2 in this volume): on the one hand there is no restriction on what may appropriately be taken into account in determining the truth value of a belief; on the other hand all beliefs are part of a belief system, for which the degree of confirmation can only be assessed as a whole.

Fodor thus proposes certain answers to questions that Gall's theory raised, and these answers raise yet further questions, some of which I will examine presently. Here are the answers:

(1F) Yes, the mind is made up of parts, and we have a relatively precise idea of what these parts are and how they are to be characterized. Nonetheless, this division into parts only concerns one sector of mental activity, and leaves out of account an important part of it. (Naturally, the modules whose existence is conjectured by Fodor have not much to do with the 27 faculties of Gall; the very notion of a faculty, which for Gall includes instincts and character traits as well as particular intellectual talents or different forms of memory, has for Fodor a precise meaning, which is related to the other postulates of his psychology.)[3]

(2F) The parts of the mind identified by Fodor, whether they are modular or not, are described as information processing systems. It is at least conceivable that the brain is the material system that carries out this handling of information, and that the modules of the mind are associated with (or have as their seat, or as their "neural substrate" as one tends to say today) subsystems of the brain that are dedicated to the execution of specialized tasks which are taken on by the corresponding module. This is made explicit by Fodor in the first part of the book in which he reviews the general framework that the cognitive sciences have used since they were founded—something that will be examined in section 2.

(3F) The ability of the mind to handle an indefinite multiplicity of situations, most of which cannot logically belong to a single faculty, is a mystery that the cognitive sciences are not currently able to explain.

[3] Fodor's modules are distinguished in general terms from the components that "faculty psychology" searched for after Gall during the entire 19th century: those were "horizontal," that is, they designated "operations" such as attention, memory, observation, clarity, quickness, sensory awareness, etc., that applied to all areas; Fodor's modules are, in contrast, "vertical": each of them has a limited competence that does not interfere with the others. Faculty psychology, which had important consequences as regards pedagogy, was definitively discredited in the early 20th century (Thorndike & Woodworth, 1901).

1.2 THE NOTION OF GENERAL INTELLIGENCE AND ITS DIFFICULTIES

When Fodor published his book, one of the most influential works in the history of cognitive science, he crossed up one of the basic intuitions that launched the whole enterprise, while remaining, in other respects, in the direct line of that research tradition. In a seminal article which appeared in 1950, the logician Alan Turing, inventor of the abstract concept of a computer, put forth the hypothesis that certain machines might be capable of thought, meaning that they might be able to accomplish all the tasks that human beings are able to accomplish in virtue of their intelligence. This hypothesis was clarified and amplified by Herbert Simon, Alan Newell and others (Newell & Simon, 1972),[4] and the project soon was given the name "artificial intelligence" [AI]. This theoretical framework gave life to the first epoch of what we now call cognitive science.[5] That which Fodor takes over from the AI framework, and which he helps clarify, is the idea that mental processes are essentially rule-governed transformations of information. What Fodor rejects, on the other hand, is the conclusion that AI drew on the basis of an —undeniably striking— logical fact, namely, the existence of a *Turing machine* (a symbolic calculator) that possesses the property of *universality*: such a machine is capable of calculating, based on the construction schema (technically: the table) of any other Turing machine, that which the other Turing machine calculates (Turing, 1936–1937). Thus Turingian neo-mechanism appears capable of getting around the essential limitation of the classical concept of mechanism, which cannot go beyond *dedicated* machines: one task, one machine.[6] A universal Turing machine (UTM) can accomplish in its proprietary domain (information processing) any conceivable task.[7] Our third question thus received an adequate response: if our minds have the functionality of a UTM, then we can explain how it is, that the mind is able to accomplish any cognitive task; and so, to the extent that it is "realized" in our organ, the brain, we can understand the feeling we have of the unity of the mind, somewhat in the manner in which we understand intuitively that our hands can execute, within certain limits, any manual gesture that we can conceive, and our lived experience of

---

[4] This publication date is misleading; the birth of AI actually took place in the mid-1950s (see Buchanan, 2005; McCorduck, 2004; Hook 1960).

[5] In the present context, "intelligence" tends to be equated with "mind" (or at least with "cognitive capacity"), and artificial intelligence is then seen as an abstract model of human intelligence. This involves a number of choices that are partly terminological, partly doctrinal; I will address the latter a little further. There is another construal of the term "intelligence," associated with a different concept of general intelligence, one linked to the attempt to measure and compare the *quality* of cognitive performances. This concerns a different area altogether, that of IQ, which only partially intersects (in the current state of knowledge) cognitive science, although eventually the question of IQ must become a fully integrated part of cognitive science. Intelligence in the sense of IQ raises problems of the most interesting kind for the philosophy of science (see for example Sternberg, 1988; Flynn, 2009; Nisbett, 2009), but which we cannot go into here.

[6] Let us remind ourselves that for Aristotle, it is because the mind knows how to take in all possible forms (that is, is able to think any object whatsoever) that it cannot be material (*De Anima* III, 4; 429a10–b9).

[7] Whatever may be its exact significance for cognitive science, the general conceptual reach of the notion of UTM is considerable (Herken, 1988).

"hand-ness" is homogeneous—it doesn't feel to us that our hands click into different "gears" when we wash the dishes, change the baby's diapers, type a paper on our laptop or play the clarinet.

Why did Fodor and the supporters of modularity give up this solution? For two main reasons. The first involves the argument from combinatory explosion. The number of operations that have to be performed in the course of a cognitive task is an exponential function of the number of items of information that might possibly be relevant. If the latter number is very great, the number of necessary operations "explodes" and exceeds the necessarily finite capacity of any material system. A "general intelligence," that is, a universal system of cognition, would by definition have a database of a nearly infinite size, something that would prevent it from executing most of its tasks, at least within a reasonable time frame (a favorite example of the modularist literature involves a tiger: in the presence of a sign that there is probably a tiger around, such as a visual perception that looks like a tiger, it is crucial to be able to make decisions quickly). The hypothesis of modularity, by drastically limiting the database for certain families of tasks, renders them quickly feasible in a physical information processing system.

The second reason to give up the UTM model is the argument from poverty of stimulus. The first case of modularity was defended by Chomsky (Chomsky, 1957, 1975; Piatelli-Palmarini, 1979): learning one's mother tongue or first language is a particularly important and complex task, yet one which all normal babies accomplish adequately. If this accomplishment were (as many thought for a long time) the work of a general capacity for learning applied to the linguistic environment of a young child, this success would be (according to Chomsky) impossible, essentially for logical reasons: that which experience provides for the child (the "stimulus")[8] turns out to be much too thin, too "poor" to allow the child to identify the grammar of his native language, meaning by that the articulated group of items of (tacit) knowledge that make it possible for him to understand and speak. The induction that makes it possible for the child to go from certain items of information that are picked up from his environment to the mastery of grammar (in an extended sense that goes far beyond traditional grammar) can only be successful in a constrained framework, comparable to the developmental sequence followed by an organ or the limb of an animal. The "language acquisition system," according to Chomsky, would therefore be a module that is essentially independent of the general faculties of the mind. Still hotly debated, the argument draws on a variety of sources: linguistics, logic, psychology, biology. More importantly still, the case of language appears to have the status of a paradigm for the entire range of cognitive processes: the Chomskyan model, as we have seen in examining Fodor's position, can be applied to other cognitive aptitudes and by the same token raises the

[8] The terminology comes from behaviorist psychology, whose theory of language drew a critique from Chomsky that is often considered to have been decisive (see his review of *Verbal Behavior* by B. F. Skinner; Chomsky, 1959).

same series of questions—to the clarification of which, philosophers have contributed a great deal. We shall now examine several aspects of this model.

1.3 DEVELOPMENT AND NATIVISM

1.3 1 The Mystery of the Infant

Since Plato, philosophers have asked themselves questions about the origin of our knowledge. The *infans,* he who does not speak (and who was long believed to think even less), develops physically and mentally. However, while it is easy to observe with the naked eye, and to harbor the impression that one understands many features of the transformation of the body, what can be observed of the transformation of the mind is deeply puzzling. The concept of growth has since Aristotle provided us with a reassuring grip on the phenomenon of organs gaining size, making room for both commonsense understanding and a very successful scientific research program. But we have remained for a long time at a loss regarding mental development.

That this mystery should have been mostly ignored for a very long time, or at least relegated to a subordinate place behind the mysteries of the origin of the cosmos, the nature of matter, or the essence of life, constitutes a philosophical mystery by itself. It may be, I venture to suggest, a case of rational renunciation, as in the fable of the fox and the sour grapes: people hit quite early on ways of addressing the last three problems, and while they may not be fully resolved today, at least a great deal of progress has been made. But until quite recently the first mystery has had the appearance of a perfectly smooth wall. We have remained in a state of paralysis, caught between a naturalistic and organic conception of mental development (the child grows mentally "as" he grows physically) and a metaphor of mental "inscription," according to which the mind is progressively "informed" through being "written on," as it were, this inscription being that which gradually renders the mind capable of carrying out operations of the sort that characterize adult cognition. To inform the mind is to train it—that is, to furnish it with what one called ideas in the 17th century, and would later be called representations. Now, either these inscriptions are present (in their totality, or in part) from birth, as nativism holds (nativism that is sometimes also called rationalism in this context), or the inscriptions all come from experience, as empiricism holds, and this experience begins with the first days of a child's life. For one camp (which has Descartes as a member) as for the other (represented by Locke), the mind is without structure (without "architecture" in the sense explained earlier): it is an essentially passive recipient, but in contrast to all other natural systems it is endowed with an aptitude for receiving "impressions" in an infinity of ways, and this aptitude is called learning or memory. The child develops mentally because it acquires knowledge, just as it develops physically because it acquires organic material in the form of muscle and bone and other kinds of tissue, all of which only help strengthen structures that already are present (with few exceptions, the organs and visible segments of an adult body are present in the body of a newborn).

1.3.2 The Modern Idea of Development

The founders of the modern conception of cognitive development (Piaget, 1926; Vygotsky, 1930; Chomsky, 1968; Bruner; 1966, 1968; Carey, 1985, 2011) were strongly at odds with one another about certain central questions, but they were all able to break away from these traditional conceptions, while retaining parts of them. They shared the belief that the architecture of the mind may be complex and differentiated; this architecture might vary over the course of development; the evolution of the cognitive capacity of an infant is the joint result of an organic development of the architecture of the mind and of the modification (through acquisition and revision) of elements of knowledge (e.g., ideas, representations, beliefs) which the mind possesses, it being understood that these elements of knowledge do not necessarily have (and ordinarily do not have) the explicit and conscious character of the knowledge possessed by an adult in a deliberative situation (the paradigm being a scientist at work).

What is carried over from tradition is the idea that while the acquisition of knowledge (in a sense that gradually diverges from both the ordinary meaning and from 17th century philosophical conceptions) plays a role in the epigenesis of cognitive capacities, these may be innate, that is, present at birth (an acquisition for the species, rather than for the individual), or acquired in the course of the development of the individual. What is rejected is, first, the axiom of homogeneity (the thought that the mind is initially undifferentiated), second, the axiom of structural or organic invariance over the course of development, and finally the idea that cognitive development is a product of the accumulation of knowledge alone.

From there on, the framework for the study of development is based on three main tenets:

- Development is the middle term in a 3-stage sequence: initial state (the newborn)/development/final state (the adult).
- Priority is given to invariant characteristics that are common to all (normal) individuals.
- An important goal is to distinguish processes of structural change (also called maturation) and processes of acquisition of knowledge (also called learning), and to understand how these two kinds of processes interact.

Although essentially independent, these three tenets taken together form a coherent theoretical framework that has been judged to be productive by many researchers. None of them is particularly obvious. To the contrary, they are quite risky, and somewhat enigmatic. The challenge will be to reduce this obscurity by combining empirical investigation and conceptual analysis. As I briefly suggested earlier (and will discuss further), the notion of an *architecture* of the mind is not straightforward. As long as it has not been sufficiently clarified, the idea of an evolution of the architecture of the mind itself remains obscure. Regarding architecture, provisionally we may content ourselves with the Gallian device of taking the anatomical map of the brain as the model for the mental structure. But this strategy, as we shall see, raises objections. These difficulties also hang over the distinction between maturation and learning, or between the evolution of mental architecture and the acquisition of knowledge. Nonetheless we shall see that there are ways to lift these barriers at the theoretical level; then we will have to ask whether the proposed framework is generally adequate to the set of empirical data.

1.3.3 What Is an Innate Capacity?

The hypothesis that is most urgently in need of clarification has to do with capacities (elements of knowledge or aptitudes) that are *innate.* Nativism (also known as innatism) plays a crucial role in cognitive science, because language is but one of the many faculties, or (to use the most inclusive term possible) cognitive structures that are claimed by one or another group of scientists to be innate, and these claims are all controversial.

The initial thought is to defer to anatomy and physiology: aren't organs and organic functions good examples of innate structures? They are certainly a plausible starting point, but they do raise a series of questions that are central for the philosophy of biology. And when it comes to mental "organs" and functions, the difficulty is compounded.

The first observation is that the most natural definition of what is innate is privative: what is innate is that which has not been acquired, whether for empirical or conceptual reasons. One may certainly think that certain concepts or capacities could be acquired, when in fact they are not; others may appear to be difficult or impossible to acquire for reasons of principle.

But what does possession of an innate cognitive structure consist in? Does the answer depend on the structure in question? The capacity to smile, to swallow, or to blink your eyes is innate: these are motor reflexes. The capacity to serve the ball in tennis is acquired: this is an ability that one learns little by little through intelligent imitation. But how can we understand that the concept of time is innate, or that the concept of a solid object is innate, while on the other hand the concept of a morganatic marriage or of a limited corporation is (one tends to think) acquired?

And what do we mean, exactly, when we deny that something might have been acquired? Are we saying that the environment played no role whatsoever? Evidently this is too much to ask: many anatomical and functional traits of the adult organism depend on the environment in order to develop, and often in order to take on one specific form among several that had at some point been possible. One might at least speak, as does Paul Griffiths (2002), a philosopher of biology, about developmental invariance, which means that the structure in question emerges over the course of development independently from environmental differences, within the limits of a broad spectrum of natural environments.

Or do we mean that the structure in question remains essentially the same over the course of the life of an organism? Relevant examples might then be gender (in mammals, barring human intervention and certain forms of hermaphroditism), eye color or the number of fingers. That is a different property from the one just discussed. Or are we saying, third possibility, that the structure was present at birth?

Second observation: what is innate is not something that belongs to the species, as the example of gender shows. Nonetheless the concept of innateness and its use come closer to the related ideas of heredity and universality within one species—in other words, we would often understand by "innate" that which is encoded in the genetic heritage of the species. That is undoubtedly what many people are thinking when they claim for example that language is a "property" of the human being, or when they observe that certain species of animals (but not all) possess elementary mathematical ability, or are capable of altruistic conduct. A difficulty that comes along with this conception: the notion of coding in or by the genome of a species gives rise to difficulties that are well-known by philosophers of biology.

A third observation: what is innate appears to give material form to a *norm* that is the property of the species. What is innate is that which normally leads to a trait that is universally shared by the normal members of the species. Women's breasts are innate in this sense, although they are not present at birth. And the same is true of innumerable metabolic systems, cerebral structures, and so forth. These traits are normative as well, inasmuch as they are functional, and thus probably directly or indirectly result from natural selection.

Apart from the questions that are raised by these characterizations, taken one by one, we could ask ourselves if they are conceptually or empirically coextensive, or at least coincide to a large extent. On the conceptual level, a first analysis yields a clearly negative response: definitions based on independence with regard to the environment or on non-learnability, definitions based on the genome of a species and intraspecific universality, and definitions based on functional and adaptive normativity are not conceptually equivalent. In fact, from an empirical standpoint, biologists have unearthed a large number of counterexamples that go against the hypothesis of even an approximate overlap. Some authors have reached the point of recommending that the notion be simply dropped. Others recommend using it in different ways according to particular contexts and theoretical purposes (a solution that is also often proposed for the concept of a gene). Most continue to entertain the idea that these different characterizations refer to properties that in fact are often associated, and that it is useful to consider the structures that possess all of them. In other words, innateness would be a *cluster property* made up of traits that come generally together, but which are not necessarily all present; we know that life (as the property a material system has of being alive) is often considered today as a property of this type.

In the case of cognitive structures, as I have noted, the difficulty is compounded by our uncertainty with regard to the nature of these structures. Take the case of language. We first observe that an important argument that is invoked by nativists is that the progression followed during language acquisition is largely independent of the individual and his mother tongue, that it takes place quickly, and that it does not require any voluntary learning. These are indications of an organic development that is comparable to the development of an organ or of a part of the body. It is also a sign that the rhythm of the developmental process is determined by maturation rather than by the acquisition of information (for in the latter were the case, one would expect significant variations from one individual to another and from one language to another). Next, it is quite clear that that which is innate cannot be a particular language that is spoken by a child: children not only learn different languages, but in fact *any* child that is immersed in a given linguistic environment learns the language that is spoken in that environment, independently of the child's origin, in exactly the same way (same stages, same pace, same final result) as all other children. That which is innate, therefore, can only be the ability to learn the language, which ability, by virtue of the argument from poverty of the stimulus, must be dedicated to language, in the sense that this capacity or ability cannot be used to learn something else. Depending on the context, Chomsky called this *universal grammar* or a *language acquisition device* (LAD). Learning consists therefore in the determination of the particular grammar of a language used in a particular environment, on the basis of available indications within that environment. To say that universal grammar is innate would be to say that it is a "primitive" cognitive capacity, according to an interpretation that is still disputed. In other words, this capacity does not belong to psychology but to biology. In this sense it would really be an organ (more precisely a functional cerebral structure) capable of receiving and processing linguistic information, and in the end producing an information-based or psychological structure constituted by representations that in combination give rise to the totality of the sentences in a language, that is, sentences that are acceptable to one's interlocutors when they hear them.

The same questions are asked every time someone puts forward the hypothesis that a particular cognitive structure or capacity is innate, and most of the time they are thinking of one or another of the three main families of properties listed previously. We might for example be led to suppose that a concept (such as time, space, the natural numbers, iteration (repeated application of a function to an argument) or (the more sophisticated) recursion, material object, movement, cause, relation, logical consequence, and even the concept of concept itself) is innate. At that point we would seek to understand what that means, in other words to pass from a diagnostic property (the concept apparently has not been learned or cannot be) to an intrinsic characterization (what does it mean, to say that a concept is innate?; Samuels, 2002; Carruthers, Laurence, & Stich, 2005; Khalidi, 2007; Margolis & Laurence, 2012).

1.3.4 The Empirical Question: Which Capacities Are Innate?

Even supposing the ontological uncertainty of the concept of innateness to have been resolved, and even if we can agree provisionally on an operational characterization of the innate character of a given cognitive structure, there are still arguments for and against that must be examined. In the case of language, apart from the properties already mentioned, the study of infants who are born blind or deaf, and thus do not benefit from all of the information to which other infants, who can see and hear, are exposed, considerably strengthens the innatist hypothesis. In the case of concepts, what speaks in favor of nativism is the apparent impossibility of inferring the extension of a concept by induction on set of examples (Fodor, 1975, 1981). The skeptics (Elman et al., 1996, Cowie, 1999) dispute the argument from poverty of stimulus; on the one hand the stimulus is not as poor as some have said, since part of the necessary information might be obtained from nonlinguistic sources. On the other hand, though they admit that the identification of a grammar requires further constraints, they deny that these constraints must necessarily take the form of tacit knowledge, understood in general as rules or parameters for universal rules. They also doubt that the myriad regularities that belong to every language can all be deduced from a reasonable number of rules or parameters. Connectionist (neural net) models (see section 3.a), which are *prima facie* incompatible with Chomskyan conceptions of linguistic competence, appear to show that the inductive impossibilities postulated by the nativists are actually the result of a lack of imagination on their part; not being able to see how a system S could learn X based on a certain body of information does not imply that X is innate in S, but only that the researcher has not found a solution (assuming one exists; Elman et al., 1996). A formal (logical) theory of learning was developed in order to overcome this kind of objection. This theory makes it possible to state and mathematically prove impossibility results of the following form: under certain assumptions, an information-processing system S that is provided with certain resources cannot identify the grammar of a language on the basis of empirical information presenting certain characteristics (Jain et al., 1999). These results must nonetheless be judged in the light of the idealizations made to give the phenomenon a mathematical formulation, and the plausibility of the substantial assumptions regarding the system; this explains why this dispute continues (Stainton, 2006, pp. 57–112). And the debate over the question of the innateness of concepts (among the skeptics: Prinz, 2002; Laurence & Margolis, 2002) is no closer to being resolved. It is fair to say that nativism has of late become a minority view within cognitive science, yet retains many advocates (for a recent defense, see Margolis & Lawrence, 2012).

1.4 THE IDEA OF A NEURAL BASE

Let's go back to modularity (without straying far from innateness). For Gall, as we have seen, the faculties possess distinct "seats," limited areas of the brain (most of the time but not always, parts of the cortex). Fodor is much more cautious, and supposes that the modules are not necessarily *anatomically* localized, but may only be functionally localized (so as to correspond with modes of neurophysiological functioning that cannot be reduced to the sum of activities in a particular area of the brain); on the other hand, localization is not strictly necessary for modularity, at least at the conceptual level. It remains true that a neural or neuro-dynamical interpretation seems to be a natural way to make concrete sense of the modularist hypothesis. Neuropsychology came from discoveries by neurologists such as Broca (Broca, 1861) and Vernicke, and took on the task of establishing correspondences between cognitive deficits and cerebral lesions. The existence of patients afflicted with very specific deficits has constituted the main empirical argument in favor of the general idea of a differentiated (as opposed to homogeneous or equipotential) central nervous system; modularity is the modern formulation of that idea, one more precisely adapted to the information based framework of today's cognitive science.

Neuropsychology is now part of cognitive neuroscience, whose goal is to uncover the so-called neural base of cognitive functions primarily in *well-functioning* human beings. The specific contribution of neuropsychology consists in comparing clinical profiles in order to formulate hypotheses about the cerebral organization that is "responsible" for certain cognitive functions. The typical situation in this regard is a "double dissociation," involving

- On the one hand, some patient X with a serious deficit with regard to capacity A (for example, *identifying* ordinary objects like a comb, a hammer, a pair of scissors; or, to take another example, reading words for *concrete* things), but having no problem with capacity B (such as *using* ordinary objects; or, in the other example, reading words for *abstract* things)
- On the other hand, some patient Y deficient with regard to B, but not at all with regard to A

Such a pair of clinical profiles, in the absence of contrary indications, suggests the modular hypothesis that attributes distinct neural bases to A and B. Of course, this is not a deduction, but at best an inference to the best explanation (or abduction, to use the Peircean term): if the neural basis of A and B are in fact localized in distinct components, that would be a very direct explanation of the possibility of clinical profiles such as those presented by A and B. By the same token, the fact that two deficits are invariably associated speaks in favor of (but doesn't firmly establish) the hypothesis according to which the neural bases of A and B largely overlap.

This approach raises a bunch of conceptual, methodological, and empirical questions. We need to clarify the notion of difference involved in speaking of cognitive functions or processes. In one sense, every difference does count: everybody accepts that different cognitive processes are "executed" by cerebral circuits that are different from one another, even if only slightly (by virtue of a principle of supervenience according to which every difference at the mental level implies a difference at the cerebral level). Yet not all differences have a theoretical interest. We would learn a lot from a relation of dependence between certain functions that appear quite different from one another (for example, spatial navigation and autobiographical memory, or perception of the direction in which someone is looking, and the understanding of another person's motives), or conversely from the mutual independence of two functions that common sense tends to identify (the pronunciation of names for concrete things as opposed to the pronunciation of names for abstract things). By contrast, nothing interesting would seem to follow the discovery of a clinical correlation, or disconnection, between the memorization of car models and that of washing machine brands. The danger that appears to threaten research into dissociations in neuropsychology, apart from fragmentation, is triviality. Cerebral lesions are never "pure" (in the sense of affecting only one functional system); double dissociations may end up being revealed as pairs of processes that exhibit minor differences with no theoretical bearing. In practice, good clinical sense and a solid theoretical framework may allow these difficulties to be avoided.

But other difficulties arise. The simplest way one neural base can be distinguished from another, as we have said, is spatially. Beyond that, one can imagine distinct circuits that are not necessarily completely separate. But a third kind of relation, much more exotic, is conceivable. Connectionist models, and in general models derived from the theory of dynamic systems, prove that distinct functions can be produced by a single complex system functioning under different dynamical regimes. This would appear to undermine the key intuition of modularity—to explain the structure of thought with reference to the organization of the material system from which it proceeds, causally or metaphysically.

Another question involves the relationship between stability and plasticity in cerebral architecture. No one denies that the central nervous system is capable of reorganizing itself on several different temporal and spatial scales. London taxi drivers are known to have a detectable overdevelopment of the hippocampus, a structure that is involved in spatial navigation (Maguire et al., 1997). Some infants who suffered from extremely serious cases of epilepsy have had one entire cerebral hemisphere removed, and later on present a cognitive profile that is essentially normal (Battro, 2001). But the question is to know to what extent the brain "constructs itself" over the course of its existence, as a result of the experience that it lives through and the tasks it accomplishes. For supporters of "neuronal constructivism," cerebral plasticity makes it impossible to determine an architecture that would belong at one and the same time to the brain and to the mind (Quartz & Sejnowski, 1997; for a recent review, Forest, 2014).

Thus it is the very concept of a neural base that is in question, at least in the version that appears to sit best with the idea of a term to term correspondence of cognitive primitives and fundamental neural structures. The importance of that idea turns on the justification it provides for the simple methodological principle according to which a single cognitive phenomenon (memory, reasoning, face recognition, planning, ...) can be studied at two levels, the level of information and the cerebral or neural level, where the two approaches are assumed to be directly connected and to provide support for one another.

1.5 THE DISTINCTION BETWEEN HIGHER AND LOWER FUNCTIONS AND THE MASSIVE MODULARITY HYPOTHESIS

Let's return to modularity according to Fodor. His conception of modular processes and modular organization of cognition was in the direct line of over a hundred years of research, but the clear break he introduced between modular systems and central systems, involving the inaccessibility (on principle) of central systems to scientific inquiry, directly challenged the presuppositions and the hopes of many researchers.

Modular processes, as we have seen, are essentially linked to perception and motor capacity (an exception being made—language being regarded as par excellence "superior"—for certain "lower" linguistic operations). These are therefore the repertory of "lower" processes, almost all of which have analogues among nonhuman animals. We should note in passing that although Fodor takes up the traditional distinction between higher and lower processes, he gives it a very different twist, in line with the framework of contemporary cognitive science. The ontological difference between psychophysical systems, sensors or effectors, pure biological machines on the one hand, and intellectual processes, purely mental or ideational, on the other, disappears in the contemporary framework and is replaced by a structural distinction between two main categories of biological systems of information processing.

Understanding these "lower" processes in human beings and in animals is no minor undertaking. In fact it involves deep problems, both scientific and philosophical; it offers a comparative perspective that proves to be essential for the understanding of these processes in human beings; it also provides essential guidance, in the guise of concepts, methods and models, for the understanding of the so-called higher processes. Still, the cognitive sciences' highest ambition is to give an account of the whole of cognition, so that an exclusion on principle of the "higher" processes would constitute, if it were really inevitable, a tremendous disappointment (and would provide confirmation of a skeptical point of view in regard to the pretensions of the psychological sciences, something many philosophers and specialists in the human sciences still uphold).

A possible response to Fodor's prognosis would be to reject all or part of his fundamental hypotheses: the existence of modules, their essentially innate character, the distinction between higher and lower processes, and so on. We will not discuss that possibility at this time, but we will say something about a different possible reaction, which consists in accepting Fodor's modular framework, but rejecting one of his two main conclusions, namely, the non-modularity of higher processes. The supporters of "massive modularity" (Tooby & Cosmides, 1992, 2005; Hirschfeld & Gelman, 1994; Sperber, 2005; Carruthers, 2006) defend the position that these processes are also modular, completely or partially. But the modularity that they possess is understood in a slightly more flexible sense than that of Fodor. Emphasis is placed on (i) domain specificity: a higher module only processes information that is related to a well-defined segment of the natural, conceptual, or social world; (ii) informational isolation or encapsulation: a module can only access a limited store of information, which is out of the reach of other modules; (iii) innateness; (iv) adaptive character. The general arguments in favor of massive modularity are exactly the same as the general arguments in favor of simple modularity. To these may be added arguments that are related specifically to various higher modules whose existence is conjectured, especially by developmental psychologists. Important examples are the so-called naïve theories: bodies of specialized tacit knowledge, present very early in development, exhibiting few interindividual differences, universal in all cultures, and that have a functionality which can be conjectured to have been very important in the adaptive environment of Homo sapiens, and may still be today. Examples of such corpora, which together constitute that which is sometimes called core knowledge (Spelke, 2000; Barner & Baron, 2016), include: one or more numeric systems, a naïve physics, a naïve psychology, a naïve biology, a naïve sociology, and a system of management of cooperation.

Corpora, core knowledge of particular domains—might suggest that what allows an infant, and later on an adult, to act in a quick and well-adapted manner in situations in each such domain, are bunches of facts. If that were the case, massive modularity would amount to a pair of claims : (i) certain kinds of empirical knowledge are domain-specific; (ii) the capacity involved is something close to (albeit obviously different from) propositional knowledge. The first claim is trivial: dealing with numeracy involves a specific competence not involved in, say, social interactions. The second claim is contentious: the sense in which an infant "knows that 1 + 1 = 2" or a toddler "knows that a painted mule is not a real zebra" is not likely to be that they possess the corresponding propositional beliefs. However, if we think of these higher modules as particular mechanisms that are activated only when the agent is confronted with a domain-specific task, then modularity takes on a genuine "architectural" meaning and opens up a fruitful line of inquiry.

There does remain a crucial question, one we mentioned at the beginning of the discussion. If you take away the modules, higher and lower, is there anything left of the mind? Both possible answers are given by different defenders of massive modularity. The positive response risks losing part of what makes the hypothesis interesting, because it plays on the possibility, quite plausible as we shall see presently, that an essential part of the properties of the human mind, especially as regards its exceptional virtues among living beings, is found in the non-modular part. Still, even a partially modular architecture of higher processes would have important theoretical and practical consequences (e.g., in education), so we should be careful not to go too far in the opposite direction.

As for the negative response, which is more daring, it raises a host of objections. One of the sources of the power of the human mind certainly appears to reside in its capacity to handle an indefinitely large number of situations, including situations that are entirely new, by using a certain number of general procedures that do not belong to any domain in particular. Next, if the modules are only competent in their own domain, how does one react to a situation which straddles the domains of two or more modules? More generally, is it not the case that flexibility and inventiveness are the marks of intelligence, and do they not confer on the mind a part of its stunning effectiveness? Wouldn't a mind that is entirely modular be reduced to reacting in a reflexive way to problems that it encounters, characterizing them according to the module, if there is one, that is competent for that situation? Is this not exactly the way that a sclerotic bureaucratic society works, with now well-recognized and well-understood effects? Our ability to implement new strategies with speed and suppleness is certainly limited in practice by habits, but in contrast to the situation with massively modular architecture, these habits do not seem to prevent it altogether. The latter argument refers us to the persistently problematic notion of *general* intelligence, which we already encountered in the context of early AI, and in relation to the existence of syndromes involving a *general* mental handicap.

The supporters of massive modularity answer in two ways. To begin with, they don't take their adversaries' arguments entirely seriously; after all, are we talking about anything other than the observations of common sense, based on nothing more than our intuitions? This feeling of flexibility, of fluidity, of mobility, accompanied by an introspective conviction of the homogeneity of higher processes—is all this going to stand up under scientific investigation any better than the feeling we have of the homogeneity of our own vision, the isotropy of our visual field or the connectedness of our retinal images (all of which can be considered today as having been definitively refuted)? For higher processes as for perception, these questions are of an empirical nature, and introspective evidence carries no weight. A second answer, more specific, was proposed by Dan Sperber (Sperber, 2001, 2005; a book-length defense of massive modularity is provided in Carruthers, 2006). First he reminds us modules should be conceived in terms of Chomsky's notion of a language acquisition system: as specialized learning systems that allow the organism to build modular components that are adapted to the environment, and which are in this sense acquired (universal grammar is an innate module, one which allows the acquisition, in contact with a particular linguistic environment, of mastery of a particular language among the five or six thousand that still exist today). Finally Sperber posits the existence of a particular higher module called "meta-representational" whose domain is made up of representations coming from all the other modules. This module could therefore "interweave" and combine information collected by various modules, and thus perform functions of transfer, generalization, and so forth that confer on the cognitive system the virtues emphasized by the opponents of massive modularity. This meta-representational hypothesis recalls in some ways an ancient conception according to which language is that which allows the human mind to reach the highest level of cognitive performance: to work with terms and phrases is no longer to deal directly with objects and matters of fact in the world, but with their linguistic representations. However, there is a chasm between the traditional conception and Sperber's hypothesis: the first treats the mind as something given, and the other claims to explain it with reference to a principle of reflection, through which a property of the mind is reflected at the level of its internal functioning. This principle is something to which we will return.

1.6 THE EVOLUTIONARY PERSPECTIVE IN COGNITIVE SCIENCE

To the defenders of massive modularity, the biological nature of the mind is of prime importance. The theory of evolution is for them an essential theoretical resource: it commands, as does for biology as a whole, a specific level of explanation that is in some sense fundamental. Nor is this a matter of mere principle, as remains the case with regard to many areas of biology that in fact have no real use for evolutionary data or explanations. Modularists have no choice but to adopt the evolutionary perspective. Here as well, they are opposed to Fodor (Fodor, 2000, 2008b), and are in agreement with Daniel Dennett (Dennett, 1995), who was the first philosopher of cognitive science to take seriously the evolutionary origin of cognition, and consequently to make the theory of evolution the very foundation of cognitive science.

Historically, the emergence of the evolutionary theme is a striking fact; it is difficult today to realize that the cognitive sciences arose, and for a long time developed in complete ignorance of the theory of evolution. Chomsky himself, one of the founders of the "cognitive revolution," early on emphasized the fundamentally biological nature of cognition, but even he resisted for quite a long time the idea that the theory of evolution could contribute to the scientific account of it. (This course of development illustrates in an ironic way one of the principal reasons, put forth by Fodor, for denying that higher processes might one day become the objects of natural science. As we have seen, according to him, an important lesson in the philosophy of science is that we can never be sure that a fact, no matter how distant from the phenomenon at hand it may appear, may not turn out to be relevant in the evaluation of a belief.)

The "evolutionary turn" in cognitive science has by now permeated the entire domain; even when no one can tell precisely *how* the phenomenon under scrutiny came to exist in the course of evolution, everyone agrees that in the best of all scientific worlds, we should be able to explain it, because this phenomenon was at first lacking, and then emerged over the course of evolution; so in order to give a complete account of it, we must at least show that its emergence is theoretically possible (Bickhard, 2002).

But there is another, more constructive and more targeted way evolutionary theory impacts cognitive science, by giving birth to two new branches, evolutionary psychology and evolutionary anthropology (also known as the evolutionary theory of culture), which are so closely interlocked as to form a single field. Many questions are raised by this field, which we can group into three main families.

First there are questions of *method*. To the general difficulties involved in the application of the theory of evolution, we must add in the case of cognition (i) the nearly total absence of a fossil record, since the essential part of the structure in question was composed of soft tissue, and the hard part (cranial anatomy, pharyngeal cavity, etc.) only provides very tenuous indications that are difficult to interpret; (ii) the lack of solid information concerning the evolutionary environment of adaptation (EEA) within which our species emerged; (iii) the still very fragmentary nature of our hypotheses concerning the architecture of the mind: the elementary components of the cognitive system are far from having been identified with the same degree of certainty and precision as the bodily organs, systems, or structures with which they are compared. The situation does improve somewhat with the development of paleogenetics and cognitive ethology, and as a result of progress in cognitive neuroscience. But the methodological problems are nonetheless many and complex.

Questions about *foundations* are no less pressing. One may perfectly well admit that the material basis of the mind, its "seat," is a biological system, comparable in this respect to the cardiovascular system, the digestive system or the locomotor system, whose present form and functions were jointly shaped by evolution. Still the mind has this essential feature, that it is endowed with dispositions that go far beyond the initial specifications that the mind might satisfy thanks to natural selection. In contrast to other biological systems, the human central nervous system supports not only specialized or dedicated functions, but also "meta-functions" capable of producing processes and entities that retain no trace (or almost none) of the evolved mechanisms[9] present in the system.[10] Culture, taken in the largest possible sense, includes processes and entities of this type, but in such a proportion that it could be the case that the sum of the explanatory resources of the theory of evolution would turn out to be of only marginal utility for a science of culture. To the extent that the mind, among its "meta-functions," possesses the capacity to absorb and to incorporate a vast quantity of external material furnished by individual experience, and even more by culture, psychology itself is "contaminated" by culture; biological determinations, particularly evolutionary ones, might take second place behind cultural determinations.

Such questions are grist to the mill of the culturalist, historicist, interpretativist currents within the sciences of man, whose hostility to naturalism is deeply rooted. (Taylor, 1985). It is generally agreed that cognitive science must pursue its own path without paying too much attention to that mistrust; as long as one is not denying any sort of relevance at all to cognitive science (thus questioning its right to exist), it must continue to pursue its goal, which is to bring to light the natural constraints within which specific human processes and events, individual and social, ephemeral and long-lasting, unfold. In the opinion of most researchers in this area, the respective importance of such constraints on one hand and of cultural determinations and historical contingencies on the other, will be settled later. It would be all the more premature to propose a settlement as its exact terms remain to be determined, inasmuch as the mode or modes of interaction between "nature" and "culture" are the subject of a great deal of current research. One of the main themes of evolutionary anthropology is the "co-evolution" of genes and culture: as many examples show, culture contributes to the selection of genes, favoring those who carry them through custom-based, institutional and material arrangements

[9] In the present context this term has a technical meaning: what is "evolved" is a mechanism, system, or process that is the result of biological evolution.

[10] The locomotor system occupies in this respect an intermediary position: it is not selected "for" dancing or acrobatics, but its "meta-functions" are very limited, and the traces of evolution are visible in all its productions.

(Richerson & Boyd, 2005; Diamond, 1997; Sterelny, 2006, 2012). This is but one among a set of important novel ideas that are gradually transforming our scientific understanding and bringing it to bear on a foundational problem that has long befuddled philosophers.

The third group of questions has to do with the actual reach of evolutionary approaches. For a long time, these approaches emphasized the reproductive functions, most directly linked to natural selection: mate selection, reproductive control, and so on (Buss, 2008). Now they extend as far as higher cognitive functions, especially language, in the framework provided by massive modularity, and even as far as the most general cognitive structures which make human sociality (and that of other species) and culture possible, including normative systems on which every human society is based. In the light of this new phase of research, the controversies of the preceding phase appear less worrisome than before. On the other hand, the higher ambitions of the new evolutionary sciences of man raise problems of their own: How will they fit in the landscape of the established, non-naturalist and pre-evolutionary disciplines, including practical philosophy, the branch dealing with ethics, action, rationality? Will they settle age-old disputes? How far will they reach? These matters are hotly debated today, but cannot be more than mentioned here (Gintis, 2007 Bowles & Gintis, 2011; Sterelny et al., 2013).

* * *

Before concluding this first section, it should be noted that modularity has served two purposes. As a characteristic example of a question in the philosophy of cognitive science, and as a guiding theme. This has led us to consider a series of questions and hypotheses that are probably more central and more fruitful than modularity itself. It may well turn out that modularity as such will cease to be the focus of discussions, after having been on the list of "live" questions in the discipline for some 30 years, a favorite subject of philosophers of cognitive science, while some of the other questions retain an enduring interest calling for further conceptual and empirical investigation. Such developments are already in progress. Today researchers are arguing about "dual" theories of cognition, so called *dual process theories*, more than about modularity: originally a theme in the theory of reasoning (Evans, 2003), it has come to serve as a more general framework for higher cognitive functions (for a recent appraisal, Evans & Stanovich, 2013; for the large picture, Kahneman, 2011). The idea is that two kinds of process are concurrently or successively at work in many cognitive processes: automatic processes, which are not under voluntary control, function rapidly and rigidly, and generally are not conscious; and on the other hand voluntary, deliberative, conscious processes that operate slowly, and are subject to error. We still find here some of the properties mentioned in the debate over modularity, but the theme of mental faculties has been displaced from center stage by a very different idea of the organization of mental functioning. What still remains of the modularist problematic is the idea of an architecture of the mind, composed of stable components.

2. The Mind as an Object for Science: The Foundations and Domain of Cognitive Science

2.1 PROVIDING A FOUNDATION FOR COGNITIVE SCIENCE

A traditional mission for philosophy of science, recognized in most schools of thought, is to uncover the foundations, those of science as a whole, or of particular disciplines. But what sort of foundations are these, how can they be uncovered, and what is the contribution of philosophy, in view of the fact that science, in the course of its development, appears itself to take responsibility for this task? Various views are held with regard to this issue.

Limiting ourselves to the present context and to the foundations of a particular discipline, we may discern two main attitudes. The aim of the philosopher, for some, must be to construct a coherent and complete metaphysical framework in which that discipline has its place. For others, the objective must be to ascertain the internal coherence of a discipline, by exhibiting its presuppositions and by displaying the logical structure of its fundamental concepts. In a word, there's a contrast between a global or external conception of the kind of intelligibility which is sought for, and a local or internal conception. A philosopher, of course, can decline to choose, and retain both goals; she may also decline to draw clear boundary between them.

This distinction crosses over another, which bears on the third question, that of the respective roles of philosophy and science. For the naturalistic philosopher, the two investigations stand in a relation of continuity, with philosophy situated at the boundary of science, in the area of its greatest abstraction. The question of the distribution of roles does not arise (or at least does not admit a stable response, because the product of philosophical activity is quickly integrated into scientific activity). According to the naturalistic philosopher, if the aim is to constitute a metaphysical picture, science contributes to this just as philosophy does, and in the same process of gradual articulation. When the goal is the conceptual "grammar" of the discipline, then the interweaving of philosophy and science is complete.

Things look differently to a philosopher who does not wholly espouse a naturalistic position. She will tend to reject the idea that science can make a major contribution to a metaphysical picture; yet she might also hold that this task no longer concerns philosophy of science, whose sole mission (one which is not on the agenda of science itself) is to make explicit the conceptual framework of the scientific discipline under scrutiny.

These questions become particularly tricky in the case of cognitive science, whose central object has remained in the purview of philosophy. On the metaphysical option, the major topics are the mind-body problem, the problem of intentionality, the nature of mental representations and perception, consciousness, mental causation, free will, and so forth. And according to whether one holds a naturalistic position or not, one will consider these matters to be on the agenda of cognitive science itself, or as the constituent parts of a general philosophical framework whose compatibility with scientific results must be established.

I will return at the end of the chapter to the distribution of labor, within philosophy itself, between the different branches that are concerned with cognitive science. The core of philosophy of cognitive science, as I will regard it for present purposes, consists in the examination of the most general concepts of the field. Take for example the mind-body problem, which actually refers to a number of distinct, though connected, enigmas (Warner & Szubka, 1994), and let's keep to a simple formulation: to account for the place of mental entities in the material order. Some believe that this question will be answered *by* cognitive science (which would in fact take this to be its primary aim), in the manner in which biology has resolved (one may believe) the life-matter problem, or in the manner in which physics overthrew Zeus and his thunderbolts in favor of electromagnetism. Others think that it is necessary to find a solution to the mind-body problem in order for cognitive science to be established on a solid foundation. But a "modest" philosopher of science will observe that cognitive science has developed a strategy which allows it precisely to get around this problem.[11] We mentioned early on the "structuralism" that is inherent in the project of cognitive science. It is time now to be more specific.

2.2 REPRESENTATION AND COMPUTATION: THE FUNCTIONALIST FRAMEWORK AND THE LANGUAGE OF THOUGHT

2.2.1 Functionalism

As cognitive science took off, it was provided with a theoretical framework that was relatively precise, and which not only historically gave it a conceptual mooring, but which also remains to this day —despite the heavy criticism is has been subjected to, and the adjustments that are proposed in the hope of saving it—the starting point for all foundational concern. This framework I will label "functionalism," in accordance with well-established usage, despite the ambiguity of the term.[12]

Functionalism is a form of structuralism applied to mental entities. It consists in substituting for the question of the nature of these entities, a description of their mutual relations. Specifically, all we need to know about mental states such as pains, beliefs, desires, memories, regrets, intentions, projects, and so on are on the one hand the relations that exist between them, on the other the relations between them and sensory stimulations and movements. These states are unobservable, theoretical entities; their role in cognitive science is that played by forces in Newtonian dynamics, or quarks in particle physics, or expected utility in economics, or the pressure

[11] This possibility was envisioned by certain psychologists as early as the 18th century (cf. Hatfield 1995).

[12] We will see that in the context of cognitive science, there are several conceptions of functionalism. But the term also refers to positions adopted in other fields, such as linguistics, anthropology, sociology, the life sciences, etc. These other uses of the term are without relation (at least without direct relation) to the functionalism of cognitive science and philosophy of mind.

of selection in evolution. Their relations to stimuli and movements, which are observable, are the analog of boundary conditions in other sciences.

The relations which internal mental states have with each other and with stimuli and motor capacities are causal relations, and they give rise to a mental dynamic (which has physical antecedents and consequences).[13] The cognitive system thus passes from one complex state to another, under the influence of forces that are a function of the constant relations that exist between different types of mental states. For example, my belief that I have had a headache for the last few minutes is grasped on a theoretical level by means of a relationship between this belief and certain sensory stimuli (the stimuli have contributed to causing this belief, and by the same token this kind of stimulus tends to cause beliefs of the type "I have had a headache for a little while"), along with desires such as doing something about my headache, which combines with another belief about the effectiveness of aspirin, which tends to cause the intention of taking aspirin, an intention which in turn, and along with other beliefs, intentions, and desires, gives rise to a plan for getting some aspirin from the medicine cabinet, and so on.

So the fundamental intuition of functionalism is this: if the object is to reveal the determinations of a mental dynamic, or, to use an old phrase, to exhibit certain "laws of thought," it is not necessary to say anything about the material out of which mental states, thoughts and so on, are cut; it is sufficient to display the constant connections that exist between them. These connections are dispositions: in the presence of certain conditions, a specific causal chain is triggered (recall the typical example of a dispositional property: when you put sugar in water, it dissolves—barring exceptional circumstances; the solubility of sugar is a dispositional property). But this causality must be discovered. It calls in fact for two explanations: one explains the general phenomenon, and the other, its distribution. The point is to understand, on the one hand, how thought can cause any event whatsoever; and on the other, why the thought of having a headache, as opposed, say, to putting an end to my life, leads me to form the intention of swallowing aspirin rather than strychnine.

To that end, we need to say a little more about mental states. Their "operationalization" remains an abstraction as long as we have not specified how they are individuated. This is where several conceptions of functionalism diverge. For *analytic* functionalism (Lewis,

[13] We run up against a well-known terminological difficulty here. Every supporter of naturalism, even if only methodological and not metaphysical, attributes to mental states and processes a physical nature; a *particular* belief or pain is not considered to be less physical than a retinal stimulation or a movement of the hand. The pertinent difference is that belief is understood as possessing a semantic content; it is certainly a physical event, but it is grasped in accordance with a particular description that is not such an event. We will return to this point in a moment, but an example from another domain may help the reader: when I talk about a 20-euro bill, I'm referring to a material object, but I speak of it in terms of its nominal value, and I select this description because it is the one I need to account for what happens at the bakery when I pay for my bread. This example raises some problems itself, but it should help a reader who had trouble with the point made.

1966, 1980), every mental state is defined by its place in a network of dispositions that are expressed through the platitudes of common sense in which it occurs. So, for example, the belief that one has a headache is related to the belief that aspirin would help in such a way as to trigger, in the presence of the desire to cease having a headache and in the absence of a number of beliefs such that one is allergic to aspirin, the intention of taking some. The belief that one has a headache is nothing over and above the functional role of a particular place within the network of such platitudes (and of course, similarly for all the *relata* that occur, such as the belief that aspirin tends to relieve headaches, the desire to relieve the headache and so on). For *empirical* functionalism (also called psychofunctionalism: Fodor 1968; for influential critical assessments: Cummins, 1985; Block, 1980b), the network of common sense only serves to designate mental entities, and it is up to science to determine their actual properties; in similar fashion, common sense *designates* water (giving the *meaning* of the word or concept), while physics and chemistry *discover* what water really is[14] (thus determining the *extension* of the concept). Finally, *Turing* functionalism, or *machine* (sometimes: *machine state)* functionalism, likens mental states to internal states of a Turing machine (or more generally, of any computational system).

2.2.2 The Computational Theory of the Mind

Turing functionalism is not on the same level with analytic or empirical functionalism, though it is compatible with either. It is an extremely abstract hypothesis, one which is admittedly hard to grasp outside the more general context of the psychological theory in which it is embedded. That theory goes by the name of the computational theory of the mind (CTM).[15]

Analytic and empirical functionalisms strive in the first place to produce a conceptual analysis of mental *states*. In other respects they are derived from "logical behaviorism"; they have rejected part of the legacy of that school, but they have retained its preference for ontological economy, and its lively sense of the difficulty of giving an essentialist definition of mental entities. This philosophical form of behaviorism was the product of a reflection for which Wittgenstein was chiefly responsible, concerning the abusive reification to which excessive confidence in the superficial form of the expressions of ordinary, everyday language might lead (Ryle, 1949).

Turing functionalism, on the other hand, first proposed by Putnam (Putnam, 1960, 1975) originates in a concern for mental *processes*, drawing on a long reflection, begun with Frege, which led in the 1930s to consideration of the notion of a *formal* language (or system). Arithmetic gives us several characteristic examples of such languages;[16] we create symbols for particular numbers such as 0 or 1, and symbols for certain operations

---

[14] We should observe in passing that the answer is not "H_2O"; it is much more complex (see e.g. Weisberg 2006). But this is an answer of the kind that science is supposed to provide.

[15] It is this more complete theory that certain authors (for example Putnam himself: Putnam, 1988) call "functionalism."

[16] There are several formal languages that can accommodate arithmetic quite naturally.

such as the passage from one natural number to the next, addition or multiplication, symbols for certain particular numbers, logical symbols (for negation, conjunction etc.), and morphological rules for the combination of all these symbols. A language of this type can float in a sphere of ideal entities or abstract concepts, and it can also be "realized" in material terms in various ways. Every calculator, from Pascal's calculator to the analogical calculators of Konrad Zuse and the mechanical and electromechanical devices that preceded electronics, and from there to present-day personal calculators and computers, realizes or "implements" a formal language for arithmetic. There are countless such instantiations, involving causal chains that have little in common, so that there is no isomorphism of any sort between them all that can be expressed in the language of physics.[17] That which they have in common is only visible from an abstract point of view, the point of view of the formal specifications that guided their construction. The fundamental intuition of Turing functionalism is that mental operations are formal, and can be physically instantiated in different ways, so that the theory of these operations is not part of physics, but rather part of a formal science that one could call the science of information (although this expression is not normally used in this sense). In a more concrete sense, a law of thought such as *modus ponens* (going from the thoughts that *A is true*, and that *A implies B*, to the thought that *B is true*) must be understood as an abstract form of *calculation*, and a material system obeys this law to the extent that it concretely carries out this calculation (like the student who *writes* "B" on the blackboard underneath the assumptions "A implies B" and "A"). It is the same, *mutatis mutandis*,[18] with the passage from the thought of a migraine and a belief about the effectiveness of aspirin to the intention of taking aspirin. This sort of abstract operation can be carried out, as a matter of fact, through mechanisms which differ at the physical level. This argument, called the argument from "multiple realizations," is the basis of Turing functionalism and the computational theory of the mind which is developed from it.

A formal language ***L*** has two faces: on one hand it is a combinatorial system of symbols, and on the other, it is the basis for an "interpretation" that attributes a meaning or value to the symbols, terms and sentences of ***L***. Thus interpreted, ***L*** designates objects, relations and states of affairs in a "universe" or "domain of interpretation" that may be abstract (the group of integers, for example, or a chessboard with its pieces, understood not as a material object but as a system of relations) or concrete, either real (a real chessboard, the real Swiss electric grid), or imaginary (the characters in a TV series). These two faces are correlated in the following way: certain strings of symbols correspond to certain states of affairs in the domain of interpretation, and by manipulating the symbols one represents changes in the relations between the

---

[17] In order to make this idea clearer, it is sometimes suggested that one imagine calculators made up of chickens that lay eggs connected by tubes, or of children yelling at one another (shouts, not words) on a playground, or again of beer cans tied together by strings, etc. Clearly as *physical events* or phenomena these have nothing in common.

[18] The two cases differ in important ways; I will return to this point presently.

interpreted entities, such that the changing state of symbolic configurations reflects relevant aspects of, or events within the domain of interpretation. Thus an air traffic controller follows the track of airplanes and guides them with symbols that indicate their identity, position and destination; the controller's operations act on the symbols, but the correspondence insures that these operations refer in a dependable manner to the actual trajectories of airplanes, such that, barring accident, the planes land safely according to the intentions of the controller.[19]

Two elements are lacking in order for this schema to constitute (even if only as a sketch) a theory of the mind. The first has to do with the interpretation of symbols: by what means do they represent what they represent, and what does it mean in concrete terms that they represent anything at all? The CTM is a *representational* theory, in a sense that has been familiar in the theory of knowledge since the 17th century. The mind contains representations, which Descartes and Locke generally call *ideas*. This is why the theory is sometimes called the computational-representational theory of mind. But it is not enough simply to give the theory another name. It is necessary to show how a representational theory of mind can also be a naturalistic theory of mind.

The example of the air traffic controller sets us on the path (without leading us to the goal): that which confers representative value on the inscriptions read by the controller on his screens and her "flight progress strips" are complex causal connections that run from the represented entities (for example, a plane identified as AF26 at location (x, y, h) somewhere between Paris and Washington at instant t) to the inscriptions that do the representing (here, the positioning of point labeled "AF26" at a certain point on the screen, associated with a coordinate pair (x,y) plus the value h of the parameter *altitude*). The symbols postulated by the CTM are similarly supposed to be naturally endowed with meaning, but what is to be understood by this is far from obvious, and I'll have more to say presently under the heading of "intentionality." We can note at this point that unlike the air traffic control system, the human cognitive system does not have a "controller" sitting in its center, equipped with the principal attributes of the mind. The internal symbols cannot be "read" in any literal sense. The solution to this difficulty is to be sought on the side of the functionalist idea: the meaning of the symbol could perhaps be defined functionally by the entire set of effects this symbol may (dispositionally) have on the system.

[19] For the sake of simplification, but at the risk of some confusion, I do not distinguish here between two types of transformation that are in fact quite different. In one case, the universe is fixed, and representations of that universe are modified (e.g., when certain new conclusions are drawn from information that is already present). In the other case the universe itself changes, especially through the intervention of an agent. The two processes are often at work simultaneously; this is the case with air traffic controllers: based on information that is valid at an instant *t*, the controller is led to deduce (to calculate) certain other items of information at the same instant; but he also infers, based on information at time *t* and on knowledge about the evolution of the system (as influenced by causes that are endogenous or exogenous, including his own intervention), information that is valid for at some later time *t′*.

The second gap that must be filled has to do with the different categories of thought. Up till now we have focused on just one, belief or assertion. But the mind, as we already noted, can involve other kinds of states, for example, desires that are anything but beliefs about the state of things in the world. If I want to buy a car, in other words, if I want the world to be one in which I am the owner of a car, this normally means that the world is not yet in that state. Besides beliefs and desires, the mind also forms all kinds of thoughts, such as intentions, hopes, fears, regrets. Yet another kind is mere hypotheses: if *the weather had been good yesterday*, we would have been able to get the hay in; if *the weather is good tomorrow*, we will get the hay in. It seems that the mind must therefore maintain separate lists for its beliefs, its desires, and so forth. But now we must understand how these lists are connected. As we have seen, certain conjunctions of desires and beliefs, for example, produce intentions; but not every desire can be conjoined with just any belief in order to produce an intention. The mind can therefore function only if certain very specific connections can be established between these various lists.

2.2.3 The Language of Thought

The CTM can in turn be embedded in a richer theory. In order to present it, I have used the example of the formal languages of logic, that come with operations or logical calculations. But nothing in the CTM forces us to postulate that the symbolic system which is at the center of the system's operations is actually a formal language, or that the operations are in effect syntactical calculations in the sense of logic. One can perfectly well imagine other systems, and other notions of computation than those of logic;[20] we will soon see (section 2.3.1) that such conceptions have in fact been put forward.

The hypothesis of a language of thought (LOTH, *language of thought hypothesis*; Fodor 1975, 2008a), for a mind trained in modern logic, is nonetheless an apparently natural extension of the CTM. The hypothesis states that the representational medium is precisely constituted by a formal language of the same type that logic constructs. This medium is called the "language of thought," or sometimes "mentalese." This hypothesis has a whole slew of consequences, which are so many arguments in its favor:

1. It gives a perfectly precise form to the dual nature of mental states and processes. The sentences in mentalese have a material form, and this form confers on them the disposition to be transformed under the effect of causal processes whose form is given by syntax. They also have a semantics, that is, they refer to entities, relations and states of affairs in a universe of interpretation (which in general is the material world to which the organism has access through perception, and on which it is able to act

[20] This claim may surprise the reader who has learned that there is in fact essentially only one mathematical notion of computation, known as computability in the technical sense (an assertion that can be disputed but is correct to a first approximation). In the present context, however, the concept is more flexible and can designate in reality almost any mechanizable procedure, even if it involves operations or dispositions that do not respect the specifications of computation in a strict logical sense.

through motor capacities). Syntax and semantics are independent, but they mirror each other. This conformity explains compositionality, a property that many attribute to thought, namely the fact that a complex thought is entirely characterized by its structure and by its constituent thoughts. This mirroring also explains how syntactical transformations are truth-tracking: a thought that is formally deduced from true thoughts (thoughts which are verified in the universe of interpretation) is true—to say it briefly, when we follow syntax, we don't stray from the path of truth.

2. It offers an elegant solution to the necessity of separating thoughts into distinct lists, in conformity with what has just been said, while rendering possible certain combinations. The belief that P may be seen as a relation of the form ***Bel***(<P>), where ***Bel*** is a predicate symbol associated with belief, and <P> is a phrase in mentalese that expresses P. The belief that P is what philosophers, after (Russell, 1918), call a *propositional attitude*, the relational account of which offered by LOTH is very natural. Similarly, the desire that P is a relation ***Des***(<P>), ***Des*** being another predicate symbol. Schematically, the fact that an individual believes that P, is realized by the presence of <P> in a sector of his mind (or of his brain) dedicated to beliefs (his "belief box," in a colorful expression coined by Schiffer, 1981); and to desire that P would consist, for the individual, in having <P> in his "desire box." This way of realizing beliefs and desires (as well as other propositional attitudes) makes possible specific pairings. If I believe that P entails Q and I desire Q, I form the intention to act so that P becomes true: the crucial point is that an unintelligent, "unminded" device can simply spot the common symbol <Q> as a (syntactic) constituent of the complex symbol <<P> → <Q>> in the belief box, and as a token in the desire box, and can mechanically extract <P> and put it in the intention box, provided it has the appropriate rule in its operating table.

3. It allows us to explain what appears to be the partial independence of thought with respect to language (natural language, that of the person harboring the cognitive system). In other words, if LOTH is true, we can understand that thought without language is possible (for example, the thought of pre-verbal infants, which turns out to be stunningly rich, and that of various animal species). By the same token, language acquisition can be seen as a process anchored in an already structured thought: the threat of circularity is deflected if a store of mentalese-based thought is available to initialize the process by which, as accords with intuition, we simultaneously acquire new linguistic and intellectual resources.

4. It accounts naturally for the intuition that different linguistic expressions express the same thought. "It's raining," "Il pleut," "Piove" mean the same thing. LOTH accounts for this very simply: it is one and the same sentence in mentalese which is, in all three cases, tokened or activated. Similarly, within a given language, sentences such as "Marie killed Pierre" and "Pierre was killed by Marie" are synonymous by virtue of being linked to one and the same mentalese sentence.[21] One may hope to be able to explain in the same

[21] The example is obviously only valid at the price of a high degree of idealization: it is clear that there are contexts in which one could not substitute one sentence for another.

way the universal character of certain mental schemas (such as rules of inference), which are translated in very different ways in different natural languages, and even in different idiolects of a given language.

5. Thought appears at first sight to be endowed with two properties deemed crucially important by students of language, *viz.* productivity and systematicity. Just as infinitely many sentences can be generated from a finite initial stock of basic sentences (productivity in this technical sense), infinitely many thoughts can be generated from an initial finite stock of thoughts. And just as any speaker who can competently utter the sentence "Marie hit Pierre" can also utter the sentence "Pierre hit Marie" (systematicity), anyone who can entertain the thought that Marie hit Pierre can entertain the thought that Pierre hit Marie, as well as such thoughts as "Someone hit Pierre," "Marie hit someone," and so forth.[22] These properties are shared by both natural languages (at least ideally) and formal languages, and their mental counterparts are direct consequences of LOTH.

And yet, LOTH is not particularly evident on its own, and it faces some serious objections. Its near-obviousness is the product of an illusion. Thought *as a product* can certainly be described with the help of a formal language—granting, for the sake of the argument, that the well-known objections against the idea that natural language, suitably idealized, has the structure of a formal language, can be gotten around by supposing that thought corresponds to the content, or to the deep structure of sentences of natural language, and not to their surface structure. But why would *that which produces* thought, that is, the mind, have precisely the same structure? It's one thing to describe the structure of thought, which is the object of logic (understood in a very wide sense); it's something else again to describe the genesis of thought, which is the object of psychology. So far from being the pedantic formulation of a truism, LOTH is thus a bold hypothesis. It claims that the mind, whatever it does on any occasion, proceeds like a "self-propelled" formal system, one that runs without external intervention. It applies rules of composition and formal inference to sets of sentences in mentalese. The truism-type version would be to explain that in order to multiply 31 × 12 (to pass from the composite thought (<multiply>, <31>, <12>) to the thought <372>, the mind applies an internal table of multiplication to symbols that in mentalese signify 31 and 12, and produces the mentalese symbol that means 372. This interpretation actually leads to a regress: how can we account for this internal operation? Must we postulate, within the mind, a sub-mind that allows it to carry out the operation?

This is the homunculus fallacy. How can LOTH escape it? It postulates that when *I* multiply 31 times 12, my cognitive system follows a trajectory that can be described

[22] In order to explain this idea, Fodor, who suggested it, put forward a parallel with a conversation guide for tourists that might happen to contain the phrase, "The London tube is more expensive than the New York underground," and not the phrase, "The New York underground is more expensive than the London tube." For a French tourist who has no idea about English syntax, the first phrase, thanks to the guide, becomes sayable, but the second remains unsayable. If we replace "sayable" by "thinkable," we get an illustration of what the non-systematicity of thought would amount to, something lacking in plausibility.

as the application of certain operations to certain complex symbols in mentalese. But what distinguishes the cognitive system from me, the conscious being whose flow of thought is what we are trying to explain, is the fact that the cognitive system is a "blind" mechanism without thought, intelligence or consciousness. On the one hand, like a robot in an assembly line, it only moves material entities around; what is for me part of the order of reasons is for the system something of the order of causes; on the other hand, that which proceeds in me from grasping the meaning of symbols corresponds in the system to a nodal position within a network of dispositions.

This explanation calls forth three remarks. The first is of a pedagogical nature: there is something misleading about this particular example; it so happens that multiplying 31 times 12 is a formal, rule-governed operation, and that most of us apply an algorithm to find the result, somewhat as a calculator or a computer does (and this is no accident; in a case such as this, machines *imitate* the mind of the person who calculates).[23] But this is a limiting case; the vast majority of cognitive processes do not have this character. What makes LOTH such a strong claim is that it accounts for all of them by *underlying* processes that are formal. Perception, memory, the understanding of the motives of other persons, linguistic communication, learning to play the piano, scientific research, finding one's way on the Tokyo subway—all these tasks, that do not have the appearance of algorithmic procedures carried out by a conscious subject, are accomplished thanks to cognitive processes of the same kind as those which underlie the multiplication of 31 × 12. Contrary to what one often reads, LOTH does not claim that mental processes are formal, but rather that the mechanisms that account for them are.

Second, it must be admitted that the grasp of meaning *by the cognitive system* remains unclear. The difficulty is twofold. On the one hand, the goal is to understand intentionality *in general* as a natural phenomenon; but it is widely believed that this problem remains unsolved for the most part. On the other hand, we need to understand where the various concepts come from, that, according to LOTH, are the meanings (interpretations) of the symbols in mentalese—in our example, the concept of multiplication and the concepts of 31, 12 and 372, not to mention the logical symbols. For reasons that we cannot explain at length here, LOTH is strongly tilted toward nativism: the primitive concepts of mentalese would be innate. Any reason for rejecting nativism places the LOTH at risk, so that someone who has misgivings about nativism needs to ask to what extent she can retain LOTH.

Yet another important question concerns the relation between the primitive concepts of mentalese (or more generally, the basic semantic units) and the concepts

---

[23] Which Turing calls the *computor*, in his seminal 1936–1937 article in which he sets forth the basis of a theory of computers. Another example that is often chosen is that of a game of chess, in which we see a human player on one side and a computer program on the other. It presents the same type of misleading obviousness.

of common sense, and more generally those that are expressible in a natural language. We shall see that this is a bone of contention.

2.2.4 Beneath Conscious Mental Life

As we have seen, when one sets out to introduce CTM and LOTH, the first examples of mental processes that come to mind understandably involve everyday phenomena, showing how these can be accounted for as processes in the underlying cognitive system: such processes are argued to be operations on certain "pre-concepts" which are terms of mentalese. This is supposed to be explanatory insofar as such terms faithfully reflect the concepts that are present and consciously deployed during a given episode of the mental life of the subject.

This choice of examples is unfortunate in two ways. I mentioned the first problem: it reawakens the homunculus fallacy. That is a conceptual problem. The second problem is more of an empirical one: it distracts our attention away from a crucial possibility. Nor is this simply an expository problem: the early phase of AI and cognitive psychology have done much to license the project of an explication of mental life based on processes situated at the same semantic level as those of William James's "conscious mental life," or as we would tend to say today, of folk-psychological concepts. Within cognitive science there remains a tension between a "homo-semantic" and a "hetero-semantic" conception. Let me unpack that non-standard terminology.

The idea is an old one, and it is periodically forgotten and then rediscovered in philosophy and in psychology. Already for Leibniz, for example, the physical movements of the mind could be explained with reference to a dynamic of "tiny perceptions" (Leibniz, 1714/2006: "Our big confused perceptions are the outcome of the infinity of tiny impressions that the whole universe makes on us"); the Scottish philosophers William Hamilton and Alexander Bain, the great German physicist, physiologist and psychologist Helmholtz, and the American neuropsychologist Karl Lashley each in his own way understood that most cognitive processes are neither conscious, nor easily described in an ordinary conceptual vocabulary, even one that has been refined to suit scientific purposes. As Bain wrote in 1893: "Outward expression, however close and consecutive, is still hop, skip and jump. It does not supply the full sequence of mental movements."[24] Whether we consider a given train of thoughts as a temporal and causal, or a rational sequence, we must admit that it is incomplete: we need to postulate, it would seem, a gapless, fully connected trajectory running at a deeper level, that includes certain "peaks" that emerge and form the "manifest expression" Bain referred to.

This intuition is no analysis, much less an empirically based theory: it is a metaphor. Yet it is at the root of what I see as the third fundamental idea of cognitive science—the first two being: information or representation as a relational

[24] I owe these references to Hamilton (1859) and to Bain, as well as the quote from the latter, to a chapter by Martin Davies with the same title as the present one (Davies, 2005).

property of the components of a material system, and computation as an abstract mechanism. Of the three, it may be the most original and the most fruitful. In the work of contemporary theorists it takes different forms, which may not be mutually compatible and are in fact defended by schools that are at strong odds in other respects. But across these differences, we may discern a common theoretical kernel, which comprises two claims. The first is that the level at which these actual causal sequences unfold, that are responsible for cognition, is separate from consciousness. The second is that the entities and processes that belong to this level are of a finer semantic grain than ordinary meanings, those present in consciousness and in language. The first thesis generalizes Chomsky's "tacit knowledge" of grammar, a set of rules and representations. The second thesis refers at once to the "sub-personal" level identified by Dennett (1978), to the "subdoxastic" processes of Stich (1978, 1983), and to the "microstructure" of cognition that theorists of connectionism (which will be introduced presently) mean to uncover (Rumelhart & McClelland, 1986; Smolensky, 1988).

Though different, these approaches share three core hypotheses:

- Thoughts and conscious mental acts can only be explained as resulting from processes that occur at an underlying level.
- These processes involve entities that are radically different from ordinary propositional attitudes.
- These entities in turn are of an essentially informational or representational or semantic nature: what goes on at this level is not *directly* or *essentially* physical—more specifically, neurophysiological: although present or happening *in* the brain, these states and processes are not most perspicuously rendered as brain states and processes.[25]

2.3 THE CRUCIAL YET LIMITED ROLE OF MODELS IN THE SEARCH FOR FOUNDATIONS

To a reader familiar with current research in cognitive science, or to someone who would accidentally wander into a busy laboratory in this area, what I have been discussing in the last few pages might sound very removed from the issues that are foremost in the minds of the scientists at work. That is a quite reasonable concern, for several reasons. The first, very general, is that the philosophical search for foundations is seldom directly relevant to scientific research. The second is that things change quickly, and a great deal of research has already escaped the framework that philosophers have been slowly constructing for the enterprise as a whole. These more recent undertakings are followed by some philosophers intent on sketching alternative frameworks, faithful

[25] We again run across the terminological problem mentioned in note 12. All particular states or processes *are* made of physical (neurophysiological) stuff; but their significant properties are those of a class of entities that are functionally similar, and these are stated using a different vocabulary.

to the new approaches, though as we shall see, these efforts remain scattered at this time, and even suggest that the unitary project that any such framework is aiming at articulating may be about to be abandoned. Finally, as we said from the first, the cognitive sciences (the plural is apt at this juncture) remain very uncertain with regard to the nature and the extension of their object, and the persistence of this uncertainty gives philosophy a larger role than usual in the clearing of the scientific ground. This role endows philosophy, for the time being, with an unusual degree of autonomy, on par with that of the positive disciplines involved, so that it develops its own ideas without always referring back to ongoing research programs; these programs in turn carry on without worrying about the framework they are supposed to fit in.

It is thus of paramount importance to construct actual connections between the philosophical perspectives on cognition and the positive sciences of cognition. This is what *models* can achieve. I don't propose to discuss the role of models in science generally, and the fact that they refer to very different things will be left aside. In cognitive science, as in other areas, different kinds of models are used, and the term itself is quite elastic. On the other hand, it also has a very specific use, and the theoretical arrangements in which it participates are of decisive importance.

2.3.1 Classical, Connectionist, Dynamical Models

Had computers not been invented, it is very hard to imagine the theoretical horizon whence the cognitive sciences might have arisen, or what they might be today. The role of the computer, as things stand, is often poorly understood, and this gives rise to critiques that are as lazy as they are unjustified. The computer was first conceived of by Turing as a model of a person calculating (the "computor" mentioned in note 22). Turing identified key aspects of the actual process and created a formal structure made up of elements and relations representing these aspects and their interactions. At this stage it was an abstract model, what a set of differential equations might be in physics. Then the first material computers were built; they incorporated Turing's schema, thus confirming its consistency, and providing support for its modeling hypotheses. But they also reflected further theoretical choices, inspired by technological or logico-mathematical, not psychological considerations, and these suggested in turn important supplementary hypotheses concerning the "computor." These are the choices that led to what is known as von Neumann architecture, still the standard model for computers of all kinds. Soon, Turing and others suggested seeing in the computer a model of human thought in general, this time in the practical and not the theoretical sense of "model," something like a mock-up or a scale model. Experiments with such models, and a re-examination of their principles of construction led to a considerable enrichment and modification of the initial theoretical model.

Thus through a complex process of model building in which models alternate between abstract and concrete, cognitive science has jointly produced, or co-constructed, a general framework and a collection of physical systems both embodying that framework and putting it to the test. In a moment (section 2.3.2) we will get a better grip on this double movement.

The reason I now refer to a collection of systems rather than to *the* computer in the singular is two-fold. First, as everyone knows, there is more than one kind of computer, in fact there is a whole slew of them, and they do not only differ with regard to the parameters that are known to the general public (e.g., CPU speed, RAM memory, size of hard drive); they also differ with regard to their architecture in the informatics sense of the term. Second, a computer has to be equipped with a basic language, or operating system, and this is what makes it a particular computer, different from another computer using a different operating system (and in fact each additional specification, given in the form of a higher-order language, adds yet another difference). It is true that all these machines have so much in common that it often makes perfect sense to put them all in the same category. One could even argue that they are only so many different ways of constructing a physical system that performs computation, in the logico-mathematical sense of the term,[26] thus showing that they are essentially the same. But it suffices to recall that real computers (often generically referred to as von Neumann computers) are finite entities to see how different they are in fact from the ideal model of a Turing machine, and to suspect that the manner in which they differ from the Turing machine involves differences between them that have a theoretical significance. Quite generally any given real-life computer operates under boundary conditions that result from many architectural decisions made by its designers, as well as from the way it is used and how its results are interpreted, and these constraints as a whole may be as important as its primary calculating function.[27]

We come now to a second framework for cognitive science, the result of a process of co-construction very similar to that which led to the framework associated with the Turing machine. Although it emerged at roughly the same time, and from the same source, it matured more slowly. That is why the Turing framework is often referred to as "classical." It is also sometimes called "symbolic," a reference to the symbols postulated by LOTH. The second framework is generally referred to as "connectionist," and I will now explain why.

Connectionism arose from an attempt to model the basic functional unity of the brain, such as this was conceived in the early 1940s: a network of neurons (what the Canadian psychologist Donald Hebb called "cell assemblies"; Hebb, 1949). Each neuron in the network may be electrically charged; the charge is a signal that is carried via synaptic connections to the neuron's neighbors. (To get a grasp on the model—in fact, again, a collection of models—the reader should consult any one of the numerous treatises available, such as Hinton & Anderson, 1981; Rumelhart & McClelland, 1986; Smolensky, 1987; Amit, 1989; Anderson, Pellionisz & Rosenfeld, 1990; Dayan & Abbott,

[26] See note 19.

[27] An amusing (but superficial) illustration of this is provided by the episode of the "Y2K bug."

2001; Kriesel, 2011.) The creators of the model, Warren McCulloch and Walter Pitts were part of the group that created cybernetics.[28] Their starting point was a schematic conception of the neuron (the so-called formal neuron) and of networks formed of such neurons, and their goal was to show that these networks were capable of carrying out basic logical calculations, and ultimately any kind of calculation (McCulloch & Pitts, 1943; Anderson & Rosenfeld, 1988). This movement is in one sense symmetrical with the Turing movement, since it also begins with a schematic conception of calculation and then conceives of a machine capable of carrying out the schema.

Today, networks of formal neurons, usually called neural nets, constitute a family of physical systems that stand to connectionism as von Neumann computers to classicism. They incorporate certain basic assumptions regarding the nature of cognition, which constitute a framework within which more specific hypotheses can be formulated, and to some extent tested on neural nets. In turn, those nets suggest modifications of the hypotheses, or entirely new hypotheses. Conversely, theories coming from cognitive science suggest architectural principles for the conception of networks. The variety of possible suggestions is greater here than it is for the classical framework, partly because there is a great variety of possible network architectures, and partly because neuroscientific hypotheses may play a role, just like psychological hypotheses, in the process of co-evolution of psychological theory and computational models. Connectionism has since its inception followed two paths, according to whether it has emphasized the brain or the mind.

As I mentioned it isn't possible to give even a summary description of the connectionist framework (informal presentations, in a philosophical perspective, can be found in Clark, 1989; Andler, 1992, and many other publications). We may nonetheless begin to situate it in relation to the classical framework, with the help of a series of oppositions. Information is processed, in the classical framework, sequentially; in the connectionist framework, they are massively parallel. The basic classical operation is inference, the paradigm of a process governed by an explicit rule. The basic connectionist operation is association, a process driven by continuous measures of distance. Internal representations in the classical context are symbolic and local (that is, each symbol represents one and only one concept); connectionist representations tend to be sub-symbolic and distributed (each representational item represents nothing on its own, and concepts are represented by groups of such items, such that each one has only a "micro-representational" value that is capable of being part of more than one representation). Classicism rests on a sharp distinction between items of knowledge (the values of a variable, in a program) and operations (the series of instructions making up the program); connectionism blends the two together. Finally, learning, in the classical framework, is reduced to the piecemeal acquisition of new knowledge, while in the connectionist framework it takes on the much more natural form of a gradual adaptation to the environment.

---

[28] A group that made Alan Turing an "honorary member"; see Heims (1991).

But these are only very general contrasts, that provide a simplistic picture of the situation. How the two frameworks stand with respect to one another is no simple matter. Since neither of them is very constraining, and since each one admits a great variety of interpretations, several ways of conceiving of this relationship have been worked out, ranging from complete incompatibility to full compatibility, passing through various intermediary positions. Among the latter, the conceptually simplest scheme combines the two kinds of systems, distributing each task to a component of the kind more naturally suited for it (inference-like tasks go to classical subsystems, perception-like tasks go to connectionist subsystems). More complex schemes take the classical architecture to be a limiting case of the connectionist one (along Bohr's principle of correspondence),[29] or again to emerge from it in one or another sense.

The fact that I got to connectionism rather late in this chapter, and that I have given it short shrift, might lead the reader to suppose that it plays a secondary role today in cognitive science, or that I myself don't regard it as of sufficient interest. Both of these conclusions would be quite wrong: connectionism is very much part of the mainstream today, and I for one have always taken it very seriously. In fact, in a field closely related to cognitive science, and very much in the news nowadays, *viz.* artificial intelligence, connectionist principles and models are at the root of a recent breakthrough, deep learning. There are three reasons connectionism hasn't come up earlier and been exposed at greater length: one is simply that choices must be made; the second is that it is difficult to talk about connectionism without referring to classicism (while the reverse isn't true); the third is that one can rely on the readers' acquaintance with computers, making it possible to forego a technical exposition of classicism, whereas connectionism would require more explanations than can be provided in the available space.

At any rate, we must now speak of a more recent contender, often called "dynamicism" (Thelen & Smith, 1994; Port & van Gelder, 1995; Kelso, 1997; Ward, 2001). The family of physical systems of reference is constituted here by dynamical systems, understood in the same sense as in the mathematical theory of the same name: physical systems that change over time, whose state is characterized at any given moment by the values, generally real numbers, of a set of variables, and whose trajectories are determined by a system of equations, most often differential. This is an immense category which includes all kinds of systems, from the solar system, the terrestrial meteorological system, and the world economy all the way down to gyroscopes, computers and connectionist networks, under a suitable description. Dynamicism however has certain particular systems in view, those which have been emphasized by cybernetics, especially those possessing properties of autonomy or self-regulation which are supported by feedback loops. These are typically control systems: a thermostat is a particularly rudimentary example, Watt's regulator a richer example. Certain robots, constructed according to

---

[29] A principle according to which a new theory (such as special relativity) must subsume an approximation of the old one (such as Newtonian dynamics), which appears in turn, a posteriori, as an approximation of a particular case of the new theory.

the principles of the dynamicist framework, are more explicit illustrations of cognitive systems considered as dynamic; they can be seen as control systems when they are placed in an environment on which they are able to act.

The dynamicist framework is by far the least developed of the three, and it is not certain, regarding its current state of development and its theoretical choices, that it will last long in its current form. Its main points of difference from the classical framework are as follows. It rejects all recourse to internal representations. In a similar manner, it conceives of the relationship between cognitive systems and the environment as a matter of coupling and control, not in terms of representation and action.

Second, it gives the temporality of processes a crucial importance, while the classical framework only sees this as an effect of the succession of operations, entailing constraints that may be important, but are not constitutive of the cognitive system. A central characteristic of the temporality in dynamicism is that it is continuous: the system interacts continually with the environment, whereas the classical system receives information at discrete moments, changes according to a discrete protocol, and executes a discrete sequence of instructions.

Finally, dynamicism espouses a radical holism, one inspired by *Gestalttheorie* (Koffka, 1935; Köhler, 1945; Kanizsa, 1979, Smith & Ehrenfels, 1988). From this point of view, only configurations of the system and system-environment connections are significant, not this element or that element, or any distinguishable aspect; taken in isolation, no simple element has a meaning. The very notion of a simple or basic element contains the germ of a fundamental error.

With respect to this contrast, connectionism occupies an intermediary position. In some of its most interesting versions, it links up with what is in my view the most solid part of the dynamicist program, without accepting it completely (something that would force it to give up a large part of that which makes it a fruitful hypothesis): it calls into question the classical conception of representation, without rejecting the idea that representation is an essential part of cognition. It adopts the perspective of dynamical systems, making time an essential dimension; it also favors a certain degree of holism.

Not everyone agrees with my sober assessment of dynamicism. To some, connectionism does not go far enough in its repudiation of classical hypotheses, and only dynamicism offers a real possibility of escaping what they see as the inacceptable limitations, even perhaps the inconsistencies of classicism.

2.3.2 Clarifying and Diversifying Theoretical Options

But in what way do these "grand models" actually contribute to research in cognitive science? The question might appear otiose; isn't that what we have been talking about all along? It's worth taking a closer look.

Let's begin with the classical framework. It is often presented (by critics within as well as outside the field) as a development of the "computer metaphor," whose relevance is claimed to be at best marginal, given that the central nervous system is really not comparable to a computer in any reasonable sense. But this is a severe misunderstanding regarding the role played in cognitive science by the "grand model" of the computer.

In fact this role is threefold. *First*, the computer provided a precise and concrete specification of the theoretical concepts used in the nascent cognitive psychology: the computer served as a proof of existence (or what comes down to the same thing, of consistency), it helped set down these abstract, novel ideas. Let's take the very general idea of a formal system, taking as our starting point the Leibnizian notion of blind thought. Can we conceive of a "syntactic machine" that delivers the services provided by an ideal "semantic machine," that is, something capable of avoiding the many pitfalls produced by ordinary language, thought and perception? The answer is *Yes*, although it comes at the price of a long development, from Aristotle to Turing, passing by way of Frege, Russell, and Gödel. . . . But can we be assured that the theoretical proposition with which we have ended up is free of contradictions? (Haven't contradictions appeared in theories whose abstract rigor and apparent simplicity seemed to guarantee consistency?) Are we sure that this proposition can be realized as a real, physical system, in the material world as we know it? Isn't it the case that a physical system is fated to produce only reflex reactions, perhaps extending as far as the operations of elementary arithmetic, but no further? It is quite remarkable that Turing succeeded in putting an end to these doubts, and that his successful attempt to determine the *limits* of formal or mechanizable thought allowed him to prove its *unlimited* extension. To take another example, the general idea that our reaction to a given situation depends on our state at that moment, is somewhat foggy. If we link it to the precise technical notion of the internal state of a Turing machine (a notion Turing himself clarified by making a comparison with the keyboard of a typewriter that can "shift" from capital to lowercase letters),[30] we get a firmer grasp on things, which allows us to make progress in conceptual reflection without being a slave to the model.

Second, the model of the computer is a source of concepts, distinctions and hypotheses that psychology, and more broadly cognitive science can make use of. There are many examples. The notion of a "default value" comes from computer science and now belongs to the basic vocabulary of cognitive science. The same is true for "active memory" (which gives rise to the notions of short-term memory and working memory), for "content addressable memory" that is "addressable by the content" or for "central control." Or again the concept of "heuristics" introduced by Herbert Simon (Simon, 1957) in the context of decision-making, and transferred by him into the domain of AI, where it takes on a precise meaning and can at that point migrate toward cognitive science. We should observe that many of these notions have

[30] The mechanical (and then electromechanical) ancestor of today's word processors; its remaining traces are the computer's keyboard, and a great deal of nostalgia on the part of several older generations.

also invaded ordinary language; the omnipresence of computers impacts the "naïve theory" of mental processes (the notion of a naïve theory is a generalization from "naïve physics," another concept created by AI). Other transfers to cognitive science are more local and more technical, for example in vision science, which space does not allow me to present. The contribution of computer science, at this more concrete level, is nonetheless disputed; for some it is of little value, but for others it is of prime importance.

It is in its third role that the usefulness of the model is the least doubtful. The computer is taken as an experimental laboratory. What is performed are either experiments in the literal sense, as the founders of AI insisted there would be, although they are experiments of a particular kind, or thought experiments, also of a special kind, much practiced by cognitive science today.

Let's see, to begin with, in what sense the computer allows cognitive science to do genuine experiments. On the early version of AI, a computer program that made the computer accomplish a cognitive task which if performed by a human would result in the exercise of cognitive capacity C literally constituted a theory of C belonging by right to scientific psychology—by way of example, let C be the capacity to read a text out loud, or the capacity to solve problems from a certain family of geometrical problems, or the capacity to stack blocks of different sizes such that the stack stays upright. Considering then a psychologist who comes up with a conjecture T relative to capacity C, she can (and according to some, she must) translate T into a program P and measure the degree of success that P achieves in the accomplishment of C; non-success can lead the psychologist to reject T, or if there is partial success, to modify T into T', and then to translate T' into a program C' which would be tested in turn. So much for the computer as a "laboratory" of cognitive science. For various reasons this procedure has been practically abandoned, except in certain particular areas, but it does retain some heuristic value, and constitutes a schema that will be taken up again in other frameworks.

Today the computer is most useful to cognitive science as a testing ground for thought experiments. When a scientist seeks to explain a cognitive capacity, if he is working within the classical framework, he will propose to analyze that capacity (seen as a particular kind of information processing) into more elementary capacities, which in turn are to be analyzed into still simpler capacities, until he reaches the point where the original capacity is reduced to a combination of capacities that he is confident can be mechanically realized. It is generally impractical to translate this analytic procedure into a complete and explicit mechanical model. The thought experiment consists in asking oneself if a computer could be programmed in conformity with the proposed analysis, and if so, whether the program would achieve the desired result. Like any thought experiment, a procedure of this kind can only have probative value when pursued by an experienced researcher. The computer serves as a discipline that guards the scientist from spurious solutions.

But a very different type of thought experiment can also be envisaged. Let C be again a cognitive capacity for which one seeks an account. Let us suppose that we succeed, through a convergent group of arguments, in convincing ourselves that all possible analyses into component parts that can be carried out using a computer with a given architecture, would present characteristics that are not observed in connection with C. In that case we would have an argument in favor of rejecting that architecture as a model of the mind (or perhaps only as a model of that kind of capacity). If one arrives at a stronger conclusion, namely, that no such analysis that can be carried out on a (classical, von Neumann) computer, regardless of its architecture, can model the main observable characteristics of C, then one would have an argument against the classical or symbolic framework itself.

And it is at this point that the theoretical usefulness of the grand models is most clear. If capacity C cannot be modeled using a classical architecture, and if there are other conceivable architectures, one may seek to realize C using these other architectures (and to re-conceptualize C as a result). Connectionism as well as dynamicism, despite its somewhat shaky foundations, then appear as alternatives to classicism. Thus in fact a wide variety of psychological theories postulate a connectionist realization, without going as far as offering an actual neural-net model, nor necessarily presenting the result as a schema of neural functioning. That several grand models exist at the same time allows psychology to formulate with greater precision than ever a whole set of questions, ranging from the most local to the most general level. Among the local questions, the classical and connectionist frameworks lead to conceptions of memory, pattern recognition, learning morphological rules in natural languages (a famous controversy arose over the manner in which children learn the past tense of verbs), concept acquisition, and so forth, that are—or appear to be—radically at odds. At the intermediate level, it is the format for the representation of items of knowledge, the role of rules in cognition, and the nature of learning that are at stake. At the highest level, different conceptions of cognition confront each other. Where classicism puts logic at the heart of cognition; connectionism places perception, and dynamicism, movement. In the classical framework, cognition is a matter of abstract information processing; the connectionist framework sees it as an informational function of systems that have a very particular form, that of cortical structures; in the dynamicist framework, cognition is understood as a dynamic, evolving coupling between the cognitive system and the environment.

How can we choose one framework over the others? This is one of the main questions in the philosophy of cognitive science, and it is linked to the other major issues in a number of ways. Its difficulty stems from two main sources. One is that, as I've suggested, the grand models don't wear their intrinsic, objective differences on their sleeves: discovering what they are remains an open question, whose resolution requires a combination of conceptual and empirical inquiries, which are yet to deliver a final verdict. The other source of difficulty comes from the fact that one cannot base oneself in this case (as one would be tempted to do) on the verdict of cognitive science itself, in its current state. One might assume the touchstone for assessing these frameworks to be how well they fit the domain, whose fundamental structure they claim to reveal: by putting forward very general hypotheses about what *cognition* in fact is, they appear as rational reconstructions of *cognitive science*, conceived as the sum total of local empirical work bearing on different aspects, at different levels of description, of particular cognitive functions. The framework that subsumes all this research in the most satisfactory manner might be declared the winner (as this is understood in the sciences, where conclusions can always be revised). But what can count as a result or as a research program that is acceptable for cognitive science is not something given; it is a hypothesis that is part of a set of hypotheses, including the general framework. In other words, the framework at least partially determines that which counts as a result or as a theory, so that the issue of the best framework cannot be made on the basis of results and theories. Thus in the best of circumstances, only at the end of a long cycle leading us from high-level hypotheses, to theories situated on a more local level, to empirical results, will a framework, a conception of the object of the cognitive sciences and the structure of their theories, and the corpus of their concepts and fundamental results be jointly stabilized.

Happily for cognitive science, the choice of a general framework is not a prerequisite for all work, for reasons we will now examine.

2.3.3 All That Remains to Be Decided: The Incompleteness of the Grand Models

Let's imagine a developmental psychologist trying to account for the manner in which an infant masters a particular capacity, concept, or skill. Let's imagine a neurolinguist who wants to understand why certain massive linguistic deficits, which appear following a stroke, sometimes disappear spontaneously, why other such deficits improve with the help of therapy, while others yet prove to be irreversible. Let's imagine a psychologist interested in the interdependence, suggested by some pathologies, between the ability to find one's way or navigate, and one's autobiographical awareness. Let's imagine a neurophysiologist who wonders how the visual system can follow the trajectory of more than one object simultaneously. Let's imagine a psychophysicist who wants to improve the hearing of profoundly deaf people with the help of better cochlear implants. Let's imagine a linguist trying to determine which indications allow us to attribute the proper referential values of certain pronouns in phrases of a certain type ("The cat ate the meat because *it* was hungry" versus "The cat ate the meat because *it* looked good"/ "Pierre asks John if *he* really believes that *he* loves Julia" versus "Pierre tells John that *he* really believes that *he* loves Julia."). Let's imagine a computer scientist who has the task of writing a software program to assist with their decision-making those in charge of security at nuclear power plants. Let's imagine an anthropologist who is studying supernatural beliefs and the way in which they coexist with common beliefs. Let's imagine an economist who is trying to correct the cognitive biases of an average subject in order to help him or her adopt behaviors that are favorable to his or her long-term interests, for example, with regard to retirement, health or highway safety. Let's imagine a philosopher wondering whether a perceived image, an imagined image, or a remembered image are the same kind of thing.

How will these researchers proceed? They can expect nothing, at least when they start out, from the grand models or their associated frameworks, for the simple reason that these models and frameworks have absolutely nothing to say regarding the questions with which the researchers are occupied. In their roles as psychologists, linguists, computer scientists, neurobiologists, anthropologists, economists and philosophers, they can only focus on the phenomenon at hand, and pursue all the paths of inquiry their disciplinary traditions suggest, while also taking advantage of indications provided by other disciplines, according to the basic organizational principle of cognitive science. The grand models bear primarily on mental processes. As Fodor has been at pains to emphasize, the issue of processes, while long neglected in the philosophical and psychological traditions, is important; but the emerging cognitive science tended to sin in the opposed direction, and to underestimate the difficulty of the question of mental states and their specific content. As they have matured, they have become interested in increasingly specific or domain-specific capacities, concerning numbers or other persons, the notion of an object or anaphoras, dyslexia or the perception of movement, and so forth, and mental states have again taken center stage, relegating processes to the background, and with them the grand models.

However it be, most cognitive scientists are generally indifferent to questions of framework, and are inclined to say what Newton said with regard to gravity: *Hypothesis non fingo*. The questions that occupy them are not completely unrelated to the general hypotheses that constitute the grand models. But the connections are usually loose ones, and if they are tightened, this is only gradually and in a reversible manner. In a word, cognitive scientists usually reject, explicitly but more often implicitly, any form of dependence with respect to any overarching framework, and more often than not abstain from any commitment in the area of foundations.

What this stance amounts to in practice will appear more clearly on a couple of examples, chosen from among those that are interesting in themselves to the philosopher of science. Nearly 40 years ago, two psychologists who were specialists in the study of the great apes asked the following question: Do chimpanzees have a theory of mind (ToM; Premack & Woodruff, 1978)? In other words, are they capable, like we are, of attributing to another chimpanzee beliefs, desires, or intentions that belong to that chimpanzee, and which may be different from their own? This question gave birth to a research program involving human beings. What is this capacity? What are its psychological mechanisms? In particular, does it depend on our capacity to understand our own propositional attitudes, or are those attitudes only accessible to us in the same manner that the attitudes of others are accessible? What are the neural bases of ToM? Is it embedded in a more general capacity, such as a "naïve psychology,"[31] conceived as a theory, more or less tacit, of mental functioning, or is it limited to the identification of the propositional attitudes of another member of our species? Do they possess

[31] Or "folk psychology"; there is no consensus about the acceptability of this locution, employed by some as a synonym for "theory of mind" in a technical sense (ToM), but also employed in a wider sense.

the characteristics of a module in the sense of massive modularity? At what age, and how is this capacity acquired by an infant? Is autism marked by the absence or deficiency of a ToM, and is this deficiency the cause or the effect of other aspects of the syndrome? In particular, is "mindblindness" (the presumed incapacity of the autistic person to see in another person an entity with a mind, as opposed to rocks or trucks) the cause of the person's inability to establish social relations? In normal infants, conversely, is a theory of mind necessary, is it sufficient to allow the child to develop his or her "social intelligence"? What is the basis of social cognition in adults? (Two recent assessments: Slors & Macdonald, 2008; Hutto et al., 2011.)

These questions concern philosophers as much as they do psychologists, and if one had the time to examine their contributions, one would be crossing many domains that belong to the philosophy of cognitive science in the broadest sense, and in all its diversity. One cannot but be struck by the character, or at least the formulation of the initial question, which a philosopher could have asked if psychologists had not already done so; the thinker who first became aware of its importance, Piaget, was equally philosopher and psychologist (Piaget, 1926). But to return to the question of the role of the grand models, we can see that they offer no assistance in this regard; they have no resources to enable us to formulate questions, or to suggest responses, or even to recommend a method of inquiry. And if one happened to say to such a researcher that his or her explanation of ToM, or his or her answer to one of the many questions that it raises, was incompatible with one or another grand model, he or she probably would not care, if only because the supposed indication of incompatibility would seem more doubtful than his or her own theory.

A second example, linked to the first, illustrates even more clearly this poverty of the grand models. More or less by chance, some 20 years ago, it was discovered that macaque monkeys had neurons that fired in the same way under two different conditions:[32] either when the animal carried out an intentional movement (such as reaching out its hand for some peanuts someone was offering to it), or else when the monkey saw another monkey (or a human) doing the same thing (Rizzolatti et al., 1996). These "mirror neurons," according to some researchers, allow the monkey to identify the intention of another monkey, which intention is expressed by gesture; Bobby "understands" my intention of giving him some peanuts because a mirror neuron fires when I extend my hand with that intention, a neuron that would also fire if Bobby extended his hand and with the same intention. Thus Bobby can relate to his observation to his own intention, and so identify my intention. These observations and this interpretation gave rise to a "motor" theory of human cognition (Rizzolatti et al., 2004; Gallese & al., 2004), particularly intended to address the question of human social cognition, which is the object of lively debates that here again involve philosophers and psychologists, and in this case, neurobiologists (Jacob & Jeannerod, 2005; Jacob,

[32] The effect of chance is never pure. See especially the real history of the « accidental » discovery of penicillin in 1928 by Fleming.

2008). These debates have nothing to do with the question of grand models; indeed, the discovery that gave rise to them is situated outside the widest possible context in which the grand models can be compared: a behavioristic—neither mentalist nor information-based—interpretation seems permissible. Only (perhaps) dynamicism (which can also be seen as a form of behaviorism) has the flexibility necessary to try to accommodate a motor theory in the strong sense in which it can be argued that each supports the other. Classicism, like connectionism, can also accommodate such a theory, but only in a weak sense: the validation of these frameworks is not linked to a validation of the motor theory. However, those who are interested in mirror neurons usually cannot be bothered with this kind of question.

2.3.4 Cognitive Science: Existence and Unity

There is a tension that exists between the last two sections. The first underlined the central character and the heuristic virtues of the grand models, the second emphasized their lack of relevance with regard to many areas of contemporary research. What does this tell us?

The first role filled by the grand models—particularly the classical model, but also the other models whose precursors did play an important role—is historical. This function, as we have seen, was to furnish the emerging cognitive science with a perspective within which it has been able to take shape, construct its first concepts, attain its first results, accommodate that which had been established by research programs which preceded it, in psychology and in other domains, and to attract a sufficient number of researchers, thus attaining a critical mass. This function, which is at once sociological and methodological, could only be performed by virtue of a relatively precise conceptualization (although one of limited applicability), which took the form of theses about the nature of the object under study and the complex methodology that was appropriate for it. Taken at face value, the grand models uphold the ontological and methodological unity of cognitive science, each in proprietary terms and for reasons of its own. Within the frameworks that they propose, cognitive science has a specific object, which constitutes a domain with natural and stable boundaries; and this domain must be studied at several levels, between which there exists an articulation that allows them to be subsumed as aspects of a single phenomenon.

Thus the grand models help obtain the conditions for the pragmatic viability of cognitive science, based on a theoretical perspective. What is at issue today is this theoretical perspective, but the conditions of viability are not necessarily affected.

We can account for this apparent paradox. Cognitive science does not require a guarantee of the ontological unity of its domain in order to get to work. At bottom, it only needs to be able to presume that this unity is thinkable, and that no decisive argument disproves this. As was the case with physics and biology, such unity may not appear until a later stage of the development of the discipline. Nor does cognitive science need to interpret literally the methodological prescriptions of one or another of the grand models. A methodological *modus vivendi* is enough for its various components, one based on the rejection of fixed boundaries, on common references, on the practice of dialogue, and on a shared goal of convergence, operating for all as a regulative ideal. Once these intellectual conditions are present, a research community emerges, and shows that the task is feasible by getting on with it. At that point, reflection on the grand models retreats to the study of foundations, as is the case in mature scientific disciplines. We may not be there yet, but one can interpret the changes that are taking place as a transition toward that stage.

If the grand models have been put in their proper place, it is not only because cognitive science has begun to mature, pursuing its course of development with minimal help from the grand models. It is also because the models have their own problems.

These problems are of two kinds. On the one hand, grand models are in search of answers to a whole group of questions of an ontological order, and in the absence of these answers they remain partly obscure. On the other hand, they are the target of criticisms that are nothing less than destructive, aimed at them but also, beyond them, at the very project of cognitive science, as it develops today. The two sets of problems are in fact connected, ranging on a scale of radicalness from philosophical conundrum to outright rejection. However the distinction does reflect a certain institutional reality: there are two fairly separate groups of authors, who communicate among themselves quite often, but do not exchange much from group to group, and have distinct perspectives on the matter.

The first group of authors are oriented toward naturalism, and actively seek naturalist solutions to the problem of the foundation of cognitive science. They may be pessimistic (in the sense in which a Borges character once said that a gentleman is only interested in lost causes), but they work side by side with optimists, accepting the terms in which the questions are presented. Such is not the case with the second group of authors, who without necessarily rejecting all forms of naturalism, nonetheless reject the kind that the first group takes for granted.

The two groups (who are not numerically equal) in practice work on different themes. The first group places at the center of its inquiries three main questions: the question of intentionality, the question of mental causality, and the question of consciousness.

The first of the three was long considered as the most important, or at least the one that had to be dealt with first. How can we make sense of the idea that a natural process be described, in the psychological idiom, as one by which a physical entity is endowed with meaning, that is stands for something (object, class, relation, state of affairs) that is situated outside it? In the LOTH framework, for example, as we have seen, the question is to know in what sense, and how the symbols of mentalese possess or acquire their reference or denotation, that is, the entities that they point to. This question splits into two. The first has to do with reference in general, and the second with the assignment of a particular reference to a given symbol. It is one thing to understand what it means, for a symbol to have something that it refers to; it is something else to know what makes a particular symbol designate trucks rather than Julius Caesar, or the equilateral triangle I am now drawing on a blackboard. Intentionality, circumscribed in this way, opens up a heady perspective: it appears to insert the world into the mind, threatening the image of the fortress of the inner mind, or the control tower. "Externalism" is the label usually given to this perspective. Externalism comes into various flavors, more or less radical, and each one offers a different conception of the way in which the world barges in the mind (Clark & Chalmers, 1998; Hutchins, 1995; Rowlands, 2003; Wilson, 2004; Menary, 2011; Shapiro, 2014).

The second main question for naturalistic philosophers is a modern version of the problem that Descartes thought he could resolve by way of the pineal gland. It is known as the problem of mental causality, and is formulated in the following way.[33] The physical world changes according to laws of physics. In principle, these laws are complete: physics may not be completely finished, but it includes the totality of the laws of nature. Thus in principle, if not in fact, it has everything it needs to give an account of any process or causal sequence. There is no place in this picture for a cause that physics cannot account for. But on another hand, we are tempted to think that our thoughts do have a causal effect. Isn't it my intention to open the door that which causes the door to open? Must we set aside this intuition, at the risk of seeing the psychology of common sense and a large portion of the scientific psychology of today disappear?

The third main question has to do with consciousness. Does it have its own reality, or is it an epiphenomenon? Does it have several forms or modalities, or is it a unitary thing? Does it play a particular role in cognition, and what would that be? If it is real, how does and how did it initially find its place in nature? Several other issues are connected to this group of questions: the question of phenomenal properties, also known as qualia, that is, those that do not seemingly intervene in the processing of information, but only accompany certain cognitive processes (the taste of a pear: the general effect on me of feeling it in my mouth); the question of the nature and role of emotions; and the question of the self.

It is more difficult to draw up a list of themes around which the reflections of the philosophers who criticize the naturalistic orientation of the first group are organized. I will venture to put forward three of them. The first two are closely linked together: (1) Can one think about the mind, even in a preliminary way, independently of society? Can one *make sense of* the mind (or thoughts) of a single human being cut off from society, Robinson style, from day one? (2) Isn't the mind *shaped by* culture, to the point where its natural and biological structure practically disappears in description and explanation? If the answer is that it is, as the philosophers in the group, together with some scientists who ask these questions tend to believe, then it is conceivable that the mind, such as it is conceived by contemporary cognitive science (the science of *cognition*), does not constitute an authentic object of science. (We recall that existence in the material world is not a sufficient condition for constituting an object of science: there is no science of objects weighing less than 350 grams, nor a science of texts lacking the letter x; there is

[33] The reader will find a much more complete exposition in chapter 3 of this volume.

no science of prestidigitation, no science of misery, and no science of faces.) The third theme concerns the body (Bermudez et al., 1995; Kelly, 2000): Is it legitimate to think that the mind is lodged, as it were, within the body, or that it is connected to the body? Is it not the case that it *is* a body, or a constitutive part of the body?

I may have given the impression that these disputes, whether they are taking place in one camp or the other, or somewhere between the two camps, are without effect on cognitive science. That is clearly false. The radical critiques of the second camp give rise to "heterodox" research programs in cognitive science, and these programs in return give life to philosophy's re-opening of certain questions. The work of naturalistic philosophers may "resonate" with scientific issues (in conformity with one of the main theses of naturalism, which affirms the continuity of science and philosophy). These are as much problems of the first order—as when a connectionist solution is proposed for the problem of the origin of language, or when the neurosciences propose a model of consciousness (Dehaene, 2001; Koch, 2004; Tononi, 2012)—as of the second order, and not less important: like the question of knowing how far psychology, linguistics and anthropology can continue their inquiries independently of the data and ongoing inquiries of neuroscience (Ravenscroft, 1998; Gold & Stolja, 1999; Bennett & Hacker, 2003; Andler, 2016, chap. 3).

The ontological questions of philosophers, as we can see, have therefore a bearing on the question of the existence and the unity of cognitive science, taken in its present state or considered with an eye toward its future development. The proactive reader will have followed this path throughout this chapter. But he or she will have to look elsewhere for a more complete presentation of ontological questions and their potential impact on cognitive science, for it is high time for the present chapter to reach an end.

* * *

And so this chapter comes to a conclusion precisely where other authors would have chosen to begin. I have raised a number of ontological questions that not only belong to the philosophy of cognitive science, according to them, but lie at their very heart, and I have done no more, almost as an afterthought, than to formulate them, and then left them hanging. I would therefore like to say a few words about the technical division of labor among philosophers who are interested in cognition.

Several terms exist that designate their area of activity: philosophy of cognitive science, philosophy of psychology, philosophical psychology, cognitive philosophy, philosophy of mind, philosophy of cognition (Guttenplan, 1994; Dretske, 1995; Warfield & Stich, 2003). Two things hardly need emphasizing: first, terminology varies from one philosopher to another and from one book to another, so there is nothing dependable that we can draw from the study of terminological choices; second, no classification should aim at eliminating all overlaps,[34] which are not only inevitable, but also play a role in the circulation of concepts and ideas and prevent doctrinal ossification and the formation of academic cliques.

[34] They are, in fact, so large that certain philosophers refuse to recognize the distinctions I propose, maintaining that they are mere terminological quirks or nuances without any theoretical bearing.

Let us focus our attention rather on the objectives that these philosophers set for themselves, and on their positions with regard to the sciences. Philosopher A asks questions about cognitive science in a way that is at one and the same time descriptive and normative or critical; she feels close to the field, but it is not her sole objective to assist it in its tasks, nor does she claim to contribute directly to it. Her attitude is similar to the one adopted by most philosophers of physics, mathematics, or biology. Philosopher B, in contrast, wants to contribute to cognitive science in any way he can: by conceptual analysis, or by participation in interdisciplinary research, necessitating on his part the acquisition of scientific competence, even if this is only for a particular reason at a particular time. Philosopher C asks direct questions about the object of cognitive science, but in a way that does not depend entirely on its productions or on its methodological choices, and leans on one or the other philosophical tradition. Philosopher D is interested in psychology in all its extent and diversity. The objectives of D are both narrower and wider than those of A: he tends to leave aside certain questions about the domain of A (questions relative for example to language, to cultural evolution, to artificial intelligence, or to the methodology of the neurosciences), and conversely he may include in his focus schools or branches of psychology that are not (at least not at this moment) within the sole competence of cognitive science (such as clinical psychology and psychoanalysis, differential psychology, educational psychology, social psychology, industrial psychology, and so on). On another hand, he or she does pay attention to specific methodological issues of scientific psychology, ranging from chronometry or priming to the significance of gaze duration or of non-nutritive sucking in infants, or again the inheritability of character traits or intelligence.[35] Similarly, C's domain is at one and the same time more restricted and wider than that of B: C, for example, in contradistinction to B, can defend dualism or take up a phenomenological point of view or a Wittgensteinian one without feeling the obligation, as B does, to join up with cognitive science in one way or another.[36]

These ideal-types (in Weber's sense) are representative of what I would call, respectively, the philosophy of cognitive science (A); cognitive philosophy or philosophical psychology, with a cognitive orientation (B); philosophy of mind (C); and the philosophy of psychology (D). Cognitive philosophy and philosophical psychology are close to cognitive science, in the sense that they share the same direct objectives; the philosophy of cognitive science and the philosophy of psychology are further away: their

[35] To the extent that linguistics, the neurosciences, and anthropology are also partly immersed in the cognitive sciences, they come up with a division of tasks that is somewhat comparable. The philosophy of cognitive sciences emphasizes the relationships between the disciplines that are included in it and on their convergences, while the philosopher of linguistics or the neurosciences or anthropology, on one hand, accepts all currents by definition, including the "non-cognitive" version of linguistics, etc., and on another hand concentrates on problems specific to his or her discipline.

[36] A movement inspired by phenomenology has appeared recently, which would like to contribute very directly to the cognitive sciences (see Dreyfus, 1982; McClamrock, 1995; Kelly, 2001/2013, Petitot et al., 1999; Smith & Thomasson, 2005; Andler, 2006; Gallagher & Zahavi, 2008, Gallagher & Schmicking, 2010; and the journal *Phenomenology and the Cognitive Sciences*).

objectives do not necessarily coincide in every way or at all times with those of cognitive science. The philosophy of psychology and philosophical psychology are obviously close to psychology as a separate and autonomous discipline; the philosophy of cognitive science and cognitive philosophy are further away because they are interested, precisely, in an approach that suggests immersing (or even dissolving) psychology in a much larger theoretical framework. Finally, philosophy of mind overlaps to a great extent with all the other branches, but it is less straightforwardly committed to a scientific perspective.

The division of labor is not the only explanation of this geography of specialties. There are also doctrinal disagreements, whether of the first order (for example, on the question of naturalism) or the second (bearing on a normative conception of the role of the philosopher). But this is a subject which we cannot go into here.

This chapter was written from the perspective of philosopher A. I have not tried to avoid the company of B, C, or D. But I have not followed the pathways they would have pursued. I have also left aside many questions that are beyond doubt relevant from A's perspective. The aim, once again, was to discuss cognitive science in the way a philosopher of biology would talk about biology, a philosopher of economics about economics, and so on. If as I fear the aim has not been fully attained, the reason lies partly with the nature of the domain, as I warned the reader, and partly of course with the author.

References

Amit, D. J. (1989), *Modeling Brain Function: The World of Attractor Neural Networks*. Cambridge: Cambridge University Press.

Anderson, J. A., & Rosenfeld E., eds. (1988), *Neurocomputing. Foundations of Research*. Cambridge, MA: MIT Press.

Anderson, J. A., Pellionisz, A., & Rosenfeld, E., eds. (1990), *Neurocomputing II*. Cambridge, MA: MIT Press.

Andler, D. (1992), "From paleo to neo-connectionism," *in* van der Vijver, G., ed., *New Perspectives on Cybernetics*. Dordrecht: Kluwer, 125–146.

Andler, D. (2006), "Phenomenology and existentialism in cognitive science and artificial intelligence," *in* Wrathall, M. & Dreyfus, H., eds., *Blackwell Companion of Phenomenology and Existentialism*. Oxford: Blackwell, 377–393.

Andler, D. (2016), *La Silhouette de l'humain. Quelle place pour le naturalisme dans le monde d'aujourd'hui?* Paris: Gallimard.

Bain, A. (1893), "The respective spheres and mutual helps of introspection and psychophysical experiment in psychology," *Mind*, 2: 42–53.

Barner, D., & Baron, A.S., eds. (2016.), *Core Knowledge and Conceptual Change*. New York: Oxford University Press.

Battro, A. (2001), *Half a Brain Is Enough: The Story of Nico*. Cambridge: Cambridge University Press.

Bennett, M. R., & Hacker, P. M. S. (2003), *Philosophical Foundations of Neuroscience*. Oxford: Blackwell.

Bermúdez, J. L., Marcel, A., & Eilan, N., eds. (1995), *The Body and the Self*. Cambridge, MA: MIT Press.

Bickhard, Mark H. (2002) "Critical principles: On the negative side of rationality," *New Ideas in Psychology* 20 (1): 1–34.

Block, N., ed. (1980a), *Readings in the Philosophy of Psychology*, vols. 1 and 2. Cambridge, MA: Harvard University Press.

Block, N. (1980b). "Troubles with Functionalism," in Block 1980a, 268–305.

Bowles, S. & Gintis, H. (2011), *A Cooperative Species: Human Reciprocity and Its Evolution*. Princeton, NJ: Princeton University Press.

Broca, P. (1861), "Remarques sur le siège de la faculté du langage articulé, suivies d'une observation d'aphémie (perte de la parole)," *Bulletin de la Société Anatomique* 6, 330–357 (online: http://psychclassics.asu.edu/Broca/aphemie.htm).

Bruner, J. (1966), *Toward a Theory of Instruction*. Cambridge, MA: Harvard University Press.

Bruner, J. (1968), *Processes of Cognitive Growth: Infancy*. Worcester, MA: Clark University Press.

Buchanan, G. (2005), "A (very) brief history of artificial intelligence," *AI Magazine, 25th Anniversary Issue,* Winter 2005: 53–60.

Buss, D. M. (2008), *Evolutionary Psychology*, 3rd ed. Boston: Pearson Education.

Carey, S. (1985), *Conceptual Change in Childhood*. Cambridge, MA: MIT Press.

Carey, S. (2011), *The Origin of Concepts*. Oxford and New York: Oxford University Press

Carruthers, P. (2006), *The Architecture of the Mind*. New York: Oxford University Press.

Carruthers, P., Laurence, S., & Stich, S., eds. (2005), *The Innate Mind, vol. 1: Structure and Contents*. Oxford: Oxford University Press.

Chomsky, N. (1957), *Syntactic Structures*. The Hague: Mouton.

Chomsky, N. (1959), "Review of B.F. Skinner's *Verbal Behavior*," *Language* 35: 26–58.

Chomsky, N. (1968), *Language and Mind*. New York: Harcourt Brace Jovanovich Inc., 1968.

Chomsky, N. (1975), *Reflections on Language*. New York: Pantheon Books.

Clark, A. (1989), *Microcognition. Philosophy, Cognitive Science, and Parallel Distributed Processing*. Cambridge, MA: MIT Press.

Clark, A., & Chalmers, D. (1998), "The extended mind," *Analysis* 58: 7–19; repr. in Menary, R. (2011).

Comte, A. (1848/1998), *Discours sur l'ensemble du positivisme*. Paris: Flammarion; available at http://classiques.uqac.ca/classiques/Comte_auguste/la_science_sociale_extraits/4_discours_ensemble_positivisme/discours_ensemble_pos.html. Engl. transl. *A General View of Positivism,* Paris, 1848.

Copeland, B. J., ed. (2004), *The Essential Turing*. Oxford: Oxford University Press.

Cowie, F. (1999), *What's Within? Nativism Reconsidered*. New York: Oxford University Press.

Cummins, R. (1985), *The Nature of Psychological Explanation*. Cambridge, MA: MIT Press.

Davies, Martin (2005), "Cognitive science," *in* Jackson, F., & Smith, M. (eds.), *The Oxford Handbook of Contemporary Philosophy*. Oxford: Oxford University Press, 358–394. (A more complete version is available on the author's website.)

Dayan, P., & Abbott, L. (2001), *Theoretical Neuroscience: Computational and Mathematical Modeling of Neural Systems*. Cambridge, MA: MIT Press.

Dehaene, S. (2001), *The Cognitive Neuroscience of Consciousness*. Cambridge, MA: MIT Press.

Dennett, D. C. (1978), *Brainstorms. Philosophical Essays on Mind and Psychology*. Cambridge, MA: MIT Press.

Dennett, D. C. (1991), *Consciousness Explained*, Boston: Little, Brown.

Dennett, D. C. (1995), *Darwin's Dangerous Idea. Evolution and the Meanings of Life*. New York: Simon & Schuster.

Diamond, J. (1997), *Guns, Germs and Steel*. New York: Norton.

Dretske, F. (1995), *Naturalizing the Mind*. Cambridge, MA: MIT Press.

Dreyfus, H. L., ed. (1982), *Husserl, Intentionality and Cognitive Science*. Cambridge, MA: MIT Press.

Elman, J., Bates L. E., Johnson, M.H., Karmiloff-Smith, A., Parisi, D., & Plunkett, D. (1996), *Rethinking Innateness. A Connectionist Perspective on Development*. Cambridge, MA: MIT Press.

Evans, J. St. B. T. (2003), "In two minds: Dual-process accounts of reasoning," *Trends in Cognitive Sciences* 7 (10): 454–459.

Evans, J. S. B. T., & Stanovich, K. E. (2013), "Dual-process theories of higher cognition: Advancing the debate," *Perspectives on Psychological Science* 8 (3): 223–241.

Flynn, James R. (2007), *What Is Intelligence? Beyond the Flynn Effect*, expanded edition (2009). Cambridge: Cambridge University Press.

Flynn, J. R. (2009), *What Is Intelligence?: Beyond the Flynn Effect*, Expanded edition. Cambridge: Cambridge University Press.

Fodor, J. (1968), *Psychological Explanation*. New York: Random House.

Fodor, J. (1975), *The Language of Thought*. New York, Thos. Crowell; reprint: Cambridge, MA, Harvard University Press.

Fodor, J. (1981a), "The present status of the innateness controversy," in Fodor, J. (1981b), chap. 10.

Fodor, J. (1981b), *Representations: Philosophical Essays on the Foundations of Cognitive Science*. Cambridge, MA: MIT Press.

Fodor, J. (1983), *The Modularity of Mind*. Cambridge, MA: MIT Press.

Fodor, J. (2000), *The Mind Doesn't Work That Way: The Scope and Limits of Computational Psychology*. Cambridge, MA: MIT Press.

Fodor, J. (2008a), "Against Darwinism," *Mind & Language* 23 (1): 1–24.

Fodor, J. (2008b), *LOT2: The Language of Thought Revisited*. Oxford: Clarendon Press.

Forest, D. (2014), "Neuroconstructivism: A developmental turn in cognitive neuroscience?," *in* Wolfe, C., ed. (2014), *Brain Theory: Essays in Critical Neurophilosophy*. London and New York: Palgrave Macmillan, 68–87.

Gall, F. J., & Spurzheim, J. C. (1810–1819), *Anatomie et physiologie du système nerveux en général, et du cerveau en particulier; avec des observations sur la possibilité de reconnoitre plusieurs dispositions intellectuelles et morales de l'homme et des animaux, par la configuration de leurs têtes, etc.*, 4 vol. Paris: F. Schoell, then J. B. Baillère.

Gallagher, S., & Schmicking, D., eds. (2010), *Handbook of Phenomenology and Cognitive Science*. Dordrecht: Springer

Gallagher, S., & Zahavi, D. (2008). *The Phenomenological Mind—An Introduction to Philosophy of Mind and Cognitive Science*. London and New York: Routledge.

Gallese, V., Keysers, C. & Rizzolatti, G. (2004), "A unifying view of the basis of social cognition," *Trends in Cognitive Sciences* 8: 396–403.

Gintis, H. (2007), "A framework for the unification of the behavioral sciences," *Behavioral and Brain Sciences* 30: 1–61.

Gold, I., & Stolja, D. (1999), "A neuron doctrine in the philosophy of neuroscience," *Behavioral and Brain Sciences* 22: 809–830.

Griffiths, P. (2002), "What is innateness?," *Monist* 85: 70–85.

Guttenplan, S., ed. (1994), *A Companion to the Philosophy of Mind*. Oxford: Blackwell.
Hatfield, G. (1995), "Remaking the science of the mind. Psychology as natural science," *in* Fox, C., Porter, R., & Wokler, R., eds. (1995), *Inventing Human Science. Eighteenth-Century Domains*. Berkeley & Los Angeles: University of California Press, 184–231.
Hebb, D. O. (1949), *The Organization of Behavior*. New York: Wiley.
Heims, S. (1991), *Constructing a Social Science for Postwar America. The Cybernetics Group 1946–1953*. Cambridge, MA: MIT Press.
Herken, R., ed., (1988), *The Universal Turing Machine. A Half-Century Survey*. Oxford: Oxford University Press.
Hinton, G. E., & Anderson, J. A. (1981), *Parallel Models of Associative Memory*. Hillsdale, NJ: Erlbaum.
Hirschfeld, L. A., & Gelman, S. A., eds. (1994), *Mapping the Mind: Domain Specificity in Cognition and Culture*. Cambridge: Cambridge University Press.
Hook, S., ed. (1960), *Dimensions of Mind*. New York: Collier Books.
Hutchins, E. (1995), *Cognition in the Wild*. Cambridge, MA: MIT Press.
Hutto, D. H., Herschbach, M., & Southgate, V., eds. (2011), "Social cognition: Mindreading and alternatives," special issue, *Review of Philosophy and Psychology*.
Jacob, P. (2008), "What do mirror neurons contribute to human social cognition?," *Mind and Language* 23 (2): 190–223.
Jacob, P., & Jeannerod, M. (2005) "The motor theory of social cognition: A critique," *Trends in Cognitive Sciences* 9 (1): 21–25.
Jain, S., Osherson, D, Royer, J.S., & Sharma, A. (1999), *Systems That Learn: An Introduction to Learning Theory*, 2nd ed. Cambridge, MA: MIT Press.
James, W. (1890) *Principles of Psychology*, vol. 1. New York: Holt.
Kahneman, D. (2011), *Thinking Fast and Slow*. New York: Farrar, Strauss & Giroux.
Kanizsa, G. (1979), *Organization in Vision: Essays on Gestalt Perception*. New York: Praeger.
Kelly, S. (2000), "Grasping at straws: Motor intentionality and the cognitive science of skilled behaviour," *in* Wrathall, M. & Malpas, J., eds. (2000), *Heidegger, Coping, and Cognitive Science, Essays in honor of Hubert L. Dreyfus*, vol. 2. Cambridge, MA: MIT Press, 161–177.
Kelly, S. D. (2001/2013), *The Relevance of Phenomenology to the Philosophy of Language and Mind*. London and New York: Routledge
Kelso, J.A.S. (1997), *Dynamic Patterns*. Cambridge, MA: MIT Press.
Khalidi, M. (2007), "Innate cognitive capacities," *Mind and Language* 22 (1): 92–115.
Koch, C. (2004), *The Quest for Consciousness. A Neurobiological Approach*. Englewood, CO: Roberts & Co.
Koffka, K. (1935), *Principles of Gestalt Psychology*. New York: Harcourt.
Köhler, W. (1945), *Gestalt Psychology*. New York: Liveright; rev. ed., 1947, New York: Mentor Books.
Kriesel, D. (2011), *A Brief Introduction to Neural Networks*, downloadable at http://www.dkriesel.com/_media/science/neuronalenetze-en-zeta2-1col-dkrieselcom.pdf.
Laurence, S., & Margolis, E. (2002), "Radical concept nativism," *Cognition* 86 (1): 22–55.
Leibniz, G. W. (1714/2006), *Principles of Nature and Grace Based on Reason*, edited by J. Bennett, http://www.earlymoderntexts.com/pdfs/leibniz1714a.pdf.
Lewis, D. (1966), "An argument for the identity theory," *Journal of Philosophy* 63: 17–25.
Lewis, D. (1980), "Mad pain and martian pain," *in* Block (1980a), 216–222.
Maguire, E. A., Frackowiak, R. S. J., & Frith, C. D. (1997), "Recalling routes around London: Activation of the right hippocampus in taxi drivers," *The Journal of Neuroscience* 17 (18): 7103–7110.

Margolis, E., & Laurence, S. (2012), "In defense of nativism," *Philosophical Studies* 165 (2): 693–718.

McClamrock, R. (1995), *Existential Cognition*. Chicago: Chicago University Press.

McCorduck, P. (2004), *Machines Who Think: Twenty-Fifth Anniversary Edition*. Natick, MA: A. K. Peters.

McCulloch, W. S. (1965/1988), *Embodiments of Mind*. Cambridge, MA: MIT Press.

McCulloch, W. S., Pitts, W. (1943), "A logical calculus of the ideas immanent in nervous activity," *The Bulletin of Mathematical Biophysics* 5: 115–133; repr. in McCulloch (1965/1988) and in Anderson & Rosenfeld (1988).

Menary, R., ed. (2011), *The Extended Mind*. Cambridge, MA: MIT Press

Nisbett, R. E. (2009), *Intelligence and How to Get It*. New York: Norton.

Petitot, J., Varela, F.J., Pachoud, B., & Roy, J.-M., eds. (1999), *Naturalizing Phenomenology: Issues in Contemporary Phenomenology and Cognitive Science*. Stanford: Stanford University Press.

Piaget, J. (1926), *La représentation du monde chez l'enfant, etc*. Paris: Alcan; Engl. transl. (1926), *The Language and Thought of the Child, etc*. London: Kegan Paul.

Piattelli-Palmarini, M. dir. (1979), *Théories du langage, théories de l'apprentissage, le débat Chomsky/Piaget*. Paris: Seuil. Engl. transl. (1980), *Language and Learning: The Debate between Jean Piaget and Noam Chomsky*. Cambridge, MA: Harvard University Press.

Poincaré, H. (1905), *La Science et l'hypothèse*. Engl. transl. (1952), *Science and Hypothesis*. New York: Dover.

Port, R.F., & van Gelder, T., eds. (1995), *Mind as Motion: Explorations in the Dynamics of Cognition*. Cambridge, MA: MIT Press.

Premack, D. & Woodruff, G. (1978), "Does the chimpanzee have a theory of mind?," *Behavioral and Brain Sciences* 4: 515–526.

Prinz, J. (2002), *Furnishing the Mind: Concepts and Their Perceptual Basis*. Cambridge, MA: MIT Press.

Putnam, H. (1960), "Minds and Machines," repr. in Putnam (1975), chap. 18, and in many anthologies.

Putnam, H. (1975), *Mind, Language, and Reality*. Cambridge: Cambridge University Press.

Putnam, H. (1988), *Representation and Reality*. Cambridge, MA: MIT Press, 1988.

Quartz, S., & Sejnowski, T. J. (1997), "The neural basis of cognitive development: A constructivist manifesto," *Behavioral and Brain Sciences* 20 (4): 537–596.

Ravenscroft, I. (1998), "Neuroscience and the mind," *Mind and Language* 13: 132–137.

Richerson, P. J., & Boyd, R. (2005), *Not by Genes Alone: How Culture Transformed Human Evolution*, Chicago: University of Chicago Press.

Rizzolatti, G., et al. (1996), "Premotor cortex and the recognition of motor actions," *Cognitive Brain Research* 3 (2): 131–141.

Rizzolatti, G., et al. (2004), "A unifying view of the basis of social cognition," *Trends in Cognitive Science* 8: 396–403.

Rowlands, M. (2003), *Externalism, Putting Mind and World Back Together Again*. Montreal and Kingston: McGill-Queens University Press.

Rumelhart, D., McClelland, J., & the PDP Research Group (1986), *Parallel Distributed Processing. The Microstructure of Cognition*, vols. 1 and 2. Cambridge, MA: MIT Press.

Russell, B. (1918), "The Philosophy of Logical Atomism," repr. *in The Philosophy of Logical Atomism and Other Essays: 1914-1919*. London: Allen & Unwin, 1986, 160–244.

Ryle, G. (1949), *The Concept of Mind*. London: Hutchinson.

Samuels, R. (2002), "Nativism in cognitive science," *Mind & Language* 17 (3): 233–265.

Schiffer, S. (1981), "Truth and the theory of content," *in* H. Parret & J. Bouveresse, eds., *Meaning and Understanding*. Berlin: Walter de Gruyter, 204–222.

Shapiro, L., ed. (2014), *The Routledge Handbook of Embodied Cognition*. London, New York: Routledge.

Simon, H. A. (1957), *Models of Man, Social and Rational: Mathematical Essays on Rational Human Behavior in a Social Setting*. New York: John Wiley & Sons.

Slors, M., Macdonald, C. (eds.) (2008), "Rethinking folk-psychology: alternatives to theories of mind," special issue, *Philosophical Explorations* 11 (3): 153–161.

Smith, B., & Ehrenfels, C., eds. (1988), *Foundations of Gestalt Theory*. Munich: Philosophia Verlag.

Smith, D. W., & Thomasson, A.L., eds., (2005), *Phenomenology and Philosophy of Mind*. New York: Oxford University Press.

Smolensky, P. (1988), "On the proper treatment of connectionism," *Behavioral and Brain Sciences* 11: 1–74; repr. in P. Smolensky & G. Legendre (2005), *The Harmonic Mind: From neural computation to optimality-theoretic grammar*. Cambridge, MA: MIT Press.

Spelke, E. S. (2000), "Core knowledge," *American Psychologist* 55: 1233–1243.

Sperber, D. (2001), "In Defense of massive modularity," in E. Dupoux (ed.), *Language, Brain and Cognitive Development: Essays in Honor of Jacques Mehler*. Cambridge, MA: MIT Press, 47–57.

Sperber, D. (2005), "Modularity and relevance: How can a massively modular mind be flexible and context-sensitive?," *in* Carruthers et al. (2005), chap. 4.

Stainton, R. J., ed. (2006), *Contemporary Debates in Cognitive Science*. Oxford: Blackwell.

Sterelny, K. (2006), "Language, evolution and modularity." *in* Macdonald, G., & Papineau, D., eds. (2006), *Teleosemantics*. Oxford: Oxford University Press, 23–41.

Sterelny, K. (2012), *The Evolved Apprentice*. Cambridge, MA: MIT Press.

Sterelny, K., Joyce, R., Calcott, B., & Fraser, B., eds. (2013), *Cooperation and Its Evolution*. Cambridge, MA: MIT Press.

Sternberg, R. J. (1988), *The Triarchic Mind: A New Theory of Human Intelligence*. New York: Penguin.

Stich, S. (1978), "Beliefs and subdoxastic states," *Philosophy of Science* 44: 589–622.

Stich, S. (1983), *From Folk Psychology to Cognitive Science*. Cambridge, MA: MIT Press.

Taylor, C. (1985), *Philosophy and the Sciences of Man*. Cambridge: Cambridge University Press.

Thelen, E., & Smith, L.B., eds. (1994), *A Dynamic System Approach to the Development of Cognition and Action*. Cambridge, MA: MIT Press.

Thorndike, E. L., & Woodworth, R. S. (1901), "The influence of improvement in one mental function upon the efficiency of other functions," *Psychological Review* 8: 247–261.

Tononi, G. (2012), "The integrated information theory of consciousness: an updated account", *Archives italiennes de biologie* 150: 290–326.

Tooby, J., & Cosmides, L. (1992), "The psychological foundations of culture," *in* Barkow, J. H., Cosmides, L., & Tooby, J., eds. (1992), *The Adapted Mind*. New York: Oxford University Press, 19–136.

Tooby, J., & Cosmides, L. (2005), "Conceptual Foundations of Evolutionary Psychology," *in* Buss, D., ed. (2005), *Handbook of Evolutionary Psychology*. Hoboken, NJ: Wiley, 5–67.

Turing, A. M. (1936–1937), "On computable numbers, with an application to the Entscheidungsproblem," *Proceedings of the London Mathematical Society, Ser.* 2, 42: 230–265. "A correction," ibid., 43 (1937): 544–546; repr. in many anthologies, including Copeland (2004).

Turing, A. M. (1950), "Computing machinery and intelligence," *Mind* 59: 433–460; repr. in many anthologies, including Copeland (2004).

Vygotsky, L. S. (1930/1978), *Mind and Society. The Development of Higher Psychological Processes*. Cambridge, MA: Harvard University Press, trans. from Russian. (Almost all of Vygotsky's writings translated into English are available online: http://www.marxists.org/archive/vygotsky/index.htm.)

Ward, L. M. (2001), *Dynamical Cognitive Science*. Cambridge, MA: MIT Press.

Warfield, T., & S. Stich, S., eds. (2003), *The Blackwell Guide to the Philosophy of Mind*. Oxford: Blackwell.

Warner, R., & Szubka, T., eds., (1994), *The Mind-Body Problem*. Oxford: Blackwell.

Weisberg, M. (2006). "Water is not H2O," *in* Baird, D., McIntyre, L., & Scerri, E., eds., *Philosophy of Chemistry: Synthesis of a New Discipline*. New York: Springer, 337–345.

Wilson, R. A. (2004), *Boundaries of the Mind. The Individual in the Fragile Sciences*. Cambridge: Cambridge University Press.

Worrall, J. (1989), "Structural realism: The best of both worlds?," *Dialectica* 43: 99–124; repr. in D. Papineau, ed., *The Philosophy of Science*. Oxford: Oxford University Press, 139–165.

17

PHILOSOPHY OF LINGUISTICS

Paul Égré (CNRS, Institut Jean-Nicod / École Normale Supérieure, PSL Research University)

1. Introduction: What Is Linguistics?

1.1 LANGUAGES AND LANGUAGE

Linguistics is the scientific study of languages and of language.[1] The distinction between languages, in the plural, and language, in the singular, exposes from the outset a duality which lies at the very heart of linguistics, and is also found in the opposition between general or theoretical linguistics on the one hand, and specialized grammar on the other, that is, the study of a specific language.

[1] This chapter is an English translation of the chapter originally written for the French edition of this handbook, with minor revisions and a few updates. I am extremely grateful to Michael Murez, who did the translation, for his meticulous work and help in adapting the text. Any infelicity should be blamed on me and on the fact that the text would have been written differently if it had been written directly into English. For the original version of this paper, I wish to thank in particular Sylvain Bromberger, Morris Halle, Philippe Schlenker, Benjamin Spector, and Noam Chomsky for the numerous suggestions, detailed comments, and criticisms they made as I was writing this chapter: without their advice and encouragements, the horizon of my research would have been considerably narrowed (any remaining shortcomings or errors are entirely my own responsibility). My thanks also extend to the MIT France Program and to the MIT Linguistics Department, where I spent the spring semester in 2007, which allowed me to start working on my research in the epistemology of linguistics and to deepen my knowledge of linguistics itself. I am also grateful to the Paris III-Censier linguistics library and to its staff. My thanks also to L. Rizzi, D. Steriade, N. Richards, D. Blitman, S. Peperkamp, C. Beyssade, D. Sportiche, A. Bachrach, and finally B. Gillon for very useful reading advice, and to M. Cozic, D. Bonnay, and F. Rivenc for their attentive proofreading. I also thank S. Hartmann, M. Nilsenova, R. Muskens, I. Douven, and J-W. Romeijn for their questions during a presentation in Tilburg, as well as H. Galinon, H. Ba, and

The first thing a linguist notices is actually the same thing as anyone who speaks a particular language: it is the diversity of languages, and the difficulty of understanding and speaking any language different from one's mother tongue. From a traditional viewpoint, linguistics thus starts with grammar, understood as the study of the rules underlying the proper formulation and usage of utterances belonging to a particular language (such as the grammar of ancient Greek, the grammar of modern Portuguese, the grammar of Moroccan Arabic, etc.). The task of the linguist, as a grammarian, is therefore to give a rational description of the relevant units and formation rules of each of the languages he has chosen to study.

For the linguist, however, noticing the diversity of languages soon leads to another observation, which, first of all, concerns the *intertranslatability* of different languages, and then, perhaps more fundamentally, the capacity of newborn infants to acquire and speak the language of the community in which they are raised (Chomsky 1965). People often marvel naively at how difficult it is to translate certain specific words from one language into another: *saudade*, in Portuguese, probably does not have a strict equivalent in French or in English.[2] Similarly, a poem by Goethe loses its poetical force in the attempt to translate it from German into some other language. But by focusing on differences in poetic value between words in different languages, one loses sight of the far more fundamental fact that it is possible to translate everyday utterances of one language into ordinary utterances in another.[3] The existence of a principled correspondence between different languages, and a young child's capacity to acquire any language whatsoever, suggest that there is a "common denominator" among the different languages (Baker, 2001).[4] According to this perspective, the linguist should not only aim to study the rules specific to a particular language, but also the more general rules, which are liable to govern language as a faculty, and they should seek to uncover linguistic invariants across languages. As Postal (1964, p. 137) sums things up, following Chomsky:[5]

> Linguistics is interested both in individual natural languages and in Language. This involves the grammarian in the two distinct but interrelated tasks of

the participants of the Lemming seminar in Paris. Finally, thanks to A. Barberousse, D. Bonnay, and M. Cozic for their encouragement and patience during preparation of this chapter. This research was done with the support of Programs ANR-10-LABX-0087 IEC and ANR-10-IDEX-0001-02 PSL*.

[2] The most common English equivalent of *saudade* is "nostalgia."

[3] As Baker (2001) very rightly points out, the problem with translating poetical texts derives from the difficulty in jointly satisfying a great number of constraints (equivalence in lexical meaning, preservation of meter, preservation of rhyme, preservation of assonances and alliterations, etc.). For precisely this reason, poetical discourse does not constitute a suitable starting point for the study of language.

[4] Of these two aspects, intertranslatability in principle between languages, and infants' capacity to learn how to speak, it is fundamentally the second which underlies the generative project and the idea of universal grammar. It is actually not obvious that infants' linguistic capacity necessarily implies the intertranslatability of different languages (I am grateful to S. Bromberger, P. Schlenker, and N. Chomsky for bringing up this point, independently from one another.)

[5] See Chomsky (1957, p. 14), who writes: "we are interested not only in particular languages, but also in the general nature of Language."

> constructing grammars for particular languages and constructing a general theory of linguistic structure which will correctly characterize the universal grammatical features of all human languages.

As a general theory of the language faculty, linguistics should therefore be distinguished from grammar in the traditional sense of that term, even if it is rooted in the work of the grammarians and comparatists of the nineteenth century (especially Schleicher, Grimm, Bopp, Verner, cf. the overview in Saussure, 1916), and even if it crucially relies on the comparative study of different languages, present or past. Furthermore, traditional grammars are essentially *normative* grammars. They are supposed to teach proper usage, which usually depends itself on the written form of the language. In contemporary linguistics, the term *grammar* is now used in a descriptive sense to refer to the implicit rules of spoken language in a way that allows for integration of different language registers.

1.2 THE SCIENCES OF LANGUAGE

Like mathematics or physics, modern linguistics is not so much a single, indivisible science, but rather a set of interrelated disciplines. Each of these disciplines corresponds to a different aspect of the study of language, and some of the sub-disciplines that make up contemporary linguistics have developed at different moments in its history. Five main sub-disciplines in the contemporary study of language are worth mentioning, presented here by their successive degrees of integration: *phonology, morphology, syntax, semantics*, and finally *pragmatics*.[6]

To give a much-simplified overview, one could say that while phonology deals with the sounds of language and how they are combined, morphology deals with the composition of words, syntax with the composition of sentences, semantics with the composition of meanings, and pragmatics with discourse and communication. In many respects, however, it is fair to say that syntax, understood as the study of the combination of units of language, is the cement that binds together each of these various subdisciplines (with the possible exception of pragmatics, although that is a topic of ongoing debate). As we will see in the following section, this view of syntax as playing a central and foundational role is inherited from the methodology advocated by Noam Chomsky in his seminal book, *Syntactic Structures*, which gave birth to generative grammar.[7] In fact, the methodology advocated by Chomsky has so radically

[6] Apart from these various fields, several cross-cutting disciplines are worth mentioning, such as historical linguistics, sociolinguistics, psycholinguistics (which includes neurolinguistics), and computational linguistics. Nevertheless, the five disciplines that we distinguished are the most fundamental areas of study, regardless of which methods are used or which of their aspects are focused on (research in historical linguistics, sociolinguistics, psycholinguistics, or computational linguistics is therefore distinguished according to whether it deals with phonology, or rather with syntax, etc.)

[7] More precisely, the great founding text of generative grammar is *The Logical Structure of Linguistic Theory*, which Chomsky wrote in 1955, but which was published 20 years later. *Syntactic Structures*, published in 1957, was the real starting point of the generative enterprise within the linguistics community. Some

transformed the way language is conceived that it seems difficult to outline the general purpose of linguistics without calling attention to its importance and heritage from the outset.

To illustrate each of the main aspects of the study of language just mentioned, consider an English sentence, such as

(1) John has talked to his mother.

The sentence is composed of six words. Each of these words corresponds to a sequence of sounds, the concatenation of which would be phonologically transcribed as the sequence /ǰɑn#hæz#tɒkt#tu#hɪz#məðər/.[8] The same six words, in different orders, make up different sequences. Some of these sequences are grammatical, such as for example (2), whereas others are not, such as (3) (marked by an asterisk, which we use to indicate that the sequence is incorrect):

(2) his mother has talked to John.

(3) *talked to John his has mother.

The theoretical goal of syntax, as Chomsky contributed to defining it in *Syntactic Structures*, is first and foremost to explain why certain combinations of the same words, such as (1) or (2), are grammatical, whereas others, such as (3), are not. More fundamentally, as we will see, its aim is to account for the structure of well-formed expressions so as to bring to light the mechanisms governing their interpretation. Clearly, the problem also arises from a theoretical point of view for all languages. Indeed, all languages are made up of discrete units, words, the combination of which produces sentences. Words, viewed as sequences of sounds, are in turn made up of discrete units, phonemes, of which a finite stock exists in each language.

Like words, phonemes obey specific combination rules in each language. To use an example from Halle (1978), a speaker of English presented with the following sequence of words:

(4) ptak thole hlad plast sram mgla vlas flitch dnom rtut

and who had never previously encountered these words, will admit that *thole, plast* and *flitch* are possible sequences of phonemes in English, whereas none of the others are (Halle, 1978). The theoretical task of phonology is to explain, more generally, why

of the ideas contained in this treatise were already present in Chomsky's master's thesis, entitled *The Morphophonemics of Modern Hebrew*.

[8] The # symbol indicates boundaries between words, the transcription relies on the American Phonetic Alphabet.

a speaker of English recognizes certain sequences of phonemes to be admissible, and others not.

Finally, in the same way, if we consider an English word such as "anticonstitutional," we know that it is composed of several more elementary units, and that it is analyzable as "anti-constitution-al," that is, as being composed of a prefix, a root, and a suffix. If we consider each of these units, which we will call *morphemes*, as so many elementary units, we can ask ourselves why the sequence "anti-constitution-al" is morphologically well-formed in English, whereas its permutations "constitution-anti-al" and "anti-al-constitution" are not. The aim of morphology, which can be defined in an analogous manner, is to account for the formation rules of words in each language, and for the constraints which govern the acceptability of given sequences of morphemes in opposition to others.

As Chomsky puts it in 1957,

> the fundamental aim in the linguistic analysis of a language L is to separate the grammatical sequences which are the sentences of L from the ungrammatical sequences which are not sentences of L and to study the structure of the grammatical sequences. (1957, p. 13)

Formulated in these terms, it becomes clear that an analogous problem arises with respect to the phonological analysis of a given language, as well as with respect to its morphological and syntactic analysis. Whether we take phonemes, morphemes, or words as basic units, the combinatorial problem which Chomsky raises is formally the same from one language to another. It is primarily in this sense that we can say the syntactic view of language, that is, the problem of discovering the fundamental combination laws of language units, presides over the linguistic enterprise as a whole.

Moreover, the notion of grammaticality that Chomsky uses is in a sense inseparable from the semantic notion of interpretability, even if Chomsky himself at first sought to dissociate the two notions. In *Syntactic Structures*, Chomsky is careful to distinguish between the notion of *grammatical* and that of *meaningful* or *significant*. To this day, the example Chomsky gives to illustrate such a difference remains among the most famous in the entire history of linguistics:

(5) Colorless green ideas sleep furiously.

For Chomsky, the point of this example is to suggest that a sentence like (5) is grammatical, even though it obviously expresses an incoherent proposition (we would no more say of an idea that it sleeps, than of a green thing that it is colorless, whereas "to sleep furiously" amounts to an oxymoron). Chomsky's gloss of the example is somewhat questionable, however, because a sentence like (5) remains potentially *interpretable*, so in this sense it carries a meaning (even if incoherent or poetic), unlike other combinations of the same words. Although the sentence violates certain rules of so-called *subcategorization* from a lexical point of view (Chomsky, 1965), the sentence does have a structure which is grammatical and which allows it to be interpretable in principle.[9]

---

[9] This issue is clarified by Chomsky in *Aspects of the Theory of Syntax*. It may be added that many utterances which would be considered deviant by competent speakers are interpretable. For example, "John like

More generally, we have seen that sentences (1) and (2) are each grammatical in the sense that they are liable to be produced and interpreted by competent speakers of English. We nonetheless observe that when the word order is changed from (1) to (2), new meanings are produced, since "John has talked to his mother" and "his mother has talked to John" are not synonymous sentences ("John has talked to his mother" implies that John has pronounced some words, but not necessarily that his mother has done the same.) The meaning of a sentence therefore depends on its syntactic structure. The purpose of natural language semantics is to account for the meaning of sentences. In particular, an adequate semantic theory for English must be able to explain such differences in meaning.

Historically, natural language semantics first developed in reaction to certain positions taken by Chomsky concerning the *autonomy of syntax* with respect to semantics,[10] mainly under the influence of Richard Montague's work (Montague 1968, 1970), and building on the logical work of Frege, Carnap and Tarski.[11] However, every semantics is actually the semantics of some syntax. In this sense, no semanticist would dream of giving an adequate theory of the meaning of sentences in a language without relying on a theory of syntax, as Montague was fully aware. The foundation of modern semantics is based on a principle first formulated by Frege, which was later elaborated by Montague, known as the *compositionality* of meanings, which states that the meaning of a complex sentence is a function of the meanings of the sentence's constituents. Adoption of the compositionality principle in semantics is consistent with the basic tenet of the generative approach to language which Chomsky initiated, according to which mastery of a language consists in mastery of the rules which allow one to generate and understand the sentences of that language in a systematic manner, without being obliged to memorize the overall meaning of arbitrary sequences of words.[12]

Before delving into detailed issues concerning semantics and the other disciplines previously mentioned, let us say a word about pragmatics. Pragmatics can be defined as a theory of discourse, as a theory of conversation, or as a theory of the use of language and communication in the broadest sense. For example, someone who utters a sentence like "John has talked to his mother" can mean that John has talked to his

---

potatoes" is no doubt interpretable in English, though deviant (missing 3rd singular form on the verb) and non-grammatical in this sense. In order to properly understand the scope of the example, it should be put in its immediate historical context, namely Quine's view according to which the grammaticality of an utterance supposedly derives from its meaningfulness. See for example Quine (1960). We are grateful to N. Chomsky for this remark.

[10] See Chomsky 1957, p. 17: "I think we are forced to conclude that grammar is autonomous and independent of meaning." The autonomy thesis, again, should be understood in the context of the debate with Quine's behaviorism as well as with structural linguistics, particularly regarding the idea that the notion of phonemic contrast should be backed by an independent notion of lexical meaning. In reality, however, the autonomy thesis does *not* imply, from Chomsky's point of view, that "the study of meaning, of reference and of language use is outside the scope of linguistics" (cf. Chomsky 1977 [1979], pp. 144–145, which clears up the misunderstanding).

[11] See especially Frege (1892), Tarksi (1933), Carnap (1947).

[12] On the definition of the principle of compositionality, cf. Partee (2004, chap. 7), Janssen (1997), Hodges (1998), and section 4.1.

mother about *the problem that he was faced with*, thereby referring to something which was supposedly present in the mind of his hearer. Similarly, when saying "John has talked to his mother," the speaker presupposes that John has a mother, that John is known to the hearer, and so on. A complete theory of the meaning of the sentence "John talked to his mother" must take into account the conversational context of the sentence. According to this point of view, pragmatics can be considered the theory of the contextual parameters which govern the use and interpretation of sentences (cf. for example the definition in Montague 1968 which refers to the so-called *indexical* elements of a sentence, such as "I," "tomorrow," whose reference varies according to the speaker, the time of the utterance, etc.). This definition, however, might seem equally appropriate to semantics *lato sensu*, understood as the theory of the truth-conditions of a given sentence.

A second conception, closer this time to the theory of speech acts (Austin, 1962; Searle, 1969), would view pragmatics as a theory of the elements which govern the *illocutionary force* of the utterance. For example, "John has talked to his mother [yeah, sure]," according to the context and intonation used in the sentence, could mean ironically that John has as a matter of fact *not* talked to his mother. More generally, and this time following Grice (1989), pragmatics can be defined as a theory of the interaction between general principles of rationality and interpretative constraints internal to grammar. Pragmatics in this sense aims to account for the inferences which allow us to detect the intentions of the speaker, including the parts of a sentence's meaning which go beyond its literal meaning and which contribute to its interpretation in context (the implied meanings, and all the indirect elements of meaning which Grice calls *implicatures*). Frequently, pragmatics has been depicted as "the wastebasket of semantics," conforming to the idea that any phenomenon relevant to meaning which cannot be explained strictly on the basis of the compositionality principle falls *de facto* under the scope of pragmatics. In fact, a precise definition of the purpose of pragmatics is a far more delicate matter, for it poses deep methodological problems concerning the boundary between meaning and linguistic usage.[13]

To give some idea of the interaction between syntax, semantics, and pragmatics, let us consider a classic example of ambiguity, such as

(6) John saw Mary with his binoculars.

It can mean that i) John saw Mary by looking through his binoculars, or that ii) John saw Mary, and saw her carrying or using his binoculars. It can be shown that these two readings correspond to structural ambiguities, or to distinct possible derivations

[13] For a historical and conceptual overview of the various definitions of pragmatics, cf. Korta and Perry (2006), who propose to distinguish between pragmatics in a narrow sense ("near-side pragmatics") and pragmatics in a broad sense ("far-side pragmatics"). They write: "*Near-side pragmatics* is concerned with the nature of certain facts that are relevant to determining what is said. *Far-side pragmatics* is focused on what happens *beyond saying*: what speech acts are performed *in* or *by* saying what is said, or *what* implicatures . . . are generated by saying what is said."

of the sentence (see below). This also explains why, in the previously quoted passage, Chomsky doesn't merely give syntax the goal of separating grammatical from ungrammatical word sequences, but also of accounting for the structure of the grammatical sequences.

These structural ambiguities, which are syntactic in nature, are correlated with distinct semantic interpretations. However, in the context of a discourse, the ambiguity of a sentence like (6) will not necessarily be present in the mind of the speaker or his interlocutor. Suppose that the speaker wishes to communicate meaning ii) to his interlocutor. He might do so without any effort on the hearer's part in a context in which it has just been said that John is a regular bird watcher who complains about his flatmate Mary, because she's constantly using cloths or equipment that belongs to him: *he saw her wearing his hat, he saw her wearing his gloves, he saw her with his binoculars* . . . Presented in a more abstract fashion, the purpose of pragmatics can thus be described as aiming to explain why a certain context favors a particular semantic choice rather than an alternate one.

To conclude this overview of the main areas of study in linguistics, it is important to note that if the boundaries between syntax and semantics, as well as between semantics and pragmatics, are sometimes difficult to define, the same is equally true with respect to phonology and morphology, or morphology and syntax. We will have an opportunity to return to this problem, but the reader should keep in mind that work in linguistics is largely carried out at the interface between several of the disciplines we've mentioned, just as the resolution of a given mathematical problem can call for methods that fall simultaneously within the purview of arithmetic, probability theory, and geometry.

2. Units and Rules: From Structural Linguistics to Generative Grammar

In the previous section, I tried to give a synoptic overview of the general purpose of linguistics and of its constituent sub-fields. The aim of this section will be to understand in fuller detail the goals of linguistics. I will consider the conflict between structuralist methodology, inherited from the work of Saussure, which dominated linguistics from the beginning of the twentieth century up until the 1950s, and the generative approach, initiated by Chomsky, starting at that time, which radically challenged the structuralist framework. The conflict between the structuralist and generative approaches is highly instructive. Even nowadays, it is little known or even ignored by the philosophical public. Yet I think this conflict offers a very concrete example of a scientific paradigm shift, in Kuhn's sense.

The main differences between the structuralist and generativist conceptions of language are the following: the structuralist tradition adopts a perspective which is primarily *analytic* and *descriptive;* it seeks to discover the elementary units of language (phonemes and morphemes), whereas generative grammar gives primacy to the search for rules rather than atoms, and therefore employs a more *synthetic* and *predictive* perspective.

Correlatively, structural linguistics views language above all as a corpus of utterances, whereas generative grammar views language primarily as a creative faculty, whose characteristic feature is recursion. Finally, as Chomsky was the first to formalize, the conception of syntax which underlies the Saussurean model considers language to be, in essence, a linear arrangement of discrete units, a view which is fundamentally inadequate. As we shall see, the Chomskyan conception of syntax is responsible for a profound renewal of the methods of phonology, which was the central domain of inquiry in structural linguistics until then.

2.1 THE SAUSSUREAN VIEW OF LANGUAGE

Until the 1950s, the work of reference for theoretical linguistics was the *Course in General Linguistics* by Ferdinand de Saussure, published in 1916 by his students following his death. The view of language propounded by Saussure was highly innovative for its time, particularly because Saussure offered an abstract perspective on language, and because he underlined the importance of a synchronic point of view (the study of language at a given time) relative to a diachronic point of view (the evolution of language over time). Also, he conferred a central role to phonology, the study of linguistic sounds, whose methods he partly defined, and which he distinguished from the study of meaning, which he called semiology.

A famous Saussurean distinction regarding other aspects of language, which is crucial to an understanding of the spirit of the structuralist approach, is the one between *langue* and *parole*. Saussure characterizes language (*langue*) as a "principle of classification," or again "a system of signs in which the only essential thing is the union of meaning and acoustic images." The abstract notion of *langue* is distinguished from that of *parole*, which is presented as the set of utterances which each individual produces in an autonomous fashion when she speaks.

In this characterization of language as a "system of signs," resides the basic principle of the structuralist approach, according to which language consists of a set of discrete meaningful units, *words* or *morphemes*, which, in turn, are composed of distinctive discrete units, *phonemes*. This distinction also corresponds to the principle of a so-called *double articulation* of language in morphemes and phonemes (cf. Benveniste, 1971a; Martinet, 1991).[14,15] Thus, two languages differ as much by their

[14] Cf. Benveniste (1971a, p. 104): "The word has an intermediary functional position that arises from its double nature. On the one hand, it breaks down into phonemic units, which are from the lower level; on the other, as a unit of meaning and together with other units of meaning, it enters into a unit of the level above."

[15] Martinet (1991) uses the term *moneme* instead of *morpheme*. Martinet is not strictly speaking a representative of structuralism, but of a different current, called functionalism. Like the structuralists, however, he explicitly claims to follow a Saussurean approach to language (cf. Complements C-1 to C-13 in Martinet, 1991, pp. 208–210), which states, clearly with hostility toward generative grammar, several tenets of functionalism having to do with the nature of language and the methodology of linguistics). The term functionalism applies to several currents apart from Martinet and his school, but is generally used by opposition with the so-called formalist conceptions (cf. Newmeyer, 1998, and section 4).

stock of phonemes as by their stock of morphemes (by which we mean a word or part of a word, a root, a suffix or a prefix, also referred to generically as *affixes* in morphology).

For example, in French there are nasal vowels, like the sounds [ã] and [õ] in *lent* and *long*, which don't exist in English and which native English speakers have trouble distinguishing when they learn French; conversely, in English there exists the initial consonant [θ] in *thing*, which is not a phoneme in French, and for which French people learning English often substitute a [s] ("*sing*"). In addition to these phonological differences, the same concept may typically be expressed by different words in different languages. Where a French person says *chien*, an English person says *dog*, and a German person says *Hund*. This observation is the foundation for the Saussurean principle of arbitrariness of the linguistic sign, which states that the same *signified* (or concept) can be expressed using different *signifiers* (sequences of phonemes; Saussure, 1916). The Saussurean principle is not surprising when we consider that the phonological repertories of two different language vary, but in principle we could imagine two languages with exactly the same phonemes, that would systematically employ different words to express the exact same concepts.[16]

According to Saussure, language can be viewed abstractly as a system of signs (words or morphemes), each of which can be analyzed as a sequence of phonemes. Sentences can be considered as concatenations of *signs* (sequences of words), and signs, in turn, as the concatenations of elementary linguistic *sounds* (sequences of phonemes). Furthermore, a noteworthy aspect of the organization of phonemes is that in each language they are finite in number, which means that the words of each language are constructed using a finite number of elementary sounds. Contemporary French, for example, consists of approximately 30 specific phonemes (whose exact number varies by a few units according to the various dialects and theories under consideration, Martinet, 1991; Dell, 1985). As discrete units which are finite in number, phonemes appear most likely to be the elementary units of language.[17] What this means is not so much that they are unanalyzable, as I will soon show in greater detail (see section 2.3), but rather that they are the basic linguistic units from which more complex units can be constituted.

A characteristic of phonemes that is central to understanding the structuralist conception of language is the fact that they are defined in a contrastive manner with respect to each other. In English, for example, the words *bet* and *get* have different meanings. These meanings are indecomposable from a morphological point of view. However, from a phonetic point of view, the words *bet* and *pet* can be analyzed

[16] This occurs, to a certain degree, in dialects internal to one language—think for example of Pig Latin with respect to ordinary English.

[17] See for example Jakobson's description of N. Troubetzkoy's work: "Among a series of brilliant discoveries we owe to him especially the first attempt at a phonological classification of the vowels and consequently a typology of the vocalic systems of the whole worlds. These are extremely far-reaching discoveries, and it is quite appropriate that they have been compared with the famous periodic table of chemical elements established by Mendeleev" (Jakobson 1978, p. 50).

as sequences of several sounds, which can be transcribed using the International Phonetic Alphabet as [bɛt] and [pɛt], respectively. These two sequences differ only in the phonetic contrast that exists between the voiceless initial occlusive consonant [p] in the one, and the voiced initial occlusive consonant [b] in the other. The contrast between these two sounds is not only acoustic or phonetic, but also has a functional value, inasmuch as the substitution of one sound for another, in the same environment (before the sequence of sounds [ɛt]), and in other similar environments (*pack* vs. *back, tap* vs. *tab*, etc.), correlates with a difference in meaning.

The two sounds [p] and [b] do not in themselves have semantic value of their own. Their semantic value is essentially contrastive, as Saussure emphasizes, who characterizes phonemes as "relative, oppositive and negative entities" (Saussure, 1916; Jakobson, 1978). In this perspective, the value of the phoneme [p] is oppositive and negative because it is defined only according to its difference from those other phonemes to which it stands in opposition. Its value is also relative, because a phoneme, in the structuralist view, may have contextual variants, called *allophones:* these variants are not contrastive and are generally predictable based on context (see the historical preamble in Steriade, 2007). In English, for example, the [p^h] sound in *pin* at the onset of a word must be distinguished from the unaspirated [p] in *spin* following the consonant [s] (more on this later). Despite this, as actually indicated in this case by the spelling of these two words, the two sounds are identified, despite the fact they are different, as combinatorial variants of the same phoneme /p/. As a result of its functional value, a phoneme is thus a more abstract entity than a phonetic sound.

Such a conception of phonemes' essentially relational and contrastive character sheds light on Saussure's view of language as a "classification system" or a "system of signs." In Saussure's view, each language has a corresponding class of specific phonemes, which it is phonology's task to inventory. According to this approach, morphemes themselves, and in particular words, also have an essentially contrastive and differential semantic value. For example, Saussure writes that "synonyms like French *redouter* 'dread,' *craindre* 'fear,' and *avoir peur* 'be afraid' have value only through their opposition: if *redouter* did not exist, all its content would go to its competitors" (Saussure, 1916, p. 160, 1959, p. 116). This purely differential conception of morphemes' value, by analogy with that of phonemes, was criticized relatively early on by certain proponents of structural linguistics, including Jakobson, who lucidly reproached Saussure for having "overhastily generalised this characterisation and sought to apply it to all linguistic entities" (Jakobson, 1978, p. 64). The example Jakobson gives is that of the morphological category of the plural, which is defined in relation and opposition to the singular, but whose value is positive according to him, namely "the designation of a plurality." This disagreement is of some importance, especially since it lay bare a limitation of the purely structural conception of meaning. But it does not call into question the core of the structuralist approach in either morphology or semantics. Thus, Jakobson admits that "Grammatical categories are relative entities, and their meanings are determined by the whole system of categories of a given language, and by the play of oppositions within this system" (1978, p. 64).

Consequently, according to the structuralist approach pioneered by Saussure, the task of language is simultaneously analytical and descriptive in essence. As Ruwet (1968, p. 50 [1973]) sums up the structuralist view of syntax:

> For Saussure (. . .) language is essentially an inventory, a taxonomy of elements. In that perspective, grammar seems to have to consist in a classification of minimal elements (corresponding to the morphemes of structuralists), of paradigmatic classes, and perhaps, of phrases.[18]

Moreover, viewed as a system of classification, language is considered by Saussure and his intellectual heirs to be a closed system, analogous in that respect to the repertory of phonemes. Admittedly, a linguist such as Martinet, for example, is careful to distinguish between the "closed list" of phonemes and the "open list" of morphemes in a language, insisting on the fact that each language creates new words (1991, p. 20).[19] Although this list of words is open, it remains essentially *finite*. As a result, Saussure's view of language as a system of signs gives to linguistics the task of describing vast corpora, and of detecting relevant systems of opposition within them.[20] As I shall show in what follows, this conception of language, despite its analytical virtues, neglects an essential dimension of language and grammar, namely its creative or productive aspect, which manifests itself, from a syntactic point of view, as recursion.

---

[18] "Pour Saussure (. . .) la langue est essentiellement un inventaire, une taxinomie d'éléments. Dans cette perspective, la grammaire semble devoir se ramener à une classification d'éléments minimaux (correspondant aux morphèmes des structuralistes), de classes paradigmatiques, et, peut-être, de syntagmes" (1968, p. 50).

[19] Furthermore, certain classes of morphemes are clearly closed, such as prepositions. Words which are routinely introduced into the language are *non-functional* or *non-logical* words, nouns, verbs, or adjectives.

[20] On the influence of the Saussurean conception of the notion of phoneme beyond linguistics, via Jakobson's teaching, in particular in anthropology, see for example the analysis of myths proposed by Lévi-Strauss. Lévi-Strauss writes in the prefaces to Jakobson's lessons (1978, p. xxii): "For we must always distinguish the meaning or meanings which a word has in the language from the mytheme which this word can denote in whole or in part. . . . In fact nobody, coming across 'sun' in a myth, would be able to say in advance just what its specific content, nature or functions were in that myth. Its meaning could only be identified from the relations of correlation and opposition in which it stands to other mythemes within this myth." Notice that Lévi-Strauss is careful to distinguish "the meaning or meanings which a word has in the language," that is, its meaning in ordinary language, from the meaning of the word in a particular discursive or symbolic context (myth, poem, song, etc.). A point worth noting is that the structuralist approach to the concept of symbolic meaning is fundamentally holistic and differential (the value of an item depends on its relationship to other items within a system or corpus). The conception of the meaning of terms in ordinary language that governs contemporary model-theoretic semantics is, by contrast, fundamentally atomistic and referential (the meaning of a word depends crucially on its reference in a given context), particularly in the idea that computation of the meaning of a sentence occurs "bottom-up" rather than "top-down" (under the principle of compositionality; see section 4).

2.2 LINGUISTIC PRODUCTIVITY, COMPETENCE, AND PERFORMANCE

In presenting Saussure's view of language and its legacy in the structuralist movement in this way, I did not attempt to give a precise and differentiated picture of structural linguistics itself, because in order to do so I would have had to consider historical issues too far removed from the methodological ones that are the focus of this chapter. However, some important points to remember from the preceding section are: recognition of the discrete nature of linguistic units, and the fact that, under the influence of phonological analysis, which largely dominated linguistics up until the 1950s because of its success, the linguistics enterprise initially focused on the segmentation and classification of linguistic units.

In this context, Chomsky's principal innovation lies in an observation which Chomsky credits to the nineteenth-century German grammarian Wilhelm von Humboldt, namely that "language makes an infinite use of finite means." Thus, at the beginning of *Syntactic Structures*, Chomsky defines language in an abstract manner as "a set (finite or infinite) of sentences, each finite in length and constructed out of a finite set of elements." (Chomsky, 1957, p. 13). Chomsky goes on to write

> All natural languages in their spoken or written form are languages in this sense, since each natural language has a finite number of phonemes (or letters in its alphabet) and each sentence is representable as a finite sequence of these phonemes (or letters), though there are infinitely many sentences.

Although the first part of this quote borrows directly from structuralist observations on language, the latter part introduces an essentially new element, namely consideration of the level of *sentences* (as opposed to the level of more basic units, such as phonemes or words). Most importantly, it introduces the observation that language potentially allows us to produce an infinite number of different sentences.[21] Contrary to phonemes, which are finite in number, the possible sentences of a given language are theoretically infinite in number. To illustrate this, consider the following set of six words *{Peter, John, man, is, believes, a, that}*, in which each word in turn uses a finite

[21] The stark oppositions drawn between generative linguistics and structural linguistics should be nuanced somewhat, especially as far as the immediate context in which generative grammar was born is concerned. For example, although reading Hockett (1954) allows us to measure the gap between pre-generative views and the generative model which was being developed contemporaneously by Chomsky, it is interesting to note that the article concludes by asserting several theses concerning the relationship between language description and prediction, which already point in the direction of the generative program. Hockett (1954, p. 232) thus writes: "the [grammatical] description must also be prescriptive, not of course in the Fidditch sense, but in the sense that by following the statements one must be able to generate any number of utterances in the language, above and beyond those observed in advance by the analyst—new utterances most, if not all, of which will pass the test of casual acceptance by a native speaker." For a more detailed overview of the work of the American school of linguistics in syntax during Chomsky's formative years, cf. especially Harris (1951). Ruwet (1967) offers a very well informed overview of the state of theoretical syntax in the early 1950s. Chomsky (1958) contains an illuminating discussion of the legacy of Harris in his own theory.

alphabet of letters. From this finite set of words, it is possible to construct an infinity of possible sentences in English:

John is a man
Peter believes that John is a man
John believes that Peter believes that John is a man
. . . .

In order to do so, it is only necessary to prefix each sentence that was previously obtained in the hierarchy with the sequence "John believes that" or "Pierre believes that." Although it is not possible to pronounce all of these sentences (a lifetime would not suffice), there is no doubt that each of them is grammatical, and that, in principle, they are all capable of being understood. For Chomsky, therefore, the real challenge facing linguistics is no longer to merely inventory the basic units of language, but rather, on the contrary, to account for the creative nature of language, and for the fact that on the basis of a finite inventory, competent users of a given language are able to produce and understand a theoretically infinite number of sentences.

The limited language just described enables the production of an infinite number of grammatical sentences on a finite basis. Technically, this means that the language in question is recursive, that is, that it contains one or more *rules* for constructing a first sentence, which can then be reapplied to that sentence in order to produce a new sentence. The notion of recursion, which was originally studied by logicians and computability theorists starting in the 1930s, is at the heart of the Chomskyan conception of langage.[22] Indeed, the essential characteristic of language, according to Chomsky, is the productive nature of syntax, that is to say the fact that there is no "longest sentence," or in other words, that it is not possible to non-arbitrarily assign a limit to the length of sentences in any given language (Hauser, Chomsky and Fitch, 2002).

In addition, the notion of recursion constitutes a link between the notion of language as a faculty and the notion of language as a "set of sentences" as Chomsky first defined it in *Syntactic Structures*. Viewed extensionally as a set of grammatical sentences, a language is what Chomsky, in his subsequent writings, calls an E-language, that is, an "externalised" or "extensional" language. This is the set of grammatical sentences that are the product of the speaker's language faculty, strictly understood. However, the language faculty itself is associated with what Chomsky calls the speaker's I-language (which stands for "internal," "individual" or "intensional" language), that is, the set of rules and principles which allow the speaker to produce and understand sentences of the language she speaks (see Chomsky and Lasnik, 1995), without the speaker necessarily being aware of those rules.

[22] It is probably because Chomsky was aware of the possibility of studying formal languages using mathematical methods that he became interested in extending this approach to natural language. On Chomsky's work in computational linguistics, cf. for example, Chomsky (1956), Chomsky (1962), Chomsky (1963), and Chomsky and Miller (1963).

By emphasizing the fact that human languages allow for the construction of a potentially infinite set of sentences, and correspondingly that recursive procedures for generating sentences exist, Chomsky overturns another aspect of Saussure's conception of language, which has to do with language learning. In his *Course in General Linguistics*, Saussure asserts that "Language is not a function of the speaker; it is a product that is passively assimilated by the individual" (1916, p. 30; 1959, p. 14). According to Saussure, language is thus fundamentally a social rather than an individual entity. However, Saussure acknowledges that a sentence "is the essence of speech," which, according to Saussure, is the level where the speaking subject's freedom is manifested (1916, p. 31), as in conjunction with the freedom of grammatical combination (1916, p. 172). But due to this fact, as Chomsky (1968, p. 37) underscores, Saussure tends to relegate syntax to a realm outside of linguistics, whose primary object of study is defined as that of *langue* rather than *parole*. To a large extent, the view that language is "registered passively" paves the way for a behaviorist conception of language learning, which Chomsky set out to refute about the time that he published *Syntactic Structures*, most famously in a review of a book by the American psychologist B. Skinner, *Verbal Behavior* (Chomsky, 1959).

Indeed, for Chomsky, what characterizes language, contrary to what behaviorists maintain, is the fact that it is free from control by external stimuli, and that it cannot be reduced to the association of sound patterns with characteristic stimuli. One of the most famous, and also most controversial, arguments Chomsky advanced in this regard is the so-called *poverty of the stimulus* argument (Chomsky, 1980), which states that no child could possibly learn a language simply by repeating previously heard sentences or patterns. One of the reasons Chomsky gives has to do with the *productive* nature of language. Children are rapidly capable of producing as well as understanding sentences which they have never heard before. Of course, it is because children hear sentences in English that they come to speak English rather than Japanese, and it is in this sense that Saussure can say that language is not "a function of the speaking subject." Nonetheless, for Chomsky, exposure to verbal stimuli largely under-determines the inferences thanks to which, within a few years, children become capable of producing sentences that they have never heard before.[23]

As a result, in the Chomskyan perspective, the fundamental purpose of linguistics is not what was ascribed to it by Saussure, that is, description of the units of a language and the systems of opposition relevant to it. Of course, by operating on the assumption that language is composed of discrete units, Chomsky assimilated the structuralist legacy of the preceding generation. But the task which Chomsky assigned to linguistics was no longer primarily that of analyzing and segmenting linguistic data in order to arrive at basic units. If the job of segmentation and analysis remains a necessary one, as I shall show in various examples, it becomes subordinate to the task of

[23] For a historical synthesis and a detailed critical evaluation of this and other arguments known as "poverty of the stimulus," see Pullum and Scholz (2002).

searching for those rules which govern the organization of the units themselves, and through which speakers express their linguistic creativity.

In this regard, as Ruwet (1968) rightly notes, Chomsky was careful to distinguish "creativity which changes the rules" from "creativity governed by rules." The first type of creativity is linked to the *performance* of subjects when they speak, and to the gradual changes that they are liable to bring about in a given language (on the lexical, phonological or syntactic levels). The second type of creativity is related to subjects' grammatical competence, that is, the mastery which they possess, without necessarily being aware of it, of recursive procedures that enable them to produce and understand an infinite number of possible sentences on the basis of a finite number of morphological and phonological units. Moreover, the distinction between competence and performance, introduced by Chomsky (1963, 1965), has a central methodological significance, in that, from Chomsky's point of view, the theory of grammar which a linguist aims to elaborate should be a theory of subjects' competence (the internal grammar of a subject), and not of subjects' performance (all the utterances they actually produce verbally). A further reason for this is the idea that there is "noise" associated with speakers' performance, noise which can be due to subjects' occasional fatigue and sometimes leads to errors, as well as noise resulting from any kind of concrete discursive situation, which may cause a sentence to be incomplete, interrupted, and so on. A theory of grammatical competence is therefore a theory that abstracts away from such noise, in keeping with the idea that children themselves, when they learn a language are capable, without realizing it, of similarly separating generative rules on the one hand from irregularities deriving from language use on the other.

2.3 A NEW APPROACH TO SYNTAX AND PHONOLOGY

To underscore the novelty of the Chomskyan approach to language, it is useful to note the interest that it evoked among certain members of the structuralist movement, particularly the French. In 1962, Benveniste, in an article on levels of linguistic analysis, concluded that the sentential level is radically different from that of phonemes and morphemes:

> Phonemes, morphemes, and words (lexemes) can be counted; there is a finite number of them. Not so with sentences. Phonemes, morphemes, and words (lexemes) have a distribution at their respective levels and a use at higher levels. Sentences have neither distribution nor use. An inventory of the uses of a word might have no end; an inventory of the uses of a sentence could not even be begun. (Benveniste, 1971a, p. 109)

The conclusion of Benveniste's article largely goes against Saussure's approach, since Benveniste ends his article with a Latin expression which translates as: "there is nothing in language which is not first in discourse." Clearly, the Benvenistian notion of discourse ("discours") is related to Saussure's concept of speech ("parole").[24] Nevertheless, as Ruwet (1967, p. 165) notes, structuralists relatively seldom make explicit the idea that linguistic creativity is governed by rules. In order to explain this lacuna, I find it useful to now describe two aspects of generative grammar responsible for its dissociation from structural linguistics. The first aspect concerns the conception of sentences' grammatical structure. The second concerns the definition of the notion of a phoneme. In both cases, Chomsky made profound and in some ways decisive objections, which it is helpful to consider jointly.

2.3.1 The Inadequacy of Finite State Grammars

I will first consider the structure of sentences. One of the principles defended by Saussure in the *Course in General Linguistics* is the principle of the "linear nature of the signifier" (1959, p.70; 1916, p. 103), by which Saussure intends to say that words, like sentences, are concatenations of signs along a linear temporal axis (the time it takes to pronounce a word or sentence). A sentence like "Peter observes a very old cat" can be seen as the concatenation of the signs: *Peter—observes—a—very—old—cat*. A second principle put forth by Saussure is the opposition between "syntagmatic relations" and "associative relations" (or paradigmatic relations) within a word or sentence. An example given by Saussure is the word *défaire* in French [undo] (Saussure 1916, p. 178). From a syntagmatic point of view, the word is a concatenation or combination of a prefix, *dé-* [un], and a root, *faire* [do]. From an associative perspective, however, each morpheme is in competition with other possible morphemes. Instead of the prefix *dé-*, there could be *re-* or *contre-*, and we would have *refaire* (redo), *contrefaire* (counterfeit). Conversely, one could substitute other verbs for the root *faire*, such as *coller* (glue), *coudre* (stitch) so as to obtain *décoller* (unglue), *découdre* (unstitch), and so on.

Similarly, each sentence can be regarded as a combination of units along the syntagmatic axis, each of which can be subject to certain substitutions along the paradigmatic axis. As an example of some possible substitutions along the paradigmatic axis, take the sentence "Pierre observes a very old cat"

Pierre	*observes*	*a*	*very*	*old*	*cat*
Mary	*eats*	*the*	*rather*	*fat*	*chicken*
Susan	*paints*	. . .	. . .	*beautiful*	*dog*
. . .	. . .	. . .	. . .	. . .	. . .

[24] Benveniste's phrase, which is modeled on that of Locke, is "*nihil est in lingua quod non prius fuerit in oratione.*" This remark must be qualified, as Benveniste recognizes, "But we must realize that in the syntagm there is no clear-cut boundary between the language fact, which is a sign of collective usage, and the fact that belongs to speaking and depends on individual freedom" (1959, p. 125). But Saussure concludes from this that this lack of clear boundaries just makes the job of linguistic classification more complicated, and not that it would actually make it impossible to draw up such an inventory in the case of sentences.

The Saussurean opposition between combination along a syntagmatic axis and selection along a paradigmatic axis can also be found in Jakobson, who suggested linking it to various linguistic disorders in aphasic patients (which Jakobson called "the contiguity disorder" and "the similarity disorder"; Jakobson, 1956). More generally, it has had an impact even outside theoretical linguistics, especially in literary theory, but also apparently in the teaching of foreign languages.

However, Chomsky, in an early chapter of *Syntactic Structures*, proposed a more abstract version of this syntactic model, calling it a *finite state grammar*, in order to show that the grammar of a language like English (or French) could not be adequately described this way. Chomsky's idea was to describe the grammar underlying the linear model as a sentence-production system, an *automaton* comprising a finite set of states, that would move from an initial state to a final state and produce a word with each transition from one state to the next. An equivalent manner of representing some of the possible combinations of sentences just given is by using the diagram in Figure 1. The diagram represents an automaton with six states, in which the state *qo* is input state, and *q5* is output state:

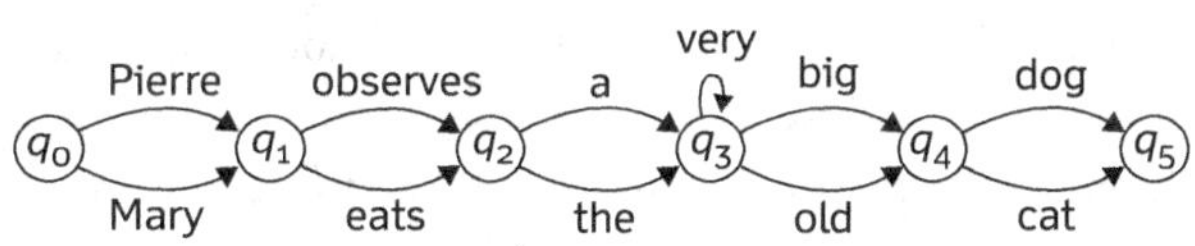

FIGURE 1 A finite state automaton

The grammar described by the automaton is not entirely trivial, since it can generate an infinite number of possible sentences based on a finite set of words—for example "Peter observes the very old dog," "Peter eats the very very fat cat," and so on—thanks to a loop that allows the automaton to produce the word *very* and then return to the same state.

At first glance, a finite state grammar of this type provides a plausible description of the type of procedure that allows a speaker to produce sentences. Nonetheless, it is possible to demonstrate mathematically, as Chomsky has done, that a finite state grammar does not allow for the production of all sentences of English and only those sentences. To show this, Chomsky first proved that a very simple language such as the formal language constructed from the alphabet {*a*, *b*} (containing only those two words), which consists of all sequences of letters of the form $a^n b^n$ (a sequence of *a*s followed by a sequence of *b*s of the same length, such as *ab*, *aabb*, *aaabbb*, etc.) cannot be generated by a finite automaton. On this basis, the argument Chomsky gave essentially amounts to showing that in the case of English or French, there are certain structures of dependency between syntactic constituents that obey this very pattern.[25] In English, for example, all sentences of the form "wolfs ate," "wolfs wolfs ate ate" (wolves that wolves ate have eaten), and so on. A finite state grammar cannot generate

[25] The target of Chomsky's demonstration was English, but it was meant to apply to any language which shared with English the relevant pattern of syntactic dependency (called *center embedding*).

the fragment of English which contains all sentences of this type, and nothing but those sentences.[26]

More fundamentally, the argument presented by Chomsky in *Syntactic Structures* depends on the mechanism of structure *embedding*, omnipresent in all languages, and analogous to so-called palindrome or "mirror" languages, which also fall outside the scope of finite state grammars (for example, based on the alphabet {*a, b*}, the language containing all sequences of the form *aa, bb, abba, aabbaa*, etc.). Consider for example the sentence-schema "the man who says that S is standing," in which the verb phrase "is standing" agrees with the subject "man." In this sentence, it is possible to substitute for S a conditional sentence of the form "if A then B." Within this conditional sentence, one can also embed, in place of A, a conjunction of the form "P and Q," and so on. Thus, a sentence such as "the man$_1$ who says that if$_2$ Peter comes$_3$ or Mary leaves$_3$, then$_2$ Julie will be happy is standing$_1$" follows a pattern of mirror-like dependencies of this type (which I summarily represent here using indices, which mark the syntactic relationships between the underlined expressions).[27]

In general, what Chomsky's argument shows, is that a finite state grammar does not adequately account for the relations of syntactic dependency between certain constituents. In *Syntactic Structures*, Chomsky therefore opposed this model to a second model, that of so-called *phrase structure grammars*, or *constituency grammars*.[28] This model, it is important to point out, is itself a direct result of the work of American linguists on so-called *immediate constituent analysis*, outlined by Bloomfield, and developed in various ways by Wells, Harris, Bloch, Nida, and Hockett in the 1940s and 1950s (see Ruwet, 1967). Unlike its predecessor, this model describes the hierarchical structure of a sentence by decomposing each of its immediate constituents in turn into other constituents (phrases, which themselves decompose into phrases). As the tree in Figure 2 shows, the syntactic structure of a sentence such as "Peter observes a very old dog" is not linear in this case, but treelike. If the tree-representation is due to Chomsky, the concept of hierarchical organization of sentences is not novel, and should be credited to the linguists who preceded him. However, Chomsky's originality resides in the fact that he proposed a unified, abstract framework for the representation

[26] The argument outlined here, though basically correct, is not conclusive on one point. In reality, it is not enough to show that a fragment L′ of a language L is not describable by a finite automaton to show that any language L itself is not. However, it suffices to show that L′ can be obtained as the intersection of L with a language L* which can be generated by a finite automaton. If L could be generated by a finite automaton, then the intersection L′ and L-L* should be possible to generate by a finite automaton. For a detailed demonstration of the fact that English is not describable by a finite state grammar, see Partee et al. (1990).

[27] Note that the same is fundamentally also true, in fact of, a sentence of the form *aabb* such as "wolfs1 wolfs2 ate2 ate1," considered this time in terms of the structural dependencies between subject and verb. From this point of view, the argument from embedding proposed by Chomsky goes beyond the inability to *weakly* generate all sequences of the form a^nb^n. One point on which N. Chomsky calls my attention (personal communication) is also that languages of the type a^nb^n can be generated by finite automata *with counters*, unlike embedded structures.

[28] Chomsky uses the term *phrase structure grammars* (1957), which is the most common in English, and occasionally that of constituent-structure grammars (1963, p. 292).

of such grammars, in the form of *rewriting systems*, and that he showed the irreducibility of the phrase-structure grammar model to the finite-state automaton model. In so doing, Chomsky helped to generalize and refine the grammatical models outlined by his predecessors, by demonstrating the equivalence of models which were previously presented as distinct (see, e.g., Hockett, 1954), or on the contrary, by establishing the principled mutual irreducibility of models that at first seemed similar (see, e.g., Hockett, 1955). More fundamentally, the framework which Chomsky elaborated made it possible for him to examine the expressive power of different grammars comparatively, according to the form of their rewriting rules for intra-sentential constituents.[29]

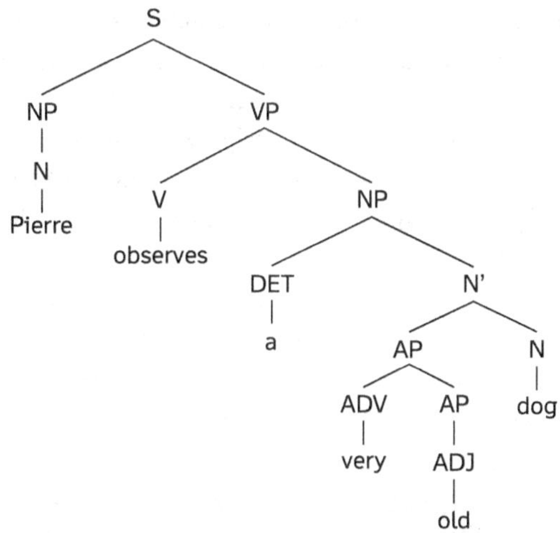

FIGURE 2 A derivation tree based on a context-free grammar

Consider the grammar underlying the derivation tree in Figure 2, which presents a particular example of a rewriting system (in this case, a context-free grammar). The system in question comprises several rewriting rules of the form: X → Y + Z, where X and Y are so-called intermediate symbols (the grammatical categories in the diagram), and Z is either an intermediate symbol or a word in the lexicon (Y may be null, in which case the rule can be written X → Z).[30] For example, the rule VP → V + NP states that a verb phrase is composed of a verb and a noun phrase. The grammar is once again recursive, since the rule AP → ADV + AP implies that an adjectival phrase may contain

[29] Strictly speaking, the finite state automaton of Figure 1, viewed as a rewriting system, also produces a tree for the sentences it generates, but the structure of these trees is trivial: the fact that a node dominates another only means that the word associated with the first comes before the word associated with the second in the sentence.

[30] We use the nomenclature of International Syntax. S is for the category of sentence ("Sentence"), VP for the 'Verb Phrase," NP for the "Noun Phrase," AP for adjectival phrase, and so forth (the term *phrase* refers to a level of sentential grammatical constituency). The reader may be surprised to find a category N′ between N and NP: the idea is that the phrase "very old dog" is the component of a broader phrase than the noun, but that it needs a determiner in order to form a complete noun phrase.

an adjectival phrase as a constituent, which in this case accounts for the generation of phrases like "very very old dog." Finally, for each basic category, such as ADJ, N, DET or V in this example, in principle one finds the specification of all the terms in the lexicon which belong to that category. For example, we would have ADJ → *old, fat, beautiful*.

The model of phrase structure grammars is more adequate than that of finite state grammars in three main respects. First, as Chomsky has shown, phrase structure grammars are strictly more expressive than finite state grammars. In particular, a context-free grammar allows for the derivation of all sequences of the form $a^n b^n$, and is thus immediately a better candidate for representing the embedded syntactic structures mentioned earlier. Furthermore, as can be seen immediately by comparing Figures 1 and 2, a phrase structure grammar takes into account the distribution of words in the lexicon into different grammatical categories, whereas the model in Figure 1 indiscriminately puts all the terms in the lexicon on equal footing. The contrast between the two models brings to the fore the fact that, underlying the linear order of the words in a sentence, as we write it from left to right, our understanding of language depends on a deeper level of representation. Finally, the derivation shown in Figure 2 lays bare basic grammatical rules, in this case the rules for composition or generation of sentences. For example, this derivation contains a rule governing the structure of a verbal group, which consists of a noun and a noun phrase, as well as a rule governing the structure of a sentence, which consists of a noun phrase and a verb phrase.

As a result, the phrase structure grammar-model is also more appropriate in another respect, that which concerns language learning. The finite state grammar-model would be plausible if we learned language by committing to memory the sentences we hear, in order to repeat them verbatim. However, the finite state model also purports to account for the fact that we make substitutions based on heard lexical patterns, in order to produce new sentences. But as it happens, nothing in the model in Figure 1 explains why we can substitute the word *the* for *a* in such a sentence, rather than any other word. In the case of a phrase structure grammar, what explains that *a* and *the* can occur in the same position, is the fact that they belong to the same grammatical category, unlike other words in the lexicon. So if children learn language on the basis of heard patterns, they must at least make inferences that allow them to operate adequate substitutions, or otherwise be able to infer the underlying grammatical structure of the sentences they hear, which renders the model of phrase structure grammars immediately superior.

It should be noted that for Chomsky, the model of phrase structure grammars also remains inadequate in several respects, particularly because it cannot account for certain specific dependencies between constituents that are distant from each other in a sentence, or can only do so by introducing a great deal of redundancy in the rules. It is this inadequacy which is responsible for Chomsky's introduction of a third model, the *transformational* model, to which I shall return.[31] Nevertheless,

[31] Chomsky's arguments concerning the limits of phrase structure grammars are presented in chapter 5 of *Syntactic Structures*. See also Chomsky and Miller (1963, pp. 296 ff.) The notion of transformation comes from the work of Z. Harris, cf. Harris (1957) and Chomsky (1955, 1958). One of the first applications of

it is important to keep in mind that the phrase structure model has in common with the more complex models that Chomsky subsequently considers, the existence of a clear distinction between the linear order of words heard or pronounced and the grammatical constituent-structure which underlies it. To any reader with even a passing knowledge of traditional grammatical analysis, the superiority of the derivation given in Figure 2 over that in Figure 1 should come as no surprise. But it is important to realize that it refutes in a precise manner a naive conception of the structure of language.

At the time when Chomsky published *Syntactic Structures* and demonstrated the inadequacy of the finite state model, his intended target was not so much the Saussurean conception of syntax, which was not very fully articulated even by Saussure himself, as much as a model inspired by the mathematical theory of communication, developed in particular by Shannon in the 1940s, in which several postwar linguists (including Jakobson and Hockett) had placed great hopes.[32] It could therefore be argued that Saussure's examples of morphological opposition, such as *dé-faire* and *contre-faire*, are compatible with a correct view of the constituent-structure of the lexicon, and do not necessarily imply a general conception of syntax similar to the one underlying Figure 1. I am willing to grant this point (see also Ruwet 1967, p. 165). But it is important to realize that Saussure's distinction between syntagmatic and paradigmatic axes, and his emphasis on the linear nature of the signifier, naively generalized to sentential structure, led to an inadequate vision of language. In refuting such a conception, Chomsky made clear that sentences are much more than mere concatenations of words or elementary units.

2.3.2 Structural Phonology and Generative Phonology

In Chomsky's view, language is first and foremost a "system of rules," rather than simply a "system of elements" (Chomsky and Halle 1965, p. 459).[33] As I have just shown, the notion of rule first appears with respect to syntax, in the very idea of the derivation of a sentence from rewriting rules. Another example of the primacy of rules over elements is provided by phonology, and the renewal of structural phonology within

the concept of transformation by Chomsky was to the case of the auxiliaries *have* and *be* and the dependency between auxiliary and past-tense forms of the verb in English (Chomsky 1957, pp. 39 ff.) See Rivenc and Sandu (2009, pp. 69–70) for a brief presentation in French, and Lasnik (2000) for more details. Let us add that other approaches have been proposed to deal with long distance dependencies; for instance, the model of generalized phrase structure grammars (GSPG and HPSG), which appeal to principles of sub-categorization in the rewrite rules. For a discussion of these grammars, and a presentation of the history of syntactic models since 1957, see Sag, Wasow, and Bender (2003).

[32] The case of Jakobson is reported by M. Halle (personal communication) in particular, see for example Jakobson (1952). Chomsky cites precisely Hockett's (1955) phonological model as an adaptation of Shannon's model.

[33] Chomsky and Halle (1965, p. 458) thus write: "We assume, with no further discussion, the distinction of *langue-parole* (except that we do not accept the Saussurean limitation of *langue* to 'system of elements,' but regard it also as a system of rules."

the generative approach, in the work of Chomsky and Halle starting in the late 1950s and throughout the 1960s.

As I emphasized, up until 1950 phonology was the flagship discipline of theoretical linguistics. Pre-war phonologists devoted a large part of their efforts to drawing up inventories of phonemes in particular languages. The basic method employed for doing so, as I previously mentioned, consisted in establishing contrasts or *minimal pairs* in order to isolate phonemes, a method which is also known as the *commutation test* (see also section 3.3.1). In English, sounds [p] and [b] in "pin" and "bin" stand in this relation of contrastive opposition. Furthermore, the same phoneme can be realized differently at the phonetic level depending on the environment in which it appears. Thus, the unaspirated [p] in *spin* in English is in fact distinct from the aspirated [p^h] sound in *pin*. However, these two sounds occur in *complementary distribution*, that is to say, never in the same environments, the aspirated [p^h] appears at the onset of an unstressed syllable, whereas the unaspirated [p] occurs in other environments. The opposition between these two sounds is therefore never contrastive in English: there are not two different words [spin] and [Sphin] for example, or [pin] and [phin], with different meanings. However, in other languages the opposition between these sounds is contrastive, for example, in Bengali (Radford et al., 1999). In the case of English, according to Bloomfield's traditional approach, the sounds [p] and [p^h] are considered to be two *allophones* of the same phoneme, written /p/, and a phoneme is defined as a class of sounds or phonetic segments in complementary distribution.

In order to properly understand the distinction between phonemes and sounds, as well as Chomsky and Halle's criticism of structural phonology, it is important to recall Troubetzkoy and Jakobson's conception of the nature of linguistic sounds. One of the significant contributions of their approach was to consider the sounds of language as sets of distinctive articulatory features, rather than as indecomposable units. According to this approach, the sound of English written [p] in *spin* is actually an abbreviation used to designate the following matrix of articulatory features [bilabial, plosive, unvoiced, oral, non-aspirated, . . .], whereas [b] abbreviates the matrix [bilabial, plosive, voiced, oral, non-aspirated, . . .]. Thus, the sounds [p] and [b] differ mainly in the feature voiced vs. unvoiced. Another of the key theses put forward by Troubetzkoy and Jakobson in phonology is also the idea that the sounds of all possible spoken languages are distributed within a space of common articulatory features, a universal set of features. From this perspective, a linguistic sound is much more than merely a sound that is heard. Instead, it must be conceived of as a set of articulatory or motor instructions, defined on the basis of a universal set of elementary articulatory gestures.

However, one of the difficulties with the Bloomfieldian definition of the notion of phoneme is that it is too broad. For example, sounds [t] and [p^h] are also in complementary distribution, but one would hesitate to say that they are combinatorial variants of the same phoneme (Halle, 1959). Another difficulty, originally highlighted by Bloch and more extensively discussed by Chomsky (1964), concerns the fact that the commutation test itself tells in favor of the existence of phonemes that are not yet accepted as such. Thus, in American English the word *writer* is commonly pronounced [rayDər], which means that the sound [t] in *write* is pronounced [D], a sound close to a [d], which is called *flap*. The word *rider* is pronounced [ra: yDər], the difference in pronunciation is based on the lengthening of the vowel [a] which is pronounced [a:], with the [d] also being transformed into [D]. If the pair *writer-rider* is considered from a phonetic point of view, one should therefore conclude that this is a contrastive difference, and that the segments [a] and [a:] are two different phonemes of English. This poses a problem, however, as soon as one considers that the verbs *write* and *ride* from which *writer* and *rider* respectively derive are given the phonological representations /rayt/ and /Rayd/ respectively: in this case, the contrastive difference concerns phonemes /d/ and /t/ and does not involve elongation of the vowel.

A radical way of viewing the problem is to question the relevance of the Bloomfieldian notion of phoneme. Thus, in the approach advocated by Chomsky and Halle, there are basically two levels of representation in phonology: a morpho-phonological (or morpho-phonemic) level of representation, which takes into account both the sound and morphological structure of words, as well as perhaps the structure of the overall context of the sentence, and another level of phonetic representation derived from the first one.[34] The task of generative phonology is to connect these two levels of representation by using derivation rules: starting from the phonological structure of a sentence, the goal is to derive its actual phonetic pronunciation, in the same way that the word-order of a sentence is derived in a bottom-up manner from rewriting rules. In so doing, Chomsky and Halle dispute the existence of an intermediary "phonemic" level of representation in between the level of phonological representations which reflect morphology, and the level of phonetic representations which are derived from them using syntactic rules.[35]

In order to fully understand what is at stake, consider how Chomsky proposes to account for the pronunciation of the words *writer* and *rider* in American

[34] In this respect, as they themselves point out, Chomsky and Halle follow an approach to phonology pioneered by Sapir, who is still regarded today as one of the most lucid and brilliant linguists of the period between the wars. The distinction between two levels of representation, phonological and phonetic, connected by derivation rules, is also fully consistent with the distinction Chomsky draws at the same time in syntax, thanks to the notion of transformation, between deep structure and surface structure (see Chomsky 1968, chap. 2). For more details on the ins and outs of generative phonology, see Anderson (1985) and Kenstowicz (2004).

[35] Bloomfield is also the author of an article entitled *Menomini Morphophonemics* in which he anticipates the generative approach, by emphasizing the order of derivation rules. See Bromberger and Halle (1989), who report that Chomsky was not aware of this article when writing his master's thesis in 1951 (which Chomsky confirms, personal communication). This factual point was controversially contested by Encrevé (1997), who emphasizes the continuity between Bloomfieldian phonology and the subsequent contributions of generative phonology (although, as Encrevé admits, Halle and Chomsky systematically give credit to Bloomfield for the originality of his 1939 article, as early as in their 1960s joint work in phonology, but precisely to emphasize its heterogeneity with other work by Bloomfield on the topic). In any event, an important element of Bromberger and Halle's testimony concerns the fact that after World War II, phonology was taught in the United States following a tripartite division between morphophonemic, phonemic and phonetic levels. Even if, as Encrevé claims, Chomsky were to have had knowledge of Bloomfield's treatise in the early 1950s (allegation that Chomsky explicitly denies, personal communication), he and Halle have drawn the consequences of problems that arose in canonical Bloomfieldian analysis in a way that revolutionized structural phonology, by refuting the relevance of the phonemic level.

English. The proposed derivation involves two rules (Chomsky, 1964, reprinted in Kenstowicz, 2004):

(i) Lengthening rule: [a] becomes [a:] before a voiced obstruent consonant.
(ii) Flapping rule: [t] and [d] become [D] between two vowels, the first stressed and the second unstressed.

TABLE 1

Two phonological derivations, *write-writer* vs. *ride-rider*

/rayt/	/rajt+ər/	/rayd/	/rayd+ər/	morpho-phonological representation
—	—	ra:yd	ra:ydər	lengthening rule
—	rayDər	—	ra:yDər	flapping rule
[rayt]	[rayDər]	[ra:yd]	[ra:yDər]	phonetic representation
write	writer	ride	rider	written form

As can be seen, the statement of the rules refers not only to the distinctive features of the units postulated in the phonological representation, but also to prosodic information: for example, the flapping rule refers to word stress; the lengthening rule does not apply to *writer* at the first stage because /t/ is not a voiced consonant, by contrast with /d/. Another crucial point in Halle and Chomsky's theory of the notion of derivation in phonology, which I will return to later, is that the order of the rules is also decisive. In principle, rules such as (i) and (ii) are supposed to apply to the language under consideration with full generality, and so reversing their order should lead to different predictions about American English pronunciation.

From this example, Chomsky draws two lessons for linguistic theory in general. The first, which is well-known among phonologists, but less well-known among philosophers of science, concerns the relativity of the concept of a minimal pair, which is an indispensable tool for the production of linguistic data. As the *writer/rider* case shows, the contrast in meaning between the two words is actually derived rather than primitive in the generative approach, contrary to what would be the case in a conventional structural analysis. For Chomsky, it follows that the concept of a minimal pair is relative, and depends not only on the phonetic level, but also on phonological analysis, which becomes part of syntactic analysis in a broader sense. Furthermore, Chomsky argues against the structuralist approach that

> it seems that no inventory (not even that of phonemes) can be determined without reference to the principles by which sentences are constructed in the language.

Thus, Chomsky argues for the primacy of syntax at all levels of linguistic analysis, including that which up until then would have seemed the most independent from its successors, that is, the level of phonology. Another important point is that by relinquishing a definition of the concept of phoneme based on the notion of contrastive alternation between sound segments, Chomsky and Halle proposed a unified account of contrastive (of the [p] vs [b] kind) and non-contrastive (of the [p] vs. [p^h] kind) alternations. Thus, Chomsky and Halle help reduce the gap which appeared considerable, within the structuralist-inherited perspective, between phonology and phonetics.

2.4 THE CHOMSKYAN REVOLUTION

At the beginning of this section, I called attention to the unprecedented impact of the Chomskyan view of language, starting with the publication of *Syntactic Structures*. Several linguists in the 1960s did not hesitate to refer to a "Chomskyan revolution" to characterize the importance of Chomsky's contribution to the study of linguistics. Before addressing issues pertaining to the methodology of linguistics as a whole, it seems useful to conclude this section with some more general considerations from philosophy of science concerning the schism caused by the Chomskyan view of language with respect to the structuralist era that preceded it, and whether or not employing the term of "revolution" is justified.

As previously explained, the Chomskyan view of language profoundly modified the structuralist view in three ways: language is seen as a cognitive faculty and as a system of rules rather than as a corpus of utterances or as a system of elements; work in linguistics is to be carried out in a synthetic and predictive perspective, rather than in a merely descriptive and analytic one; this synthetic and predictive perspective is closely tied to the methodology adopted by Chomsky, which consists, first of all, in drawing a parallel between the grammar of natural languages and that of formal languages, and second of all in seeking to determine which kind of grammar is best suited to exactly generating all the sentences of a particular language.

One of the features which, in my opinion, best highlights the radical change brought about by Chomsky's view, is the fact just mentioned that syntax overturned and largely dethroned phonology as the flagship discipline of linguistics from the 1950s onward. Of course, phonology continues to develop to this day, but the goals and methods of phonology changed profoundly, and the book *The Sound Pattern of English*, published by Chomsky and Halle in 1968, marked a new stage in the revolution brought about by the generative approach. Similarly, the reader should be aware that linguistics started to be taught in an entirely different way in the 1960s, in particular in the United States: until then, linguistics departments were mainly departments of "linguistics and philology," or of "linguistics and Slavic languages" (as at Columbia in the 1940s), and so forth. Starting in the 1960s, theoretical linguistics departments were created and separated from departments specialized in languages. Syntax, understood as the study of the structure of sentences, expanded remarkably, thanks to several generations of students, many from MIT, where Chomsky and Halle exerted considerable influence through their teaching (up until the 1990s and beyond).[36]

[36] For a more detailed discussion of the sociological evolution of linguistics from the 1950s to the 1980s, see F. Newmeyer (1986). See also Anderson (1985, pp. 315 ff.).

Apart from such sociological change in the organization of the field of linguistics, it should also be noted that the Chomskyan approach called into question the scientific methodology which was considered appropriate for studying language. As early as 1968, in an article published in French in the *Diogenes* collection, E. Bach highlighted the contrast between the "Keplerian" perspective that guides Chomsky's approach, and the "Baconian" one that underlies Bloomfield's work in particular. In referring to Bacon, Bach intends to underscore the priority given by Bloomfield to induction and observation in science, which is manifested by the statement that "the only useful generalizations about language are inductive generalizations" (Bloomfield, 1933). By alluding to Kepler, Bach targets, by contrast, the hypothetical-deductive approach which proceeds by formulating general hypotheses, and then investigating their consequences in order to explain observable phenomena. Thus, as we have seen, one of the central generalizations of Chomsky's approach lies in the affirmation of the infinitary nature of natural languages, which is inseparable from the concept of recursion. However, at any given time, or even in the space of a lifetime, we can only observe, utter or hear a finite number of actual sentences. In this respect, the emphasis Chomsky places on recursion is comparable to the emphasis Galileo places on the principle of inertia: recursion is no more directly observable than the principle of inertia. Chomsky's approach implies the relation of language to possible sentences rather than to sentences that are really or actually produced.[37] Indeed, Chomsky himself explicitly opposes the rationalist character of his approach to the empiricism and behaviorism which characterized the dominant conception of language in the 1940s and 1950s.

Bach did not hesitate to write that "Chomsky's revolution has certain analogies with both that of Copernicus and that of Kant."[38] One of these analogies has to do with the rationalism of Chomsky's approach, and with his internalist conception of language as a faculty rather than as a body of utterances. A second analogy, no less important, is to be found in the universalism of Chomsky's methodology. One of the postulates of Chomsky's approach is indeed that natural languages have a common cognitive basis. This postulate is by no means self-evident, far from it. To this day, in the eyes of many linguists, the opposite seems true. Sapir and Joos are thus often cited as claiming that "languages can differ from each other without limit and unpredictable ways" (Joos, 1957). Joos's point of view appears to be amply confirmed by experience, if one considers the syntactic, morphological and phonological variability between different languages. However, Chomsky's view goes against the idea that languages could differ

---

[37] No comparison is made in Bach's article between recursion and the principle of inertia; I note the parallel. However, Bach does conclude his article by referring to Koyré's work on the importance of a priori knowledge in science, in a way that seems to support exactly such an analogy. Chomsky himself mentions Koyré when he speaks of the "Galilean style" in science (see Chomsky, 2000).

[38] See also Pollock (2007, p. 102), who writes that "generative linguistics is one of the rare human sciences which has adopted the methodology that the natural sciences have made their own since the scientific revolution of the 16th and 17th centuries, the "Galilean style." "The Galilean style" is explicitly endorsed and discussed by Chomsky (see Chomsky, 2000).

"without limit": as I shall show, certain universal constraints on the syntactic structure of languages plausibly exist. Bach, in his article, emphasizes that a statement such as "all languages are similar to Latin" immediately has more predictive power than the opposing claim that languages are radically diverse, in the sense that it calls for the uncovering of principles of universal grammar which can be tested on languages that have not yet been described. In the view represented by Bach, such a claim is introduced above all as a regulatory ideal: for Chomsky, however, the assertion of the existence of a universal grammar plays more than a mere regulatory role, it is linked to a conception of the language faculty as essentially innate in character, and beginning in the 1960s the very term "universal grammar" is no longer used by Chomsky to designate a grammar strictly speaking, but rather a theory of the genetic component of the language faculty.

3. Description, Explanation, and Prediction in Linguistics

In the previous section I gave an overview of the evolution of linguistics over the course of the 20th century. In showing the opposition between the structural linguistics that Saussure inspired, and the generative linguistics that grew out of Chomsky's works, I presented the idea of an epistemological and methodological break between Chomsky's view of language and that which underlies the structuralist movement that preceded the generative approach. In this section, I will now discuss epistemological issues relating to description, explanation and prediction in linguistics. From now on, I will adopt a perspective that focuses more on methodological problems, and less on the historical aspects of the development of linguistics. The first issue I will examine concerns the analogy between the explanatory schema used in generative linguistics, and the explanatory schema employed by the other natural sciences. Then, using examples, I will discuss in greater detail the generation of linguistic data, the formulation of explanatory hypotheses, and the problem of their confirmation. The following section will be devoted to a broader discussion of the status of the very notion of a universal grammar and, in particular, of what deserves to be called a linguistic universal.

3.1 Chomsky's Three Levels of Adequacy

As a result of its dual cognitive and mathematical orientation, generative linguistics aims to deal with language in the same way as the other natural sciences deal with natural phenomena, by providing an explanatory and predictive model of linguistic phenomena. As previously mentioned, there is first of all a theoretical sense in which the generativist enterprise is simultaneously descriptive and predictive. This sense is related to a parallel drawn by Chomsky between natural languages and formal languages. According to Chomsky's view in 1957, a generative grammar is a recursive system of rules from which it must be possible to generate all the sentences of a given natural language and nothing but these sentences, as well as an adequate description of their structure.

For example, a context-free grammar like the one just described is a rewriting system from which it is possible to generate sentences like "Peter watches a dog," but also other sentences of the same type, such as "Mary drives a car" and certain slightly more complex sentences, like "Peter watches a very old dog." The generative power of a grammar of this type is similar to the *predictive* capacity of a hypothetical-deductive system, or also to the *expressive* power of a system of axioms. Consider for example Newton's laws of dynamics. In theory, these laws are used to describe and predict the movement of any moving object whose initial position and acceleration are fixed in a Galilean reference frame. Newton's laws of dynamics describe and predict what specific path is possible. This view is comparable to the one according to which the purpose of an adequate grammar is to separate those sequences of words in a given language which are grammatical, and would be accepted by a competent speaker, from those that would not. Like physicists who seek a set of laws that would enable them to characterize the various possible states of a system over time, linguists seek a set of rules that would enable them to derive the various possible sentences a competent speaker is liable to say or accept.

For example, if one were to fully specify the rewriting rules underlying the construction of the tree in Figure 2, one could see that the system in question is not trivial, in the sense that it provides for the generation of other grammatical sentences than those previously listed, such as "a very old dog watches Peter" or "a very old dog watches a fat cat." Similarly, one could, without changing the rules, extend the final lexicon so as to account for a large number of transitive constructions of the same type (via Rule V → *watches, hits, loves, directs, . . .*). However, the *descriptive* power of this grammar is obviously very limited. Suppose, to take a simple example, that one wanted to extend the lexicon by admitting plural forms for the nouns already present, with *dogs, cats,* and likewise by enriching the lexicon for determiners like *some*. In that case, some new rules are needed concerning agreement. Without such rules, the grammar would immediately over-generate (*some dogs observes a dog* could be produced). But even with the incorporation of agreement mechanisms, the grammar would still under-generate relative to a whole set of constructions: how should one derive interrogative sentences such as "Does Peter watch a dog?," negative sentences like "Peter did not watch a dog," and so on.

I have given these examples in order to show how difficult it is to extend a specific grammar, which appears adequate for a fragment of natural language, to the entire language. At first glance, the grammar underlying the tree in Figure 2 makes a necessary and adequate distinction between grammatical categories, for example between the determiner *a* and the noun *cat*. A grammar of the same type would allow one to generate in a precisely analogous fashion the sentence "Peter watches two cows," provided one had chosen an appropriate lexicon. But it is clear that unifying these two grammars in such a way as to integrate the singular and plural based on a common set of rules is not straightforward. This problem shows that generative syntax cannot immediately aim to generate in an adequate manner from the outset all the possible sentences in a given language and only those sentences. To achieve such a goal, correct hypotheses must be formulated about the structure of sentences and that of the lexicon itself, which can easily be generalized.

Chomsky (1964, 1965) thus distinguishes three levels of adequacy or success for grammatical description: *observational* adequacy, *descriptive* adequacy, and finally, *explanatory* adequacy. The first and most basic level, consists in having an adequate inventory of the units required for the purpose of description, of the constructions that are acceptable and those that are deviant. According to Chomsky, the second level, that of descriptive adequacy, aims to give a correct theory of the intuition of a native speaker; formally, this means formulating a grammar that can generate all the grammatical sentences of a given language (or fragment thereof), as well as providing what Chomsky calls a correct *structural description* of each sentence.[39] For instance, what my example suggests is that the grammar underlying Figure 2 provides at best a first approximation of the correct structural description underlying the sentence "Peter watches a very old dog," because it does not account for the markers of gender and number in particular, nor for verb tense and mood, or many other subtler aspects of the structure of the lexicon that are used by competent English speakers in order to interpret the sentence in question. Evidently, in order to obtain a correct structural description of a sentence like "Peter watches a very old dog," one must be able to account for the differences and similarities in structure between a potentially large number of sentences that superficially share the same structure.

Explanatory adequacy, the third level that Chomsky distinguishes, is more abstract than the previous two. Chomsky imagines that, in principle, two different grammars could generate the same adequate set of sentences, and also provide structural descriptions that are equally compatible with the intuitions of a given speaker, but nonetheless remain distinct. At this stage, the comparison between the explanatory

[39] The difference between sentences and structural descriptions is the same as that between the sequence of words in a sentence and its syntactic derivation tree. (See Chomsky, 1965, chapter 1, Section 9). Chomsky distinguishes the *weak* generative capacity of a grammar (all the sentences it generates) from its *strong* generative capacity (the set of structural descriptions it generates). According to Chomsky, a grammar is descriptively adequate if it strongly generates all its correct structural descriptions. According to Chomsky, the only one of these two concepts that is fundamentally relevant from the point of view of linguistic inquiry is the concept of strong generation. On this subject, N. Chomsky provides the following historical clarification (personal communication, December 2009): "*Syntactic Structures* is, basically, undergraduate course notes, and it formulated the problem at the outset in terms of weak generation, for one reason, because one pedagogical goal was to undermine the near-universal view at the time among engineers and psychologists that Markovian sources and information-theoretic notions sufficed to account for language, and these kept to weak generation (in fact very special cases of weak generation, even weaker than finite automata). One of the early footnotes points this out, and the rest of the monograph goes on to deal with strong generation, the only really linguistically interesting (or even clear) concept. The exposition has been misleading for this reason. In fact, almost all of *Syntactic Structures* and *LSLT* is devoted to strong generation and, furthermore, to semantic interpretation. Many people have been misled because they did not go beyond the first few pages of *SS*."

power of the two grammars depends on different criteria. The simplicity of one grammar relative to the other is one of these criteria, but in itself, the definition of the concept of simplicity is problematic. However, Chomsky emphasizes two aspects which seem essential to the characterization of the notion of explanatory adequacy. On the one hand, Chomsky considers that a grammar would be more appropriate than another from an explanatory point of view if, for example, it were more compatible with certain data related to language acquisition, and to how a child learning the language internally constructs correct generalizations about the language he speaks.[40] Chomsky also puts forward the idea that a grammar is more explanatory if it formulates more significant generalizations (1965, pp. 63–64). Once again, however, the notion of significant generalization is presented as a problem rather than as a primitive notion:

> The major problem in constructing an evaluation measure for grammars is that of determining which generalizations about a language are significant ones; an evaluation measure must be selected in such a way as to favor these. We have a generalization when a set of rules about distinct items can be replaced by a single rule (or, more generally, partially identical rules) about the whole set.

In order to briefly illustrate the different levels of adequacy that have been distinguished, I will consider an example from syntactic theory, that pertains to Chomsky's introduction of the concept of *transformation*. Chomsky argues in particular that a transformational grammar would be more explanatory than a context-free grammar, even if both were to have the same descriptive power.

3.2 THE EXAMPLE OF MOVEMENT

To illustrate the three levels of adequacy distinguished by Chomsky, I will first reproduce an example of syntactic contrast discussed by Chomsky himself in Chomsky (1964, p. 34). Consider the following pair:

(7) John is easy to please.

(8) John is eager to please.

Inasmuch as these two sentences are accepted as well-formed by a competent speaker, a grammar would achieve the level of observational adequacy if it included the sentences in question on the list of those sentences in the language being considered that are grammatical. Superficially, the two sentences differ only by the substitution of the two adjectives "easy" and "eager." So one might think that the two sentences have the

[40] As Pesetsky eloquently emphasizes (1995, p. 1) at the opening of his book: "Although linguists struggle to make sense of the grammatical patterns of human languages, children take a mere two years or less to discover most of the grammar and much of the basic vocabulary of their native language."

same syntactic structure. However, if a grammar were to give these two sentences the same structural description (the same syntactic representation in the form of a tree), it would fail to achieve the level of descriptive adequacy. Indeed, in (7) "John" is actually the direct object of the verb "please," whereas in (8) it is the subject. To convince oneself of the difference between these two constructions, one needs only to compare other occurrences of the words "easy" and "eager" in distinct environments:

(9) It is easy to please John.

(10) *It is eager to please John.

(11) Pleasing John is easy.

(12) *Pleasing John is eager.

(13) *John is easy to please those around him.

(14) John is eager to please those around him.

(15) *Who is John easy to please?

(16) Who is John eager to please?

To be descriptively adequate, a grammar must then assign separate structural descriptions to (7) and (8), capable of deriving the fact that in (7) "John" is logically the object of the verb "please," whereas in (8) it is the subject. Moreover, in order to be adequate from the point of view of explanation, a grammar should at least provide an explanation of the relation between the structural description of (7) and that of (8) and of the contrasts observed in (9)–(16)—as far as our judgments of the sentences' grammaticality or incorrectness are concerned. As Chomsky explains, in order to achieve this, a grammar must include principles which make it possible, for example, to derive the acceptability of (9) and the impropriety of (10) based on the structural descriptions assigned to (7) and (8). In this way, the grammatical theory under consideration would be able to provide an explanation for speakers' linguistic intuition. A grammar that could predict in a unified manner the contrasts observed in (7)–(16) would be more appropriate from an explanatory point of view, *ceteris paribus*, than a grammar that was only able to derive some of the contrasts in question, or that failed to give a unified explanation of them.

Logically, explanatory adequacy presupposes descriptive adequacy, which in turn presupposes observational adequacy, but as these examples show, in practice the different levels of adequacy Chomsky distinguishes prove to be interdependent. In order to obtain an adequate structural description of sentences (7) and (8), it is necessary to supplement observation with the consideration of other sentences, so as to elucidate the intuition that the syntactic position of the word "John" differs from one sentence to the other. Simultaneously, this example suggests that a grammar will only be perfectly adequate from a descriptive point of view, according to Chomsky's definition, if it is based on a set of explanatory generalizations that are sufficient, from the explanatory point of view, to unify the description of a large number of sentences.

In addition, Chomsky's example illustrates a key aspect of the generative approach, which has to do with the notion of transformation. In theory, it is conceivable to derive sentences (7) and (8) using separate rewriting rules in a context-free grammar. However, in doing so, the end result would be a system of rules that would fail to account for the semantic and syntactic kinship between the two sentences. Yet one of the purposes of syntactic theory, as Chomsky has emphasized, is not only to generate all the sentences of a given language and nothing but those sentences, but also to account for the systematic dependencies which exist between certain classes of grammatical structures. It is such a perspective that motivates the introduction of the concept of transformation.

Thus, one way to account for the underlying structure of (7) is to assume that (7) is derived from the underlying structure of sentence (9) together with a certain transformation. Consider the following schematic structural description, intended as a first approximation: [$_{\text{TP}}$ it [is [easy [$_{\text{CP}}$ ~~for~~ [PRO$_j$ [$_{\text{VP}}$ to please$_j$ John]]]]]], and compare it to the description [$_{\text{TP}}$ John$_i$ [is [easy [$_{\text{CP}}$ ~~for~~ [PRO$_j$ [$_{\text{VP}}$ to please __ $_i$]]]]]]. One way to describe the relation between these two structures would be to consider that the word "John," which in the first description appears in the *complement* position of the verb "please," *moves* into the *subject* position of the verb "is" in the second description.[41] To symbolize this displacement or movement, I co-indexed the word "John" and its initial position as complement to the verb "please."[42]

The term of "movement" or "transformation" should naturally be regarded with circumspection: the connection between the two structures is best understood as the expression of a rule that allows for the production of a new syntactic structure based on a more basic syntactic structure, rather than as the product of a mental operation. The notion of transformation plays a critical role in Chomsky's theory of syntax as a result of its ubiquity. For example, the relation between a sentence in the active mode, such as "John loves Mary" and the passive sentence "Mary is

---

[41] The type of syntactic dependency this passage illustrates, known as "tough-movement" (with reference to sentences of the type "This problem is easy / tough to solve"), is the subject of an extensive literature and of rival analyses since the 1960s. The transformational analysis of the phenomenon that we have outlined is no longer considered adequate today. See especially Lasnik and Fiengo (1974) for criticism, and Rezac (2006) for a recent presentation and a detailed overview of the literature. Regardless of the details of this example, the reader should keep in mind that the notion of movement remains central in syntax more generally, as soon as it comes to accounting for dependencies between syntactic constituents that are distant from one another in a sentence. For a detailed discussion of the concept of movement, see Fox (2002).

[42] Next, we discuss the significance of null subject "PRO." Here, crossed-out words indicate that the transformation of a sentence into another involves making silent some syntactic element. Recall that "VP" means "verb phrase," and "CP" means "complement phrase."

loved by John" corresponds to a specific transformation rule. The same goes for the affirmative sentence "John loves Mary" and the interrogative sentence "Who does John love?"

The concept of transformation does not play merely a descriptive role, inasmuch as it does not aim merely to simplify the rules of a given generative grammar. It also plays an explanatory role. Take for an example the occurrence of the expletive pronoun "it" in (9). The occurrence of this type of pronoun is predicted in the theory of government and binding (Chomsky, 1981), thanks to a postulate, called the extended projection principle, otherwise known as EPP, which states:

(EPP) The subject position of a tensed phrase (TP) must be filled.

A tensed phrase (or TP) is a proposition whose main verb is in a finite mode (other than the infinitive). For instance, consider the sentence: "It is easy for Mary to please John." In this sentence, the word "Mary" is in the subject position of the infinitive verb "to please." As the representation shows, the subject of a verb in the infinitive can be null or not expressed, as in "It's easy to please John." However, the extended projection principle precludes one from saying "*is easy to please John," because in this case the verb "is," which is the present tense, has no subject. In this case there are at least two ways to satisfy the EPP principle: either by using the expletive pronoun "it" or by *moving* the noun "John" into the subject position.

So that the reader does not become confused at this stage of the explanation, I should add that the EPP principle is not sufficient to explain all of the data listed earlier. Consider the case of (10). "Eager" is part of a family of predicates known as "control." The underlying structure of "John is inclined to seduce Mary" is in this case $[_{TP}$ John$_i$ [is [eager $[_{CP}$ PRO$_i$ $[_{VP}$ to please[Mary]]]]]], where PRO is a null subject, unexpressed phonetically, whose reference is controlled by an antecedent in the main sentence (in this case, by "John," which is represented by co-indexation, the description being understandable as: "John is eager that *John* please Mary"). To account for the anomalousness of (10), however, that is, "*it is eager to please John," the EPP principle is not sufficient. An explanation of this phenomenon follows from the theory of Case, which governs the distribution of noun phrases according to the Cases assigned to them, whose details I will not go into (see Bobalijk and Wurmbrand (2008) for an overview, and Vergnaud (1977) for the source).[43]

The main point to remember from this series of examples concerns the articulation between the three levels of observation, of description and of explanation discussed by Chomsky. From an abstract perspective, which Chomsky adopts in the first pages of *Syntactic Structures,* a grammar is a hypothetico-deductive system on the basis of which it should be possible to reconstruct an entire language. In this respect, the perspective

[43] The relevant notion of Case corresponds to a generalization of the morphological notion of case (nominative, accusative, oblique, etc.).

adopted by Chomsky is very close to the deductive-nomological model proposed by Hempel and Oppenheim to account for explanation in science (see Hempel, 1965). Yet before arriving at such a system of rules, the task of the linguist is to formulate hypotheses or significant "generalizations" from which, given a lexicon, it becomes possible to predict the order of words in a given language.

The example of sentences (7) and (8) is emblematic of linguistics' approach for several reasons. In particular, it highlights the fact that linguists must first formulate sufficiently general hypotheses about the syntactic structure of sentences they consider. It is only on the basis of sophisticated syntactic analyses that linguists can attempt to infer the rules that allow the generation of sentences. Moreover, the job of linguists is not merely to find rules for deriving sentences individually. Their aim is to connect different classes of structures to each other, and from that point, to try and explain why certain structures are illicit.

3.3 THEORY COMPARISON AND HYPOTHESIS CONFIRMATION IN LINGUISTICS

3.3.1 The Method of Minimal Pairs

In previous sections, I have already given a significant overview of the topic of the constitution of linguistic data. Whether phonology, morphology, syntax or semantics are concerned, the starting point for the vast majority of linguistic theories resides in the constitution of *minimal pairs*. For example, the two sentences "John is eager to please Mary" and "*John is easy to please Mary" form a minimal pair: the two sentences differ only in a change of one parameter (switching "easy" for "eager"), a variation that changes the status of the sentence (from acceptable to inacceptable). This variation exposes a structural difference. As shown, it also serves to corroborate the grammatical intuition that the sentences "John is eager to please" and "John is easy to please" have different structures.

As Chomsky's quote about the switch test in phonology indicated, the notion of a minimal pair is not absolute, in the sense that it is necessarily relative to a theory (to a preliminary hypothesis, to another set of data pairs, etc.). However, the production of a minimal pair is the first step that must be taken to control available linguistic data. This remark may seem self-evident, but a minimal pair is the linguistic equivalent of a controlled experiment in which the linguist is trying to confirm or refute such and such a hypothesis concerning the structure of a sentence. Sometimes the production of a minimal pair is the *explanandum* of a theory, while in other cases it acts as *explanans*, along with other general hypotheses: for example, one may ask why (9) and (10) present a contrast, but one can also use this contrast to confirm the intuition that (7) and (8) have different underlying structures.

One aspect worth noting is that the concept of minimal pair is primarily a legacy of structural linguistics, since it is associated with a methodology that is found both in Bloomfield's phonology and in Z. Harris's work in syntax on the distribution of syntaxic constituents.[44] Nonetheless, the systematic usage of minimal pairs marks an essential departure from the methodology of accounting only for the sentences actually pronounced within a given corpus. As should be obvious from the preceding examples, a convention that is now universally adopted in linguistics is to mark with a star sequences of words that are deviant or unacceptable to a competent speaker. The method of producing such starred sentences—ungrammatical sentences—starting from grammatical ones, has been criticized by certain linguists, who believe that proper linguistics can only be practiced on already existing discourses.[45] But such criticism is based on misunderstanding and narrow empiricism, since it neglects an essential aspect of empirical investigation in linguistics: to compare grammatical sentences to ungrammatical sentences with neighboring configurations is to compare admissible sentences with inadmissible ones, in order to reveal the structure of the admissible sentences. By comparing grammatical sentences to ungrammatical ones, linguists seek to identify the constraints that govern the judgments of native speakers about their own language.

Of course, there is debate regarding the limits of theory-building in linguistics that would be based only on the kind of preliminary task that linguists routinely undertake, which is to obtain grammaticality judgments from competent speakers (who are often the linguists themselves when they are working on their own language). These discussions relate to more fundamental questions about the psychology of language, especially concerning the limitations of the introspective method in linguistics. However, there are more precise ways of controlling data-collection from a linguistic point of view, either by comparing the judgments of a number of speakers, or by comparing explicit judgments to brain and behavioral data, which can be obtained either simultaneously or independently. In any case, the rise of more complex experimental techniques does not call into question the validity of the method of minimal pairs, which remains an essential starting point for the constitution of data and hypotheses in linguistics.[46]

---

[44] Gillon (2017) emphasizes that the method of minimal pairs is already attested in the work of ancient Indian grammarians, and rightly notes that it can be seen as a special case of the so-called method of agreement and difference discussed by Mill (1843) in his analysis of causal inferences.

[45] See for example a point made about F. Newmeyer (1998, p. 96): "Certain linguists dismiss any interest in explaining judgments by native speakers about sentences that would rarely, if ever, be used in actual discourse." T. Givón is one of the linguists who Newmeyer cites in support of this remark (1998, p. 38). The use of the asterisk to mark deviant constructions or utterances dates back at least to Bloomfield (see, e.g., Bloomfield 1933, p. 167 and passim).

[46] On this point, see in particular Marantz (2005) and Sprouse and Almeida (2012). Sprouse and Almeida give a scientific comparison between informal acceptability judgments from a standard syntax textbook and judgments collected from a large sample of participants. They find an impressive rate of replication (98%) of the textbook's acceptability judgments, suggesting to them that "there is no reason to favor formal experiments over traditional methods solely out of a concern about false positives" (2012, p. 634). For a philosophical discussion of the status of linguistic "intuitions" in relation to intuitions in thought experiments, see Devitt (2006, chap. 7). For a comparison of informal semantic judgments with judgments based on systematic surveys, see in particular Chemla, Homer, and Rothschild's (2011) discussion of polarity items.

3.3.2 The Notion of Prediction in Linguistics

The purpose of a theory in linguistics, as in the other empirical sciences, is to formulate explanatory and predictive hypotheses about the nature of linguistic phenomena. A hypothesis is predictive if it can explain data not already predicted by the theory, or not readily accessible. There is some debate regarding the claims of theories in linguistics as to providing explanatory and predictive hypotheses. Some consider that linguists' claim that they formulate hypotheses with the same status as those in physical science is illusory. Givón, for example, writes in a controversial remark that:

> In essence, a formal model is nothing but a restatement of the facts at a tighter level of generalization . . . There is one thing, however, that a formal model can never do: It cannot explain a single thing . . . The history of transformational-generative linguistics boils down to nothing but a blatant attempt to represent the formalism as 'theory', to assert that it 'predicts a range of facts', that it 'makes empirical claims', and that it somehow 'explains'. (Givón 1979a, pp. 5–6; emphasis in original)

Givón's remark is not entirely unfounded. One criticism that is often made of explanatory hypotheses in linguistics is indeed that they are no more nor less than descriptive generalizations in disguise. Consider once more the extended projection principle (EPP), which states that any finite tense phrase must have an expressed subject (that is, that the specifier position of the TP must be filled). The EPP principle can be considered to be a descriptive generalization about the structure of sentences. This way of seeing the principle is well-founded, since it is a universal statement that quantifies over the class of all sentences (English or French), and in that sense it *describes* a presumed regularity in the linguistic structure of sentences.

Despite this, Givón's remark underestimates the fact that any significant linguistic generalization is necessarily based on a set of hypotheses and theoretical concepts that have an explanatory goal. Thus, the concept of a *specifier* is a theoretical concept (developed in the X-bar theory, see Jackendoff, 1972, and Radford, 1995, for an introduction), which actually generalizes the notion of subject of a verb to other syntactic categories, a point that is far from obvious. More fundamentally, one of the aspects of the EPP principle is that it is intended to account for a variety of hypotheses which concern a wide class of grammatical structures. For example, the EPP principle accounts for certain transformations in several classes of structures (the passive, the raising of the subject, or the displacement of the object in the theory which treats (7) as a case of movement), which is to say that it is a generalization which unifies the description of a wide range of phenomena. As Newmeyer rightly emphasizes, contra Givón (Newmeyer, 1986, 1998), the relationship between formal hypotheses and facts in generative grammar is often indirect, and therefore does not justify the remark that such a theory would be a mere "reformulation of the facts."

To illustrate the idea that linguistic hypotheses in generative grammar really do have a predictive dimension, consider an example discussed by Morris Halle, which involves the formulation of hypotheses in phonology (Halle, 1978). Halle's example concerns the phonological rule governing the formation of plural nouns in English (see, e.g., Bloomfield 1933, pp. 210–211, where this generalization has already been formulated). The rule is based on a preliminary inventory of the various ways to form plural nouns from singular ones in English. There are three main classes of words, regarding the pronunciation of the morphological mark of the plural in English, certain representatives of which are as follows:

(17) a) *bus, bush, batch, buzz, garage, badge*, . . . the plural of which is pronounced / iz / (pronounced as in *nozzles, bushes*, etc.).
b) *lick, pit, pick, cough, sixth*, . . . the plural of which is pronounced with the sound / s /
(pronounced as in *licks, pits*, etc.).
c) *cab, lid, rogue, cove, cam, can, cal,l* . . . the plural of which is pronounced with the / z /
(pronounced as in *cabs, lids, rogues*, etc.).

Based on this observation, the question Halle raises is the following: "In what form does the English speaker internalize his knowledge of the plural rule?" Several hypotheses are compatible with the data: one of them would be that, for each word of English, the speaker memorizes the singular and the plural form. This hypothesis is unconvincing if one considers that the rule which underlies the formation of the plural is a productive rule: a competent speaker is capable of forming plurals from singular words he has never heard before. The second hypothesis envisaged by Halle is that the rule could be formulated in terms of sounds. According to this hypothesis, the rule could be stated as follows:

(18) a) if the noun ends with /s, z, š, ž, č, θ, ǰ/, add /iz/
b) if the noun ends with /p, t k, f, θ/, add /s/
c) otherwise, add /z/

As the reader can verify, this hypothesis is consistent with the data collected in (17). Halle notes, however, that rule (18) is formulated in terms of sounds, not in terms of articulatory features. But a more fundamental assumption in phonology, already mentioned here in the discussion of Jakobson's work, is that "features rather than sounds are the ultimate constituents of language." A rival manner of formulating the rule would therefore be in terms of features, as follows:

(19) a) if the word ends with a sound that is [coronal, strident], add /iz/
b) if the word ends with a sound that is [unvoiced], add /s/
c) otherwise, add /z/

This second version of the rule is also consistent with the data available in (17). At first glance, one could say that the two rules are therefore only "reformulations at a narrower level of generality" of the observations obtained in (17). However, Halle points out that the two rules (18) and (19) are predictive, insofar as they are also supposed to apply to words that were not part of the initial inventory. According to Chomsky's typology, it would seem that the two rules nonetheless have the same level of descriptive adequacy. Nevertheless, Halle notes that the two rules make different predictions. One way to test these two hypotheses, as Lise Menn suggested to Halle, would be to ask a native speaker of English to form the plural of words involving foreign sounds that are not used in English. The proposed test concerns the /x/ sound in the German word "Bach" (in its German pronunciation). If the speaker uses the rule, formulated in terms of sound, then the prediction is that the plural would be pronounced /z/ (case c) of the rule). But if the rule is formulated in terms of features as in (19), the plural of "Bach" should be pronounced /s/, since the sound /x/ is not [coronal, strident] but [unvoiced] (case b)). By testing English speakers (using this word and other similar cases), it is observed that they form the plural by adding /s/ rather than /z/.

Halle's example is indicative of the fact that an "interesting" descriptive generalization, as soon as it reaches a sufficient level of generality, necessarily has a predictive or ampliative dimension. By comparing (18) and (19), we also observe that not only is the formulation of the rule in terms of features more economical, but it also makes better predictions than the version in terms of sounds, in those cases not previously considered by the theory. With respect to the avowed purpose of trying to account for the mechanisms according to which a competent speaker internalizes the rules of plural formation in English, the rule given in (19) is thus more explanatory than the one given in (18). Contrary to what Givón maintains, Halle's example shows that no clear opposition can be made between the level of description and the level of explanation in linguistics. To achieve an adequate description of the pluralization rule, one that is faithful to the intuitions of speakers, it is necessary to invoke the phonological theory of decomposition of sounds in terms of articulatory features. Contra Givón, it follows that the statement of a rule can actually have a predictive dimension, in the sense in which I have defined that notion.

3.3.3 Confirmation and Refutation of Linguistic Hypotheses

Nevertheless, the example advanced by Halle is subject to a standard objection in philosophy of science, which was originally raised by Duhem (1906): the reason why we prefer (19) to (18) cannot be grounded purely and simply on the "crucial" experiment of testing English speakers with respect to the sound /x/. Indeed, what would happen if there were independent reasons for favoring the hypothesis according to which the ultimate constituents of language were sounds rather than articulatory features? In such a case, one could imagine a way of "fixing" rule (18) by adding the /x/ sound to the list of sounds for which the plural is formed by affixation of the /s/ sound. In order to decide between rule (19) and the amended version of rule (18), new tests would then be required. In fact, the test considered by Halle is supposed to suffice, since Halle gives independent reasons to believe that articulatory features play a more fundamental functional role than sounds, from a phonological point of view, and also because of the hypothesis that the /x/ sound is not an English phoneme, but rather is borrowed from German phonology.

However, this situation precisely reflects the fact that an isolated language test is not sufficient to refute or confirm a given hypothesis, except in trivial cases. To further illustrate this point, now consider an example taken from the semantics of natural language, an area I have not discussed very much thus far. A general problem in linguistics is to explain the limited distribution of certain classes of lexical items. In English, expressions like "any" or "ever" are called *negative polarity items* or NPIs. These expressions are so called because their occurrence seems to require the presence of a "negative" environment. For example, compare the following sentences:

(20) John has not met any students.

(21) *John has met any students

(22) I do not think there will ever be a new Aristotle.

(23) *I think there will ever be a new Aristotle.

A first hypothesis to consider is that words like "any" or "ever" need to be preceded syntactically by a negation. The situation is actually more complex however, since one can say:

(24) I doubt that John has met any students.

(25) Every student who has ever been to Rome has returned amazed.

Obviously, a verb like "doubt" has a "negative meaning," but assuming as much already goes against the hypothesis of a purely syntactic constraint governing the distribution of NPIs. A more detailed hypothesis, originally formulated by Fauconnier (1975), and developed by Ladusaw (1979), is based on a semantic generalization of the intuition according to which NPIs need to be preceded by a negation. The generalization is as follows:

Fauconnier-Ladusaw Generalization: an NPI is grammatical only if it appears in a *monotone decreasing* environment.

An environment is said to be monotone decreasing if it behaves like a monotonically decreasing function in terms of its arguments. A function f is monotonically decreasing if it reverses the order of its arguments, for example, if it is such that $f(y) < f(x)$ when $x < y$. By extension, a function from sets to sets is monotonically decreasing if it reverses the inclusion relationship between the sets. Semantically, however, determiners such as "a," "no," "every" can be treated as expressing relations between two sets.[47] For example, "every student smokes" is true if the set of all students is *included* in the set of all smokers, "a student smokes" is true if the set of smoking students is *non-empty*, "no students smoke" is true if the set of smoking students is *empty*. A determiner is said to be *monotone decreasing* (resp. increasing) for one of its arguments if, when it takes as an argument a subset (resp. superset) of a given set, the relation of logical consequence gets inversed (resp. preserved). For example, "no" is monotone *decreasing* for each of its arguments. Thus, "to smoke cigars" entails "to smoke" (but not vice versa), yet we have

(26) a) No student smokes. => No student smokes cigars.
b) No smoker is a student. => No cigar smoker is a student.

By contrast, the determiner "a" is monotone *increasing* for each of its arguments, whereas "every" is monotone *decreasing* on its first argument, and monotone *increasing* on its second argument:

(27) a) Every smoker is a student. => Every cigar smoker is a student.
b) Every student smokes cigars. => Every student smokes.

(28) a) A student smokes cigars. => A student smokes.
b) A cigar smoker is a student. => A smoker is a student.

As von Fintel (1999) writes on the subject of determiners, "Quite spectacularly, we find that NPI licensing exactly mirrors these entailment properties." For example, we have

(29) a. A student (*who has ever been to Rome) (*bought any postcards there).
b. Every (student who has ever been to Rome) (*bought any postcards there).
c. No (student who has ever been to Rome) (bought any postcards there).

As it once again becomes apparent, the Fauconnier-Ladusaw generalization is far from being a mere redescription of the facts at a higher level of generality, since it establishes a correlation between a syntactic property (i.e., occurrence of NPIs) and a semantic property (i.e., occurrence in a monotone decreasing environment). However, and this is the relevant point in this section, there are many counter-examples to the Fauconnier-Ladusaw generalization, that is, there are cases where NPIs are allowed but where the monotone-decreasing consequence relation is not valid. In such cases, one

[47] Such an account, inspired by Boolean logic, is based on the work of R. Montague (1974) and is the subject of the theory of generalized quantifiers. See the source article by Barwise and Cooper (1981) for a classic reference and the volume by Westerstahl and Peters (2006) for an encyclopedic presentation. On NPIs and the relation between grammar and logic, see the overview of Spector (2003) and, more recently, the experimental work of Chemla, Homer, and Rothschild (2011).

can say that the generalization *under-generates*, in the sense that it is too restrictive with respect to all environments in which NPI are licensed. But the generalization can just as well be seen as *over-generating*, in the sense that taken literally, it incorrectly implies that certain environments that are not monotone decreasing should be so in principle. A counter-example envisioned by von Fintel concerns the word "only":

(30) Only John has ever met any students.

(31) Only John smokes. ⇏ Only John smokes cigars.

As (30) shows, "only" licenses NPIs. However, the inference in (31) is not valid: it could be the case that John is the only smoker, yet he only smokes cigarettes, in which case the premise of (31) is true, but not its conclusion. As von Fintel discusses, there are other counter-examples to the generalization, which include superlatives (cf. "the greatest man I've ever met . . ."), and the antecedents of conditionals ("if John ever met any student . . .").

In spite of these counter-examples, there have been many attempts to amend the Fauconnier-Ladusaw hypothesis. One of the reasons for this, stressed by Linebarger (cited by von Fintel, 1999, p. 101) is the "surprisingly algorithmic" character of the hypothesis, which, according to von Fintel, is "worth defending against challenges." The meaning of this remark is that the hypothesis also has an explanatory dimension (in Chomsky's sense): one way of envisaging the hypothesis is indeed to consider that it is because speakers are logically capable of recognizing monotone decreasing environments that they infer from this the rule that NPIs are allowed in such environments.

The point of von Fintel's article is thus to reformulate the Fauconnier-Ladusaw generalization. Von Fintel shows that, if a notion of logical entailment that is sensitive to the presuppositions present in the premises and conclusion of the argument is adopted, then recalcitrant examples can be dealt with (such a concept is called *Strawson Entailment* by von Fintel, in reference to Strawson's work on presuppositions). For example, "Only John smokes cigars" semantically presupposes that "John smokes cigars." Assuming this presupposition is satisfied (by virtue of the lexical semantics of the word "only"), then supposing that "only John smokes" is true, its monotonically decreasing entailment "only John smokes cigars" holds this time. A rough reformulation of the Fauconnier-Ladusaw generalization is thus:

Fauconnier-Ladusaw-Fintel Generalization: an NPI is grammatical only if it appears in a *monotone decreasing* environment under Strawson Entailment.

I chose von Fintel's discussion of negative polarity items because it provides a realistic and easy to explain example of how hypotheses are refined. As this case demonstrates, examples which at first appear to go against a hypothesis can become new confirming instances, once the hypothesis has been properly refined. Few significant linguistic generalizations are immediately descriptively adequate. Most often, a unifying hypothesis *under-generates* or *over-generates* when it is applied to a sufficiently large data set. Hypotheses in linguistics, as in other empirical sciences, are largely under-determined by the available data. Linguists mainly give priority to a hypothesis' unifying and explanatory value. If the hypothesis is interesting, it will most likely be revised rather than being considered to be refuted.

3.4 HISTORICAL EXPLANATIONS AND THEIR LIMIT

Explanations can be seen as answering questions such as "why does this phenomenon occur?," but also "how does this phenomenon occur?." For example, the Fauconnier-Ladusaw generalization is supposed to answer the question as to why a particular class of lexical items has a limited distribution. The answer to this question lies partly in the generalization itself. If the reader were to ask this question to a linguist today, she would be very likely to receive the following answer: "It is because the items in question may only appear in monotone decreasing environments." In other words, the answer to this question would be a statement of the Ladusaw-Fauconnier generalization. As already shown, this generalization is explanatory in the sense that it establishes a correlation between a distributional property and a semantic property, and it realizes a deductive-nomological schema such as: "Any expression of the NPI type can only appear in a monotone-decreasing environment; expressions such as *ever, the least*, . . . are NPIs; so *ever, the least* can only appear in monotone-decreasing environments." If one were to reiterate the question, and ask why NPIs can only appear in monotone decreasing environments, there would be two possible responses. One would be an attempt to derive the generalization from a more basic set of rules or constraints that involve the lexical items in question. The other would consist in supposing that the generalization itself was the expression of a primitive rule of the grammar.

In principle, the same also applies to the other linguistic generalizations I mentioned in previous examples. For example, if one asks "why is the plural of the word [bəs] in English [bəsiz] (rather than [bəss] or [bəsz]))?" The best explanation we would have is: "because the final consonant of *bus* is [coronal strident]." In this case, the explanation is an enthymeme, which involves the pluralization rule formulated earlier as an implicit premise. Once again, it would be natural to think of deriving the rule from more general constraints, or else considering it to be primitive. Many such examples could be found, but they are indicative of the approach inspired by Chomsky in *Syntactic Structures*, which consists in assuming that language is the expression of internal rules governing the order and distribution of linguistic elements.

However, the deductive-nomological perspective that is adopted in generative grammar may seem too narrowly focused on synchronous phenomena. For instance, if we consider the bulk of linguistic research carried out in the nineteenth century, its main objective was to account for the evolution of languages, especially with respect to pronunciation and morphology.[48] The perspective of such research was essentially historical and diachronic, and explanation was thought to consist primarily in asking *how* certain linguistic forms had been arrived at. The importance as well as the enduring influence of such an approach should not be underestimated.[49] Consider a question such as, "Why is the future of *je chante* in French *je chanterai*, whereas in English the future of *I sing* is *I will sing*? ." In a more apt formulation, why is the future in French formed by suffixation, while English employs a periphrastic form? To give an explanation of the genetic type, in the case of French, one would observe that the future is formed from the infinitive of the verb, to which the verb *avoir* (to have) is postfixed (*je chanter-ai, tu chanter-as, il chanter-a, nous chanter-(av)ons, vous chanter-(av)ez, il chanter-ont*) [I to-sing-have, you to-sing have, he to-sing has, . . .]. In other words, the future in French is formed by grammaticalization of what was originally a periphrastic form (je chanter-ai = « *j'ai à chanter* » [I have to sing]).[50] This genetic hypothesis is confirmed by comparison with the way in which the future is expressed in the other Romance languages.[51]

As Lightfoot points out, however, although the phenomenon of grammaticalization is real enough, it is not obvious that it has "explanatory force" (Lightfoot 2006).[52] The reason Lightfoot gives is threefold: first, grammaticalization corresponds to reanalysis of the units of the language, but it is a local phenomenon. What is truly interesting is whether or not this phenomenon is correlated with the reorganization of other elements in the structure of the language. On the other hand, if grammaticalization is one phenomenon among many, then it calls for a theory: it should be taken as *explanandum* rather than as *explanans*, and one should enquire as to why an evolution took place in this direction rather than another. Finally, and this is a point first made by Chomsky and Halle (1968, pp. 249–252), language change can be viewed as the addition of new rules to the grammar of a given language. This was first illustrated by Chomsky and Halle with respect to phonetic change, but an even more striking illustration is provided by the syntactic evolution of the verbal systems of English and French.

To show this, I will briefly summarize the main elements of Pollock's analysis of the verb phrase, as well as his examples (see Pollock, 1997; Pollock, 2007; and also

---

[48] See Lightfoot (2006), chap. 2, for a very clear and informative overview of historical linguistics in the 19th century, which also explains the emergence of structuralism as a reaction to historicism.

[49] The principal achievement of the comparative-historical method lies in the various laws of phonetic change formulated in the 19th century for Germanic languages, including Grimm's law and Verner's law (see following section). It is interesting to note that the posterity of Grimm and Verner's laws has reached generative grammar (see Halle 2002, and Halle personal communication) in which phonological rules can be seen as "laws," synchronic laws having to do with changes in sounds, as explained in section 2.

[50] The notion of grammaticalization is due to Meillet (1937) and describes, according to Lightfoot's definition (2006, p. 37): "the semantic tendency for an item with a full lexical meaning to be bleached over time and to come to be used as a grammatical function."

[51] See Teyssier (2004) and Benveniste (1974, p. 131) for a description of the stages of the Latin future's transformation into the Romance one.

[52] Cf. Lightfoot (2006), p. 38 and p. 177: "Grammaticalisation, interesting as a PHENOMENON, is not an explanatory force."

Lightfoot, 2006). In English, the negation of verbs in the present tense are constructed thanks to the auxiliary *do*, and the same is true for the interrogative:

(32) I do not sing.

(33) Do you sing?

Up until the 16th century, however, negation and the interrogative could be constructed directly, like in French:

(34) *I sing not.

(35) *Sing you?

Note that the negation is to the right of the verb in Old English and modern French, while in contemporary English it appears to the left of the verb. This contrast is correlated with two others that involve the position of adverbs and quantifiers in English and French. In French, adverbs and quantifiers appear to the right of the verb:

(36) J'embrasse souvent Marie.

(37) Ils embrassent tous Marie

In contemporary English, however, analogous sentences are incorrect, and adverbs and quantifiers must be to the left to the verb:

(38) *I kiss often Mary.

(39) I often kiss Mary.

(40) *They kiss all Mary.

(41) They all kiss Mary.

But as several studies have shown, sentences (34)–(35) and (38) and (40) disappeared simultaneously from the grammar of English, at the same time as the verbal morphology of English became more impoverished as well (English lost most verbal markers for person, such as *thou singst* vs. *you sing*). As Pollock notes, this covariation suggests that a single property governs all of these phenomena. In order to explain word-order in contemporary English, certainly one could just say: "It is because at the turn of the sixteenth century, the rules changed." But in this case, what are the rules? A deeper explanation might be provided by attributing to English and French a level of shared structure, and by seeking to discover which rules are commonly used in one language and not in the other. Pollock's explanation postulates that French and

English sentences have a common structure, in which syntactic categories are hierarchically organized (INFL stands for the auxiliary or modal or temporal inflection, NEG for negation, ADV for adverb, Qnf for quantification, V for verb):[53]

$[_S\ NP_{subject}\ [_{INFL} \ldots [_{NEG}\ pas/not\ [_{ADV}\ souvent/often\ [_{QNF}\ tous/all\ [_{VP}\ \mathrm{V}]]]]]$

What becomes apparent is that in French, the verb *chante* in *je (ne) chante pas* appears in the INFL position, which is where the auxiliary *do* appears in English in *I do not sing*. We can account for this contrast if we suppose that the verb really does occupy the V position in principle, but is attracted by the INFL position, according to the rule:

$[_S\ NP_{subject}\ [_{INFL}\ \varnothing\ [_{NEG}\ pas/not\ [_{ADV}\ souvent/often\ [_{QNF}\ tous/all\ [_{VP}\ \mathrm{V\ X}]]]]] \rightarrow [_S\ GN_{sujet}$
$[_{INFL}\ \mathrm{V}\ [_{NEG}\ pas/not\ [_{ADV}\ souvent/often\ [_{QNF}\ tous/all\ [_{VP}\ \mathrm{X}]]]]]$

This is another example of a rule of transformation or syntactic movement. In this case, the rule states that V moves to the INFL position in French but not in English. Thanks to the hierarchy of categories, the principle simultaneously accounts for the other contrasts established previously. One way to describe the evolution of the English language would be to say that the rule of displacement from V to INFL was once active in Old English, but ceased to be, in correlation with the evolution of verbal morphology.

As the explanation I have just outlined suggests, it makes good sense to account for linguistic evolution by appealing to the addition or subtraction of rules that are supposed to hold synchronously. The type of explanation given by Pollock, in keeping with Chomsky's approach, is an internal and formal explanation, not of the causes of linguistic change, but of the connection that can be established between the grammars that underlie the two states of English. This type of explanation is in opposition to approaches that would seek to explain the nature of a certain rule first and foremost because of the occurrence of some external change in the way language is used. Explanations of this latter kind are usually called *functionalist* or *external*, since they depend on the idea that rules change essentially by virtue of pragmatic constraints having to do with language use.

I will return to this debate in the next section, but for now, suffice it to say that in principle the two modes of explanation are not necessarily mutually exclusive (see Newmeyer, 1998, 2005; Baker 2001, and Lightfoot, 2006, who argue extensively for such a conclusion). However, substantial differences remain concerning the issue of the goals of linguistics: as Pollock's examples convincingly demonstrate, explanation of a given linguistic phenomenon cannot be limited to purely historical considerations, such as when a new construction appeared or when another fell into disuse, or else all of linguistics would be reduced to an inventory of such changes. As Chomsky and Halle

[53] $NP_{subject}$ refers not to a syntactic category but to a noun phrase that is subject of the sentence.

(1968, p. 251) stressed concerning this subject, the rules that are found in a synchronic grammar cannot all merely be reduced to the pure and simple expression of changes that affected previous rules. Not only is this not the case, but if it were the case, it would lead to an infinite regress, which would in any case require one to look to psychology in order to uncover the bases for the first rules to which historical investigation could lead.[54]

3.5 PARTIAL CONCLUSION

As I have shown, Chomsky proceeded from a deductive-nomological ideal: a grammar is descriptively adequate if it is capable of weakly generating all and only those sentences of a given language, as well as of strongly generating the structural descriptions of the sentences in question. As Chomsky points out, the major part of linguists' work takes place precisely at the level of giving an adequate structural description of the sentences of a given language. To accomplish such a task, linguists must make generalizations that are capable of accounting for the distribution of lexical items of the language, so as to derive underlying constraints on word order. As a result, in practice the deductive-nomological ideal on the basis of which Chomsky founded modern linguistics is inevitably confronted with the inductive problem of formulating descriptive generalizations and explanatory hypotheses. The second point that I have emphasized was that the notion of linguistic prediction plays an important role in linguistics, and that, in this respect, linguistics is no different from other empirical sciences. Problems concerning confirmation and refutation have the same status in linguistics as they do elsewhere.

But for contemporary linguistics, a nagging question remains concerning the unification of its various explanatory hypotheses. Whether it's a matter of syntax or semantics, even moderately attentive readers may be surprised by the proliferation of a large number of explanatory hypotheses in linguistics. Such readers may wonder, what organic link is there between a syntactic and semantic generalization such as the Fauconnier-Ladusaw generalization and a syntactic principle such as the extended projection principle? Are these each time only local generalizations, or might one expect that they will all fit into a unified structure? An even more radical way of posing such a question is as follows: are there linguistics rules that have the same degree of generality or the same unifying character as Newton's laws in relation to physics, for example? To answer these questions, in the following section, I propose to examine

[54] For further details on this point I refer to Lightfoot (2006), chap. 7, which deals with the emergence of new grammars. See also Pinker (1994) and Senghas et al. (2004) on the emergence of structures in Nicaraguan Sign Language, a recent and spectacular example of creolization (transition from pidgin to an articulated language). Incidentally, as we shall see, Greenberg, arguably the most prominent historical linguistic in the 20th century (cf. Greenberg, 2005), himself puts forward the idea that the existence of a rule cannot be purely a matter of survival, but rather involves autonomous psychological constraints (see Greenberg, 1957, p. 89, who mentions Sapir as his source of inspiration on this point).

the problem of universality in language, and the status of the concept of a universal grammar.

4. The Concept of a Linguistic Universal

As I noted in the introduction to this chapter, the goal of theoretical linguistics is to account not only for linguistic diversity, but also for the faculty of language, insofar as it remains invariant from one language to another. One of the postulates of the generative enterprise launched by Chomsky is that

> the grammar of a particular language . . . is to be supplemented by a universal grammar that accommodates the creative aspect of language use and expresses the deep-seated regularities which, being universal, are omitted from the grammar itself. (1965, p. 17)

The idea of a universal grammar is ancient, and Chomsky explicitly associates it with the rationalist tradition in philosophy (Descartes, Leibniz) and with the philosophical grammar of the seventeenth and eighteenth centuries, such as the Port-Royal grammar or Du Marsais's grammar (cf. Chomsky, 1966). The postulation of a universal grammar is also based on the idea that robust regularities exist across languages, which reveal the very nature of the language faculty. However, the concept of a universal grammar raises several problems.

The first problem concerns the extent to which such a concept is consistent with the fact of linguistic diversity and that of the evolution of languages. A related issue concerns the degree to which the form of particular grammars depends on individual and social uses of language. Chomsky's conception of grammar is essentially nativist, internalist and formalist, and thus goes against more social, externalist or functionalist conceptions of the nature of language, which leave open the possibility that language is more authentically the product of culture than of nature.

A second problem concerns the very definition of what deserves to be called a cross-linguistic regularity, and at which level of abstraction such a concept should be deployed. In speaking of a linguistic universal, one sometimes refers to an architectural *principle*, and at other times to the occurrence of *elements, structures* or grammatical *categories* which are identical across languages. The level of abstraction involved is not the same in each case. The purpose of this section will be to clarify these issues. I will begin by discussing the central role played by the compositionality principle and by the notion of recursion in defining the concept of universal grammar in syntax and semantics. The second part of this section will be devoted to distinguishing different ways in which the concept of linguistic regularity can be characterized. In the last section, I will discuss in greater detail the phenomenon of the diversity of languages and the question of the relationship between diversity and singularity, especially in the model that is most influential nowadays, the so-called Principles and Parameters theory.

4.1 Universal Grammar, Recursion, and Compositionality

Before describing in detail the various meanings of the concepts of invariance and universality in linguistics, it seems important to recall that in the recent history of linguistics the notion of "universal grammar" is closely associated with the work of Chomsky in syntax, and that of R. Montague and several of his colleagues in semantics.[55] It is mainly from 1965 onward, with the publication of *Aspects of the Theory of Syntax*, that Chomsky discusses the concept. Some years later, in 1970, Montague gave the title "Universal Grammar" to one of his pioneering articles in formal semantics. An important point is that Montague, like Chomsky in *Syntactic Structures*, approached language from the perspective of a logician.[56] From this point of view, it can be said that both Montague and Chomsky generalize the idea that natural language functions in a manner essentially analogous to formal language. In particular, Montague writes (1970, p. 223):[57]

> There is in my opinion no important theoretical difference between natural languages and the artificial languages of logicians; indeed, I consider it possible to comprehend the syntax and semantics of both kinds of languages within a single natural and mathematically precise theory. On this point I differ from a number of philosophers, but agree, I believe, with Chomsky and his associates.

Although Chomsky's approach and that of Montague have much in common in terms of style as well as inspiration, each was guided by distinct aspects of the functioning of formal languages.[58] For Chomsky, as I already pointed out, it is the

[55] I would like to mention, in particular, D. Lewis (1970), T. Parsons, E. Keenan, and B. Partee, who contributed greatly to making formal semantics a discipline of linguistics in its own right. For a detailed historical overview, see Partee (2004, chap. 1).

[56] Note, however, that for Chomsky, the study of formal languages, as useful as it is, throws only some light on human language, considered as a biological object (it is limited, in particular, to those aspects which relate to recursion). On this topic, N. Chomsky adds the following (personal communication, Dec. 2009): "*Morphophonemics of Modern Hebrew* had nothing to do with formal languages, and in *Logical Structure of Linguistic Theory*, formal language theory is not mentioned at all. Clarification of the notions of computability were surely influential, but that is a separate matter. Formal language theory is mentioned at the beginning of *Syntactic Structures*, for pedagogic reasons, since the MIT undergrad students, engineers and mathematicians had been taught about the alleged universality of information-theoretic Markov source models. But even SS goes on pretty soon to what always seemed to me the central issues. The study of automata theory and formal languages is an interesting topic, but the implications for linguistics always seemed to me slight, even when I was working on these topics in the 50s and early 60s." See also note 39.

[57] Church (1951, p. 106) defends views that largely prefigure Montague's famous claim, when he writes: "Although all the foregoing account has been concerned with the case of a formalized language, I would go on to say that in my opinion there is no difference in principle between this case and that of one of the natural languages."

[58] The syntactic framework used by Montague is that of categorical grammar, which was first developed by Ajdukiewicz and Bar-Hillel. See Rivenc and Sandu (2009, chap.1) for more details on the relationship between these different formalisms.

concept of *recursion* that unites formal and natural languages, that is, the existence of a finite number of rules which enable an infinite number of sentences to be generated from a finite set of symbols. For Montague, Lewis and those whose research program was mainly driven by the goal of elaborating a recursive theory of meaning and interpretation (in line with the work of Tarski and Davidson), the central notion is the related concept of *compositionality*, or the idea that the meaning of a complex expression is a function of the meanings of its parts and the way they are combined. The concepts of recursion and compositionality, although distinct, are closely linked, because they are associated in varying degrees to other characteristics specific to human speech and language as a faculty.[59] In fact, they are both introduced side by side in one of Frege's pioneering logical texts on the composition of thoughts (Frege [1923] 1963, p. 1):

> It is astonishing what language can do. With a few syllables it can express an incalculable number of thoughts, so that even a thought grasped by a human being for the very first time can be put into a form of words which will be understood by someone to whom the thought is entirely new. This would be impossible, were we not able to distinguish parts in the thought corresponding to the parts of a sentence, so that the structure of the sentence serves as an image of the structure of the thought.

Already found in the quotation from Frege, and among the features of language which are adduced as evidence in favor of both the recursion and compositionality hypotheses, it is worth mentioning *productivity*, *learnability*, and *systematicity*. Recall that productivity refers to the ability to produce a potentially infinite number of new sentences; learnability is the capacity for language to be learned and mastered in a finite period of time; systematicity commonly refers to the possibility of recombining the units of language, that is, the idea that if an individual is able to interpret a given sequence of words (like "John loves Mary"), then in principle she is also able to interpret any sequence which is obtainable from it as a result of substituting expressions of the same category (such as "Mary loves John"). Productivity, learnability and systematicity are partially pre-theoretical concepts whose precise definition is still hotly debated.[60] If this is also true, to some extent, for the notions of compositionality and recursion (see for examples Hodges, 1998, on the distinction between different forms of compositionality), it seems fair to say that sufficiently precise definitions

[59] The concepts are distinct because compositionality implies a notion of interpretation for a language. The syntax of a language can be recursive without necessarily being associated with a compositional semantics. For examples of non-compositional semantics for recursive languages, see Janssen (1997) and Hodges (1998). Typically, a semantics for a given language is said to be non-compositional if it is possible to assign a semantic value to a complex expression without certain component expressions receiving semantic values of their own. In this sense, lack of compositionality corresponds to a kind of breakdown of the parallelism between syntax and semantics.

[60] On systematicity, see for example Fodor and Pylyshyn (1998) and Pullum and Scholz (2007).

of these two concepts exist in the case of formal languages.[61] In addition, the two concepts are so closely connected to the very definition of the syntax and semantics of formal languages that the question of their relevance to natural languages seems almost obvious.

I should add that each of the notions mentioned previously can be considered strong candidates for the status of constitutive properties of language as a faculty. Therefore, from among these various properties it is possible to select an equal number of properties which are presumed to be universal to human languages. The American linguist Hockett, for example, already included in the list of universal properties of human languages (which Hockett called "design features" of language) the fact that "all human languages are productive" or that "all human languages are learnable" (see Hockett, 1960, 1963). Hockett's perspective was nevertheless distinct from that of Frege, Chomsky or Montague. Hockett sought to uncover a set of features related to communication, which were such that no other animal communication system possessed them all, whereas Chomsky proposed to characterize human language internally, in relation to its structure and expressive power, that is to say, independently from the problem of communication.

In fact, formal syntax as well as formal semantics are both based on two postulates that can be stated as follows: "all human languages are recursive" (meaning that the syntax of all human languages is recursive) and "all human languages are compositional" (meaning that the process of semantic interpretation in all human languages obeys the principle of compositionality). In particular, when Montague wrote that the purpose of his theory was "to develop a universal syntax and semantics." (Montague, 1970, p. 223), his goal was to generalize and rigorously articulate the Fregean postulate that a sentence fundamentally possesses a function-argument type structure (cf. Frege, 1891/1960). Thus, in Montague grammar, a complex expression is constructed through concatenation of two or more constituent sub-expressions, and their interpretation is treated as a function which associates a resulting interpretation to the interpretation of those constituent expressions.[62] Montague's hypothesis, which continues to drive semantics to this day, is that the expressions of natural language each have different functional types, whose combination ultimately enables the process of interpretation.[63]

---

[61] In particular, strong and weak compositionality should be distinguished, cf. Hodges (1998) and Schlenker (2008). Strong compositionality means that the meaning of an expression is a function of the meaning of its *immediate* constituents and of their mode of combination.

[62] To give a simple example: a sentence such as " [[a man] sleeps]" will be dealt with in such a way that "a" denotes a function of an appropriate type, which takes "man" as argument, and produces as its value a function of another type; this latter function, associated to the complex phrase "a man," will in turn take "sleeps" as its argument, and produce as its value a truth-value (such as True or False).

[63] For a contemporary introduction to Montague grammar, cf. Gamut (1991, vol. 2) and Heim and Kratzer (1998). Schlenker (2008) gives an overview of the different fields of application of compositional semantics. Lapierre and Lepage (2000) and Rivenc and Sandu (2009) provide a detailed introduction to pioneering articles by Montague, including Montague (1970a).

What is one to think of the idea that all human languages are compositional and recursive? As to be expected, each of the corresponding assertions raises objections and has its opponents. First of all, with regard to recursion, the idea that there could exist non-recursive human languages is difficult to articulate precisely. The main reason for this is that any finite mathematical language is trivially recursive. Therefore, a counter-example cannot rely solely on the consideration of a finite corpus of utterances, but must take structure into consideration. Hence, when the concept of recursion is criticized, what is claimed is that there are non-productive human languages, or languages that might eventually be sufficient to express new thoughts, yet possess no recursion rule as properly understood (rules that would allow phrases to appear as constituents within phrases of the same grammatical category).[64] Recently, Everett (2005), a linguist and anthropologist, has argued that Pirahã, an Amazonian language of Brazil, lacks certain rules of embedding, which are common in English. He infers that "productivity (. . .) is severely restricted by the Pirahã culture." As things stand, however, few linguists seem to give credence to this hypothesis, not only because the data that has been advanced is very partial and largely inconclusive (Everett is one of the only non-native speakers to understand Pirahã, making expert assessment of his data difficult), but also because Everett does not analyze the embedding structures systematically.[65]

However, the thesis that languages are fully compositional has more commonly been challenged on the grounds that the meaning of expressions is sensitive to the context of their use, or because of the existence of idioms whose meaning seems fixed and non-functional with respect to their constituents (cf. Partee, 2004; Travis, 1997; Szabo,

---

[64] The latter possibility should not immediately be discounted, since one could imagine certain general mechanisms taking the place of recursion, such as rich mechanisms for anaphora. A sentence like "A man says that a dog barks," which is usually dealt with thanks to a recursive rule, for example, could be expressed as "A dog barks. A man said that," which would not involve any embedding rule (I thank P. Schlenker for this observation and for the example).

[65] Everett gives the example of the apparent limitation of the number of genitive embeddings in Pirahã (e.g., "the son of the sister of Jean") but offers no systematic test. However, he suggests that in some cases, Pirahãs nevertheless use circumlocutions. The type of causal argument that Everett gives for the restrictions imposed by culture on grammar goes in the direction explicitly opposite to that of the so-called Sapir-Whorf thesis (cf. Whorf, 1956), according to which the grammar of each language would have an influence on culture. An example of the kind of causal reasoning proposed by Everett is as follows: "No more than one possessor per noun phrase is ever allowed. Removing one of the possessors in either sentence makes it grammatical. A cultural observation here is, I believe, important for understanding this restriction. Every Pirahã knows every other Pirahã, and they add the knowledge of newborns very quickly. Therefore one level of possessor is all that is ever needed." On productivity, see the response by Pawley (2005, p. 638). On the Sapir-Whorf thesis, see the commentary by Levinson (2005, pp. 637–638). Potentially the strongest argument that is available to Everett is the fact that Pirahãs have a limited number system, which could be correlated to certain restrictions on recursive structures more generally. However, one may wonder if he does not make the mistake of reducing (grammatical and logical) competence to limitations that have to do with performance. Wierzbicka (2005, p. 641) argues in particular that "many languages have no numerals, yet as the Australian experience shows, their speakers can easily borrow or develop them when they feel the need to do so." See also Nevins. Pesetsky, and Rodrigues (2009) for a thorough critical analysis.

2007). The question of the interaction between compositionality and context dependence remains an open topic of exploration in linguistics, of which I will speak no further at this time. However, it should be noted that in this specific case, there would be little sense in imagining that certain human languages are "more" compositional than others, especially because the concept of compositionality is relative to a semantics and a grammar, and also because, as argued by Hodges (1998), to declare that a given language poses a challenge for compositionality implies that one already possesses a sufficiently complete and precise definition of its semantics as well as its grammar.

More generally, theoretical issues pertaining to the status of compositionality, and to that of recursion, are not so much about whether language is recursive or compositional in itself, but have more to do with specifying more precisely the complexity of the recursive grammars underlying natural languages, or the minimal syntactic constraints consistent with the compositionality hypothesis.[66] In other words, the recursion hypothesis, like the compositionality hypothesis, vastly underdetermine the form that a grammar should take (be it universal, or that of a given language). If then, there is a debate concerning the comparison between languages, it mostly touches on the issue of whether different languages have the same structure or not, or if certain syntactic constraints observed in a given language have analogs in others. Although there is some similarity between the Montagovian conception of universal grammar and that of Chomsky's, I must emphasize that a Montague grammar provides only a general framework for the formal description of compositional and recursive grammars, and remains neutral as to the nature of the universal constraints liable to govern word order across languages.

4.2 DIFFERENT TYPES OF LINGUISTIC UNIVERSALS

Recursion and compositionality belong to what I have called the *architectonic principles* of universal grammar. In this section, I will now examine the question of whether there are *units, categories* or *structures* that are invariant across languages. Consideration of this question will lead me to distinguish different aspects of the concepts of a linguistic universal and of a cross-linguistic regularity.

4.2.1 Laws and Rules

For a philosopher of science with an interest in comparing linguistics to the other natural sciences, including physics, an important issue that I have already mentioned

[66] For a recent discussion of the status of the compositionality hypothesis, see D. Dowty's (2007) article, which also discusses the issue of whether compositionality can be direct or transparent (in many formalisms, for example, a semantics is compositional only if there are rules for changing the type of expressions, making it the case that compositionality is not direct or clear). On the logical problem of whether any semantics for a given recursive grammar can be made compositional, cf. Janssen (1997, section 9) and Hodges (1998). The answer to this question depends on the precise way in which the problem is formulated.

is whether or not there are laws of language, analogous to those of nature in physics. The concept of a law is not often used in linguistics, where the notion of a rule is preeminent. However, the concept of a rule is in large part relative to the grammar of a particular language. When one refers to *laws*, what one generally has in mind are statements that would describe cross-linguistic regularities, or which would formulate general constraints on the form that a *system of rules* should take.

Tellingly, the concept of a law was first used in historical linguistics to describe certain systematic correspondences between the phonetic systems of several languages. This is the case of the so-called Grimm and Verner laws. Grimm's law, which was later complemented by Verner's law, established a correspondence between the sounds of several ancient languages such as Sanskrit, Greek and Latin, and the Germanic languages of Gothic and English. The law states that certain unvoiced occlusives in Latin and Greek become fricatives in Gothic and English (e.g., the Latin root *ped-* of the French word *pied* corresponds to *fetus* in Gothic and to *foot* in English, so that [p] corresponds to [f]; similarly, voiced occlusives become unvoiced (*decem*, which corresponds to *dix* in French, corresponds to *ten* in English, so [d] becomes [t]), and so forth. In speaking of laws, Grimm and Verner therefore mainly had in mind rules of phonetic evolution or change. The sense in which all this is a matter of laws, is that these principles of correspondence are systematic and especially "exceptionless" (a point emphasized by Verner himself) when they are brought to bear on the entire lexicon of the languages in question. However, as rightly pointed out by Lightfoot (2006, p. 29), these laws are essentially descriptions of changes specific to a small group of languages, changes that may be contingent, so that these laws cannot claim the same universality as the laws of Boyle and Newton. Pursuing this analogy further, one could say that the "laws" of Grimm and Verner have, at best, the same epistemological position as Kepler's laws for the motion of planets in the solar system: of course, they are laws of evolution, but essentially descriptive ones, relative to a limited domain, which call for a more general explanation.

4.2.2 Substantive Universals and Formal Universals

Therefore, if I mention the concept of a law, understood in this sense, it is actually in order to better clarify what deserves to be called a linguistic universal, or a linguistic invariant. From the logical point of view, a linguistic universal presents itself as a universal statement that quantifies over the class of all human languages. I have already discussed some examples, such as "all human languages are recursive." There are many other universal statements of this type that range over the class of all human languages, but they do not all have the same epistemological status. For example, there are universal statements about the phonology of human languages, such as "All languages have syllables" or "all languages have consonants and vowels," or more subtly, "all languages have at least two of the three voiceless plosives [p, t, k]" (Gussenhoven and Jacobs, 1998, pp. 28–29). However, *universal facts* of this kind are not necessarily significant from a theoretical point of view.

For a deeper understanding of this point, it is useful to distinguish several dimensions of the concept of a linguistic universal. Chomsky (1965) proposes to distinguish two types of universals, *formal* universals and *substantive* universals. A different concept of a universal, is the notion of a typological universal, associated with the work of Greenberg (1963a; 1963b), whose relation to Chomsky's distinction remains to be clarified. Finally, some linguistic universals, particularly those found in certain areas of formal semantics, amount to *logical* universals. Their status appears to be hybrid vis-à-vis the Chomskyan distinction.

Chomsky's notion of a substantive universal concerns generalizations about phonological, morphological, syntactic, or semantic units, that are supposed to be the constitutive elements of any human language. Chomsky's first example concerns Jakobson's theory of distinctive features in phonology, which states that the sounds of each language can be characterized in terms of a finite and universal inventory of articulatory features. The corresponding universal statement would be in this case: "The phonology of all human languages can be represented from the same universal set of features." The examples Chomsky gives in the case of syntax and semantics relate to grammatical categories or to the realization of certain semantic functions by specific lexical items. For example, Hockett (1963) writes that in all languages, there is a distinction similar to the one we have in English between nouns and verbs, or that all languages have deictics (pronouns such as "he," "her," "this," "that," etc.) or again, that all languages have proper names. Most if not all universals proposed by Hockett count as substantive universals in Chomsky's sense.

Unlike substantive universals, the universals Chomsky calls *formal* designate universal constraints on the form of grammars. The example Chomsky gives in the case of syntax involves the very concept of transformation, and the corresponding statement would be that the grammar of all human languages includes transformational rules. Another example of an allegedly universal constraint on the grammar of human languages is provided by the so-called X-bar theory of syntax, which states that the items of each grammatical category are organized according to the "X-bar" schema, that is to say that words are organized hierarchically in ordered phrases by projection of certain functional heads (for example: a noun phrase NP is the maximal projection of a noun of type N, see the diagram in Figure 2), so that for each category one can distinguish a notion of *complement, adjunct,* and *specifier* (see Radford, 1995; and Chomsky and Lasnik, 1995).[67] Just like recursion or compositionality, formal universals such as the X-bar schema are theoretical hypotheses about the nature of the computational system to which universal grammar corresponds. Hence, the assumption that all languages can be described using X-bar theory, or that all languages have a level of deep structure which allows for transformations, is more informative about the structure of language than the simple assertion that all languages are compositional.

[67] For example, in the noun phrase "a tall student of physics," "student" is of level N, "of physics" is its complement, "tall" is its adjunct, and the determiner "a" can be considered a specifier.

4.2.3 Typological Universals

An aspect shared by what Chomsky calls substantive universals and by what he calls formal universals is that each of the postulated universals is supposed to play an explanatory role in linguistic analysis. In this respect, they must be distinguished, for example, from Greenberg's *typological* universals, which give descriptive generalizations about the *surface order* of words across languages. Most of Greenberg's universals are "implicational universals," that is to say, restricted universal statements, for example: "Languages with dominant Verb Subject Object (VSO) order are always prepositional" (Greenberg's Universal 3). For example, Gaelic is a VSO-type language, unlike French, which is Subject Verb Object (SVO). In Gaelic and English, words like "of," "to" "toward" are ante-posed with respect to the nouns that govern them (in English one says "to the city" and not "the city to," as would be the case in a postpositional language such as Basque). Greenberg's text includes forty-five alleged universals of this kind established on the basis of a corpus of thirty different languages belonging to diverse language groups. As shown in the example of Greenberg's Universal 3, the universals in question are "tendential" and actually describe regularities of a statistical nature.

In the literature the issue concerning the relevance of typological universals to the update of the strictly formal universals of universal grammar remains widely discussed. According to Baker (2001), for example, certain typological regularities must result from the principles of universal grammar. For example, it seems that there is no OSV type language, as is shown by the corpus of more than 600 languages established by Dryer (see Baker, 2001, p. 128).[68] According to Baker, this lacuna must follow from a general constraint of universal grammar, namely that the verb and its complement must combine as soon as possible (which Baker calls the *verb-object constraint*, p. 93). As Baker argues, this constraint does not explain everything, since there are also VSO languages such as Gaelic, in which, precisely, the subject seems to intervene between the verb and its complement. However, Baker suggests that if one takes auxiliaries into account, VSO type languages are actually languages in which the order is Aux SVO, so that the verb-object constraint is violated only in appearance, in this case.

Contrary to Baker, Newmeyer (2005) defends the thesis that typological patterns do not come from universal grammar, but rather call for functional explanations related to performance. For Newmeyer, as indeed originally for Chomsky, "Typological generalizations belong to the domain of E-language," and not to I-language: in other words, these generalizations may have to do with linguistic conventions (in a broad sense), rather than with internal constraints of the computational system specific to the language faculty. Newmeyer's main argument is that most of Greenberg's universals seem to face significant exceptions, which therefore invalidate the idea that universal grammar would directly encode these typological constraints.

---

[68] A controversial case for such a generalization, cited by Baker, is that of Warao, a South-American language.

However, Newmeyer's argument is based in part on those of Greenberg's universals which only capture imperfect statistical trends. Some facts seem absolutely universal. As noted by Pinker (1994) or by Comrie (2003), for example, no language forms questions by palindrome starting from the words of the affirmative sentence (such as from: *Marie is at the beach* to *beach the at is Mary?*): this universal fact in itself provides little information, but it is at least indicative of the fact that a semantic structure must be realized according to some minimal constraints across languages. In this sense, typological universals, although they may not provide direct access to universal grammar, may nonetheless be indicative of constraints on the deep structure of utterances across languages.[69]

4.2.4 Semantic Universals

In addition to typological universals, I would like to mention in closing what I will call *logical* or *semantic* universals. Such universals were brought to light in the 1980s in research on quantification and generalized quantifiers. The corresponding generalizations naturally involve the syntax of natural languages, but what makes it appropriate to refer to them as semantic universals is that the discriminant properties (such as monotonicity) primarily concern the entities used to *interpret* a particular class of syntactic objects.

A pioneering article by Barwise and Cooper (1981) provides an example of a *substantial* universal in Chomsky's sense, which states that any natural language contains syntactic elements whose function is to express generalized quantifiers over the domain of discourse. In particular, this universal predicts that there should be no language which cannot express universal quantification (such as "all the men have arrived").[70] The rest of the article, however, is devoted to the statement of finer-grained generalizations concerning the form of determiners in all natural languages. One of these universals, for example, is the monotonicity constraint, according to which simple noun phrases in natural language express monotonic quantifiers or conjunctions of monotonic quantifiers (see section 3.3.3). This constraint predicts that no language will grammaticalize an expression like "an even number of X" under the form "Q X," where Q is a simple determiner, for the reason that the quantifier "an even number of X" is not monotonic.[71] Like the syntactic universals postulated by Chomsky, this type of semantic universal is a formal universal, which may account for a typological regularity.

Significantly, the inductive basis for the universals proposed by Barwise and Cooper is essentially limited to English, the main arguments used to generalize being precisely hypotheses concerning the logicality of quantifiers. However, the ambition of the research program started by Barwise and Cooper is to account for the form of possible

[69] See e.g. Cinque (2005) for an attempt to derive Greenberg's Universal 20 in generative grammar.

[70] Such a point may seem self-evident, but it is precisely contested by Everett in the case of Pirahã.

[71] Cf. the definitions given in section 3.

grammars, so as, for example, to try to characterize in genuinely semantic terms the grammatical categories that are cross-linguistically robust. This research program, of course, by no means rules out more empirical research on the properties of particular grammars.[72]

4.3 THE EXPLANATION OF LINGUISTIC UNIVERSALS

If one compares the different types of universal generalizations just reviewed, one sees that they do not all exist on the same plane. Typological generalizations, regardless of the level of the language concerned, are rather signs of specific constraints on universal grammar than the direct expression of such constraints. An important aspect of the classification just provided is that it is in fact indicative of the nature of what is meant by "universal grammar." Universal grammar is not simply a catalog of descriptive generalizations that are robust across languages.[73] By universal grammar one must understand the computational constraints on the system thanks to which we produce and interpret sentences. The examples of syntactic or semantic universals I have given are supposed to correspond to properties of this complex computational system. However, this characterization raises a new problem: how is one to explain the emergence and robustness of a semantic or syntactic property, from a cross-linguistic point of view? More precisely, does the datum of a presumed universal of natural language involve mechanisms specific to language, or on the contrary, general mechanisms of the human mind?

Pinker (1994) underlines two important points concerning the derivation of formal linguistic universals: the first is that these universals are distinct from any universal conventions that would be passed down from generation to generation. Pinker writes, that "children could learn that English is SVO *and* has prepositions, but nothing could show them that *if a* language is SVO, *then* it must have prepositions." In this sense, the underlying typological generalization, if indeed it is universal, must reflect a constraint on the computational system itself. The second thesis that Pinker defends is that the constraints of universal grammar should not be confused with the constraints that govern other cognitive systems. For example, a lexical universal seems to be that any language that contains the word "purple" also has the word "red," but this universal would appear to relate to constraints that concern the visual system.[74]

Pinker's remarks raise a difficult and still largely unresolved problem in linguistics, which concerns the delimitation of the language system and its relation to other cognitive systems. For example, consider a Jakobsonian phonological universal such as

---

[72] Cf. Keenan and Stabler (2003) for a review of recent research on the relation between grammatical invariants and semantic invariants.

[73] See Pinker (1994, p. 237), who writes: "In any case, Greenbergisms are not the best place to look for a neurologically given Universal Grammar that existed before Babel. It is the organization of grammar as a whole, not some laundry list of facts, that we should be looking at."

[74] Cf. Berlin and Kay (1969) for a study of color terms across languages.

"all phonetic features have binary representations."[75] Is this the expression of a computational constraint strictly speaking (what Hauser et al., 2002, call the faculty of language *in the narrow sense*), or rather a constraint related to the auditory and articulatory system (what Hauser et al., 2002, call the faculty of language *in the broad sense*)? Clearly, such questions concern cognitive science well beyond purely internal or formal research into the nature of particular grammars.

Pinker's position on these issues is opposed to a position that can be described as functionalist in a broad sense. The term functionalism covers a wide range of schools of thought, but in recent years it has been commonly associated with the idea that the properties of language are not necessarily within the purview of an autonomous and innate linguistic system, but rather of general properties of the cognitive system, or of pragmatic constraints on the use of language and communication. First of all, it should be noted that several points of convergence exist between functionalism and formalism. A functionalist linguist such as Comrie (2003), for example, agrees with Pinker on the idea that linguistic universals cannot be explained simply by the survival of properties of a primitive universal language (what Comrie calls the *Hypothesis of monogenesis*). In addition, Comrie also agrees with Pinker that the rules of grammar of any human language obey constraints of "structure-dependence" (see Chomsky 1979, to whom the concept is due).[76] A rule of question formation that would function by palindrome, for example, would alleviate the need for an analysis of sentences into differentiated phrases and would not involve structure-dependence.

However, Comrie argues that "this property of structure-dependence is not a specific property of language, but rather a general property of human cognition." (2003, p. 200). Comrie proposes two arguments in support of this thesis: the first is that when it comes to memorizing sequences of digits (e.g., phone numbers) we typically segment or "chunk" the sequence into sub-sequences, seemingly in virtue of constraints related to the faculty of memory rather than language. The other argument is that the task of reciting the alphabet backward, a learned and unstructured sequence of letters, is itself very difficult to do successfully. In this case, the fact that one does not form questions by palindrome should therefore follow from the fact that the very process of forming palindromes is cognitively demanding.

The arguments advanced by Comrie are challenging, but none of them is quite conclusive: in particular, it could be that the difficulty of carrying out certain operations on unstructured word sequences comes precisely from the fact that we memorize arbitrary sequences of words or letters by using principles of organization which are in themselves linguistic.[77] Moreover, even if a general cognitive principle could explain

[75] See e.g. Kenstowicz and Kisserberth (1979, p. 23), who write: "languages such as French make a distinction between whether a vowel is round (like lune, [lün]) or non-round (like ligne [liN]). But so far as is known, no language makes distinctions between three degrees of rounding."

[76] Chomsky puts forward such a concept in the context of a debate with Piaget, who could easily be classified as a functionalist. Cf. Piatelli-Palmarini (1979).

[77] Think e.g. of how the alphabet is learned thanks to the Alphabet Song.

that some syntactic operations are illicit across languages, such principles do not necessarily explain why licit operations obey certain particular positive constraints.

This point can be illustrated by a second example of functional explanation advanced by Comrie, this time to explain a Greenberg-style typological universal. The universal in question concerns the distribution of reflexive pronouns across languages. Comrie observes that languages are clearly divided into three types. Some languages, such as contemporary English, distinguish morphologically between reflexive pronouns and non-reflexive pronouns for all persons (*myself* vs. *me*, *yourself* vs. *you*, *himself* vs. *him*, etc.). Other languages, such as French, do not distinguish between reflexive and non-reflexive for the first and second persons (*me*, *te*), but distinguish them in the third person (*se* vs. *le/la/les*). Take the following sentences:

(42) Pierre se voit dans le miroir / Pierre sees himself in the mirror.

(43) Pierre le voit dans le miroir / Pierre sees him in the mirror.

(44) Je me vois dans le miroir / I see myself in the mirror.

(45) Pierre me voit dans le miroir / Pierre sees me in the mirror.

A sentence such as (43), in particular, whether in French or English, cannot be construed so that the pronoun "him" or "le" is coreferential with the subject "Peter/Pierre." Coreference in this case is prohibited, a phenomenon that constitutes one of the basic principles of the theory of binding.[78] A third group, however, includes language that do not distinguish morphologically between reflexives and non-reflexives, for any person (Comrie gives the example of Old English). A universal fact that Comrie notes, however, is that there seems to be no language symmetrical to French, namely a language that distinguishes reflexives and non-reflexives in the first and second persons, but not the third. The implicational universal that Comrie draws from this is that if a language distinguishes reflexive pronouns and non-reflexive ones, it must distinguish them in the third person. According to Comrie, this fact cannot be explained in a purely internal manner. The asymmetry between the first and second person on the one hand, and the third person on the other, must instead be explained, according to him, by the observation that both the first and the second person serve to designate the speaker or interlocutor, reference to whom is usually unambiguous. This does not hold true for the third person. It would be very uneconomical if a language distinguished reflexive and non-reflexives in cases where reference is unambiguous, but did not make this distinction in cases where there is ambiguity.

Thus, the explanation suggests that the morphological distinction between reflexive and non-reflexive is only useful where the pronoun's reference is potentially

[78] The principle in question is "condition B," which states that a non-reflexive pronoun cannot be c-commanded by a coreferential antecedent. For a presentation of binding theory, cf. Büring (2005).

ambiguous. But as can be seen, this explanation does not explain everything. In particular, it does not explain why some languages, such as Old English, can do without the morphological distinction in the third person. This gap in the explanation is not necessarily invalidating, since it is likely that other principles will explain why this possibility can obtain, but clearly it is less satisfactory than if there were no language such as old English.

A feature functional explanations have in common is that they seek to account for linguistic regularities on the basis of principles that have to do either with cognition in general, or with the use of language in general and therefore its pragmatic dimension. The conversational maxims of Grice (1967), which play a central role in explanations of a pragmatic nature, undeniably have a functional dimension, insofar as they embody principles of rationality supposed to hold universally, regardless of the language used, while being liable to interact with morphology and syntax.[79] Horn (1989, pp. 254–255), for example, proposes to explain the absence of lexicalization across languages of a simple quantifier equivalent to "not all" based on Grice's maxim of quantity and a theory of scalar implicatures.[80] More generally, the theory of optimality that is used in phonology and more recently in pragmatics, proposes to account for the exclusion of certain phonetic and syntactic forms by postulating systems of lexicographically ordered constraints (rather than derivational systems of rules), that are intended to account not only for the categorical exclusion of certain forms, but also for the relative preference given to certain realizations rather than others. A review of optimality theory would lead us too far afield, but what should be retained from this brief discussion of the derivation of linguistic universals is that they are conceived of in conflicting ways, either as the expression of autonomous rules of the faculty of language, or as the expression of more general cognitive and pragmatic constraints, not necessarily specific to language.[81]

4.4. LINGUISTIC DIVERSITY, PRINCIPLES, AND PARAMETERS

To close this chapter and in order to further clarify the opposition just mentioned between functional explanations and formal explanations, I propose to conclude with a brief discussion of the problem of linguistic diversity. There are several aspects of the problem of diversity. One is the issue of the evolution of languages and of their differentiation: how are languages born, how do they evolve, and how do they come to differ from one another? Another issue is the compatibility of the hypothesis of universal grammar with the very observation of linguistic diversity.

---

[79] See Grice (1989).

[80] Horn's theory, roughly summarized, is based on the observation that usage of the quantifier "some" causes in positive environments the pragmatic inference (or implicature) "some but not all." For example: "Some students have arrived" is usually taken to mean "some students have arrived, but not all." Such a systematic strengthening, which can be explained by appealing to Grice's maxim of quantity (make your contribution as informative as possible), is supposed to account, according to Horn, for the absence of lexicalization of a determiner such as "not all."

[81] On optimality theory, see e.g. Prince and Smolensky (1997).

Before considering these questions, it is useful to recall certain salient facts concerning the phenomenon of linguistic diversity. It is estimated that there are currently between 5,000 and 8,000 languages spoken in the world (see Evans and Levinson, 2009). An exact count of languages at any given moment in time is problematic, because if one chooses to define a language based on the concept of mutual understanding between speakers, this is a relative concept, which does not allows one to draw sharp boundaries between given idioms (see Picq et al., 2008). So when one counts 5000 to 8000 languages, this is done on the basis of multiple criteria, which take into account geographical location, and also the perception of the users of the language community to which they belong. A second aspect of linguistic diversity is the fact that in addition to spoken languages, there is also a wide variety of signed languages. As Emmorey (2002, p. 1) points out, it is necessary to avoid the prejudice that there exists a universal sign language:

> There are many distinct sign languages that have evolved independently of each other. Just as spoken languages differ in their lexicon, in the types of grammatical rules they contain and in historical relationships, signed languages also differ along these parameters.

Thus a census of the number of signed languages is as a matter of principle subject to exactly the same limits as that of spoken languages, even if to date more than one hundred sign languages have been documented (Evans and Levinson, 2009). To this dual synchronic diversity between spoken and signed languages, one should naturally add diachronic diversity, which is implied by the fact that Latin and ancient Greek, for example, are no longer spoken by a living community, so that we know them through writing. As a result the evolution of languages over time makes the project of counting human languages as arduous and difficult in principle as that of counting living species.

The analogy between languages and living species brings us to the heart of the problem at hand. By putting emphasis, in the previous sections, on the hypothesis of universal grammar, or on the notion of linguistic prediction, it might seem like I have exaggerated the importance of these concepts and missed a more illuminating analogy, which would be to picture the linguist as a naturalist or biologist engaged in the description of languages similarly to that of living species. However, it is important to be very careful about what such an analogy is worth, in this case. A language can certainly be seen as a complex organism, the product of a large number of factors and constraints. These constraints have to do with communication and with the conventions specific to certain communities of individuals. Such conventions can evolve in accidental and contingent ways, particularly as is the case insofar as the lexicon of each language is concerned, as well as its pronunciation, or morphology. By extension, it may seem as if none of the architectural dimensions of language were free of change and variation. Seen in such a light, the "predictive" dimension of linguistic research might seem entirely illusory.

Nevertheless, as I have repeatedly stressed, the constraints that account for language use are not only a historical and collective social product: every individual is born predisposed to speak, and as Chomsky points out, for this reason language must also be considered internally, and ultimately as dependent on a mental, neurological and genetic architecture. If, therefore, language is to be compared to biology, it is paramount to remember that the linguist is in as complex a position as the biologist vis-à-vis the living: just as the study of life cannot be reduced to a simple taxonomy of life forms, but is bound up with chemistry, physics and ethology, the study of language is interconnected with neurology, biology, psychology, as well as with studies of a historical nature of facts concerning the evolution of spoken forms. Viewed in this way, the phenomenon of linguistic diversity is hardly easier to explain than the diversity of life itself.

I will put aside, at this time, the question of the origins of language, or that of the driving forces of the evolution of a language,[82] all hotly debated issues, in order to focus on the relationship between linguistic diversity and the hypothesis of universal grammar. The dominant model in generative grammar since the late 1970s is the one known as "Principles and Parameters" (Chomsky, 1981; Rizzi, 1978). During the 1950s and 1960s, as Rizzi (2007) explains, universal grammar was considered by Chomsky and by the generativists essentially as "a kind of grammatical metatheory, a theory explaining the form of the rules and expressing general conditions on their application." Particular grammars were themselves viewed as "systems of rules specific to the language and to its constructions." Beginning in the late 1970s, this view of the relationship between universal grammar and particular grammars changed. From then on universal grammar was seen as a system of principles and parameters, and particular grammars were conceived of as so many realizations of universal grammar in which these parameters are set in a specific way.

One of the most eloquent examples of the notion of a parameter is probably the one which concerns word order in different languages, or more precisely constituent structure. English or French, for example, are so-called head-initial languages, in the sense that the functional head of a phrase precedes the phrase. But Japanese, for example, as well as Lakhota, the language of the Sioux Indians (Baker 2001, p. 61) are head-final languages, where the functional head of a phrase now comes at the end of phrase. This means that a sentence of English such as "John found that letter under his bed," whose analysis in constituents is approximately: $[_{IP}$John$[_{VP}$found $[_{DP}$that letter] $[_{PP}$under $[_{DP}$his bed]]]], would be in Lakhota or Japanese "John letter that bed his under found," that is $[_{IP}$John $[_{VP}[_{DP}$letter that] $[_{PP}[_{DP}$bed his] under] found]]] (cf. Baker 2001, p. 61). For example, within the DP "that letter," the determiner precedes the noun in English, whereas in Japanese or Lakhota the determiner follows the noun in the phrase. Similarly, the verb comes first in the VP in English, but last in Japanese or Lakhota. This example is significant, because at the same time as it shows the

[82] See e.g. Pinker (1994), Chomsky (2000), Baker (2001), Hauser et al. (2002), Lightfoot (2006), and more recently Chomsky (2010) on the relevance and limits of Darwinian explanations of the evolution of language and languages.

difference between English and Japanese, it suggests that in each language sentences have a common constituent structure, which follows the same principle of functional head projection. The underlying principle of universal grammar is thus that in all languages, every sentence is the projection of a functional head, but the parameter related to this principle is that the functional head can be to the left or to the right of its complement within the phrase.

According to Baker (2001, p. 45), parameters can hence more generally be seen as "the atoms of linguistic diversity." For example, it will probably not have escaped the reader's attention, given the previous example, that in Japanese as in French, the subject of a finite temporal phrase is at the beginning of the sentence. But there are other head-initial languages in which the subject comes last (languages like Malagasy, see Baker 2001, p. 166). This suggests that the positioning of the subject can in turn be treated as a parameter. More abstractly, taking the so-called Principles and Parameters vision of language to its limit, one could therefore represent each language as a vector in a multidimensional space, each coordinate of which would indicate the value of the corresponding parameter.

However, the "Principles and Parameters" model does not merely aim to unify linguistic diversity and universality in the abstract. In the view initially defended by Chomsky, the notion of a parameter is also relevant to account for the acquisition of language, since one may consider that the child, when he learns language, basically aims to gradually set the parametric values of his parents' language (cf. Rizzi, 2007). Finally, as I showed earlier, the parametric view also serves to account for the diversity of languages from a diachronic perspective, in the sense that a morphological or syntactic change is often indicative of a level of shared structure (see Pollock, 1997, and Baker, 2001, p. 136, who proposes to speak in terms of a *verb attraction parameter* in relation to the distinction between French and English concerning order of the verb, auxiliary and adverbs, see section 3.4).

The so-called Principles and Parameters approach remains to this the day the frame of reference for generativists, but it too has opponents and critics. One of the problems with such a view concerns the question of whether the number of parameters is actually finite or not, and the question of whether the parameters are prioritized (logically, but also in terms of learning). Baker is probably one of the most committed defenders of this view, since he has sketched out a hierarchy of parameters, aimed at connecting language groups to each other which at first glance seem very heterogeneous (see Baker, 2001). Baker does not hesitate to compare the task of the linguist in this respect to the effort it required in order to establish a periodic table of the chemical elements.

Among the opponents of the parametric model, there are some theorists who could be described as "moderate," such as Newmeyer (2005), for whom the term parameter is simply less explanatory than the notion of a rule specific to a given language. According to Newmeyer, an explanation of linguistic diversity must take into account how linguistic performance may interact with certain sociolinguistic conventions.[83] Newmeyer

[83] On the concept of a linguistic convention, and for an attempt to reconcile a "formal" definition of language with a "social" one, cf. also Lewis (1968).

can be described as moderate in his criticism, however, inasmuch as he remains a supporter of the idea of universal grammar, although his approach is closer to the metatheoretical conception of early generative grammar. Other critics, however, are more radical, such as Evans and Levinson (2009). According to them, even the notion of constituent structure should be counted among the dogmas of modern linguistics which need to be revised.[84] One of the theses they put forward is in fact that "language diversity is characterized not by sharp boundaries between possible and impossible languages, between sharply parameterized variables, or by selection from a finite set of types." Their hypothesis is that "instead, [linguistic diversity] is characterized by clusters around alternative architectural solutions, by prototypes (like "subject") with unexpected outliers, and by family-resemblance relations between structures ("words," "noun phrases") and inventories ("adjectives")." In this respect, Evans and Levinson belong to the functionalist tradition that I have discussed, and underlying diversity, they are willing to see certain statistical regularities or "recurrent clustering of solutions" to given constraints, rather than the expression of invariant mechanisms. In this respect, more than Newmeyer, Evans and Levinson emphasize the need to reassess the initial Chomskyan opposition between competence and performance.

It would be foolhardy and out of our jurisdiction to adjudicate this debate. One point that should be emphasized, however, is that this debate illustrates the vitality of the opposition between performance models and competence models, that has existed ever since the beginning of generative grammar and the methodological primacy given by Chomsky to the notion of competence over that of performance. As previously noted, one of the issues still open in this debate is not so much to determine whether language involves innate mechanisms or not (this is clearly the case), but rather the extent to which language involves autonomous computational constraints rather than functional constraints involving a large number of systems (communication, phonation, hearing, memory, etc..)

5. Conclusion and Avenues for Further Research

In closing this chapter, let me summarize the main steps of our journey. I sought to clarify four groups of questions:

(i) What is linguistic theory and what are its goals?

[84] Evans and Levinson suggest in particular that the concept of constituent structure is too closely linked to the grammatical model of languages such as English, in which word order is relatively rigid, as opposed to certain morphologically rich languages in which the order of words is very free (they give the example of Latin). In transformational grammar, however, it is recognized that the so-called free word order languages are just languages in which word order is relatively less constrained, certain syntactic operations (e.g., the formation of questions) remaining subject to strong syntactic constraints. An even more extreme case than Latin is Warlpiri, an aboriginal language of Australia, where the word order was considered entirely free up until the work of Ken Hale, among others, starting in the 1960s, on free-order languages.

(ii) What does the evolution of linguistics from the structuralist to the generativist framework represent from the perspective of history and philosophy of science?
(iii) What do the concepts of generalization, explanation, and prediction signify in linguistics?
(iv) What is the status of the notion of linguistic universal in linguistics?

My goal will have been achieved if, concerning each of these questions, I have given the reader a fairly accurate idea, albeit brief, of the methods of contemporary linguistics, and of the similarity in style between linguistics and the other natural sciences, as well as of the key methodological debates within the discipline.

In conclusion, it seems important to once again situate linguistics with respect to the other sciences and to highlight some of the opportunities that are open to linguistics in the years to come. For a long time, especially during the structuralist period, theoretical linguistics was ranked alongside social anthropology, especially because of the view that language is a reflection of a society and a culture (cf. Jakobson, 1952) or conversely because of the idea that language in turn affects the way people see the world (see especially Whorf, 1956). Since the inception of generative grammar, and under the influence of Chomsky, linguistics gradually made its place alongside cognitive psychology and the other cognitive sciences, whose goals it helped to define. This is in large part a reflection of Chomsky's hypothesis that language should be considered above all as an internal tool for the individual expression of thought, rather than as a social and external instrument of communication between individuals. In this respect, Chomsky's opposition to structuralism, as to behaviorism or certain varieties of functionalism, is surely a form of methodological individualism. For Chomsky, of course, the point is not to deny that language is an instrument of communication, but to argue that the parameters that govern communication are secondary relative to those that govern the expression of thoughts. This view, as we have emphasized, remains controversial, but it must be recognized that it has significantly renewed the study of language for over half a century.

If the merits of methodological individualism are granted, many difficult questions still remain unanswered for linguistics. One of these questions concerns the nature of the biological and genetic basis of the faculty of language: what is the biological material that distinguishes man from other animals, including apes, from the linguistic point of view? (See Pinker, 1994; Hauser et al., 2002). A precise answer to this question should help clarify the extent of the actual innate component of language. Another series of questions concerns the nature of the brain processes underlying the acquisition and the processing of language as well as of meaning. Since the 1960s formal syntax and semantics have led to the development of analytical tools for some fragments of natural languages (Montague, 1973), and even of computer regimenting such fragments (see Blackburn and Bos, 2005). However, there is clearly a significant difference between these computational models of meaning and the description of the psychological and neurological processes of verbal production and comprehension.

This does not mean, of course, that current mathematical models of meaning are lacking in value or useless. As pointed out by Poeppel and Embick (2005), a central and still unresolved epistemological problem for neurolinguistics concerns in particular the establishment of a plausible functional correspondence between the phonological, morphological and syntactic units and operations postulated by linguists, and the relevant units and operations from the point of view of brain imaging. To date, as argued convincingly by Poeppel (2005), or also Grodzinsky (2007), the study of syntactic structures and grammatical analysis remains the most reliable guide for a theory of the units and underlying neurolinguistic processes, rather than vice versa, contrary to what a naively reductionist view might suggest. Ultimately, however, there is hope that a harmonious integration of formal theories of meaning and of the computational processes involved in the brain will eventually take place.

References

Anderson, S. R. (1985), *Phonology in the Twentieth Century: Theories of Rules and Theories of Representation*, Chicago: University of Chicago Press.

Austin, J. L. (1962), *How to Do Things with Words*. Oxford: Clarendon.

Bach, E. (1965) "Linguistique structurelle et philosophie des sciences," in E. Benveniste, N. Chomsky, & R. Jakobson (1966), *Problèmes du Langage*, Collection Diogène (n° 51), Paris: Gallimard, pp. 117–136.

Baker, M. (2001) *The Atoms of Language*, Oxford: Oxford University Press.

Barwise, J. & Cooper, J. (1981) "Generalized Quantifiers and Language," *Linguistics and Philosophy* 4, pp. 159–219.

Benveniste, E. (1971a) "The Levels of Linguistic Analysis," in *Problems in General Linguistics*, Coral Gables, FL: University of Miami Press, pp. 101–112. Originally published in French as E. Benveniste (1962) "Les niveaux de l'analyse linguistique," in *Problèmes de linguistique générale*, tome 1, Paris: Tel Gallimard, chap. 10, pp. 126–136.

Benveniste, E. (1974) "La transformation des catégories linguistiques," in *Problèmes de linguistique générale*, tome 2, Tel Gallimard, chap. 9, pp. 126–136.

Berlin, B., & Kay, P. (1969) *Basic Color Terms: Their Universality and Evolution*, rev. ed., 1999, Stanford, CA: CSLI Publications.

Blackburn, P., & Bosn, J. (2005) *Representation and Inference for Natural Language: A First Course in Computational Semantics*, Stanford, CA: CSLI Publications.

Bloch, B. (1941) "Phonemic Overlapping," *American Speech* 16, pp. 278–284.

Bloomfield, L. (1933) *Language*, New York: Holt.

Bloomfield, L. (1939) "Menomini Morphophonemics," in *Études dédiées à la mémoire de M. le Prince N. S. Trubetzkoy, Travaux du Cercle linguistique de Prague* 8, pp. 105–115.

Bobalijk, J. D., & Wurmbrand, S. (2008) "Case in GB/Minimalism," in A. Malchukov and A. Spencer (eds.), *Handbook of Case*, Oxford: Oxford University Press, pp. 44–58.

Bromberger, S., & Halle, M. (1989) "Why Phonology Is Different," *Linguistic Inquiry* 20, pp. 51–70, repr. in Halle (2002).

Büring, D. (2005) *Binding Theory*, Cambridge: Cambridge University Press.

Carnap, R. (1947), *Meaning and Necessity: A Study in Semantics and Modal Logic*. Chicago: University of Chicago Press.

Chemla, E., Homer, V., & Rothschild, D. (2011) "Modularity and Intuitions in Formal Semantics: The Case of Polarity Items," *Linguistics and Philosophy* 34(6), pp. 537–566.

Chomsky, N. (1955) *The Logical Structure of Linguistic Theory*, Den Haag: Mouton.

Chomsky, N. (1956) "Three Models for the Description of Language," *IRE, Transactions on Information Theory* IT-2, pp. 113–124.

Chomsky, N. (1957) *Syntactic Structures*, Den Haag: Mouton.

Chomsky, N. (1958) "A Transformational Approach to Syntax," repr. in J. A. Fodor & J. J. Katz (1964), pp. 211–245.

Chomsky, N. (1959), "A Review of BF Skinner's Verbal Behavior." *Language* 35(1), pp. 26–58.

Chomsky, N. (1961) "On the Notion 'Rule of Grammar,'" repr. in J. A. Fodor & J. J. Katz (1964), pp. 119–136.

Chomsky, N. (1962) "Explanatory Models in Linguistics:," in E. Nagel, P. Suppes, & A. Tarski (eds.), *Logic, Methodology and Philosophy of Science: Proceedings of the 1960 International Congress*, Stanford, CA: Stanford University Press, pp. 528–550.

Chomsky, N. (1963) "Formal Properties of Grammars," in R. D. Luce, R. Bush, & E. Galanter (eds.), *Handbook of Mathematical Psychology*, vol. 2, New York: Wiley, pp. 323–418.

Chomsky, N. (1964) *Current Issues in Linguistic Theory*, Den Haag: Mouton.

Chomsky, N. (1965) *Aspects of the Theory of Syntax*, Cambridge, MA: MIT Press.

Chomsky, N. (1966) *Cartesian Linguistics: A Chapter in the History of Rationalist Thought*, New York: Harper and Row.

Chomsky, N. (1968) *Language and Mind*, Boston: Harcourt.

Chomsky, N. (1979) *Language and Responsibility: Based on Conversations with Mitsou Ronat*, New York: Pantheon Books. Originally published in French as N. Chomsky & M. Ronat (1977), *Langue, Linguistique, Politique—Dialogues avec Mitsou Ronat*, Paris: Flammarion.

Chomsky, N. (1979), "A propos des structures cognitives et de leur développement: une réponse à Piaget," in Piatelli-Palmarini (1979), pp. 65–87.

Chomsky, N. (1980) *Rules and Representations*, Oxford: Basil Blackwell.

Chomsky, N. (1981) *Lectures on Government and Binding*, Dordrecht: Floris Publications.

Chomsky, N. (1988) *Language and Problems of Knowledge*, the Managua Lectures, Cambridge, MA: MIT Press.

Chomsky, N. (1995) *The Minimalist Program*, Cambridge, MA: MIT Press.

Chomsky, N. (2000) *On Nature and Language*, Cambridge: Cambridge University Press.

Chomsky, N. (2010) "Some Simple Evo Devo Theses: How True Might They Be for Language?" in R. Larson, V. Deprez, & H. Yamakido (eds.) (2011), *The Evolution of Human Language: Biolinguistic Perspectives*, Cambridge: Cambridge University Press, chapter 2.

Chomsky, N., & Halle, M. (1965) "Some Controversial Questions in Phonological Theory," *Journal of Linguistics* (1), pp. 97–138.

Chomsky, N., & Halle, M. (1968) *The Sound Pattern of English*, Cambridge, MA: MIT Press.

Chomsky, N., & Miller G. (1963) "Introduction to the Formal Analysis of Natural Languages," in Luce R.D., Bush R., & Galanter E. (eds.) (1963), *Handbook of Mathematical Psychology*, vol. 2, New York: Wiley, pp. 269–322.

Church, A. (1951) "The Need for Abstract Entities in Semantic Analysis," *Proceedings of the American Academy of Arts and Sciences* 80(1), pp. 100–112.

Cinque, G. (2005), "Deriving Greenberg's Universal 20 and Its Exceptions," *Linguistic Inquiry* 36(3), pp. 315–332.

Comrie, B. (2003) "On Explaining Language Universals," in Tomasello 2(003), pp. 195–209.

Dell, F. (1985) *Les règles et les sons, Introduction à la phonologie générative*, Paris: Hermann.

Devitt, M. (2006) *Ignorance of Language*, Oxford: Oxford University Press.

Dowty, D. (2007) "Compositionality as an Empirical Problem," in C. Barker & P. Jacobson (eds.) (2007), *Direct Compositionality*, Oxford: Oxford Studies in Theoretical Linguistics 14, pp. 23–101.

Duhem, P. (1906) *La théorie physique, Son objet—Sa Structure*, 2nd ed., 1997, Paris: Vrin.

Emmorey, K. (2002) *Language, Cognition and the Brain: Insights from Sign Language Research*, Wahwah, NJ: Lawrence Erlbaum Associates.

Encrevé, P. (1997) "L'ancien et le nouveau: quelques remarques sur la phonologie et son histoire," *Langages* 125(31), pp. 100–123.

Everett, D. (2005) "Cultural Constraints on Grammar and Cognition in Pirahã," *Current Anthropology* 46(4), pp. 621–646.

Fauconnier, G. (1975) "Polarity and the Scale Principle," *Chicago Linguistics Society* 11, pp. 188–199.

Fodor, J. A., & Katz, J. J. (1964) *The Structure of Language, Readings in the Philosophy of Language*, Upper Saddle River, NJ: Prentice-Hall, Inc.

Fodor, J. A., & Pylyshyn, Z. (1998) "Connectionism and Cognitive Architecture: A Critical Analysis," *Cognition* 28, 3–71.

Fox, D. (2002) "On Logical Form," in R. Hendrick (ed.) (2002), *Minimalist Syntax*, Oxford: Blackwell, pp. 82–123.

Frege, G. ([1891] 1960) "Function and Concept," in *Translations from the Philosophical Writings of Gottlob Frege*, Oxford: Blackwell, pp. 21–41.

Frege, G. ([1892] 1960), "On Sense and Reference," in *Translations from the Philosophical Writings of Gottlob Frege*, Oxford: Blackwell, pp. 56–78.

Frege, G. ([1923] 1963) "Compound Thoughts," English translation in *Mind* 72(285), pp. 1–17.

Gamut, L.T.F. (1991) *Logic, Language and Meaning*, vol. 2, *Intensional Logic and Logical Grammar*, Chicago: University of Chicago Press.

Gillon, B. (2017) "Language, Linguistics, Semantics," chapter 1 of *Grammatical Structure and Its Interpretation: An Introduction to Natural Language Semantics*, manuscript available on the author's webpage, McGill University.

Givón, T. (1979) *On Understanding Grammar*, New York: Academic Press.

Greenberg, J. (1957) *Essays in Linguistics*, Chicago: University of Chicago Press.

Greenberg, J. (ed.) (1963a) *Universals of Language*, Cambridge, MA: MIT Press.

Greenberg, J. (1963b) "Some Universals of Grammar with Particular Reference to the Order of Meaningful Elements," in Greenberg (ed.) (1963a), pp. 58–90.

Greenberg, J. H. (2005) *Genetic Linguistics, Essays on Theory and Method*, ed. and introduced by W. Croft, Oxford: Oxford Linguistics.

Grice, H. P. (1967) "Logic and Conversation," reprinted in Grice (1989), pp. 22–40.

Grice, H. P. (1989) *Studies in the Way of Words*, Cambridge, MA: Harvard University Press.

Grodzinski, Y. (2007) "La syntaxe générative dans le cerveau," in *L'Herne 88, Chomsky*. Paris: L'Herne, pp. 169–178.

Gussenhoven, C., & Jacobs, H. (1998). *Understanding Phonology*. London: Hodder Arnold [3rd ed., Routledge, 2013].

Halle, M. (1954) "Why and How Do We Study the Sounds of Speech?," repr. in Halle (2002), pp. 18–23.

Halle, M. (1978) "Knowledge Unlearned and Untaught: What Speakers Know about the Sounds of Their Language," repr. in Halle (2002), pp. 95–104.

Halle, M. (2002) *From Memory to Speech and Back, Papers on Phonetics and Phonology, 1954–2002*, Boston: Mouton De Gruyter.

Harris, Z. (1951) *Methods in Structural Linguistics*, Chicago: University of Chicago Press.

Harris, Z. (1957) "Co-occurrence and Transformation in Linguistic Structure," *Language* 33(3), pp. 283–340.

Hauser, M., Chomsky, N., & Fitch, W.T. (2002) "The Faculty of Language: What Is It, Who Has It, and How Did It Evolve?" *Science* 298, no. 5598, pp. 1569–1579, November. doi: 10.1126/science.298.5598.1569

Heim, I., & Kratzer, A. (1998) *Semantics in Generative Grammar*, Oxford: Blackwell.

Hempel, C. G. (1965) *Aspects of Scientific Explanation and other Essays in the Philosophy of Science*, New York: The Free Press.

Hempel, C. G. (1966) *Philosophy of Natural Science*, Englewood Cliffs, NJ: Prentice-Hall.

Hockett, C. (1954) "Two Models of Grammatical Description," *Word* X, 210–231, repr. in Katamba (2003).

Hockett, C. (1955) *A Manual of Phonology*, Baltimore: Indiana University Publications in Anthropology and Linguistics.

Hockett, C. F. (1958) *A Course in Modern Linguistics*, New York: McMillan.

Hockett, C. F. (1960) "The Origin of Speech," *Scientific American* 203, pp. 88–96.

Hockett, C. F. (1963) "The Problem of Universals in Language," in Greenberg (1963a), pp. 1–22.

Hodges, W. (1998) "Compositionality Is Not the Problem," *Logic and Logical Philosophy* 6, pp. 7–33.

Horn, L. (1989) *A Natural History of Negation*, Stanford, CA: CSLI Publication.

Humboldt, W. von (1836), *Über die Verschiedenheit des Menschlichen Sprachbaues*, Berlin.

Jackendoff, R. (1972) *Semantic Interpretation in Generative Grammar*, Cambridge, MA: MIT Press.

Jakobson, R. (1952) "Results of a Joint Conference of Anthropologists and Linguists," chap. 1 of *Fundamentals of Language*, Berlin: Mouton De Gruyter.

Jakobson, R. (1956) "Two Aspects of Language and Two Types of Aphasic Disturbances" chap. 2 of *Fundamentals of Language*, Berlin: Mouton De Gruyter.

Jakobson, R. (1978) *Six Lectures on Sound and Meaning*, Hassocks: Harvester Press.

Janssen, T. M. V. (1997) "Compositionality," in J. van Benthem & A. ter Meulen (eds.) (1997), *Handbook of Logic and Language*. Amsterdam: Elsevier, pp. 417–473.

Joos, M. (ed.), (1957) *Readings in Linguistic*, Chicago: University of Chicago Press.

Katamba, F. (ed.), (2003) *Morphology, Critical Concepts in Linguistics*, London: Routledge.

Keenan, E. & Stabler, E. (2003) *Bare Grammar, Lectures of Linguistic Invariants*, Stanford Monographs in Linguistics, Stanford, CA: CSLI Publications.

Kenstowicz, M. (2004) "Generative Phonology," in *Encyclopedia of Language and Linguistics*, 2nd ed., Amsterdam: Elsevier.

Kenstowicz, M., & Kisseberth, M. (1979) *Generative Phonology, Description and Theory*, New York: Academic Press.

Korta, K., & Perry, J. (2006) "Pragmatics," *The Stanford Encyclopedia of Philosophy* (Fall 2008 Edition), Edward N. Zalta (ed.), http://plato.stanford.edu/archives/fall2008/entries/pragmatics/.

Ladusaw, W. (1979) "Polarity Sensitivity as Inherent Scope Relations," PhD, University of Texas–Austin.

Lasnik, H. (2000) *Syntactic Structures Revisited: Contemporary Lectures on Classic Transformational Theory*, Cambridge, MA: MIT Press

Lasnik, H., & Fiengo, R. (1974) "Complement Object Deletion," *Linguistic Inquiry* 5 (4), pp. 535–571.

Lepage, F., & Lapierre, S. (2000) *Logique partielle et savoir*, Paris: Vrin, Collection Analytiques 11.

Lévi-Strauss, C. (1978) preface to R. Jakobson (1978) *Six Lectures on Sound and Meaning*, Hassocks: Harvester Press, pp. xi–xxvi.

Levinson, S. C. (2005) Comment to Everett (2005), in Everett (2005), pp. 637–638.

Lewis, D. (1968) "Languages and Language," repr. in D. Lewis, *Philosophical Papers*, vol. 1, Oxford: Oxford University Press, pp. 163–188.

Lewis, D. (1970) "General Semantics," repr. in D. Lewis, *Philosophical Papers*, vol. 1, Oxford: Oxford University Press, pp. 189–232.

Lightfoot, D. (2006) *How New Languages Emerge*, Cambridge: Cambridge University Press.

Marantz, A. (2005) "Generative Linguistics within the Cognitive Neuroscience of Language," *The Linguistic Review* 22, pp. 429–445.

Martinet, A. (1991) *Éléments de linguistique générale*, 3rd ed., Paris: Armand Colin.

Meillet, A. (1937) *Introduction à l'étude comparative des langues indo-européennes*, Paris: Hachette.

Mill, J. S. (1843) *A System of Logic, Ratiocinative and Inductive*, London: Parker.

Montague, R. (1968) "Pragmatics," repr. in R. Thomason (ed.) (1974), *Formal Philosophy: Selected Papers of Richard Montague*. New Haven, CT: Yale University Press, pp. 95–118.

Montague, R. (1970a) "English as a Formal Language," repr. in R. Thomason (ed.) (1974), *Formal Philosophy: Selected Papers of Richard Montague*. New Haven: Yale University Press, pp. 188–221.

Montague, R. (1970b) "Universal Grammar," repr. in R. Thomason (ed.) (1974), *Formal Philosophy: Selected Papers of Richard Montague*. New Haven, CT: Yale University Press, pp. 222–246.

Montague, R. (1973) "The Proper Treatment of Quantification in Ordinary English," reprinted in R. Thomason (ed.) (1974), *Formal Philosophy: Selected Papers of Richard Montague*. New Haven, CT: Yale University Press, pp. 247–270.

Nevins, A., Pesetsky, D., & Rodrigues, C. (2009) "Pirahã Exceptionality: A Reassessment," *Language* 85(2), pp. 355–404.

Newmeyer, F. (1986) "Has There Been a 'Chomskyan Revolution' in Linguistics?" *Language* 62(1), pp. 1–18.

Newmeyer, F. (1998) *Language Form and Language Function*, Cambridge, MA: MIT Press.

Newmeyer, F. (2005) *Possible and Probable Languages, A Generative Perspective on Linguistic Typology*, Oxford: Oxford Linguistics.

Ngyen, N., Wauquier-Gravelines, S., Durand, J. (eds.) (2005) *Phonologie et Phonétique, Forme et Substance*, Paris: Hermès Science Publications.

Partee, B. (2004) *Compositionality in Formal Semantics: Selected Papers of Barbara Partee*, Oxford: Blackwell Publishers.

Partee, B., & ter Meulen, A., & Wall, R. (1990) *Mathematical Methods in Linguistics*, Studies in Linguistics and Philosophy, vol. 30, Dordrecht, Boston, and London: Kluwer Academic Publishers.

Pawley, A. (2005) Comment on Everett (2005), in Everett (2005), pp. 638–639.

Peters, S., & Westerståhl, D. (2006) *Quantifiers in Language and Logic*, Oxford: Oxford University Press.

Pesetsky, D. (1995) *Zero Syntax*, Cambridge, MA: MIT Press.

Piatelli-Palmarini, M. (ed.) (1979) *Théories du langage, Théories de l'apprentissage—Le débat entre Jean Piaget et Noam Chomsky*, Paris: Le Seuil.

Picq, P., Sagart, L., Dehaene, G., & Lestienne, C. (2008) *La plus belle histoire du langage*, Paris: le Seuil.

Pinker, S. (1994) *The Language Instinct*, New York: Harper.

Poeppel, D. (2005) "Interdisciplinary Cross-Fertilization or Cross-Sterilization? Challenges at the Interface of Research on Brain and Language," manuscript.

Poeppel, D., & Embick, D. (2005) "Defining the Relation between Linguistics and Neuroscience," in Anne Cutler (ed.) (2005), *Twenty-First Century Psycholinguistics: Four Cornerstones*, Mahwah: Lawrence Erlbaum Associates, pp. 103–118.

Pollock, J-Y (1997) *Langage et cognition—Introduction au programme minimaliste de la grammaire générative*, Paris: PUF.

Pollock, J-Y (2007) "La grammaire générative et le programme minimaliste," in *L'Herne* 88, *Chomsky*. Paris: L'Herne, pp. 94–119.

Postal, P. (1964) "Limitations of Phrase Structure Grammars," in Fodor & Katz (1964), pp. 135–154.

Prince, A., & Smolensky, P. (1997) "Optimality: From Neural Networks to Universal Grammar," *Science* 275, pp. 1604–1610.

Pullum, G., & Scholz, B. (2002) "Empirical Assessment of Stimulus Poverty Arguments," *The Linguistic Review* 19 (2002), pp. 9–50.

Pullum, G., & Scholz, B. (2007) "Systematicity and Natural Language Syntax," *Croatian Journal of Philosophy* 7(21), pp. 375–402.

Quine, W. V. O. (1960) *Word and Object*, Cambridge, MA: MIT Press.

Radford, A. (1995) *Transformational Grammar, A First Course*, Cambridge: Cambridge University Press.

Radford, A., Atkinson, R.M., Britain, D., Clahsen, H., & Spencer, A. J. (1999) *Linguistics: An Introduction*, Cambridge: Cambridge University Press.

Rezac, M. (2006) "On Tough-Movement," in C. Boeckx (ed.) (2006), *Minimalist Essays*, Amsterdam: John Benjamins, pp. 288–325.

Rivenc, F., & Sandu, G. (2009) *Entre logique et langage*, Paris: Vrin.

Rizzi, L. (1980) "Violations of the Wh-Island Constraint in Italian and the Subjacency Condition," *Journal of Italian Linguistics* 5(1), pp. 157–191.

Rizzi, L. (2007) "L'acquisition de la langue et la faculté de langage," in *L'Herne* 88, *Chomsky*, Paris: L'Herne, pp. 147–157.

Ruwet, N. (1967) *Introduction à la grammaire générative*, Paris: Plon. Translated in English as N. Ruwet (1973), *Introduction to generative grammar*, Amsterdam: North Holland.

Sag, I., Wasow, T., & Bender, E.M. (2003) *Syntactic Theory: A Formal Introduction*, Stanford, CA: CSLI Publications.

Sapir, E. (1921) *Language: An Introduction to the Study of Speech*, Boston: Mariner Books.

Sapir, E. (1925) "Sound Patterns in Language," *Language* 1(2), 37–51.

Sapir, E. ([1933] 1985) "The Psychological Reality of Phonemes" in *Selected Writings of Edward Sapir in Language, Culture and Personality*, Berkeley: University of California Press, pp. 46–60.

Saussure, F. ([1916] 1959) *Course in General Linguistics*, New York: Philosophical Library.

Schlenker, P. (2008) "Semantics," in K. Malmkjaer (ed.) (2008), *Linguistics Encyclopedia*, London: Routledge.

Searle, J. (1969) *Speech Acts: An Essay in the Philosophy of Language*. Cambridge: Cambridge University Press.

Senghas, A., Kita, S., & Özyürek, A., (2004) "Children Creating Core Properties of Language: Evidence from an Emerging Sign Language in Nicaragua," *Science* 305(5691), pp. 1779–1782.

Spector, B. (2003) "Grammaire et logique," *Labyrinthe* 14, pp. 25–46.

Sprouse, J., & Almeida, D. (2012) "Assessing the Reliability of Textbook Data in Syntax: Adger's *Core Syntax*," *Journal of Linguistics* 48 (2012), pp. 609–652.

Steriade, D. (2007) "Contrast," in Paul de Lacy (ed.) (2007), *The Cambridge Handbook of Phonology*, Cambridge: Cambridge University Press, pp. 139–157.

Szabo, Z. (2007) "Compositionality," *The Stanford Encyclopedia of Philosophy*, E. Zalta (ed.), URL= http://plato.stanford.edu/entries/compositionality/.

Tarski, A. (1933) "The Concept of Truth in Formalized Languages," repr. *in* A. Tarski (ed.) (1983), *Logic, Semantics, Metamathematics*, Oxford: Oxford University Press, pp. 152–278.

Teyssier, P. (2004) *Comprendre les langues romanes*, Paris: Chandeigne.

Tomasello, M. (ed.) (2003) *The New Psychology of Language—Cognitive and Functional Approaches to Language Structure*, 2 vols., Mahwah: Lawrence Erlbaum Associates.

Travis, C. (1997) "Pragmatics," in B. Hale and C. Wright (eds.) (1997) *A Companion to the Philosophy of Language*, Oxford: Blackwell, pp. 87–106.

Vergnaud, J-R. (1977) "Letter to Noam Chomsky and Howard Lasnik on 'Filters and Control,'" April 17, 1977, repr. in *Foundational Issues in Linguistic Theory, Essays in Honor of Jean-Roger Vergnaud*, Cambridge, MA: MIT Press 2008, 3–16.

von Fintel, K. (1999) "NPI Licensing, Strawson Entailment and Context-Dependency," *Journal of Semantics* (16), pp. 97–148.

Whorf, B. L. (1956) *Language, Thought and Reality*, Cambridge, MA: MIT Press.

Wierzbicka, A. (2005) Comment on Everett 2005, in Everett (2005), p. 641.

Index